LOGARITHMS

$y = \log_a x$ means $a^y = x$

$\log_a a^x = x$ $\qquad a^{\log_a x} = x$

$\log_a 1 = 0$ $\qquad \log_a a = 1$

$\log x = \log_{10} x$ $\qquad \ln x = \log_e x$

$\log_a xy = \log_a x + \log_a y$

$\log_a\left(\dfrac{x}{y}\right) = \log_a x - \log_a y$

$\log_a x^b = b\,\log_a x$ $\qquad \log_b x = \dfrac{\log_a x}{\log_a b}$

EXPONENTIAL AND LOGARITHMIC FUNCTIONS

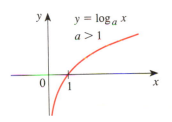

$y = a^x$
$a > 1$

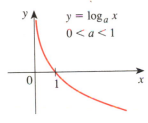

$y = a^x$
$0 < a < 1$

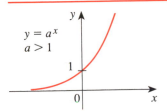

$y = \log_a x$
$a > 1$

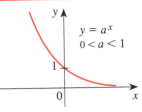

$y = \log_a x$
$0 < a < 1$

ANGLE MEASUREMENT

π radians $= 180°$

$1° = \dfrac{\pi}{180}$ rad $\qquad 1$ rad $= \dfrac{180°}{\pi}$

$s = r\theta$
(θ in radians)

TRIGONOMETRIC FUNCTIONS OF REAL NUMBERS

$\sin t = y$ $\qquad \csc t = \dfrac{1}{y}$

$\cos t = x$ $\qquad \sec t = \dfrac{1}{x}$

$\tan t = \dfrac{y}{x}$ $\qquad \cot t = \dfrac{x}{y}$

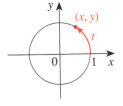

TRIGONOMETRIC FUNCTIONS OF ANGLES

$\sin \theta = \dfrac{y}{r}$ $\qquad \csc \theta = \dfrac{r}{y}$

$\cos \theta = \dfrac{x}{r}$ $\qquad \sec \theta = \dfrac{r}{x}$

$\tan \theta = \dfrac{y}{x}$ $\qquad \cot \theta = \dfrac{x}{y}$

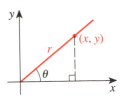

RIGHT ANGLE TRIGONOMETRY

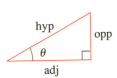

$\sin \theta = \dfrac{\text{opp}}{\text{hyp}}$ $\qquad \csc \theta = \dfrac{\text{hyp}}{\text{opp}}$

$\cos \theta = \dfrac{\text{adj}}{\text{hyp}}$ $\qquad \sec \theta = \dfrac{\text{hyp}}{\text{adj}}$

$\tan \theta = \dfrac{\text{opp}}{\text{adj}}$ $\qquad \cot \theta = \dfrac{\text{adj}}{\text{opp}}$

SPECIAL VALUES OF THE TRIGONOMETRIC FUNCTIONS

θ	radians	$\sin \theta$	$\cos \theta$	$\tan \theta$
$0°$	0	0	1	0
$30°$	$\pi/6$	$1/2$	$\sqrt{3}/2$	$\sqrt{3}/3$
$45°$	$\pi/4$	$\sqrt{2}/2$	$\sqrt{2}/2$	1
$60°$	$\pi/3$	$\sqrt{3}/2$	$1/2$	$\sqrt{3}$
$90°$	$\pi/2$	1	0	—
$180°$	π	0	-1	0
$270°$	$3\pi/2$	-1	0	—

SPECIAL TRIANGLES

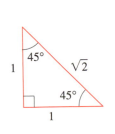

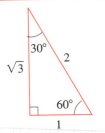

TO THE STUDENT

Many students experience difficulty with *precalculus* mathematics in calculus courses. Calculus requires that you understand and remember precalculus topics. For this reason, it may be helpful to retain this text as a reference in your calculus course. It has been written with this purpose in mind.

SOME REFERENCES TO CALCULUS TOPICS IN THIS TEXT

MATHEMATICS FOR CALCULUS

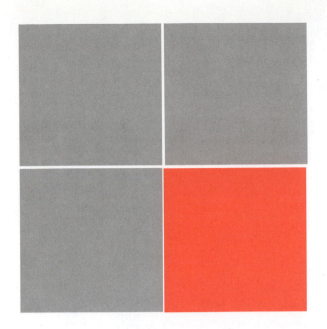

MATHEMATICS FOR CALCULUS

SECOND EDITION

JAMES STEWART
McMaster University

LOTHAR REDLIN
The Pennsylvania State University

SALEEM WATSON
California State University, Long Beach

BROOKS/COLE PUBLISHING COMPANY
Pacific Grove, California

To our students

Brooks/Cole Publishing Company
A Division of Wadsworth, Inc.
© 1993, 1989 by Wadsworth, Inc., Belmont, California 94002.

Printed in the United States of America
10 9 8 8 7 6 5 4 3 2

Library of Congress Cataloging-in-Publication Data
  Stewart, James.
  Mathematics for calculus / James Stewart, Lothar Redlin, Saleem Watson .—2nd ed.
    Includes index.
  ISBN 0-534-20250-0
  1. Mathematics. I. Redlin, L. II. Watson, Saleem. III. Title.
  QA39.2.S75 1993
  512′.1—dc20
         92-42810
         CIP

Illustration and photo credits
page 40 Granger Collection / **52** Bettman Archives / **101** Stanford University / **123** Stanford University /
541 California Institute of Technology, Pasadena

Sponsoring Editor: *Jeremy Hayhurst*
Editorial Assistant: *Nancy Champlin*
Production Coordination: *Kathi Townes*
Production Services Manager: *Joan Marsh*
Manuscript Editor: *Kathi Townes*
Permissions Editor: *Carline Haga*
Interior Design: *John Edeen*
Cover Design: *Katherine Minerva*
Cover Photo: *Ed Young Photography*
Cover Artwork: Marriage of Opposites (1992) *by Anthony Marsh; courtesy of the artist*
Interior Illustration: *TECHarts, Scientific Illustrators, Eric Bosch*
Typesetting: *TECHarts, Jonathan Peck Typographer Ltd.*
Cover Printing: *The Lehigh Press*
Printing and Binding: *Arcata Graphics/Fairfield*

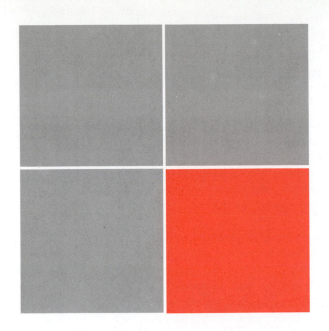

PREFACE

What does a student really need to know to prepare for calculus? This question has motivated the writing of this text.

Students often begin their study of calculus unprepared for the task that lies ahead. What they need is not only technical skill but a correct concept of the nature of mathematics. Accordingly, we present precalculus mathematics as a problem-solving endeavor. The intention throughout the book is to encourage the student to organize his or her thoughts when tackling a mathematical problem.

A clear understanding of the idea of a function and its graph are fundamental to a student beginning calculus. We introduce functions early and we emphasize this concept throughout the text. Inexpensive graphing calculators, now readily available, provide a powerful tool for developing the student's intuition for functions and their graphs. In this edition, we have added a number of optional sections and numerous exercises devoted to the use of these tools. These sections and exercises are identified with a special icon: ▨. They should be useful to students even if they don't have a calculator in hand while reading these sections.

This text is a thorough revision of the first edition. We have rewritten more than half of the sections to sharpen the clarity of the exposition. Much of the material, particularly the trigonometry, has been reorganized to make the structure of the book more flexible to serve a wider variety of precalculus courses. In addition, we have rewritten many of the exercise sets to provide a more consistent gradation of routine to more challenging problems.

Our aim throughout is to provide a thoroughly modern text that will prepare students for the calculus courses of today and of the future.

SPECIAL FEATURES OF THE SECOND EDITION

Graphing Calculators and Computers

The invention of the Cartesian plane allowed mathematicians to draw graphs so they could "see" the relationships they were studying. The philosopher John Stuart Mill called this invention "the greatest single step in the progress of science." It can be argued that the ability to draw—in an instant—the most complex graphs is the next great step. Calculator and computer technology provides completely new ways of visualizing mathematics and is affecting not only how a topic is taught but also what is emphasized.

We have integrated the technology carefully in six sections devoted to the fundamentals of graphing devices. In other sections we have included examples and exercises to show how the graphing calculator can be used; see, for instance, examples on pages 268, 441, and 641 and exercises 39–54 on page 270. These sections, examples, and exercises, all marked with the special logo ⬚ , are optional and may be skipped without loss of continuity, but we encourage you to try them. You may in fact wish to integrate graphing devices more fully into your course by using them for teaching the "regular" sections of the text as well. It has been our experience that students enjoy working with the graphing calculator, and learn the material better and with more enthusiasm when we require or suggest its use in our classes.

Of course, graphing via technology is useless if students don't understand what the device is doing. We don't believe that recipe-oriented button pushing will yield much insight without the proper conceptual framework. We stress that graphing devices can't be used without thought, and we point out possible pitfalls in using these devices.

Flexible Approach to Trigonometry

The trigonometric functions can be defined in two different, although equivalent, ways—one based on right triangles and the other based on arc length on the unit circle. One way to teach the subject is to begin with the right triangle approach, building on the foundation of a conventional high school course in trigonometry. The other way is to begin with the unit circle approach, thus emphasizing that the trigonometric functions are functions of real numbers, just like polynomial, logarithmic, and exponential functions. Each way of teaching trigonometry has its merits, and so the trigonometry chapters of this text have been carefully written so that *either approach may be studied first*. In both cases, the applications appropriate to the approach are presented together with it. This provides early motivation for studying these functions and also makes it clear why we need to study two different ways of looking at the same functions.

Chapter 5 gives the unit circle treatment and the applications for which it is most appropriate, and Chapter 6 gives the right triangle treatment and its applications. Either Chapter 5 *or* Chapter 6 may be studied first. Marginal notes in each chapter relate the two approaches to each other, but the chapters are otherwise independent, so there is absolutely no loss of continuity if the chapters are studied in either order. Chapter 7, on analytic trigonometry, may be studied after either or both of the preceding chapters.

Focus on Problem Solving

We are committed to helping students develop mathematical thinking rather than "memorizing all the rules," an endless and pointless task that many students are not taught to avoid. We believe that mathematical thinking is best taught by careful exposition and examples and well-graded exercise sets, but also by giving guidelines for problem solving. Thus an emphasis on problem solving is integrated throughout the text. In addition, we have concluded each chapter with a section entitled *Focus on Problem Solving,* which highlights a particular problem-solving technique. The first of these sections, on pages 100–105, gives a general introduction to the principles of problem solving. Each *Focus* section includes problems of a varied nature that encourage students to apply their problem-solving skills. In selecting these problems we have kept the following advice from David Hilbert in mind: "A mathematical problem should be difficult in order to entice us, yet not completely inaccessible, lest it mock our efforts."

Mathematical "Vignettes"

Throughout this book we make use of the margins to provide short biographies of interesting mathematicians, and applications of precalculus mathematics to the "real world." The biographies often include a key insight that the mathematician discovered and which is relevant to precalculus. (See, for instance, the vignettes on Viète, page 40; Salt Lake City, page 73; and radio-carbon dating, page 273.) These serve to enliven the material and show the student that mathematics is an important, vital activity, and that even at this elementary level, it is fundamental in everyday life.

Review Sections

Each chapter ends with an extensive review section, including a Chapter Test designed to help the students gauge their progress. Brief answers to the odd-numbered exercises in each section and to all questions on the Chapter Tests are given in the back of the book.

■ MAJOR CHANGES FOR THE SECOND EDITION

- ■ Chapter 1, the review of basic algebra, has been expanded. As much or as little of this chapter as the students require may be taught; for particularly well prepared classes, the chapter may be omitted entirely.

- ■ Sections 2.3, 2.7, 3.2, 3.6, 3.10, and 5.5, which deal with the use of the graphing calculator, have been added. In addition, many examples and exercises involving graphing devices have been added throughout the text where appropriate.

- Chapter 3, on polynomial and rational functions, has been reorganized, with the theory of equations now coming after the material on graphing polynomials. In addition, the exercise sets in Sections 3.1, 3.4, 3.8, and 3.9 have been rewritten to provide a greater variety of simple as well as more challenging problems.

- Chapters 5, 6, and 7, on trigonometry, have been completely rewritten to make it possible to teach either the unit circle approach or the right triangle approach first.

- Section 8.3 now includes a discussion of *echelon form*, to clarify the process of Gaussian elimination.

- Extensive changes have been made in the text, the figures, and the examples in every chapter to increase the clarity of the exposition.

- Interesting marginal notes containing mathematical biographies and descriptions of applications of mathematics have been added throughout the text.

SUPPLEMENTS

The following supplements are available from Brooks/Cole Publishing Company:

Complete Solutions Manual by Stewart, Redlin, and Watson
Student Solutions Manual by Stewart, Redlin, and Watson
Study Guide by John Banks, *University of California, Santa Cruz*
EXP-Test, computerized testing software for the IBM and PC-compatibles
Exam Builder, computerized testing software for the Macintosh

ACKNOWLEDGMENTS

We thank the following reviewers for their thoughtful and constructive comments:

Paul Beem, Indiana University
Franklin Cheek, University of Wisconsin
Estela Llinas, University of Pittsburgh, Greensburg
John Sarli, California State University, San Bernardino
Kenneth Shabell, Riverside Community College

We thank the talented staff at Brooks/Cole Publishing Company: Joan Marsh, production services manager; Katherine Minerva, cover designer; Arthur Minsberg, product manager; Margaret Parks, director of marketing communications; Nancy Champlin, editorial assistant. We also thank our designer, John Edeen, as well as Anthony Marsh, whose sculpture appears on the cover.

We are especially grateful to Kathi Townes, production coordinator, and her staff at TECHarts, who worked tirelessly to ensure the completion and accuracy of every aspect of this book. They have done an outstanding job.

Finally, we would like to express our heartfelt thanks to Jeremy Hayhurst, mathematics editor at Brooks/Cole, without whose guidance this project could not have been completed. His insight in many crucial editorial decisions has been invaluable.

TO THE STUDENT

This textbook was written for you to use as a guide to mastering precalculus mathematics. Here are some suggestions to help you get the most out of your course.

First of all, you should read the appropriate section of text *before* you attempt your homework problems. Reading a mathematics text is quite different from reading a novel, a newspaper, or even another textbook. You may find that you have to reread a passage several times before you understand it. Pay special attention to the examples, and work them out yourself with pencil and paper as you read. With this kind of preparation you will be able to do your homework much more quickly and with more understanding.

Don't make the mistake of trying to memorize every single rule or fact you may come across. Mathematics does not consist simply of memorization. Mathematics is a problem-solving art, not just a collection of facts. To master the subject you must solve problems—lots of problems. Do as many of the exercises as you can. Be sure to write your solutions in a logical, step-by-step fashion. Don't give up on a problem if you can't solve it right away. Try to understand the problem more clearly—reread it thoughtfully and relate it to what you have learned from your teacher and from the examples in the text. Struggle with it until you solve it. Once you have done this a few times you will begin to understand what mathematics is really all about.

Answers to the odd-numbered exercises, as well as all the answers to each chapter test, appear in the back of the book. If your answer differs from the one given, don't immediately assume that you are wrong. There may be a calculation that connects the two answers and makes both correct. For example, if you get $1/(\sqrt{2} - 1)$ but the answers given is $1 + \sqrt{2}$, your answer *is* correct, because you can multiply both numerator and denominator of your answers by $\sqrt{2} + 1$ to change it to the given answer.

The symbol 🚫 is used to warn against committing an error. We have placed this symbol in the margin to point out situations where we have found that many of our students make the same mistake.

CALCULATORS AND CALCULATIONS

A calculator is essential in most mathematics and science subjects. Precalculus courses are no exception. Although calculators are powerful tools, they need to be used with care. Here are some guidelines to help you obtain meaningful answers from your calculator.

For this course you will need a *scientific* calculator—one that has, as a minimum, the usual arithmetic operations ($+$, $-$, $\times$, $\div$, $\sqrt{}$, x^y), exponential and logarithm functions (e^x, 10^x, $\ln x$, $\log x$), and the trigonometric functions (sin, cos, tan, $\sin^{-1}$, $\cos^{-1}$, $\tan^{-1}$). In addition, a memory and at least some degree of programmability will be useful.

Your instructor may recommend or require that you purchase a *graphing* calculator for your work in this course. Several different kinds are now on the market, including the Texas Instruments TI81 and TI85, the Casio FX8500, the Hewlett-Parkard 48SX, and the Sharp EL-5200. This book has optional sections and exercises that are intended to be used only by students who own, or have access to, a graphing calculator or a computer with graphing software. These special sections and exercises are indicated by the symbol ⨎

Whatever kind of calculator you have, you should learn how to use it by reading the owner's manual carefully. Work through the examples given in it to become familiar with how your calculator performs each operation. When using graphing calculators, it is particularly important to choose carefully the range of values for x and y. We refer you to Section 2.3 for advice on how to choose an appropriate viewing rectangle.

It is important to realize that, because of limited resolution, a graphing calculator gives only an *approximation* to the graph of a function. It can plot only a finite number of points and then connect them to form a *representation* of a graph. In Chapters 2, 3, and 5 we point out that you sometimes have to be careful when interpreting graphs produced by calculators.

Most of the applied examples and exercises in this book involve approximate values. For example, one exercise states that the moon has a radius of 1074 miles. This does not mean that the moon's radius is *exactly* 1074 miles but simply that this is the radius rounded to the nearest mile.

One simple method for specifying the accuracy of a number is to state how many **significant digits** it has. The significant digits in a number are the ones from the first nonzero digit to the last nonzero digit (reading from left to right). Thus 1074 has four significant digits, 1070 has three, 1100 has two, and 1000 has one significant digit.

This rule may sometimes lead to ambiguities. For example, if a distance is 200 km to the nearest kilometer, then the number 200 really has three significant digits, not just one. This ambiguity is avoided if we use *scientific notation*—that is, if we express the number as a multiple of a power of 10:

$$2.00 \times 10^2$$

Number	Significant Digits
12,300	3
3000	1
3000.58	6
0.03416	4
4200	2
4.2×10^3	2
4.20×10^3	3

When working with approximate values, students often make the mistake of giving a final answer with *more* significant digits than the original data. This is incorrect because you cannot "create" precision by using a calculator. The final result can be no more accurate than the measurements given in the problem. For example, suppose we are told that the two shorter sides of a right triangle are measured to be 1.25 and 2.33 inches long. By the Pythagorean Theorem, we find, using a calculator, that the hypotenuse has length

$$\sqrt{1.25^2 + 2.33^2} \approx 2.644125564 \text{ in.}$$

But since the given lengths were expressed to three significant digits, the answer cannot be any more accurate. We can therefore say only that the hypotenuse is 2.64 in. long, rounding to the nearest hundredth.

In general, the final answer should be expressed with the same accuracy as the *least*-accurate measurement given in the statement of the problem. The following rules make this principle more precise.

**RULES FOR WORKING
WITH APPROXIMATE DATA**

1. When multiplying or dividing, round off the final result so that it has as many *significant digits* as the given value with the fewest number of significant digits.
2. When adding or subtracting, round off the final result so that it has its last significant digit in the *decimal place* in which the least-accurate given value has its last significant digit.
3. When taking powers or roots, round off the final result so that it has the same number of *significant digits* as the given value.

As an example, suppose that a rectangular table top is measured to be 122.64 in. by 37.3 in. We express its area and perimeter as follows:

Area = length × width = 122.64 × 37.3 ≈ 4570 in^2 (*three significant digits*)

Perimeter = 2(length + width) = 2(122.64 + 37.3) ≈ 319.9 in.   (*tenths digit*)

Note that in the formula for the perimeter, the value 2 is an exact value, not an approximate measurement. It therefore does not affect the accuracy of the final result. In general, if a problem involves only exact values, we may express the final answer with as many significant digits as we wish.

Note also that to make the final result as accurate as possible, you should wait until the last step to round off your answer. If necessary, use the memory feature of your calculator to retain the results of intermediate calculations.

ABBREVIATIONS

cm	centimeter	**MHz**	megahertz
dB	decibel	**mi**	mile
F	farad	**min**	minute
ft	foot	**mL**	milliliter
g	gram	**mm**	millimeter
gal	gallon	**N**	Newton
h	hour	**qt**	quart
H	henry	**oz**	ounce
Hz	Hertz	**s**	second
in.	inch	**Ω**	ohm
J	Joule	**V**	volt
kcal	kilocalorie	**W**	watt
kg	kilogram	**yd**	yard
km	kilometer	**yr**	year
kPa	kilopascal	**°C**	degree Centigrade
L	liter	**°F**	degree Fahrenheit
lb	pound	**°K**	degree Kelvin
M	mole of solute per liter of solution	$\Rightarrow$	implies
m	meter	$\Leftrightarrow$	is equivalent to

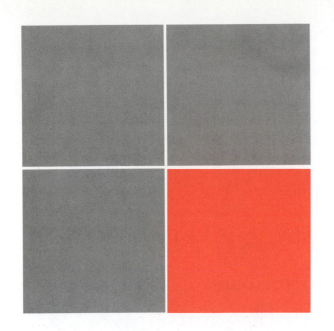

CONTENTS

FOCUS ON PROBLEM SOLVING: Pattern Recognition   **173**

3   POLYNOMIALS AND RATIONAL FUNCTIONS   176

FOCUS ON PROBLEM SOLVING: Introducing Something Extra   **260**

4   EXPONENTIAL AND LOGARITHMIC FUNCTIONS   262

FOCUS ON PROBLEM SOLVING: Indirect Reasoning   **303**

These sections are optional and require graphing calculators or computers

Either Chapter 5 or Chapter 6 may be studied first.

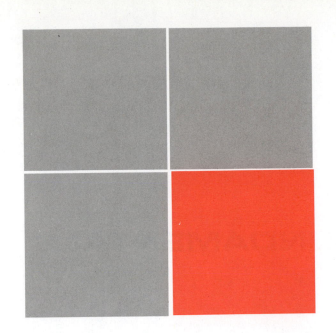

MATHEMATICS
FOR CALCULUS

1 FUNDAMENTALS

One learns by doing the thing; for though you think you know it, you have no certainty until you try.

SOPHOCLES

A great discovery solves a great problem but there is a grain of discovery in the solution of any problem. Your problem may be modest; but if it challenges your curiosity and brings into play your inventive faculties, and if you solve it by your own means, you may experience the tension and enjoy the triumph of discovery.

GEORGE POLYA

In this first chapter we review the basic ideas from algebra and coordinate geometry that will be needed throughout this book.

Let us recall the types of numbers that make up the real number system. We start with the **natural numbers**:

$$1, 2, 3, 4, \ldots$$

The **integers** consist of the natural numbers together with their negatives and 0:

$$\ldots, -3, -2, -1, 0, 1, 2, 3, 4, \ldots$$

We construct the **rational numbers** by taking ratios of integers. Thus any rational number r can be expressed as

$$r = \frac{m}{n} \qquad \text{where } m \text{ and } n \text{ are integers and } n \neq 0$$

Examples of rational numbers are

$$\tfrac{1}{2} \qquad -\tfrac{3}{7} \qquad 46 = \tfrac{46}{1} \qquad 0.17 = \tfrac{17}{100}$$

(Recall that division by 0 is always ruled out, so expressions like $\frac{3}{0}$ and $\frac{0}{0}$ are undefined.) There are also real numbers, such as $\sqrt{2}$, that cannot be expressed as a ratio of integers and are therefore called **irrational numbers**. It can be shown, with varying degrees of difficulty, that the following are also examples of irrational numbers:

$$\sqrt{3} \qquad \sqrt{5} \qquad \sqrt[3]{2} \qquad \pi \qquad \frac{3}{\pi^2}$$

The set of all real numbers is usually denoted by the symbol R. When we use the word *number* without qualification, we mean "real number."

Every number has a decimal representation. If the number is rational, then the corresponding decimal is repeating. For example,

$$\tfrac{1}{2} = 0.5000\ldots = 0.5\overline{0} \qquad\qquad \tfrac{2}{3} = 0.66666\ldots = 0.\overline{6}$$

$$\tfrac{157}{495} = 0.3171717\ldots = 0.3\overline{17} \qquad \tfrac{9}{7} = 1.285714285714\ldots = 1.\overline{285714}$$

(The bar indicates that the sequence of digits repeats forever.) On the other hand, if the number is irrational, the decimal representation is nonrepeating:

$$\sqrt{2} = 1.414213562373095. . . \qquad\qquad \pi = 3.141592653589793. . .$$

If we stop the decimal expansion of any number at a certain place, we get an approximation to the number. For instance, we can write

$$\pi \approx 3.14159265$$

where the symbol $\approx$ is read "is approximately equal to." The more decimal places we retain, the better the approximation we get.

The real numbers can be represented by points on a line as in Figure 1. The positive direction (to the right) is indicated by an arrow. We choose an arbitrary reference point O, called the **origin**, which corresponds to the real number 0. Given any convenient unit of measurement, each positive number x is represented by the point on the line a distance of x units to the right of the origin, and the corresponding negative number $-x$ is represented by the point x units to the left of the origin. Thus every real number is represented by a point on the line, and every point P on the line corresponds to exactly one real number. The number associated with the point P is called the **coordinate** of P and the line is then called a **coordinate line**, or a **real number line**, or simply a **real line**. Often we identify the point with its coordinate and think of a number as a point on the real line.

Figure 1
The real line

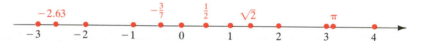

PROPERTIES OF REAL NUMBERS

In combining real numbers using the basic operations of addition and multiplication, we need to use the following properties of real numbers:

$a + b = b + a \qquad ab = ba$		(Commutative Laws)
$(a + b) + c = a + (b + c) \qquad (ab)c = a(bc)$		(Associative Laws)
$a(b + c) = ab + ac$		(Distributive Law)

The Commutative Law says that, when we add or multiply two real numbers a and b, the order of the numbers does not matter.

According to the Associative Law, when we add or multiply three numbers, it does not matter which pair of numbers we start with. For instance, $(2 + 4) + 7 = 6 + 7$ gives the same sum as $2 + (4 + 7) = 2 + 11$. For this reason, we usually just write $2 + 4 + 7$.

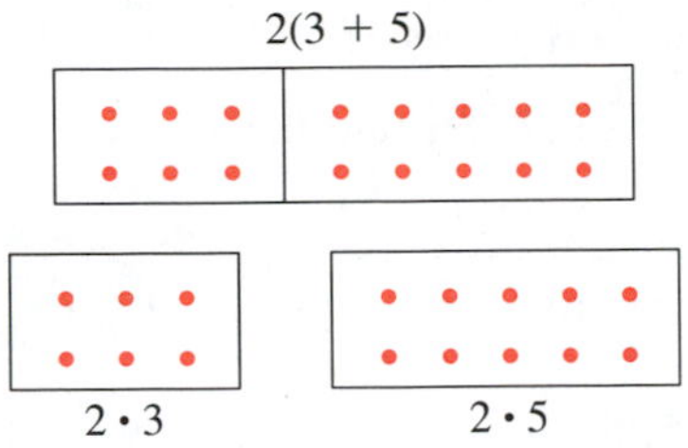

2(3 + 5)

2·3   2·5

Figure 2

The Distributive Law applies when we multiply a number by a sum. It is illustrated in Figure 2 for the case where the numbers are positive integers but it is true for any real numbers a, b, and c. For instance,

$$4\left(\tfrac{1}{2} + \pi\right) = 4 \cdot \tfrac{1}{2} + 4 \cdot \pi = 2 + 4\pi$$

We will be using this law frequently when we manipulate algebraic expressions in Section 1.3.

Subtraction is the operation that undoes addition. The difference $a - b$ is the number that, when added to b, gives a:

$$a - b = c \qquad \text{means} \qquad b + c = a$$

Similarly, division is the operation that undoes multiplication. If $b \neq 0$, the quotient a/b is the number that, when multiplied by b, gives a:

$$\frac{a}{b} = c \qquad \text{means} \qquad bc = a$$

The real numbers are ordered. We say a *is less than* b and write $a < b$ if $b - a$ is a positive number. Geometrically this means that a lies to the left of b on the number line. (Equivalently, we say b *is greater than* a and write $b > a$.) The symbol $a \leq b$ (or $b \geq a$) means that either $a < b$ or $a = b$ and is read "a is less than or equal to b." For instance, the following are true inequalities:

$$7 < 7.4 < 7.5 \qquad -3 > -\pi \qquad \sqrt{2} < 2 \qquad 2 \leq 2$$

SET NOTATION

In what follows we need to use set notation. A **set** is a collection of objects, and these objects are called the **elements** of the set. If S is a set, the notation $a \in S$ means that a is an element of S, and $a \notin S$ means that a is not an element of S. For example, if Z represents the set of integers, then $-3 \in Z$ but $\pi \notin Z$.

Some sets can be described by listing their elements between braces. For instance, the set A consisting of all positive integers less than 7 can be written as

$$A = \{1, 2, 3, 4, 5, 6\}$$

We could also write A in set-builder notation as

$$A = \{x \mid x \text{ is an integer and } 0 < x < 7\}$$

which is read "A is the set of x such that x is an integer and $0 < x < 7$."

If S and T are sets, then their **union** $S \cup T$ is the set consisting of all elements that are in S *or* T (or in both S and T). The **intersection** of S and T is the set $S \cap T$ consisting of all elements that are both in S *and* T. In other words, $S \cap T$ is the common part of S and T. The **empty set**, denoted by $\varnothing$, is the set that contains no elements.

If $S = \{1, 2, 3, 4, 5\}$, $T = \{4, 5, 6, 7\}$, and $V = \{6, 7, 8\}$, find the sets $S \cup T$, $S \cap T$, and $S \cap V$.

SOLUTION

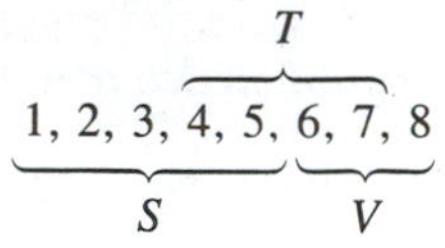

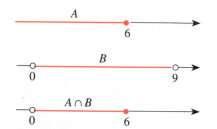

$$S \cup T = \{1, 2, 3, 4, 5, 6, 7\}$$
$$S \cap T = \{4, 5\}$$
$$S \cap V = \varnothing$$

■

EXAMPLE 2

If $A = \{x \mid x \leq 6\}$ and $B = \{x \mid 0 < x < 9\}$, find $A \cap B$.

SOLUTION

Let $x \in A \cap B$. Since $x \in A$, we have $x \leq 6$. Since $x \in B$, we also have $0 < x < 9$. Therefore,

$$0 < x \leq 6$$

so the intersection of the sets A and B is

$$A \cap B = \{x \mid 0 < x \leq 6\}$$

Figure 3

This is illustrated in Figure 3.

■

■ **INTERVALS**

There are certain sets of real numbers, called **intervals**, that occur frequently in calculus and correspond geometrically to line segments. For example, if $a < b$, the **open interval** from a to b consists of all numbers between a and b and is denoted by the symbol (a, b). Using set-builder notation, we can write

$$(a, b) = \{x \mid a < x < b\}$$

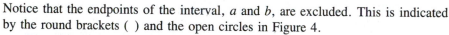

Figure 4
The open interval (a, b)

Notice that the endpoints of the interval, a and b, are excluded. This is indicated by the round brackets ( ) and the open circles in Figure 4.

The **closed interval** from a to b is the set

$$[a, b] = \{x \mid a \leq x \leq b\}$$

Figure 5
The closed interval $[a, b]$

Here the endpoints of the interval are included. This is indicated by the square brackets [ ] and the solid circles in Figure 5. It is also possible to include only one endpoint in an interval, as shown in the table of intervals preceding Example 3.

We also need to consider infinite intervals such as

$$(a, \infty) = \{x \mid x > a\}$$

This does not mean that ∞ ("infinity") is a number. The notation (a, ∞) stands for the set of all numbers that are greater than a, so the symbol ∞ simply indicates that the interval extends indefinitely far in the positive direction.

The following table lists the nine possible types of intervals. When these intervals are discussed, it will always be assumed that $a < b$.

Notation	Set Description	Picture
(a, b)	$\{x \mid a < x < b\}$	
$[a, b]$	$\{x \mid a \leq x \leq b\}$	
$[a, b)$	$\{x \mid a \leq x < b\}$	
$(a, b]$	$\{x \mid a < x \leq b\}$	
(a, ∞)	$\{x \mid x > a\}$	
$[a, \infty)$	$\{x \mid x \geq a\}$	
$(-\infty, b)$	$\{x \mid x < b\}$	
$(-\infty, b]$	$\{x \mid x \leq b\}$	
$(-\infty, \infty)$	R (set of all real numbers)	

EXAMPLE 3

Express the following intervals in terms of inequalities and graph the intervals:

(a) $[-1, 2)$ (b) $[1.5, 4]$ (c) $(-3, \infty)$

SOLUTION

(a) $[-1, 2) = \{x \mid -1 \leq x < 2\}$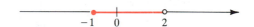

(b) $[1.5, 4] = \{x \mid 1.5 \leq x \leq 4\}$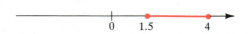

(c) $(-3, \infty) = \{x \mid x > -3\}$

EXAMPLE 4

Graph the following sets: (a) $(1, 3) \cap [2, 7]$ (b) $(-2, -1) \cup (1, 2)$

SOLUTION

(a) The intersection of two intervals consists of the numbers that are in both intervals. Therefore

$$(1, 3) \cap [2, 7] = \{x \mid 1 < x < 3 \quad \text{and} \quad 2 \leqslant x \leqslant 7\}$$
$$= \{x \mid 2 \leqslant x < 3\}$$
$$= [2, 3)$$

This interval is shown in Figure 6(a).

(b) The union of the intervals $(-2, -1)$ and $(1, 2)$ consists of the numbers that are in either $(-2, -1)$ or $(1, 2)$, so

$$(-2, -1) \cup (1, 2) = \{x \mid -2 < x < -1 \quad \text{or} \quad 1 < x < 2\}$$

This set is illustrated in Figure 6(b).

Figure 6

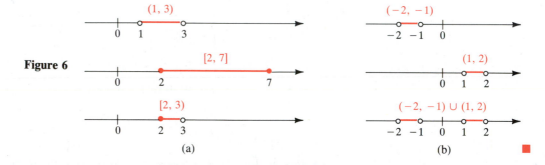

(a)   (b)

ABSOLUTE VALUE

The **absolute value** of a number a, denoted by $|a|$, is the distance from a to 0 on the real number line. (See Figure 7.) Distances are always positive or zero, so we have

$$|a| \geqslant 0 \quad \text{for every number } a$$

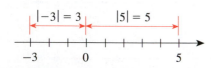

Figure 7

Remembering that $-a$ is positive when a is negative, we have

$$|a| = a \qquad \text{if } a \geqslant 0$$
$$|a| = -a \qquad \text{if } a < 0$$

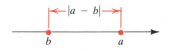

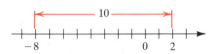

Figure 8
Length of a line segment = $|a - b|$

Figure 9

EXAMPLE 5

(a) $|3| = 3$

(b) $|-3| = -(-3) = 3$

(c) $|0| = 0$

(d) $|\sqrt{2} - 1| = \sqrt{2} - 1$ (since $\sqrt{2} > 1 \;\Rightarrow\; \sqrt{2} - 1 > 0$)

(e) $|3 - \pi| = -(3 - \pi) = \pi - 3$ (since $\pi > 3 \;\Rightarrow\; 3 - \pi < 0$)   ■

If a and b are any real numbers, then the distance between a and b is the absolute value of the difference, $|a - b|$, which is also equal to $|b - a|$. See Figure 8.

EXAMPLE 6

The distance between the numbers -8 and 2 is

$$|-8 - 2| = |-10| = 10$$

This calculation can be checked by looking at Figure 9.   ■

EXERCISES 1.1

In Exercises 1–8 use properties of real numbers to write the given expression without parentheses.

1. $3(x + y)$

2. $8(a - b)$

3. $4(2m)$

4. $\frac{1}{2}(10z)$

5. $\frac{4}{3}(-6y)$

6. $-2(r + s)$

7. $-\frac{5}{2}(2x - 4y)$

8. $(3a)(b + c - 2d)$

In Exercises 9–18 state whether the given inequality is true or false.

9. $-6 < -10$

10. $-3 < -1$

11. $0.66 < \frac{2}{3}$

12. $\sqrt{2} > 1.41$

13. $\frac{10}{11} < \frac{12}{13}$

14. $-\pi > -3$

15. $8 \leq 8$

16. $8 \leq 9$

17. $\pi \geq 3$

18. $1.1 > 1.\overline{1}$

In Exercises 19–28 write the statement in terms of inequalities.

19. x is positive

20. y is negative

21. t is less than 4

22. z is greater than 1

23. a is greater than or equal to π

24. b is at most 8

25. x is less than $\frac{1}{3}$ and is greater than -5

26. w is positive and is less than or equal to 17

27. The distance from p to 3 is at most 5

28. y is at least 2 units from π

In Exercises 29–34 find the given set if $A = \{1, 2, 3, 4, 5, 6\}$, $B = \{2, 4, 6, 8\}$, and $C = \{7, 8, 9, 10\}$.

29. $A \cup B$

30. $A \cap B$

31. $B \cap C$

32. $B \cup C$

33. $A \cup C$

34. $A \cap C$

In Exercises 35–38 find the given set if $A = \{x \mid x \geq -2\}$, $B = \{x \mid x < 4\}$, and $C = \{x \mid -1 < x \leq 5\}$.

35. $A \cap B$

36. $B \cap C$

37. $A \cap C$

38. $B \cup C$

In Exercises 39–48 express the interval in terms of inequalities and graph the interval.

39. $(-3, 0)$ **40.** $(2, 8]$

41. $[2, 8)$ **42.** $\left[-6, -\frac{1}{2}\right]$

43. $[-1, 1]$ **44.** $(-4, \infty)$

45. $[2, \infty)$ **46.** $(-\infty, 1)$

47. $(-\infty, -2]$ **48.** $(0, \pi)$

In Exercises 49–56, express the inequality in interval notation and graph the corresponding interval.

49. $x \leq 1$ **50.** $0 < x < 8$ **51.** $1 \leq x \leq 2$

52. $x < 3$ **53.** $-2 < x \leq 1$ **54.** $x \geq -5$

55. $x > -1$ **56.** $-5 < x < 2$

In Exercises 57–64 graph the set.

57. $(-2, 0) \cup (-1, 1)$ **58.** $(-2, 0) \cap (-1, 1)$

59. $[-4, 6] \cap [0, 8)$ **60.** $[-4, 6] \cup [0, 8)$

61. $(-\infty, -4) \cup (4, \infty)$ **62.** $(-\infty, 6] \cap (2, 10)$

63. $(0, 7) \cap [2, \infty)$ **64.** $(-1, 0] \cup [1, 2)$

In Exercises 65–76 evaluate the expression.

65. $|100|$ **66.** $|-73|$

67. $|2 - 6|$ **68.** $|-8 - (-23)|$

69. $|-\pi|$ **70.** $|\pi - 10|$

71. $|\sqrt{3} - 3|$ **72.** $||-6| - |-4||$

73. $\dfrac{-1}{|-1|}$ **74.** $|-26 - 14|$

75. $\left|2 - |-12|\right|$ **76.** $-1 - \left|1 - |-1|\right|$

In Exercises 77–82 find the distance between the given numbers.

77. 2 and 17   **78.** -3 and 21

79. -14 and 12   **80.** 100 and -150

81. -38 and -57 **82.** -2.6 and -1.8

83. Show that the sum, difference, and product of rational numbers are rational numbers.

84. (a) Is the sum of two irrational numbers always an irrational number?
   (b) Is the product of two irrational numbers always an irrational number?

SECTION 1.2
EXPONENTS AND RADICALS

INTEGER EXPONENTS

A product of identical numbers is usually written in exponential notation. For example, $5 \cdot 5 \cdot 5$ is written as 5^3. In general, if a is any real number and n is a positive integer, then the ***n*th power of *a*** is

$$a^n = \underbrace{a \cdot a \cdot \cdots \cdot a}_{n \text{ factors}}$$

The number a is called the **base** and n is called the **exponent.**

EXAMPLE 1

(a) $\left(\dfrac{1}{2}\right)^5 = \left(\dfrac{1}{2}\right)\left(\dfrac{1}{2}\right)\left(\dfrac{1}{2}\right)\left(\dfrac{1}{2}\right)\left(\dfrac{1}{2}\right) = \dfrac{1}{32}$

(b) $(-3)^4 = (-3) \cdot (-3) \cdot (-3) \cdot (-3) = 81$

(c) $-3^4 = -(3 \cdot 3 \cdot 3 \cdot 3) = -81$ ∎

In Example 1, note the distinction between $(-3)^4$ and -3^4. In part (b), $(-3)^4$ means that the exponent 4 applies to the negative sign as well as to the 3. In part (c), -3^4 means that the exponent applies only to the 3. Similarly, in $(3x)^n$ the n applies to both 3 and x, but in $3x^n = 3(x^n)$ the n applies only to x.

There are several useful rules for working with exponential notation. To discover the rule for multiplication we multiply 5^4 by 5^2:

$$5^4 \cdot 5^2 = \underbrace{(5 \cdot 5 \cdot 5 \cdot 5)}_{4 \text{ factors}}\underbrace{(5 \cdot 5)}_{2 \text{ factors}} = \underbrace{5 \cdot 5 \cdot 5 \cdot 5 \cdot 5 \cdot 5}_{6 \text{ factors}} = 5^6 = 5^{4+2}$$

It appears that *to multiply two powers of the same base we add exponents*. In general, we can confirm this by considering any real number a and any positive integers m and n:

$$a^m a^n = \underbrace{(a \cdot a \cdot \cdots \cdot a)}_{m \text{ factors}}\underbrace{(a \cdot a \cdot \cdots \cdot a)}_{n \text{ factors}} = \underbrace{a \cdot a \cdot a \cdot \cdots \cdot a}_{m + n \text{ factors}} = a^{m+n}$$

Thus we have shown that

$$\boxed{a^m a^n = a^{m+n}}$$

when m and n are positive integers.

We would like this rule to be true even when m and n are 0 or negative integers. Thus, for instance, we must have

$$2^0 \cdot 2^3 = 2^{0+3} = 2^3$$

But this can be true only if $2^0 = 1$. Therefore, for any number $a \neq 0$, we define

$$\boxed{a^0 = 1}$$

Likewise, we want to have

$$5^4 \cdot 5^{-4} = 5^{4+(-4)} = 5^{4-4} = 5^0 = 1$$

and this will be true if $5^{-4} = 1/5^4$. In general, when $a \neq 0$ and n is a positive integer, we define

$$a^{-n} = \frac{1}{a^n}$$

EXAMPLE 2

(a) $\left(\dfrac{4}{7}\right)^0 = 1$ 　　　　　　　　　(b) $x^{-1} = \dfrac{1}{x^1} = \dfrac{1}{x}$

(c) $(-2)^{-3} = \dfrac{1}{(-2)^3} = \dfrac{1}{-8} = -\dfrac{1}{8}$ 　　　　　　　　　■

It is essential to be familiar with the following rules for working with exponents and bases.

LAWS OF EXPONENTS

Let a and b be real numbers and m and n be integers. Then

1. $a^m a^n = a^{m+n}$ 　　　　To multiply two powers of the same number, we add the exponents.

2. $\dfrac{a^m}{a^n} = a^{m-n}$ $(a \neq 0)$ 　　　To divide two powers of the same number, we subtract the exponents.

3. $(a^m)^n = a^{mn}$ 　　　　To raise a power to a new power, we multiply the exponents.

4. $(ab)^n = a^n b^n$ 　　　　To raise a product to a power, we raise each factor to the power.

5. $\left(\dfrac{a}{b}\right)^n = \dfrac{a^n}{b^n}$ $(b \neq 0)$ 　　　To raise a quotient to a power, we raise both numerator and denominator to the power.

We have already seen how to prove Law 1 when m and n are positive integers, but it can be proved for any integer exponents using the definition of negative exponents. We now give the proof of Law 3. The proofs of Laws 2, 4, and 5 are requested in Exercise 128.

Proof of Law 3   If m and n are positive integers, we have

$$(a^m)^n = (\underbrace{a \cdot a \cdot \cdots \cdot a}_{m \text{ factors}})^n$$

$$= \underbrace{(\underbrace{a \cdot a \cdots \cdot a}_{m \text{ factors}})(\underbrace{a \cdot a \cdots \cdot a}_{m \text{ factors}}) \cdots (\underbrace{a \cdot a \cdots \cdot a}_{m \text{ factors}})}_{n \text{ groups of factors}}$$

$$= \underbrace{a \cdot a \cdot \cdots \cdot a}_{mn \text{ factors}} = a^{mn}$$

The cases where $m \leq 0$ or $n \leq 0$ can be proved using the definition of negative exponents. ■

EXAMPLE 3

(a) $x^4 x^7 = x^{4+7} = x^{11}$ (b) $y^4 y^{-7} = y^{4-7} = y^{-3} = \dfrac{1}{y^3}$

(c) $\dfrac{c^9}{c^5} = c^{9-5} = c^4$ (d) $\dfrac{d^2}{d^{10}} = d^{2-10} = d^{-8} = \dfrac{1}{d^8}$

(e) $(xy)^3 = x^3 y^3$ (f) $\left(\dfrac{x}{2}\right)^5 = \dfrac{x^5}{2^5} = \dfrac{x^5}{32}$

(g) $(b^4)^5 = b^{4 \cdot 5} = b^{20}$ (h) $(2^m)^3 = 2^{3m}$

EXAMPLE 4

Write the following numbers as powers of 2:

(a) 4^7 (b) $\dfrac{2^8}{8^2}$

SOLUTION

(a) $4^7 = (2^2)^7 = 2^{14}$

(b) $\dfrac{2^8}{8^2} = \dfrac{2^8}{(2^3)^2} = \dfrac{2^8}{2^6} = 2^{8-6} = 2^2$

EXAMPLE 5

Simplify: (a) $(2a^3 b^2)(3ab^4)^3$ (b) $\left(\dfrac{x}{y}\right)^3 \left(\dfrac{y^2 x}{z}\right)^4$

SOLUTION

(a) $(2a^3 b^2)(3ab^4)^3 = (2a^3 b^2)[3^3 a^3 (b^4)^3] = (2a^3 b^2)(27a^3 b^{12})$
$$= (2)(27)a^3 a^3 b^2 b^{12} = 54a^6 b^{14}$$

(b) $\left(\dfrac{x}{y}\right)^3 \left(\dfrac{y^2 x}{z}\right)^4 = \dfrac{x^3}{y^3} \dfrac{y^8 x^4}{z^4} = (x^3 x^4)\left(\dfrac{y^8}{y^3}\right)\dfrac{1}{z^4} = \dfrac{x^7 y^5}{z^4}$

EXAMPLE 6

Eliminate negative exponents and simplify: (a) $\dfrac{6st^{-4}}{2s^{-2}t^2}$ (b) $\left(\dfrac{y}{3z^2}\right)^{-2}$

SOLUTION

(a) $\dfrac{6st^{-4}}{2s^{-2}t^2} = \dfrac{6}{2} \cdot \dfrac{s}{s^{-2}} \cdot \dfrac{t^{-4}}{t^2} = 3s^{1-(-2)}t^{-4-2} = 3s^3t^{-6} = \dfrac{3s^3}{t^6}$

(b) $\left(\dfrac{y}{3z^2}\right)^{-2} = \dfrac{1}{\left(\dfrac{y}{3z^2}\right)^2} = \dfrac{1}{\dfrac{y^2}{3^2z^4}} = \dfrac{9z^4}{y^2}$

$\dfrac{a^{-m}}{b^{-n}} = \dfrac{b^n}{a^m}$

It is worthwhile to note that in simplifying expressions like the one in Example 6(a) we can move a factor from numerator to denominator (or vice versa) by changing the sign of its exponent. So another way of simplifying that expression is as follows:

$$\frac{6st^{-4}}{2s^{-2}t^2} = \frac{6ss^2}{2t^4t^2} = \frac{3s^3}{t^6}$$

$\left(\dfrac{a}{b}\right)^{-n} = \left(\dfrac{b}{a}\right)^n$

Likewise, when working with an expression like the one in Example 6(b), where a fraction has a negative exponent, a shortcut is to switch the numerator and denominator and make the exponent positive:

$$\left(\frac{y}{3z^2}\right)^{-2} = \left(\frac{3z^2}{y}\right)^2 = \frac{9z^4}{y^2}$$

■ RADICALS

The symbol $\sqrt{}$ means *the positive square root of*. Thus

DEFINITION OF SQUARE ROOT

$$\boxed{\sqrt{a} = b \qquad \text{means} \qquad b^2 = a \quad \text{and} \quad b \geq 0}$$

Since $a = b^2 \geq 0$, the symbol $\sqrt{a}$ makes sense only when $a \geq 0$.
   For instance,

$$\sqrt{9} = 3 \qquad \text{because} \qquad 3^2 = 9 \qquad \text{and} \qquad 3 \geq 0$$

It is true that the number 9 has two square roots, 3 and -3, but the notation $\sqrt{9}$ is reserved for the *positive* square root of 9 (sometimes called the *principal square root* of 9). If we want the negative root, we *must* write $-\sqrt{9}$, which is -3.

Here are two rules for working with square roots:

PROPERTIES OF SQUARE ROOTS

If $a \geq 0$ and $b > 0$, then

$$\sqrt{ab} = \sqrt{a}\sqrt{b} \qquad \sqrt{\frac{a}{b}} = \frac{\sqrt{a}}{\sqrt{b}}$$

There is no similar rule for the square root of a sum. In fact you should avoid making the following common error:

$$\sqrt{a + b} \neq \sqrt{a} + \sqrt{b}$$

For instance, if we take $a = 9$ and $b = 16$ we see the error:

$$\sqrt{9 + 16} \stackrel{?}{=} \sqrt{9} + \sqrt{16}$$
$$\sqrt{25} \stackrel{?}{=} 3 + 4$$
$$5 \stackrel{?}{=} 7 \qquad \text{(Wrong!)}$$

EXAMPLE 7

Simplify the following expressions:

(a) $\sqrt{72}$ (b) $\sqrt{6}\sqrt{24}$ (c) $\dfrac{\sqrt{18}}{\sqrt{2}}$

SOLUTION

(a) $\sqrt{72} = \sqrt{36 \cdot 2} = \sqrt{36} \cdot \sqrt{2} = 6\sqrt{2}$

(b) $\sqrt{6}\sqrt{24} = \sqrt{6 \cdot 24} = \sqrt{144} = 12$

(c) $\dfrac{\sqrt{18}}{\sqrt{2}} = \sqrt{\dfrac{18}{2}} = \sqrt{9} = 3$ ∎

Notice that

$$\sqrt{4^2} = \sqrt{16} = 4 \qquad \text{but} \qquad \sqrt{(-4)^2} = \sqrt{16} = 4 = |-4|$$

Thus the equation $\sqrt{a^2} = a$ is not always true; it is true only when $a \geq 0$. If $a < 0$, then $-a > 0$, so we have $\sqrt{a^2} = -a$. In terms of absolute values, however, we can always write

$$\sqrt{a^2} = |a|$$

EXAMPLE 8

(a) $\sqrt{7^2} = 7$ (b) $\sqrt{(-3)^2} = |-3| = 3$

(c) If x is any real number and y is a positive number, then

$$\sqrt{x^2 y} = \sqrt{x^2}\sqrt{y} = |x|\sqrt{y}$$ ∎

◼ RATIONALIZING THE DENOMINATOR

When a denominator contains a radical, it is often useful to eliminate the radical by multiplying both the numerator and denominator by an appropriate expression. This procedure is called **rationalizing the denominator**. If the denominator is of the form $\sqrt{a}$, we multiply numerator and denominator by $\sqrt{a}$. Since in effect we are multiplying the given quantity by 1, we do not change its value. For instance,

$$\frac{1}{\sqrt{a}} = \frac{1}{\sqrt{a}} \cdot 1 = \frac{1}{\sqrt{a}} \cdot \frac{\sqrt{a}}{\sqrt{a}} = \frac{\sqrt{a}}{\left(\sqrt{a}\right)^2} = \frac{\sqrt{a}}{a}$$

Notice that the denominator in the last fraction contains no radical.

EXAMPLE 9

$$\frac{2}{\sqrt{3}} = \frac{2}{\sqrt{3}} \cdot \frac{\sqrt{3}}{\sqrt{3}} = \frac{2\sqrt{3}}{\left(\sqrt{3}\right)^2} = \frac{2\sqrt{3}}{3}$$ ∎

If the denominator is of the form $a + b\sqrt{c}$, we multiply numerator and denominator by the **conjugate radical** $a - b\sqrt{c}$. This is effective because the product of the denominator and its conjugate radical does not contain a radical:

$$\left(a + b\sqrt{c}\right)\left(a - b\sqrt{c}\right) = a^2 - ab\sqrt{c} + ab\sqrt{c} - b^2\left(\sqrt{c}\right)^2 = a^2 - b^2 c$$

EXAMPLE 10

Rationalize the denominator in the following expressions:

(a) $\dfrac{1}{1 + \sqrt{2}}$ (b) $\dfrac{2}{\sqrt{5} - \sqrt{3}}$

SOLUTION

(a) We multiply both the numerator and the denominator by the conjugate radical of $1 + \sqrt{2}$, which is $1 - \sqrt{2}$:

$(a + b)(a - b) = a^2 - b^2$

$$\frac{1}{1 + \sqrt{2}} = \frac{1}{1 + \sqrt{2}} \cdot \frac{1 - \sqrt{2}}{1 - \sqrt{2}} = \frac{1 - \sqrt{2}}{1^2 - \left(\sqrt{2}\right)^2}$$

$$= \frac{1 - \sqrt{2}}{1 - 2} = \frac{1 - \sqrt{2}}{-1} = \sqrt{2} - 1$$

Notice that we did not change the value of the given fraction because we multiplied by 1.

(b) The conjugate radical of $\sqrt{5} - \sqrt{3}$ is $\sqrt{5} + \sqrt{3}$, so we have:

$$\frac{2}{\sqrt{5} - \sqrt{3}} = \frac{2}{\sqrt{5} - \sqrt{3}} \cdot \frac{\sqrt{5} + \sqrt{3}}{\sqrt{5} + \sqrt{3}} = \frac{2(\sqrt{5} + \sqrt{3})}{5 - 3}$$

$$= \frac{2(\sqrt{5} + \sqrt{3})}{2} = \sqrt{5} + \sqrt{3} \qquad ■$$

nTH ROOTS

If n is any positive integer, then the principal nth roots are defined as follows:

DEFINITION OF nTH ROOT

$$\sqrt[n]{a} = b \qquad \text{means} \qquad b^n = a$$

where, if n is even, we require that $a \geq 0$ and $b \geq 0$.

Thus

$$\sqrt[4]{81} = 3 \qquad \text{because} \qquad 3^4 = 81 \quad \text{and} \quad 3 \geq 0$$

$$\sqrt[3]{-8} = -2 \qquad \text{because} \qquad (-2)^3 = -8$$

But $\sqrt{-8}$, $\sqrt[4]{-8}$, and $\sqrt[6]{-8}$ are not defined. (For instance, $\sqrt{-8}$ is not defined because the square of every real number is nonnegative.) Notice also that odd roots are unique but even roots are not.

The equation $x^5 = 31$ has only one real solution: $\quad x = \sqrt[5]{31}$.

The equation $x^4 = 31$ has two real solutions: $\quad x = \pm\sqrt[4]{31}$.

The following rules are used in working with nth roots. In each case we assume that all the roots exist.

PROPERTIES OF nTH ROOTS

$$\sqrt[n]{ab} = \sqrt[n]{a}\sqrt[n]{b} \qquad\qquad \sqrt[n]{\frac{a}{b}} = \frac{\sqrt[n]{a}}{\sqrt[n]{b}}$$

$$\sqrt[m]{\sqrt[n]{a}} = \sqrt[mn]{a} \qquad\qquad \sqrt[n]{a^n} = \begin{cases} a & \text{if } n \text{ is odd} \\ |a| & \text{if } n \text{ is even} \end{cases}$$

EXAMPLE 11

(a) $\dfrac{\sqrt[5]{64}}{\sqrt[5]{2}} = \sqrt[5]{\dfrac{64}{2}} = \sqrt[5]{32} = 2$

(b) $\sqrt[3]{32} + \sqrt[3]{108} = \sqrt[3]{8 \cdot 4} + \sqrt[3]{27 \cdot 4} = \sqrt[3]{8}\sqrt[3]{4} + \sqrt[3]{27}\sqrt[3]{4}$

$\qquad\qquad\qquad = 2\sqrt[3]{4} + 3\sqrt[3]{4} = 5\sqrt[3]{4}$

(c) If $t \geq 0$, we have $\sqrt[3]{\sqrt[4]{t}} = \sqrt[12]{t}$.   ■

EXAMPLE 12

(a) $\sqrt[4]{81x^8y^4} = \sqrt[4]{81}\sqrt[4]{x^8}\sqrt[4]{y^4} = 3\sqrt[4]{(x^2)^4}\,|y| = 3x^2|y|$

(b) $\sqrt[3]{x^4} = \sqrt[3]{x^3x} = \sqrt[3]{x^3}\sqrt[3]{x} = x\sqrt[3]{x}$ ■

RATIONAL EXPONENTS

To define what is meant by fractional exponents we need to use radicals. This is because to give meaning to the symbol $a^{1/n}$ in a way that is consistent with the Laws of Exponents, we require that

$$(a^{1/n})^n = a^{(1/n)n} = a^1 = a$$

For this reason we define

$$a^{1/n} = \sqrt[n]{a}$$

Notice that if n is even, we require that $a \geq 0$.

EXAMPLE 13

(a) $64^{1/3} = \sqrt[3]{64} = 4$ (b) $4^{1/2} = \sqrt{4} = 2$ ■

Finally we define powers for any rational exponent m/n in lowest terms, where m and n are integers and $n > 0$. If $\sqrt[n]{a}$ exists as a real number (and this will be the case for any number a if n is odd and for $a > 0$ if n is even), we define

DEFINITION OF RATIONAL EXPONENTS

$$a^{m/n} = \left(\sqrt[n]{a}\right)^m \quad \text{or equivalently} \quad a^{m/n} = \sqrt[n]{a^m}$$

With this definition it can be proved that the Laws of Exponents are still true for rational exponents.

EXAMPLE 14

(a) $4^{3/2} = \left(\sqrt{4}\right)^3 = 2^3 = 8$
Alternate solution: $4^{3/2} = \sqrt{4^3} = \sqrt{64} = 8$

(b) $(125)^{-1/3} = \dfrac{1}{(125)^{1/3}} = \dfrac{1}{\sqrt[3]{125}} = \dfrac{1}{5}$

(c) $\dfrac{1}{\sqrt[3]{x^4}} = \dfrac{1}{x^{4/3}} = x^{-4/3}$ ■

EXAMPLE 15

(a) $\left(2\sqrt{x}\right)\left(3\sqrt[3]{x}\right) = (2x^{1/2})(3x^{1/3}) = 6x^{1/2+1/3} = 6x^{5/6}$

(b) $(2a^3b^4)^{3/2} = 2^{3/2}(a^3)^{3/2}(b^4)^{3/2} = \left(\sqrt{2}\right)^3 a^{3(3/2)}b^{4(3/2)} = 2\sqrt{2}\,a^{9/2}b^6$

(c) $\left(\dfrac{2x^{3/4}}{y^{1/3}}\right)^3\left(\dfrac{y^4}{x^{-1/2}}\right) = \dfrac{2^3(x^{3/4})^3}{(y^{1/3})^3}\cdot(y^4x^{1/2}) = \dfrac{8x^{9/4}}{y}\cdot y^4x^{1/2} = 8x^{11/4}y^3$ ■

EXAMPLE 16

$$\sqrt{x\sqrt{x\sqrt{x}}} = [x(xx^{1/2})^{1/2}]^{1/2} = [x(x^{3/2})^{1/2}]^{1/2}$$
$$= (x\cdot x^{3/4})^{1/2} = (x^{7/4})^{1/2} = x^{7/8}$$ ■

EXERCISES 1.2

In Exercises 1–30 evaluate the number.

1. $(-2)^6$ **2.** -2^6 **3.** $2^{-3}5^4$

4. $\left(\dfrac{1}{3}\right)^4$ $\frac{1}{81}$ **5.** $\left(\dfrac{1}{4}\right)^{-2}$ **6.** $(2^4\cdot 2^2)^2$ 4096

7. $\dfrac{10^9}{10^4}$ **8.** $16\cdot\left(\dfrac{4}{9}\right)^0\cdot 2^{-3}$ **9.** $\sqrt[5]{-32}$

10. $\sqrt[4]{256}$ 4 **11.** $\sqrt[6]{\dfrac{1}{64}}$ **12.** $\sqrt[3]{-64}$ -4

13. $\sqrt{0.0004}$ **14.** $\sqrt[3]{0.000001}$ **15.** $\sqrt{2}\sqrt{50}$

16. $\sqrt{7}\sqrt{28}$ 14 **17.** $\dfrac{\sqrt{72}}{\sqrt{2}}$ **18.** $\dfrac{\sqrt{48}}{\sqrt{3}}$ $=4$

19. $\sqrt[3]{3}\sqrt[3]{9}$ **20.** $\sqrt[4]{24}\sqrt[4]{54}$ $=6$ **21.** $9^{7/2}$

22. $(-32)^{2/5}$ 4 **23.** $(-125)^{-1/3}$ **24.** $81^{3/4}$ 27

25. $128^{8/7}$ **26.** $64^{-4/3}$ $\frac{1}{256}$ **27.** $\left(\dfrac{25}{64}\right)^{3/2}$

28. $\left(-\dfrac{27}{8}\right)^{2/3}$ $\frac{9}{4}$ **29.** $3^{1/2}9^{1/4}$ **30.** $1024^{-0.1}$ $\frac{1}{2}$

In Exercises 31–36 state whether the equation is true or false.

31. $2^3\cdot 2^5 = 2^{15}$ **32.** $2^3\cdot 2^5 = 4^8$

33. $2^3 + 2^5 = 2^{15}$ **34.** $(2^3)^5 = 2^{15}$

35. $(2\cdot 3)^5 = 2^5\cdot 3^5$ **36.** $2^3\cdot 5^2 = 10^5$

In Exercises 37–44 write the number as a power of 2.

37. 128 **38.** $2^6\cdot 8^4$ 2^{18}

39. $(2^9)^4$ **40.** $\dfrac{1}{4}$ 2^{-2}

41. $\dfrac{2^{12}}{2^{18}}$ **42.** 1 2^0

43. $2^8\cdot 4^{-7}$ **44.** $4\sqrt{2}$ $2^{\frac{5}{2}}$

In Exercises 45–62 simplify the expression and eliminate any negative exponents.

45. t^7t^{-2} **46.** $(4x^2)(6x^7)$ $24x^9$

47. $(12x^2y^4)\left(\tfrac{1}{2}x^5y\right)$ **48.** $(6y)^3$ $216y^3$

49. $\dfrac{x^9(2x)^4}{x^3}$

50. $\dfrac{a^{-3}b^4}{a^{-5}b^5}$

51. $b^4\left(\tfrac{1}{3}b^2\right)(12b^{-8})$

52. $(2s^3t^{-1})\left(\tfrac{1}{4}s^6\right)(16t^4)$

53. $(rs)^3(2s)^{-2}(4r)^4$

54. $(2u^2v^3)^3(3u^3v)^{-2}$

55. $\dfrac{(6y^3)^4}{2y^5}$

56. $\dfrac{(2x^3)^2(3x^4)}{(x^3)^4}$

57. $\dfrac{(x^2y^3)^4(xy^4)^{-3}}{x^2y}$

58. $\left(\dfrac{c^4d^3}{cd^2}\right)\left(\dfrac{d^2}{c^3}\right)^3$

59. $\dfrac{(xy^2z^3)^4}{(x^3y^2z)^3}$

60. $\left(\dfrac{xy^{-2}z^{-3}}{x^2y^3z^{-4}}\right)^{-3}$

61. $\left(\dfrac{q^{-1}rs^{-2}}{r^{-5}sq^{-8}}\right)^{-1}$

62. $(3ab^2c)\left(\dfrac{2a^2b}{c^3}\right)^{-2}$

In Exercises 63–78 simplify the expression and eliminate any negative exponents. Assume that all letters denote positive numbers.

63. $x^{2/3}x^{1/5}$

64. $(-2a^{3/4})(5a^{3/2})$

65. $(4b)^{1/2}(8b^{2/5})$

66. $(8x^6)^{-2/3}$

67. $(c^2d^3)^{-1/3}$

68. $(4x^6y^8)^{3/2}$

69. $(y^{3/4})^{2/3}$

70. $(a^{2/5})^{-3/4}$

71. $(2x^4y^{-4/5})^3(8y^2)^{2/3}$

72. $(x^{-5}y^3z^{10})^{-3/5}$

73. $\left(\dfrac{x^6y}{y^4}\right)^{5/2}$

74. $\left(\dfrac{-2x^{1/3}}{y^{1/2}z^{1/6}}\right)^4$

75. $\left(\dfrac{3a^{-2}}{4b^{-1/3}}\right)^{-1}$

76. $\dfrac{(y^{10}z^{-5})^{1/5}}{(y^{-2}z^3)^{1/3}}$

77. $\dfrac{(9st)^{3/2}}{(27s^3t^{-4})^{2/3}}$

78. $\left(\dfrac{a^2b^{-3}}{x^{-1}y^2}\right)^3\left(\dfrac{x^{-2}b^{-1}}{a^{3/2}y^{1/3}}\right)$

In Exercises 79–92 simplify the expression. Here the letters denote any real numbers.

79. $\sqrt{75}$

80. $\sqrt{8}+\sqrt{50}$

81. $\sqrt{245}-\sqrt{125}$

82. $\sqrt[3]{54}-\sqrt[3]{16}$

83. $\sqrt[4]{x^4}$

84. $\sqrt[3]{x^3y^6}$

85. $\sqrt[3]{x^3y}$

86. $\sqrt[3]{x^4y^3}$

87. $\sqrt[5]{a^6b^7}$

88. $\sqrt[3]{a^2b}\,\sqrt[3]{a^4b}$

89. $\sqrt{x^2y^6}$

90. $\sqrt[4]{x^4y^2z}$

91. $\sqrt[3]{\sqrt{64x}}$

92. $\sqrt[4]{r^{2n+1}}\,\sqrt[4]{r^{-1}}$

In Exercises 93–102 write the given expression as a power of x.

93. $x^ax^bx^c$

94. $((x^a)^b)^c$

95. $\sqrt[3]{x^5}$

96. $\dfrac{1}{\sqrt[7]{x^3}}$

97. $x^2\sqrt{x}$

98. $x\sqrt[3]{x}$

99. $\dfrac{(x^2)^nx^5}{x^n}$

100. $\dfrac{(x^2)^m(x^3)^n}{x^{m+n}x^{m-n}}$

101. $\sqrt{x\sqrt{x}}$

102. $\sqrt{x\sqrt{x\sqrt{x\sqrt{x}}}}$

In Exercises 103–112 rationalize the denominator.

103. $\dfrac{1}{\sqrt{6}}$

104. $\sqrt{\dfrac{2}{3}}$

105. $\sqrt{\dfrac{3}{20}}$

106. $\sqrt{\dfrac{x^5}{2}}$

107. $\sqrt{\dfrac{x}{3y}}$

108. $\sqrt{\dfrac{1}{2x^3y^5}}$

109. $\dfrac{1}{3-\sqrt{5}}$

110. $\dfrac{3}{\sqrt{2}+\sqrt{5}}$

111. $\dfrac{2}{\sqrt{a}+1}$

112. $\dfrac{1}{\sqrt{x}-\sqrt{y}}$

In Exercises 113–116 rewrite the expression by rationalizing the numerator; that is, multiply numerator and denominator by the conjugate radical of the numerator and simplify.

113. $\dfrac{1+\sqrt{2}}{2}$

114. $\dfrac{1-\sqrt{5}}{3}$

115. $7-\sqrt{3}$

116. $\sqrt{x}+\sqrt{2}$

In Exercises 117–122 state whether the given equation is true for all nonzero values of x.

117. $\sqrt[6]{x^6}=x$

118. $\sqrt[3]{x^3}=x$

119. $x^3x^{-1/3}=x^{8/3}$

120. $(x^3)^4=x^7$

121. $2^x\cdot2^{3x}=2^{3x^2}$

122. $\sqrt{x^2+4}=|x|+2$

123. It follows from Kepler's Third Law of planetary motion that the average distance from a planet to the sun, in meters, is

$$d=\left(\dfrac{GM}{4\pi^2}\right)^{1/3}T^{2/3}$$

where $M=1.99\times10^{30}$ kg is the mass of the sun,

$G = 6.67 \times 10^{-11}$ Nm^2/kg^2 is the gravitational constant, and T is the period of the orbit (in seconds). Use the fact that the period of the earth's orbit is about 365.25 days to find the distance from the earth to the sun.

124. A certain type of glass allows only 85% of the light to pass through a pane of glass that is 1 cm thick.
  (a) What percentage of the light will pass through a pane that is 2 cm thick?
  (b) What percentage of the light will pass through a pane that is t centimeters thick?
  (c) What percentage of the light will pass through a pane that is 0.75 cm thick?

125. Without using a calculator, determine which of the numbers $7^{1/4}$ and $4^{1/3}$ is larger.

126. Without using a calculator, determine which of the numbers $\sqrt[3]{5}$ and $\sqrt{3}$ is larger.

127. In which of the following pairs are the numbers closer together?
  (a) 10^{10} and 10^{53}
  (b) 10^{100} and 10^{101}

128. (a) Prove Law 2 of exponents for the case where m and n are positive integers and $m > n$. (The proof is similar to the proof of Law 1.)
  (b) Prove Law 4 of exponents for the case where m and n are positive integers.
  (c) Prove Law 5 of exponents for the case where m and n are positive integers.

SECTION 1.3
ALGEBRAIC EXPRESSIONS

Algebraic expressions such as

$$2x^2 - 3x + 4 \qquad ax + b \qquad \frac{y - 1}{y^2 + 2} \qquad \frac{cx^2y + dy^2z}{\sqrt{x^2 + y^2 + z^2}} \qquad (1)$$

are obtained by starting with variables such as x, y, and z and constants such as 2, -3, a, b, c, and d, and combining them using addition, subtraction, multiplication, division, and roots. A **variable** is a letter that can represent any number in a given set of numbers whereas a **constant** represents a fixed number. Usually, we use letters near the end of the alphabet (like t, u, x, y, z) for variables, and letters near the beginning of the alphabet (like a, b, c) for constants. The **domain** of a variable is the set of numbers that the variable is permitted to represent. For instance, in the expression $\sqrt{x}$ the domain of x is $\{x \mid x \geq 0\}$ whereas in the expression $2/(x - 3)$ the domain of x is $\{x \mid x \neq 3\}$.

The simplest types of algebraic expressions use only addition, subtraction, and multiplication, and are called **polynomials**. The first two expressions in (1) are polynomials of degree 2 and 1, respectively. The general form of a polynomial of degree n in the variable x is

The degree is the highest power in the expression.

$$a_n x^n + a_{n-1} x^{n-1} + \cdots + a_1 x + a_0$$

where $a_0, a_1, \ldots, a_n$ are constants and $a_n \neq 0$. Notice that any polynomial is a sum of **terms** of the form ax^k (called **monomials**) where a is a constant and k is a nonnegative integer.

For instance,

$$2x^3 - x^2 + 4x - \sqrt{3}$$

is a polynomial of degree 3 in x and

$$y^5 + 8y^2 - 17$$

is a polynomial of degree 5 in y.

OPERATIONS ON ALGEBRAIC EXPRESSIONS

We **add** and **subtract** polynomials using the properties of real numbers that were discussed in Section 1.1. The idea is to combine **like terms** (that is, terms with the same variable raised to the same powers) using the Distributive Law. For instance,

$ac + bc = (a + b)c$

$$5x^7 + 3x^7 = (5 + 3)x^7 = 8x^7$$

EXAMPLE 1

Find the sum $(x^3 - 6x^2 + 2x + 4) + (3x^3 + 5x^2 - 4x)$.

SOLUTION

$$(x^3 - 6x^2 + 2x + 4) + (3x^3 + 5x^2 - 4x)$$

Group like terms
$$= (x^3 + 3x^3) + (-6x^2 + 5x^2) + (2x - 4x) + 4$$

Combine like terms
$$= 4x^3 - x^2 - 2x + 4$$ ∎

In subtracting polynomials we must remember that if a minus sign precedes parentheses, then the sign of every term inside the parentheses is changed:

$$-(b + c) = -b - c$$

[This is simply the Distributive Law, $a(b + c) = ab + ac$, with $a = -1$.]

EXAMPLE 2

Find the difference $(x^3 - 6x^2 + 2x + 4) - (3x^3 + 5x^2 - 4x)$.

SOLUTION

$$(x^3 - 6x^2 + 2x + 4) - (3x^3 + 5x^2 - 4x)$$

$$= x^3 - 6x^2 + 2x + 4 - 3x^3 - 5x^2 + 4x$$

Group like terms
$$= (x^3 - 3x^3) + (-6x^2 - 5x^2) + (2x + 4x) + 4$$

Combine like terms
$$= -2x^3 - 11x^2 + 6x + 4$$ ∎

To find the **product** of polynomials or other algebraic expressions, we need to use the Distributive Law repeatedly. In particular, using it three times, we get

$$(a + b)(c + d) = (a + b)c + (a + b)d = ac + bc + ad + bd$$

This says that we multiply two factors by multiplying each term in one factor by each term in the other factor and adding the products. Schematically, we have

$$(a + b)(c + d)$$

EXAMPLE 3

(a) $(2x + 1)(3x - 5) = 6x^2 + 3x - 10x - 5 = 6x^2 - 7x - 5$

(b) $3(x - 1)(4x + 3) - 2(x + 6) = 3(4x^2 - x - 3) - 2x - 12$
$$= 12x^2 - 3x - 9 - 2x - 12$$
$$= 12x^2 - 5x - 21 \qquad \blacksquare$$

In general, we can multiply any two polynomials by using the Distributive Law and the Laws of Exponents.

EXAMPLE 4

Find the product $(x^2 - 3)(x^3 + 2x + 1)$.

SOLUTION

We start by regarding the second factor as a single number:

$$(x^2 - 3)(x^3 + 2x + 1) = x^2(x^3 + 2x + 1) - 3(x^3 + 2x + 1)$$
$$= x^5 + 2x^3 + x^2 - 3x^3 - 6x - 3$$
$$= x^5 - x^3 + x^2 - 6x - 3 \qquad \blacksquare$$

The next example shows that the methods we have used for multiplying polynomials also apply to other algebraic expressions.

EXAMPLE 5

(a) $\sqrt{x}(x^2 + 2x + \sqrt{x}) = x^2\sqrt{x} + 2x\sqrt{x} + \sqrt{x}\sqrt{x}$
$$= x^{5/2} + 2x^{3/2} + x$$

(b) $(1 + \sqrt{x})(2 - 3\sqrt{x}) = 2 - 3\sqrt{x} + 2\sqrt{x} - 3(\sqrt{x})^2$
$$= 2 - \sqrt{x} - 3x \qquad \blacksquare$$

Certain types of products occur so frequently that you should memorize them. You can verify the following formulas by performing the multiplications.

SPECIAL PRODUCT
FORMULAS

> **1.** $(a - b)(a + b) = a^2 - b^2$
> **2.** $(a + b)^2 = a^2 + 2ab + b^2$
> **3.** $(a - b)^2 = a^2 - 2ab + b^2$
> **4.** $(a + b)^3 = a^3 + 3a^2b + 3ab^2 + b^3$
> **5.** $(a - b)^3 = a^3 - 3a^2b + 3ab^2 - b^3$

EXAMPLE 6

Use the special product formulas to find the following products:

(a) $(2x + 5)^2$ (b) $\left(2\sqrt{y} - \dfrac{1}{\sqrt{x}}\right)\left(2\sqrt{y} + \dfrac{1}{\sqrt{x}}\right)$ (c) $(x^2 - 2)^3$

SOLUTION

(a) Product Formula 2, with $a = 2x$ and $b = 5$, gives

$$(2x + 5)^2 = (2x)^2 + 2(2x)(5) + 5^2 = 4x^2 + 20x + 25$$

(b) Using Product Formula 1 with $a = 2\sqrt{y}$ and $b = 1/\sqrt{x}$, we have

$$\left(2\sqrt{y} - \frac{1}{\sqrt{x}}\right)\left(2\sqrt{y} + \frac{1}{\sqrt{x}}\right) = \left(2\sqrt{y}\right)^2 - \left(\frac{1}{\sqrt{x}}\right)^2 = 4y - \frac{1}{x}$$

(c) Putting $a = x^2$ and $b = 2$ in Product Formula 5, we get

$$(x^2 - 2)^3 = (x^2)^3 - 3(x^2)^2(2) + 3x^2(2)^2 - 2^3$$
$$= x^6 - 6x^4 + 12x^2 - 8$$

■

■ FACTORING

We have used the Distributive Law to expand algebraic expressions. We sometimes need to reverse this process (again using the Distributive Law) by factoring an expression as a product of simpler ones. The easiest case is when there is a *common factor* as follows:

$$\overset{\longleftarrow \text{ expanding } \longrightarrow}{3x^2 - 6x = 3x(x - 2)}$$
$$\underset{\longleftarrow \text{ factoring } \longrightarrow}{}$$

Factoring quadratics

To factor a second-degree polynomial, or **quadratic**, of the form $x^2 + bx + c$, we note that

$$(x + r)(x + s) = x^2 + (r + s)x + rs$$

so we need to choose numbers r and s so that $r + s = b$ and $rs = c$.

Factoring Numbers

To factor a whole number completely means to write it as a product of smaller whole numbers that cannot themselves be factored, that is, as a product of primes. For example, $420 = 2 \cdot 2 \cdot 3 \cdot 5 \cdot 7$. To factor a very large number can be a difficult task. High-speed computers employing the fastest known methods would take about a day to factor an arbitrary 30-digit number and about a million years to factor a 40-digit number. Ted Rivest, Adi Shamir, and Leonard Adleman used this fact in the 1970s to devise the RSA code for sending secret messages. This code uses an extremely large number to encode a message, but requires knowledge of the factors to decode it. Since multiplying numbers is easy but factoring the result is hard, this code is very difficult to break. It was at first thought that a carefully selected 80-digit number would provide an unbreakable code, but recent advances in the study of factoring have made numbers as large as 200 digits necessary to assure complete security.

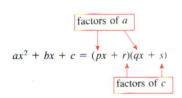

EXAMPLE 7

Factor $x^2 + 7x + 12$.

SOLUTION

In this case $rs = 12$, so r and s must be factors of 12 and their sum must be 7. We enumerate the possibilities as follows:

Trial factors:	$1 \cdot 12$	$2 \cdot 6$	$3 \cdot 4$
Corresponding sum:	13	8	7

Therefore, taking $r = 3$ and $s = 4$, we have the desired factorization:

$$x^2 + 7x + 12 = (x + 3)(x + 4)$$ ∎

EXAMPLE 8

Factor $2x^2 - 7x - 4$.

SOLUTION

Even though the coefficient of x^2 is not 1, we can still look for factors of the form $2x + r$ and $x + s$, where $rs = -4$. After trial and error we find that

$$2x^2 - 7x - 4 = (2x + 1)(x - 4)$$ ∎

In general, to factor a quadratic of the form $ax^2 + bx + c$, where $a \neq 1$, we look for factors of the form $px + r$ and $qx + s$:

$$ax^2 + bx + c = (px + r)(qx + s) = pqx^2 + (ps + qr)x + rs$$

Therefore we try to find numbers p, q, r, and s such that

$$pq = a \qquad rs = c \qquad ps + qr = b$$

If these numbers are all integers, there will be a limited number of possibilities to try.

Some special quadratics can be factored using the following formulas. They are just the first three product formulas, but written backward. The first one is referred to as the formula for a **difference of squares**.

FACTORING FORMULAS

1. $a^2 - b^2 = (a - b)(a + b)$ difference of squares

2. $a^2 + 2ab + b^2 = (a + b)^2$

3. $a^2 - 2ab + b^2 = (a - b)^2$

EXAMPLE 9

Factor: (a) $4x^2 - 25$ (b) $x^2 + 6x + 9$ (c) $x^2 - xy + \frac{1}{4}y^2$

SOLUTION

(a) Using the formula for a difference of squares with $a = 2x$ and $b = 5$, we have

$$4x^2 - 25 = (2x)^2 - 5^2 = (2x - 5)(2x + 5)$$

(b) Using Formula 2 with $a = x$ and $b = 3$, we get

$$x^2 + 6x + 9 = x^2 + 2(3x) + 3^2 = (x + 3)^2$$

(c) Here we take $a = x$ and $b = \frac{1}{2}y$ in Formula 3:

$$x^2 - xy + \tfrac{1}{4}y^2 = x^2 - 2x\left(\tfrac{1}{2}y\right) + \left(\tfrac{1}{2}y\right)^2$$
$$= \left(x - \tfrac{1}{2}y\right)^2 \quad\blacksquare$$

Factoring other algebraic expressions

In general it is not easy to factor a polynomial of degree higher than two. We will give a more systematic approach in Chapter 3, but some of these polynomials can be factored by means of the following formulas for the sum or difference of cubes. The formulas can be verified by multiplying out the right sides.

FACTORING FORMULAS

4. $a^3 - b^3 = (a - b)(a^2 + ab + b^2)$ difference of cubes
5. $a^3 + b^3 = (a + b)(a^2 - ab + b^2)$ sum of cubes

EXAMPLE 10

Factor: (a) $27x^3 - 1$ (b) $x^6 + 8$

SOLUTION

(a) Using the formula for a difference of cubes with $a = 3x$ and $b = 1$, we get

$$27x^3 - 1 = (3x)^3 - 1^3 = (3x - 1)[(3x)^2 + (3x)(1) + 1^2]$$
$$= (3x - 1)(9x^2 + 3x + 1)$$

(b) Using the formula for a sum of cubes with $a = x^2$ and $b = 2$, we have

$$x^6 + 8 = (x^2)^3 + 2^3 = (x^2 + 2)(x^4 - 2x^2 + 4) \quad\blacksquare$$

As the following example shows, some higher-degree polynomials can be treated by first removing a common factor.

EXAMPLE 11

Factor $2x^4 - 8x^2$.

SOLUTION

$$2x^4 - 8x^2 = 2x^2(x^2 - 4) \qquad \text{(common factor)}$$
$$= 2x^2(x - 2)(x + 2) \qquad \text{(difference of squares)} \quad \blacksquare$$

Notice that when factoring out a common factor, we look for the term with the smallest exponent. When factoring expressions that involve fractional powers, we also need to use the Laws of Exponents.

EXAMPLE 12

Factor $3x^{3/2} - 9x^{1/2} + 6x^{-1/2}$.

SOLUTION

We factor out the power of x with the *smallest exponent*:

$$3x^{3/2} - 9x^{1/2} + 6x^{-1/2} = 3x^{-1/2}(x^2 - 3x + 2)$$
$$= 3x^{-1/2}(x - 1)(x - 2) \quad \blacksquare$$

Grouping

Polynomials with at least four terms can sometimes be factored by grouping terms. The following example illustrates the technique.

EXAMPLE 13

Factor $x^3 + x^2 + 4x + 4$.

SOLUTION

$$x^3 + x^2 + 4x + 4 = (x^3 + x^2) + (4x + 4) \qquad \text{(grouping)}$$
$$= x^2(x + 1) + 4(x + 1) \qquad \text{(common factors)}$$
$$= (x^2 + 4)(x + 1) \quad \blacksquare$$

EXERCISES 1.3

In Exercises 1–46 perform the indicated operations and simplify.

1. $3(2 - x) - 5(x + 2)$

2. $4(2x + 3) - 3(3x - 1)$

3. $(x^2 + 2x + 3) + (2x^2 - 3x + 4)$

4. $(x^2 + 2x + 3) - (2x^2 - 3x + 4)$

5. $(x^3 + 6x^2 - 4x + 7) - (3x^2 + 2x - 4)$

6. $4(x^2 - x + 2) - 5(x^2 - 2x + 1)$

7. $2(2 - 5t) + t^2(t - 1) - (t^4 - 1)$

8. $5(3t - 4) - (t^2 + 2) - 2t(t - 3)$

9. $\sqrt{x}(x - \sqrt{x})$

10. $x^{3/2}(\sqrt{x} - 1/\sqrt{x})$

11. $\sqrt[3]{y}(y^2 - 1)$

12. $\sqrt{x}(1 + \sqrt{x} - 2x)$

13. $(x + 6)(2x - 3)$

14. $(4x - 1)(3x + 7)$

15. $(3t - 2)(7t - 5)$

16. $(t + 6)(t + 5) - 3(t + 4)$

17. $(x + 2y)(3x - y)$

18. $(4x - 3y)(2x + 5y)$

19. $(1 - 2y)^2$

20. $(3u - v)(3u + v)$

21. $(\sqrt{x} + \sqrt{y})(\sqrt{x} - \sqrt{y})$

22. $(3x + 4)^2$

23. $(2x - 5)(x^2 - x + 1)$

24. $(x^2 + 3)(5x - 6)$

25. $x(x - 1)(x + 2)$

26. $(1 + 2x)(x^2 - 3x + 1)$

27. $(x^2 + 1)(x^2 - x + 1)$

28. $(2u^2 + u - 3)(u^3 + 1)$

29. $y^4(6 - y)(5 + y)$

30. $(t - 5)^2 - 2(t + 3)(8t - 1)$

31. $(2x^2 + 3y^2)^2$

32. $(\sqrt{r} - 2\sqrt{s})^2$

33. $(x^2 - a^2)(x^2 + a^2)$

34. $(3x - 4)^3$

35. $(1 + a^3)^3$

36. $(x - 1)(x^2 + x + 1)$

37. $\left(\sqrt{a} - \dfrac{1}{b}\right)\left(\sqrt{a} + \dfrac{1}{b}\right)$

38. $\left(c + \dfrac{1}{c}\right)^2$

39. $(x^2 + x - 2)(x^3 - x + 1)$

40. $(1 + x + x^2)(1 - x + x^2)$

41. $(1 + x^{4/3})(1 - x^{2/3})$

42. $(x^{3/2} - x + 1)(x^2 + x^{1/2} - 2)$

43. $(1 - b)^2(1 + b)^2$

44. $(1 + x - x^2)^2$

45. $(3x^2y + 7xy^2)(x^2y^3 - 2y^2)$

46. $(x^4y - y^5)(x^2 + xy + y^2)$

In Exercises 47–94 factor the expression.

47. $2x + 12x^3$

48. $8x^5 + 4x^3$

49. $6y^4 - 15y^3$

50. $5ab - 8abc$

51. $x^2 + 7x + 6$

52. $x^2 - x - 6$

53. $x^2 - 2x - 8$

54. $x^2 - 14x + 48$

55. $y^2 - 8y + 15$

56. $z^2 + 6z - 16$

57. $2x^2 + 5x + 3$

58. $2x^2 + 7x - 4$

59. $9x^2 - 36$

60. $8x^2 + 10x + 3$

61. $6x^2 - 5x - 6$

62. $x^2 + 10x + 25$

63. $8x^2 + 14x - 15$

64. $6 + 5t - 6t^2$

65. $t^3 + 1$

66. $4t^2 - 9s^2$

67. $4t^2 - 12t + 9$

68. $x^3 - 27$

69. $x^3 + 2x^2 + x$

70. $3x^3 - 27x$

71. $4x^2 + 4xy + y^2$

72. $4r^2 - 12rs + 9s^2$

73. $x^4 + 2x^3 - 3x^2$

74. $x^6 + 64$

75. $8x^3 - 125$

76. $x^4 + 2x^2 + 1$

77. $x^4 + x^2 - 2$

78. $x^3 + 3x^2 - x - 3$

79. $y^3 - 3y^2 - 4y + 12$

80. $y^3 - y^2 + y - 1$

81. $2x^3 + 4x^2 + x + 2$

82. $3x^3 + 5x^2 - 6x - 10$

83. $x^6 - y^6$

84. $x^8 - 1$

85. $s^3 + t^9$

86. $t^6 - 5t^4 + 4t^2$

87. $x^6 - x^4 + x^2 - 1$

88. $x^5 + 2x^4 + x^2 + 2x$

89. $x^{5/2} - x^{1/2}$

90. $3x^{-1/2} + 4x^{1/2} + x^{3/2}$

91. $x^{-3/2} + 2x^{-1/2} + x^{1/2}$

92. $(x - 1)^{7/2} - (x - 1)^{3/2}$

93. $(x^2 + 1)^{1/2} + 2(x^2 + 1)^{-1/2}$

94. $x^{1/2}(y - 1)^{-1/2} + (y - 1)^{1/2} + 3x^{-1/2}(y - 1)^{3/2}$

95. Verify Product Formulas 1, 2, and 4.

96. Verify Product Formulas 3 and 5.

97. Verify the formula for a difference of cubes.

98. Verify the formula for a sum of cubes.

99. Factor $x^4 + 3x^2 + 4$. [*Hint:* Write the expression as $(x^4 + 4x^2 + 4) - x^2$ and note that this is a difference of squares.]

100. Factor $y^4 + 1$.

SECTION 1.4
FRACTIONAL EXPRESSIONS

A quotient of two algebraic expressions is called a **fractional expression**. We assume that all fractions are defined; that is, *we deal only with values of the variables such that the denominators are not zero*.

A common type of fractional expression occurs when both the numerator and denominator are polynomials. This is called a **rational expression**. For instance,

$$\frac{4x^3 + 2x + 5}{x + 3}$$

is a rational expression whose denominator is 0 when $x = -3$. So in dealing with this expression we implicitly assume that $x \neq -3$.

In simplifying rational expressions, we factor both numerator and denominator and cancel any common factors. In doing so, we are using the following property of real numbers:

SIMPLIFYING FRACTIONS

$$\frac{ac}{bc} = \frac{a}{b} \qquad (b \neq 0, \; c \neq 0)$$

This says that we can divide both numerator and denominator by the number c.

EXAMPLE 1

Simplify: (a) $\dfrac{x^2 - 1}{x^2 + x - 2}$ (b) $\dfrac{2x^3 + 5x^2 - 3x}{6 - x - x^2}$

SOLUTION

We cannot cancel the x^2's in $\dfrac{x^2 - 1}{x^2 + x - 2}$ because they are not factors.

(a) $\dfrac{x^2 - 1}{x^2 + x - 2} = \dfrac{(x - 1)(x + 1)}{(x - 1)(x + 2)}$ (factor)

$= \dfrac{x + 1}{x + 2}$ (cancel)

(b) $\dfrac{2x^3 + 5x^2 - 3x}{6 - x - x^2} = \dfrac{x(2x^2 + 5x - 3)}{-(x^2 + x - 6)} = \dfrac{x(2x - 1)(x + 3)}{-(x - 2)(x + 3)}$ (factor)

$= -\dfrac{x(2x - 1)}{x - 2}$ (cancel) ■

When multiplying fractional expressions we use the following property of real numbers. It says that to multiply two fractions we multiply the numerators and multiply the denominators.

MULTIPLYING FRACTIONS

$$\frac{a}{b} \cdot \frac{c}{d} = \frac{ac}{bd}$$

EXAMPLE 2

Perform the indicated multiplication and simplify:

$$\frac{x^2 + 2x - 3}{x^2 + 8x + 16} \cdot \frac{3x + 12}{x - 1}$$

SOLUTION

We first factor:

$$\frac{x^2 + 2x - 3}{x^2 + 8x + 16} \cdot \frac{3x + 12}{x - 1} = \frac{(x - 1)(x + 3)}{(x + 4)^2} \cdot \frac{3(x + 4)}{x - 1}$$

$$= \frac{3(x - 1)(x + 3)(x + 4)}{(x - 1)(x + 4)^2}$$

$$= \frac{3(x + 3)}{x + 4}$$

When dividing fractional expressions we use the following property of real numbers. It says that to divide two fractions we invert the denominator and multiply.

DIVIDING FRACTIONS

$$\frac{\dfrac{a}{b}}{\dfrac{c}{d}} = \frac{a}{b} \cdot \frac{d}{c} = \frac{ad}{bc}$$

EXAMPLE 3

Perform the indicated division and simplify:

$$\frac{x - 4}{x^2 - 4} \div \frac{x^2 - 3x - 4}{x^2 + 5x + 6}$$

SOLUTION

$$\frac{x - 4}{x^2 - 4} \div \frac{x^2 - 3x - 4}{x^2 + 5x + 6} = \frac{x - 4}{x^2 - 4} \cdot \frac{x^2 + 5x + 6}{x^2 - 3x - 4}$$

$$= \frac{(x - 4)(x + 2)(x + 3)}{(x - 2)(x + 2)(x - 4)(x + 1)}$$

$$= \frac{x + 3}{(x - 2)(x + 1)}$$

In adding and subtracting rational expressions, we first find a common denominator and then use the following property of real numbers.

ADDING FRACTIONS

$$\frac{a}{b} + \frac{c}{b} = \frac{a + c}{b}$$

(2)

Although any common denominator will work, it is best to use the **least common denominator** (LCD). This is found by factoring each denominator and taking the product of the distinct factors, using the largest exponents that appear in any of the factors.

EXAMPLE 4

Combine and simplify: $\dfrac{3}{x - 1} + \dfrac{x}{x + 2}$

SOLUTION

Here the LCD is simply the product $(x - 1)(x + 2)$, so we have

$$\frac{3}{x - 1} + \frac{x}{x + 2} = \frac{3(x + 2)}{(x - 1)(x + 2)} + \frac{x(x - 1)}{(x - 1)(x + 2)}$$

$$= \frac{3x + 6 + x^2 - x}{(x - 1)(x + 2)}$$

$$= \frac{x^2 + 2x + 6}{(x - 1)(x + 2)}$$ ■

EXAMPLE 5

Combine and simplify: $\dfrac{1}{x^2 - 1} - \dfrac{2}{(x + 1)^2}$

SOLUTION

The LCD of $x^2 - 1 = (x - 1)(x + 1)$ and $(x + 1)^2$ is $(x - 1)(x + 1)^2$, so we have

$$\frac{1}{x^2 - 1} - \frac{2}{(x + 1)^2} = \frac{1}{(x - 1)(x + 1)} - \frac{2}{(x + 1)^2}$$

$$= \frac{(x + 1) - 2(x - 1)}{(x - 1)(x + 1)^2}$$

$$= \frac{x + 1 - 2x + 2}{(x - 1)(x + 1)^2}$$

$$= \frac{3 - x}{(x - 1)(x + 1)^2}$$ ■

EXAMPLE 6

Combine and simplify: $\dfrac{1}{x^2 + 4x - 5} - \dfrac{1}{2x} + \dfrac{x + 1}{x^2 - x}$

SOLUTION

We start by factoring the denominators:

$$\frac{1}{x^2 + 4x - 5} - \frac{1}{2x} + \frac{x + 1}{x^2 - x} = \frac{1}{(x - 1)(x + 5)} - \frac{1}{2x} + \frac{x + 1}{x(x - 1)}$$

The LCD of the denominators is $2x(x - 1)(x + 5)$ and so

$$\frac{1}{(x - 1)(x + 5)} - \frac{1}{2x} + \frac{x + 1}{x(x - 1)} = \frac{2x - (x - 1)(x + 5) + (x + 1)(2)(x + 5)}{2x(x - 1)(x + 5)}$$

$$= \frac{2x - (x^2 + 4x - 5) + (2x^2 + 12x + 10)}{2x(x - 1)(x + 5)}$$

$$= \frac{x^2 + 10x + 15}{2x(x - 1)(x + 5)} \qquad \blacksquare$$

In the next example we simplify a **compound fraction** in which there are fractions in both the numerator and the denominator.

EXAMPLE 7

Simplify:

$$\frac{\dfrac{x}{y} + 1}{1 - \dfrac{y}{x}}$$

SOLUTION

We combine the terms in the numerator into a single fraction. We do the same in the denominator. Then we invert and multiply.

$$\frac{\dfrac{x}{y} + 1}{1 - \dfrac{y}{x}} = \frac{\dfrac{x + y}{y}}{\dfrac{x - y}{x}} = \frac{x + y}{y} \cdot \frac{x}{x - y}$$

$$= \frac{x(x + y)}{y(x - y)} \qquad \blacksquare$$

The remaining examples show situations in calculus where facility with fractional expressions is required.

EXAMPLE 8

Simplify:

$$\frac{\dfrac{1}{(a + h)^2} - \dfrac{1}{a^2}}{h}$$

As in Example 7, we begin by taking the numerator to a common denominator:

To divide by h, multiply by its reciprocal.

$$\frac{\dfrac{1}{(a+h)^2} - \dfrac{1}{a^2}}{h} = \frac{\dfrac{a^2 - (a+h)^2}{(a+h)^2 a^2}}{h}$$

$$= \frac{a^2 - (a^2 + 2ah + h^2)}{(a+h)^2 a^2} \cdot \frac{1}{h}$$

$$= \frac{-2ah - h^2}{h(a+h)^2 a^2}$$

$$= \frac{h(-2a - h)}{h(a+h)^2 a^2}$$

$$= -\frac{2a + h}{(a+h)^2 a^2}$$

■

EXAMPLE 9

Rationalize the numerator: $\dfrac{\sqrt{4+h} - 2}{h}$

SOLUTION

We multiply numerator and denominator by the conjugate radical $\sqrt{4+h} + 2$. The advantage of doing so is that we can use Product Formula 1 and the roots disappear from the numerator.

$$\frac{\sqrt{4+h} - 2}{h} = \frac{\sqrt{4+h} - 2}{h} \cdot \frac{\sqrt{4+h} + 2}{\sqrt{4+h} + 2}$$

$$= \frac{\left(\sqrt{4+h}\right)^2 - 2^2}{h\left(\sqrt{4+h} + 2\right)}$$

$$= \frac{4 + h - 4}{h\left(\sqrt{4+h} + 2\right)}$$

$$= \frac{h}{h\left(\sqrt{4+h} + 2\right)} = \frac{1}{\sqrt{4+h} + 2}$$

■

Note   It is sometimes useful to use Equation 2 backward, that is, in the form

$$\frac{a + c}{b} = \frac{a}{b} + \frac{c}{b}$$

For example, in certain calculus problems it is advantageous to write

$$\frac{x + 3}{x} = \frac{x}{x} + \frac{3}{x} = 1 + \frac{3}{x}$$

But remember to avoid the following common error:

$$\frac{a}{b + c} \neq \frac{a}{b} + \frac{a}{c}$$

(For instance, take $a = b = c = 1$ to see the error.)

EXERCISES 1.4

In Exercises 1–48 simplify the expression.

1. $\dfrac{x^2 + 3x + 2}{x^2 + 5x + 6}$

2. $\dfrac{x^2 + x - 6}{x^2 - 4}$

3. $\dfrac{y - y^2}{y^2 - 1}$

4. $\dfrac{2y^2 - 9y - 18}{4y^2 + 16y + 15}$

5. $\dfrac{2x^3 - x^2 - 6x}{2x^2 - 7x + 6}$

6. $\dfrac{1 - x^2}{x^3 - 1}$

7. $\dfrac{t - 3}{t^2 + 9} \cdot \dfrac{t + 3}{t^2 - 9}$

8. $\dfrac{x^2 - x - 6}{x^2 + 2x} \cdot \dfrac{x^3 + x^2}{x^2 - 2x - 3}$

9. $\dfrac{x^2 + 7x + 12}{x^2 + 3x + 2} \cdot \dfrac{x^2 + 5x + 6}{x^2 + 6x + 9}$

10. $\dfrac{x^2 + 2xy + y^2}{x^2 - y^2} \cdot \dfrac{2x^2 - xy - y^2}{x^2 - xy - 2y^2}$

11. $\dfrac{2x^2 + 3x + 1}{x^2 + 2x - 15} \div \dfrac{x^2 + 6x + 5}{2x^2 - 7x + 3}$

12. $\dfrac{4y^2 - 9}{2y^2 + 9y - 18} \div \dfrac{2y^2 + y - 3}{y^2 + 5y - 6}$

13. $\dfrac{\dfrac{x^3}{x + 1}}{\dfrac{x}{x^2 + 2x + 1}}$

14. $\dfrac{\dfrac{2x^2 - 3x - 2}{x^2 - 1}}{\dfrac{2x^2 + 5x + 2}{x^2 + x - 2}}$

15. $\dfrac{x/y}{z}$

16. $\dfrac{x}{y/z}$

17. $\dfrac{1}{x + 5} + \dfrac{2}{x - 3}$

18. $\dfrac{1}{x + 1} + \dfrac{1}{x - 1}$

19. $\dfrac{1}{x + 1} - \dfrac{1}{x + 2}$

20. $\dfrac{x}{x - 4} - \dfrac{3}{x + 6}$

21. $\dfrac{x}{(x + 1)^2} + \dfrac{2}{x + 1}$

22. $\dfrac{5}{2x - 3} - \dfrac{3}{(2x - 3)^2}$

23. $u + 1 + \dfrac{u}{u + 1}$

24. $\dfrac{2}{a^2} - \dfrac{3}{ab} + \dfrac{4}{b^2}$

25. $\dfrac{1}{x^2} + \dfrac{1}{x^2 + x}$

26. $\dfrac{1}{x} + \dfrac{1}{x^2} + \dfrac{1}{x^3}$

27. $\dfrac{2}{x + 3} - \dfrac{1}{x^2 + 7x + 12}$

28. $\dfrac{x}{x^2 - 4} + \dfrac{1}{x - 2}$

29. $\dfrac{1}{x + 3} + \dfrac{1}{x^2 - 9}$

30. $\dfrac{x}{x^2 + x - 2} - \dfrac{2}{x^2 - 5x + 4}$

31. $\dfrac{2}{x} + \dfrac{3}{x - 1} - \dfrac{4}{x^2 - x}$

32. $\dfrac{x}{x^2 - x - 6} - \dfrac{1}{x + 2} - \dfrac{2}{x - 3}$

33. $\dfrac{1}{x^2 + 3x + 2} - \dfrac{1}{x^2 - 2x - 3}$

34. $\dfrac{1}{x + 1} - \dfrac{2}{(x + 1)^2} + \dfrac{3}{x^2 - 1}$

35. $\dfrac{\dfrac{x}{y} - \dfrac{y}{x}}{\dfrac{1}{x^2} - \dfrac{1}{y^2}}$

36. $x - \dfrac{y}{\dfrac{x}{y} + \dfrac{y}{x}}$

37. $\dfrac{1 + \dfrac{1}{c - 1}}{1 - \dfrac{1}{c - 1}}$

38. $1 + \dfrac{1}{1 + \dfrac{1}{1 + x}}$

39. $\dfrac{\dfrac{5}{x - 1} - \dfrac{2}{x + 1}}{\dfrac{x}{x - 1} + \dfrac{1}{x + 1}}$

40. $\dfrac{\dfrac{a - b}{a} - \dfrac{a + b}{b}}{\dfrac{a - b}{b} + \dfrac{a + b}{a}}$

41. $\dfrac{x^{-2} - y^{-2}}{x^{-1} + y^{-1}}$

42. $\dfrac{x^{-1} + y^{-1}}{(x + y)^{-1}}$

43. $\dfrac{\dfrac{1}{a + h} - \dfrac{1}{a}}{h}$

44. $\dfrac{(x + h)^{-3} - x^{-3}}{h}$

45. $\dfrac{\dfrac{1 - (x + h)}{2 + (x + h)} - \dfrac{1 - x}{2 + x}}{h}$

46. $\dfrac{(x + h)^3 - 7(x + h) - (x^3 - 7x)}{h}$

47. $\sqrt{1 + \left(\dfrac{x}{\sqrt{1 - x^2}}\right)^2}$

48. $\sqrt{1 + \left(x^3 - \dfrac{1}{4x^3}\right)^2}$

In Exercises 49–54 rationalize the numerator.

49. $\dfrac{\sqrt{r} + \sqrt{s}}{t}$

50. $\dfrac{\sqrt{3(x + h) + 5} - \sqrt{3x + 5}}{h}$

51. $\dfrac{\sqrt{x} - \sqrt{x + h}}{h\sqrt{x}\sqrt{x + h}}$

52. $\sqrt{x^2 + 1} - x$

53. $\sqrt{x^2 + x + 1} + x$

54. $\sqrt{x + 1} - \sqrt{x}$

In Exercises 55–64 state whether the given equation is true for all values of the variables. (Disregard values that make denominators 0.)

55. $\dfrac{16 + a}{16} = 1 + \dfrac{a}{16}$

56. $\dfrac{b}{b - c} = 1 - \dfrac{b}{c}$

57. $\dfrac{2}{4 + x} = \dfrac{1}{2} + \dfrac{2}{x}$

58. $\dfrac{x + 1}{y + 1} = \dfrac{x}{y}$

59. $\dfrac{x}{x + y} = \dfrac{1}{1 + y}$

60. $2\left(\dfrac{a}{b}\right) = \dfrac{2a}{2b}$

61. $\dfrac{-a}{b} = -\dfrac{a}{b}$

62. $\dfrac{1 + x + x^2}{x} = \dfrac{1}{x} + 1 + x$

63. $\dfrac{x^2 + 1}{x^2 + x - 1} = \dfrac{1}{x - 1}$

64. $\dfrac{x^2 - 1}{x - 1} = x + 1$

SECTION 1.5
EQUATIONS

Examples of Equations

$6 = 2 + 4$

$4x + 8 = 0$

$y^2 - 16 = (y - 4)(y + 4)$

Many problems involving numerical quantities can be solved by changing the statement of the problem to an equation. In fact, solving equations is what most people think of when they hear the word "algebra." An **equation** is a statement that two mathematical expressions are equal. Some examples of equations can be found at the left. The first one is a simple statement of fact, but the second and third involve **variables**, which are letters that represent numbers. If we substitute the value -2 for the variable x in the second equation, we get the true statement $4(-2) + 8 = 0$, so we say that $x = -2$ is a solution of this equation. A **solution** or **root** of an equation is a value of the variable that makes the equation true. In the equation $y^2 - 16 = (y - 4)(y + 4)$, any value substituted for y will make the equation true. (Do you see why this is the case?) An equation that is true for every value of the variable is called an **identity**.

To **solve** an equation means to find all roots of the equation. Two equations are **equivalent** if they have exactly the same solutions. We solve an equation by transforming it into simpler equivalent equations with solutions that are obvious. We do this by adding or subtracting a quantity to both sides of the equation and by multiplying or dividing both sides by a nonzero quantity. For example, the equations in the steps below are all equivalent:

$$4x + 8 = 0$$
$$4x = -8 \quad \text{(subtracting 8 from both sides)}$$
$$x = -2 \quad \text{(dividing both sides by 4)}$$

This proves that -2 is the only solution of the equation $4x + 8 = 0$.

TRANSFORMING EQUATIONS

1. Adding or subtracting the same quantity from both sides of an equation does not change the solutions of the equation.

2. Multiplying or dividing both sides of an equation by the same nonzero quantity does not change the solutions of the equation.

■ LINEAR EQUATIONS

Linear Equations

$4x - 5 = 3$

$2x = \frac{1}{2}x - 5$

Nonlinear Equations

$x^2 + 2x = 8$

$\sqrt{x} - \frac{3}{x} = 6x - 1$

The simplest type of equation is a **linear equation**, or first-degree equation, which is an equation in which each term is either a constant or a nonzero multiple of the variable. That means it is equivalent to an equation of the form $ax + b = 0$. Here a and b represent real numbers with $a \neq 0$, and x is the unknown variable that we are solving for. The equation in the following example is linear.

EXAMPLE 1

Solve the equation $7x - 4 = 3x + 8$.

SOLUTION

We solve this equation by changing it to an equivalent equation with all terms involving x on one side and all constant terms on the other.

$$7x - 4 = 3x + 8$$
$$7x = 3x + 12 \quad \text{(adding 4 to both sides)}$$
$$4x = 12 \quad \text{(subtracting } 3x \text{ from both sides)}$$
$$x = 3 \quad \text{(dividing both sides by 4)}$$

The solution is 3. This can be checked by substituting $x = 3$ into the original equation.

$$7(3) - 4 \stackrel{?}{=} 3(3) + 8$$
$$17 = 17$$

The last statement is true, so $x = 3$ is the solution.   ■

In the next example we solve an equation that doesn't look as if it is linear to begin with, but that can be simplified to an equivalent linear equation.

EXAMPLE 2

Solve the equation $\dfrac{x}{x + 1} = \dfrac{2x + 1}{2x - 3}$.

SOLUTION

If $x \neq -1$ and $x \neq \frac{3}{2}$, we can multiply both sides of the equation by the LCD, which is $(x + 1)(2x - 3)$:

$$(x + 1)(2x - 3)\left(\frac{x}{x + 1}\right) = (x + 1)(2x - 3)\left(\frac{2x + 1}{2x - 3}\right)$$

$$(2x - 3)x = (x + 1)(2x + 1)$$

$$2x^2 - 3x = 2x^2 + 3x + 1$$

$$-3x = 3x + 1 \qquad \text{(subtracting } 2x^2 \text{ from both sides)}$$

$$-6x = 1 \qquad \text{(subtracting } 3x \text{ from both sides)}$$

$$x = -\frac{1}{6} \qquad \text{(dividing both sides by } -6)$$

The solution is $-\frac{1}{6}$, as you can verify. ∎

QUADRATIC EQUATIONS

Linear equations are first-degree equations of the form $ax + b = 0$. Quadratic equations are second-degree equations, containing an additional term involving the square of the variable.

QUADRATIC EQUATIONS

> A quadratic equation is an equation equivalent to one of the form
>
> $$ax^2 + bx + c = 0$$
>
> where a, b, and c are real numbers with $a \neq 0$.

Some quadratic equations can be solved by factoring and using the following basic property of real numbers.

ZERO-PRODUCT PROPERTY

> $ab = 0$ if and only if $a = 0$ or $b = 0$

EXAMPLE 3

Solve the equation $x^2 + 5x = 24$.

SOLUTION

We first subtract 24 from both sides of the equation and then factor the left side:

$$x^2 + 5x = 24$$
$$x^2 + 5x - 24 = 0$$
$$(x - 3)(x + 8) = 0$$
$$x - 3 = 0 \quad \text{or} \quad x + 8 = 0$$
$$x = 3 \qquad\qquad x = -8$$

The solutions are $x = 3$ and $x = -8$. ∎

Note that to use factoring to solve an equation, one side of the equation must be 0. For example, factoring the equation in the preceding example by writing $x(x + 5) = 24$ does not help in finding the solutions, since there are infinitely many ways to factor 24 (for example, $6 \cdot 4$, $\frac{1}{2} \cdot 48$, $\left(-\frac{2}{5}\right) \cdot (-60)$, and so on).

EXAMPLE 4

Solve the equation $x^2 = 12$.

SOLUTION

We solve by subtracting 12 from both sides and factoring as follows:

$$x^2 - 12 = 0 \qquad \text{(difference of squares)}$$
$$\left(x - \sqrt{12}\right)\left(x + \sqrt{12}\right) = 0$$
$$x = \sqrt{12} \quad \text{or} \quad x = -\sqrt{12}$$
$$x = 2\sqrt{3} \qquad\qquad x = -2\sqrt{3}$$

The solutions are $x = \pm 2\sqrt{3}$. ∎

Remember that every positive real number has two square roots. If $c > 0$, then $\sqrt{c}$ represents the positive (or *principal*) square root of c, but both $\sqrt{c}$ and $-\sqrt{c}$ are solutions of the equation $x^2 - c = 0$. We can thus solve simple equations like the one in Example 4 by using the following fact.

> The solutions of the equation $x^2 = c$ are $x = \sqrt{c}$ and $x = -\sqrt{c}$.

Euclid (circa 300 B.C.) *taught in Alexandria. His* Elements *is the most widely influential scientific book in history. For 2000 years it was the standard introduction to geometry in the schools and for many generations was considered the best way to develop logical reasoning. Abraham Lincoln, for instance, studied the* Ele- *ments as a way to sharpen his mind. The story is told that King Ptolemy once asked Euclid if there was a faster way to learn geometry than through the* Elements. *Euclid replied that there is "no royal road to geome- try"—meaning by this that mathemat- ics does not respect wealth or social status. Euclid was revered in his own time and was referred to by the title "The Geometer" or "The Writer of the Elements." The greatness of the* Elements *stems from its precise, logi- cal, and systematic treatment of geometry. For dealing with equality, Euclid lists the following rules he calls "common notions":*

1. *Things that are equal to the same thing are equal to each other.*
2. *If equals are added to equals, the sums are equal.*
3. *If equals are subtracted from equals, the remainders are equal.*
4. *Things which coincide with one another are equal.*

If a quadratic equation does not factor readily, we can solve it using the technique of **completing the square**, which means to add a constant to an expression to make it a perfect square. For example, to make $x^2 - 6x$ into a perfect square, we must add 9, since $x^2 - 6x + 9 = (x - 3)^2$. In general, from the identity

$$x^2 + bx + \left(\frac{b}{2}\right)^2 = \left(x + \frac{b}{2}\right)^2$$

it follows that to make $x^2 + bx$ a perfect square, we must add the square of half the coefficient of x. This procedure is summarized in the following box.

COMPLETING THE SQUARE

> To make $x^2 + bx$ a perfect square, add $\left(\frac{b}{2}\right)^2$.

EXAMPLE 5

Solve the following equations by completing the square:
(a) $x^2 - 8x + 13 = 0$ (b) $3x^2 - 12x + 6 = 0$

SOLUTION

(a) We first subtract 13 from both sides of the equation. Then we add the square of half the coefficient of x, namely, $(-4)^2 = 16$, to both sides in order to complete the square.

$$x^2 - 8x + 13 = 0$$
$$x^2 - 8x \quad\quad = -13$$
$$x^2 - 8x + 16 = -13 + 16$$
$$(x - 4)^2 = 3$$
$$x - 4 = \sqrt{3} \quad\quad \text{or} \quad\quad x - 4 = -\sqrt{3}$$
$$x = 4 + \sqrt{3} \quad\quad\quad\quad x = 4 - \sqrt{3}$$

Completing the Square

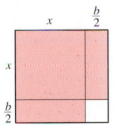

Area of shaded region is

$$x^2 + 2\left(\frac{b}{2}\right)x = x^2 + bx$$

Add small square $\left[area \left(\frac{b}{2}\right)^2 \right]$ to "complete" the square.

(b) After subtracting 6 from both sides of the equation, we must factor out the coefficient of x (the 3) from the left side of the equation to put it in the correct form for completing the square.

$$3x^2 - 12x + 6 = 0$$
$$3x^2 - 12x \quad\quad = -6$$
$$3(x^2 - 4x \quad\quad) = -6$$

Now we complete the square by adding $(-2)^2 = 4$ inside the parentheses.

Since everything inside the parentheses is multiplied by 3, this means that we are really adding $3 \cdot 4 = 12$ to the left side of the equation. Thus we must add 12 to the right side as well.

$$3(x^2 - 4x + 4) = -6 + 12$$
$$3(x - 2)^2 = 6$$
$$(x - 2)^2 = 2$$
$$x - 2 = \pm\sqrt{2}$$
$$x = 2 \pm\sqrt{2}$$

We can use the technique of completing the square to derive a formula for the roots of the general quadratic equation $ax^2 + bx + c = 0$. First, we divide both sides of the equation by a and move the constant to the right, giving

$$x^2 + \frac{b}{a}x = -\frac{c}{a}$$

We now complete the square by adding $[b/(2a)]^2$ to both sides of the equation:

$$x^2 + \frac{b}{a}x + \left(\frac{b}{2a}\right)^2 = -\frac{c}{a} + \left(\frac{b}{2a}\right)^2$$

and so

$$\left(x + \frac{b}{2a}\right)^2 = \frac{-4ac + b^2}{4a^2}$$

Taking square roots of both sides, we see that

$$x + \frac{b}{2a} = \pm\sqrt{\frac{-4ac + b^2}{4a^2}} = \pm\frac{\sqrt{b^2 - 4ac}}{2a}$$

Solving for x, we get

$$x = \frac{-b \pm \sqrt{b^2 - 4ac}}{2a}$$

François Viète (1540–1603) was a French mathematician, sometimes known by the Latin form of his name, Vieta. He introduced a new level of abstraction in algebra by using letters to stand for known quantities in an equation. Before Viète's time, each equation had to be solved on its own. For instance, the quadratic equations

$$3x^2 + 2x + 8 = 0$$
and $$5x^2 - 6x + 11 = 0$$

had to be solved separately for the unknown x. Viète's idea was to consider all quadratic equations at once by writing $ax^2 + bx + c = 0$ where a, b, c are known quantities. Thus it is possible to write a formula (in this case the quadratic formula) involving a, b, and c that can be used to solve all such equations in one fell swoop.

THE QUADRATIC FORMULA

> The roots of the quadratic equation $ax^2 + bx + c = 0$, where $a \neq 0$, are
>
> $$x = \frac{-b \pm \sqrt{b^2 - 4ac}}{2a}$$

The quadratic formula could be used to solve the equations in Examples 3, 4, and 5. You should carry out the details of these calculations.

EXAMPLE 6

Find all solutions of each of the following equations:

(a) $3x^2 - 5x - 1 = 0$

(b) $4x^2 + 12x + 9 = 0$

(c) $x^2 + 2x + 2 = 0$

SOLUTION

(a) Using the quadratic formula with $a = 3$, $b = -5$, and $c = -1$, we get

$$x = \frac{-(-5) \pm \sqrt{(-5)^2 - 4(3)(-1)}}{2(3)} = \frac{5 \pm \sqrt{37}}{6}$$

If approximations are desired, we use a calculator and obtain

$$x = \frac{5 + \sqrt{37}}{6} \approx 1.8471 \qquad \text{or} \qquad x = \frac{5 - \sqrt{37}}{6} \approx -0.1805$$

(b) Using the quadratic formula with $a = 4$, $b = 12$, and $c = 9$ gives

Another Method

$4x^2 + 12x + 9 = 0$

$(2x + 3)^2 = 0$

$2x + 3 = 0$

$x = -\frac{3}{2}$

$$x = \frac{-12 \pm \sqrt{12^2 - 4 \cdot 4 \cdot 9}}{2 \cdot 4} = \frac{-12 \pm 0}{8} = -\frac{3}{2}$$

In this case there is only one solution, $x = -\frac{3}{2}$.

(c) $$x = \frac{-2 \pm \sqrt{2^2 - 4 \cdot 2}}{2} = \frac{-2 \pm \sqrt{-4}}{2}$$

$$= \frac{-2 \pm 2\sqrt{-1}}{2} = -1 \pm \sqrt{-1}$$

Since the square of any real number is nonnegative, $\sqrt{-1}$ is undefined in the real number system. The equation has no real solutions. ■

In the next section we will study the complex number system, in which the square roots of negative numbers do exist.

The quantity $b^2 - 4ac$ that appears under the square root sign in the quadratic formula is called the **discriminant** of the equation $ax^2 + bx + c = 0$, and is given the symbol D. If $D < 0$, then $\sqrt{b^2 - 4ac}$ is undefined, so the quadratic equation has no real solutions, as in Example 6(c). If $D = 0$, then there is only one solution, as in Example 6(b). Finally, if $D > 0$, then the equation will have two different solutions, as in Example 6(a).

The following box summarizes what we have observed about the discriminant.

THE DISCRIMINANT

> The discriminant of the general quadratic $ax^2 + bx + c = 0$ $(a \neq 0)$ is $D = b^2 - 4ac$.
>
> **1.** If $D > 0$, then the equation has two different real solutions.
> **2.** If $D = 0$, then the equation has exactly one real solution.
> **3.** If $D < 0$, then the equation has no real solutions.

EXAMPLE 7

Use the discriminant to determine how many real solutions the following equations have:

(a) $x^2 + 2x + 8 = 0$ (b) $3x^2 - 5x + \frac{3}{2} = 0$

SOLUTION

(a) The discriminant is

$$D = 2^2 - 4 \cdot 1 \cdot 8$$
$$= -28 < 0$$

Thus the equation has no real solutions.

(b) $D = (-5)^2 - 4 \cdot 3 \cdot \frac{3}{2} = 25 - 18 = 7 > 0$

The equation has two distinct real solutions. ■

OTHER EQUATIONS

In our final four examples we look at other types of equations, including those that involve higher powers or radicals.

EXAMPLE 8

Find all real solutions of the following equations:

(a) $x^3 = -8$ (b) $16x^4 = 81$

SOLUTION

(a) Since every real number has exactly one real cube root, we can solve this equation by taking cube roots of both sides.

$$(x^3)^{1/3} = (-8)^{1/3}$$
$$x = -2$$

(b) Here we must remember that if n is even, then every positive real number has *two* real nth roots, a positive one and a negative one.

$$x^4 = \frac{81}{16}$$

$$(x^4)^{1/4} = \pm\left(\frac{81}{16}\right)^{1/4}$$

$$x = \pm\frac{3}{2}$$ ∎

Some equations that at first glance may not appear to be quadratic can be changed into quadratic equations by performing simple algebraic operations on them, such as multiplying both sides by a common denominator or squaring both sides. Special care needs to be taken in solving such equations, as we see in the following example.

EXAMPLE 9

Solve the following equations:

(a) $x + 3 = \dfrac{2x^2 - 7x + 3}{3 - x}$ (b) $x = 1 - \sqrt{2 - \dfrac{x}{2}}$

SOLUTION

(a) Multiplying both sides of the equation by $x - 3$ to clear the denominator (which we can do as long as $x \neq 3$), we get $x^2 - 9 = -2x^2 + 7x - 3$, so

$$3x^2 - 7x - 6 = 0$$

Another Method

To solve $3x^2 - 7x - 6 = 0$, use the quadratic formula:

$$x = \frac{7 \pm \sqrt{7^2 - 4 \cdot 3 \cdot (-6)}}{2 \cdot 3}$$

$$= \frac{7 \pm 11}{6} = 3 \text{ or } -\frac{2}{3}$$

Factoring, we get $\qquad (3x + 2)(x - 3) = 0$

Thus $x = -\frac{2}{3}$ or $x = 3$. But $x = 3$ does not satisfy the original equation (since division by 0 is impossible), so the only solution is $x = -\frac{2}{3}$.

(b) To eliminate the square root, we first write the equation as

$$x - 1 = -\sqrt{2 - \frac{x}{2}}$$

and then square both sides, giving

$$(x - 1)^2 = 2 - \frac{x}{2}$$

so $\qquad x^2 - 2x + 1 = 2 - \dfrac{x}{2}$

and $\qquad 2x^2 - 3x - 2 = 0$

$$(x - 2)(2x + 1) = 0$$

Thus $x = 2$ and $x = -\frac{1}{2}$ are potential solutions. Substituting these into the original equation, we see that $x = -\frac{1}{2}$ is a solution, since this leads to

$$-\tfrac{1}{2} = 1 - \sqrt{2 + \tfrac{1}{4}}$$

which is true, but that $x = 2$ is not a solution (since $2 \neq 1 - \sqrt{2 - 1}$). The only solution is $x = -\frac{1}{2}$. ■

Extraneous solutions often are introduced when we square both sides of an equation since the operation of squaring can turn an inequality into an equality. For example, $-1 \neq 1$, but $(-1)^2 = 1^2$. We saw another instance in Example 9(b) in the case where $x = 2$. When both sides of an equation have been squared, our initial solutions are *potential* solutions that must be checked by substituting them into the original equation.

In some cases, a polynomial equation of degree 4 or higher can be changed into a quadratic equation by performing an algebraic substitution, as in the following example.

EXAMPLE 10

Find the real solutions of $x^4 - 2x^2 - 2 = 0$.

SOLUTION

If we set $w = x^2$, then

$$x^4 - 2x^2 - 2 = (x^2)^2 - 2x^2 - 2$$
$$= w^2 - 2w - 2 = 0$$

and the solutions of the equation in w are

$$w = \frac{2 \pm \sqrt{2^2 + 4 \cdot 2}}{2} = 1 \pm \sqrt{3}$$

Since $x^2 = w$, $x = \pm\sqrt{w}$, and since $1 - \sqrt{3}$ is a negative number, we get

$$x = \pm\sqrt{1 + \sqrt{3}}$$

as the only real solutions of the original equation. ■

EXAMPLE 11

Find all solutions of the equation $x^{5/6} + x^{2/3} = 2x^{1/2}$.

SOLUTION

The terms in this equation have the common factor $x^{1/2}$, so after subtracting $2x^{1/2}$

from both sides, we factor as follows:

$$x^{5/6} + x^{2/3} - 2x^{1/2} = 0$$
$$x^{1/2}(x^{1/3} + x^{1/6} - 2) = 0$$

We can factor the expression in parentheses further by noting that $(x^{1/6})^2 = x^{1/3}$, so the expression becomes a quadratic if we let $w = x^{1/6}$.

$$w^3(w^2 + w - 2) = 0$$
$$w^3(w - 1)(w + 2) = 0$$

So $\quad w = 0 \quad$ or $\quad w = 1 \quad$ or $\quad w = -2$

$\qquad x^{1/6} = 0 \qquad\quad x^{1/6} = 1 \qquad\quad x^{1/6} = -2$

$\qquad x = 0 \qquad\quad x = 1^6 = 1 \qquad\quad x = (-2)^6 = 64$

Substituting these values into the original equation, we see that $x = 0$ and $x = 1$ are solutions, but $x = 64$ is not, since the left side is

$$64^{5/6} + 64^{2/3} = 32 + 16 = 48$$

but the right-hand side is $2(64)^{1/2} = 16$. The only solutions are 0 and 1. ■

Note that it would have been wrong to divide both sides of the original equation in this example by the common factor $x^{1/2}$, since we would have lost the solution $x = 0$ by doing so. Never divide both sides of an equation by an expression containing the variable (unless you know that expression can never equal 0).

EXERCISES 1.5

In Exercises 1–14 determine whether the given equation is an identity.

1. $\dfrac{x^3 - 8}{x - 2} = x^2 + 2x + 4 \quad (x \neq 2)$

2. $\dfrac{1}{x} + \dfrac{1}{2} = \dfrac{1}{x + 2}$

3. $\dfrac{y + 16}{2} - \dfrac{3}{2}y = 8 - y$

4. $\sqrt{4 - x^2} = 2 - x \quad (-2 \leq x \leq 2)$

In Exercises 5–8 state whether the given equation is equivalent to a linear equation. If it is, solve the equation.

5. $4x - 7 = 2x + 11$

6. $\dfrac{x}{4} + 12 = \dfrac{3x}{10} + \dfrac{2}{5}$

7. $y^2 - 7y + 4 = 2y^2 + 11$

8. $2x - 1 = \sqrt{x^2 - 3}$

In Exercises 9–26 solve the equation.

9. $2x + 4 = 12$

10. $x - 3 = 4x + 33$

11. $\dfrac{1}{2}y - 2 = \dfrac{1}{3}y$

12. $\dfrac{z}{5} = \dfrac{3}{10}z + 7$

13. $2(1 - x) = 3(1 + 2x) + 5$

14. $5(x + 3) + 9 = -2(x - 2) - 1$

15. $4\left(y - \tfrac{1}{2}\right) - y = 6(5 - y)$

16. $\dfrac{2}{3}y + \dfrac{1}{2}(y - 3) = \dfrac{y + 1}{4}$

17. $\dfrac{1}{x} = \dfrac{4}{3x} + 1$

18. $\dfrac{2x - 1}{x + 2} = \dfrac{4}{5}$

19. $\dfrac{2}{t + 6} = \dfrac{3}{t - 1}$

20. $\dfrac{1}{t - 1} + \dfrac{t}{3t - 2} = \dfrac{1}{3}$

21. $\dfrac{2x - 7}{2x + 4} = \dfrac{2}{3}$

22. $\dfrac{1}{3 - t} + \dfrac{4}{3 + t} + \dfrac{16}{9 - t^2} = 0$

23. $\sqrt{x - 4} = \sqrt{2x}$

24. $\sqrt{2x - 3} = \sqrt{6x}$

25. $\dfrac{3}{x + 4} = \dfrac{1}{x} + \dfrac{6x + 12}{x^2 + 4x}$

26. $\dfrac{1}{x} - \dfrac{2}{2x + 1} = \dfrac{1}{2x^2 + x}$

In Exercises 27–30 find all solutions of the equation.

27. $x^2 = 121$ **28.** $3x^2 = 72$

29. $8x^2 - 64 = 0$ **30.** $5x^2 - 125 = 0$

In Exercises 31–34 solve the equation by factoring.

31. $x^2 - x - 12 = 0$ **32.** $x^2 + 2x = 15$

33. $3y^2 + 17y + 10 = 0$ **34.** $10z^2 = z + 2$

In Exercises 35–38 solve the equation by completing the square.

35. $x^2 + 2x - 2 = 0$ **36.** $x^2 - 4x + 2 = 0$

37. $x^2 + x - 1 = 0$ **38.** $2x^2 + 8x + 1 = 0$

In Exercises 39–78 find all real solutions of the equation.

39. $x^2 - 2x - 15 = 0$ **40.** $6x^2 + 7x - 3 = 0$

41. $x^2 + 36x - 1440 = 0$ **42.** $12x^2 + 140x + 375 = 0$

43. $3x^2 + 2x - 2 = 0$ **44.** $x^2 - 6x + 1 = 0$

45. $2y^2 - y - \tfrac{1}{2} = 0$ **46.** $\theta^2 - \tfrac{3}{2}\theta + \tfrac{9}{16} = 0$

47. $3x^2 + 6x + 4 = 0$ **48.** $4x = x^2 + 1$

49. $3 + 5z + z^2 = 0$ **50.** $w^2 = 3(w - 1)$

51. $x^2 - \sqrt{5}x + 1 = 0$ **52.** $\sqrt{6}x^2 + 2x - \sqrt{\tfrac{3}{2}} = 0$

53. $\dfrac{x^2}{x + 100} = 50$

54. $1 + \dfrac{2x}{(x + 3)(x + 4)} = \dfrac{2}{x + 3} + \dfrac{4}{x + 4}$

55. $\dfrac{x + 5}{x - 2} = \dfrac{5}{x + 2} + \dfrac{28}{x^2 - 4}$

56. $\dfrac{x}{2x + 7} - \dfrac{x + 1}{x + 3} = 1$

57. $x^3 = 125$ **58.** $x^5 - 32 = 0$

59. $x^4 - 9 = 0$ **60.** $8x^6 = 27$

61. $x^4 - 3x^3 + 2x^2 = 0$ **62.** $(x + 1)^5 - 4(x + 1)^3 = 0$

63. $x^3 - 4x^2 - 2x + 8 = 0$ **64.** $2x^3 + x^2 - 4x - 2 = 0$

65. $\sqrt{2x + 1} + 1 = x$ **66.** $x - \sqrt{9 - 3x} = 0$

67. $\sqrt{5 - x} + 1 = \sqrt{x}$

68. $\sqrt{2x + 1} + \sqrt{x + 1} = 2$

69. $\sqrt{\sqrt{x - 5} + x} = 5$ **70.** $x^4 - 5x^2 + 4 = 0$

71. $2x^4 + 4x^2 + 1 = 0$ **72.** $x^6 - 2x^3 - 3 = 0$

73. $4(x + 1)^{1/2} - 5(x + 1)^{3/2} + (x + 1)^{5/2} = 0$

74. $x^{1/2} + 3x^{-1/2} = 10x^{-3/2}$

75. $x^{4/3} - 5x^{2/3} + 6 = 0$ **76.** $\sqrt{x} - 3\sqrt[4]{x} - 4 = 0$

77. $\dfrac{1}{x^3} + \dfrac{4}{x^2} + \dfrac{4}{x} = 0$ **78.** $4x^{-4} - 16x^{-2} + 4 = 0$

In Exercises 79–82 use a calculator to find all real solutions, correct to three decimals.

79. $2.15x - 4.63 = x + 1.19$

80. $3.95 - x = 2.32x + 2.00$

81. $x^2 - 0.011x - 0.064 = 0$

82. $2.232x^2 - 4.112x = 6.219$

In Exercises 83–90 solve the equation for the indicated variable.

83. $PV = nRT;$ for R **84.** $F = G\dfrac{mM}{r^2};$ for m

85. $\dfrac{1}{R} = \dfrac{1}{R_1} + \dfrac{1}{R_2}$;   for R_1

86. $\dfrac{a+1}{b} = \dfrac{a-1}{b} + \dfrac{b+1}{a}$;   for a

87. $V = \frac{1}{3}\pi r^2 h$;   for r

88. $F = G\dfrac{mM}{r^2}$;   for r

89. $a^2 + b^2 = c^2$;   for b

90. $S = \dfrac{n(n+1)}{2}$;   for n

In Exercises 91 and 92 find the value(s) of k that will ensure that the indicated value(s) of x are the solutions of the given equation.

91. $3x + k - 5 = kx - k + 1$;   solution $x = 2$

92. $kx^2 + x - 4 = 0$;   solutions $x = -\frac{4}{3},\ 1$

In Exercises 93–98, use the discriminant to determine the number of real solutions of the equation. Do not solve the equation.

93. $5x^2 + 3x + 1 = 0$

94. $x^2 = 6x + 6$

95. $x^2 + 2.20x + 1.21 = 0$

96. $x^2 + 2.21x + 1.21 = 0$

97. $x^2 - rx + s = 0$ $\left(s > 0,\, r > 2\sqrt{s}\right)$

98. $x^2 + rx - s = 0$ $(s > 0)$

In Exercises 99 and 100 find all values for k that ensure that the given equation has exactly one solution.

99. $4x^2 + kx + 25 = 0$

100. $kx^2 + 36x + k = 0$

101. The formula for the volume of a right circular cone is $V = \frac{1}{3}\pi r^2 h$, where r is the radius of the base and h is the height. Find the height (correct to two decimals) of a cone with volume 30.00 in^2 and radius 4.00 in.

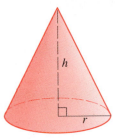

SECTION 1.6
PROBLEM SOLVING WITH EQUATIONS

Many problems in the sciences, economics, finance, medicine, and numerous other fields can be translated into algebra problems, which is one reason that algebra is so useful. Students often find it difficult at first to translate a word problem into an equation. This skill is developed through practice. After considering a few examples, we will list some general principles to guide your thinking in this translation process.

EXAMPLE 1

Mary inherits $100,000 and invests it in two certificates of deposit. One pays $10\frac{1}{2}\%$ and the other pays 9% simple interest annually. If her total interest is $9525 per year, how much is invested at each rate?

SOLUTION

The key to solving any word problem is to translate the information given into the language of mathematics; that is, into an equation. First, we need to identify the variables. This can usually be done by a careful reading of the question asked in the problem. Here we are asked to find out how much is invested at each rate. So we let

$$x = \text{amount invested at } 10\tfrac{1}{2}\%$$

The rest of Mary's money is invested at 9%, so

$$100,000 - x = \text{amount invested at } 9\%$$

Now we translate the fact that her total interest is $9525 into an equation:

$$\left(\text{interest at } 10\tfrac{1}{2}\%\right) + \left(\text{interest at } 9\%\right) = 9525 \qquad (1)$$

Simple Interest

Interest = Principal × rate × time

$$I = Prt$$

Since x dollars are invested at $10\tfrac{1}{2}\%$, the annual interest from this certificate is $10\tfrac{1}{2}\%$ of x, or $0.105x$. Similarly the interest received from the 9% certificate will be $0.09(100,000 - x)$, and so Equation (1) is expressed in terms of x as

$$0.105x + 0.09(100,000 - x) = 9525$$

Now we solve for x:

$$0.105x + 9000 - 0.09x = 9525$$
$$0.015x + 9000 = 9525$$
$$0.015x = 525$$
$$x = \frac{525}{0.015} = 35,000$$

Thus Mary has invested $35,000 at $10\tfrac{1}{2}\%$ and the remaining $65,000 at 9%.   ■

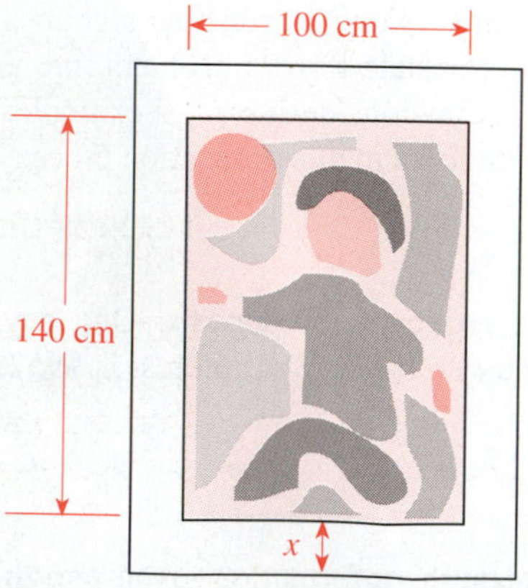

EXAMPLE 2

A poster has on it a rectangular printed area 100 cm by 140 cm, with a blank strip of uniform width around the edges. The perimeter of the poster is $1\tfrac{1}{2}$ times the perimeter of the printed area. What is the width of the blank strip, and what are the dimensions of the poster?

SOLUTION

In a problem such as this that involves geometry, it is essential to draw a diagram as in Figure 1. Let

$$x = \text{width of the blank strip}$$

From the picture we see that the poster is $(100 + 2x)$ cm by $(140 + 2x)$ cm, so its perimeter is $2(100 + 2x) + 2(140 + 2x)$. The perimeter of the printed area is

Figure 1

$2(100) + 2(140) = 480$ cm. We are told that

$$(\text{perimeter of poster}) = \tfrac{3}{2} \times (\text{perimeter of printed part})$$

or

$$2(100 + 2x) + 2(140 + 2x) = \tfrac{3}{2} \cdot 480$$
$$480 + 8x = 720$$
$$8x = 240$$
$$x = 30$$

The blank strip is 30 cm wide, so the poster is 160 cm by 200 cm.   ■

We adapt here the problem-solving principles presented after this chapter to this situation. To solve any problem the first step is to read it carefully to make sure you understand it. In particular, you must understand exactly what you are being asked to find. The following guidelines should help you to set up the equation that expresses the English statement of the problem in the language of algebra.

Guidelines for Solving Word Problems

Identify the variable

1. Identify the quantity that the problem asks you to find. This can usually be determined by a careful reading of the question posed at the end of the problem. Give this quantity a name (x or some other variable). Make sure to write down precisely what the variable represents.

Express all unknown quantities in terms of the variable

2. Read each sentence in the problem again, and express all the quantities mentioned in the problem in terms of the variable you defined in Step 1. For instance, in Example 1 we expressed the principal invested at each interest rate and the annual interest earned by each certificate in terms of the variable x. Sometimes it is helpful to organize this information into a chart or diagram, as in Example 2.

Relate the quantities

3. Find the crucial fact in the problem that relates two or more of the expressions you listed in Step 2. For instance, in Example 1 we were told that the total interest from the two certificates was $9525. This fact produced the equation for that problem. A statement that one quantity "is," "equals," or "is the same as" another often signals the type of relationship we are looking for in this step.

Set up an equation

4. Set up an equation that expresses the crucial fact you found in Step 3 in algebraic form. You will often need to use a formula to translate from English to algebra. For instance, in Example 2 you need to know that the perimeter of a rectangle is twice the length plus twice the width. This can be expressed as $P = 2L + 2W$.

Solve the equation

5. Solve the equation, and check to make sure your answer satisfies the original word problem.

EXAMPLE 3

The sum of three consecutive even integers is 288. What are the integers?

SOLUTION

Identify the variable

If we call the first integer x, then the others must be $x + 2$ and $x + 4$, since consecutive even integers are two units apart. This means that we must solve the equation

Express all unknown quantities in terms of the variable

Set up an equation

Solve

$$x + (x + 2) + (x + 4) = 288$$
$$3x + 6 = 288$$
$$3x = 282$$
$$x = 94$$

The integers are 94, 96, and 98. ∎

EXAMPLE 4

A manufacturer of soft drinks makes a type of orange soda that he advertises as "naturally flavored," although it contains only 5% orange juice. A new federal regulation stipulates that to be called natural, a drink must contain at least 10% juice. How much pure orange juice must he add to 900 gal of his soda to conform to the regulations?

SOLUTION

In any problem of this type where two different substances are to be mixed, a diagram helps to set up the required equation (see Figure 2).

Figure 2

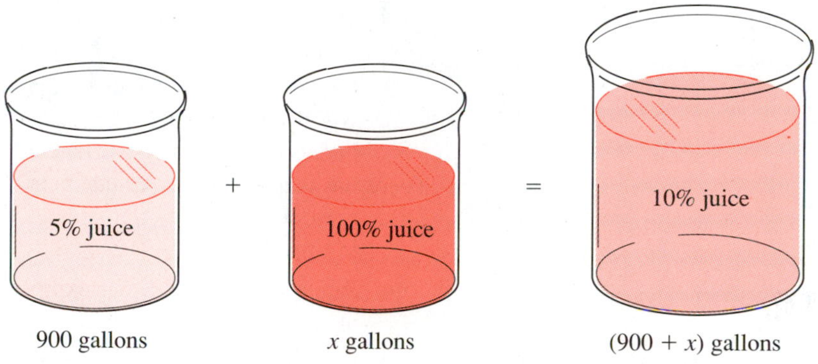

| 5% juice | + | 100% juice | = | 10% juice |
| 900 gallons | | x gallons | | $(900 + x)$ gallons |

Identify the variable

Express all unknown quantities in terms of the variable

Relate the quantities

Let x be the amount (in gallons) of pure juice to be used. Then $900 + x$ gallons of 10% orange juice mixture will result. The key idea to turn the picture into an equation here is to notice that the total amount of juice on both sides of the equal sign is the same. The orange juice in the first vat is 5% of 900 gal, or 45 gal. The second vat contains x gallons of juice, and the third contains $0.1(900 + x)$ gal. So we write the equation

$$45 + x = 0.1(900 + x)$$
$$45 + x = 90 + 0.1x$$
$$0.9x = 45$$
$$x = \frac{45}{0.9} = 50$$

The manufacturer should add 50 gal of pure orange juice to the soda. ∎

EXAMPLE 5

Because of an anticipated heavy rainstorm, the water level in a reservoir must be lowered by 1 ft. Opening spillway A lowers the level by this amount in 4 h, while opening the smaller spillway B does the job in 6 h. How long will it take if both spillways are opened?

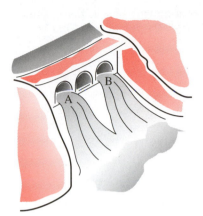

Figure 3

SOLUTION

As usual, we represent the number we are seeking by x:

Identify the variable

$$x = \text{number of hours it takes to lower the water}$$
$$\text{level by 1 ft if both spillways are open}$$

Finding an equation relating x to the other quantities in this problem is not easy. Certainly x is not simply $4 + 6$, since that would mean it takes longer to lower the water level if both spillways are open, which doesn't make sense. Instead, *we look at the fraction of the job that can be done in one hour by each spillway.*

Spillway A lowers the water by $\frac{1}{4}$ ft in 1 h

Relate the quantities

Spillway B lowers the water by $\frac{1}{6}$ ft in 1 h

Both spillways lower the water by $\frac{1}{x}$ ft in 1 h

Therefore we get the equation

Set up an equation

$$\frac{1}{4} + \frac{1}{6} = \frac{1}{x}$$
$$3x + 2x = 12$$
$$5x = 12$$

Solve

$$x = \frac{12}{5}$$

It will take $2\frac{2}{5}$ h, or 2 h 24 min, to lower the water level by 1 ft if both spillways are open. ∎

The next example deals with distance, rate (speed), and time. The formula to keep in mind here is

$$\text{distance} = \text{rate} \times \text{time}$$

where the rate is the constant or average speed of a moving object. For example, driving at 60 mi/h for 4 h takes you a distance of $60 \cdot 4 = 240$ mi.

EXAMPLE 6

A jet flew from New York to Los Angeles, a distance of 4200 km. On the return trip the speed was increased by 100 km/h. If the total trip took 13 h, what was the speed from New York to Los Angeles?

SOLUTION

Let $\qquad\qquad\qquad s = $ speed from New York to Los Angeles

Then $\qquad\qquad s + 100 = $ speed from Los Angeles to New York

We organize the given data into the following table.

	Distance (km)	Speed (km/h)	Time (h)
N.Y. to L.A.	4200	s	$\dfrac{4200}{s}$
L.A. to N.Y.	4200	$s + 100$	$\dfrac{4200}{s + 100}$

The trip took a total of 13 h, so we have the equation

$$\frac{4200}{s} + \frac{4200}{s + 100} = 13$$

Multiplying by the common denominator $s(s + 100)$ we get

$$4200(s + 100) + 4200s = 13s(s + 100)$$

$$8400s + 420{,}000 = 13s^2 + 1300s$$

$$13s^2 - 7100s - 420{,}000 = 0$$

Although this equation does factor, with numbers this large it is probably quickest

Pierre de Fermat (*1601–1665*) *was a French lawyer who became interested in mathematics at the age of 30. Because of his job as a magistrate Fermat had little time to write complete proofs for his discoveries and often wrote them in the margin of whatever book he was reading at the time. After his death his copy of Diophantus'* Arithmetica (*see p. 23*) *was found to contain a particularly tantalizing comment. Where Diophantus discusses the solutions of* $x^2 + y^2 = z^2$ (*for example,* $x = 3$, $y = 4$, $z = 5$; *see Review Exercise 138*) *Fermat states in the margin that for* $n \geqslant 3$ *there are no natural number solutions to the equation* $x^n + y^n = z^n$. *In other words, it is impossible for a cube to equal the sum of two cubes, a fourth power to equal the sum of two fourth powers, and so on. Fermat writes "I have discovered a truly wonderful proof for this but the margin is too small to contain it." All the other margin comments in Fermat's copy of the* Arithmetica *have proved to be true but no one has been able to prove this one, in spite of attempts by very great mathematicians. It has come to be called "Fermat's Last Theorem."*

to use the quadratic formula and a calculator.

$$s = \frac{7100 \pm \sqrt{7100^2 - 4(13)(-420{,}000)}}{2(13)}$$

$$= \frac{7100 \pm 8500}{26}$$

$$s = 600 \qquad \text{or} \qquad s = \frac{-700}{13}$$

Since s represents speed, we reject the negative answer and conclude that the jet's speed from New York to Los Angeles was 600 km/h.   ■

EXAMPLE 7

An object thrown or fired straight up at v_0 ft/s will reach a height of h feet after t seconds, where h and t are related by the formula

$$h = -16t^2 + v_0 t$$

(This formula is derived in elementary physics courses, and depends on the fact that the acceleration due to gravity is constant near the surface of the earth. Here we are neglecting the effect of air resistance.)

Suppose that a bullet is shot straight up with a muzzle speed of 800 ft/s.
(a) When does the bullet fall back to the ground?
(b) When does it reach a height of 6400 ft?
(c) When does it reach a height of 2 mi?
(d) How high is the highest point it reaches?

SOLUTION

The formula in this case is $h = -16t^2 + 800t$.
(a) Ground level corresponds to $h = 0$, so we must solve the equation

$$0 = -16t^2 + 800t$$
$$= -16t(t - 50)$$

Thus $t = 0$ or $t = 50$. This means the bullet starts ($t = 0$) at ground level, and returns to ground level after 50 s.
(b) Setting $h = 6400$ gives the equation

$$6400 = -16t^2 + 800t$$
$$16t^2 - 800t + 6400 = 0$$
$$t^2 - 50t + 400 = 0$$
$$(t - 10)(t - 40) = 0$$
$$t = 10 \qquad \text{or} \qquad t = 40$$

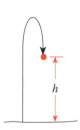

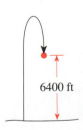

6400 ft

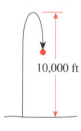

2 mi

The bullet reaches 6400 ft after 10 s (on the way up) and again after 40 s (on the way down).

(c) Two miles is $2 \times 5280 = 10,560$ ft.

$$10,560 = -16t^2 + 800t$$
$$16t^2 - 800t + 10,560 = 0$$
$$t^2 - 50t + 660 = 0$$

The discriminant of this equation is $D = 50^2 - 4(660) = -140$, which is negative. Thus the equation has no real solution. The bullet never reaches a height of 2 mi.

(d) Each height the bullet reaches is attained twice, once on the way up and once on the way down. The only exception is the highest point of its path, which is only reached once. This means that for the highest value of h, the equation

$$h = -16t^2 + 800t$$

or
$$16t^2 - 800t + h = 0$$

has only one solution for t. This in turn means that the discriminant of the equation is 0, and so

$$D = (-800)^2 - 4(16)h = 0$$
$$640,000 - 64h = 0$$
$$h = 10,000$$

The maximum height reached is 10,000 ft.

10,000 ft

EXAMPLE 8

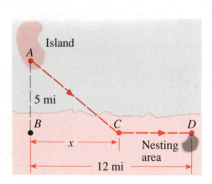

Island

5 mi

B C D

x Nesting area

12 mi

Figure 4

Ornithologists have determined that some species of birds tend to avoid flights over large bodies of water during daylight hours, since air generally rises over land and falls over water in the daytime, so that flying over water requires more energy. A bird is released from point A on an island that is 5 mi from the nearest point B on a straight shoreline. It flies to a point C on the shore, and then flies along the shoreline to its nesting area D, as shown in Figure 4. Suppose the bird requires 10 kcal/mi of energy to fly over land and 14 kcal/mi to fly over water. It has 170 kcal of energy reserves left.

(a) Where should the point C be located so that the bird uses exactly 170 kcal in its flight?

(b) Does the bird have enough energy to fly directly from A to D?

SOLUTION

(a) Let x be the distance in miles from B to C. Then

$$\text{distance flown over water} = \sqrt{x^2 + 25} \qquad \text{(Pythagorean Theorem)}$$
$$\text{distance flown over land} = 12 - x$$

Pythagoras (*circa 580–500* B.C.) *founded a school in Croton in southern Italy devoted to the study of arithmetic, geometry, music, and astronomy. The Pythagoreans, as they were called, were a secret society with peculiar rules and initiation rites. Because of their secretive rules nothing was written down and they were not to reveal to anyone what they had learned from the Master. They were vegetarians but also did not eat beans, which they considered sacred. Although women were barred by law from attending public meetings, Pythagoras allowed women in his school, and his most accomplished student was Theano (whom Pythagoras later married). The outstanding contribution of Pythagoras is the theorem that bears his name: In a right triangle, the area of the square on the hypotenuse is equal to the sum of the areas of the squares on the other two sides.*

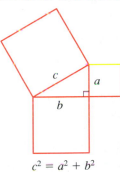

$$c^2 = a^2 + b^2$$

Since

$$\text{energy used} = (\text{energy per mile}) \times (\text{miles flown})$$

and since

$$\text{total energy} = (\text{energy over water}) + (\text{energy over land})$$

we get the following equation for the energy that the bird uses on its trip:

$$170 = 14\sqrt{x^2 + 25} + 10(12 - x)$$

To solve this equation, we eliminate the square root by first bringing all other terms to the left of the equal sign, and then squaring both sides.

$$170 - 10(12 - x) = 14\sqrt{x^2 + 25}$$
$$50 + 10x = 14\sqrt{x^2 + 25}$$
$$(50 + 10x)^2 = (14)^2(x^2 + 25)$$
$$2500 + 1000x + 100x^2 = 196x^2 + 4900$$
$$0 = 96x^2 - 1000x + 2400$$

This equation can be factored, but because the numbers are so large it is easier to use the quadratic formula and a calculator:

$$x = \frac{1000 \pm \sqrt{(1000)^2 - 4(96)(2400)}}{2(96)}$$
$$= \frac{1000 \pm 280}{192} = 6\frac{2}{3} \text{ or } 3\frac{3}{4}$$

Point C should be either $6\frac{2}{3}$ mi or $3\frac{3}{4}$ mi from B for the bird to use exactly 170 kcal of energy during its flight.

(b) By the Pythagorean Theorem, the length of the route directly from A to D is $\sqrt{5^2 + 12^2} = 13$ mi, so the energy the bird requires for that route is $14 \times 13 = 182$ kcal. This is more energy than the bird has, so that route is impossible. ■

EXERCISES 1.6

In Exercises 1–8 express the given quantity in terms of the indicated variable.

1. The sum of three consecutive integers; x = the first integer of the three

2. The sum of four consecutive odd integers; n = the first integer of the four

3. The perimeter (in cm) of a rectangle that is 3 cm longer than it is wide; w = width of the rectangle, in cm

4. The perimeter of a rectangle that is 4 times as long as it is wide; w = width of the rectangle

5. The time (in hours) it takes to travel a given distance at 55 mi/h; d = the given distance, in mi

6. The distance (in miles) traveled when driving at a certain speed for 2 h, then driving 15 mi/h faster for another $1\frac{1}{2}$ h; s = initial speed, in mi/h

(handwritten: $d = 2s + 1.5(s+15) = 3.5s + 22.5$)

7. The amount of alcohol (in mL) in a 750-mL bottle of vodka; c = concentration of alcohol in the vodka, in percent

8. The value (in cents) of the change in a purse that contains twice as many nickels as pennies, four more dimes than nickels, and as many quarters as dimes and nickels combined; p = number of pennies

(handwritten: $N = 2P$, $D = 2P+4$, $Q = 4P+4$; $P + 10P + 10(2P+4) + 25(4P+4) = 131P + 140$)

9. The length of a rectangular garden is 7 m greater than its width. If its perimeter is 210 m, what are the dimensions of the garden?

10. During his major league career, Hank Aaron hit 31 more home runs than Babe Ruth. Together they hit 1459 home runs. How many home runs did Babe Ruth hit?

11. A cash box contains $525 in $5 and $10 bills. The total number of bills is 74. How many bills of each denomination are there?

12. A change purse contains an equal number of pennies, nickels, and dimes. The total value of the coins is $1.44. How many coins of each type does the purse contain?

13. Mary has $3 made up of nickels, dimes, and quarters. If she has twice as many dimes as quarters, and five more nickels than dimes, how many of each type of coin does she have?

14. One-half John's age two years from now plus one-third his age three years ago is 20 years. How old is John?

(handwritten: $\frac{1}{2}(x+2) + \frac{1}{3}(x-3) = 20$; $\frac{1}{2}x + 1 + \frac{1}{3}x - 1 = 20$; $\frac{5}{6}x = 20$; $x = 24$)

15. A movie star, unwilling to give his age, posed the following riddle to a gossip columnist. "Seven years ago, I was eleven times as old as my daughter. Now I am four times as old as she is." How old is he?

(handwritten: x = daughter now; $4x$ = he; $4x-7 = 11(x-7)$; $4x-7 = 11x-77$)

16. A merchant mixes tea that sells for $3.00 per pound with tea that sells for $2.75 per pound to get 80 lb of a mixture that sells for $2.90 per pound. How many pounds of each type of tea does he use?

17. Steve invested $6000, part at an interest rate of 9% per year and the rest at a rate of 8% per year. After a year the total interest on these investments was $525. How much did he invest at each rate?

18. If Ravi invests $8000 at 8% per year, how much additional money must he invest at 11% to ensure that the interest he receives each year is 9% of the total invested?

(handwritten: $8000(.08) + .11x = .09(8000 + x)$; $x = 4000$)

19. An airliner took off from Kansas City for San Francisco, a distance of 2550 km, at a speed of 800 km/h. At the same time a private jet, traveling at 900 km/h, left San Francisco for Kansas City. How long after takeoff will they pass each other?

(handwritten: $\frac{1}{6}$ of an hour)

20. After robbing a bank in Dodge City, the robber gallops off at 14 mi/h. Ten minutes later the marshall leaves to pursue him at 16 mi/h. How long does it take the marshall to catch up to the bank robber?

(handwritten: t = marshall; $t + \frac{1}{6}$ = robber; $d = 16t$; $d = 14(t + \frac{1}{6})$; $16t = 14(t + \frac{1}{6})$; $t = \frac{7}{6}$)

21. Wilma drove at an average speed of 50 mi/h from her home in Boston to visit her sister in Buffalo. She stayed in Buffalo 10 h, and on the trip back averaged 45 mi/h. She returned home 29 h after leaving. How far is it from Boston to Buffalo?

(handwritten: t = to Buff; $19-t$ = to Boston; $d = 50t$; $d = 45(19-t)$)

22. A coast guard boat that patrols a river separating two countries cruises at 20 knots (nautical miles per hour) in still water. The river flows at 4 knots. For each patrol the captain of the boat is ordered to travel upstream and then return. His trip takes exactly 6 h. How far upstream does he travel?

23. Wendy took a trip from Davenport to Omaha, a distance of 300 mi. She traveled part of the way by bus and arrived at the train station just in time to complete her journey by train. The bus averaged 40 mi/h and the train 60 mi/h. The entire trip took $5\frac{1}{2}$ h. How long did she spend on the train?

24. Craig drove from Fresno to Bakersfield at 50 mi/h. Mike left at the same time but drove at 56 mi/h. When Mike arrived in Bakersfield, Craig had another 12 mi to drive. How far is it from Fresno to Bakersfield?

(handwritten: t = time; $50t + 12 = 56t$; $112mi$; $d = 50t$; $d = 56t$)

25. Two cyclists, 90 mi apart, start riding toward each other at the same time. One cycles twice as fast as the other. If they meet 2 h later, at what speeds are they traveling?

(handwritten: $r = x$; $r = 2x$; $t = 2$; $2x + 4x = 90$; $d = 2x$; $d = 4x$)

26. A woman driving a car 14 ft long is passing a bus 30 ft long. The bus is traveling at 50 mi/h. How fast must the woman drive her car so that she can pass the bus completely in 6 s, from the position shown in (a) to the position in (b)? [*Hint:* Use feet and seconds instead of miles and hours.]

(handwritten bottom, problem 16:)
x = lbs at 3.00
$80-x$ = lbs at 2.75
$300x + 2.75(80-x) = 2.90(80)$
$300x + 220 - 2.75x = 232$
$.25x = 12$
$x = 48$ at 3.00
52 at 2.75

[handwritten top-left: 1 mix / 1 dist. / 1 work   41 / 55]

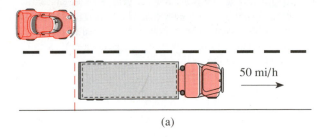

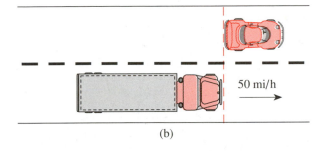

(a)

(b)

50 mi/h

50 mi/h

27. Find three consecutive odd integers with a sum of 219.

28. Dr. Plath is raising fruit flies for use in a biology experiment. She notices that every day, the number of new baby flies in her colony is always one-half the previous day's total population, but that 200 adult flies die each day. If after three days she has 3100 fruit flies, how many did she start with?

29. What quantity of a 60% acid solution must be mixed with a 30% solution to produce 300 mL of a 50% solution?

30. A health clinic uses a solution of bleach to disinfect petri dishes in which cultures are grown. They have a 100-gal tank containing a solution made up of 2% ordinary household bleach mixed with pure distilled water. New research indicates that the concentration of bleach should be 5% for complete effectiveness in killing the cultures. How much of their solution should they drain and replace with bleach to increase the bleach content to the recommended level?

[handwritten: x = amount drained + added. 3.06 gal]
$$.02(100 - x) + x = .05(100)$$

31. The radiator in a car is filled with a solution of 60% antifreeze and 40% water. The manufacturer of the antifreeze suggests that for summer driving, optimal cooling of the engine is obtained with only 50% antifreeze. If the capacity of the radiator is 3.6 L, how much coolant should be drained and replaced with water to reduce the antifreeze concentration to 50%?

[handwritten: x = amount drained + water added.]
[handwritten: work with water]
$$.4(3.6 - x) + x = .5(3.6)$$

[handwritten at top of right column: x = time together $\frac{x}{70} + \frac{x}{80} = 1$ 150x = 5600   x = 37.3]

32. Candy and Tim share a paper route. It takes Candy 70 min to deliver all the papers, while Tim takes 80 min. How long would it take them if they work together?

33. Betty and Karen have been hired to paint the houses in a new development. Working together they can paint a house in two-thirds the time that it takes Karen working alone. Betty takes 6 h to paint a house alone. How long does it take Karen to do the job alone?

[handwritten: x = Karen   together = $\frac{2}{3}x$ $\frac{2x}{3}$ $\frac{\frac{2}{3}x}{x} + \frac{\frac{2}{3}x}{6} = 1$]

34. A man is walking away from a lamppost with a light source 6 m above the ground. The man is 2 m tall. How far from the lamppost is he when his shadow is 5 m long? [*Hint:* Use similar triangles.]

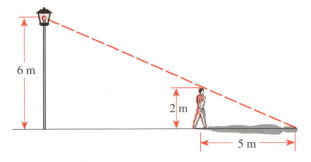

6 m

2 m

5 m

35. A rectangular building lot is 50 ft wide. The length of a diagonal between opposite corners is 10 ft more than the length of the lot. What is the length of the lot?

36. This problem is taken from a Chinese mathematics textbook called *Chui-chang suan-shu*, or *Nine Chapters on the Mathematical Art*, which was written about 250 B.C. A 10-ft-long bamboo is broken in such a way that the tip touches the ground 3 ft from the base of the stem (as shown in the figure). What is the height of the break?

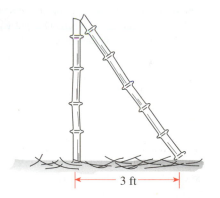

3 ft

37. Find two numbers whose sum is 55 and whose product is 684.

$x(55-x) = 684 \quad 0 = x^2 + 55x + 684$
$55x - x^2 = 684$

38. The sum of the squares of two consecutive even integers is 1252. Find the integers.

39. Helen has two posters on her bedroom door. The smaller one is twice as long as it is wide. The larger one is 4 in. wider and 12 in. longer than the smaller one, and it has twice the area of the smaller one. What are the dimensions of the smaller poster?

40. A circle has a radius of 4 cm. By how much should the radius be increased so that the area is increased by 10 cm^2?

41. A box with a square base and no top is to be made from a square piece of cardboard by cutting 4-in. squares from each corner and folding up the sides as in the figure. The box is to hold 100 in^3. How big a piece of cardboard is needed?

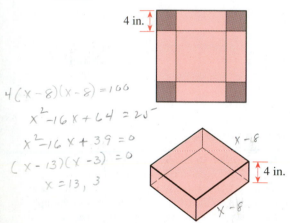

4 in.

$4(x-8)(x-8) = 100$

$x^2 - 16x + 64 = 25$

$x^2 - 16x + 39 = 0$

$(x-13)(x-3) = 0$

$x = 13, 3$

$x-8$ 4 in.

$x-8$

42. A cylindrical tin can has a volume of 40π cm^3 and is 10 cm tall. What is its diameter? [*Hint:* Use the volume formulas inside the front cover of this book.]

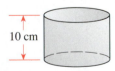

10 cm

43. A building lot is 6 ft longer than it is wide. Each diagonal from one corner to the opposite corner is 174 ft long. What are the dimensions of the lot?

44. A flagpole is secured on opposite sides by two guy wires, each of which is 5 ft longer than the pole. The distance between the points where the wires are fixed to the ground is equal to the length of one guy wire. How tall is the flagpole (to the nearest inch)?

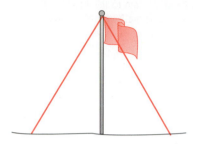

In Exercises 45 and 46 use the formula

$$h = -16t^2 + v_0 t$$

discussed in Example 7.

45. A ball is thrown straight upward at 40 ft/s.
   (a) When does it reach a height of 24 ft?
   (b) When does it reach a height of 48 ft?
   (c) What is the greatest height reached by the ball?
   (d) When does the ball reach the highest point of its path?
   (e) When does the ball hit the ground?

46. How fast would a ball have to be thrown upward to reach a maximum height of 100 ft? [*Hint:* Use the discriminant of the equation $16t^2 - v_0 t + h = 0$.]

47. The fish population in a certain lake rises and falls according to the formula

$$F = 1000(30 + 17t - t^2)$$

Here F is the number of fish at time t, where t is measured in years since January 1, 1992, when the fish population was first estimated.
   (a) When will the fish population again be the same as it was on January 1, 1992?
   (b) When will all the fish in the lake have died out?

48. A salesman drives from Ajax to Barrington, a distance of 120 mi, at a steady speed. He then increases his speed by 10 mi/h to drive the 150 mi from Barrington to Collins. If the second leg of his trip took 6 min more time than the first, how fast was he driving between Ajax and Barrington?

49. It took a crew 2 h 40 min to row 6 km upstream and back again. If the rate of flow of the stream was 3 km/h, what was the rowing rate of the crew in still water?

50. A factory is to be built on a lot measuring 180 ft by 240 ft. A local building code specifies that a lawn of uniform width and equal in area to the factory must surround the factory. What must the width of this lawn be, and what are the dimensions of the factory?

51. Henry and Irene, working together, can wash all the windows of their house in 1 h 48 min. Working alone, it takes Henry $1\frac{1}{2}$ h more than Irene to do the job. How long does it take each person alone to wash all the windows?

52. Jack, Kay, and Lynn deliver advertising flyers in a small town. Working alone, it takes Jack 4 h to deliver all the flyers, and it takes Lynn 1 h longer than it takes Kay. Working together, they can deliver all the flyers in 40% of the time it takes Kay working alone. How long does it take Kay to do the delivery alone?

53. If an imaginary line segment is drawn between the centers of the earth and the moon, then the net gravitational force F acting on an object situated on this line segment is

$$F = \frac{-K}{x^2} + \frac{0.012K}{(239 - x)^2}$$

where $K > 0$ is a constant and x is the distance of the object from the center of the earth, measured in thousands of miles. How far from the center of the earth is the "dead spot" where there is no net gravitational force acting on the object? (Express your answer to the nearest thousand miles.)

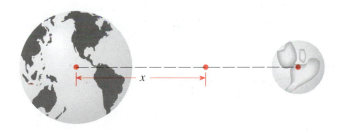

54. A man wishes to determine the water level in a deep well. He drops a stone into the well and hears the splash 3 s later. If the stone drops $16t^2$ feet after t seconds, and the speed of sound is 1090 ft/s, how far down is the surface of the water (to the nearest foot)?

55. The town of Foxton lies 10 mi north of an abandoned east-west road that runs through Grimley, as shown in the figure. The point on the abandoned road closest to Foxton is 40 mi from Grimley. County officials wish to build a new road connecting Foxton and Grimley. They have determined that restoring the old road would cost $100,000 per mile, whereas building a completely new road would cost $200,000 per mile. How much of the abandoned road should they use (as indicated in the figure) if they intend to spend exactly $6.8 million? Would it be cheaper to build a completely new road directly between the two cities?

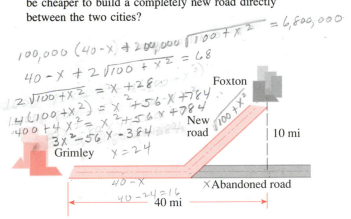

56. A boardwalk is parallel to and 210 ft inland from a straight shoreline. There is a sandy beach between the boardwalk and the shoreline. A man is standing on the boardwalk, exactly 750 ft across the sand from his beach umbrella, which is right at the water-line. The man walks 4 ft/s on the boardwalk and 2 ft/s on the sand. How far should he go on the boardwalk before veering off onto the sand if he wishes to reach his umbrella in exactly 4 min 45 s?

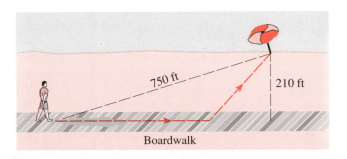

SECTION 1.7
INEQUALITIES

Some problems in algebra lead to **inequalities** instead of equations. An inequality looks just like an equation, except that in the place of the equal sign it has one of the symbols $<, >, \leq, \geq$. Here are some examples of inequalities:

$$5 < 7 \qquad (x - 1)^2 \geq 0 \qquad 3x - 1 \leq 5$$

The second inequality in this list is true for all values of the variable x, since the square of any real number is nonnegative. But the third is true only for some values of x. For example, it is true for $x = 1$, since substituting 1 for x produces the true statement $2 \leq 5$, but it is not true for $x = 3$, since $3(3) - 1 = 8$, and 8 is *not* less than or equal to 5. To **solve** an inequality that contains a variable means to find all values of the variable that make the inequality true. The methods we use for solving inequalities are very similar to the methods we use for solving equations.

When simplifying inequalities, we use the following rules.

RULES FOR INEQUALITIES

1. If $a \leq b$, then $a + c \leq b + c$.
2. If $a \leq b$ and $c \leq d$, then $a + c \leq b + d$.
3. If $a \leq b$ and $c > 0$, then $ac \leq bc$.
4. If $a \leq b$ and $c < 0$, then $ac \geq bc$.
5. If $0 < a < b$, then $1/a > 1/b$.

Rules 1–4 apply more generally if the relation $\leq$ is replaced by any of the other order relations $\geq$, $<$, and $>$. Rule 1 says that we can add (or subtract) any number to (or from) both sides of an inequality, and Rule 2 says that two inequalities can be added. However, we have to be careful with multiplication. Rule 3 says that we can multiply (or divide) both sides of an inequality by a *positive* number, but Rule 4 says that if we multiply both sides of an inequality by a negative number, then we reverse the direction of the inequality. For example, if we take the inequality

$$3 < 5$$

and multiply by 2, we get

$$6 < 10$$

but if we multiply by -2, we get

$$-6 > -10$$

Finally, Rule 5 says that if we take reciprocals, then we reverse the direction of an inequality (provided the numbers are positive).

EXAMPLE 1

Solve the inequality $1 + x < 7x + 5$ and sketch the solution set.

SOLUTION

We wish to simplify this to an equivalent inequality in which the variable x stands alone on one side of the inequality symbol. First we subtract 1 from each side of the inequality (using Rule 1 with $c = -1$):

$$x < 7x + 4$$

Then we subtract $7x$ from both sides (Rule 1 with $c = -7x$), to combine the terms involving x:

$$-6x < 4$$

Now we divide both sides by -6 $\left(\text{Rule 4 with } c = -\frac{1}{6}\right)$:

$$x > -\frac{4}{6} = -\frac{2}{3}$$

The solution set consists of all numbers greater than $-\frac{2}{3}$. In other words, the solution of the inequality is the interval $\left(-\frac{2}{3}, \infty\right)$. It is graphed in Figure 1. ■

Figure 1

EXAMPLE 2

Solve the inequalities $4 \leqslant 3x - 2 < 13$.

SOLUTION

The solution set consists of all values of x that satisfy both inequalities. Using Rules 1 and 3, we see that the following inequalities are equivalent:

$$4 \leqslant 3x - 2 < 13$$
$$6 \leqslant 3x < 15 \qquad \text{(adding 2)}$$
$$2 \leqslant x < 5 \qquad \text{(dividing by 3)}$$

Therefore the solution set is $[2, 5)$. It is shown in Figure 2. ■

Figure 2

EXAMPLE 3

Solve $2x + 1 \leqslant 4x - 3 \leqslant x + 7$.

SOLUTION

Since all three expressions in these inequalities involve the variable x, it is impossible to deal with both inequalities simultaneously as we did in Example 2. So here we first solve the inequalities separately:

$$2x + 1 \leqslant 4x - 3 \qquad\qquad 4x - 3 \leqslant x + 7$$
$$4 \leqslant 2x \qquad\qquad\qquad 3x \leqslant 10$$
$$2 \leqslant x \qquad\qquad\qquad\quad x \leqslant \tfrac{10}{3}$$

Since x must satisfy both inequalities, we have

$$2 \leqslant x \leqslant \tfrac{10}{3}$$

Figure 3

Thus the solution set is the closed interval $\left[2, \tfrac{10}{3}\right]$ and is shown in Figure 3.   ■

EXAMPLE 4

The instructions on a box of film indicate that it should be stored at a temperature between 5° and 30° Celsius. What range of temperatures does this correspond to on the Fahrenheit scale?

SOLUTION

The relationship between degrees Celsius (C) and degrees Fahrenheit (F) is given by the equation $C = \tfrac{5}{9}(F - 32)$. Expressing the statement on the film box in terms of inequalities, we have

$$5 < C < 30$$

so that the corresponding Fahrenheit temperatures satisfy

$$5 < \tfrac{5}{9}(F - 32) < 30$$
$$\tfrac{9}{5} \cdot 5 < F - 32 < \tfrac{9}{5} \cdot 30$$
$$9 < F - 32 < 54$$
$$9 + 32 < F < 54 + 32$$
$$41 < F < 86$$

The film should be stored at a temperature between 41 °F and 86 °F.   ■

So far we have learned how to solve linear inequalities. To solve inequalities involving squares and other powers, we use factoring and the following fact: The product of two expressions is positive if both are positive or both are negative, while the product is negative if one expression is positive and the other is negative.

EXAMPLE 5

Solve the inequality $x^2 - 5x + 6 \leqslant 0$.

SOLUTION

First we factor the left side: $(x - 2)(x - 3) \leqslant 0$

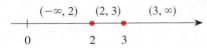

Figure 4

We know that the corresponding equation $(x - 2)(x - 3) = 0$ has the solutions 2 and 3. These are the only points at which the expression $(x - 2)(x - 3)$ can change sign. As shown in Figure 4, the numbers 2 and 3 divide the real line into three intervals: $(-\infty, 2)$, $(2, 3)$, and $(3, \infty)$. On each of these intervals we determine the signs of the factors. For instance,

$$x \in (-\infty, 2) \;\Rightarrow\; x < 2 \;\Rightarrow\; x - 2 < 0$$

and $\qquad\qquad\qquad x < 2 \;\Rightarrow\; x < 3 \;\Rightarrow\; x - 3 < 0$

We then record these signs in the following chart. We get the sign in the third column by multiplying the entries of the first two columns.

Interval	$x - 2$	$x - 3$	$(x - 2)(x - 3)$
$x < 2$	$-$	$-$	$+$
$2 < x < 3$	$+$	$-$	$-$
$x > 3$	$+$	$+$	$+$

Another method for obtaining the information in the chart is to use **test values**. We choose a value in each of the intervals, and check the sign of the factors $x - 2$ and $x - 3$ at the values selected. For instance, if we use the test value $x = 1$ for the interval $(-\infty, 2)$, then substitution in $x - 2$ and $x - 3$ gives

$$x - 2 = 1 - 2 = -1 < 0$$

and $\qquad\qquad\qquad x - 3 = 1 - 3 = -2 < 0$

Since neither of the factors $x - 2$ and $x - 3$ changes sign on the interval $(-\infty, 2)$, we get the signs indicated in the first line of the chart. Similarly, we can use the test value $x = \frac{5}{2}$ for the interval $(2, 3)$ and $x = 4$ for the interval $(3, \infty)$ to obtain the remaining signs in the chart.

We read from the completed chart that $(x - 2)(x - 3)$ is negative when $2 < x < 3$. Thus the solution of the inequality $(x - 2)(x - 3) \leq 0$ is

$$\{x \mid 2 \leq x \leq 3\} = [2, 3]$$

Notice that we have included the endpoints 2 and 3 because we seek values of x such that the product is either negative or zero. The solution is illustrated in Figure 5.

Figure 5

When using the factoring technique illustrated in the preceding example, it is important that all nonzero terms appear on one side of the inequality symbol. If this is not the case, the inequality first must be rewritten, as we see in the next example. We also illustrate a second solution technique that can be applied to problems of this type.

EXAMPLE 6

Solve the inequality $x^2 + 3x > 4$.

SOLUTION 1

First we take all nonzero terms to one side of the inequality sign and factor the resulting expression.

$$x^2 + 3x - 4 > 0 \quad \text{or} \quad (x - 1)(x + 4) > 0$$

As in Example 5, we solve the corresponding equation $(x - 1)(x + 4) = 0$ and use the solutions $x = 1$ and $x = -4$ to divide the real line into the three intervals $(-\infty, -4)$, $(-4, 1)$, and $(1, \infty)$. On each interval the product keeps a constant sign as shown in the following chart. The signs in the chart could be obtained by using, for example, the test values -5, 0, and 2, respectively, for the three intervals.

Interval	$x - 1$	$x + 4$	$(x - 1)(x + 4)$
$x < -4$	$-$	$-$	$+$
$-4 < x < 1$	$-$	$+$	$-$
$x > 1$	$+$	$+$	$+$

We read from the chart that the solution set is

$$\{x \mid x < -4 \text{ or } x > 1\} = (-\infty, -4) \cup (1, \infty)$$

The solution is illustrated in Figure 6.

Figure 6

SOLUTION 2

As in the first solution we write the inequality in the form

$$(x - 1)(x + 4) > 0$$

Then we use the fact that the product of two numbers is positive when both factors are positive or when both factors are negative. We examine each of these two cases separately.

Case I   When $x - 1 > 0$ and $x + 4 > 0$, we have $x > 1$ and $x > -4$. But if $x > 1$, then $x > -4$ holds automatically. So the solution in this case is given by $x > 1$.

Case II   When $x - 1 < 0$ and $x + 4 < 0$, we have $x < 1$ and $x < -4$. But if $x < -4$, then $x < 1$ holds automatically. So the solution in this case is given by $x < -4$.

Combining both cases, we see, as before, that the solution set is

$$\{x \mid x < -4 \text{ or } x > 1\} = (-\infty, -4) \cup (1, \infty) \qquad \blacksquare$$

In the remaining examples in this section we will use the method of the first solution in the preceding example. In the next example, we solve an inequality involving a quotient instead of a product.

EXAMPLE 7

Solve $\dfrac{1 + x}{1 - x} \geqslant 1$.

SOLUTION

It is tempting to simplify the inequality by multiplying both sides by $1 - x$. But this does not work, since we do not know whether $1 - x$ is positive or negative. If $1 - x$ turns out to be negative for some values of x in the solution set, then multiplying by $1 - x$ would reverse the direction of the inequality. Therefore we must use the following technique in solving inequalities of this type. First we take all nonzero terms to the left side, and then we simplify using a common denominator:

$$\frac{1 + x}{1 - x} \geqslant 1$$

$$\frac{1 + x}{1 - x} - 1 \geqslant 0$$

$$\frac{1 + x}{1 - x} - \frac{1 - x}{1 - x} \geqslant 0$$

$$\frac{1 + x - 1 + x}{1 - x} \geqslant 0$$

$$\frac{2x}{1 - x} \geqslant 0$$

The numerator is zero when $x = 0$ and the denominator is zero when $x = 1$. As before, we set up a chart to determine the sign on each of the intervals $(-\infty, 0)$, $(0, 1)$, and $(1, \infty)$.

Interval	$2x$	$1 - x$	$2x/(1 - x)$
$x < 0$	−	+	−
$0 < x < 1$	+	+	+
$x > 1$	+	−	−

From the chart we see that the solution set is $\{x \mid 0 \leqslant x < 1\} = [0, 1)$. We include the endpoint 0 since the original inequality requires the quotient to be greater than *or equal to* 1. However, we do not include the endpoint 1, since the quotient in the inequality is not defined at 1. Always check the endpoints of the solution intervals to see whether they satisfy the original inequality.

The solution set $[0, 1)$ is illustrated in Figure 7.   ■

Figure 7

Example 7 shows why we must never multiply both sides of an inequality by an expression unless we know the sign of that expression.

EXAMPLE 8

A stone thrown straight up into the air at 96 ft/s reaches a height of h feet after t seconds, where h and t are related by the formula

$$h = 96t - 32t^2$$

During what time interval will the stone be at least 64 ft above the ground?

SOLUTION

We wish to determine for what times t it is true that $h \geq 64$. Thus we must solve the inequality $96t - 32t^2 \geq 64$. After first dividing by the common factor 32, we get

$$3t - t^2 \geq 2$$
$$0 \geq t^2 - 3t + 2$$
$$0 \geq (t - 1)(t - 2)$$

This leads to the following chart.

Interval	$t - 1$	$t - 2$	$(t - 1)(t - 2)$
$t < 1$	$-$	$-$	$+$
$1 < t < 2$	$+$	$-$	$-$
$t > 2$	$+$	$+$	$+$

Thus the solution of the inequality is the interval $[1, 2]$. The stone is at least 64 ft high between 1 and 2 seconds (inclusive) after it is thrown. ■

EXERCISES 1.7

In Exercises 1–48 solve the given inequality. Express each solution in interval form and illustrate the solution set on the real number line.

1. $2x \leq 10$

2. $-3x > 5$

3. $2x + 7 > 3$

4. $3x - 11 < 4$

5. $1 - x \leq 2$

6. $4 - 3x \geq 6$

7. $2x + 1 < 5x - 8$

8. $1 + 5x > 5 - 3x$

9. $4 - 3x \leq -(1 + 8x)$

10. $2(7x - 3) \leq 12x + 16$

11. $-1 < 2x - 5 < 7$

12. $1 < 3x + 4 \leq 16$

13. $0 \leq 1 - x < 1$

14. $-5 \leq 3 - 2x \leq 9$

15. $4x < 2x + 1 \leq 3x + 2$

16. $2x + 3 < x + 4 < 3x - 2$

17. $1 - x \geq 3 - 2x \geq x - 6$

18. $x > 1 - x \geq 3 + 2x$

19. $\dfrac{x-1}{2} \geqslant \dfrac{2}{3} \geqslant 1 - \dfrac{x}{6}$ **20.** $4x < \dfrac{1-x}{12} < \dfrac{2x+1}{3}$

21. $\dfrac{1}{x} < 4$ **22.** $-3 < \dfrac{1}{x} \leqslant 1$

23. $(x-1)(x-2) > 0$ **24.** $(2x+3)(x-1) \geqslant 0$

25. $x^2 - 2x - 8 \leqslant 0$ **26.** $x^2 + 5x + 6 > 0$

27. $2x^2 + x \leqslant 1$ **28.** $x^2 < 2x + 8$

29. $3x^2 - 6x > 2x^2 - 3x + 4$

30. $5x^2 + 3x \geqslant 3x^2 + 2$

31. $x^2 > 3(x+6)$ **32.** $3x^2 + 2x > 1$

33. $x^2 < 4$ **34.** $x^2 \geqslant 5$

35. $-6x^2 \leqslant 12$ **36.** $-x^3 > x^2$

37. $\dfrac{x-1}{x+2} \geqslant 0$ **38.** $\dfrac{2x+6}{x-6} < 0$

39. $\dfrac{x}{x+3} > 4$ **40.** $-2 < \dfrac{x+1}{x-3}$

41. $\dfrac{2x+1}{x-5} \leqslant 3$ **42.** $\dfrac{2+x}{3-x} \geqslant 1$

43. $\dfrac{4}{x} < x$ **44.** $\dfrac{x}{x+1} > 3x$

45. $\dfrac{x^2-1}{x^2+1} \geqslant 0$ **46.** $x^3 > x$

47. $1 + \dfrac{2}{x+1} \leqslant \dfrac{2}{x}$ **48.** $\dfrac{3}{x-1} - \dfrac{x}{x+1} \geqslant 1$

49. Use the relationship between C and F given in Example 4 to find the interval on the Fahrenheit scale corresponding to the temperature range $20 \leqslant C \leqslant 30$.

50. What interval on the Celsius scale corresponds to the temperature range $50 \leqslant F \leqslant 95$?

51. A charter airline finds that on its Saturday morning flights from Philadelphia to London, it is able to sell all 120 seats if it prices its tickets at $200. However, for each $3 increase in ticket price, the number of seats sold decreases by 1.
(a) Find a formula for the number of seats sold if the ticket price is P dollars.
(b) Over a certain period, the number of seats sold ranged between 90 and 115. What was the corresponding range of ticket prices?

52. As dry air moves upward, it expands and in so doing cools at a rate of about 1 °C for each 100 m rise, up to about 12 km.
(a) If the ground temperature is 20 °C, write a formula for the temperature at height h.
(b) What range of temperatures can be expected if a plane takes off and reaches a maximum height of 5 km?

53. Using calculus it can be proved that if a ball is thrown upward from the top of a building 128 ft high with an initial velocity of 16 ft/s, then the height h above the ground t seconds later will be

$$h = 128 + 16t - 16t^2$$

During what time interval will the ball be at least 32 ft above the ground?

54. The gravitational force F exerted by the earth on an object having a mass of 100 kg is given by the equation

$$F = \dfrac{4{,}000{,}000}{d^2}$$

where d is the distance (in km) of the object from the center of the earth, and the force F is measured in newtons. For what distances will the gravitational force exerted by the earth on this object be between 0.0004 N and 0.01 N?

In Exercises 55–58 determine the values of the variable for which the expression is defined as a real number.

55. $\sqrt{16 - 9x^2}$ **56.** $\sqrt{3x^2 - 5x + 2}$

57. $\left(\dfrac{1}{x^2 - 5x - 14}\right)^{1/2}$ **58.** $\sqrt[4]{\dfrac{1-x}{2+x}}$

59. Prove Rule 1. [*Hint:* Remember that $a < b$ means that $b - a > 0$.]

60. Prove Rule 3. [See the hint for Exercise 59 and remember that the product of two positive numbers is positive.]

61. Prove Rule 4. [*Hint:* Use the fact that the product of a positive number and a negative number is negative.]

62. Prove Rule 5.

In Exercises 63 and 64 solve the inequality for x assuming that a, b, and c are positive constants.

63. $a(bx - c) \geqslant bc$ **64.** $a \leqslant bx + c < 2a$

In Exercises 65 and 66 solve the inequality for x assuming that a, b, and c are negative constants.

65. $ax + b < c$

66. $\dfrac{ax + b}{c} \le b$

67. Show that if $a < b$, then $a < \dfrac{a + b}{2} < b$.

68. Show that if $0 < a < b$, then $a^2 < b^2$.

69. Suppose that a, b, c, and d are positive numbers such that

$$\frac{a}{b} < \frac{c}{d}$$

Show that

$$\frac{a}{b} < \frac{a + c}{b + d} < \frac{c}{d}$$

SECTION 1.8
ABSOLUTE VALUE

Recall from Section 1.1 that the absolute value of a number a is given by

$$|a| = \begin{cases} a & \text{if } a \ge 0 \\ -a & \text{if } a < 0 \end{cases}$$

and that it represents the distance from a to the origin on the real number line (see Figure 1).

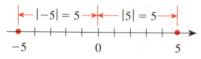

Figure 1

$\int$ **EXAMPLE 1**

Express $|2x - 1|$ without using the absolute value symbol.

SOLUTION

$$|2x - 1| = \begin{cases} 2x - 1 & \text{if } 2x - 1 \ge 0 \\ -(2x - 1) & \text{if } 2x - 1 < 0 \end{cases}$$

$$= \begin{cases} 2x - 1 & \text{if } x \ge \frac{1}{2} \\ 1 - 2x & \text{if } x < \frac{1}{2} \end{cases}$$
■

The following properties can be proved using the definition of absolute value.

PROPERTIES OF ABSOLUTE VALUE

Suppose a and b are any real numbers and n is an integer. Then

1. $|ab| = |a|\,|b|$ **2.** $\left|\dfrac{a}{b}\right| = \dfrac{|a|}{|b|}$ $(b \ne 0)$ **3.** $|a^n| = |a|^n$

EXAMPLE 2

Simplify the expression $|2x - 6|$.

SOLUTION

Using Property 1 we have

$$|2x - 6| = |2(x - 3)| = |2| \, |x - 3| = 2|x - 3|$$ ■

When solving equations or inequalities involving absolute values, it is often very helpful to use the following properties:

FURTHER PROPERTIES OF ABSOLUTE VALUE

Suppose $a > 0$. Then

4. $|x| = a$ if and only if $x = \pm a$
5. $|x| < a$ if and only if $-a < x < a$
6. $|x| > a$ if and only if $x > a$ or $x < -a$.

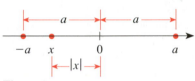

Figure 2

For instance, the inequality $|x| < a$ says that the distance from x to the origin is less than a, and you can see from Figure 2 that this is true if and only if x lies between $-a$ and a.

EXAMPLE 3

Solve the equation $|2x - 5| = 3$.

SOLUTION

By Property 4, $|2x - 5| = 3$ is equivalent to

$$2x - 5 = 3 \quad \text{or} \quad 2x - 5 = -3$$

So $\qquad\qquad\qquad 2x = 8 \quad \text{or} \quad 2x = 2$

Thus $\qquad\qquad\qquad x = 4 \quad \text{or} \quad x = 1$ ■

EXAMPLE 4

Solve the inequality $|x - 5| < 2$.

SOLUTION 1

By Property 5, $|x - 5| < 2$ is equivalent to

$$-2 < x - 5 < 2$$

Therefore, adding 5 to each side, we have

$$3 < x < 7$$

and the solution set is the open interval (3, 7).

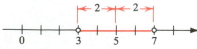

Figure 3

SOLUTION 2

Geometrically, the solution set consists of all numbers x whose distance from 5 is less than 2. From Figure 3 we see that this is the interval (3, 7). ∎

EXAMPLE 5

Solve the inequality $|3x + 2| \geq 4$.

SOLUTION

By Properties 4 and 6, $|3x + 2| \geq 4$ is equivalent to

$$3x + 2 \geq 4 \qquad \text{or} \qquad 3x + 2 \leq -4$$

In the first case $3x \geq 2$, which gives $x \geq \frac{2}{3}$. In the second case $3x \leq -6$, which gives $x \leq -2$. So the solution set is

$$\left\{x \mid x \leq -2 \quad \text{or} \quad x \geq \tfrac{2}{3}\right\} = (-\infty, -2] \cup \left[\tfrac{2}{3}, \infty\right)$$

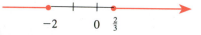

Figure 4

It is graphed in Figure 4. ∎

EXAMPLE 6

Solve the inequality $\dfrac{3}{|2x - 5|} \geq 1$.

SOLUTION

Since the absolute value of any number is always nonnegative, if we multiply both sides of the inequality by $|2x - 5|$, we will not reverse the direction of the inequality. So, as long as $2x - 5 \neq 0$, or $x \neq \frac{5}{2}$, the following inequality is equivalent to the one we are solving:

$$3 \geq |2x - 5|$$
$$3 \geq 2x - 5 \geq -3 \qquad \text{(Property 5)}$$
$$8 \geq 2x \geq 2 \qquad \text{(adding 5)}$$
$$4 \geq x \geq 1$$

Figure 5

Thus the solution consists of all numbers in the interval [1, 4] except the number $\frac{5}{2}$. We write this set as $\left[1, \frac{5}{2}\right) \cup \left(\frac{5}{2}, 4\right]$. It is graphed in Figure 5. ∎

Another important property of absolute value, called the Triangle Inequality, is used frequently not only in calculus but throughout mathematics in general.

THE TRIANGLE INEQUALITY

> If a and b are any real numbers, then
>
> $$|a + b| \leqslant |a| + |b|$$

Observe that if the numbers a and b are both positive or negative, then the two sides in the Triangle Inequality are actually equal. But if a and b have opposite signs, the left side involves a subtraction and the right side does not. This makes the Triangle Inequality seem reasonable, but we can prove it as follows.

Notice that

$$-|a| \leqslant a \leqslant |a|$$

is always true because a equals either $|a|$ or $-|a|$. If we write the corresponding statement for b, we have

$$-|b| \leqslant b \leqslant |b|$$

and adding these inequalities we get

$$-\big(|a| + |b|\big) \leqslant a + b \leqslant |a| + |b|$$

If we now apply Properties 4 and 5 $\big($with x replaced by $a + b$ and a by $|a| + |b|\big)$ we obtain

$$|a + b| \leqslant |a| + |b|$$

which is what we wanted to show.

EXAMPLE 7

If $|x - 4| < 0.1$ and $|y - 7| < 0.2$, use the Triangle Inequality to estimate $|(x + y) - 11|$.

SOLUTION

In order to use the given information, we use the Triangle Inequality with $a = x - 4$ and $b = y - 7$:

$$
\begin{aligned}
|(x + y) - 11| &= |(x - 4) + (y - 7)| \\
&\leqslant |x - 4| + |y - 7| \\
&< 0.1 + 0.2 = 0.3
\end{aligned}
$$

Thus $|(x + y) - 11| < 0.3$ ■

EXERCISES 1.8

In Exercises 1–6 use the definition of absolute value to write the given expression without the absolute value symbol, as in Example 1.

1. $|x - 3|$

2. $|4x - 7|$

3. $|3x - 10|$

4. $|8 - 5x|$

5. $|x^2 + 1|$

6. $|x^2 - 1|$

In Exercises 7–12 simplify the expression, as in Example 2.

7. $|3x + 9|$

8. $|4x - 16|$

9. $\left|\frac{1}{2}x - \frac{5}{2}\right|$

10. $|-2x - 10|$

11. $|-x^2 - 9|$

12. $\left|\frac{x - 1}{1 - x}\right|$

In Exercises 13–20 solve the equation.

13. $|2x| = 3$

14. $|3x + 5| = 1$

15. $|x - 4| = 0.01$

16. $|x - 6| = -1$

17. $\left|\frac{2x - 1}{x + 1}\right| = 3$

18. $6 + 3|x + 5| = 5$

19. $|x - 1| = |3x + 2|$

20. $|x + 3| = |2x + 1|$

In Exercises 21–40 solve the given inequality. Express your answer using interval notation.

21. $|x| < 3$

22. $|x| \geq 3$

23. $|x - 4| < 1$

24. $|x - 6| < 0.1$

25. $|x + 5| \geq 2$

26. $|x + 1| \geq 3$

27. $|2x - 3| \leq 0.4$

28. $|5x - 2| < 6$

29. $\left|\frac{x - 2}{3}\right| < 2$

30. $\left|\frac{x + 1}{2}\right| \geq 4$

31. $|x + 6| < 0.001$

32. $|x - a| < d$

33. $1 \leq |x| \leq 4$

34. $0 < |x - 5| \leq \frac{1}{2}$

35. $\frac{1}{|x + 7|} > 2$

36. $\left|1 - \frac{1}{x}\right| \leq 2$

37. $|x| > |x - 1|$

38. $|2x - 5| \leq |x + 4|$

39. $\left|\frac{x}{2 + x}\right| < 1$

40. $\left|\frac{2 - 3x}{1 + 2x}\right| \leq 4$

41. Prove that $|ab| = |a|\,|b|$.

42. Prove that $\left|\frac{a}{b}\right| = \frac{|a|}{|b|}$.

43. Suppose that $|x - 2| < 0.01$ and $|y - 3| < 0.04$. Show that $|(x + y) - 5| < 0.05$.

44. If $|a - 1| < 2$ and $|b - 1| < 3$, show that $|a + b - 2| < 5$.

45. Show that $|x - y| \leq |x| + |y|$ for all real numbers x and y.

46. Show that $|x - y| \geq |x| - |y|$ for all real numbers x and y. [*Hint:* Use the Triangle Inequality with $a = x - y$ and $b = y$.]

47. Prove that $|a| = \sqrt{a^2}$ for any real number a.

SECTION 1.9
COORDINATE GEOMETRY

Just as the points on a line can be identified with real numbers by assigning them coordinates, as described in Section 1.1, so the points in a plane can be identified with ordered pairs of real numbers. We start by drawing two perpendicular coordinate lines that intersect at the origin O on each line. Usually one line is horizontal with positive direction to the right and is called the **x-axis**; the other line is vertical with positive direction upward and is called the **y-axis**.

Any point P in the plane can be located by a unique ordered pair of numbers as follows. Draw lines through P perpendicular to the x- and y-axes. These lines will intersect the axes in points with coordinates a and b as shown in Figure 1. Then the point P is assigned the ordered pair (a, b). The first number a is called the **x-coordinate** (or **abscissa**) of P; the second number is called the **y-coordinate** (or **ordinate**) of P. We say that P is the point with coordinates (a, b) and we denote the point by the symbol $P(a, b)$. We can think of the coordinates of a point as its "address" in the plane, since the coordinates specify the location of the point. Several points are labeled with their coordinates in Figure 2.

The coordinates of a point in the xy-plane uniquely determine its location. We can think of the coordinates as the "address" of the point. In Salt Lake City, Utah, the addresses of most buildings are in fact expressed as coordinates. The city is divided into quadrants with Main Street as the vertical (North-South) axis and S. Temple Street as the horizontal (East-West) axis. An address such as

1760 W   2100 S

indicates a location 17.6 blocks west of Main Street and 21 blocks south of S. Temple Street. (This is the address of the main post office in Salt Lake City.) With this logical system it is possible for someone unfamiliar with the city to locate any address immediately, as easily as one can locate a point on the coordinate plane.

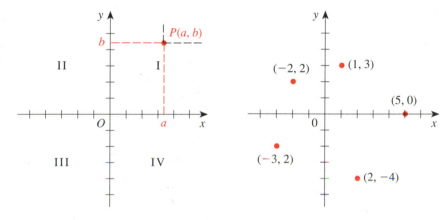

Figure 1   **Figure 2**

Post Office
1760 W 2100 S

By reversing the preceding process we can start with an ordered pair (a, b) and arrive at the corresponding point P. Often we identify the point P with the ordered pair (a, b) and refer to "the point (a, b)." [Although the notation used for an open interval (a, b) is the same as the notation used for a point (a, b), you will be able to tell from the context which meaning is intended.]

This coordinate system is called the **rectangular coordinate system** or the **Cartesian coordinate system** in honor of the French mathematician René Descartes (1596–1650). [Another Frenchman, Pierre de Fermat (1601–1665), also invented the principles of analytic geometry at about the same time as Descartes.] The plane, supplied with this coordinate system, is called the **coordinate plane** or the **Cartesian plane** and is denoted by R^2.

The x- and y-axes are called **coordinate axes** and divide the Cartesian plane into four quadrants, which are labeled I, II, III, and IV in Figure 1. Notice that the first quadrant consists of those points whose x- and y-coordinates are both positive. (The points *on* the coordinate axes are not assigned to any quadrant.)

René Descartes (1596–1650) was born at the town of La Haye in southern France, the son of a provincial judge. At an early age Descartes liked mathematics because of the certainty of its results. His interest in truth and falsehood led him to study philosophy, a subject he was to revolutionize. His most important work is his book Discourse on the Method of Rightly Conducting the Reason in the Search for Truth in the Sciences. In a famous appendix to this book, Descartes invented what is now called the Cartesian plane. This idea of combining algebra and geometry allowed mathematicians for the first time to "see" the functions they were studying. The philosopher John Stuart Mill called the invention of the Cartesian plane "the greatest single step ever made in the progress of the exact sciences." Descartes liked to get up late and spend the morning in bed thinking and writing. However, in 1649 he became the tutor of Queen Christina of Sweden and she liked her lessons at 5 A.M., when, she said, her mind was sharpest. Descartes, of course, conformed to the Queen's schedule. But the change from his habits combined with the ice-cold library where they studied proved too much for him. In February 1650, after just two months of this, he caught pneumonia and died.

EXAMPLE 1

Describe and sketch the regions given by the following sets:
(a) $\{(x, y) \mid x \geq 0\}$ (b) $\{(x, y) \mid y = 1\}$ (c) $\{(x, y) \mid |y| < 1\}$

SOLUTION

(a) The points whose x-coordinates are zero or positive lie on the y-axis or to the right of it [see Figure 3(a)].
(b) The set of all points with y-coordinate 1 is a horizontal line one unit above the x-axis [see Figure 3(b)].
(c) Recall from Section 1.8 that

$$|y| < 1 \qquad \text{if and only if} \qquad -1 < y < 1$$

The given region consists of those points on the plane whose y-coordinates lie between -1 and 1. Thus the region consists of all points that lie between (but not on) the horizontal lines $y = 1$ and $y = -1$. These lines are shown as broken lines in Figure 3(c) to indicate that the points on these lines do not lie in the set.

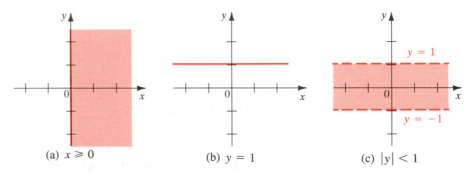

(a) $x \geq 0$ (b) $y = 1$ (c) $|y| < 1$

Figure 3

Recall from Section 1.1 that the distance between points a and b on a number line is $|a - b| = |b - a|$. Thus the distance between points $P_1(x_1, y_1)$ and $P_3(x_2, y_1)$ on a horizontal line must be $|x_2 - x_1|$ and the distance between $P_2(x_2, y_2)$ and $P_3(x_2, y_1)$ on a vertical line must be $|y_2 - y_1|$ (see Figure 4).

To find the distance $|P_1P_2|$ between any two points $P_1(x_1, y_1)$ and $P_2(x_2, y_2)$, we note that the triangle $P_1P_2P_3$ in Figure 4 is a right triangle, and so by the Pythagorean Theorem we have

$$|P_1P_2| = \sqrt{|P_1P_3|^2 + |P_2P_3|^2} = \sqrt{|x_2 - x_1|^2 + |y_2 - y_1|^2}$$
$$= \sqrt{(x_2 - x_1)^2 + (y_2 - y_1)^2}$$

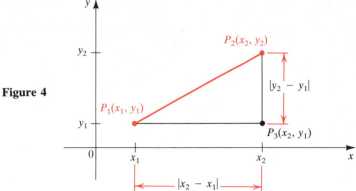

Figure 4

DISTANCE FORMULA

The distance between the points $P_1(x_1, y_1)$ and $P_2(x_2, y_2)$ is

$$|P_1P_2| = \sqrt{(x_2 - x_1)^2 + (y_2 - y_1)^2}$$

EXAMPLE 2

Which of the points $P(1, -2)$ or $Q(8, 9)$ is closer to the point $A(5, 3)$?

SOLUTION

The distance from P to A is

$$|PA| = \sqrt{(5 - 1)^2 + [3 - (-2)]^2} = \sqrt{4^2 + 5^2} = \sqrt{41}$$

Similarly $|QA| = \sqrt{(5 - 8)^2 + (3 - 9)^2} = \sqrt{(-3)^2 + (-6)^2} = \sqrt{45}$

This shows that $|PA| < |QA|$, so P is closer to A. ■

Let us now find the coordinates (x, y) of the midpoint M of the line segment that joins the point $P_1(x_1, y_1)$ to the point $P_2(x_2, y_2)$. Notice that triangles P_1AM and MBP_2 in Figure 5 are congruent because $|P_1M| = |MP_2|$ and corresponding angles are equal. It follows that $|P_1A| = |MB|$ and so

$$x - x_1 = x_2 - x$$

Solving this equation for x_1, we get

$$2x = x_1 + x_2 \qquad x = \frac{x_1 + x_2}{2}$$

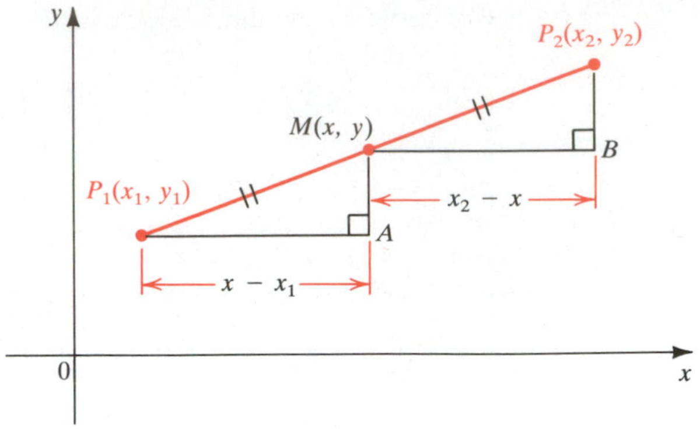

Figure 5

Similarly, $y = (y_1 + y_2)/2$.

MIDPOINT FORMULA

The midpoint of the line segment from $P_1(x_1, y_1)$ to $P_2(x_2, y_2)$ is

$$\left(\frac{x_1 + x_2}{2}, \frac{y_1 + y_2}{2}\right)$$

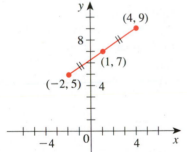

Figure 6

EXAMPLE 3

The midpoint of the line segment that joins $(-2, 5)$ and $(4, 9)$ is

$$\left(\frac{-2 + 4}{2}, \frac{5 + 9}{2}\right) = (1, 7)$$

(See Figure 6.) ∎

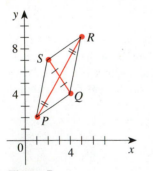

Figure 7

EXAMPLE 4

Show that the quadrilateral with vertices $P(1, 2)$, $Q(4, 4)$, $R(5, 9)$, and $S(2, 7)$ is a parallelogram, by proving that its two diagonals bisect each other.

SOLUTION

If the two diagonals have the same midpoint, then they must bisect each other. The midpoint of the diagonal PR is

$$\left(\frac{1 + 5}{2}, \frac{2 + 9}{2}\right) = \left(3, \frac{11}{2}\right)$$

and the midpoint of the diagonal QS is

$$\left(\frac{4 + 2}{2}, \frac{4 + 7}{2}\right) = \left(3, \frac{11}{2}\right)$$

so each diagonal bisects the other (see Figure 7). (It is a theorem from elementary geometry that the quadrilateral is therefore a parallelogram.) ∎

EQUATIONS AND GRAPHS

Suppose we have an equation involving the variables x and y, such as

$$x^2 + y^2 = 25 \quad \text{or} \quad x = y^2 \quad \text{or} \quad y = \frac{2}{x}$$

A point (x, y) **satisfies** the equation if the equation is true when the coordinates of the point are substituted into the equation. For example, the point $(3, 4)$ satisfies the equation $x^2 + y^2 = 25$, since $3^2 + 4^2 = 25$, but the point $(2, -3)$ does not, since $2^2 + (-3)^2 = 13 \neq 25$. The **graph** of such an equation in x and y is the set of all points (x, y) that satisfy the equation. The graphs of most of the equations that we will encounter are curves. For example, we will see later in this section that the first equation represents a circle; we sketch the graph of the second equation in Example 5; we investigate the graph of the third equation later in the book. Graphs are important in the study of equations because they give visual representations of equations.

EXAMPLE 5

Sketch the graph of the equation $x = y^2$.

SOLUTION

There are infinitely many points on the graph and it is not possible to plot all of them. But the more points we plot, the better we can imagine what the curve represented by the equation must look like. We list some of the points on this graph in the following table and plot them in Figure 8. Then we connect the points by a smooth curve as shown. A curve with this shape is called a *parabola*.

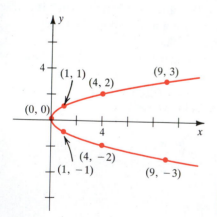

Figure 8

y	0	1	2	3	-1	-2	-3
$x = y^2$	0	1	4	9	1	4	9
(x, y)	$(0, 0)$	$(1, 1)$	$(4, 2)$	$(9, 3)$	$(1, -1)$	$(4, -2)$	$(9, -3)$

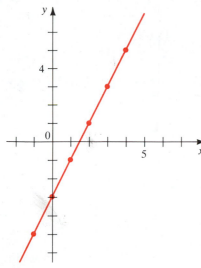

Figure 9

EXAMPLE 6

Sketch the graph of the equation $2x - y = 3$.

SOLUTION

If we solve the given equation for y, we get $y = 2x - 3$. This helps us calculate the y-coordinates in the following table.

x	-1	0	1	2	3	4
y	-5	-3	-1	1	3	5

When we plot the points with these coordinates it appears that they lie on a line, and we sketch the graph by joining the points as in Figure 9. (In the next section we verify that the graph is indeed a line.) ∎

The x-coordinates of the points where a graph intersects the x-axis are called the **x-intercepts** of the graph and are obtained by setting $y = 0$ in the equation of the graph. The y-coordinates of the points where a graph intersects the y-axis are called the **y-intercepts** of the graph and are obtained by setting $x = 0$ in the equation of the graph. In Example 6, the y-intercept is -3, and to find the x-intercept we set $y = 0$, which gives $2x = 3$ or $x = \frac{3}{2}$.

So far we have discussed how to find the graph of an equation in x and y. The converse problem is to find an *equation of a graph*, that is, an equation that represents a given curve in the xy-plane. Such an equation is satisfied by the coordinates of the points on the curve and by no other points. This is the other half of the basic principle of analytic geometry as formulated by Descartes and Fermat. The idea is that if a geometric curve can be represented by an algebraic equation, then the rules of algebra can be used to analyze the geometric problem.

As an example of this type of problem, let us find the equation of a circle with radius r and center (h, k). By definition, the circle is the set of all points $P(x, y)$ whose distance from the center $C(h, k)$ is r (see Figure 10). Thus P is on the circle if and only if $|PC| = r$. From the Distance Formula we have

$$\sqrt{(x - h)^2 + (y - k)^2} = r$$

or equivalently, squaring both sides,

$$(x - h)^2 + (y - k)^2 = r^2$$

This is the desired equation.

Fundamental Principle of Analytic Geometry

A point (x, y) lies on the graph of an equation if and only if its coordinates satisfy the equation.

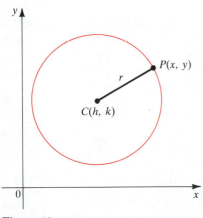

Figure 10

EQUATION OF A CIRCLE

An equation of a circle with center (h, k) and radius r is

$$(x - h)^2 + (y - k)^2 = r^2$$

In particular, if the center is the origin $(0, 0)$, the equation is

$$x^2 + y^2 = r^2$$

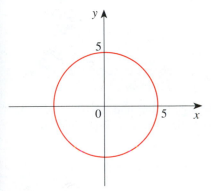

Figure 11

EXAMPLE 7

Determine the graph of the equation $x^2 + y^2 = 25$.

SOLUTION

Rewriting the equation as $x^2 + y^2 = 5^2$, we see that this is an equation of the circle of radius 5 centered at the origin. Its graph is sketched in Figure 11. ■

EXAMPLE 8

Find an equation of the circle with radius 3 and center $(2, -5)$.

SOLUTION

Using the equation of a circle with $r = 3$, $h = 2$, and $k = -5$, we obtain

$$(x - 2)^2 + (y + 5)^2 = 9$$

Its graph is sketched in Figure 12. ■

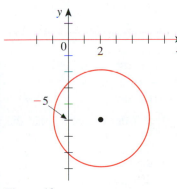

Figure 12

EXAMPLE 9

Find an equation of the circle that has the points $P(1, 8)$ and $Q(5, -6)$ as the endpoints of a diameter.

SOLUTION

We first observe that the center is the midpoint of the diameter PQ, so by the Midpoint Formula the center is

$$\left(\frac{1 + 5}{2}, \frac{8 - 6}{2}\right) = (3, 1)$$

The radius r is the distance from P to the center, so

$$r^2 = (3 - 1)^2 + (1 - 8)^2 = 2^2 + (-7)^2 = 53$$

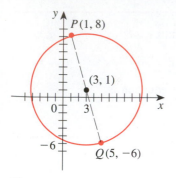

Figure 13

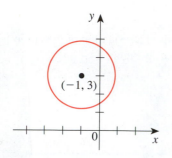

Figure 14
$x^2 + y^2 + 2x - 6y + 7 = 0$

Therefore the equation of the circle is

$$(x - 3)^2 + (y - 1)^2 = 53$$

Its graph is shown in Figure 13.   ■

EXAMPLE 10

Sketch the graph of the equation $x^2 + y^2 + 2x - 6y + 7 = 0$ by first showing that it represents a circle and then finding its center and radius.

SOLUTION

We first group the x-terms and y-terms as follows:

$$(x^2 + 2x) + (y^2 - 6y) = -7$$

Then we complete the square within each grouping, adding the square of half the coefficient of x and the square of half the coefficient of y to both sides of the equation:

$$(x^2 + 2x + 1) + (y^2 - 6y + 9) = -7 + 1 + 9$$

or

$$(x + 1)^2 + (y - 3)^2 = 3$$

Comparing this equation to the standard equation of a circle, we see that $h = -1$, $k = 3$, and $r = \sqrt{3}$, so the given equation represents a circle with center $(-1, 3)$ and radius $\sqrt{3}$. It is sketched in Figure 14.   ■

■ SYMMETRY

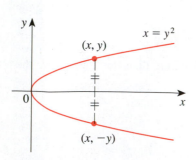

Figure 15

In Example 5 we sketched the graph of $x = y^2$. Notice from Figure 15 that the part of the curve in the fourth quadrant is a mirror image of the part of the curve in the first quadrant. This is because whenever (x, y) is on the curve, so is the point $(x, -y)$, and these points are reflections of each other about the x-axis. In this situation we say the curve is **symmetric with respect to the x-axis**.

Similarly we say the curve is **symmetric with respect to the y-axis** if whenever the point (x, y) is on the curve, so is $(-x, y)$. Figure 16 shows the graph of the curve $y = x^2$. We see from the graph that the part of the curve in the second quadrant (the left half) is the mirror image of the part in the first quadrant (the right half). The two halves are reflections of each other about the y-axis.

A curve is called **symmetric with respect to the origin** if whenever (x, y) is on the curve, so is $(-x, -y)$. The circle $x^2 + y^2 = 1$, graphed in Figure 17, exhibits this type of symmetry. Each point (x, y) on the circle is the reflection about the origin of the point $(-x, -y)$, which also lies on the graph.

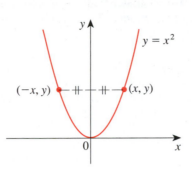

Figure 16

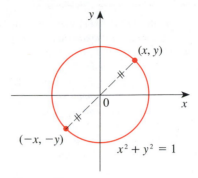

Figure 17

The three types of symmetry that we have discussed are illustrated more generally in Figure 18.

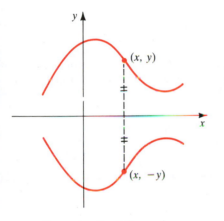

Symmetry about the x-axis

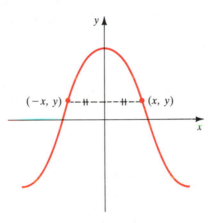

Symmetry about the y-axis

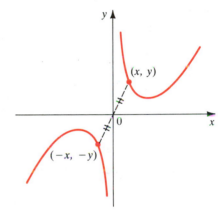

Symmetry about the origin

Figure 18

The tests described in the following box can be used to check whether the graph of any given equation has one of these three types of symmetry.

TESTS FOR SYMMETRY

1. A curve is symmetric with respect to the x-axis if its equation is unchanged when y is replaced by $-y$.
2. A curve is symmetric with respect to the y-axis if its equation is unchanged when x is replaced by $-x$.
3. A curve is symmetric with respect to the origin if its equation is unchanged when x is replaced by $-x$ and y is replaced by $-y$.

One reason that symmetry is important is that we can use it to help us sketch the graphs of equations. The remaining examples in this section illustrate the use of symmetry in graphing.

EXAMPLE 11

Find the x- and y-intercepts of the graph of the equation $y = x^3 - x$. Test the equation for symmetry and sketch its graph.

SOLUTION

Setting $x = 0$ in the equation gives $y = 0$, so the y-intercept is $(0, 0)$. Setting $y = 0$ gives $x^3 - x = 0$, or $x(x^2 - 1) = x(x - 1)(x + 1) = 0$. Thus the x-intercepts are $(0, 0)$, $(1, 0)$, and $(-1, 0)$.

If we replace x by $-x$ and y by $-y$ in the equation, we get

$$-y = (-x)^3 - (-x)$$
$$-y = -x^3 + x$$
$$y = x^3 - x$$

and so the equation is unchanged. This means that the curve is symmetric with respect to the origin. We sketch it by first plotting the points for $x > 0$ and then reflecting in the origin, as shown in Figure 19.

x	y
0	0
$\frac{1}{2}$	$-\frac{3}{8}$
1	0
2	6
3	24

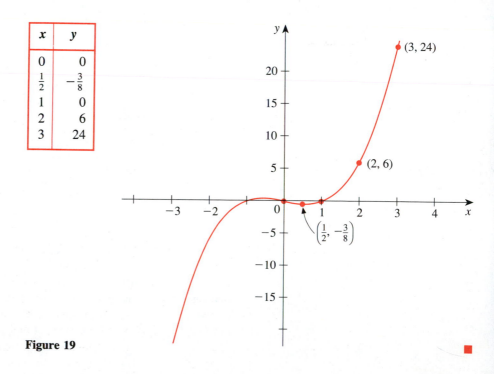

Figure 19

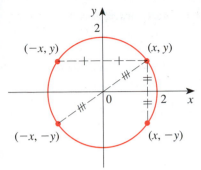

Figure 20

EXAMPLE 12

The equation of the circle $x^2 + y^2 = 4$ remains unchanged when x is replaced by $-x$ and when y is replaced by $-y$, since $(-x)^2 = x^2$ and $(-y)^2 = y^2$, so the circle exhibits all three types of symmetry: It is symmetric with respect to the x-axis, the y-axis, and the origin (see Figure 20). ■

EXERCISES 1.9

In Exercises 1–6 (a) plot the points on a coordinate plane, (b) find the distance between them, and (c) find the midpoint of the segment that joins them.

1. $(1, 1)$, $(4, 5)$

2. $(1, -3)$, $(5, 7)$

3. $(6, -2)$, $(-1, 3)$

4. $(1, -6)$, $(-1, -3)$

5. $(2, 5)$, $(4, -7)$

6. $\left(-3, \frac{1}{2}\right)$, $\left(\frac{5}{2}, -1\right)$

7. Plot the points $A(1, 0)$, $B(5, 0)$, $C(4, 3)$, and $D(2, 3)$ on a coordinate plane. Draw the segments AB, BC, CD, and DA. What kind of quadrilateral is $ABCD$, and what is its area?

8. Plot the points $P(5, 1)$, $Q(0, 6)$, and $R(-5, 1)$ on a coordinate plane. Where must the point S be located so that the quadrilateral $PQRS$ is a square? Find the area of this square.

In Exercises 9–22 sketch the region given by the set.

9. $\{(x, y) \mid x < 0\}$

10. $\{(x, y) \mid y > 0\}$

11. $\{(x, y) \mid xy < 0\}$

12. $\{(x, y) \mid xy > 0\}$

13. $\{(x, y) \mid x = 3\}$

14. $\{(x, y) \mid y = -2\}$

15. $\{(x, y) \mid 1 < x < 2\}$

16. $\{(x, y) \mid 0 \leq y \leq 4\}$

17. $\{(x, y) \mid x \geq 1 \text{ and } y < 3\}$

18. $\{(x, y) \mid |y| > 1\}$

19. $\{(x, y) \mid |x| \leq 2\}$

20. $\{(x, y) \mid |x| < 3 \text{ and } |y| > 2\}$

21. $\{(x, y) \mid |y - 3| \leq 3\}$

22. $\{(x, y) \mid |x - 1| < 2 \text{ and } |y + 1| \leq 2\}$

23. Which of the points $A(6, 7)$ or $B(-5, 8)$ is closer to the origin?

24. Which of the points $C(-6, 3)$ or $D(3, 0)$ is closer to the point $E(-2, 1)$?

25. Show that the triangle with vertices $A(0, 2)$, $B(-3, -1)$, and $C(-4, 3)$ is isosceles.

26. Show that the triangle with vertices $A(6, -7)$, $B(11, -3)$, and $C(2, -2)$ is a right triangle using the converse of the Pythagorean Theorem. Find the area of the triangle.

27. Show that the points $A(-2, 9)$, $B(4, 6)$, $C(1, 0)$, and $D(-5, 3)$ are the vertices of a square.

28. Show that the points $A(-1, 3)$, $B(3, 11)$, and $C(5, 15)$ are collinear by showing that $|AB| + |BC| = |AC|$.

29. Find a point on the y-axis that is equidistant from the points $(5, -5)$ and $(1, 1)$.

30. Find the lengths of the medians of the triangle with vertices $A(1, 0)$, $B(3, 6)$, and $C(8, 2)$. (A median is a line segment from a vertex to the midpoint of the opposite side.)

31. Find the point that is one-fourth of the way from the point $P(-1, 3)$ to the point $Q(7, 5)$ along the segment PQ.

32. Plot the points $P(-2, 1)$ and $Q(12, -1)$. Which (if any) of the points $A(5, -7)$ or $B(6, 7)$ lies on the perpendicular bisector of the segment PQ?

In Exercises 33–54, make a table of values and sketch the graph of the given equation. Test each curve for symmetry. Find x- and y-intercepts for each curve.

33. $y = x - 1$

34. $y = 2x + 5$

35. $3x - y = 5$

36. $x + y = 3$

37. $y = 1 - x^2$

38. $y = x^2 + 2x$

39. $4y = x^2$

40. $8y = x^3$

41. $xy = 2$

42. $x + y^2 = 4$

43. $y = \sqrt{x}$

44. $x + \sqrt{y} = 4$

45. $x^2 + y^2 = 9$

46. $9x^2 + 9y^2 = 49$

47. $y = \sqrt{4 - x^2}$

48. $x = -\sqrt{25 - y^2}$

49. $y = |x|$

50. $x = |y|$

51. $y = 4 - |x|$

52. $y = |4 - x|$

53. $x = y^3$

54. $y = x^3 - 4x$

In Exercises 55–60 sketch a graph of the equation without making a table of values.

55. $x = 3$

56. $y = -1$

57. $x^2 + y^2 = 4$

58. $4x^2 + 4y^2 = 9$

59. $(x - 1)^2 + (y + 3)^2 = 25$

60. $\left(x + \frac{1}{2}\right)^2 + \left(y - \frac{1}{2}\right)^2 = \frac{1}{4}$

In Exercises 61–68 find an equation of the circle that satisfies the given conditions.

61. Center $(3, -1)$;   radius 5

62. Center $(-2, -8)$;   radius 10

63. Center at the origin;   passes through $(4, 7)$

64. Center $(-1, 5)$;   passes through $(-4, -6)$

65. Endpoints of a diameter are $P(-1, 1)$ and $Q(5, 5)$

66. Endpoints of a diameter are $P(-1, 3)$ and $Q(7, -5)$

67. Center $(7, -3)$;   tangent to the x-axis

68. Circle lies in the first quadrant, tangent to both x- and y-axis;   radius 5

In Exercises 69–74 show that the given equation represents a circle and find the center and radius of the circle.

69. $x^2 + y^2 - 4x + 10y + 13 = 0$

70. $x^2 + y^2 + 6y + 2 = 0$

71. $x^2 + y^2 + x = 0$

72. $x^2 + y^2 + 2x + y + 1 = 0$

73. $2x^2 + 2y^2 - x + y = 1$

74. $16x^2 + 16y^2 + 8x + 32y + 1 = 0$

In Exercises 75–78 sketch the graph of the equation.

75. $x^2 + y^2 + 4x - 10y = 21$

76. $4x^2 + 4y^2 + 2x = 0$

77. $x^2 + y^2 + 6x - 12y + 45 = 0$

78. $x^2 + y^2 - 16x + 12y + 200 = 0$

In Exercises 79–82 sketch the region given by the set.

79. $\{(x, y) \mid x^2 + y^2 \leq 1\}$

80. $\{(x, y) \mid x^2 + y^2 > 4\}$ outside ⊙

81. $\{(x, y) \mid 1 \leq x^2 + y^2 < 9\}$

82. $\{(x, y) \mid 2x < x^2 + y^2 \leq 4\}$

83. Find the area of the region that lies outside the circle $x^2 + y^2 = 4$ but inside the circle

$$x^2 + y^2 - 4y - 12 = 0$$

84. Sketch the region in the coordinate plane that satisfies both the inequalities $x^2 + y^2 \leq 9$ and $y \geq |x|$. What is the area of this region?

85. Under what conditions on the coefficients a, b, and c does the equation $x^2 + y^2 + ax + by + c = 0$ represent a circle? When that condition is satisfied, find the center and radius of the circle.

SECTION 1.10
LINES

In this section we find equations for straight lines lying in a coordinate plane. We first need to measure the steepness of a line and so we define its slope.

SLOPE OF A LINE

The **slope** of a nonvertical line that passes through the points $P_1(x_1, y_1)$ and $P_2(x_2, y_2)$ is

$$m = \frac{y_2 - y_1}{x_2 - x_1}$$

The slope of a vertical line is not defined.

Thus the slope of a line is the ratio of the change in y (the rise) to the change in x (the run) between two points on the line. From the similar triangles in Figure 1 we see that the slope is independent of which two points are chosen on the line:

$$\frac{y_2 - y_1}{x_2 - x_1} = \frac{y_2' - y_1'}{x_2' - x_1'}$$

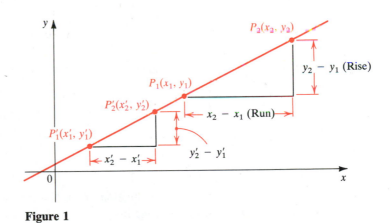

Figure 1

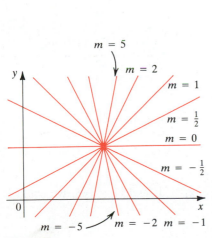

Figure 2

Figure 2 shows several lines labeled with their slopes. Notice that lines with positive slope slant upward to the right, whereas lines with negative slope slant downward to the right. Notice also that the steepest lines are the ones for which the absolute value of the slope is the largest and that a horizontal line has slope zero.

EXAMPLE 1

Find the slope of the line that passes through the points $P(2, 1)$ and $Q(8, 5)$.

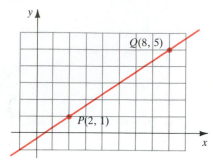

Figure 3

SOLUTION

Since any two different points determine a line, there is only one line that passes through these two points. From the definition, the slope is

$$m = \frac{y_2 - y_1}{x_2 - x_1} = \frac{5 - 1}{8 - 2} = \frac{4}{6} = \frac{2}{3}$$

This says that for every three units we move to the right, the line rises two units. The line is drawn in Figure 3. ∎

Now let us find the equation of the line that passes through a given point $P_1(x_1, y_1)$ and has slope m. A point $P(x, y)$ with $x \neq x_1$ lies on this line if and only if the slope of the line through P_1 and P is equal to m (see Figure 4); that is,

$$\frac{y - y_1}{x - x_1} = m$$

This equation can be rewritten in the form

$$y - y_1 = m(x - x_1)$$

and we observe that this equation is also satisfied when $x = x_1$ and $y = y_1$. Therefore it is an equation of the given line.

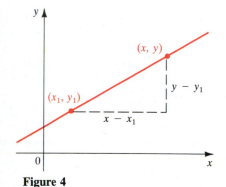

Figure 4

POINT-SLOPE FORM OF THE EQUATION OF A LINE

An equation of the line that passes through the point (x_1, y_1) and has slope m is

$$y - y_1 = m(x - x_1)$$

EXAMPLE 2

(a) Find an equation of the line through $(1, -3)$ with slope $-\frac{1}{2}$.
(b) Sketch the line.

SOLUTION

(a) Using the point-slope form with $m = -\frac{1}{2}$, $x_1 = 1$, and $y_1 = -3$, we obtain an equation of the line as

$$y + 3 = -\tfrac{1}{2}(x - 1)$$

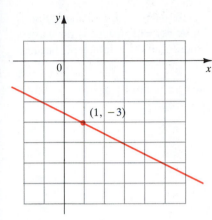

Figure 5

which we can rewrite as

$$2y + 6 = -x + 1$$

or

$$x + 2y + 5 = 0$$

(b) The fact that the slope is $-\frac{1}{2}$ tells us that when we move two units to the right, the line drops one unit. This enables us to sketch the line in Figure 5. ∎

EXAMPLE 3

Find an equation of the line through the points $(-1, 2)$ and $(3, -4)$.

SOLUTION

The slope of the line is

$$m = \frac{-4 - 2}{3 - (-1)} = -\frac{3}{2}$$

Using the point-slope form with $x_1 = -1$ and $y_1 = 2$, we obtain

$$y - 2 = -\tfrac{3}{2}(x + 1)$$

which simplifies to

$$2y - 4 = -3x - 3$$

or

$$3x + 2y = 1$$ ∎

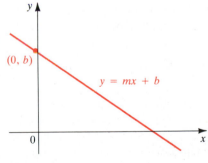

Figure 6

Suppose a nonvertical line has slope m and y-intercept b (see Figure 6). This means it intersects the y-axis at the point $(0, b)$, so the point-slope form of the equation of the line, with $x = 0$ and $y = b$, becomes

$$y - b = m(x - 0)$$

This simplifies to $y = mx + b$, which is called the **slope-intercept form** of the equation of a line.

SLOPE-INTERCEPT FORM OF THE EQUATION OF A LINE

An equation of the line that has slope m and y-intercept b is

$$y = mx + b$$

As a special case, if a line is horizontal, its slope is $m = 0$, so its equation is $y = b$, where b is the y-intercept (see Figure 7). A vertical line does not have a slope, but we can write its equation as $x = a$, where a is the x-intercept, because the x-coordinate of every point on the line is a.

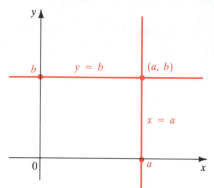

Figure 7

VERTICAL AND HORIZONTAL LINES

The equation of a vertical line through (a, b) is $x = a$.

The equation of a horizontal line through (a, b) is $y = b$.

Observe that the equation of every line can be written as $Ax + By + C = 0$, as in the solution to Example 2(a), because a nonvertical line has the equation $y = mx + b$ or $-mx + y - b = 0$ ($A = -m$, $B = 1$, $C = -b$), and a vertical line has the equation $x = a$ or $x - a = 0$ ($A = 1$, $B = 0$, $C = -a$). Conversely, if we start with a general first-degree equation, that is, an equation of the form $Ax + By + C = 0$, where A, B, and C are constants and A and B are not both 0, then we can show that it is the equation of a line. If $B = 0$, it becomes $Ax + C = 0$ or $x = -C/A$, which represents a vertical line with x-intercept $-C/A$. If $B \neq 0$, it can be rewritten by solving for y:

$$y = -\frac{A}{B}x - \frac{C}{B}$$

and we recognize this as being the slope-intercept form of the equation of a line ($m = -A/B$, $b = -C/B$). Therefore an equation of this form is called a **linear equation** or the **general equation of a line**. For brevity, we often refer to "the line $Ax + By + C = 0$" instead of "the line whose equation is $Ax + By + C = 0$."

GENERAL EQUATION OF A LINE

Every line has an equation of the form

$$Ax + By + C = 0$$

Conversely, every equation of this form in which A and B are not both zero is the equation of a line.

EXAMPLE 4

Sketch the graph of the equation $3x - 5y = 15$.

SOLUTION

Since the equation is linear, its equation is a line. To draw the graph it is enough for us to find two points on the line. It is easiest to find the intercepts. Substituting $y = 0$ (the equation of the x-axis) in the given equation, we get $3x = 15$, so $x = 5$ is the x-intercept. Substituting $x = 0$ in the equation, we find that the y-intercept is -3. This allows us to sketch the graph as in Figure 8.

Another method is to write the equation in slope-intercept form:

$$3x - 5y = 15$$
$$5y = 3x - 15$$
$$y = \tfrac{3}{5}x - 3$$

This is in the form $y = mx + b$, so the slope is $m = \tfrac{3}{5}$ and the y-intercept is $b = -3$. This information can be used to sketch the line. ■

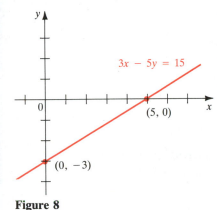

Figure 8

PARALLEL AND PERPENDICULAR LINES

Since slope measures the steepness of a line, it seems reasonable that parallel lines should have the same slope. In fact, we can prove this.

PARALLEL LINES

> Two nonvertical lines are parallel if and only if they have the same slope.

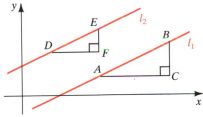

Figure 9

Proof Let the lines l_1 and l_2 in Figure 9 have slopes m_1 and m_2. If the lines are parallel, then the right triangles ABC and DEF are similar, so

$$m_1 = \frac{|BC|}{|AC|} = \frac{|EF|}{|DF|} = m_2$$

Conversely, if the slopes are equal, then the triangles will be similar and so $\angle BAC = \angle EDF$. Thus the lines are parallel. ■

EXAMPLE 5

Find an equation of the line through the point $(5, 2)$ that is parallel to the line $4x + 6y + 5 = 0$.

SOLUTION

The given line can be written in slope-intercept form as follows:

$$6y = -4x - 5$$
$$y = -\tfrac{2}{3}x - \tfrac{5}{6}$$

Thus the line has slope $m = -\frac{2}{3}$. Parallel lines have the same slope, so the required line has slope $-\frac{2}{3}$ and its equation in point-slope form is

$$y - 2 = -\tfrac{2}{3}(x - 5)$$

This simplifies to

$$3y - 6 = -2x + 10 \qquad \text{or} \qquad 2x + 3y - 16 = 0$$

The condition for perpendicular lines is not so obvious.   ■

PERPENDICULAR LINES

> Two lines with slopes m_1 and m_2 are perpendicular if and only if $m_1 m_2 = -1$; that is, their slopes are negative reciprocals:
>
> $$m_2 = -\frac{1}{m_1}$$
>
> Also, a horizontal line (slope 0) is perpendicular to a vertical line (no slope).

Proof   In Figure 10 we show two lines intersecting at the origin. (If the lines intersect at some other point, we consider lines parallel to these that intersect at the origin. These lines have the same slopes as the original lines.)

Figure 10

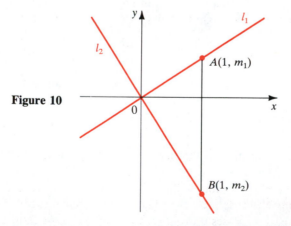

If the lines l_1 and l_2 have slopes m_1 and m_2, then their equations are $y = m_1 x$ and $y = m_2 x$. Notice that $A(1, m_1)$ lies on l_1 and $B(1, m_2)$ lies on l_2. By the

Pythagorean Theorem, and its converse, $OA \perp OB$ if and only if

$$|OA|^2 + |OB|^2 = |AB|^2$$

By the Distance Formula, this becomes

$$(1^2 + m_1^2) + (1^2 + m_2^2) = (1 - 1)^2 + (m_2 - m_1)^2$$

$$2 + m_1^2 + m_2^2 = m_2^2 - 2m_1m_2 + m_1^2$$

$$2 = -2m_1m_2$$

$$m_1m_2 = -1 \qquad\blacksquare$$

EXAMPLE 6

Show that the points $P(3, 3)$, $Q(8, 17)$, and $R(11, 5)$ are the vertices of a right triangle.

SOLUTION

The slopes of the lines PR and QR are

$$m_1 = \frac{5 - 3}{11 - 3} = \frac{1}{4}$$

and

$$m_2 = \frac{5 - 17}{11 - 8} = -4$$

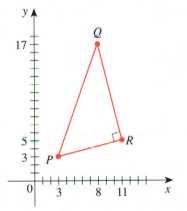

Figure 11

Since $m_1m_2 = -1$, these lines are perpendicular and so PQR is a right triangle. It is sketched in Figure 11. $\qquad\blacksquare$

EXAMPLE 7

Find the equation of a line that is perpendicular to the line $4x + 6y + 5 = 0$ and passes through the origin.

SOLUTION

In Example 5 we found that the slope of the line $4x + 6y + 5 = 0$ is $-\frac{2}{3}$. Thus the slope of a perpendicular line is the negative reciprocal, that is, $\frac{3}{2}$. Since the required line passes through $(0, 0)$, the point-slope form gives

$$y - 0 = \tfrac{3}{2}(x - 0) \qquad \text{or} \qquad y = \tfrac{3}{2}x \qquad\blacksquare$$

EXAMPLE 8

The relationship between the Fahrenheit (F) and Celsius (C) temperature scales is given by the equation $F = \frac{9}{5}C + 32$.

(a) Sketch a graph of this equation, letting the horizontal axis be the C-axis and the vertical axis be the F-axis.

(b) What is the slope of this graph and what does it represent? What is the F-intercept and what does it represent?

SOLUTION

(a) This is a linear equation, so its graph is a line. (We are calling the coordinates C and F instead of x and y, but this makes no difference in the shape of the graph.) Since two points determine a line, we first find two points that satisfy the equation and plot those. Then we draw the straight line that contains them. When $C = 0$, $F = \frac{9}{5}(0) + 32 = 32$, and when $C = 5$, $F = \frac{9}{5}(5) + 32 = 41$. Thus the points $(0, 32)$ and $(5, 41)$ lie on the line. These points and the line containing them are shown in Figure 12.

(b) Since the equation is given in the slope-intercept form, we see that the slope is $\frac{9}{5}$ and the F-intercept is 32. The slope represents the change in °F for every 1 °C change; thus a 9 °F increase in temperature corresponds to a 5 °C increase. The F-intercept is the F-coordinate of the point on the graph whose C-coordinate is 0. Thus 32 °F is the same as 0 °C (the freezing point of water). ∎

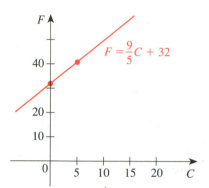

Figure 12

EXAMPLE 9

(a) As dry air moves upward, it expands and cools. If the ground temperature is 20 °C and the temperature at a height of 1 km is 10 °C, express the temperature T (in °C) in terms of the height h (in kilometers), assuming the expression is linear.

(b) Draw the graph of the linear equation. What does the slope represent?

(c) What is the temperature at a height of 2.5 km?

SOLUTION

(a) Because we are assuming a linear relationship between h and T, the equation must be of the form

$$T = mh + b$$

where m and b are constants. When $h = 0$, we are given that $T = 20$, so

$$20 = m(0) + b \qquad \text{or} \qquad b = 20$$

Thus, we have

$$T = mh + 20$$

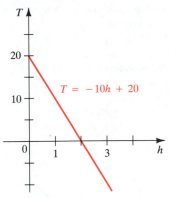

Figure 13

When $h = 1$, we have $T = 10$ and so

$$10 = m(1) + 20$$
$$m = 10 - 20 = -10$$

The required expression is

$$T = -10h + 20$$

(b) The graph is sketched in Figure 13. The slope is $m = 10$ °/km and represents the rate of change of temperature with respect to distance.

(c) At a height of $h = 2.5$ km, the temperature is

$$T = -10(2.5) + 20 = -25 + 20 = -5 \,°\text{C} \qquad \blacksquare$$

EXERCISES 1.10

In Exercises 1–4 find the slope of the line through P and Q.

1. $P(1, 5)$, $Q(4, 11)$ **2.** $P(-1, 6)$, $Q(4, -3)$

3. $P(-3, 3)$, $Q(-1, -6)$ **4.** $P(-1, -4)$, $Q(6, 0)$

In Exercises 5 and 6 find an equation for the line whose graph is sketched.

5.

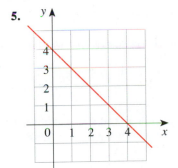

6.

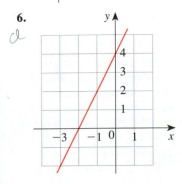

In Exercises 7–22 find an equation of the line that satisfies the given conditions.

7. Through $(2, -3)$; slope 6

8. Through $(-1, 4)$; slope -3

9. Through $(1, 7)$; slope $\frac{2}{3}$

10. Through $(-3, -5)$; slope $-\frac{7}{2}$

11. Through $(2, 1)$ and $(1, 6)$

12. Through $(-1, -2)$ and $(4, 3)$

13. Slope 3; y-intercept -2

14. Slope $\frac{2}{5}$; y-intercept 4

15. x-intercept 1; y-intercept -3

16. x-intercept -8; y-intercept 6

17. Through $(4, 5)$; parallel to the x-axis

18. Through $(4, 5)$; parallel to the y-axis

19. Through $(1, -6)$; parallel to the line $x + 2y = 6$

20. y-intercept 6; parallel to the line $2x + 3y + 4 = 0$

21. Through $(-1, -2)$; perpendicular to the line
$2x + 5y + 8 = 0$

22. Through $\left(\frac{1}{2}, -\frac{2}{3}\right)$; perpendicular to the line $4x - 8y = 1$

23. (a) Sketch the line with slope $\frac{3}{2}$ that passes through the point $(-2, 1)$.
 (b) Find an equation for this line.

24. (a) Sketch the line with slope -2 that passes through the point $(4, -1)$.
(b) Find an equation for this line.

In Exercises 25–34 find the slope and y-intercept of the line and draw its graph.

25. $x + y = 5$

26. $2x - 3y = 6$

27. $x + 3y = 0$

28. $2x - 5y = 0$

29. $\frac{1}{2}x - \frac{1}{3}y + 1 = 0$

30. $-3x - 5y + 30 = 0$

31. $3x - 4y = 12$

32. $y = 1.5$

33. $3x + 4y - 1 = 0$

34. $4x + 5y = 10$

35. Show that $A(1, 1)$, $B(7, 4)$, $C(5, 10)$, and $D(-1, 7)$ are vertices of a parallelogram.

36. Show that $A(-3, -1)$, $B(3, 3)$, and $C(-9, 8)$ are vertices of a right triangle.

37. Show that $A(1, 1)$, $B(11, 3)$, $C(10, 8)$, and $D(0, 6)$ are vertices of a rectangle. rectangle = ▱ with one rt ∠

38. Use slopes to determine whether the given points are collinear (lie on a line).
(a) $(1, 1)$, $(3, 9)$, and $(6, 21)$
(b) $(-1, 3)$, $(1, 7)$, and $(4, 15)$

$C =$

39. Find the equation of the perpendicular bisector of the line segment joining the points $A(1, 4)$ and $B(7, -2)$.

40. (a) Find equations for the sides of the triangle with vertices $P(1, 0)$, $Q(3, 4)$, and $R(-1, 6)$.
(b) Find equations for the medians of this triangle. Where do they intersect?

41. (a) Show that if the x- and y-intercepts of a line are nonzero numbers a and b, then the equation of the line can be put in the form

$$\frac{x}{a} + \frac{y}{b} = 1$$

This equation is called the **two-intercept form** of the equation of a line.
(b) Use part (a) to find an equation of the line whose x-intercept is 6 and whose y-intercept is -8.

42. (a) Find an equation for the line tangent to the circle $x^2 + y^2 = 25$ at the point $(3, -4)$.

(b) At what other point on the circle will the tangent line be parallel to the line in part (a)?

43. A small appliance manufacturer finds that if he produces x toaster ovens in a month his production costs are given by the equation $y = 6x + 3000$ (where y is measured in dollars).
(a) Sketch a graph of this linear equation.
(b) What do the slope and y-intercept of the graph represent?

44. The manager of a weekend flea market knows from past experience that if he charges x dollars for a space at the flea market, then the number y of spaces he can sell is given by the equation $y = 200 - 4x$.
(a) Sketch a graph of this linear equation. (Remember that the charge per space and the number of spaces sold must both be nonnegative quantities.)
(b) What do the slope, the y-intercept, and the x-intercept of the graph represent? for every $1 increase of fewer spaces will be sold

45. The monthly cost of driving a car depends on the number of miles driven. Lynn found that in May it cost her $380 to drive 480 mi and in June it cost $460 to drive 800 mi.
(a) Express the monthly cost C in terms of the distance driven d, assuming that a linear relationship gives a suitable model.
(b) Use part (a) to predict the cost of driving 1500 mi per month.
(c) Draw the graph of the linear equation. What does the slope of the line represent?
(d) What does the y-intercept of the graph represent?
(e) Why is a linear relationship a suitable model in this situation?

46. Jason and Debbie leave Detroit at 2:00 P.M. and drive at a constant speed west along I-90. They pass Ann Arbor, 40 mi from Detroit, at 2:50 P.M.
(a) Express the distance traveled in terms of the time elapsed.
(b) Draw the graph of the equation in part (a).
(c) What is the slope of this line? What does it represent?

47. At the surface of the ocean, the water pressure is the same as the air pressure above the water, 15 pounds per square inch. Below the surface, the water pressure increases by 4.34 lb/in^2 for every 10 ft of descent.
(a) Find an equation for the relationship between pressure and depth below the surface.
(b) At what depth is the pressure 100 lb/in^2?

48. The manager of a furniture factory finds that it costs $2200 to manufacture 100 chairs in a day, and $4800 to produce 300 chairs in a day.
  (a) Assuming that the relationship between cost and the number of chairs produced is linear, find an equation that expresses this relationship, and graph the equation.
  (b) What is the slope of the line of part (a), and what does it represent?
  (c) What is the y-intercept of this line, and what does it represent?

CHAPTER 1 REVIEW

Define, state, or discuss each of the following.

1. Integers
2. Rational and irrational numbers
3. Real number line
4. Commutative Law
5. Associative Law
6. Distributive Law
7. Union of sets
8. Intersection of sets
9. Open interval, closed interval
10. Absolute value of a number
11. Base and exponent
12. Laws of Exponents
13. Principal nth root
14. Rationalizing the denominator
15. Rational exponents
16. Variable and constant
17. Special product formulas for $(a + b)^2$, $(a - b)^2$, $(a + b)^3$, $(a - b)^3$
18. Difference of squares formula
19. Difference of cubes formula
20. Sum of cubes formula
21. Equation
22. Variable
23. Solution, root
24. Identity
25. Linear equation
26. Quadratic equation
27. Zero-product property
28. Completing the square
29. Quadratic formula
30. Discriminant of a quadratic equation
31. Inequality
32. Linear inequality
33. Solving inequalities using test values
34. Solving inequalities that involve absolute value
35. Triangle Inequality
36. Coordinate plane
37. Distance Formula
38. Midpoint Formula
39. Graph of an equation
40. x- and y-intercepts
41. Equation of a circle
42. Symmetry with respect to the x-axis, the y-axis, and the origin
43. Slope of a line
44. Point-slope equation for a line
45. Slope-intercept equation for a line
46. General equation for a line
47. Parallel lines
48. Perpendicular lines

REVIEW EXERCISES

In Exercises 1 and 2 express the interval in terms of inequalities and graph the interval.

1. $(-1, 3]$

2. $(-\infty, 4]$

In Exercises 3 and 4 express the inequality in interval notation and graph the corresponding interval.

3. $x > 2$

4. $1 \leqslant x \leqslant 6$

Evaluate the numbers in Exercises 5–12.

5. $\big|3 - |-9|\big|$

6. $1 - \big|1 - |-1|\big|$

7. $2^{-3} - 3^{-2}$

8. $\sqrt[3]{-125}$

9. $216^{-1/3}$

10. $64^{2/3}$

11. $\dfrac{\sqrt{242}}{\sqrt{2}}$

12. $\sqrt[4]{4}\,\sqrt[4]{324}$

In Exercises 13–16 write the given expression as a power of x.

13. $\dfrac{1}{x^2}$

14. $x\sqrt{x}$

15. $x^2 x^m (x^3)^m$

16. $((x^m)^2)^n$

Simplify the expressions in Exercises 17–26.

17. $(2x^3y)^2(3x^{-1}y^2)$

18. $(a^2)^{-3}(a^3b)^2(b^3)^4$

19. $\dfrac{x^4(3x)^2}{x^3}$

20. $\left(\dfrac{r^2 s^{4/3}}{r^{1/3}s}\right)^6$

21. $\sqrt[3]{(x^3y)^2 y^4}$

22. $\sqrt{x^2 y^4}$

23. $\dfrac{x}{2 + \sqrt{x}}$

24. $\dfrac{\sqrt{x} + 1}{\sqrt{x} - 1}$

25. $\dfrac{8r^{1/2}s^{-3}}{2r^{-2}s^4}$

26. $\left(\dfrac{ab^2c^{-3}}{2a^3b^{-4}}\right)^{-2}$

In Exercises 27–40 factor the expression.

27. $x^2 + 3x - 10$

28. $6x^2 + x - 12$

29. $4t^2 - 13t - 12$

30. $x^4 - 2x^2 + 1$

31. $25 - 16t^2$

32. $2y^6 - 32y^2$

33. $x^6 - 1$

34. $y^3 - 2y^2 - y + 2$

35. $x^{-1/2} - 2x^{1/2} + x^{3/2}$

36. $a^4 b^2 + ab^5$

37. $4x^3 - 8x^2 + 3x - 6$

38. $8x^3 + y^6$

39. $(x^2 + 2)^{5/2} + 2x(x^2 + 2)^{3/2} + x^2\sqrt{x^2 + 2}$

40. $3x^3 - 2x^2 + 18x - 12$

In Exercises 41–62 perform the indicated operations.

41. $(2x + 1)(3x - 2) - 5(4x - 1)$

42. $(2y - 7)(2y + 7)$

43. $(2a^2 - b)^2$

44. $(1 + x)(2 - x) - (3 - x)(3 + x)$

45. $(x - 1)(x - 2)(x - 3)$ **46.** $(2x + 1)^3$

47. $\sqrt{x}\big(\sqrt{x} + 1\big)\big(2\sqrt{x} - 1\big)$

48. $\dfrac{x^3 + 2x^2 + 3x}{x}$

49. $\dfrac{x^2 - 2x - 3}{2x^2 + 5x + 3}$ **50.** $\dfrac{t^3 - 1}{t^2 - 1}$

51. $\dfrac{x^2 + 2x - 3}{x^2 + 8x + 16} \cdot \dfrac{3x + 12}{x - 1}$

52. $\dfrac{x^3/(x - 1)}{x^2/(x^3 - 1)}$

53. $\dfrac{x^2 - 2x - 15}{x^2 - 6x + 5} \div \dfrac{x^2 - x - 12}{x^2 - 1}$

54. $x - \dfrac{1}{x + 1}$

55. $\dfrac{1}{x - 1} - \dfrac{x}{x^2 + 1}$ **56.** $\dfrac{2}{x} + \dfrac{1}{x - 2} + \dfrac{3}{(x - 2)^2}$

57. $\dfrac{1}{x - 1} - \dfrac{2}{x^2 - 1}$

58. $\dfrac{1}{x + 2} + \dfrac{1}{x^2 - 4} - \dfrac{2}{x^2 - x - 2}$

59. $\dfrac{\dfrac{1}{x} - \dfrac{1}{2}}{x - 2}$ **60.** $\dfrac{\dfrac{1}{x} - \dfrac{1}{x + 1}}{\dfrac{1}{x} + \dfrac{1}{x + 1}}$

61. $\dfrac{3(x + h)^2 - 5(x + h) - (3x^2 - 5x)}{h}$

62. $\dfrac{\sqrt{x + h} - \sqrt{x}}{h}$ (rationalize the numerator)

In Exercises 63–68 state whether the given equation is true for all values of the variables. (Disregard values that make denominators 0.)

63. $(x + y)^3 = x^3 + y^3$

64. $\dfrac{1 + \sqrt{a}}{1 - a} = \dfrac{1}{1 - \sqrt{a}}$

65. $\dfrac{12 + y}{y} = \dfrac{12}{y} + 1$

66. $\sqrt[3]{a + b} = \sqrt[3]{a} + \sqrt[3]{b}$

67. $\sqrt{a^2} = a$

68. $\dfrac{1}{x + 4} = \dfrac{1}{x} + \dfrac{1}{4}$

In Exercises 69–84 find all real solutions of the equation.

69. $2x - 3 = 15$

70. $6x + 10 = 9 - 4x$

71. $\frac{1}{2}x + 4(1 - 3x) = 5x + 3$

72. $\dfrac{x + 1}{x - 1} = \dfrac{2x - 1}{2x + 1}$

73. $2x^2 + x = 1$

74. $3x^2 + 5x - 2 = 0$

75. $4x^3 - 25x = 0$

76. $x^3 - 2x^2 - 5x + 10 = 0$

77. $3x^2 + 4x - 1 = 0$

78. $x^2 - 3x + 9 = 0$

79. $\dfrac{1}{x} + \dfrac{2}{x - 1} = 3$

80. $\dfrac{x}{x - 2} + \dfrac{1}{x + 2} = \dfrac{8}{x^2 - 4}$

81. $x^4 - 8x^2 - 9 = 0$

82. $x - 4\sqrt{x} = 32$

83. $x^{-1/2} - 2x^{1/2} + x^{3/2} = 0$

84. $|2x - 5| = 9$

85. The owner of a store sells raisins for $3.20 per pound and nuts for $2.40 per pound. He decides to mix the raisins and nuts and sell 50 lb of the mixture for $2.72 per pound. What quantities of raisins and nuts should he use?

86. Anthony leaves Kingstown at 2:00 P.M. and drives to Queensville, 160 mi distant, at 45 mi/h. At 2:15 P.M. Hortense leaves Queensville and drives to Kingstown at 40 mi/h. At what time do they pass each other on the road?

87. A woman cycles 8 mi/h faster than she runs. Every morning she cycles 4 mi and runs $2\frac{1}{2}$ mi, for a total of 1 hour of exercise. How fast does she run?

88. The approximate distance d in feet that a driver travels after noticing that he or she must come to a sudden stop is given by the following formula (where x is the speed of the car in miles per hour):

$$d = x + \frac{x^2}{20}$$

If a car travels 75 ft before stopping, what must its speed have been before the brakes were applied?

89. The hypotenuse of a right triangle has length 20 cm. The sum of the lengths of the other two sides is 28 cm. Find the lengths of the other two sides of the triangle.

90. Abbie paints twice as fast as Beth and three times as fast as Cathie. If it takes 60 min to paint a living room if all three work together, how long would it take Abbie if she works alone?

In Exercises 91–102 solve the inequality. Express the solution using interval notation and graph the solution set on the real number line.

91. $-1 < 2x + 5 \leq 3$

92. $3 - x \leq 2x - 7$

93. $x^2 + 4x - 12 > 0$

94. $x^2 \leq 1$

95. $\dfrac{2x + 5}{x + 1} \leq 1$

96. $2x^2 \geq x + 3$

97. $\dfrac{x - 4}{x^2 - 4} \leq 0$

98. $\dfrac{5}{x^3 - x^2 - 4x + 4} < 0$

99. $|x - 5| \leq 3$

100. $|x - 4| < 0.02$

101. $|2x + 1| \geq 1$

102. $|x - 1| < |x - 3|$ [*Hint:* Interpret the quantities as distances.]

103. For what values of x are the following algebraic expressions defined as real numbers?

(a) $\sqrt{24 - x - 3x^2}$

(b) $\dfrac{1}{\sqrt[4]{x - x^4}}$

104. The volume of a sphere is given by $V = \frac{4}{3}\pi r^3$, where r is the radius. Find the interval of values of the radius so that the volume is between 8 ft^3 and 12 ft^3, inclusive.

In Exercises 105–108 (a) plot the points P and Q on a coordinate plane. Then find (b) the distance from P to Q, and (c) the midpoint of the segment PQ. (d) Sketch the line determined by P and Q, and find its equation in slope-intercept form. (e) Sketch the circle that has P at its center and that contains the point Q, and find the equation of this circle.

105. $P(0, 0)$, $Q(3, 4)$

106. $P(1, 1)$, $Q(-6, 13)$

107. $P(5, -3)$, $Q(-2, -1)$

108. $P(0, 7)$, $Q(-5, 0)$

In Exercises 109 and 110 sketch the region in the plane whose points (x, y) satisfy the given condition.

109. $|x| < 4$ and $|y| < 2$ **110.** $|x| \geqslant 4$ or $|y| \geqslant 2$

111. Which of the points $A(4, 4)$ or $B(5, 3)$ is closer to the point $C(-1, -3)$?

112. Show that the points $A(1, 2)$, $B(2, 6)$, and $C(9, 0)$ are the vertices of a right triangle. Find the area of the triangle.

113. Find an equation of the circle that has center $(2, -5)$ and radius $\sqrt{2}$.

114. Find an equation of the circle that has center $(-5, -1)$ and passes through the origin.

115. Find an equation for the circle that contains the points $P(2, 3)$ and $Q(-1, 8)$ and that has the midpoint of the segment PQ as its center.

116. Find an equation for the circle that contains the points $P(1, -3)$ and $Q(-1, 2)$ as the endpoints of a diameter.

In Exercises 117–120 determine whether the equation represents a circle or a point, or has no graph. Where the equation represents a circle, find its center and radius.

117. $x^2 + y^2 + 2x - 6y + 9 = 0$

118. $2x^2 + 2y^2 - 2x + 8y = \frac{1}{2}$

119. $x^2 + y^2 + 72 = 12x$

120. $x^2 + y^2 - 6x - 10y + 34 = 0$

121. Find an equation for the line that passes through the points $(-1, -6)$ and $(2, -4)$.

122. Find an equation for the line that has x-intercept 4 and y-intercept 12.

123. Find an equation for the line that passes through the origin and is parallel to the line $3x + 15y = 22$.

124. Find an equation for the line that passes through the point $(1, 7)$ and is perpendicular to the line $x - 3y + 16 = 0$.

In Exercises 125–136 test the equation for symmetry and sketch its graph.

125. $y = 2 - 3x$ **126.** $2x - y + 1 = 0$

127. $x + 3y = 21$ **128.** $\frac{x}{4} + \frac{y}{5} = 0$

129. $y = 16 - x^2$ **130.** $8x + y^2 = 0$

131. $x = \sqrt{y}$ **132.** $y = -\sqrt{1 - x^2}$

133. $2x^2 + 2y^2 = 5$ **134.** $xy = 4$

135. $y = x^3 - 4x$ **136.** $x = y^4 - 4y^2$

137. If $t = \dfrac{1}{2}\left(x^3 - \dfrac{1}{x^3}\right)$, show that

$$\sqrt{1 + t^2} = \frac{1}{2}\left(x^3 + \frac{1}{x^3}\right)$$

138. If $m > n > 0$ and $a = 2mn$, $b = m^2 - n^2$, $c = m^2 + n^2$, show that $a^2 + b^2 = c^2$.

139. Assume $a = b$. What is wrong with the following argument?

$$a = b$$
$$a^2 = ab$$
$$a^2 - b^2 = ab - b^2$$
$$(a + b)(a - b) = b(a - b)$$
$$a + b = b$$

Now put $a = b = 1$: $2 = 1$

CHAPTER 1 TEST

1. Graph the intervals $[-2, 1]$ and $(6, \infty)$ on a real number line.

2. Evaluate the following:
   (a) $(-2)^4$
   (b) 2^{-4}
   (c) $\dfrac{3^{16}}{3^{13}}$
   (d) $16^{-3/4}$

3. Express $\dfrac{(x^2)^a(\sqrt{x})^b}{x^{a+b}x^{a-b}}$ as a power of x.

4. Simplify:
   (a) $\sqrt{200} - \sqrt{8}$
   (b) $(2a^3b^2)(3ab^4)^3$
   (c) $\left(\dfrac{x^2y^{-3}}{y^5}\right)^{-4}$
   (d) $\dfrac{x^2 + 3x + 2}{x^2 - x - 2}$
   (e) $\dfrac{x}{x^2 - 4} + \dfrac{1}{x + 2}$

5. Expand the following:
   (a) $4(3 - x) - 3(x + 5)$
   (b) $(x - 5)(2x + 3)$
   (c) $(\sqrt{x} + \sqrt{y})(\sqrt{x} - \sqrt{y})$
   (d) $(3t + 4)^2$
   (e) $(2 - x^2)^3$

6. Factor the following:
   (a) $9x^2 - 25$
   (b) $6x^2 + 7x - 5$
   (c) $x^3 - 4x^2 - 3x + 12$
   (d) $x^4 + 27x$
   (e) $3x^{3/2} - 9x^{1/2} + 6x^{-1/2}$

7. Rationalize the denominator: $\dfrac{x}{\sqrt{x} - 2}$

8. Solve for x:
   (a) $3x + 9 = 7 + \frac{3}{2}x$
   (b) $\dfrac{x + 1}{x - 1} = \dfrac{2x - 1}{2x + 1}$

9. Bill drove from Ajax to Bixby at an average speed of 50 mi/h. On the way back, he drove at 60 mi/h. The total trip took $4\frac{2}{5}$ hours of driving time. How far is it between these two cities?

10. Find all solutions, real and imaginary, of the following equations:
    (a) $x^2 - x - 12 = 0$
    (b) $x^2 + x - 1 = 0$
    (c) $\sqrt{3 - \sqrt{x + 5}} = 2$
    (d) $x^{1/2} - 3x^{3/2} + 2x^{5/2} = 0$

11. A rectangular building lot is 70 ft longer than it is wide. Each diagonal between opposite corners is 130 ft. What are the dimensions of this lot?

12. Solve the inequalities. Sketch the solutions on the real number line, and write your final answers using interval notation.
    (a) $-1 \le 5 - 2x < 10$
    (b) $x(x - 1)(x - 2) > 0$
    (c) $|x - 3| < 2$
    (d) $\dfrac{2x + 5}{x + 1} \le 1$

13. A bottle of medicine is to be stored at a temperature between 5 °C and 10 °C. What range does this correspond to on the Fahrenheit scale? [*Note:* The Fahrenheit (F) and the Celsius (C) scales satisfy the relation $C = \frac{5}{9}(F - 32)$.]

14. Graph the following set in the coordinate plane:
$$\{(x, y) \mid x \ge 3 \quad \text{and} \quad |y| < 2\}$$

15. For the points $A(9, -4)$ and $B(-1, 20)$, find the length and the midpoint of the segment AB. Then find an equation of the circle for which AB is a diameter.

16. Sketch graphs of the following equations:
    (a) $x^2 + y^2 - 6x + 10y + 9 = 0$
    (b) $2x^2 + 2y^2 + 6x + 10y + 17 = 0$

17. Find the equation of the line (in slope-intercept form) that is perpendicular to the line $2x + 6y = 13$ and that passes through the point $(5, 0)$.

PRINCIPLES OF PROBLEM SOLVING

There are no hard and fast rules that will ensure success in solving problems. However, it is possible to outline some general steps in the problem-solving process and to give some principles that may be useful in the solution of certain problems. These steps and principles are just common sense made explicit. They have been adapted from George Polya's book *How to Solve It*.

 UNDERSTAND THE PROBLEM

The first step is to read the problem and make sure that you understand it clearly. Ask yourself the following questions:

> *What is the unknown?*
> *What are the given quantities?*
> *What are the given conditions?*

For many problems it is useful to

> *draw a diagram*

and identify the given and required quantities on the diagram.
    Usually it is necessary to

> *introduce suitable notation.*

In choosing symbols for the unknown quantities we often use letters such as a, b, c, m, n, x, and y, but in some cases it helps to use initials as suggestive symbols, for instance, V for volume, or t for time.

 THINK OF A PLAN

Find a connection between the given information and the unknown that will enable you to calculate the unknown. It often helps to ask yourself explicitly: "How can I relate the given to the unknown?" If you do not see a connection immediately, the following ideas may be helpful in devising a plan.

Try to recognize something familiar

Relate the given situation to previous knowledge. Look at the unknown and try to recall a more familiar problem that has a similar unknown.

George Polya (1887–1985) *is famous among mathematicians for his ideas on problem solving. His lectures on problem solving at Stanford University attracted overflow crowds whom he held on the edges of their seats, leading them to discover solutions for themselves. He was able to do this because of his deep insight into the psychology of problem solving. His well known book* How to Solve It *has been translated into 15 languages. He said that Euler (see p. 222) was unique among great mathematicians because he explained how he found his results. Polya often said to his students and colleagues, "Yes, I see that your proof is correct—but how did you discover it?" In the preface to* How to Solve It *Polya writes, "A great discovery solves a great problem but there is a grain of discovery in the solution of any problem. Your problem may be modest; but if it challenges your curiosity and brings into play your inventive faculties, and if you solve it by your own means, you may experience the tension and enjoy the triumph of discovery."*

Try to recognize patterns

Some problems are solved by recognizing that some kind of pattern is occurring. The pattern could be geometric, or numerical, or algebraic. If you can see regularity or repetition in a problem, then you might be able to guess what the continuing pattern is and then prove it.

Use analogy

Try to think of an analogous problem, that is, a similar problem, a related problem, but one that is easier than the original problem. If you can solve the similar, simpler problem, then it might give you the clues you need to solve the original, more difficult problem. For instance, if a problem involves very large numbers, you could first try a similar problem with smaller numbers. Or if the problem is in three-dimensional geometry, you could look for a similar problem in two-dimensional geometry. Or if the problem you start with is a general one, you could first try a special case.

Introduce something extra

It may sometimes be necessary to introduce something new, an auxiliary aid, to help make the connection between the given and the unknown. For instance, in a problem where a diagram is useful the auxiliary aid could be a new line drawn in the diagram. In a more algebraic problem it could be a new unknown that is related to the original unknown.

Take cases

We may sometimes have to split a problem into several cases and give a different argument for each of the cases. For instance, we often have to use this strategy in dealing with absolute value.

Work backward

Sometimes it is useful to imagine that your problem is solved and work backward, step by step, until you arrive at the given data. Then you may be able to reverse your steps and thereby construct a solution to the original problem. This procedure is commonly used in solving equations. For instance, in solving the equation $3x - 5 = 7$, we suppose that x is a number that satisfies $3x - 5 = 7$ and work backward. We add 5 to each side of the equation and then divide each side by 3 to get $x = 4$. Since each of these steps can be reversed, we have solved the problem.

Establish subgoals

In a complex problem it is often useful to set subgoals (in which the desired situation is only partially fulfilled). If we can first reach these subgoals, then we may be able to build on them to reach our final goal.

Indirect reasoning

Sometimes it is appropriate to attack a problem indirectly. In using **proof by contradiction** to prove that P implies Q we assume that P is true and Q is false and try to see why this cannot happen. Somehow we have to use this information and arrive at a contradiction to what we absolutely know is true.

Mathematical induction

In proving statements that involve a positive integer n, it is frequently helpful to use the Principle of Mathematical Induction, which is discussed in Section 9.7.

3 CARRY OUT THE PLAN

In Step 2 a plan was devised. In carrying out that plan we have to check each stage of the plan and write the details that prove that each stage is correct.

4 LOOK BACK

Having completed our solution, it is wise to look back over it, partly to see if there are errors in the solution and partly to see if there is an easier way to solve the problem. Another reason for looking back is that it will familiarize us with the method of solution and this may be useful for solving a future problem. Descartes said, "Every problem that I solved became a rule which served afterwards to solve other problems."

We illustrate some of these principles of problem solving on an example. Further illustrations of these principles will be presented at the end of every chapter.

EXAMPLE

A driver sets out on a journey. For the first half of the distance she drives at the leisurely pace of 30 mi/h; she drives the second half at 60 mi/h. What is her average speed on this trip?

Preliminary Thoughts It is tempting to take the average of the speeds and say that the average speed for the entire trip is

$$\frac{30 + 60}{2} = 45 \text{ mi/h}$$

But is this simple-minded approach really correct?

Try a special case

Let us look at an easily calculated special case. Suppose that the total distance travelled is 120 mi. Since the first 60 mi is travelled at 30 mi/h, it takes 2 h. The second 60 mi is traveled at 60 mi/h, so it takes 1 h. Thus the total time is $2 + 1 = 3$ h and the average speed is

$$\frac{120}{3} = 40 \text{ mi/h}$$

So the guess of 45 mi/h is wrong.

SOLUTION

Understand the problem

We need to look more carefully at the meaning of average speed. It is defined as

$$\text{average speed} = \frac{\text{distance traveled}}{\text{time elapsed}}$$

Introduce notation

Let d be the distance traveled on each half of the trip. Let t_1 and t_2 be the times taken for the first and second halves of the trip. Now we are in a position to write down the information that we are given. For the first half of the trip we have

State what is given

$$30 = \frac{d}{t_1} \tag{1}$$

and for the second half we have

$$60 = \frac{d}{t_2} \tag{2}$$

Identify the unknown

Let us now identify the quantity we are asked to find:

$$\text{average speed for entire trip} = \frac{\text{total distance}}{\text{total time}} = \frac{2d}{t_1 + t_2}$$

Connect the given with the unknown

To calculate this quantity we need to know t_1 and t_2, so we solve Equations 1 and 2 for these times:

$$t_1 = \frac{d}{30} \qquad t_2 = \frac{d}{60}$$

103

Now we have the ingredients needed to calculate the desired quantity:

$$\text{average speed} = \frac{2d}{t_1 + t_2} = \frac{2d}{\dfrac{d}{30} + \dfrac{d}{60}}$$

$$= \frac{60(2d)}{60\left(\dfrac{d}{30} + \dfrac{d}{60}\right)} \qquad \text{(Multiply numerator and denominator by 60)}$$

$$= \frac{120d}{2d + d} = \frac{120d}{3d} = 40$$

The average speed for the entire trip is 40 mi/h. ∎

PROBLEMS

1. **(a)** The radius of the earth is about 3960 mi. What length of red ribbon would you need to wrap around the earth at the equator?
   (b) How much more red ribbon would you need if you raised the ribbon one foot above the earth?

2. A piece of wire is bent as shown in the figure. You can see that one cut through the wire produces four pieces and two parallel cuts produce seven pieces. How many pieces will be produced by 142 parallel cuts? Write a formula for the number of pieces if there are n parallel cuts.

3. Which is better, a 40% discount or two successive discounts of 20%?

4. A car and a van are 120 mi apart on a straight road. The drivers start driving toward each other at noon, each at a speed of 40 mi/h. A fly starts from the front bumper of the van at noon and flies to the bumper of the car, then immediately back to the bumper of the van, back to the car, and so on, until the car and the van meet. If the fly flies at a speed of 100 mi/h, how far does the fly fly?

5. Use a calculator to find the value of the expression

$$\sqrt{3 + 2\sqrt{2}} - \sqrt{3 - 2\sqrt{2}}$$

The number looks very simple. Show that the calculated value is correct.

Srinivasa Ramanujan (1887–1920) *was born into a poor family in the small town of Kumbakonam in India. He was self-taught in mathematics and worked in virtual isolation from other mathematicians. At the age of 25 he wrote a letter to Hardy, the leading British mathematician at the time, listing some of his discoveries. Hardy immediately recognized Ramanujan's genius and for the next six years the two worked together in London until Ramanujan fell ill and went back to his hometown in India, where he died a year later. Ramanujan was a genius with phenomenal ability to see hidden patterns in the properties of numbers. Most of his discoveries were written as complicated infinite series, the importance of which was not recognized till many years after his death. In the last year of his life he wrote 130 pages of mysterious formulas, many of which still defy proof. Hardy tells the story that when he visited Ramanujan in a hospital and arrived in a taxi, he remarked to Ramanujan that the cab's number, 1729, was uninteresting. Ramanujan replied "No, it is a very interesting number. It is the smallest number expressible as the sum of two cubes in two different ways." (See Problem 7.)*

6. Use a calculator to evaluate

$$\frac{\sqrt{2} + \sqrt{6}}{\sqrt{2} + \sqrt{3}}$$

Show that the calculated value is correct.

7. The number 1729 is the smallest positive integer that can be represented in two different ways as the sum of two cubes. What are the two different ways?

8. Two runners start running laps at the same time from the same starting position. George takes 50 s to run a lap; Sue takes 30 s to run a lap. When will the runners next be even with each other?

9. Player A has a higher batting average than Player B for the first half of the baseball season. Player A also has a higher batting average than Player B for the second half of the season. Is it true that Player A has a higher batting average than Player B for the entire season?

10. A person starts at a point P on the earth's surface and walks 1 mi south, then 1 mi east, then 1 mi north and finds herself back at P where she started. Describe all points P for which this is possible. (There are infinitely many.)

11. You have an 8×8 grid with two diagonally opposite corner squares removed, as shown in the figure. You also have a set of dominoes, each of which covers exactly two squares of the grid. Is it possible to cover all the remaining squares of the grid with dominoes (without overlapping the dominoes)? [*Hint:* Color the grid as a checkerboard.]

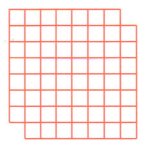

FUNCTIONS

That flower of modern mathematical thought—the notion of a function.

THOMAS J. MC CORMACK

When we cannot use the compass of mathematics or the torch of experience . . . it is certain we cannot take a single step forward.

VOLTAIRE

One of the most basic and important ideas in all of mathematics is that of a *function*. In this chapter we study functions, their graphs, and some of their applications.

The area A of a circle depends on the radius r of the circle. The rule that connects r and A is given by the equation $A = \pi r^2$. With each positive number r there is associated one value of A, and we say that A is a function of r.

The number N of bacteria in a culture depends on the time t. If the culture starts with 5000 bacteria and the population doubles every hour, then after t hours the number of bacteria will be $N = (5000)2^t$. This is the rule that connects t and N. For each value of t there is a corresponding value of N, and we say that N is a function of t.

The cost C of mailing a first-class letter depends on the weight w of the letter. Although there is no single neat formula that connects w and C, the post office has a rule for determining C when w is known.

In each of these examples there is a rule whereby, given a number (r, t, or w), another number (A, N, or C) is assigned. In each case we say that the second number is a function of the first number.

> A **function** f is a rule that assigns to each element x in a set A exactly one element, called $f(x)$, in a set B.

We usually consider functions for which the sets A and B are sets of real numbers. The set A is called the **domain** of the function. The symbol $f(x)$ is read "f of x" or "f at x" and is called the **value of f at x**, or the **image of x under f**. The **range** of f is the set of all possible values of $f(x)$ as x varies throughout the domain, that is, $\{f(x) \mid x \in A\}$.

The symbol that represents an arbitrary number in the *domain* of a function f is called an **independent variable**. The symbol that represents a number in the *range* of f is called a **dependent variable**. For instance, in the bacteria example, t is the independent variable and N is the dependent variable.

It is helpful to think of a function as a **machine** (see Figure 1). If x is in the domain of the function f, then when x enters the machine, it is accepted as an input and the machine produces an output $f(x)$ according to the rule of the function. Thus we can think of the domain as the set of all possible inputs and the range as the set of all possible outputs.

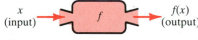

Figure 1

The preprogrammed functions in a calculator are good examples of a function as a machine. For example, the $\sqrt{x}$ key on your calculator is such a function. First you input x into the display. Then you press the key labeled $\sqrt{x}$. If $x < 0$, then x is not in the domain of this function; that is, x is not an acceptable input and the

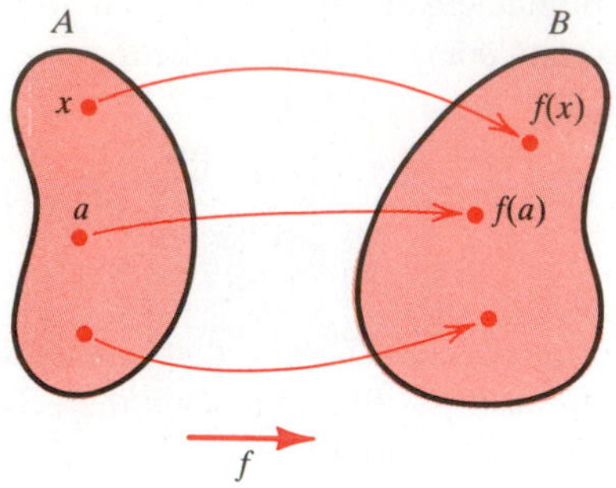

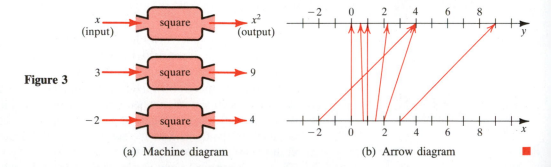

Figure 2
Arrow diagram for f

calculator will indicate an error. If $x \geqslant 0$, then an approximation to $\sqrt{x}$ will appear in the display, correct to a certain number of decimal places. [Thus the $\sqrt{x}$ key on your calculator is not quite the same as the exact mathematical function f defined by $f(x) = \sqrt{x}$.]

Another way to picture a function is by an **arrow diagram** as in Figure 2. Each arrow connects an element of A to an element of B. The arrow indicates that $f(x)$ is associated with x, $f(a)$ is associated with a, and so on.

EXAMPLE 1

The squaring function assigns to each real number x its square x^2. It is defined by the equation

$$f(x) = x^2$$

(a) Evaluate $f(3)$, $f(-2)$, and $f(\sqrt{5})$.
(b) Find the domain and range of f.
(c) Draw a machine diagram and an arrow diagram for f.

SOLUTION

(a) The values of f are found by substituting for x in the equation $f(x) = x^2$:

$$f(3) = 3^2 = 9 \qquad f(-2) = (-2)^2 = 4 \qquad f(\sqrt{5}) = (\sqrt{5})^2 = 5$$

(b) The domain of f is the set R of all real numbers. The range of f consists of all values of $f(x)$, that is, all numbers of the form x^2. But $x^2 \geqslant 0$ for all real numbers x, and any nonnegative number c is a square since $c = (\sqrt{c})^2 = f(\sqrt{c})$. Therefore the range of f is $\{y \mid y \geqslant 0\} = [0, \infty)$.

(c) Machine and arrow diagrams for this function are shown in Figure 3.

Figure 3

(a) Machine diagram   (b) Arrow diagram

Note   We previously have used letters to stand for numbers. Here we do something quite different. We use letters to represent *rules*. For instance, in Example 1 the name of the function is f and the rule is "square the number."

EXAMPLE 2

If we define a function g by

$$g(x) = x^2 \qquad 0 \leqslant x \leqslant 3$$

then the domain of g is given as the closed interval $[0, 3]$. This is different from the function f in Example 1 because in considering g we are restricting our attention to those values of x between 0 and 3. The range of g is

$$\{x^2 \mid 0 \leqslant x \leqslant 3\} = \{y \mid 0 \leqslant y \leqslant 9\} = [0, 9] \qquad \blacksquare$$

EXAMPLE 3

If $f(x) = 2x^2 + 3x - 1$, evaluate the following:

(a) $f(a)$ (b) $f(-a)$ (c) $f(a + h)$ (d) $\dfrac{f(a + h) - f(a)}{h}, \quad h \neq 0$

[Expressions like the one in part (d) occur frequently in calculus where they are called *difference quotients*.]

SOLUTION

(a) $$f(a) = 2a^2 + 3a - 1$$

(b) $$f(-a) = 2(-a)^2 + 3(-a) - 1 = 2a^2 - 3a - 1$$

(c) $$f(a + h) = 2(a + h)^2 + 3(a + h) - 1$$
$$= 2(a^2 + 2ah + h^2) + 3(a + h) - 1$$
$$= 2a^2 + 4ah + 2h^2 + 3a + 3h - 1$$

(d) $$\frac{f(a + h) - f(a)}{h} = \frac{(2a^2 + 4ah + 2h^2 + 3a + 3h - 1) - (2a^2 + 3a - 1)}{h}$$
$$= \frac{4ah + 2h^2 + 3h}{h} = 4a + 2h + 3 \qquad \blacksquare$$

In Examples 1 and 2 the domain of the function was given explicitly. But if a function is given by a formula and the domain is not stated explicitly, **the convention is that the domain is the set of all real numbers for which the formula makes sense and defines a real number.**

We should distinguish between a function f and the number $f(x)$, which is the value of f at x. Nonetheless, it is common to abbreviate an expression such as

$$\text{the function } f \text{ defined by } f(x) = x^2 + x$$

to

$$\text{the function } f(x) = x^2 + x$$

EXAMPLE 4

Find the domain of the function $f(x) = \dfrac{1}{x^2 - x}$.

SOLUTION

The function is not defined when the denominator is 0. Since

$$f(x) = \frac{1}{x^2 - x} = \frac{1}{x(x - 1)}$$

we see that $f(x)$ is not defined when $x = 0$ or $x = 1$. Thus the domain of f is

$$\{x \mid x \neq 0,\, x \neq 1\}$$

which could also be written in interval notation as

$$(-\infty, 0) \cup (0, 1) \cup (1, \infty) \qquad \blacksquare$$

EXAMPLE 5

Find the domain of the function $g(t) = \dfrac{t}{\sqrt{t + 1}}$.

SOLUTION

The square root of a negative number is not defined (as a real number), so we require that $t + 1 \geq 0$. Also the denominator cannot be 0, that is, $\sqrt{t + 1} \neq 0$. Thus $g(t)$ exists when $t + 1 > 0$, that is, $t > -1$. So the domain of g is

$$\{t \mid t > -1\} = (-1, \infty) \qquad \blacksquare$$

EXAMPLE 6

Find the domain of $h(x) = \sqrt{2 - x - x^2}$.

SOLUTION

Since the square root of a negative number is not defined (as a real number), the domain of h consists of all values of x such that $2 - x - x^2 \geq 0$. We solve this inequality using the methods of Section 1.7. Since $2 - x - x^2 = (2 + x)(1 - x)$, the product will change sign when $x = -2$ or 1 as indicated in the following chart:

Interval	$2 + x$	$1 - x$	$(2 + x)(1 - x)$
$x < -2$	$-$	$+$	$-$
$-2 < x < 1$	$+$	$+$	$+$
$x > 1$	$+$	$-$	$-$

Therefore, the domain of h is

$$\{x \mid -2 \leqslant x \leqslant 1\} = [-2, 1]$$

EXERCISES 2.1

1. If $f(x) = x^2 - 3x + 2$, find $f(1)$, $f(-2)$, $f(\frac{1}{2})$, $f(\sqrt{5})$, $f(a)$, $f(-a)$, and $f(a + b)$.

2. If $f(x) = x^3 + 2x^2 - 3$, find $f(0)$, $f(3)$, $f(-3)$, $f(-x)$, and $f(1/a)$.

3. If $g(x) = \dfrac{1 - x}{1 + x}$, find $g(2)$, $g(-2)$, $g(\pi)$, $g(a)$, $g(a - 1)$, and $g(-a)$.

4. If $h(t) = t + \dfrac{1}{t}$, find $h(1)$, $h(\pi)$, $h(t + 1)$, $h(t) + h(1)$, and $h(x)$.

5. If $f(x) = 2x^2 + 3x - 4$, find $f(0)$, $f(2)$, $f(\sqrt{2})$, $f(1 + \sqrt{2})$, $f(-x)$, $f(x + 1)$, $2f(x)$, and $f(2x)$.

6. If $f(x) = 2 - 3x$, find $f(1)$, $f(-1)$, $f(\frac{1}{3})$, $f(x/3)$, $f(3x)$, $f(x^2)$, and $[f(x)]^2$.

In Exercises 7–10, find $f(a)$, $f(a) + f(h)$, $f(a + h)$, and $\dfrac{f(a + h) - f(a)}{h}$, where a and h are real numbers and $h \neq 0$.

7. $f(x) = 1 + 2x$

8. $f(x) = x^2 - 3x + 4$

9. $f(x) = 3 - 5x + 4x^2$

10. $f(x) = x^3 + x + 1$

In Exercises 11–14 find $f(2 + h)$, $f(x + h)$, and $\dfrac{f(x + h) - f(x)}{h}$, where $h \neq 0$.

11. $f(x) = 8x - 1$

12. $f(x) = x - x^2$

13. $f(x) = \dfrac{1}{x}$

14. $f(x) = \dfrac{x}{x + 1}$

In Exercises 15–16 draw a machine diagram and an arrow diagram for the given function.

15. $f(x) = \sqrt{x}$, $\quad 0 \leqslant x \leqslant 4$

16. $f(x) = \dfrac{2}{x}$, $\quad 1 \leqslant x \leqslant 4$

In Exercises 17–26 find the domain and range of the given function.

17. $f(x) = 2x$, $-1 \leqslant x \leqslant 5$

18. $f(x) = 2x + 7$, $-1 \leqslant x \leqslant 6$

19. $f(x) = 6 - 4x$, $-2 \leqslant x \leqslant 3$

20. $f(x) = x^2 + 1$, $-1 \leqslant x \leqslant 5$

21. $g(x) = 2 - x^2$

22. $g(x) = \sqrt{7 - 3x}$

23. $h(x) = \sqrt{2x - 5}$

24. $h(x) = 1 - \sqrt{x}$

25. $F(x) = 3 + \sqrt{1 - x^2}$

26. $G(x) = \sqrt{x^2 - 9}$

In Exercises 27–48 find the domain of the given function.

27. $f(x) = \dfrac{1}{x + 4}$

28. $f(x) = \dfrac{2}{3x - 5}$

29. $f(x) = \dfrac{x + 2}{x^2 - 1}$

30. $f(x) = \dfrac{x^4}{x^2 + x - 6}$

31. $f(x) = \dfrac{x + 2}{x^2 + 1}$

32. $f(x) = \dfrac{x^3 + 1}{x^3 - x}$

33. $f(x) = \sqrt{5 - 2x}$

34. $f(x) = \sqrt[4]{x - 9}$

35. $f(t) = \sqrt[3]{t - 1}$

36. $f(t) = \sqrt{t^2 + 1}$

37. $g(x) = \dfrac{\sqrt{2 + x}}{3 - x}$

38. $g(x) = \dfrac{\sqrt{x}}{2x^2 + x - 1}$

39. $F(x) = \dfrac{x}{\sqrt{x - 10}}$

40. $F(x) = x^2 - \dfrac{x}{\sqrt{9 - 2x}}$

41. $G(x) = \sqrt{x} + \sqrt{1 - x}$

42. $G(x) = \sqrt{x + 2} - 2\sqrt{x - 3}$

43. $f(x) = \sqrt{4x^2 - 1}$

44. $f(x) = \sqrt{2 - 3x^2}$

45. $g(x) = \sqrt[4]{x^2 - 6x}$

46. $g(x) = \sqrt{x^2 - 2x - 8}$

47. $\phi(x) = \sqrt{\dfrac{x}{\pi - x}}$

48. $\phi(x) = \sqrt{\dfrac{x^2 - 2x}{x - 1}}$

SECTION 2.2
GRAPHS OF FUNCTIONS

In the preceding section we saw how to picture functions using machine diagrams and arrow diagrams. A third method for visualizing a function is its graph. If f is a function with domain A, its **graph** is the set of ordered pairs

$$\{(x, f(x)) \mid x \in A\}$$

In other words, the graph of f consists of all points (x, y) in the coordinate plane such that $y = f(x)$ and x is in the domain of f. Thus the equation of the graph of f is $y = f(x)$.

The graph of a function f gives us a useful picture of the behavior or "life history" of a function. Since the y-coordinate of any point (x, y) on the graph is $y = f(x)$, we can read the value of $f(x)$ from the graph as being the height of the graph above the point x (see Figure 1). The graph of f also allows us to picture the domain and range of f on the x-axis and y-axis as in Figure 2.

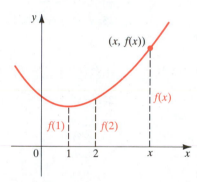

Figure 1
The graph of f

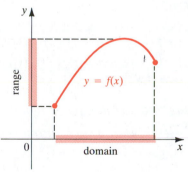

Figure 2

EXAMPLE 1

Sketch the graph of the function $f(x) = 2x - 1$.

SOLUTION

The equation of the graph is $y = 2x - 1$, and we recognize this as the equation of a line with slope 2 and y-intercept -1. This enables us to sketch the graph of f in Figure 3. ■

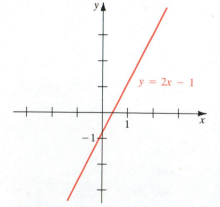

Figure 3
Graph of $f(x) = 2x - 1$

In general, a function f defined by an equation of the form $f(x) = mx + b$ is called a **linear function** because the equation of its graph is $y = mx + b$, which represents a line with slope m and y-intercept b. A special case of the linear function occurs when the slope is $m = 0$. The function $f(x) = b$, where b is a given number, is called a **constant function** because all its values are the same number, namely b. Its graph is the horizontal line $y = b$.

EXAMPLE 2

Sketch the graph of $f(x) = x^2$.

SOLUTION

The equation of the graph is $y = x^2$. We draw it in Figure 4 by setting up a table of values and plotting points as in Section 1.9. Recall from Example 1 of Section 2.1 that the domain is R and the range is $[0, \infty)$.

x	$y = x^2$
0	0
$\pm\frac{1}{2}$	$\frac{1}{4}$
± 1	1
± 2	4
± 3	9

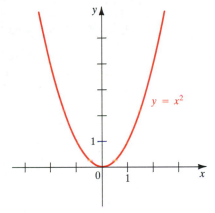

Figure 4
Graph of $f(x) = x^2$

EXAMPLE 3

Sketch the graph of $f(x) = x^3$.

SOLUTION

We list some functional values and the corresponding points on the graph in the following table:

x	0	$\frac{1}{2}$	1	2	$-\frac{1}{2}$	-1	-2
$f(x) = x^3$	0	$\frac{1}{8}$	1	8	$-\frac{1}{8}$	-1	-8
(x, x^3)	$(0, 0)$	$\left(\frac{1}{2}, \frac{1}{8}\right)$	$(1, 1)$	$(2, 8)$	$\left(-\frac{1}{2}, -\frac{1}{8}\right)$	$(-1, -1)$	$(-2, -8)$

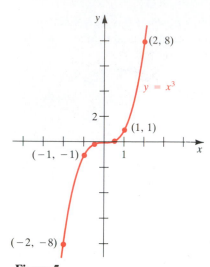

Figure 5
Graph of $f(x) = x^3$

Then we plot points to obtain the graph shown in Figure 5.   ■

EXAMPLE 4

If $f(x) = \sqrt{4 - x^2}$, sketch the graph of f and find its domain and range.

SOLUTION

The equation of the graph is $y = \sqrt{4 - x^2}$, from which it follows that $y \geq 0$. Squaring this equation, we get

$$y^2 = 4 - x^2 \qquad \text{or} \qquad x^2 + y^2 = 4$$

which we recognize as the equation of a circle with center the origin and radius 2. But, since $y \geq 0$, the graph of f consists of only the upper half of this circle. From Figure 6 we see that the domain is the closed interval $[-2, 2]$ and the range is $[0, 2]$. We could also have found the domain as follows:

$$\begin{aligned} \text{domain} &= \{x \mid 4 - x^2 \geq 0\} = \{x \mid x^2 \leq 4\} \\ &= \{x \mid |x| \leq 2\} = \{x \mid -2 \leq x \leq 2\} \\ &= [-2, 2] \end{aligned}$$

Figure 6
Graph of $f(x) = \sqrt{4 - x^2}$

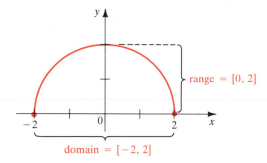

■

The graph of a function is a curve in the xy-plane. But the question arises: Which curves in the xy-plane are graphs of functions? This is answered by the following test.

THE VERTICAL LINE TEST

> A curve in the plane is the graph of a function if and only if no vertical line intersects the curve more than once.

We can see from Figure 7 why the Vertical Line Test is true. If each vertical line $x = a$ intersects a curve only once at (a, b), then exactly one functional value is defined by $f(a) = b$. But if a line $x = a$ intersects the curve twice at (a, b) and

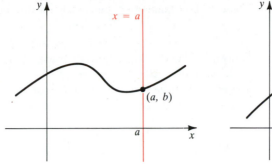

Figure 7

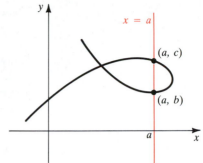

(a, c), then the curve cannot represent a function because a function cannot assign two different values to a.

EXAMPLE 5

Using the Vertical Line Test, we see that the curves in Figures 8(b) and (c) represent functions, whereas those in parts (a) and (d) do not.

Figure 8

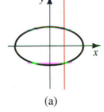 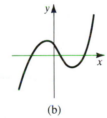 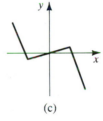 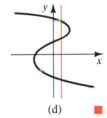

(a)   (b)   (c)   (d)   ■

EXAMPLE 6

Sketch the graph of $f(x) = \sqrt{x}$.

SOLUTION

First we note that the domain is $\{x \mid x \geq 0\} = [0, \infty)$. Then we plot the points given by the table in the margin and use them to draw the sketch in Figure 9.

x	$y = \sqrt{x}$
0	0
1	1
2	$\sqrt{2}$
3	$\sqrt{3}$
4	2
5	$\sqrt{5}$

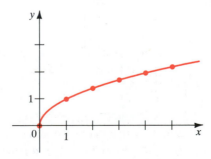

Figure 9
Graph of $f(x) = \sqrt{x}$

■

Note   The equation of the graph of the square-root function in Example 6 is $y = \sqrt{x}$. If we square this equation, we get $y^2 = x$, but we have to remember that $y \geqslant 0$. We saw in Section 1.9 that the equation $x = y^2$ represents the parabola shown in Figure 10(a). By the Vertical Line Test this parabola does not represent a function of x. But we can regard $y^2 = x$ as representing *two* functions of x; the upper and lower halves of this parabola are the graphs of the functions $f(x) = \sqrt{x}$ and $g(x) = -\sqrt{x}$. [See Figures 10(b) and (c).]

Figure 10

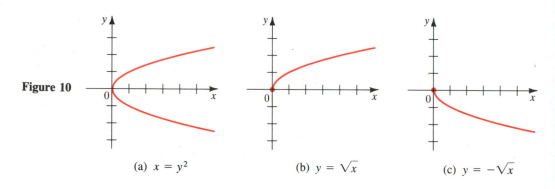

(a) $x = y^2$ (b) $y = \sqrt{x}$ (c) $y = -\sqrt{x}$

INCREASING AND DECREASING FUNCTIONS

It is very useful to know where the graph of a function rises and where it falls. The graph shown in Figure 11 rises from A to B, falls from B to C, and rises again from C to D. The function f is said to be increasing on the interval $[a, b]$, decreasing on $[b, c]$, and increasing again on $[c, d]$. Notice that if x_1 and x_2 are any two numbers between a and b with $x_1 < x_2$, then $f(x_1) < f(x_2)$. We use this as the defining property of an increasing function.

Figure 11

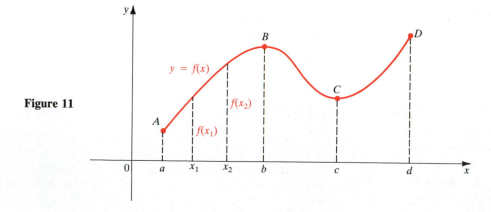

**INCREASING AND
DECREASING FUNCTIONS**

A function f is called **increasing** on an interval I if

$$f(x_1) < f(x_2) \qquad \text{whenever} \qquad x_1 < x_2 \text{ in } I$$

It is called **decreasing** on I if

$$f(x_1) > f(x_2) \qquad \text{whenever} \qquad x_1 < x_2 \text{ in } I$$

For instance, the functions in Examples 1, 3, and 6 are all increasing on their domains. In Example 2, f is decreasing on $(-\infty, 0]$ and increasing on $[0, \infty)$. In Example 4, f is increasing on $[-2, 0]$ and decreasing on $[0, 2]$.

EXAMPLE 7

State the intervals on which the function whose graph is shown in Figure 12 is increasing or decreasing.

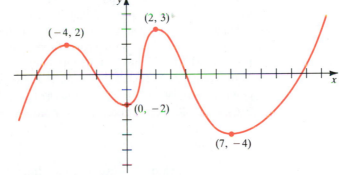

Figure 12

SOLUTION

The function is increasing on $(-\infty, -4]$, $[0, 2]$, and $[7, \infty)$. It is decreasing on $[-4, 0]$ and $[2, 7]$.   ■

■ **FUNCTIONS DEFINED BY MORE THAN ONE EQUATION**

The functions we have looked at so far have been defined by means of simple formulas. But there are many functions that are not given by such formulas. Here are some examples: the cost of mailing a first-class letter as a function of its weight, the population of New York City as a function of time, and the cost of a taxi ride as a function of distance. The following examples give more illustrations.

EXAMPLE 8

A function f is defined by

$$f(x) = \begin{cases} 1 - x & \text{if } x \le 1 \\ x^2 & \text{if } x > 1 \end{cases}$$

Evaluate $f(0)$, $f(1)$, and $f(2)$ and sketch the graph.

SOLUTION

Remember that a function is a rule. For this particular function the rule is the following: First look at the value of the input x. If it happens that $x \le 1$, then the value of $f(x)$ is $1 - x$. On the other hand, if $x > 1$, then the value of $f(x)$ is x^2.

Since $0 \le 1$, we have $f(0) = 1 - 0 = 1$.

Since $1 \le 1$, we have $f(1) = 1 - 1 = 0$.

Since $2 > 1$, we have $f(2) = 2^2 = 4$.

How do we draw the graph of f? We observe that if $x \le 1$, then $f(x) = 1 - x$, so the part of the graph of f that lies to the left of the vertical line $x = 1$ must coincide with the line $y = 1 - x$, which has slope -1 and y-intercept 1. If $x > 1$, then $f(x) = x^2$, so the part of the graph of f that lies to the right of the line $x = 1$ must coincide with the graph of $y = x^2$, which we sketched in Example 2. This enables us to sketch the graph in Figure 13. The solid dot indicates that the point is included on the graph; the open dot indicates that the point is excluded from the graph. ∎

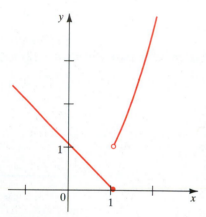

Figure 13

EXAMPLE 9

Sketch the graph of the absolute value function $y = |x|$.

SOLUTION

Recall that

$$|x| = \begin{cases} x & \text{if } x \ge 0 \\ -x & \text{if } x < 0 \end{cases}$$

Using the same method as in Example 8, we see that the graph of f coincides with the line $y = x$ to the right of the y-axis and coincides with the line $y = -x$ to the left of the y-axis (see Figure 14). ∎

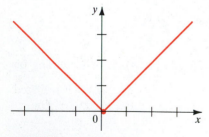

Figure 14
Graph of $f(x) = |x|$

EXAMPLE 10

The cost of a long-distance daytime phone call from Toronto to New York City is 69 cents for the first minute and 58 cents for each additional minute (or part of a minute). Draw the graph of the cost C (in dollars) of the phone call as a function of time t (in minutes).

SOLUTION

Let $C(t)$ be the cost for t minutes. Since $t > 0$, the domain of the function is $(0, \infty)$. From the given information we have

$$
\begin{aligned}
C(t) &= 0.69 & \text{if } 0 < t \le 1 \\
C(t) &= 0.69 + 0.58 = 1.27 & \text{if } 1 < t \le 2 \\
C(t) &= 0.69 + 2(0.58) = 1.85 & \text{if } 2 < t \le 3 \\
C(t) &= 0.69 + 3(0.58) = 2.43 & \text{if } 3 < t \le 4
\end{aligned}
$$

and so on. The graph is shown in Figure 15.   ■

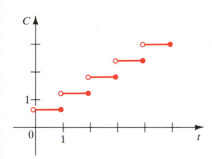

Figure 15

SYMMETRY

If a function f satisfies $f(-x) = f(x)$ for every number x in its domain, then f is called an **even function**. For instance, the function $f(x) = x^2$ of Example 2 is even because

$$ f(-x) = (-x)^2 = (-1)^2 x^2 = x^2 = f(x) $$

The geometric significance of an even function is that its graph is symmetric with respect to the y-axis (see Figure 16). This means that if we have plotted the graph of f for $x \ge 0$, we obtain the entire graph simply by reflecting in the y-axis.

If f satisfies $f(-x) = -f(x)$ for every number x in its domain, then f is called an **odd function**. For example, the function $f(x) = x^3$ of Example 3 is odd because

$$ f(-x) = (-x)^3 = -x^3 = -f(x) $$

The graph of an odd function is symmetric about the origin (see Figure 17). If we already have the graph of f for $x \ge 0$, we can obtain the entire graph by rotating through $180°$ about the origin. For instance, in Example 3 we could have plotted the graph of $y = x^3$ only for $x \ge 0$ and then rotated that part about the origin.

We summarize this discussion of symmetry as follows:

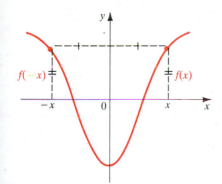

Figure 16
An even function

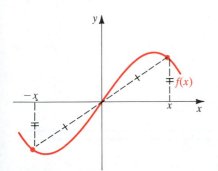

Figure 17
An odd function

An **even function** f satisfies $f(-x) = f(x)$ for all x in its domain.
The graph of an even function is symmetric about the y-axis.

An **odd function** f satisfies $f(-x) = -f(x)$ for all x in its domain.
The graph of an odd function is symmetric about the origin.

EXAMPLE 11

Determine whether each of the following functions is even, odd, or neither even nor odd:

(a) $f(x) = x^5 + x$ (b) $g(x) = 1 - x^4$ (c) $h(x) = 2x - x^2$

SOLUTION

(a)
$$f(-x) = (-x)^5 + (-x) = (-1)^5 x^5 + (-x)$$
$$= -x^5 - x = -(x^5 + x)$$
$$= -f(x)$$

Therefore f is an odd function.

(b)
$$g(-x) = 1 - (-x)^4 = 1 - x^4 = g(x)$$

So g is even.

(c)
$$h(-x) = 2(-x) - (-x)^2 = -2x - x^2$$

Since $h(-x) \neq h(x)$ and $h(-x) \neq -h(x)$, we conclude that h is neither even nor odd.   ■

The graphs of the functions in Example 11 are shown in Figure 18. The graph of f was drawn by plotting points for $x \geq 0$ and rotating about the origin. The graph of g was drawn by plotting points for $x \geq 0$ and reflecting about the y-axis. Notice that the graph of h is symmetric neither about the y-axis nor about the origin.

Figure 18

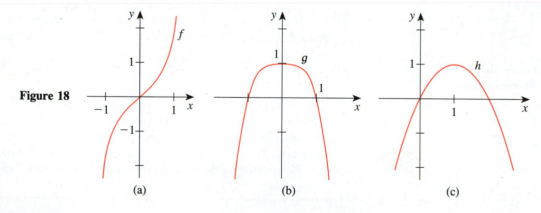

(a)   (b)   (c)

EXERCISES 2.2

1. The graph of a function is given.
  (a) State the values of $f(-2)$, $f(0)$, $f(2)$, and $f(3)$.
  (b) State the domain of f.
  (c) State the range of f.
  (d) On what interval is f decreasing? On what interval is it increasing?

odd 1–7

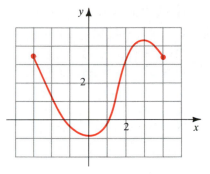

2. The graph of a function g is given.
  (a) State the values of $g(-3)$, $g(1)$, $g(2)$, and $g(3)$.
  (b) State the domain of g.
  (c) State the range of g.

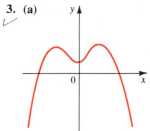

In Exercises 3 and 4 determine which of the curves are graphs of functions of x.

3. (a)

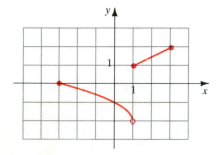

(b)

(c)

(d)

4. (a)

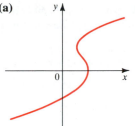

(b)

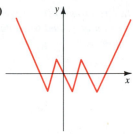

(c)

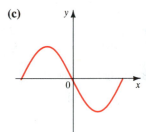

(d)

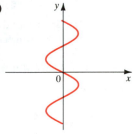

In Exercises 5–8 state whether the given curve is the graph of a function of x. If it is, state the domain and range of the function.

5.

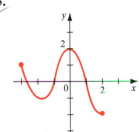

6.

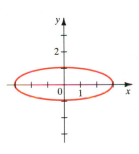

7.

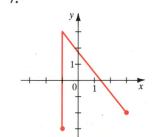

8.

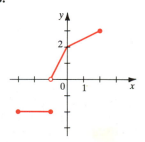

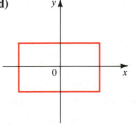

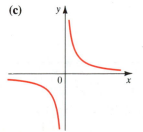

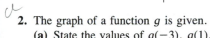

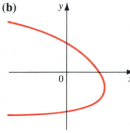

In Exercises 9–16 (a) sketch the graph of f, (b) find the domain of f, and (c) state the intervals on which f is increasing or decreasing.

9. $f(x) = 1 - x$

10. $f(x) = \frac{1}{2}(x + 1)$

11. $f(x) = x^2 - 4x$

12. $f(x) = x^2 - 4x + 1$

13. $f(x) = 1 + \sqrt{4 - x}$

14. $f(x) = \sqrt{x^2 - 4}$

15. $f(x) = -\sqrt{25 - x^2}$

16. $f(x) = x^3 - 3x + 1$

17. Draw the graphs of the functions $y = x$, $y = x^3$, and $y = x^5$ on the same coordinate axes.

18. Draw the graphs of the functions $y = x^2$, $y = x^4$, and $y = x^6$ on the same coordinate axes.

In Exercises 19–51 sketch the graph of the function.

19. $f(x) = 2$

20. $f(x) = -3$

21. $f(x) = 3 - 2x$

22. $f(x) = \dfrac{x + 3}{2}, \ -2 \le x \le 2$

23. $f(x) = -x^2$

24. $f(x) = x^2 - 4$

25. $f(x) = x^2 + 2x - 1$

26. $f(x) = -x^2 + 6x - 7$

27. $g(x) = 1 - x^3$

28. $g(x) = 2x^2 - x^4$

29. $g(x) = \sqrt{-x}$

30. $g(x) = \sqrt{6 - 2x}$

31. $F(x) = \dfrac{1}{x}$

32. $F(x) = \dfrac{2}{x + 4}$

33. $G(x) = |x| + x$

34. $G(x) = |x| - x$

35. $H(x) = |2x|$

36. $H(x) = |x + 1|$

37. $f(x) = |2x - 3|$

38. $f(x) = \dfrac{x}{|x|}$

39. $f(x) = \dfrac{x^2 - 1}{x - 1}$

40. $f(x) = \dfrac{x^2 + 5x + 6}{x + 2}$

41. $f(x) = \begin{cases} x + 1 & \text{if } x \ne 1 \\ 1 & \text{if } x = 1 \end{cases}$

42. $f(x) = \begin{cases} x + 3 & \text{if } x \ne -2 \\ 4 & \text{if } x = -2 \end{cases}$

43. $f(x) = \begin{cases} 0 & \text{if } x < 2 \\ 1 & \text{if } x \ge 2 \end{cases}$

44. $f(x) = \begin{cases} -1 & \text{if } x < -1 \\ 1 & \text{if } -1 \le x \le 1 \\ -1 & \text{if } x > 1 \end{cases}$

45. $f(x) = \begin{cases} x & \text{if } x \le 0 \\ x + 1 & \text{if } x > 0 \end{cases}$

46. $f(x) = \begin{cases} 2x + 3 & \text{if } x < -1 \\ 3 - x & \text{if } x \ge -1 \end{cases}$

47. $f(x) = \begin{cases} -1 & \text{if } x < -1 \\ x & \text{if } -1 \le x \le 1 \\ 1 & \text{if } x > 1 \end{cases}$

48. $f(x) = \begin{cases} |x| & \text{if } |x| \le 1 \\ 1 & \text{if } |x| > 1 \end{cases}$

49. $f(x) = \begin{cases} x + 2 & \text{if } x \le -1 \\ x^2 & \text{if } x > -1 \end{cases}$

50. $f(x) = \begin{cases} 1 - x^2 & \text{if } x \le 2 \\ 2x - 7 & \text{if } x > 2 \end{cases}$

51. $f(x) = \begin{cases} -1 & \text{if } x \le -1 \\ 3x + 2 & \text{if } |x| < 1 \\ 7 - 2x & \text{if } x \ge 1 \end{cases}$

52. A taxi company charges $2 for the first mile (or part of a mile) and 20 cents for each succeeding tenth of a mile (or part). Express the cost C (in dollars) of a ride as a function of the distance x traveled (in miles) for $0 < x < 2$ and sketch the graph of this function.

In Exercises 53–56 find a function whose graph is the given curve.

53. The line segment joining the points $(-2, 1)$ and $(4, -6)$

54. The line segment joining the points $(-3, -2)$ and $(6, 3)$

55. The bottom half of the parabola $x + (y - 1)^2 = 0$

56. The top half of the circle $(x - 1)^2 + y^2 = 1$

In Exercises 57–64 determine whether f is even, odd, or neither. If f is even or odd, use symmetry to sketch its graph.

57. $f(x) = x^{-2}$

58. $f(x) = x^{-3}$

59. $f(x) = x^2 + x$

60. $f(x) = x^4 - 4x^2$

61. $f(x) = x^3 - x$

62. $f(x) = 3x^3 + 2x^2 + 1$

63. $f(x) = 1 - \sqrt[3]{x}$

64. $f(x) = x + \dfrac{1}{x}$

65. (a) Sketch the graph of the function $f(x) = x^2 - 1$
(b) Sketch the graph of the function $g(x) = |x^2 - 1|$.

66. (a) Sketch the graph of the function $f(x) = |x| - 1$.
(b) Sketch the graph of the function $g(x) = ||x| - 1|$.

SECTION 2.3
GRAPHING CALCULATORS AND COMPUTERS

In this section we assume that you have access to a graphing calculator or a computer with graphing software. We will see that the use of such a device enables us to graph more complicated functions and to solve more complex problems than would otherwise be possible. We also point out some of the pitfalls that can occur with these machines.

A graphing calculator or computer displays a rectangular portion of the graph of a function in a **display window** or **viewing screen**, which we refer to as a **viewing rectangle**. The default screen often gives an incomplete or misleading picture, so it is important to choose the viewing rectangle with care. If we choose the x-values to range from a minimum value of $Xmin = a$ to a maximum value of $Xmax = b$ and the y-values to range from a minimum of $Ymin = c$ to a maximum of $Ymax = d$, then the portion of the graph lies in the rectangle

$$[a, b] \times [c, d] = \{(x, y) \mid a \leqslant x \leqslant b, c \leqslant y \leqslant d\}$$

shown in Figure 1. We refer to this rectangle as the $[a, b]$ by $[c, d]$ *viewing rectangle*.

The machine draws the graph of a function f much as you would. It plots points of the form $(x, f(x))$ for a certain number of equally spaced values of x between a and b. If an x-value is not in the domain of f, or if $f(x)$ lies outside the viewing rectangle, it moves on to the next x-value. The machine connects each point to the preceding plotted point to form a representation of the graph of f.

Figure 1
The viewing rectangle $[a, b]$ by $[c, d]$

(a, d) $y = d$ (b, d)
$x = a$ $x = b$
(a, c) $y = c$ (b, c)

EXAMPLE 1

Draw the graph of the function $f(x) = x^2 + 3$ in the following viewing rectangles:
(a) $[-2, 2]$ by $[-2, 2]$ (b) $[-4, 4]$ by $[-4, 4]$
(c) $[-10, 10]$ by $[-5, 30]$ (d) $[-50, 50]$ by $[-100, 1000]$

Donald Knuth was born in Milwaukee in 1938 and is now Professor of Computer Science at Stanford University. While still a graduate student at Caltech he started writing a monumental series of books entitled The Art of Computer Programming. *President Carter awarded him the National Medal of Science in 1979. When Knuth was a high school student he became fascinated with graphs of functions and laboriously drew many hundreds of them because* he wanted to see the behavior of a great variety of functions. (Now, with progress in computers and programming, it is far easier to use computers and graphing calculators to do this.) Knuth is also famous for his invention of T$_E$X, a system of computer-assisted typesetting. He has written a novel entitled* Surreal Numbers: How Two Ex-Students Turned on to Pure Mathematics and Found Total Happiness.

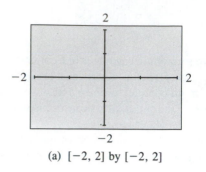

(a) [−2, 2] by [−2, 2]

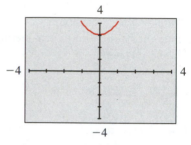

(b) [−4, 4] by [−4, 4]

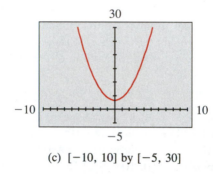

(c) [−10, 10] by [−5, 30]

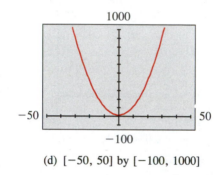

(d) [−50, 50] by [−100, 1000]

Figure 2
Graphs of $f(x) = x^2 + 3$

SOLUTION

For part (a) we select the range by setting $Xmin = -2$, $Xmax = 2$, $Ymin = -2$, and $Ymax = 2$. The resulting graph is shown in Figure 2(a). The display window is blank! A moment's thought provides the explanation: Notice that $x^2 \geq 0$ for all x, so $x^2 + 3 \geq 3$ for all x. Thus the range of the function $f(x) = x^2 + 3$ is $[3, \infty)$. This means that the graph of f lies entirely outside the viewing rectangle $[-2, 2]$ by $[-2, 2]$.

The graphs for the viewing rectangles in parts (b), (c), and (d) are shown in Figure 2, parts (b), (c), and (d). Observe that we get a more complete picture in parts (c) and (d), but in part (d) it is not clear that the y-intercept is 3.

■

We see from Example 1 that the choice of a viewing rectangle can make a big difference in the appearance of a graph. Sometimes it is necessary to change to a larger viewing rectangle to obtain a more complete picture, a more global view, of the graph. In the next example we see that knowledge of the domain and range of a function sometimes provides us with enough information to select a good viewing rectangle.

EXAMPLE 2

Determine an appropriate viewing rectangle for the function $f(x) = \sqrt{8 - 2x^2}$ and use it to graph f.

SOLUTION

The expression for $f(x)$ is defined when

$$8 - 2x^2 \geq 0 \quad \Leftrightarrow \quad 2x^2 \leq 8 \quad \Leftrightarrow \quad x^2 \leq 4$$
$$\Leftrightarrow \quad |x| \leq 2 \quad \Leftrightarrow \quad -2 \leq x \leq 2$$

Therefore the domain of f is the interval $[-2, 2]$. Also,

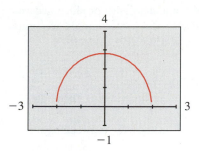

Figure 3

$$0 \leqslant \sqrt{8 - 2x^2} \leqslant \sqrt{8} = 2\sqrt{2} \approx 2.83$$

so the range of f is the interval $\left[0, 2\sqrt{2}\right]$.

We choose the viewing rectangle so that the x-interval is somewhat larger than the domain and the y-interval is larger than the range. Taking the viewing rectangle to be $[-3, 3]$ by $[-1, 4]$ we get the graph shown in Figure 3. ■

EXAMPLE 3

Graph the function $y = x^3 - 49x$.

SOLUTION

Here the domain is R, the set of all real numbers. That does not help us choose a viewing rectangle. Let us experiment. If we start with the viewing rectangle $[-5, 5]$ by $[-5, 5]$, we get the graph in Figure 4. On most calculators the screen appears to be blank, but it is not quite blank because the point $(0, 0)$ has been plotted. It turns out that for all the other x-values that the calculator chooses between -5 and 5, the values of $f(x)$ are greater than 5 or less than -5, so the corresponding points on the graph lie outside the viewing rectangle.

If we use the zoom-out feature of a graphing calculator to change the viewing rectangle to $[-10, 10]$ by $[-10, 10]$, we get the picture shown in Figure 5(a). The graph appears to consist of vertical lines, but we know that can't be correct. If we look carefully while the graph is being drawn, we see that the graph leaves the screen and reappears during the graphing process. This indicates that we need to see more in the vertical direction, so we change the viewing rectangle to $[-10, 10]$ by $[-100, 100]$. The resulting graph is shown in Figure 5(b). It still doesn't quite reveal all the main features of the function, so we try $[-10, 10]$ by $[-200, 200]$ in Figure 5(c). Now we are more confident that we have arrived at an appropriate viewing rectangle. In Chapter 3, when we discuss third-degree polynomials, we will see that the graph shown in Figure 5(c) does indeed reveal all the main features of the function.

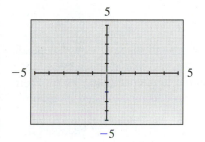

Figure 4

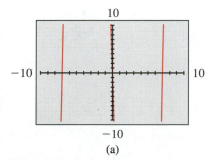

(a)

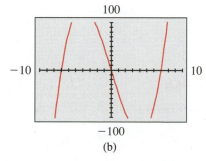

(b)

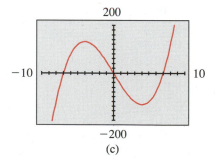

(c)

Figure 5
$f(x) = x^3 - 49x$

■

One of the uses of graphing calculators and computers is to produce accurate graphs of standard functions for future reference. In the next example we graph some of the **power functions** $f(x) = x^n$, where n is a positive integer.

EXAMPLE 4

(a) Graph the functions $y = x^2$, $y = x^4$, and $y = x^6$ in the viewing rectangle $[-2, 2]$ by $[-1, 3]$.

(b) Graph the functions $y = x$, $y = x^3$, and $y = x^5$ in the viewing rectangle $[-2, 2]$ by $[-2, 2]$.

(c) Graph all of the functions from parts (a) and (b) in the viewing rectangle $[-1, 3]$ by $[-1, 3]$.

(d) What conclusions can you draw from these graphs?

SOLUTION

The graphs are shown in Figure 6. We see that the general shape of $y = x^n$ depends on whether n is even or odd. From part (a) we see that if n is even, then $f(x) = x^n$ is an even function and its graph is similar to the parabola $y = x^2$. From part (b) we see that if n is odd, then $f(x) = x^n$ is an odd function and its graph is similar to that of $y = x^3$.

We notice from part (c) that as n increases the graph of $y = x^n$ becomes flatter near 0 and steeper when $x > 1$. When $0 < x < 1$, the lower powers of x are the bigger functions. But when $x > 1$, the higher powers of x are the dominant functions.

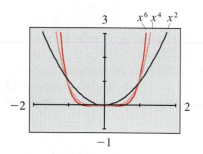

(a) Even powers of x

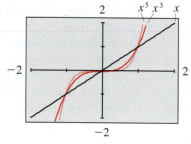

(b) Odd powers of x

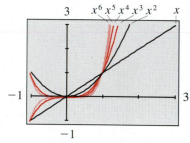

(c) Even and odd powers

Figure 6
Power functions

EXAMPLE 5

Graph the circle $x^2 + y^2 = 1$.

SOLUTION

We know from Section 1.9 how to graph this circle by hand; it is a circle with center the origin and radius 1. But to graph the circle with a graphing calculator is not straightforward because a circle is not a graph of a function (it fails the Vertical Line Test). However, if we solve the equation of the circle for y, we get

$$y^2 = 1 - x^2 \qquad y = \pm\sqrt{1 - x^2}$$

Therefore we can regard the circle as being described by the graphs of *two* functions:

$$f(x) = \sqrt{1 - x^2} \qquad \text{and} \qquad g(x) = -\sqrt{1 - x^2}$$

If we graph f with viewing rectangle $[-2, 2]$ by $[-2, 2]$, we get the semicircle shown in Figure 7(a), which is the top half of the given circle. The graph of g is the semicircle in Figure 7(b), which is the bottom half of the circle. Graphing these semicircles together on the same viewing screen, we get the full circle in Figure 7(c).

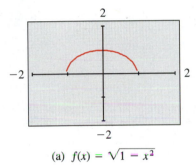
(a) $f(x) = \sqrt{1 - x^2}$

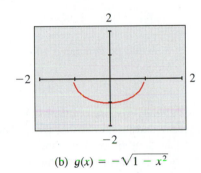
(b) $g(x) = -\sqrt{1 - x^2}$

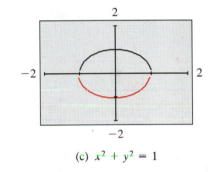
(c) $x^2 + y^2 = 1$

Figure 7

FUNCTIONS DEFINED BY MORE THAN ONE EQUATION

In Section 2.2 we looked at piecewise-defined functions, that is, functions that are defined by different formulas in different parts of their domains. The following example shows how graphing calculators can be used to help draw the graphs of such functions.

EXAMPLE 6

Draw the graph of the function

$$f(x) = \begin{cases} 1 - x & \text{if } x \leq 1 \\ x^2 & \text{if } x > 1 \end{cases}$$

SOLUTION

First we use a calculator to draw the graphs of the functions

$$g(x) = 1 - x$$

and $\qquad\qquad\quad h(x) = x^2$

On many graphing calculators the graph in Figure 8(b) can be produced by using the logical functions in the calculator. For example, on the TI-81 the following equation gives the required graph:

$$y = (x < 1)(1 - x) + (x > 1)x^2$$

in the viewing rectangle $[-1, 3]$ by $[-1, 3]$ [see Figure 8(a)]. Then we draw the graph of f (by hand) in Figure 8(b) by taking the part of the graph of g to the left of $x = 1$ and combining it with the part of the graph of h to the right of $x = 1$. Note that $f(1) = 1 - 1 = 0$, so we put a solid dot at $(1, 0)$ and an open dot at $(1, 1)$.

Figure 8

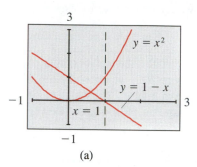

(a)

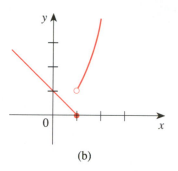

(b)

Example 6 should be compared with Example 8 in Section 2.2.

PITFALLS

We have already met one of the pitfalls in using graphing calculators and computers: Examples 1 and 3 showed that the use of an inappropriate viewing rectangle can give a misleading impression of the graph of a function. We also saw how to remedy the situation: We included the crucial parts of the function by changing to a larger viewing rectangle.

Another type of pitfall is illustrated in the next example.

EXAMPLE 7

Draw the graph of the function $y = \dfrac{1}{1 - x}$.

SOLUTION

Figure 9(a) shows the graph produced by a graphing calculator with viewing rectangle $[-9, 9]$ by $[-9, 9]$. In connecting successive points on the graph, the calculator produced a steep line segment from the top to the bottom of the screen. That

Figure 9

$$y = \frac{1}{1 - x}$$

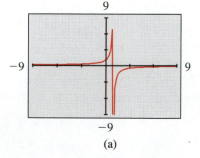

(a)

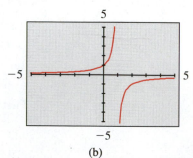

(b)

Another way to avoid the extraneous line is to change the graphing mode on the calculator so that the dots are not connected.

line segment is not truly part of the graph. Notice that the domain of the function $y = 1/(1 - x)$ is $\{x \mid x \neq 1\}$. We can get rid of the extraneous near-vertical line by experimenting with a change of scale. When we change to the smaller viewing rectangle $[-5, 5]$ by $[-5, 5]$, we obtain the much better graph in Figure 9(b).

EXERCISES 2.3

1. Use a graphing calculator or computer to determine which of the following viewing rectangles produces the most appropriate graph of the function $f(x) = x^4 + 2$.
   (a) $[-2, 2]$ by $[-2, 2]$ (b) $[0, 4]$ by $[0, 4]$
   (c) $[-4, 4]$ by $[-4, 4]$ (d) $[-8, 8]$ by $[-4, 40]$
   (e) $[-40, 40]$ by $[-80, 800]$

2. Use a graphing calculator or computer to determine which of the following viewing rectangles produces the most appropriate graph of the function $f(x) = x^2 + 7x + 6$.
   (a) $[-5, 5]$ by $[-5, 5]$ (b) $[0, 10]$ by $[-20, 100]$
   (c) $[-15, 8]$ by $[-20, 100]$
   (d) $[-10, 3]$ by $[-100, 20]$

3. Use a graphing calculator or computer to determine which of the following viewing rectangles produces the most appropriate graph of the function $f(x) = 10 + 25x - x^3$.
   (a) $[-4, 4]$ by $[-4, 4]$ (b) $[-10, 10]$ by $[-10, 10]$
   (c) $[-20, 20]$ by $[-100, 100]$
   (d) $[-100, 100]$ by $[-200, 200]$

4. Use a graphing calculator or computer to determine which of the following viewing rectangles produces the most appropriate graph of the function $f(x) = \sqrt{8x - x^2}$.
   (a) $[-4, 4]$ by $[-4, 4]$ (b) $[-5, 5]$ by $[0, 100]$
   (c) $[-10, 10]$ by $[-10, 40]$
   (d) $[-2, 10]$ by $[-2, 6]$

In Exercises 5–16 determine an appropriate viewing rectangle for the given function and use it to draw the graph.

5. $f(x) = 4 + 6x - x^2$ 6. $f(x) = 0.3x^2 + 1.7x - 3$

7. $f(x) = \sqrt[4]{256 - x^2}$ 8. $f(x) = \sqrt{12x - 17}$

9. $f(x) = 0.01x^3 - x^2 + 5$

10. $f(x) = x(x + 6)(x - 9)$

11. $y = \dfrac{1}{x^2 + 25}$ 12. $y = \dfrac{x}{x^2 + 25}$

13. $y = x^4 - 4x^3$ 14. $y = x^3 + \dfrac{1}{x}$

15. $y = 1 + |x - 1|$ 16. $y = 2x - |x^2 - 5|$

In Exercises 17–20 draw the graph of the function.

17. $f(x) = \begin{cases} x + 2 & \text{if } x \leqslant -1 \\ x^2 & \text{if } x > -1 \end{cases}$

18. $f(x) = \begin{cases} 2x - x^2 & \text{if } x > 1 \\ (x - 1)^3 & \text{if } x \leqslant 1 \end{cases}$

19. $f(x) = \begin{cases} x^3 - 2x + 1 & \text{if } x \leqslant 0 \\ x - x^2 & \text{if } 0 < x < 1 \\ \sqrt[4]{x - 1} & \text{if } x \geqslant 1 \end{cases}$

20. $f(x) = \begin{cases} \sqrt{-x} & \text{if } x < 0 \\ \sqrt{2x - x^2} & \text{if } 0 \leqslant x \leqslant 2 \\ \sqrt{x - 2} & \text{if } x > 2 \end{cases}$

21. Graph the ellipse $4x^2 + 2y^2 = 1$ by graphing the functions whose graphs are the upper and lower halves of the ellipse.

22. Graph the hyperbola $y^2 - 9x^2 = 1$ by graphing the functions whose graphs are the upper and lower halves of the hyperbola.

23. (a) Graph the root functions $y = \sqrt{x}$, $y = \sqrt[4]{x}$, and $y = \sqrt[6]{x}$ on the same screen using the viewing rectangle $[-1, 4]$ by $[-1, 3]$.
    (b) Graph the root functions $y = x$, $y = \sqrt[3]{x}$, and $y = \sqrt[5]{x}$ on the same screen using the viewing rectangle $[-3, 3]$ by $[-2, 2]$.
    (c) Graph the root functions $y = \sqrt{x}$, $y = \sqrt[3]{x}$, $y = \sqrt[4]{x}$, and $y = \sqrt[5]{x}$ on the same screen using the viewing rectangle $[-1, 3]$ by $[-1, 2]$.
    (d) What conclusions can you make from these graphs?

24. (a) Graph the functions $y = 1/x$ and $y = 1/x^3$ on the same screen using the viewing rectangle $[-3, 3]$ by $[-3, 3]$.

(b) Graph the functions $y = 1/x^2$ and $y = 1/x^4$ on the same screen using the same viewing rectangle.

(c) Graph all of the functions in parts (a) and (b) on the same screen using the viewing rectangle $[-1, 3]$ by $[-1, 3]$.

(d) What conclusions can you make from these graphs?

25. Use graphs to determine which of the functions $f(x) = 10x^2$ and $g(x) = x^3/10$ is eventually larger (that is, larger when x is very large).

26. Use graphs to determine which of the functions $f(x) = x^4 - 100x^3$ and $g(x) = x^3$ is eventually larger.

27. (a) Draw the graphs of the the functions
$$f(x) = x^2 + x - 6 \text{ and } g(x) = |x^2 + x - 6|.$$
(b) How are the graphs of f and g related?

28. (a) Draw the graphs of the functions $f(x) = x^4 - 6x^2$ and $g(x) = |x^4 - 6x^2|$.
(b) How are the graphs of f and g related?

29. Draw the graphs of the following functions on the same screen using the viewing rectangle $[-5, 5]$ by $[-10, 10]$.

(a) $y = x^2$ (b) $y = x^2 + 3$ (c) $y = x^2 - 5$
How are the graphs in parts (b) and (c) related to the one in part (a)?

30. Draw the graphs of the following functions on the same screen using the viewing rectangle $[-5, 5]$ by $[-10, 10]$.
(a) $y = x^2$ (b) $y = 2x^2$ (c) $y = \frac{1}{2}x^2$ (d) $y = -x^2$
How are the graphs in parts (b), (c), and (d) related to the one in part (a)?

31. Draw the graphs of the following functions on the same screen using the viewing rectangle $[-10, 10]$ by $[-10, 10]$.
(a) $y = x^2$ (b) $y = (x - 7)^2$ (c) $y = (x + 5)^2$
How are the graphs in parts (b) and (c) related to the one in part (a)?

32. Draw the graphs of the following functions on the same screen using the viewing rectangle $[-10, 10]$ by $[-10, 10]$.
(a) $y = x^3$ (b) $y = (x - 4)^3$ (c) $y = (x + 6)^3$
How are the graphs in parts (b) and (c) related to the one in part (a)?

SECTION 2.4
APPLIED FUNCTIONS

In this section we use functions to describe how one quantity depends on another. For instance, we say that population is a function of time and pressure is a function of temperature. When scientists or applied mathematicians talk about a **mathematical model** for a real-world phenomenon, they mean a function that describes, at least approximately, the dependence of one physical quantity on another.

Our first example shows that even when a precise formula for functional dependence is not available, it is still possible to visualize the situation from a graph of the function.

EXAMPLE 1

When you turn on a hot water tap, the temperature T of the water depends on how long the water has been running. Draw a rough graph of T as a function of the time t that has elapsed since the tap was turned on.

SOLUTION

The initial temperature of the water is close to room temperature because of the water that was in the pipes. When the water from the hot water tank starts coming out, T increases quickly. In the next phase, T is constant at the temperature of the

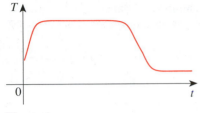

T

0

t

Figure 1

water in the tank. When the tank is drained, T decreases to the temperature of the water supply. This enables us to make the rough sketch of T as a function of t in Figure 1. ∎

A more accurate graph of the function in Example 1 could be obtained by using a thermometer to measure the temperature of the water at 10-second intervals. In general, scientists collect experimental data and use them to sketch the graphs of functions because they may not know defining formulas for the functions. In the next two examples, however, we are able to find explicit formulas for the functions.

EXAMPLE 2

Express the perimeter of a square as a function of its area.

SOLUTION

Let P, A, and s represent the perimeter, area, and side length of the square, respectively. We start with the known formulas

$$P = 4s \qquad \text{and} \qquad A = s^2$$

Substituting $s = \sqrt{A}$ from the second equation into the first, we get

$$P = 4\sqrt{A}$$

This equation expresses P as a function of A. Since A must be positive, the domain is given by $A > 0$ or $(0, \infty)$. ∎

In setting up expressions for functions in applied situations, it is useful to recall some of the problem-solving principles that were given after Chapter 1 and adapt them to the present situation.

Steps in Setting Up Applied Functions

Understand the problem

1. The first step is to read the problem carefully until it is clearly understood. Ask yourself: What is the unknown? What are the given quantities? What are the given conditions?

Draw a diagram

2. In most problems it is useful to draw a diagram and identify the given and required quantities on the diagram.

Introduce notation

3. If the problem asks you for an expression for a certain quantity, assign a symbol to that quantity (let us call it Q for now). Also select symbols $(a, b, c, \ldots, x, y)$ for other unknown quantities and label the diagram with these symbols. It may help to use initials as suggestive symbols—for example, A for area, h for height, t for time.

4. Express Q in terms of some of the other symbols from Step 3.

5. If Q has been expressed as a function of more than one variable in Step 4, use the given information to find relationships (in the form of equations) among

these variables. Then use these equations to eliminate all but one of the variables in the expression for Q. Thus Q will be expressed as a function of one variable.

EXAMPLE 3

A can holds 1 L (liter) of oil. Express the surface area of the can as a function of its radius.

SOLUTION

Introduce notation

Draw a diagram

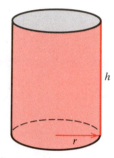

Figure 2

Let r be the radius and h the height of the can, in centimeters (see Figure 2). Then the area of the top is πr^2, the area of the bottom is also πr^2, and the area of the sides is the circumference times the height, that is, $2\pi rh$. So the total surface area is

$$A = 2\pi r^2 + 2\pi rh$$

To express A as a function of r, we need to eliminate h and we do this by using the fact that the volume is given as 1 L, which we take to be 1000 cm^3. Thus

$$\pi r^2 h = 1000$$

Substituting $h = 1000/(\pi r^2)$ into the expression for A, we have

$$A = 2\pi r^2 + 2\pi r\left(\frac{1000}{\pi r^2}\right) = 2\pi r^2 + \frac{2000}{r}$$

Therefore the equation

$$A = 2\pi r^2 + \frac{2000}{r} \qquad r > 0$$

expresses A as a function of r. ■

DIRECT AND INVERSE VARIATION

Two types of mathematical models occur so often in various sciences that they are given special names. The first is called *direct variation* and occurs when one quantity is a constant multiple of another.

DIRECT VARIATION

If the variables x and y are related by an equation

$$y = kx$$

for some constant $k \neq 0$, we say that y **varies directly as** x, or y **is directly proportional to** x, or simply y **is proportional to** x. The constant k is called the **constant of proportionality**.

Recall that a linear function is a function of the form $f(x) = mx + b$ and its graph is a line with slope m and y-intercept b. So if y varies directly as x, then the equation

$$y = kx$$

or equivalently, $f(x) = kx$, shows that y is a linear function of x. The graph of this function is a line with slope k, the constant of proportionality. The y-intercept is $b = 0$, so this line passes through the origin.

EXAMPLE 4

During a thunderstorm you see the lightning before you hear the thunder because light travels much faster than sound. The distance between you and the center of the storm varies directly as the time interval between the lightning and the thunder.
(a) If thunder from a storm whose center is 5400 ft away takes 5 s to reach you, determine the constant of proportionality and write the equation for the variation.
(b) Sketch the graph of this equation. What does the constant of proportionality represent?
(c) If the time interval is 8 s, how far away is the center of the storm?

SOLUTION

(a) Let d be the distance from you to the storm and let t be the length of the time interval. We are given that d varies directly as t, so

$$d = kt$$

where k is a constant. To find k, we use the fact that $t = 5$ when $d = 5400$. Substituting these values in the equation we get

$$5400 = k(5)$$

Therefore $$k = \frac{5400}{5} = 1080$$

Putting this value of k in the equation for d, we obtain

$$d = 1080t$$

as the equation for d as a function of t.
(b) The graph is a line through the origin with slope 1080 and is shown in Figure 3. The constant $k = 1080$ is the slope of this line and represents the approximate speed of sound.
(c) When $t = 8$, we have

$$d = 1080 \cdot 8 = 8640$$

So the storm center is 8640 ft $\approx$ 1.6 mi away. ■

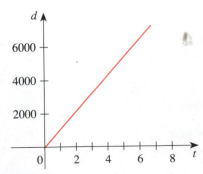

Figure 3

Another function that is used frequently in mathematical modeling is the function

$$f(x) = \frac{k}{x}$$

where k is a constant.

INVERSE VARIATION

> If the quantities x and y are related by the equation
>
> $$y = \frac{k}{x}$$
>
> for some constant $k \neq 0$, we say that y **is inversely proportional to** x, or y **varies inversely as** x.

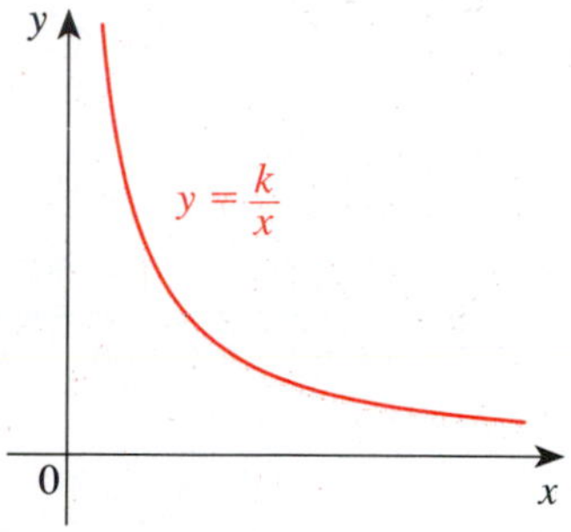

$$y = \frac{k}{x}$$

Figure 4

The graph of the function $f(x) = k/x$, $x > 0$, is shown in Figure 4 for the case $k > 0$. It gives a picture of what happens when y is inversely proportional to x.

EXAMPLE 5

Boyle's Law states that when a sample of gas is compressed at a constant temperature, the pressure of the gas is inversely proportional to the volume of the gas.
(a) Write the equation that expresses the inverse proportionality.
(b) The pressure of a sample of air that occupies 0.106 m³ at 25 °C is 50 kPa. Find the constant of proportionality.
(c) If the sample expands to a volume of 0.3 m³, find the new pressure.

SOLUTION

(a) Let P be the pressure of the sample of gas and let V be its volume. Then by the definition of inverse proportionality, we have

$$P = \frac{k}{V}$$

where k is a constant.
(b) We are given that $P = 50$ when $V = 0.106$ m³, so

$$50 = \frac{k}{0.106}$$

Thus $k = (50)(0.106) = 5.3$

(c) Using the value of k from part (b), we now have

$$P = \frac{5.3}{V}$$

When $V = 0.3$ m³, we have

$$P = \frac{5.3}{0.3} \approx 17.7$$

So the new pressure is about 17.7 kPa.   ■

A physical quantity often depends on more than one other quantity. For instance, if the quantities x, y, and z are related by the equation

$$z = kxy$$

where k is a nonzero constant, then we say that z **varies jointly as** x and y, or z **is jointly proportional to** x and y. If

$$z = k\frac{x}{y}$$

we say that z **is proportional to** x and **inversely proportional to** y.

EXAMPLE 6

Newton's Law of Gravitation says that two objects with masses m_1 and m_2 attract each other with a force F that is jointly proportional to their masses and is inversely proportional to the square of the distance r between the objects. Express Newton's Law of Gravitation as an equation.

SOLUTION

Using the definitions of joint and inverse proportionality, and using the traditional notation G for the constant of proportionality, we have

$$F = G\frac{m_1 m_2}{r^2}$$ ■

EXERCISES 2.4

1. You put some ice cubes in a glass, fill the glass with cold water, and then let the glass sit on a table. Sketch a rough graph of the temperature of the water as a function of the elapsed time.

2. Your height depends on your age. Sketch a rough graph of your height as a function of your age.

3. When a football is kicked off after a touchdown, its height depends on the time elapsed since kickoff. Sketch a rough graph of the height of the football as a function of time.

4. Sketch a rough graph of the number of hours of daylight as a function of the time of year.

5. Sketch a rough graph of the outdoor temperature as a function of time during a typical spring day.

6. You place a frozen pie in an oven and bake it for an hour. Then you take it out and let it cool before eating it. Sketch a rough graph of the temperature of the pie as a function of time.

In Exercises 7–20 find a formula for the described function and state its domain.

7. A rectangle has a perimeter of 20 ft. Express the area A of the rectangle as a function of the length x of one of its sides.

8. A rectangle has an area of 16 m². Express the perimeter P of the rectangle as a function of the length x of one of its sides.

9. Express the area A of an equilateral triangle as a function of the length x of a side.

10. Express the surface area A of a cube as a function of its volume V.

11. Express the radius r of a circle as a function of its area A.

12. Express the area A of a circle as a function of the circumference C.

13. An open rectangular box with a volume of 12 ft³ has a square base. Express the surface area A of the box as a function of the length x of a side of the base.

14. A woman 5 ft tall is standing near a street lamp that is 12 feet tall. Express the length L of her shadow as a function of her distance d from the base of the lamp.

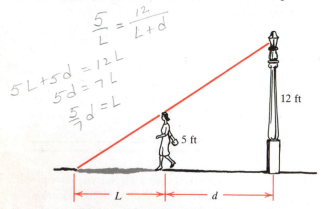

15. A Norman window has the shape of a rectangle surmounted by a semicircle as shown in the figure. If the perimeter of the window is 30 ft, express the area A of the window as a function of the width x of the window.

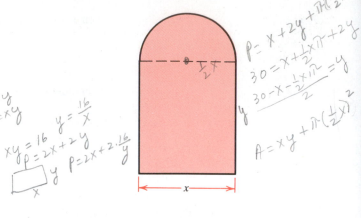

16. A box with an open top is to be constructed from a rectangular piece of cardboard with dimensions 12 in. by 20 in. by cutting out equal squares of side x at each corner and then folding up the sides as in the figure. Express the volume V of the box as a function of x.

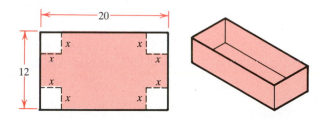

17. A farmer has 2400 ft of fencing and wants to fence off a rectangular field that borders a straight river. He needs no fence along the river. Express the area A of the field in terms of the width x of the field (see the figure).

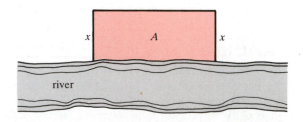

river

18. A rectangle is inscribed in a semicircle of radius r as in the figure. Express the area A of the rectangle as a function of the height h of the rectangle.

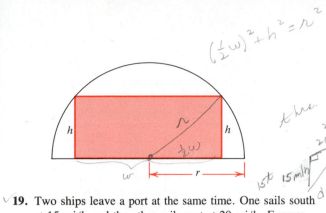

the square of the speed and the fact that 1 mi/h = 88 ft/s, she finds that the number N of cars per minute that pass a given point is a function of the speed given by

$$N(s) = \frac{88s}{17 + 17\left(\dfrac{s}{20}\right)^2}$$

Sketch a graph of this function in an appropriate viewing rectangle.

19. Two ships leave a port at the same time. One sails south at 15 mi/h and the other sails east at 20 mi/h. Express the distance d between the ships as a function of t, the time (in hours) after their departure.

20. A man is at a point A on a bank of a straight river, 2 mi wide, and wants to reach point B, 7 mi downstream on the opposite bank, by first rowing his boat to a point P on the opposite bank and then walking the remaining distance x to B. He can row at 4 mi/h and walk at 3 mi/h. Express the total time T that he takes to go from A to B as a function of x.

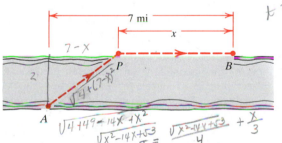

21. The weight w of an object at a height h above the surface of the earth is given by

$$w = \left(\frac{R}{R + h}\right)^2 w_0$$

where w_0 is the weight of the object at sea level and $R = 6400$ km is the radius of the earth. Suppose the weight of an object at sea level is 80 N. Using an appropriate viewing rectangle, sketch the graph of the weight of the object as a function of its height.

22. A highway engineer wants to estimate the maximum number of cars that can safely use a particular highway at a given speed. She assumes that each car is 17 ft long, travels at a speed s, and follows the car in front of it at the "safe following distance" for that speed. The safe following distance at 20 mi/h is one car length. Using the fact that the distance required to stop is proportional to

23. For a fish swimming at a speed v relative to the water, the energy expenditure per unit time is proportional to v^3. If the fish is swimming against a current with speed u miles per hour, where $u < v$, then the time required to travel a distance of L miles is $L/(v - u)$ and the total energy required to travel the distance is

$$E(v) = 2.73v^3 \frac{L}{v - u}$$

Suppose the speed of the current is $u = 5$ mi/h. Sketch the graph of the energy $E(v)$ needed to swim a distance of 10 mi as a function of the speed v of the fish.

24. In the theory of relativity the mass of an object changes as its speed changes. If m_0 is the rest mass of the object then its mass m is a function of its speed v given by

$$m(v) = \frac{m_0}{\sqrt{1 - \dfrac{v^2}{c^2}}}$$

where $c = 3.0 \times 10^5$ km/s is the speed of light. Sketch the graph of the mass m as a function of the speed v for an object with rest mass 1000 kg.

In Exercises 25–32 write an equation that expresses the statement.

25. R varies directly as t.

26. P is directly proportional to u.

27. v is inversely proportional to z.

28. w is jointly proportional to m and n.

29. y is proportional to s and inversely proportional to t.

30. P varies inversely as T.

31. z is proportional to the square root of y.

32. A is proportional to the square of t and inversely proportional to the cube of x. $A = k\dfrac{t^2}{x^3}$

In Exercises 33–38 express the statement as a formula and use the given information to find the constant of proportionality.

33. y is directly proportional to x. If $x = 4$, then $y = 72$.

34. z varies inversely as t. If $t = 3$, then $z = 5$. $z = \dfrac{k}{t}$

35. M varies directly as x and inversely as y. If $x = 2$ and $y = 6$, then $M = 5$.

36. S varies jointly as p and q. If $p = 4$ and $q = 5$, then $S = 180$.

37. W is inversely proportional to the square of r. If $r = 6$, then $W = 10$.

38. t is directly proportional to x and s and is inversely proportional to r. If $x = 2$, $s = 3$, and $r = 12$, then $t = 25$.

39. Hooke's Law states that the force needed to keep a spring stretched x units beyond its natural length is directly proportional to x. Here the constant of proportionality is called the **spring constant**.
  (a) Write Hooke's Law as an equation. $F = kx$
  (b) If a spring has a natural length of 10 cm and a force of 40 N is required to maintain the spring stretched to a length of 15 cm, find the spring constant. $40 = k \cdot 5$ $\quad 8 = k$
  (c) What force is needed to keep the spring stretched to a length of 14 cm? $F = 8(4) = 32$

40. The period of a pendulum (the time it takes for one complete swing of the pendulum) varies directly with the square root of the length of the pendulum.
  (a) Express this fact by writing an equation. $P = k\sqrt{l}$
  (b) In order to double the period, how would we have to change the length? $2P = 2k\sqrt{l}$ $\quad 2P = k\sqrt{4l}$ $\quad$ if $l = 4$ $\quad P = k\sqrt{4}$ $\quad = 2k$

41. The cost of printing a magazine is jointly proportional to the number of pages in the magazine and the number of magazines printed.
  (a) Write an equation for this joint variation if it costs $60,000 to print 4000 copies of a 120-page magazine.
  (b) How much would it cost to print 5000 copies of a 92-page magazine?

42. The pressure of a sample of gas is directly proportional to the temperature and inversely proportional to the volume.
  (a) Write an equation that expresses this fact if 100 L of gas exerts a pressure of 33.2 kPa at a temperature of 400 °K (absolute temperature measured on the Kelvin scale).
  (b) If the temperature is increased to 500 °K and the volume is decreased to 80 L, what is the pressure of the gas?

43. The resistance of a wire varies directly as its length and inversely as the square of its diameter.
  (a) A wire 1.2 m long and 0.005 m in diameter has a resistance of 140 Ω. Write an equation for this variation and find the constant of proportionality.
  (b) Find the resistance of a wire made of the same material that is 3 m long and has a diameter of 0.008 m.

44. Kepler's Third Law of Planetary Motion states that the square of the period T of a planet (the time it takes for the planet to make a complete revolution about the sun) is directly proportional to the cube of the average distance d from the sun.
  (a) Express Kepler's Third Law by writing an equation.
  (b) Find the constant of proportionality by using the fact that for our planet the period is about 365 days and the average distance is about 93 million miles.
  (c) The planet Neptune is about 2.79×10^9 mi from the sun. Find the period of Neptune.

SECTION 2.5
TRANSFORMATIONS OF FUNCTIONS

In this section we see how to reduce the amount of work required to graph certain functions by using the following transformations: shifting, reflecting, and stretching.

EXAMPLE 1

Sketch the graphs of the following functions:
  (a) $f(x) = x^2 + 3$ (b) $g(x) = x^2 - 2$

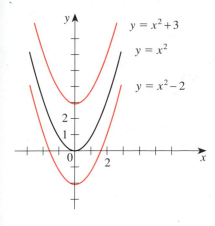

Figure 1

SOLUTION

(a) We start with the graph of the function $y = x^2$ from Example 2 of Section 2.2. Then the equation $y = x^2 + 3$ indicates that the y-coordinate of a point on the graph of f is 3 more than the y-coordinate of the corresponding point on the curve $y = x^2$. This means that we obtain the graph of $y = x^2 + 3$ simply by shifting the graph of $y = x^2$ upward by 3 units as in Figure 1.

(b) Similarly we get the graph of $g(x) = x^2 - 2$ by moving the parabola $y = x^2$ downward by 2 units. ■

In general we have the following rule for the translation (or shifting) of functions.

VERTICAL SHIFTS OF GRAPHS

Let $c > 0$. To obtain the graph of

$y = f(x) + c$ move the graph of $y = f(x)$ c units upward

$y = f(x) - c$ move the graph of $y = f(x)$ c units downward

EXAMPLE 2

Given the graph of $y = f(x)$ in Figure 2, sketch the graphs of the following functions:

(a) $g(x) = f(x - 5)$ (b) $h(x) = f(x + 8)$

SOLUTION

(a) The equation $g(x) = f(x - 5)$ says that the value of g at x is the same as the value of f at $x - 5$. Thus the value of g at a number is the same as the value of f, 5 units to the left of the number (see Figure 3). Therefore the graph of g is just the graph of f shifted 5 units to the right.

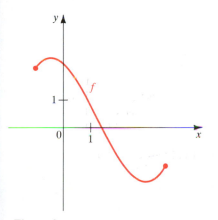

Figure 2

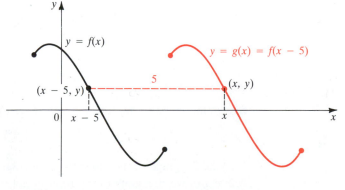

Figure 3

(b) Similar reasoning shows that the graph of $h(x) = f(x + 8)$ is the graph of $y = f(x)$ shifted 8 units to the left (see Figure 4).

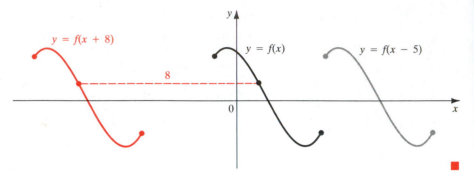

Figure 4

HORIZONTAL SHIFTS OF GRAPHS

Let $c > 0$. To obtain the graph of

$y = f(x - c)$ move the graph of $y = f(x)$ c units to the right
$y = f(x + c)$ move the graph of $y = f(x)$ c units to the left

EXAMPLE 3

Sketch the graph of the function $f(x) = (x + 4)^2$.

SOLUTION

We start with the graph of $y = x^2$ and move it 4 units to the left as in Figure 5.

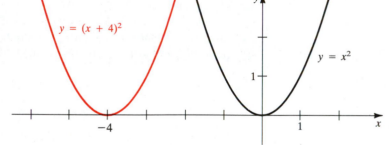

Figure 5

EXAMPLE 4

Sketch the graph of the function $y = \sqrt{x - 3} + 4$.

SOLUTION

We take the graph of the square root function $y = \sqrt{x}$ from Example 6 in Section 2.2 and move it 3 units to the right to get the graph of $y = \sqrt{x - 3}$. Then we move it 4 units upward to obtain the graph of $y = \sqrt{x - 3} + 4$ shown in Figure 6.

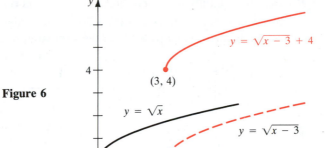

Figure 6

Vertical and horizontal shifts are illustrated in Figure 7.

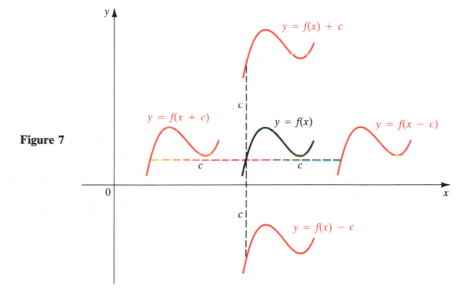

Figure 7

EXAMPLE 5

Sketch the graphs of the following functions:
(a) $y = 3x^2$ (b) $y = \frac{1}{3}x^2$

SOLUTION

(a) If we start with the parabola $y = x^2$ and multiply the y-coordinate of each point by 3, we get the curve $y = 3x^2$ shown in Figure 8(a). It is a narrower parabola than $y = x^2$ and is obtained by stretching vertically by a factor of 3.

(b) Here we multiply the y-coordinate of each point on $y = x^2$ by $\frac{1}{3}$ to get the parabola $y = \frac{1}{3}x^2$ shown in Figure 8(b). This wider parabola is obtained by shrinking the original parabola in the vertical direction.

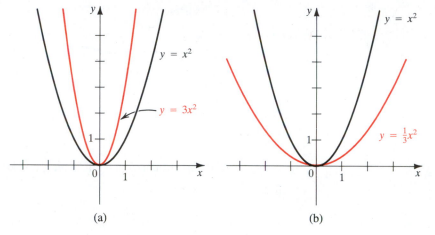

Figure 8

(a)   (b)

Figure 9

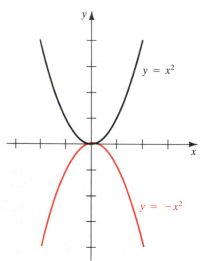

EXAMPLE 6

Sketch the graph of $y = -x^2$.

SOLUTION

We start with the parabola $y = x^2$ and multiply the y-coordinate of every point by -1. The point (x, y) is replaced by the point $(x, -y)$ and so the original graph is reflected in the x-axis (see Figure 9). ■

We summarize the effect of multiplying a function f by a constant as follows.

VERTICAL STRETCHING, SHRINKING, AND REFLECTING

Let $a > 1$. To obtain the graph of

$y = af(x)$ stretch the graph of $y = f(x)$ vertically by a factor of a

$y = \dfrac{1}{a}f(x)$ shrink the graph of $y = f(x)$ vertically by a factor of a

$y = -f(x)$ reflect the graph of f in the x-axis

Figure 10

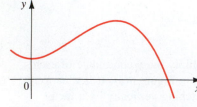

EXAMPLE 7

Given the graph of f in Figure 10, draw the graphs of the following functions:
(a) $y = 2f(x)$ (b) $y = \frac{1}{2}f(x)$ (c) $y = -f(x)$

(d) $y = -2f(x)$ (e) $y = -\frac{1}{2}f(x)$

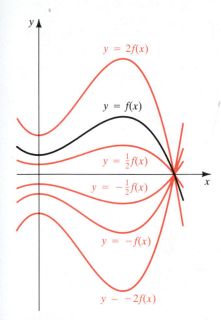

Figure 11

SOLUTION

The graphs are shown in Figure 11. Notice, for instance, that the graph of $y = -2f(x)$ is obtained by stretching by a factor of 2 and reflecting in the x-axis. ∎

We illustrate the effect of combining shifts, reflections, and stretching in the following example.

EXAMPLE 8

Sketch the graph of the function $f(x) = 1 - 2(x - 3)^2$.

SOLUTION

Starting with the graph of $y = x^2$, we first shift 3 units to the right to get the graph of $y = (x - 3)^2$. Then we reflect in the x-axis and stretch by a factor of 2 to get the graph of $y = -2(x - 3)^2$. Finally we shift upward 1 unit to get the graph of $y = 1 - 2(x - 3)^2$ shown in Figure 12.

Figure 12

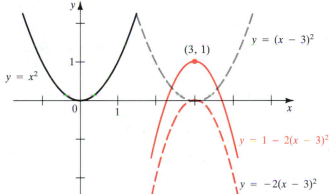

EXERCISES 2.5

In Exercises 1–11 suppose that the graph of f is given. Describe how the graphs of the following functions can be obtained from the graph of f.

1. $y = f(x) - 10$

2. $y = f(x - 10)$

3. $y = f(x + 1)$

4. $y = f(x) + 1$

5. $y = 8f(x)$

6. $y = -f(x)$

7. $y = -6f(x)$

8. $y = -\frac{1}{5}f(x)$

9. $y = f(x - 2) - 3$

10. $y = -2f(x - 3)$

11. $y = \frac{1}{2}f(x) + 9$

12. The graph of f is given. Draw the graphs of the following functions:

   (a) $y = f(x + 4)$ **(b)** $y = f(x) + 4$

   (c) $y = 2f(x)$ **(d)** $y = -\frac{1}{2}f(x) + 3$

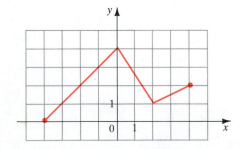

13. (a) Draw the graph of $f(x) = 1/x$ by plotting points.
   (b) Use the graph of f in part (a) to draw the graphs of the following functions:

   (i) $y = -\dfrac{1}{x}$ (ii) $y = \dfrac{1}{x-1}$

   (iii) $y = \dfrac{2}{x+2}$ (iv) $y = 1 + \dfrac{1}{x-3}$

14. (a) Draw the graph of $g(x) = \sqrt[3]{x}$ by plotting points.
   (b) Use the graph of f in part (a) to draw the graphs of the following functions:

   (i) $y = \sqrt[3]{x} - 2$ (ii) $y = \sqrt[3]{x+2} + 2$

   (iii) $y = 1 - \sqrt[3]{x}$

In Exercises 15–30 sketch the graph of the function, not by plotting points, but by starting with the graph of a standard function and applying transformations.

15. $f(x) = (x - 2)^2$ **16.** $f(x) = (x + 7)^2$

17. $f(x) = -(x + 1)^2$ **18.** $f(x) = 1 - x^2$

19. $f(x) = x^3 + 2$ **20.** $f(x) = -x^3$

21. $y = 1 + \sqrt{x}$ **22.** $y = 2 - \sqrt{x+1}$

23. $y = \frac{1}{2}\sqrt{x+4} - 3$ **24.** $y = 3 - 2(x-1)^2$

25. $y = 5 + (x + 3)^2$ **26.** $y = \frac{1}{3}x^3 - 1$

27. $y = |x| - 1$ **28.** $y = |x - 1|$

29. $y = |x + 2| + 2$ **30.** $y = 2 - |x|$

 In Exercises 31–34 graph the given functions on the same screen using the given viewing rectangle to illustrate the principles of this section.

31. Viewing rectangle $[-8, 8]$ by $[-2, 8]$
   (a) $y = \sqrt[4]{x}$ **(b)** $y = \sqrt[4]{x+5}$
   (c) $y = 2\sqrt[4]{x+5}$ **(d)** $y = 4 + 2\sqrt[4]{x+5}$

32. Viewing rectangle $[-8, 8]$ by $[-6, 6]$
   (a) $y = |x|$ **(b)** $y = -|x|$
   (c) $y = -3|x|$ **(d)** $y = -3|x - 5|$

33. Viewing rectangle $[-4, 6]$ by $[-4, 4]$
   (a) $y = x^6$ **(b)** $y = \frac{1}{3}x^6$
   (c) $y = -\frac{1}{3}x^6$ **(d)** $y = -\frac{1}{3}(x - 4)^6$

34. Viewing rectangle $[-6, 6]$ by $[-4, 4]$

   (a) $y = \dfrac{1}{\sqrt{x}}$ **(b)** $y = \dfrac{1}{\sqrt{x+3}}$

   (c) $y = \dfrac{1}{2\sqrt{x+3}}$ **(d)** $y = \dfrac{1}{2\sqrt{x+3}} - 3$

35. (a) The graph of f is given.

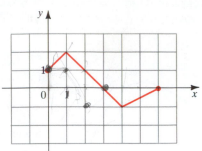

Use it to graph the following functions:

   (i) $y = f(2x)$ (ii) $y = f(\frac{1}{2}x)$
   (iii) $y = f(-x)$ (iv) $y = -f(-x)$

   (b) If $a > 1$, how are the graphs of the following functions obtained from the graph of f?

   (i) $y = f(ax)$ (ii) $y = f(x/a)$
   (iii) $y = f(-x)$

36. The graph of f is given. Use the conclusions of Exercise 35 part (b) to draw the graphs of the following functions:
   (a) $y = f(3x)$ **(b)** $y = f(x/3)$ **(c)** $y = f(-x)$

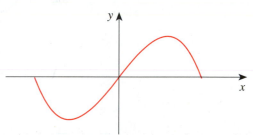

37. If $f(x) = \sqrt{2x - x^2}$, graph the following functions in the viewing rectangle $[-5, 5]$ by $[-4, 4]$.
   (a) $y = f(x)$ **(b)** $y = f(2x)$ **(c)** $y = f(\frac{1}{2}x)$

   How are the graphs in parts (b) and (c) related to the graph in part (a)?

38. If $f(x) = \sqrt{2x - x^2}$, graph the following functions in the viewing rectangle $[-5, 5]$ by $[-4, 4]$.

   (a) $y = f(x)$ **(b)** $y = f(-x)$
   (c) $y = -f(-x)$ **(d)** $y = f(-2x)$
   (e) $y = f(-\frac{1}{2}x)$

   How are the graphs in parts (b)–(e) related to the graph in part (a)?

SECTION 2.6
QUADRATIC FUNCTIONS
AND THEIR EXTREME VALUES

A **quadratic function** is a function f of the form

$$f(x) = ax^2 + bx + c$$

where a, b, and c are real numbers and $a \neq 0$.

In particular, if we take $a = 1$ and $b = c = 0$, we get the simple quadratic function $f(x) = x^2$ whose graph is the standard parabola that we drew in Example 2 of Section 2.2. In fact, the graphs of all quadratic functions have a similar shape and we will see that they can all be obtained from the graph of $y = x^2$ by the transformations given in Section 2.5.

EXAMPLE 1

Sketch the graph of the quadratic function $f(x) = -2x^2 + 3$.

SOLUTION

As in Section 2.5, we start with the graph of $y = x^2$. We stretch vertically by a factor of 2 to get the narrower parabola $y = 2x^2$. Then we reflect in the x-axis to get the graph of $y = -2x^2$. Finally we shift this graph up 3 units to obtain the graph of $f(x) = -2x^2 + 3$ as shown in Figure 1.

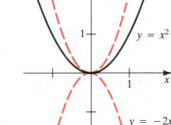
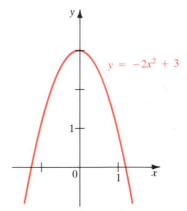

Figure 1
Steps in graphing $y = -2x^2 + 3$

If $f(x) = ax^2 + bx + c$, where $b \neq 0$, we first have to complete the square to put the function in a more convenient form for graphing.

EXAMPLE 2

Sketch the graph of $y = x^2 + 4x$.

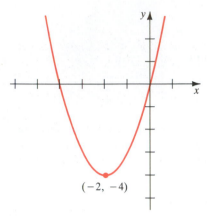

Figure 2
$y = x^2 + 4x$

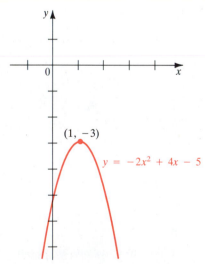

Figure 3

SOLUTION

We complete the square by adding and subtracting the square of half the coefficient of x:

$$y = x^2 + 4x$$
$$= (x^2 + 4x + 4) - 4$$
$$= (x + 2)^2 - 4$$

In this form we see that the given graph is obtained by moving the graph of $y = x^2$ two units to the left and four units downward. The vertex of the parabola is now located at the point $(-2, -4)$. For greater accuracy in graphing we could find the x-intercepts by solving the equation $x^2 + 4x = 0$. This gives $x(x + 4) = 0$, so the x-intercepts are 0 and -4. The graph is sketched in Figure 2. ■

EXAMPLE 3

Graph the function $f(x) = -2x^2 + 4x - 5$.

SOLUTION

Since the coefficient of x^2 is not 1, we must factor it from the terms involving x before we complete the square:

$$f(x) = -2x^2 + 4x - 5$$
$$= -2(x^2 - 2x) - 5$$
$$= -2(x^2 - 2x + 1) - 5 + 2$$
$$= -2(x - 1)^2 - 3$$

(Notice that we had to add 2 because the 1 inside the parentheses was multiplied by -2.) This form of the function tells us that we get the graph of f by taking the parabola $y = x^2$, shifting it 1 unit to the right, stretching it by a factor of 2, reflecting in the x-axis, and moving it 3 units downward. Notice that the vertex is at $(1, -3)$ and the parabola opens downward. We sketch the graph in Figure 3 after noting that the y-intercept is $f(0) = -5$.

From the graph we see that there is no x-intercept. Another way to reveal this information is to compute the discriminant:

$$D = b^2 - 4ac = 4^2 - 4(-2)(-5) = -24$$

From Section 1.5 we know that a negative discriminant means that there is no real solution of the equation $-2x^2 + 4x - 5 = 0$, so f has no x-intercept. ■

If we start with a general quadratic function $f(x) = ax^2 + bx + c$ and complete the square as in Examples 2 and 3, we will arrive at an expression in the standard form

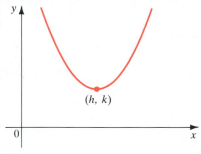

Figure 4
$y = a(x - h)^2 + k$, $a > 0$, $h > 0$,
$k > 0$

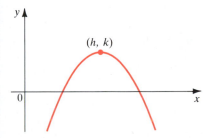

Figure 5
$y = a(x - h)^2 + k$, $a < 0$, $h > 0$,
$k > 0$

If the graph of f is produced by a graphing calculator or computer, then the approximate maximum value can be found by moving the cursor to the highest point on the parabola and observing the value of the y-coordinate. To get a more accurate reading, we can use the zoom-in feature to enlarge the region near the vertex. (For more details, see Section 2.7.)

X=.4923   Y=1.246

$y = -x^2 + x + 1$, [0, 1] by [1, 1.5]

It appears that the maximum value of the function in Example 4 is approximately 1.246. Compare this with the exact value of 1.25 found in the solution to Example 4.

$$y = a(x - h)^2 + k$$

We know from Section 2.5 that the graph of this function is obtained from the graph of $y = x^2$ by a horizontal shift, a stretch, and a vertical shift. The vertex of the resulting parabola is the point (h, k). If $a > 0$, the parabola opens upward and is as shown in Figure 4. But if $a < 0$, then a reflection in the x-axis is also required, so the parabola opens downward as in Figure 5.

Observe from Figure 4 that if $a > 0$, then the lowest point on the parabola is the vertex (h, k), so the minimum value of the function occurs when $x = h$ and this **minimum value** is $f(h) = k$. Even without the picture we could note that $(x - h)^2 \geq 0$ for all x, so $a(x - h)^2 \geq 0$ (since $a > 0$) and therefore

$$f(x) = a(x - h)^2 + k \geq k \qquad \text{for all } x$$

and $f(h) = k$. Similarly, if $a < 0$, then the highest point on the parabola is (h, k), so the **maximum value** of f is $f(h) = k$.

EXAMPLE 4

Sketch the graph of the function $f(x) = -x^2 + x + 1$ and find its maximum value and intercepts.

SOLUTION

First we complete the square:

$$\begin{aligned} y &= -x^2 + x + 1 \\ &= -(x^2 - x) + 1 \\ &= -\left(x^2 - x + \tfrac{1}{4}\right) + 1 + \tfrac{1}{4} \\ &= -\left(x - \tfrac{1}{2}\right)^2 + \tfrac{5}{4} \end{aligned}$$

From this standard form we see that the graph is a parabola that opens downward and has vertex $\left(\tfrac{1}{2}, \tfrac{5}{4}\right)$. The maximum value occurs at the vertex and is

$$f\left(\tfrac{1}{2}\right) = \tfrac{5}{4}$$

The y-intercept is $f(0) = 1$. To find the x-intercepts we use the quadratic formula to solve the equation $-x^2 + x + 1 = 0$:

$$x = \frac{-1 \pm \sqrt{1^2 - 4(-1)(1)}}{-2} = \frac{1 \pm \sqrt{5}}{2}$$

The graph of f is sketched in Figure 6.

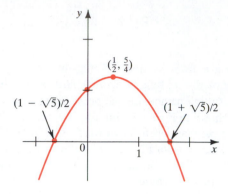

Figure 6
$f(x) = -x^2 + x + 1$

EXAMPLE 5

Find the minimum value of the function $f(t) = 5t^2 - 30t + 49$.

SOLUTION

Completing the square, we have

$$f(t) = 5t^2 - 30t + 49$$
$$= 5(t^2 - 6t) + 49$$
$$= 5(t^2 - 6t + 9) + 49 - 45$$
$$= 5(t - 3)^2 + 4$$

The minimum value of f occurs at the vertex of the graph and is

$$f(3) = 4$$

    In solving applied maximum and minimum problems, our initial task is to find an expression for an appropriate quadratic function as in Section 2.4.

EXAMPLE 6

Among all pairs of numbers whose sum is 100, find a pair whose product is as large as possible.

SOLUTION

Let the numbers be x and y. We know that $x + y = 100$, so $y = 100 - x$. Thus their product is

$$P = xy = x(100 - x)$$

and so we must find the maximum value of the quadratic function

$$P(x) = 100x - x^2$$

We do this by completing the square:

$$P(x) = -x^2 + 100x$$
$$= -(x^2 - 100x)$$
$$= -(x^2 - 100x + 2500) + 2500$$
$$= -(x - 50)^2 + 2500$$

From this standard form we see that the maximum value is 2500 and it occurs when $x = 50$. Then $y = 100 - 50 = 50$, so the two numbers are 50 and 50. ■

EXAMPLE 7

A farmer wants to enclose a rectangular field by a fence and divide it into two smaller rectangular fields by a fence parallel to one side of the field. He has 3000 yd of fencing. Find the dimensions of the field so that the total enclosed area is a maximum.

SOLUTION

We draw a diagram as in Figure 7 and let x and y be the dimensions of the field (in yards). Then the area of the field is $A = xy$, but we need to eliminate y by using the fact that the total length of the fencing is 3000. Therefore

$$x + x + x + y + y = 3000$$
$$3x + 2y = 3000$$
$$2y = 3000 - 3x$$
$$y = 1500 - \tfrac{3}{2}x$$

Figure 7

This enables us to express A in terms of x alone:

$$A = xy = x\left(1500 - \tfrac{3}{2}x\right) = 1500x - \tfrac{3}{2}x^2$$

If we now complete the square, we get

$$A = -\tfrac{3}{2}x^2 + 1500x$$
$$= -\tfrac{3}{2}(x^2 - 1000x)$$
$$= -\tfrac{3}{2}(x^2 - 1000x + 250{,}000) + 375{,}000$$
$$= -\tfrac{3}{2}(x - 500)^2 + 375{,}000$$

This expression shows that the maximum occurs when $x = 500$. Then $y = 1500 - \tfrac{3}{2}(500) = 750$. Thus the field should be 500 yd by 750 yd. ■

EXAMPLE 8

A hockey team plays in an arena with a seating capacity of 15,000 spectators. With ticket prices set at \$12, average attendance at a game has been 11,000. A market survey indicates that for each dollar that ticket prices are lowered, the average attendance will increase by 1000. How should the owners of the team set ticket prices so as to maximize their revenue from ticket sales?

SOLUTION

Let x be the selling price of the ticket. Then $12 - x$ is the amount the ticket price has been lowered, so the number of tickets sold is

$$11{,}000 + 1000(12 - x) = 23{,}000 - 1000x$$

The revenue is

$$\begin{aligned}
R(x) &= x(23{,}000 - 1000x) \\
&= -1000x^2 + 23{,}000x \\
&= -1000(x^2 - 23x) \\
&= -1000\left(x^2 - 23x + \tfrac{529}{4}\right) + 250 \cdot 529 \\
&= -1000(x - 11.5)^2 + 132{,}250
\end{aligned}$$

This shows that the revenue is maximized when $x = 11.5$. The owners should set the ticket price at \$11.50.   ■

EXERCISES 2.6

In Exercises 1–10 sketch the graph of the given parabola and state the coordinates of its vertex and its intercepts.

1. $y = \frac{1}{2}x^2 - 1$

2. $y = 4 - x^2$

3. $y = -3x^2 - 2$

4. $y = x^2 + 6x$

5. $y = x^2 - 5x$

6. $y = x^2 - 2x + 2$

7. $y = -x^2 - 6x - 8$

8. $y = x^2 + 4x - 4$

9. $y = 2x^2 - 20x + 57$

10. $y = -3x^2 + 6x - 2$

In Exercises 11–20 sketch the graph of the given quadratic function and find its maximum or minimum value.

11. $f(x) = 2x - x^2$

12. $f(x) = x + x^2$

13. $f(x) = x^2 + 2x - 1$

14. $f(x) = x^2 - 8x + 8$

15. $f(x) = -x^2 - 3x + 3$

16. $f(x) = 1 - 6x - x^2$

17. $g(x) = 3x^2 - 12x + 13$

18. $g(x) = 2x^2 + 8x + 11$

19. $h(x) = 1 - x - x^2$

20. $h(x) = 3 - 4x - 4x^2$

In Exercises 21–24 find the maximum or minimum value of the given function.

21. $f(x) = x^2 + x + 1$

22. $f(x) = 1 + 3x - x^2$

23. $f(t) = 100 - 50t - 7t^2$

24. $f(t) = 10t^2 + 40t + 113$

25. Find a function whose graph is a parabola with vertex $(1, -2)$ and that passes through the point $(4, 16)$.

26. Find a function whose graph is a parabola with vertex $(3, 4)$ and that passes through the point $(1, -8)$.

27. Find the domain and range of the function $f(x) = -x^2 + 4x - 3$.

28. Find the domain and range of the function $f(x) = x^2 - 2x - 3$.

29. If a ball is thrown directly upward into the air with a velocity of 40 ft/s, its height in feet after t seconds is given by $y = 40t - 16t^2$. What is the maximum height attained by the ball?

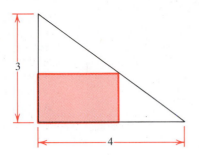

30. The effectiveness of a television commercial depends on how many times a viewer watches it. After some experiments, an advertising agency found that if the effectiveness E is measured on a scale of 0 to 10, then

$$E(n) = \tfrac{2}{3}n - \tfrac{1}{90}n^2$$

where n is the number of times a viewer watches a given commercial. For a commercial to have maximum effectiveness, how many times should a viewer watch it?

31. Find two numbers whose difference is 100 and whose product is as small as possible.

32. Find two positive numbers whose sum is 100 and the sum of whose squares is a minimum.

33. Find two numbers whose sum is -24 and whose product is a maximum.

34. Among all rectangles that have a perimeter of 20 ft, find the dimensions of the one with the largest area.

35. A farmer has 2400 ft of fencing and wants to fence off a rectangular field that borders a straight river. He needs no fence along the river. (See the figure for Exercise 17 in Section 2.4.) What are the dimensions of the field that has the largest area?

36. Find the area of the largest rectangle that can be inscribed in a right triangle with legs of lengths 3 cm and 4 cm if two sides of the rectangle lie along the legs as in the figure.

37. A farmer with 750 ft of fencing wants to enclose a rectangular area and then divide it into four pens with fencing parallel to one side of the rectangle. What is the largest possible total area of the four pens?

38. A student makes and sells necklaces on the beach during the summer months. The material for each necklace costs her $6 and she has been selling about 20 per day at $10 each. She has been wondering whether to raise her price, so she takes a survey and finds that for every dollar increase she would lose two sales a day. What should her selling price be in order to maximize profits?

39. Show, by completing the square, that the vertex of the parabola $y = ax^2 + bx + c$ is

$$\left(\frac{-b}{2a}, \frac{4ac - b^2}{4a}\right)$$

40. Find the maximum value of the function $f(x) = 3 + 4x^2 - x^4$. [*Hint:* Let $t = x^2$.]

41. Find a function whose graph is a parabola that passes through the points $(1, -1)$, $(-1, -3)$, and $(3, 9)$.

![Section 2.7 icon] **SECTION 2.7**
USING GRAPHING DEVICES TO FIND EXTREME VALUES

In Section 2.6 we were able to find maximum and minimum values of quadratic functions by completing the square to find the vertex of a parabola. This technique does not apply to other functions, so in this section we show how to locate extreme values of any function that can be graphed with a calculator or computer.

EXAMPLE 1

Draw the graph of the function $f(x) = x^3 - 8x + 1$ in the viewing rectangle $[-5, 5]$ by $[-20, 20]$. Comment on the peaks and valleys of the graph.

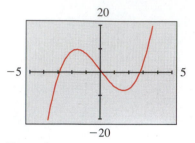

Figure 1

SOLUTION

The graph is shown in Figure 1. Notice that a high point of the graph occurs for a value of x between -3 and 0. In fact if we restrict our attention to the viewing rectangle $[-3, 0]$ by $[0, 15]$, as in Figure 2(a), we see that there is a highest point on the restricted graph. Similarly, when we confine our attention to the viewing rectangle $[0, 3]$ by $[-15, 0]$ in Figure 2(b), we see that there is a lowest point on the restricted graph.

Figure 2

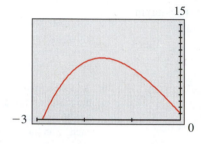

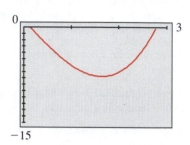

(a) f restricted to $[-3, 0]$ by $[0, 15]$ (b) f restricted to $[0, 3]$ by $[-15, 0]$ ■

In general, if there is a viewing rectangle such that the point $(a, f(a))$ is the highest point on the graph of f *within* the viewing rectangle (not on the edge), then the number $f(a)$ is called a **local maximum value** of f. (See Figure 3.) Notice that $f(a) \geq f(x)$ for all numbers x that are close to a.

Figure 3

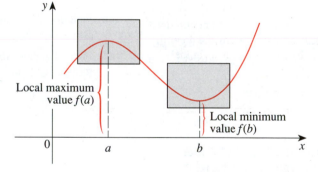

Similarly, if there is a viewing rectangle such that the point $(b, f(b))$ is the lowest point on the graph of f within the viewing rectangle, then the number $f(b)$ is called a **local minimum value** of f. In this case, $f(b) \leq f(x)$ for all numbers x that are close to b.

EXAMPLE 2

Find the approximate local maximum and minimum values of the function $f(x) = x^3 - 8x + 1$ of Example 1.

SOLUTION

Let us first find the local maximum value. Figure 2(a) shows that the maximum point lies in the rectangle $[-2, -1]$ by $[9, 10]$, so we draw the graph restricted to this viewing rectangle in Figure 4(a). By moving the cursor close to the maximum point, we see that the y-coordinates do not change very much in the vicinity of the maximum. The maximum value is about 9.7, and since the distance between the scale marks is 0.1, this estimate for the maximum value is accurate to within 0.1.

If we want an even more accurate estimate we can zoom in to a smaller rectangle containing the maximum point. Figure 4(b) shows the restriction of the graph to the viewing rectangle $[-1.7, -1.6]$ by $[9.6, 9.8]$. The curve in Figure 4(b) looks quite flat. It is easier to pinpoint the maximum value if we use a viewing rectangle that is much wider than it is high. So we use the viewing rectangle $[-1.7, -1.6]$ by $[9.7, 9.71]$ in Figure 4(c). By moving the cursor along the curve and observing how the y-coordinates change, we can estimate the maximum value quite accurately. It appears that the local maximum value is about 9.709 and occurs when x is about -1.633.

Figure 4
$f(x) = x^3 - 8x + 1$

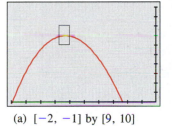 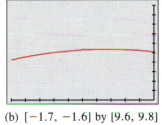 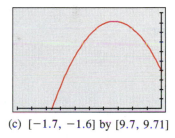

(a) $[-2, -1]$ by $[9, 10]$ (b) $[-1.7, -1.6]$ by $[9.6, 9.8]$ (c) $[-1.7, -1.6]$ by $[9.7, 9.71]$

We locate the local minimum in a similar fashion. By zooming in to the rectangles shown in Figure 5 we find that the local minimum value is about -7.709 and it occurs when $x \approx 1.633$.

Figure 5
$f(x) = x^3 - 8x + 1$

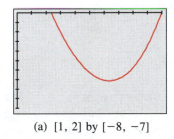 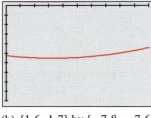 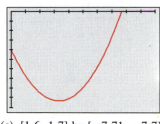

(a) $[1, 2]$ by $[-8, -7]$ (b) $[1.6, 1.7]$ by $[-7.8, -7.6]$ (c) $[1.6, 1.7]$ by $[-7.71, -7.7]$

■

EXAMPLE 3

A can is to be made to hold 1 L of oil. Find the value of the radius that minimizes the cost of the metal to make the can.

SOLUTION

In order to minimize the cost of the metal, we minimize the total surface area. In Example 3 in Section 2.4 we found the surface area to be

$$A = 2\pi r^2 + \frac{2000}{r} \qquad r > 0$$

where r is the radius of the can in centimeters. The graph of A as a function of r is shown in Figure 6(a). We see that the minimum point occurs in the rectangle $[5, 6]$ by $[500, 600]$, so we draw the graph in that viewing rectangle in Figure 6(b). But that graph is very flat, so we change to $[5, 6]$ by $[550, 560]$ in Figure 6(c). Moving the cursor near the minimum point and observing the y-coordinates (or A-coordinates in this case) we see that the minimum value of A is about 554 and occurs when the radius is about 5.4 cm.

Figure 6

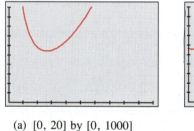   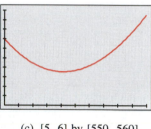

(a) $[0, 20]$ by $[0, 1000]$ (b) $[5, 6]$ by $[500, 600]$ (c) $[5, 6]$ by $[550, 560]$

EXERCISES 2.7

1. **(a)** Use a graphing calculator or computer to find the minimum value of the quadratic function $f(x) = x^2 + 1.79x - 3.21$ correct to two decimal places.

(b) Use the methods of Section 2.6 to find the exact minimum value of f and compare with your answer to part (a).

2. **(a)** Use a graphing calculator or computer to find the minimum value of the quadratic function $f(x) = 1 + x - \sqrt{2}x^2$ correct to two decimal places.

(b) Use the methods of Section 2.6 to find the exact minimum value of f and compare with your answer to part (a).

In Exercises 3–10 find the local maximum and minimum values of the function and the values of x at which they occur.

State your answers correct to two decimal places.

3. $f(x) = x^3 - x$

4. $f(x) = 3 + x + x^2 - x^3$

5. $g(x) = x^4 - 2x^3 - 11x^2$

6. $g(x) = x^5 - 8x^3 + 20x$

7. $U(x) = x\sqrt{6 - x}$

8. $U(x) = x\sqrt{x - x^2}$

9. $V(x) = \dfrac{1 - x^2}{x^3}$

10. $V(x) = \dfrac{1}{x^2 + x + 1}$

11. For a fish swimming at a speed v relative to the water, the energy expenditure per unit time is proportional to v^3. In Exercise 23 in Section 2.4 we saw that if a fish is swimming against a current of 5 mi/h, then the total energy required to travel a distance of 10 mi is

$$E(v) = 2.73v^3 \frac{10}{v - 5}$$

It is believed that migrating fish try to minimize the total energy required to swim a fixed distance. Find the value of v that minimizes energy. (*Note*: This result has been verified; migrating fish swim against a current at a speed 50% greater than the speed of the current.)

12. A highway engineer wants to estimate the maximum number of cars that can safely use a particular highway at a given speed. She assumes that each car is 17 ft long, travels at a speed s, and follows the car in front of it at the "safe following distance" for that speed. We saw in Exercise 22 in Section 2.4 that the number N of cars per minute that pass a given point is a function of the speed given by

$$N(s) = \frac{88s}{17 + 17\left(\dfrac{s}{20}\right)^2}$$

At what speed can the largest number of cars use the highway safely?

13. A box with a square base and an open top must have a volume of 12 ft^3. Find the dimensions of the box that minimize the amount of material used. (See Exercise 13 in Section 2.4.)

14. A box with an open top is to be constructed from a rectangular piece of cardboard with dimensions 12 in. by 20 in. by cutting out equal squares of side length x inches at each corner and then folding up the sides. (See the figure in Exercise 16 in Section 2.4.) Find the largest volume that such a box can have.

15. A Norman window has the shape of a rectangle surmounted by a semicircle. (See the figure in Exercise 15 in Section 2.4.) If the perimeter of the window is 30 ft, find the dimensions of the window so that the greatest possible amount of light is admitted.

16. Find the area of the largest rectangle that can be inscribed in a semicircle of radius 1 ft. (See the figure in Exercise 18 in Section 2.4.)

17. Find the dimensions of the rectangle of largest area that has its base on the x-axis and its other two vertices above the x-axis and lying on the parabola $y = 8 - x^2$.

18. A man is at a point A on a bank of a 2-mi wide straight river, and wants to reach point B, 7 mi downstream on the opposite bank, by first rowing his boat to a point P on the opposite bank and then walking the remaining distance x to B. (See the figure in Exercise 20 in Section 2.4.) He can row at a speed of 4 mi/h and walk at a speed of 3 mi/h. Where should he land in order to reach B as soon as possible?

SECTION 2.8
COMBINING FUNCTIONS

Two functions f and g can be combined to form new functions $f + g$, $f - g$, fg, and f/g in a manner similar to the way we add, subtract, multiply, and divide real numbers.

If we define the sum $f + g$ by the equation

$$(f + g)(x) = f(x) + g(x) \tag{1}$$

then the right side of Equation 1 makes sense if both $f(x)$ and $g(x)$ are defined, that is, if x belongs to the domain of f and also to the domain of g. If the domain of f is A, and the domain of g is B, then the domain of $f + g$ is the intersection of these domains, that is, $A \cap B$.

Notice that the $+$ sign on the left side of Equation 1 stands for the operation of addition of *functions*, but the $+$ sign on the right side of the equation stands for addition of the *numbers* $f(x)$ and $g(x)$.

Similarly, we can define the difference $f - g$ and the product fg, and their domains will also be $A \cap B$. But in defining the quotient f/g we must remember not to divide by zero.

ALGEBRA OF FUNCTIONS

Let f and g be functions with domains A and B. Then the functions $f + g$, $f - g$, fg, and f/g are defined as follows:

$$(f + g)(x) = f(x) + g(x) \qquad \text{domain} = A \cap B$$
$$(f - g)(x) = f(x) - g(x) \qquad \text{domain} = A \cap B$$
$$(fg)(x) = f(x)g(x) \qquad \text{domain} = A \cap B$$
$$\left(\frac{f}{g}\right)(x) = \frac{f(x)}{g(x)} \qquad \text{domain} = \{x \in A \cap B \mid g(x) \neq 0\}$$

EXAMPLE 1

If $f(x) = \sqrt{x}$ and $g(x) = \sqrt{4 - x^2}$, find the functions $f + g$, $f - g$, fg, and f/g.

SOLUTION

The domain of $f(x) = \sqrt{x}$ is $[0, \infty)$. The domain of $g(x) = \sqrt{4 - x^2}$ consists of all numbers x such that $4 - x^2 \geq 0$, that is, $x^2 \leq 4$. Taking square roots of both sides, we get $|x| \leq 2$, or $-2 \leq x \leq 2$, so the domain of g is the interval $[-2, 2]$. The intersection of the domains of f and g is

$$[0, \infty) \cap [-2, 2] = [0, 2]$$

Thus we have

$$(f + g)(x) = \sqrt{x} + \sqrt{4 - x^2} \qquad\qquad 0 \leq x \leq 2$$
$$(f - g)(x) = \sqrt{x} - \sqrt{4 - x^2} \qquad\qquad 0 \leq x \leq 2$$
$$(fg)(x) = \sqrt{x}\,\sqrt{4 - x^2} = \sqrt{4x - x^3} \qquad 0 \leq x \leq 2$$
$$\left(\frac{f}{g}\right)(x) = \frac{\sqrt{x}}{\sqrt{4 - x^2}} = \sqrt{\frac{x}{4 - x^2}} \qquad\qquad 0 \leq x < 2$$

Notice that the domain of f/g is the interval $[0, 2)$ because we must exclude the points where $g(x) = 0$, that is, $x = \pm 2$. ■

The graph of the function $f + g$ is obtained from the graphs of f and g by **graphical addition**. This means that we add corresponding y-coordinates as in Figure 1.

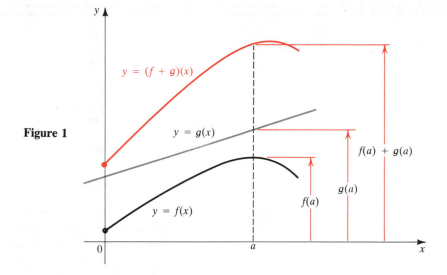

Figure 1

EXAMPLE 2

Use graphical addition to graph the function $f + g$ from Example 1.

SOLUTION

We first graph the root function $f(x) = \sqrt{x}$ and the function $g(x) = \sqrt{4 - x^2}$ (the top half of the circle $x^2 + y^2 = 4$) in Figure 2. To obtain the graph of $f + g$ we add the y-coordinates of the points on the graphs of f and g. Given the points $Q(x, f(x))$ and $R(x, g(x))$, the point S has coordinates $(x, f(x) + g(x))$. Thus $|PQ| = |RS|$.

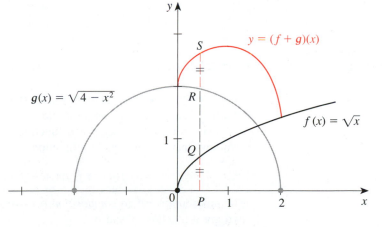

Figure 2

EXERCISES 2.8

In Exercises 1–6 find $f + g$, $f - g$, fg, and f/g and their domains.

1. $f(x) = x^2 - x$, $g(x) = x + 5$

2. $f(x) = x^3 + 2x^2$, $g(x) = 3x^2 - 1$

3. $f(x) = \sqrt{1 + x}$, $g(x) = \sqrt{1 - x}$

4. $f(x) = \sqrt{9 - x^2}$, $g(x) = \sqrt{x^2 - 1}$

5. $f(x) = \sqrt{x}$, $g(x) = \sqrt[3]{x}$

6. $f(x) = \sqrt[4]{x + 1}$, $g(x) = \sqrt{x + 2}$

In Exercises 7 and 8 find the domain of the given function.

7. $F(x) = \dfrac{\sqrt{4 - x} + \sqrt{3 + x}}{x^2 - 2}$

8. $F(x) = \sqrt{1 - x} + \sqrt{x - 2}$

In Exercises 9 and 10 copy or trace the graphs of f and g and use graphical addition to sketch the graph of $f + g$.

9.

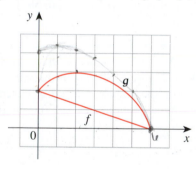

10.

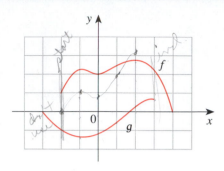

In Exercises 11–14 use the graphs of f and g and the method of graphical addition to sketch the graph of $f + g$.

11. $f(x) = x^3$, $g(x) = x$

12. $f(x) = x^2$, $g(x) = 1 - x$

13. $f(x) = x$, $g(x) = \dfrac{1}{x}$

14. $f(x) = x^3$, $g(x) = -x^2$

In Exercises 15–18 draw the graphs of f, g, and $f + g$ on a common screen to illustrate the method of graphical addition.

15. $f(x) = \sqrt{1 + x}$, $g(x) = \sqrt{1 - x}$

16. $f(x) = x^2$, $g(x) = \sqrt{x}$

17. $f(x) = x^2$, $g(x) = x^3$

18. $f(x) = \sqrt[4]{1 - x}$, $g(x) = \sqrt{1 - x^2/9}$

SECTION 2.9
COMPOSITION OF FUNCTIONS

In this section we consider a very important way of combining two functions to obtain a new function. For example, let us suppose that $y = f(u) = \sqrt{u}$ and $u = g(x) = x^2 + 1$. Since y is a function of u and u is, in turn, a function of x, it follows that y is ultimately a function of x. We compute this by substitution:

$$y = f(u) = f(g(x)) = f(x^2 + 1) = \sqrt{x^2 + 1}$$

The procedure is called *composition* because the new function is composed of the two given functions f and g.

In general, given any two functions f and g, we start with a number x in the domain of g and find its image $g(x)$. If this number $g(x)$ is in the domain of f, then we can calculate the value of $f(g(x))$. The result is a new function $h(x) = f(g(x))$ obtained by substituting g into f. It is called the *composition* (or *composite*) of f and g and is denoted by $f \circ g$ ("f circle g").

COMPOSITION OF FUNCTIONS

Given two functions f and g, the **composite function** $f \circ g$ (also called the **composition** of f and g) is defined by

$$(f \circ g)(x) = f(g(x))$$

The domain of $f \circ g$ is the set of all x in the domain of g such that $g(x)$ is in the domain of f. In other words, $(f \circ g)(x)$ is defined whenever both $g(x)$ and $f(g(x))$ are defined. The best way to picture $f \circ g$ is by a machine diagram (Figure 1) or an arrow diagram (Figure 2).

Figure 1
The $f \circ g$ machine is composed of the g machine (first) and then the f machine.

Figure 2
Arrow diagram for $f \circ g$

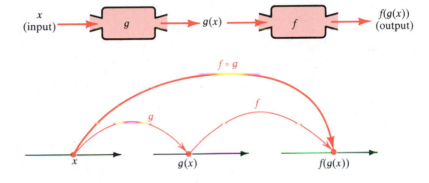

EXAMPLE 1

If $f(x) = x^2$ and $g(x) = x - 3$, find the composite functions $f \circ g$ and $g \circ f$ and their domains.

SOLUTION

We have

$$
\begin{aligned}
(f \circ g) &= f(g(x)) && \text{(definition of } f \circ g\text{)} \\
&= f(x - 3) && \text{(definition of } g\text{)} \\
&= (x - 3)^2 && \text{(definition of } f\text{)}
\end{aligned}
$$

and

$$
\begin{aligned}
(g \circ f)(x) &= g(f(x)) && \text{(definition of } g \circ f\text{)} \\
&= g(x^2) && \text{(definition of } f\text{)} \\
&= x^2 - 3 && \text{(definition of } g\text{)}
\end{aligned}
$$

The domains of both $f \circ g$ and $g \circ f$ are R (the set of all real numbers). ■

🚫 *Note*   You can see from Example 1 that, in general, $f \circ g \neq g \circ f$. Remember that the notation $f \circ g$ means that the function g is applied first and then f is applied second. In Example 1, $f \circ g$ is the function that *first* subtracts 3 and *then* squares; $g \circ f$ is the function that *first* squares and *then* subtracts 3.

EXAMPLE 2

If $f(x) = \sqrt{x}$ and $g(x) = \sqrt{2 - x}$, find the following functions and their domains:
(a) $f \circ g$ (b) $g \circ f$ (c) $f \circ f$ (d) $g \circ g$

SOLUTION

(a)
$$(f \circ g)(x) = f(g(x))$$
$$= f\left(\sqrt{2 - x}\right)$$
$$= \sqrt{\sqrt{2 - x}}$$
$$= \sqrt[4]{2 - x}$$

The domain of $f \circ g$ is

$$\{x \mid 2 - x \geq 0\} = \{x \mid x \leq 2\} = (-\infty, 2]$$

(b)
$$(g \circ f)(x) = g(f(x))$$
$$= g\left(\sqrt{x}\right)$$
$$= \sqrt{2 - \sqrt{x}}$$

For $\sqrt{x}$ to be defined we must have $x \geq 0$. For $\sqrt{2 - \sqrt{x}}$ to be defined we must have $2 - \sqrt{x} \geq 0$, that is, $\sqrt{x} \leq 2$, or $x \leq 4$. Thus we have $0 \leq x \leq 4$, so the domain of $g \circ f$ is the closed interval $[0, 4]$.

(c)
$$(f \circ f)(x) = f(f(x))$$
$$= f\left(\sqrt{x}\right)$$
$$= \sqrt{\sqrt{x}}$$
$$= \sqrt[4]{x}$$

The domain of $f \circ f$ is $[0, \infty)$.

(d)
$$(g \circ g)(x) = g(g(x))$$
$$= g\left(\sqrt{2 - x}\right)$$
$$= \sqrt{2 - \sqrt{2 - x}}$$

This expression is defined when $2 - x \geq 0$, that is, $x \leq 2$, and $2 - \sqrt{2 - x} \geq 0$. This latter inequality is equivalent to $\sqrt{2 - x} \leq 2$, or $2 - x \leq 4$, that is, $x \geq -2$. Thus $-2 \leq x \leq 2$, so the domain of $g \circ g$ is $[-2, 2]$.   ∎

It is possible to take the composition of three or more functions. For instance, the composite function $f \circ g \circ h$ is found by first applying h, then g, and then f, as follows:

$$(f \circ g \circ h)(x) = f(g(h(x)))$$

EXAMPLE 3

Find $f \circ g \circ h$ if $f(x) = x/(x + 1)$, $g(x) = x^{10}$, and $h(x) = x + 3$.

SOLUTION

$$
\begin{aligned}
(f \circ g \circ h) &= f(g(h(x))) \\
&= f(g(x + 3)) \\
&= f((x + 3)^{10}) \\
&= \frac{(x + 3)^{10}}{(x + 3)^{10} + 1}
\end{aligned}
$$

  ■

So far we have used composition to build up complicated functions from simpler ones. But in calculus it is useful to be able to decompose a complicated function into simpler ones, as in the following example.

EXAMPLE 4

Given $F(x) = \sqrt[4]{x + 9}$, find functions f and g such that $F = f \circ g$.

SOLUTION

Since the formula for F says to first add 9 and then take the fourth root, we let

$$g(x) = x + 9 \qquad \text{and} \qquad f(x) = \sqrt[4]{x}$$

Then
$$
\begin{aligned}
(f \circ g)(x) &= f(g(x)) \\
&= f(x + 9) \\
&= \sqrt[4]{x + 9} \\
&= F(x)
\end{aligned}
$$

  ■

EXERCISES 2.9

In Exercises 1–12 use $f(x) = 3x - 5$ and $g(x) = 2 - x^2$ to evaluate the expression.

1. $f(g(0))$

2. $g(f(1))$

3. $f(f(4))$

4. $g(g(3))$

5. $(f \circ g)(-2)$

6. $(g \circ f)(-2)$

7. $(f \circ f)(-1)$

8. $(g \circ g)(2)$

9. $(f \circ g)(x)$

10. $(g \circ f)(x)$

11. $(f \circ f)(x)$

12. $(g \circ g)(x)$

In Exercises 13–18 use the given graphs of f and g to evaluate the expression.

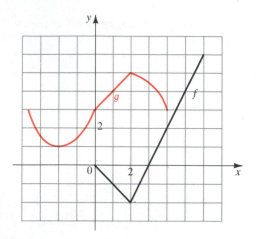

13. $f(g(2))$

14. $g(f(0))$

15. $(g \circ f)(4)$

16. $(f \circ g)(0)$

17. $(g \circ g)(-2)$

18. $(f \circ f)(4)$

In Exercises 19–28 find the functions $f \circ g$, $g \circ f$, $f \circ f$, and $g \circ g$, and their domains.

19. $f(x) = 2x + 3$, $g(x) = 4(x) - 1$

20. $f(g) = 6x - 5$, $g(x) = \dfrac{x}{2}$

21. $f(x) = 2x^2 - x$, $g(x) = 3x + 2$

22. $f(x) = \dfrac{1}{x}$, $g(x) = x^3 + 2x$

23. $f(x) = \sqrt{x - 1}$, $g(x) = x^2$

24. $f(x) = \dfrac{1}{x - 1}$, $g(x) = \dfrac{x - 1}{x + 1}$

25. $f(x) = \sqrt[3]{x}$, $g(x) = 1 - \sqrt{x}$

26. $f(x) = \sqrt{x^2 - 1}$, $g(x) = \sqrt{1 - x}$

27. $f(x) = \dfrac{x + 2}{2x + 1}$, $g(x) = \dfrac{x}{x - 2}$

28. $f(x) = \dfrac{1}{\sqrt{x}}$, $g(x) = x^2 - 4x$

In Exercises 29–32 find $f \circ g \circ h$.

29. $f(x) = x - 1$, $g(x) = \sqrt{x}$, $h(x) = x - 1$

30. $f(x) = \dfrac{1}{x}$, $g(x) = x^3$, $h(x) = x^2 + 2$

31. $f(x) = x^4 + 1$, $g(x) = x - 5$, $h(x) = \sqrt{x}$

32. $f(x) = \sqrt{x}$, $g(x) = \dfrac{x}{x - 1}$, $h(x) = \sqrt[3]{x}$

In Exercises 33–38 express the given function in the form $f \circ g$.

33. $F(x) = (x - 9)^5$

34. $F(x) = \sqrt{x} + 1$

35. $G(x) = \dfrac{x^2}{x^2 + 4}$

36. $G(x) = \dfrac{1}{x + 3}$

37. $H(x) = |1 - x^3|$

38. $H(x) = \sqrt{1 + \sqrt{x}}$

In Exercises 39–42 express the given function in the form $f \circ g \circ h$.

39. $F(x) = \dfrac{1}{x^2 + 1}$

40. $F(x) = \sqrt[3]{\sqrt{x} - 1}$

41. $G(x) = \left(4 + \sqrt[3]{x}\right)^9$

42. $G(x) = \dfrac{2}{\left(3 + \sqrt{x}\right)^2}$

43. A stone is dropped into a lake, creating a circular ripple that travels outward at a speed of 60 cm/s. Express the area of this circle as a function of time t (in seconds).

44. A spherical balloon is being inflated. If the radius of the balloon is increasing at a rate of 1 cm/s, express the volume of the balloon as a function of time t (in seconds).

45. If $f(x) = 3x + 5$ and $h(x) = 3x^2 + 3x + 2$, find a function g such that $f \circ g = h$.

46. If $f(x) = x + 4$ and $h(x) = 4x - 1$, find a function g such that $g \circ f = h$.

47. Suppose g is an even function and let $h = f \circ g$. Is h always an even function?

48. Suppose g is an odd function and let $h = f \circ g$. Is h always an odd function? What if f is odd? What if f is even?

SECTION 2.10
ONE-TO-ONE FUNCTIONS AND THEIR INVERSES

Let us compare the functions f and g whose arrow diagrams are shown in Figure 1. Note that f never takes on the same value (any two numbers in A have different images), whereas g does take on the same value twice (both 2 and 3 have the same image, 4). In symbols,

$$g(2) = g(3)$$

but $$f(x_1) \neq f(x_2) \quad \text{whenever } x_1 \neq x_2$$

Functions that have this latter property are called *one-to-one*.

Figure 1

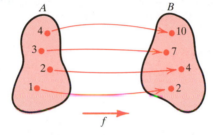 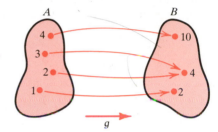

> A function with domain A is called a **one-to-one function** if no two elements of A have the same image; that is,
>
> $$f(x_1) \neq f(x_2) \quad \text{whenever } x_1 \neq x_2$$

An equivalent way of writing the condition for a one-to-one function is this:

$$\text{If } f(x_1) = f(x_2), \text{ then } x_1 = x_2.$$

If a horizontal line intersects the graph of f in more than one point, then we see from Figure 2 that there are numbers $x_1 \neq x_2$ such that $f(x_1) = f(x_2)$. This means that f is not one-to-one. Therefore we have the following geometric method for determining whether or not a function is one-to-one.

Figure 2
This function is not one-to-one because $f(x_1) = f(x_2)$

HORIZONTAL LINE TEST

> A function is one-to-one if and only if no horizontal line intersects its graph more than once.

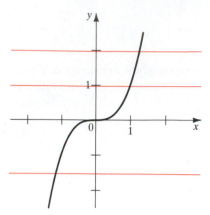

Figure 3
$f(x) = x^3$ is one-to-one

EXAMPLE 1

Is the function $f(x) = x^3$ one-to-one?

SOLUTION 1

If $x_1 \neq x_2$, then $x_1^3 \neq x_2^3$ (two different numbers cannot have the same cube). Therefore $f(x) = x^3$ is one-to-one.

SOLUTION 2

From Figure 3 we see that no horizontal line intersects the graph of $f(x) = x^3$ more than once. Therefore, by the Horizontal Line Test, f is one-to-one. ■

Notice that the function f of Example 1 is increasing and is also one-to-one. In fact it can be proved that every increasing function and every decreasing function is one-to-one (see Exercise 39).

EXAMPLE 2

Is the function $g(x) = x^2$ one-to-one?

SOLUTION 1

This function is not one-to-one because, for instance,

$$g(1) = 1 = g(-1)$$

and so 1 and -1 have the same image.

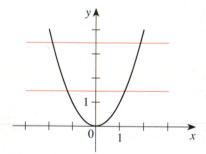

Figure 4
$g(x) = x^2$ is not one-to-one

SOLUTION 2

From Figure 4 we see that there are horizontal lines that intersect the graph of g more than once. Therefore, by the Horizontal Line Test, g is not one-to-one. ■

Although the function g in Example 2 is not one-to-one, it is possible to restrict its domain so that the resulting function is one-to-one. In fact, if we define

$$h(x) = x^2 \qquad x \geq 0$$

then h is one-to-one, as you can see from Figure 5 and the Horizontal Line Test.

EXAMPLE 3

Show that the function $f(x) = 3x + 4$ is one-to-one.

SOLUTION

Suppose there are numbers x_1 and x_2 such that $f(x_1) = f(x_2)$. Then

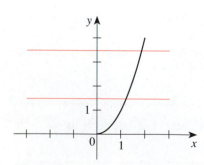

Figure 5
$h(x) = x^2$, $x \geq 0$, is one-to-one

$$3x_1 + 4 = 3x_2 + 4$$
$$3x_1 = 3x_2$$
$$x_1 = x_2$$

Therefore f is one-to-one. ■

INVERSE FUNCTIONS

One-to-one functions are important because they are precisely the functions that possess inverse functions according to the following definition.

> Let f be a one-to-one function with domain A and range B. Then its **inverse function** f^{-1} has domain B and range A and is defined by
>
> $$f^{-1}(y) = x \quad \Leftrightarrow \quad f(x) = y \qquad (1)$$
>
> for any y in B.

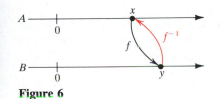

Figure 6

This definition says that if f takes x into y, then f^{-1} takes y back into x. (If f were not one-to-one, then f^{-1} would not be uniquely defined.) The arrow diagram in Figure 6 indicates that f^{-1} reverses the effect of f. Note the following:

> $$\text{domain of } f^{-1} = \text{range of } f$$
> $$\text{range of } f^{-1} = \text{domain of } f$$

For example, the inverse function of $f(x) = x^3$ is $f^{-1}(x) = x^{1/3}$ because if $y = x^3$, then

$$f^{-1}(y) = f^{-1}(x^3) = (x^3)^{1/3} = x$$

 Warning   Do not mistake the -1 in f^{-1} for an exponent. Thus

$$f^{-1}(x) \text{ does not mean } \frac{1}{f(x)}$$

The reciprocal $1/f(x)$ could, however, be written as $[f(x)]^{-1}$.

The letter x is traditionally used as the independent variable, so when we concentrate on f^{-1} rather than on f we usually reverse the roles of x and y in (1) and write

$$f^{-1}(x) = y \quad \Leftrightarrow \quad f(y) = x \qquad (2)$$

By substituting for y in (1) and substituting for x in (2) we get the following equations:

$$f^{-1}(f(x)) = x \text{ for every } x \text{ in } A$$
$$f(f^{-1}(x)) = x \text{ for every } x \text{ in } B \qquad (3)$$

The first equation says that if we start with x, apply f, and then apply f^{-1}, we arrive back at x, where we started. Thus f^{-1} undoes what f does. The second equation says that f undoes what f^{-1} does.

For example, if $f(x) = x^3$, then $f^{-1}(x) = x^{1/3}$ and the equations in (3) become

$$f^{-1}(f(x)) = (x^3)^{1/3} = x$$
$$f(f^{-1}(x)) = (x^{1/3})^3 = x$$

These equations simply say that the cube function and the cube root function cancel each other out.

Let us now see how to compute inverse functions. If we have a function $y = f(x)$ and are able to solve this equation for x in terms of y, then according to (1) we must have $x = f^{-1}(y)$. If we then interchange x and y, we have $y = f^{-1}(x)$, which is the desired equation.

HOW TO FIND THE INVERSE FUNCTION OF A ONE-TO-ONE FUNCTION f	**1.** Write $y = f(x)$. **2.** Solve this equation for x in terms of y (if possible). **3.** Interchange x and y. The resulting equation is $y = f^{-1}(x)$.

Note that Steps 2 and 3 could be reversed. In other words, it is possible to interchange x and y first and then solve for y in terms of x.

EXAMPLE 4

Find the inverse function of $f(x) = x^3 + 2$.

SOLUTION

We first write

$$y = x^3 + 2$$

Then we solve this equation for x:

$$x^3 = y - 2$$
$$x = \sqrt[3]{y - 2}$$

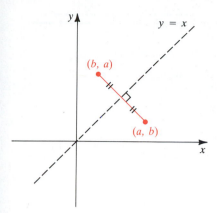

Figure 7

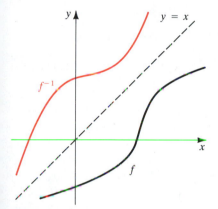

Figure 8

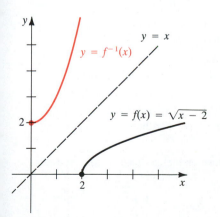

Figure 9

Finally we interchange x and y:

$$y = \sqrt[3]{x - 2}$$

Therefore the inverse function is $f^{-1}(x) = \sqrt[3]{x - 2}$.

This formula for f^{-1} seems reasonable if we state the rules for f and f^{-1} in words. The instructions for $f(x) = x^3 + 2$ are "Cube, then add 2," whereas the instructions for $f^{-1}(x) = \sqrt[3]{x - 2}$ are "Subtract 2, then take the cube root." ■

The principle of interchanging x and y to find the inverse function also gives us the method for obtaining the graph of f^{-1} from the graph of f. If $f(a) = b$, then $f^{-1}(b) = a$. Thus the point (a, b) is on the graph of f if and only if the point (b, a) is on the graph of f^{-1}. But we get the point (b, a) from the point (a, b) by reflecting in the line $y = x$ (see Figure 7). Therefore, as illustrated in Figure 8:

> The graph of f^{-1} is obtained by reflecting the graph of f in the line $y = x$.

EXAMPLE 5

(a) Sketch the graph of $f(x) = \sqrt{x - 2}$.
(b) Use the graph of f to sketch the graph of f^{-1}.
(c) Find an equation for f^{-1}.

SOLUTION

(a) As in Section 2.5 we sketch the graph of $y = \sqrt{x - 2}$ by taking the graph of the square root function (Example 6 in Section 2.2) and moving it 2 units to the right.

(b) The graph of f^{-1} is obtained by taking the graph of f from part (a) and reflecting it in the line $y = x$ as in Figure 9.

(c) Solve $y = \sqrt{x - 2}$ for x:

$$\sqrt{x - 2} = y \qquad \text{(Note that } y \geq 0\text{)}$$
$$x - 2 = y^2$$
$$x = y^2 + 2 \qquad y \geq 0$$

Interchange x and y:

$$y = x^2 + 2 \qquad x \geq 0$$

Thus
$$f^{-1}(x) = x^2 + 2 \qquad x \geq 0$$

This expression shows that the graph of f^{-1} is the right half of the parabola $y = x^2 + 2$ and, checking with Figure 9, that seems reasonable. ■

EXERCISES 2.10

In Exercises 1–6 the graph of a function f is given. Determine whether f is one-to-one.

1.

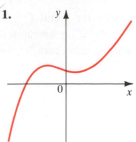

2.

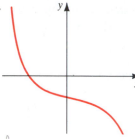

3.

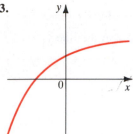

4.

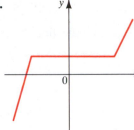

5.

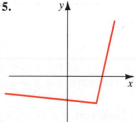

6.

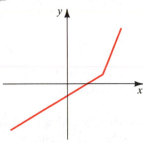

In Exercises 7–12 determine whether the given function is one-to-one.

7. $f(x) = 7x - 3$

8. $f(x) = x^2 - 2x + 5$

9. $g(x) = \sqrt{x}$

10. $g(x) = |x|$

11. $h(x) = x^4 + 5$

12. $h(x) = x^4 + 5,\ 0 \le x \le 2$

In Exercises 13–16 determine whether the given function is one-to-one.

13. $f(x) = x^3 - x$

14. $f(x) = x^3 + x$

15. $f(x) = \dfrac{x + 12}{x - 6}$

16. $f(x) = \sqrt{x^3 - 4x + 1}$

17. If f is one-to-one and $f(2) = 7$, evaluate $f^{-1}(7)$.

18. Find $f^{-1}(4)$ if f is a one-to-one function and $f(1) = 4$.

19. If $f(x) = 5 - 2x$, find $f^{-1}(3)$.

20. If $g(x) = x^2 + 4x$, $x \ge -2$, find $g^{-1}(5)$.

In Exercises 21–26 show that f is one-to-one and find its inverse function.

21. $f(x) = 4x + 7$

22. $f(x) = \dfrac{x - 2}{x + 2}$

23. $f(x) = \dfrac{1 + 3x}{5 - 2x}$

24. $f(x) = 5 - 4x^3$

25. $f(x) = \sqrt{2 + 5x}$

26. $f(x) = x^2 + x,\ x \ge -\frac{1}{2}$

In Exercises 27–32 find the inverse function of f and then verify that f^{-1} and f satisfy Equations 3.

27. $f(x) = 3 - 5x$

28. $f(x) = \dfrac{1}{x + 2},\ x > -2$

29. $f(x) = 4 - x^2,\ x \ge 0$

30. $f(x) = \sqrt{2x - 1}$

31. $f(x) = 4 + \sqrt[3]{x}$

32. $f(x) = (2 - x^3)^5$

In Exercises 33–38 (a) sketch the graph of f, (b) use the graph of f to sketch the graph of f^{-1}, and (c) find an equation for f^{-1}.

33. $f(x) = 2x + 1$

34. $f(x) = 6 - x$

35. $f(x) = 1 + \sqrt{1 + x}$

36. $f(x) = 9 - x^2,\ 0 \le x \le 3$

37. $f(x) = x^4,\ x \ge 0$

38. $f(x) = 1 - x^3$

39. Prove that if f is an increasing function on its entire domain, then f is one-to-one.

40. Show that it is possible to restrict the domain of the function $f(x) = x^2 + x + 1$ in such a way that the resulting function is one-to-one.

41. For what values of the number m is the linear function $f(x) = mx + b$ one-to-one? For those values of m, find f^{-1}.

42. Under what conditions on the constants a, b, c, and d is the function

$$f(x) = \frac{ax + b}{cx + d}$$

one-to-one?

CHAPTER 2 REVIEW

Define, state, or discuss the following.

1. Function
2. Domain of a function
3. Range of a function
4. Independent and dependent variables
5. Arrow diagram
6. Graph of a function
7. Linear function
8. Constant function
9. Vertical Line Test
10. Increasing function
11. Decreasing function
12. Even function
13. Odd function
14. Directly proportional

15. Inversely proportional
16. Jointly proportional
17. Vertical shifts of graphs
18. Horizontal shifts of graphs
19. Vertical stretching, shrinking, and reflecting
20. Quadratic functions
21. Maximum and minimum values of a function
22. Sum, difference, product, and quotient of functions
23. Composition of functions
24. One-to-one functions
25. Horizontal Line Test
26. Inverse function
27. Procedure for finding an inverse function
28. Graph of an inverse function

REVIEW EXERCISES

1. If $f(x) = 1 + \sqrt{x - 1}$, find $f(5)$, $f(9)$, $f(a + 1)$, $f(-x)$, $f(x^2)$, and $[f(x)]^2$.

2. The graph of a function is given.
   (a) State the values of $f(-2)$ and $f(2)$.
   (b) State the domain of f.
   (c) State the range of f.
   (d) On what intervals is f increasing? On what intervals is it decreasing?
   (e) Is f one-to-one?

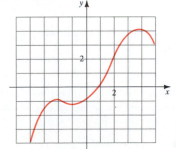

3. Which of the following figures are graphs of functions? Which of the functions are one-to-one?

(a)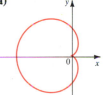

(b)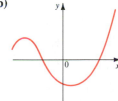

(c)

(d)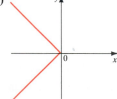

4. Find the domain and range of the function $f(x) = 2 + \sqrt{x + 3}$.

5. Find the domain and range of the function $F(t) = t^2 + 2t + 5$.

In Exercises 6–10 find the domain of the function.

6. $f(x) = \dfrac{2x + 1}{2x - 1}$ 　　　 **7.** $f(x) = 3x - \dfrac{2}{\sqrt{x + 1}}$

8. $f(x) = \dfrac{\sqrt[3]{2x + 1}}{\sqrt[3]{2x + 2}}$ 　　 **9.** $g(x) = \dfrac{2x^2 + 5x + 3}{2x^2 - 5x - 3}$

10. $h(x) = \sqrt{4 - x} + \sqrt{x^2 - 1}$

In Exercises 11–28 sketch the graph of the function.

11. $f(x) = 1 - 2x$

12. $f(x) = \frac{1}{3}(x - 5),\ 2 \le x \le 8$

13. $g(x) = \dfrac{1}{x^2}$ 　　　　 **14.** $G(x) = \dfrac{1}{(x - 3)^2}$

15. $h(x) = \sqrt[3]{x}$ 　　　　 **16.** $H(x) = x^3 - 3x + 1$

17. $f(t) = 1 - \frac{1}{2}t^2$ 　　 **18.** $g(t) = t^2 - 2t$

19. $f(x) = x^2 - 6x + 6$ 　 **20.** $f(x) = 3 - 8x - 2x^2$

21. $y = 1 - \sqrt{x}$ 　　　 **22.** $y = -|x - 5|$

23. $y = \frac{1}{2}(x + 1)^3$ 　　 **24.** $y = 2 + \sqrt{x + 3}$

25. $f(x) = \begin{cases} 1 - x & \text{if } x < 0 \\ 1 & \text{if } x \ge 0 \end{cases}$

26. $f(x) = \begin{cases} 1 - 2x & \text{if } x \le 0 \\ 2x - 1 & \text{if } x > 0 \end{cases}$

27. $f(x) = \begin{cases} x + 6 & \text{if } x < -2 \\ x^2 & \text{if } x \ge -2 \end{cases}$

28. $f(x) = \begin{cases} -x & \text{if } x < 0 \\ 2x - x^2 & \text{if } 0 \le x < 2 \\ 1 & \text{if } x \ge 2 \end{cases}$

 29. Which of the following viewing rectangles produces the most appropriate graph of the function
$f(x) = 6x^3 - 15x^2 + 4x - 1$?
(a) $[-2, 2]$ by $[-2, 2]$ 　 **(b)** $[-8, 8]$ by $[-8, 8]$
(c) $[-4, 4]$ by $[-12, 12]$
(d) $[-100, 100]$ by $[-100, 100]$

 30. Which of the following viewing rectangles produces the most appropriate graph of the function
$f(x) = \sqrt{100 - x^3}$?
(a) $[-4, 4]$ by $[-4, 4]$ 　 **(b)** $[-10, 10]$ by $[-10, 10]$
(c) $[-10, 10]$ by $[-10, 40]$
(d) $[-100, 100]$ by $[-100, 100]$

In Exercises 31–34 determine an appropriate viewing rectangle for the given function and use it to draw the graph.

31. $f(x) = x^2 + 25x + 173$

32. $f(x) = 1.1x^3 - 9.6x^2 - 1.4x + 3.2$

33. $y = \dfrac{x}{\sqrt{x^2 + 16}}$ 　　 **34.** $y = |x(x + 4)(x + 10)|$

35. Find, approximately, the domain of the function
$f(x) = \sqrt{x^3 - 4x + 1}$.

36. Find, approximately, the range of the function
$f(x) = x^4 - x^3 + x^2 + 3x - 6$.

37. Suppose that the graph of f is given. Describe how the graphs of the following functions can be obtained from the graph of f.
(a) $y = f(x) + 8$ 　　　 **(b)** $y = f(x + 8)$
(c) $y = 1 + 2f(x)$ 　　　 **(d)** $y = f(x - 2) - 2$
(e) $y = -f(x)$ 　　　　 **(f)** $y = f^{-1}(x)$

38. The graph of f is given. Draw the graphs of the following functions.
(a) $y = f(x - 8)$ 　　　 **(b)** $y = -f(x)$
(c) $y = 2 - f(x)$ 　　　 **(d)** $y = \frac{1}{2}f(x) - 1$
(e) $y = f^{-1}(x)$ 　　　 **(f)** $y = f^{-1}(x + 3)$

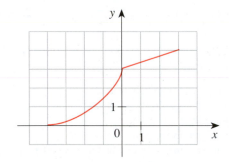

39. Find the maximum value of the function
$f(x) = 1 - x - x^2$.

40. Find the minimum value of the function
$g(x) = 2x^2 + 3x - 5$.

In Exercises 41–42 find the local maximum and minimum values of the function and the values of x at which they occur. State your answers to two decimal places.

41. $f(x) = 3.3 + 1.6x - 2.5x^3$

42. $f(x) = x^{2/3}(6 - x)^{1/3}$

43. Determine whether f is even, odd, or neither even nor odd.
  (a) $f(x) = 2x^5 - 3x^2 + 2$
  (b) $f(x) = x^3 - x^7$
  (c) $f(x) = \dfrac{1 - x^2}{1 + x^2}$ (d) $f(x) = \dfrac{1}{x + 2}$

44. If $f(x) = x^2 - 3x + 2$ and $g(x) = 4 - 3x$, find
  (a) $f + g$ (b) $f - g$ (c) fg
  (d) f/g (e) $f \circ g$ (f) $g \circ f$

In Exercises 45 and 46 find the functions $f \circ g$, $g \circ f$, $f \circ f$, $g \circ g$, and their domains.

45. $f(x) = 3x - 1$, $g(x) = 2x - x^2$

46. $f(x) = \sqrt{x}$, $g(x) = \dfrac{2}{x - 4}$

47. Find $f \circ g \circ h$, where $f(x) = \sqrt{1 - x}$, $g(x) = 1 - x^2$, and $h(x) = 1 + \sqrt{x}$.

48. If $T(x) = \dfrac{1}{\sqrt{1 + \sqrt{x}}}$, find functions f, g, and h such that $f \circ g \circ h = T$.

In Exercises 49–54 determine whether the function is one-to-one.

49. $f(x) = 3 + x^3$ **50.** $g(x) = 2 - 2x + x^2$

51. $h(x) = \dfrac{1}{x^4}$ **52.** $r(x) = 2 + \sqrt{x + 3}$

53. $p(x) = 3.3 + 1.6x - 2.5x^3$

54. $q(x) = 3.3 + 1.6x + 2.5x^3$

55. Show that $f(x) = 3x - 2$ is a one-to-one function and find its inverse function.

56. If $f(x) = 1 + \sqrt[5]{x - 2}$, find f^{-1}.

57. (a) Sketch the graph of the function $f(x) = x^2 - 4$, $x \geq 0$.
  (b) Use part (a) to sketch the graph of f^{-1}.
  (c) Find an equation for f^{-1}.

58. (a) Show that the function $f(x) = 1 + \sqrt[4]{x}$ is one-to-one.
  (b) Sketch the graph of f.
  (c) Use part (b) to sketch the graph of f^{-1}.
  (d) Find an equation for f^{-1}.

59. Suppose that M varies directly as z, and $M = 120$ when $z = 15$. Write an equation that expresses this variation.

60. Suppose that z is inversely proportional to y. If $y = 16$, then $z = 12$. Use this information to write an expression for z.

61. The intensity of illumination I from a light varies inversely as the square of the distance from the light.
  (a) Write this statement as an equation.
  (b) Determine the constant of proportionality if it is known that a lamp has an intensity of 1000 candles at a distance of 8 m.
  (c) What is the intensity of this lamp at a distance of 20 m?

62. The number of Christmas cards sold by a greeting card store depends on the time of year. Sketch a rough graph of the number of Christmas cards sold as a function of the time of year.

63. An isosceles triangle has a perimeter of 8 cm. Express the area A of the triangle as a function of the length b of the base of the triangle.

64. A rectangle is inscribed in an equilateral triangle with a perimeter of 30 cm as in the figure.
  (a) Express the area A of the rectangle as a function of the width of the rectangle.
  (b) Find the dimensions of the rectangle with the largest area.

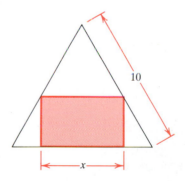

65. A piece of wire 10 m long is cut into two pieces. One piece, of length x, is bent into the shape of a square. The other piece is bent into the shape of an equilateral triangle.
  (a) Express the total area enclosed as a function of x.
  (b) For what value of x is this total area a minimum?

66. A baseball team plays in a stadium that holds 55,000 spectators. With ticket prices at \$10, the average attendance has been 27,000. A market survey indicates that for every dollar that ticket prices are lowered, attendance will increase by 3000. How should ticket prices be set to maximize revenue?

 67. A farmer wants to enclose a pen so that it has an area of 100 m². Find the dimensions of the pen that will require the minimum amount of fencing to enclose.

 68. A bird is released from point A on an island 5 mi from the nearest point B on a straight shoreline. It flies to a point C on the shore, and then flies along the shoreline to its nesting area D (see the figure). Suppose the bird requires 10 kcal/mi of energy to fly over land and 14 kcal/mi to fly over water (see Example 8 in Section

1.6). If the bird instinctively chooses a path that minimizes its energy expenditure, to what point does it fly?

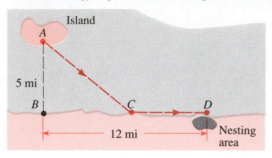

CHAPTER 2 TEST

1. State which of the following curves are graphs of functions. State which of the functions are one-to-one.

(a)

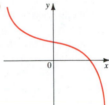

(b)

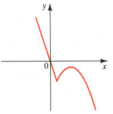

(c)

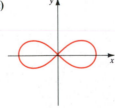

2. Find the domain of the function $f(x) = \dfrac{x^2 - 1}{\sqrt{x - 2}}$.

3. (a) Sketch the graph of the function $f(x) = x^3$.
   (b) Use part (a) to graph the function $g(x) = (x + 2)^3 - 3$.

4. (a) Sketch the graph of the function $f(x) = x^2 - 8x + 28$.
   (b) What is the minimum value of f?

5. Let $f(x) = \begin{cases} 1 - x^2 & \text{if } x \leq 0 \\ 2x + 1 & \text{if } x > 0 \end{cases}$
   (a) Evaluate $f(-2)$ and $f(1)$.

   (b) Sketch the graph of f.

6. How is the graph of $y = 2 - f(x - 3)$ obtained from the graph of f?

7. If $f(x) = x^2 + 2x - 1$ and $g(x) = 2x - 3$, find $f \circ g$ and $g \circ f$.

8. (a) If $f(x) = \sqrt{3 - x}$, find the inverse function f^{-1}.
   (b) Sketch the graphs of f and f^{-1} on the same coordinate axes.

9. (a) If 800 ft of fencing are available to build three adjacent pens as in the diagram, express the total area as a function of x.

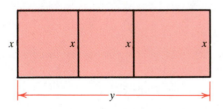

   (b) What value of x will maximize the total area?

 10. Let $f(x) = 3x^4 - 14x^2 + 5x - 3$.
   (a) Draw the graph of f in an appropriate viewing rectangle.
   (b) Is f one-to-one?
   (c) Find the local maximum and minimum values of f and the values of x at which they occur. State your answers to two decimal places.
   (d) State the approximate range of f.

FOCUS ON PROBLEM SOLVING

One of the most important principles of problem solving is the **recognition of patterns**. If we can discern that a numerical pattern or a geometrical pattern occurs in a problem, we may be able to guess what the continuing pattern is and thereby solve the problem.

This principle is often combined with the principle of **analogy**. If we are able to solve the simpler problems obtained by taking special cases or by using smaller numbers, we can often see that they lead to a pattern, which, in turn, gives us the answer to our problem.

EXAMPLE 1

Find the final digit in the number 3^{459}.

SOLUTION

Analogy

First notice that 3^{459} is a very large number—far too large for a calculator. Therefore we attack this problem by first looking at analogous problems. A similar, but simpler, problem would be to find the final digit in 3^9 or 3^{59}. In fact, let us start with the exponents 1, 2, 3, . . . and see what happens.

Number	Final digit
3^1	3
3^2	9
3^3	7
3^4	1
3^5	3
3^6	9
3^7	7
3^8	1

Pattern

By now you can see a pattern. The final digits occur in a cycle with length 4: 3, 9, 7, 1, 3, 9, 7, 1, 3, 9, 7, 1, ... Which number occurs in the 459th position? If we divide 459 by 4, the remainder is 3. So the final digit is the third number in the cycle, namely 7.

The final digit in the number 3^{459} is 7. ■

EXAMPLE 2

If $f_0(x) = 1/(1 - x)$ and $f_{n+1} = f_0 \circ f_n$ for $n = 0, 1, 2, . . .$, find $f_{1000}(1000)$.

173

Instead of computing $f_{1000}(x)$ directly, we start with the easier problem of computing $f_1(x)$, $f_2(x)$, and so on. Putting $n = 0$ in the equation

$$f_{n+1} = f_0(f_n(x)) \tag{1}$$

we get

$$f_1(x) = f_0(f_0(x)) = f_0\left(\frac{1}{1-x}\right)$$

$$= \frac{1}{1 - \dfrac{1}{1-x}} = \frac{1}{\dfrac{1-x-1}{1-x}} = \frac{-1}{\dfrac{-x}{1-x}}$$

$$= \frac{1-x}{-x} = 1 - \frac{1}{x}$$

Next we put $n = 1$ in Equation 1:

$$f_2(x) = f_0(f_1(x)) = f_0\left(1 - \frac{1}{x}\right)$$

$$= \frac{1}{1 - \left(1 - \dfrac{1}{x}\right)} = \frac{1}{\dfrac{1}{x}} = x$$

Now put $n = 2$ in Equation 1:

$$f_3(x) = f_0(f_2(x)) = f_0(x) = \frac{1}{1-x}$$

Thus we have returned to the function we started with, and we see a pattern:

$$f_0(x) = \frac{1}{1-x} \qquad f_1(x) = 1 - \frac{1}{x} \qquad f_2(x) = x$$

$$f_3(x) = \frac{1}{1-x} \qquad f_4(x) = 1 - \frac{1}{x} \qquad f_5(x) = x$$

The pattern is that successive functions repeat in a cycle of length 3.

Where does $f_{1000}(x)$ occur in the cycle? When we divide 1000 by 3, we get a remainder of 1, so

$$f_{1000}(x) = f_1(x) = 1 - \frac{1}{x}$$

Therefore

$$f_{1000}(1000) = 1 - \frac{1}{1000} = \frac{999}{1000}$$

■

The marginal notes read: "Analogy" and "Pattern".

1. Find the final digit in the number 947^{362}.

2. How many digits does the number $8^{15} \cdot 5^{37}$ have?

3. If $f_0(x) = x^2$ and $f_{n+1}(x) = f_0(f_n(x))$ for $n = 0, 1, 2, \ldots$, find a formula for $f_n(x)$.

4. If $f_0(x) = \dfrac{1}{2 - x}$ and $f_{n+1} = f_0 \circ f_n$ for $n = 0, 1, 2, \ldots$, find $f_{100}(3)$.

5. Use the techniques of solving a simpler problem and looking for a pattern to evaluate the number

$$3999999999999^2$$

6. Find the domain of the function

$$f(x) = \sqrt{1 - \sqrt{2 - \sqrt{3 - x}}}$$

7. Sketch the graph of the function $f(x) = \left| x^2 - 4|x| + 3 \right|$.

8. Sketch the graph of the function $g(x) = \left| x^2 - 1 \right| - \left| x^2 - 4 \right|$.

9. The **greatest integer function** is defined by $[\![x]\!] = $ the largest integer that is less than or equal to x. (For instance, $[\![4]\!] = 4$, $[\![4.8]\!] = 4$, $[\![\pi]\!] = 3$, $[\![-\frac{1}{2}]\!] = -1$.) Sketch the graph of the greatest integer function.

10. Sketch the graph of the function $f(x) = x - [\![x]\!]$, where $[\![x]\!]$ is the greatest integer function defined in Problem 9.

11. Sketch the region in the plane defined by the equation

$$[\![x]\!]^2 + [\![y]\!]^2 = 1$$

where $[\![x]\!]$ denotes the greatest integer function defined in Problem 9.

12. The positive integers are written in order starting with 1:

$$123456789101112131415161718192021 \ldots$$

What digit is in the 300,000th position?
[*Hint:* Using analogy, start with the simpler problem of finding the digit in the 80th position. Then find the digit in the 800th position. Armed with the clues from solving these easier problems, tackle the given problem.]

3 POLYNOMIALS AND RATIONAL FUNCTIONS

Each problem that I solved became a rule which served afterwards to solve other problems.

RENÉ DESCARTES

He thought he saw a Garden-Door
That opened with a key:
He looked again, and found it was
A Double Rule of Three:
"And all its mystery," he said,
"Is clear as day to me!"

LEWIS CARROLL

We have previously studied constant, linear, and quadratic functions, which have the equations $f(x) = c$, $f(x) = mx + b$, and $f(x) = ax^2 + bx + c$, respectively. All these functions are special cases of an important class of functions called polynomials. A **polynomial** P **of degree** n is a function of the form

$$P(x) = a_n x^n + a_{n-1} x^{n-1} + \cdots + a_1 x + a_0$$

where $a_n \neq 0$. Polynomials are constructed using the operations of addition, subtraction, and multiplication. If we introduce division as well, we obtain the set of rational functions. A **rational function** r is a function of the form

$$r(x) = \frac{P(x)}{Q(x)}$$

where P and Q are polynomials.

Virtually all the functions used in mathematics and the sciences are evaluated numerically by using polynomial approximations. In this chapter we study this important class of functions by first learning about the graphs of polynomial functions. We then learn how to find rational, irrational, and imaginary solutions of polynomial equations. Finally, we study rational functions, and learn how to solve polynomial and rational inequalities.

Working with polynomials and rational functions has numerical, algebraic, and graphical aspects, and we use all three approaches to give us a deeper understanding of these functions.

SECTION 3.1
POLYNOMIAL FUNCTIONS AND THEIR GRAPHS

Before we learn to work with polynomials, we must agree on some terminology. Let

$$P(x) = a_n x^n + a_{n-1} x^{n-1} + \cdots + a_1 x + a_0$$

be a polynomial of degree n. The numbers a_0, a_1, a_2, ... , a_n are called the **coefficients** of the polynomial. The number a_0 is called the **constant coefficient** and the number a_n, the coefficient of the highest power, is called the **leading coefficient**.

The graphs of polynomials of degree 0 or 1 are lines, and the graphs of polynomials of degree 2 are parabolas. Accurate graphing techniques for these polynomials were considered in Chapters 1 and 2. The greater the degree of a polynomial, the more complicated its graph, and to draw accurately the graph of a polynomial function of degree 3 or more requires the techniques of calculus or the use of a graphing calculator or computer (see Section 3.2). In this section we study the general features of polynomial graphs. We will see that the graphs of polynomials are smooth, continuous curves with no gaps or jumps. We begin by graphing the simplest polynomial functions, those of the form $y = x^n$.

EXAMPLE 1

Graph the functions:

(a) $y = x$ (b) $y = x^2$ (c) $y = x^3$ (d) $y = x^4$ (e) $y = x^5$

SOLUTION

We are already familiar from our previous work with some of these graphs but we include them for completeness.

x	x^2	x^3	x^4	x^5
0.1	0.01	0.001	0.0001	0.00001
0.2	0.04	0.008	0.0016	0.00032
0.5	0.25	0.125	0.0625	0.03125
0.7	0.49	0.343	0.2401	0.16807
1.0	1.0	1.0	1.0	1.0
1.2	1.44	1.728	2.0736	2.48832
1.5	2.25	3.375	5.0625	7.59375
2.0	4.0	8.0	16.0	32.0

If n is odd, $(-x)^n = -x^n$, so $y = x^n$ is an odd function, and if n is even, $(-x)^n = x^n$, so $y = x^n$ is an even function (see Section 2.2). We use these facts to find the values for negative x from the table. Plotting the points leads to the graphs in Figure 1.

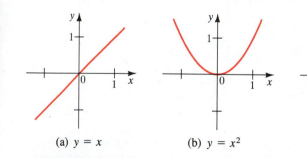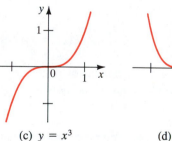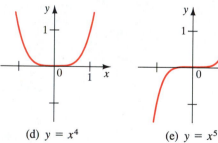

(a) $y = x$ (b) $y = x^2$ (c) $y = x^3$ (d) $y = x^4$ (e) $y = x^5$

Figure 1

As Example 1 suggests, when n is odd, the graph of $y = x^n$ has the same general shape as $y = x^3$, and when n is even, the graph of $y = x^n$ has more or less the same U-shape as $y = x^2$. However, note that as the degree n becomes larger, the graphs become flatter around the origin and steeper elsewhere.

EXAMPLE 2

Sketch the graphs of the functions:

(a) $y = -x^3$ (b) $y = (x - 2)^4$ (c) $y = -2x^5 + 4$

SOLUTION

We use the graphs in Example 1 and transform them using the techniques of Section 2.5.

(a) The function $y = -x^3$ is the negative of $y = x^3$, so we simply reflect the graph in Figure 1(c) about the x-axis to obtain Figure 2.

(b) The graph of the function $y = (x - 2)^4$ has the same shape as $y = x^4$. Replacing the x by $x - 2$ shifts the graph of $y = x^4$ in Figure 1(d) to the right by 2 units (see Figure 3).

(c) We begin with the graph of $y = x^5$ in Figure 1(e). Multiplying the function by 2 stretches the graph vertically. The negative sign reflects the graph about the x-axis, so at this stage we have the graph in Figure 4. Finally, adding 4 to the function shifts the graph up 4 units along the y-axis (see Figure 5). Since $-2x^5 + 4 = 0$ when $x^5 = 2$, the graph crosses the x-axis where $x = \sqrt[5]{2}$.

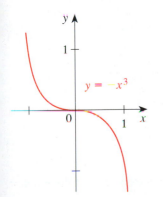

Figure 2

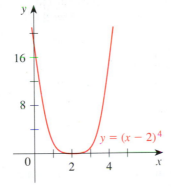

Figure 3

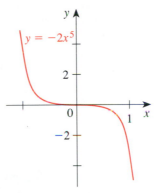

Figure 4

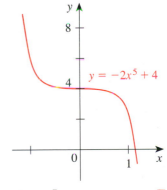

Figure 5

If $y = P(x)$ is a polynomial function and if c is a number such that $P(c) = 0$, then we say that c is a **zero** of P. In other words, the zeros of P are the solutions of the polynomial equation $P(x) = 0$. In the next example we use the zeros of the polynomial to help us sketch its graph. Note that if $P(c) = 0$, then the graph of $y = P(x)$ touches the x-axis where $x = c$; that is, c is an x-intercept of the graph. Between successive zeros, the values of the polynomial are either all positive or all negative. Thus the graph will lie entirely above or entirely below the x-axis between any two successive zeros.

EXAMPLE 3

Sketch the graph of the function $y = P(x) = x^3 - x^2 - 4x + 4$.

SOLUTION

We first find the zeros of P by factoring.

$$
\begin{aligned}
P(x) &= x^3 - x^2 - 4x + 4 \\
&= x^2(x - 1) - 4(x - 1) \\
&= (x^2 - 4)(x - 1) \\
&= (x + 2)(x - 2)(x - 1)
\end{aligned}
$$

Thus $P(x) = 0$ when $x + 2 = 0$, $x - 2 = 0$, and when $x - 1 = 0$. It follows that the zeros of P are -2, 2, and 1, so these are the x-intercepts of its graph. Since $P(0) = 4$, the y-intercept is 4.

To sketch the graph, we must calculate $P(x)$ for other values of x as well. We choose values between (and to the right and left of) successive zeros to determine if $P(x)$ is positive or negative on the intervals determined by the zeros. Note that if $P(x)$ is positive at any x between successive zeros, then the graph of $y = P(x)$ lies above the x-axis on the interval between those zeros, and if $P(x)$ is negative on such an interval, then the graph lies below the x-axis on that interval.

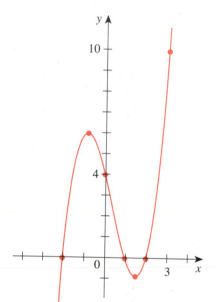

Figure 6

$y = x^3 - x^2 - 4x + 4$

x	-3	-2	-1	0	1	$\frac{3}{2}$	2	3
$P(x)$	-20	0	6	4	0	$-\frac{7}{8}$	0	10

Plotting the points given in the table and completing the sketch, we get the graph in Figure 6. ■

The graph in Example 3 has two **turning points** (or **local extrema**), which are points at which a particle moving along the graph would change from moving upward to moving downward, or vice versa. The turning points are thus "peaks" and "valleys" on the graph. Without using calculus we cannot find the exact locations of such points. We cannot even prove that we have in fact found all the turning points that a graph may have, but it can be shown (again, using calculus) that a polynomial function of degree n can have at most $n - 1$ turning points (although it may have fewer than $n - 1$). This means that in this case we have indeed found all of them, and by plotting sufficiently many points one can locate them approximately, as we did in Example 3.

EXAMPLE 4

Sketch the graph of the function $P(x) = -2x^4 - x^3 + 3x^2$.

SOLUTION

We factor to obtain

$$P(x) = -x^2(2x^2 + x - 3)$$
$$= -x^2(2x + 3)(x - 1)$$

so the zeros of P are 0, $-\frac{3}{2}$, and 1. The y-intercept is $P(0) = 0$. In the following table we give the values of P at a number of other points. (Such values are most easily obtained using a programmable calculator.) Remember that we must choose points between successive zeros to determine whether the polynomial is positive or negative between these zeros.

x	-2	-1.5	-1	-0.5	0	0.5	1	1.5
$P(x)$	-12	0	2	0.75	0	0.5	0	-6.75

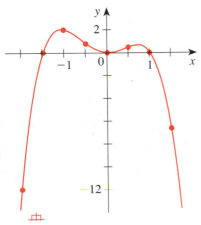

Figure 7
$y = -2x^4 - x^3 + 3x^2$

Plotting these points and completing the graph, we obtain the curve shown in Figure 7. The graph cannot have any more turning points than the three we have found, since the degree of P is 4.

Note that although $x = 0$ is a zero of the polynomial, the graph does not cross the x-axis at the x-intercept 0. It just touches the x-axis there and then goes back up again. We say that the graph is *tangent* to the x-axis at $x = 0$.   ■

As a further aid in graphing polynomials, note that for large $|x|$, the values of

$$a_n x^n + a_{n-1} x^{n-1} + \cdots + a_1 x + a_0$$

are close to $a_n x^n$ because

$$a_n x^n + a_{n-1} x^{n-1} + \cdots + a_1 x + a_0 = a_n x^n \left[1 + \frac{a_{n-1}}{a_n x} + \frac{a_{n-2}}{a_n x^2} + \cdots + \frac{a_0}{a_n x^n} \right]$$

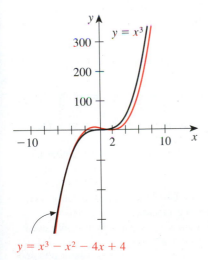

$y = x^3 - x^2 - 4x + 4$

Figure 8

and if $|x|$ is large, the quantity inside the brackets is close to 1 in value (since the reciprocals of large numbers are close to 0). For example, for large $|x|$ the graph of $y = P(x) = x^3 - x^2 - 4x + 4$ sketched in Figure 6 looks very much like the graph of $y = x^3$ (see Figure 8). Of course, when $|x|$ is small (say, $-5 < x < 5$), the two graphs are quite different, since $y = P(x)$ crosses the x-axis three times, at $x = -2$, 1, and 2, whereas $y = x^3$ crosses the x-axis only at the origin.

EXERCISES 3.1

In Exercises 1–8 sketch the graph of the function by transforming the graph of the appropriate function of the form $y = x^n$. Indicate all x- and y-intercepts on your graph.

1. $y = -3x^4$

2. $y = (x + 3)^3$

3. $y = x^3 + 5$

4. $y = -x^6 + 8$

5. $y = -(x - 1)^4 + 1$

6. $y = 3x^5 - 9$

7. $y = 4(x - 2)^5 - 4$

8. $y = 3x^4 - 27$

In Exercises 9–26 sketch the graph of the function by first plotting all x-intercepts, the y-intercept, and sufficiently many other points to detect the shape of the curve, and then filling in the rest of the graph.

9. $y = (x - 2)(x + 3)(x - 4)$

10. $y = (2x - 3)(x - 5)(x + 1)$

11. $y = x(x - 2)(x + 1)$ **12.** $y = \frac{1}{5}x(x - 5)^2$

13. $y = (x - 1)^2(x - 3)$

14. $y = \frac{1}{4}(x + 1)^3(x - 3)$

15. $y = \frac{1}{12}(x + 2)^2(x - 3)^2$

16. $y = x^3 + x^2 - x - 1$

17. $y = x^3 + 3x^2 - 4x - 12$

18. $y = 2x^3 - x^2 - 18x + 9$

19. $y = x^3 - x^2 - 6x$ **20.** $y = (x^2 - 2x - 3)^2$

21. $y = \frac{1}{8}(2x^4 + 3x^3 - 16x - 24)$

22. $y = x^4 - 3x^2 - 4$

23. $y = x^4 - 2x^3 - 8x + 16$

24. $y = x^4 - 2x^3 + 8x - 16$

25. $y = x^5 - 9x^3$

26. $y = x^6 - 2x^3 + 1$

27. (a) On the same coordinate axes, draw graphs (as accurately as possible) of the functions
$y = x^3 - 2x^2 - x + 2$ and $y = -x^2 + 5x + 2$.
(b) Based on your drawing, at how many points do the two graphs appear to intersect?
(c) Find the coordinates of all the intersection points.

28. Portions of the graphs of $y = x^2$, $y = x^3$, $y = x^4$, $y = x^5$, and $y = x^6$ are plotted in the accompanying figures. Determine which function belongs to which graph.

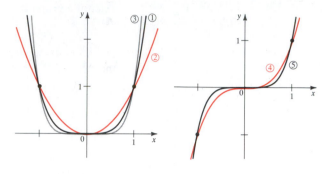

29. A cardboard box has a square base, with each of the four edges of the base having length x inches, as shown in the figure. The total length of all 12 edges of the box is 144 inches.

(a) Express the volume V of the box as a function of x.
(b) Sketch the graph of the function V.
(c) Since both x and V represent positive quantities (length and volume, respectively), what is the domain of V?

30. Recall that a function f is **odd** if $f(-x) = -f(x)$ and is **even** if $f(-x) = f(x)$, for all real x.
(a) Show that if f and g are both odd, then so is the function $f + g$.
(b) Show that if f and g are both even, then so is the function $f + g$.
(c) Show that if f is odd and g is even, and neither has constant value 0, then the function $f + g$ is neither even nor odd.
(d) Show that a polynomial $P(x)$ that contains only odd powers of x is an odd function.
(e) Show that a polynomial $P(x)$ that contains only even powers of x is an even function.
(f) Show that if a polynomial $P(x)$ contains both odd and even powers of x, then it is neither an odd nor an even function.
(g) Express the function

$$P(x) = x^5 + 6x^3 - x^2 - 2x + 5$$

as the sum of an odd function and an even function.

SECTION 3.2
USING GRAPHING DEVICES
TO GRAPH POLYNOMIALS

In the preceding section we studied some basic properties of the graphs of polynomial functions. In this section we use graphing devices to explore more features of their graphs.

Recall from Section 2.7 that if the point $(a, f(a))$ is the highest point on the graph of f within some viewing rectangle, then $f(a)$ is a local maximum value of f, and if $(b, f(b))$ is the lowest point on the graph of f within a viewing rectangle, then $f(b)$ is a local minimum value. (See Figure 1.) We will also say that such a point $(a, f(a))$ is a **local maximum point** on the graph and that $(b, f(b))$ is a **local minimum point**. The set of all local maximum and minimum points on the graph of a function are called its **local extrema**.

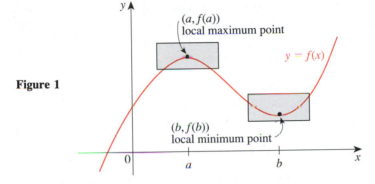

Figure 1

EXAMPLE 1

Graph the polynomial function

$$P(x) = x^3 - 3x^2 - x + 15$$

Determine the x- and y-intercepts and the coordinates of the local extrema of this graph.

SOLUTION

By setting $x = 0$, we see that the y-intercept of the graph is 15, so we must choose a viewing rectangle that extends at least this far upward. The graph of P in the viewing rectangle $[-5, 5]$ by $[-10, 20]$ is shown in Figure 2. The graph has one x-intercept, $x = -2$, and two extrema, a local maximum and a local minimum. By zooming in on each of these and tracing along the graph with the cursor (as described in Section 2.7), we find that the local maximum point is $(-0.16, 15.08)$ and the local minimum point is $(2.15, 8.92)$, rounded to two decimal places.

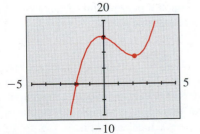

Figure 2
$P(x) = x^3 - 3x^2 - x + 15$

EXAMPLE 2

Using the viewing rectangle $[-5, 5]$ by $[-100, 100]$, graph each of the following polynomials, and determine how many local extrema each function has.
(a) $P_1(x) = x^4 + x^3 - 16x^2 - 4x + 48$
(b) $P_2(x) = x^5 + 3x^4 - 5x^3 - 15x^2 + 4x - 15$

SOLUTION

The graphs are shown in Figure 3. We see that P_1 has two local minimum points and one local maximum point, for a total of three local extrema. The polynomial P_2 has two local minimum and two local maximum points—a total of four local extrema.

Figure 3

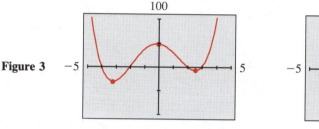 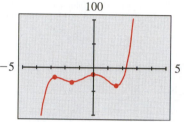

(a) $P_1(x) = x^4 + x^3 - 16x^2 - 4x + 48$ (b) $P_2(x) = x^5 + 3x^4 - 5x^3 - 15x^2 + 4x - 15$

Notice that the polynomial P in Example 1 is of third degree and has 2 local extrema. In Example 2, P_1 is a fourth-degree polynomial and has 3 local extrema, while P_2 is of fifth degree and has 4 local extrema. The number of extrema in each case is one less than the degree. This is not a coincidence, as the principle in the following box indicates. (A proof of this principle requires the methods of calculus.)

LOCAL EXTREMA OF POLYNOMIALS

> If $P(x) = a_nx^n + a_{n-1}x^{n-1} + \cdots + a_1x + a_0$ is a polynomial of degree n, then the graph of P has at most $n - 1$ local extrema.

A polynomial of degree n may in fact have less than $n - 1$ local extrema, as we see in the following example. All that the preceding box tells us is that a polynomial of degree n can have *no more* than $n - 1$ local extrema.

EXAMPLE 3

Graph the following polynomials using a graphing device, and determine the approximate coordinates of all local extrema.

(a) $F(x) = -2x^3 + 3x^2 - 5x + 2$

(b) $G(x) = 7x^4 + 3x^2 - 10x$

SOLUTION

(a) Using the viewing rectangle $[-5, 5]$ by $[-100, 100]$ gives the graph of F shown in Figure 4(a). As the graph is plotted, we see that the values of $F(x)$ appear to decline continuously, so there is no local maximum or minimum. If we select other viewing rectangles, we observe the same behavior, so the function has no local extrema.

(b) In the viewing rectangle $[-5, 5]$ by $[-100, 100]$, the graph of G has the shape shown in Figure 4(b). It appears to have just a single local minimum in the fourth quadrant. To confirm this, we zoom in on the portion of the graph near this minimum point. In the viewing rectangle $[0, 1]$ by $[-5, 0]$ we obtain the graph in Figure 4(c). Using the cursor, we locate the local minimum point at $(0.61, -4.01)$, correct to two decimals.

Figure 4

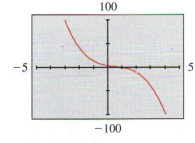

(a) $F(x) = -2x^3 + 3x^2 - 5x + 2$

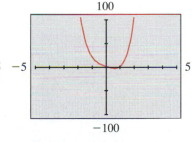

(b) $G(x) = 7x^4 + 3x^2 - 10x$

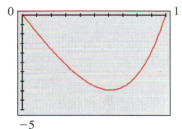

(c) $G(x) = 7x^4 + 3x^2 - 10x$

END BEHAVIOR

All the polynomial functions that we have graphed so far have had the property that, if the viewing rectangle is wide enough, the graph of the function eventually disappears off the top or bottom of the screen. This means that $|y|$ can be made as large as we wish on the graph of the polynomial $y = P(x)$ if we choose $|x|$ sufficiently large. By the **end behavior** of a function, we mean a description of what happens to the values of a function as $|x|$ becomes large. In the next example we show how end behavior is described for polynomials.

EXAMPLE 4

Graph the polynomials

$$P(x) = 3x^5 - 5x^3 + 2x \qquad \text{and} \qquad Q(x) = 3x^5$$

on the same screen, first using the viewing rectangle $[-2, 2]$ by $[-2, 2]$ and then changing the rectangle to $[-10, 10]$ by $[-10{,}000, 10{,}000]$. Describe and compare the end behavior of the two functions.

SOLUTION

On the smaller viewing rectangle the two functions look quite different [see Figure 5(a)]. The polynomial P has four local extrema, while Q has none. However, on the large viewing rectangle, the two functions look almost the same [Figure 5(b)]. This means that P and Q have the same end behavior. As x becomes large in the positive direction, the y-coordinates of points on the graphs of both P and Q become large in the positive direction as well, and as x becomes large in the negative direction, so does y. In symbols, we write this as follows:

$$y \to \infty \quad \text{as} \quad x \to \infty \qquad \text{and} \qquad y \to -\infty \quad \text{as} \quad x \to -\infty$$

Figure 5
$P(x) = 3x^5 - 5x^3 + 2x$
$Q(x) = 3x^5$

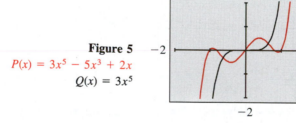

(a)

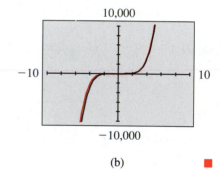

(b)

In the preceding example we saw that the end behavior of a polynomial P is determined by the term containing the highest power of the variable (in this case, the term $3x^5$). This is because when $|x|$ is large, most of the value of $P(x)$ is determined by the highest power term. For example, substituting $x = 10$ into the polynomials of Example 4 gives

$$P(10) = 3(10)^5 - 5(10)^3 + 2(10) = 300{,}000 - 5000 + 20 = 295{,}020$$
$$Q(10) = 3(10)^5 = 300{,}000$$

The two values are, relatively speaking, very close to each other—they are less than $\frac{1}{2}\%$ apart.

EXAMPLE 5

Describe the end behavior of the polynomial

$$P(x) = -x^4 + 3x^3 - 6x + 100$$

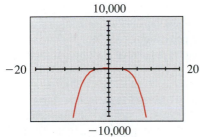

Figure 6
$P(x) = -x^4 + 3x^3 - 6x + 100$

SOLUTION

We graph P on a large viewing rectangle, since we are only interested in end behavior. From Figure 6, we see that

$$y \to -\infty \quad \text{as} \quad x \to \infty \quad \text{and} \quad y \to -\infty \quad \text{as} \quad x \to -\infty$$

As Examples 4 and 5 suggest, if $y = P(x)$ is a polynomial function of *odd* degree, the values of y go in opposite directions as $x \to \infty$ and as $x \to -\infty$, but if P is of *even* degree, then the values of y go in the same direction as $x \to \infty$ and as $x \to -\infty$. (See Figure 7.)

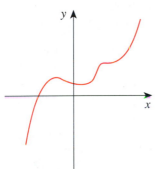

degree of P odd,
leading coefficient positive

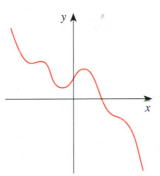

degree of P odd,
leading coefficient negative

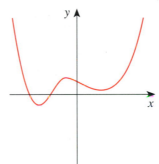

degree of P even,
leading coefficient positive

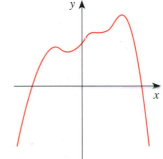

degree of P even,
leading coefficient negative

Figure 7

■

EXERCISES 3.2

In Exercises 1–6 graph the polynomials in the given viewing rectangle. Find the x-intercept(s), if any, the y-intercept, and the coordinates of all local extrema. State your answers correct to two decimals.

1. $y = x^3 - 12x + 9$, $[-5, 5]$ by $[-30, 30]$

2. $y = 2x^3 - 3x^2 - 12x - 32$, $[-5, 5]$ by $[-60, 30]$

3. $y = x^4 + 4x^3$, $[-5, 5]$ by $[-30, 30]$

4. $y = x^4 - 18x^2 + 32$, $[-5, 5]$ by $[-100, 100]$

5. $y = 3x^5 - 5x^3 + 3$, $[-3, 3]$ by $[-5, 10]$

6. $y = x^5 - 5x^2 + 6$, $[-3, 3]$ by $[-5, 10]$

In Exercises 7–16 find all local extrema of the polynomial correct to two decimals.

7. $y = x^3 - x^2 - x$ **8.** $y = 6x^3 + 3x + 1$

9. $y = x^4 - 5x^2 + 4$

10. $y = 1.2x^5 + 3.75x^4 - 7x^3 - 15x^2 + 18x$

11. $y = (x - 2)^5 + 32$ **12.** $y = (x^2 - 2)^3$

13. $y = (x - 4)(2x + 1)^2$ **14.** $y = (x + 2)^2(x - 3)^2$

15. $y = x^8 - 3x^4 + x$

16. $y = \frac{1}{3}x^7 - 17x^6 + x^3 + 7$

In Exercises 17–24 graph the polynomial and describe its end behavior.

17. $y = 3x^3 - x^2 + 5x + 1$

18. $y = -\frac{1}{8}x^3 + \frac{1}{4}x^2 + 12x$

19. $y = x^4 - 7x^2 + 5x + 5$

20. $y = (1 - x)^5$

21. $y = x^{11} - 9x^9$

22. $y = 2x^2 - x^{12}$

23. $y = 200x^2 - 0.001x^5$

24. $y = x^{20}$

25. A market analyst working for a small-appliance manufacturer finds that if the firm produces x blenders annually, the total profit in dollars is

$$P(x) = 8x + 0.3x^2 - 0.0013x^3 - 372$$

Graph the function P in an appropriate viewing rectangle and use the graph to answer the following questions.

(a) When just a few blenders are manufactured, the firm suffers a loss, since profit is negative. [For example, $P(10) = -263.3$, so the firm loses $263.30 if it produces only 10 blenders.] How many blenders must be produced so that the firm breaks even?

(b) Does profit increase indefinitely as more blenders are produced? If not, what is the largest possible profit the firm can have?

26. The rabbit population on a small island is observed to be given by the function

$$P(t) = 120t - 0.4t^4 + 1000$$

where t is the time (in months) since observations of the island began.

(a) When is the maximum population attained, and what is that maximum population?

(b) When does the rabbit population disappear from the island?

27. (a) Graph the function $P(x) = (x - 1)(x - 3)(x - 4)$ and find all local extrema, correct to the nearest tenth.

(b) Graph the function

$$Q(x) = (x - 1)(x - 3)(x - 4) + 5$$

and use your answers to part (a) to find all local extrema, correct to the nearest tenth.

(c) If $a < b < c$, explain why the function $P(x) = (x - a)(x - b)(x - c)$ must have two local extrema.

(d) If $a < b < c$ and d is any real number, explain why the function $Q(x) = (x - a)(x - b)(x - c) + d$ must have two local extrema.

28. Give an example of a polynomial that has six local extrema.

29. (a) How many x-intercepts and how many local extrema does the polynomial $P(x) = x^3 - 4x$ have?

(b) How many x-intercepts and how many local extrema does the polynomial $Q(x) = x^3 + 4x$ have?

(c) If $a > 0$, how many x-intercepts and how many local extrema do the polynomials $P(x) = x^3 - ax$ and $Q(x) = x^3 + ax$ have? Explain your answer.

30. Is it possible for a third-degree polynomial to have exactly one local extremum? Explain your answer. [*Hint:* Think about the end behavior of such a polynomial.]

SECTION 3.3
DIVIDING POLYNOMIALS

So far in this chapter we have been studying polynomial functions *graphically*. In this section we begin to study polynomials *algebraically*. Most of our work will be concerned with finding the solutions of polynomial equations, but first we consider division of polynomials.

Long division for polynomials is very much like the familiar process of long division for numbers. For example, to divide $6x^2 - 26x + 12$ (the **dividend**) by $x - 4$ (the **divisor**), we arrange our work as follows.

$$
\begin{array}{r}
6x - 2 \quad \text{quotient} \\
x - 4 \overline{)6x^2 - 26x + 12} \quad \text{dividend} \\
\underline{6x^2 - 24x} \\
-2x + 12 \\
\underline{-2x + 8} \\
4 \quad \text{remainder}
\end{array}
$$

The **quotient** at the top is obtained by first dividing the initial term of the dividend by the initial term of the divisor $[6x^2/x = 6x]$, which gives the first term of the quotient. This is then multiplied by the divisor $[6x(x - 4) = 6x^2 - 24x]$ and the result is written below the corresponding terms of the dividend and subtracted from those terms. After the next term of the dividend [the $+12$] is brought down, the whole process is repeated with the new line $[-2x + 12]$ now being treated as the dividend. The process stops if, after the subtraction has been performed, the result has degree *smaller* than the degree of the divisor and there is nothing in the original dividend left to bring down. The last line of the process will contain the **remainder**, and the result of the division can be interpreted in either of the following two ways:

$$\frac{6x^2 - 26x + 12}{x - 4} = 6x - 2 + \frac{4}{x - 4}$$

or

$$6x^2 - 26x + 12 = (x - 4)(6x - 2) + 4$$

We summarize what happens in this or any such long division problem with the following theorem.

THE DIVISION ALGORITHM

If $P(x)$ and $D(x)$ are polynomials, with $D(x) \neq 0$, then there exist unique polynomials $Q(x)$ and $R(x)$ such that

$$P(x) = D(x) \cdot Q(x) + R(x)$$

where $R(x)$ is either 0 or of degree less than the degree of $D(x)$. $P(x)$ and $D(x)$ are called the **dividend** and **divisor**, respectively, $Q(x)$ is the **quotient**, and $R(x)$ is the **remainder**.

EXAMPLE 1

Let $P(x) = 8x^4 + 6x^2 - 3x + 1$ and $D(x) = 2x^2 - x + 2$. Find polynomials $Q(x)$ and $R(x)$ such that $P(x) = D(x) \cdot Q(x) + R(x)$.

SOLUTION

We use long division after first inserting the term $0x^3$ into the dividend to ensure that the columns will line up correctly in the long division process.

$$
\begin{array}{r}
4x^2 + 2x \\
2x^2 - x + 2 \overline{\smash{)}8x^4 + 0x^3 + 6x^2 - 3x + 1} \\
\underline{8x^4 - 4x^3 + 8x^2} \\
4x^3 - 2x^2 - 3x \\
\underline{4x^3 - 2x^2 + 4x} \\
-7x + 1
\end{array}
$$

We stop the process at this point because $-7x + 1$ is of lesser degree than the divisor $2x^2 - x + 2$. From the long division table, we see that $Q(x) = 4x^2 + 2x$ and $R(x) = -7x + 1$, so

$$8x^4 + 6x^2 - 3x + 1 = (2x^2 - x + 2)(4x^2 + 2x) + (-7x + 1) \qquad ■$$

■ THE REMAINDER THEOREM

If the divisor in the Division Algorithm is of the form $x - c$ for some real number c, then the remainder must be a constant (since the degree of the remainder is less than the degree of the divisor). If we call this constant r, then

$$P(x) = (x - c) \cdot Q(x) + r$$

Setting $x = c$ in this equation, we get $P(c) = (c - c) \cdot Q(x) + r = 0 + r = r$. This proves the next theorem.

REMAINDER THEOREM

> If the polynomial $P(x)$ is divided by $x - c$, then the remainder is the value $P(c)$.

$$\begin{array}{r} x^2 - 2x\ -\ 1 \\ x - 2\overline{)x^3 - 4x^2 + 3x + 5} \\ \underline{x^3 - 2x^2} \\ -2x^2 + 3x \\ \underline{-2x^2 + 4x} \\ -x + 5 \\ \underline{-x + 2} \\ 3 \end{array}$$

EXAMPLE 2

Let $P(x) = x^3 - 4x^2 + 3x + 5$. If we divide $P(x)$ by $x - 2$ using long division, we obtain a quotient of $x^2 - 2x - 1$ and a remainder of 3. Thus by the Remainder Theorem, the value of $P(2)$ should be 3. To verify this, we calculate

$$P(2) = 2^3 - 4 \cdot 2^2 + 3 \cdot 2 + 5 = 8 - 16 + 6 + 5 = 3 \qquad \blacksquare$$

■ FACTORS AND ZEROS OF POLYNOMIALS

If $R(x) = 0$ in the Division Algorithm, then $P(x) = D(x) \cdot Q(x)$ and we say that $D(x)$ and $Q(x)$ are **factors** of $P(x)$. If $D(x) = x - c$ is a factor of $P(x)$, then obviously $P(c) = (c - c) \cdot Q(c) = 0$. On the other hand, if $P(c) = 0$, then by the Remainder Theorem

$$P(x) = (x - c) \cdot Q(x) + 0 = (x - c) \cdot Q(x)$$

so that $x - c$ is a factor of $P(x)$. Thus we have proved the following theorem.

FACTOR THEOREM

> $P(c) = 0$ if and only if $x - c$ is a factor of $P(x)$.

EXAMPLE 3

Let $P(x) = x^3 - 7x + 6$. Show that $P(1) = 0$, and use this fact to factor $P(x)$ completely.

SOLUTION

$$\begin{array}{r} x^2 + \ x\ -\ 6 \\ x - 1\overline{)x^3 + 0x^2 - 7x + 6} \\ \underline{x^3 - \ x^2} \\ x^2 - 7x \\ \underline{x^2 - \ x} \\ -6x + 6 \\ \underline{-6x + 6} \\ 0 \end{array}$$

Substituting, we see that $P(1) = 1^3 - 7 \cdot 1 + 6 = 0$. By the Factor Theorem, this means that $x - 1$ is a factor of $P(x)$. Using long division, we see that

$$P(x) = (x - 1)(x^2 + x - 6)$$

$$= (x - 1)(x - 2)(x + 3) \qquad \text{(factoring the quadratic } x^2 + x - 6\text{)} \qquad \blacksquare$$

Recall that if $P(c) = 0$, then the number c is a **zero** of the polynomial P. We also express this by saying that c is a **root** of the polynomial equation $P(x) = 0$. Thus the zeros of the polynomial of Example 3 are 1, 2, and -3. With this terminology, the Factor Theorem tells us that $x - c$ is a factor of $P(x)$ if and only if c is a zero of $P(x)$.

EXAMPLE 4

Find a polynomial $F(x)$ of degree 4 that has zeros -3, 0, 1, and 5.

SOLUTION

By the Factor Theorem, $x - (-3)$, $x - 0$, $x - 1$, and $x - 5$ must all be factors of the desired polynomial, so let

$$F(x) = (x + 3)(x - 0)(x - 1)(x - 5) = x^4 - 3x^3 - 13x^2 + 15x$$

Since $F(x)$ is to have degree 4, any other solution of the problem must be a constant multiple of the polynomial we have chosen (because multiplication by any polynomial other than a constant will increase the degree).   ■

In Section 1.3 we developed some techniques and formulas that allowed us to factor certain special kinds of polynomials. We have seen in Example 3 that the Factor Theorem can help us factor polynomials for which those techniques and formulas do not work. We will exploit this use of the Factor Theorem more fully in the next section.

■ SYNTHETIC DIVISION

The Remainder Theorem does not seem at first glance to provide a particularly useful way to evaluate polynomials. After all, to find the remainder when $x - c$ is divided into a polynomial requires division, and the long division process appears much more complicated than simply substituting c in place of x and evaluating the polynomial. However, it turns out that if the divisor is of the form $x - c$, the long division process can be substantially simplified. For example, let us divide the polynomial $2x^3 - 7x^2 + 5$ by $x - 3$:

$$
\begin{array}{r}
2x^2 - x - 3 \\
x - 3 \overline{)2x^3 - 7x^2 + 0x + 5} \\
\underline{2x^3 - 6x^2} \\
-x^2 + 0x \\
\underline{-x^2 + 3x} \\
-3x + 5 \\
\underline{-3x + 9} \\
-4
\end{array}
$$

Notice that the powers of x act simply as place-holders in this format, and we can certainly omit writing them down. Moreover, all the coefficients in the lines underneath the dividend except those in boxes are repeated either in the dividend or in the quotient. (For example, the 0 and the 5 were brought down from the dividend, and the 2, -1, and -3 are repeated in the quotient line, in the same column.) This means we can omit these as well. If we now write the quotient line, together with the remainder, at the bottom instead of at the top, the long division format simplifies to

$$
\begin{array}{r|rrrr}
-3 & 2 & -7 & 0 & 5 \\
& & -6 & 3 & 9 \\
\hline
& 2 & -1 & -3 & -4
\end{array}
$$

Compare this carefully with the long division. The top line contains the coefficients of the divisor and dividend, while the bottom line contains the coefficients of the quotient and remainder. The middle line contains the remaining numbers in boxes in the long division format. Note that each entry in the bottom line is the difference of the two numbers immediately above it.

We can simplify the format even further by changing the sign of the entry that represents the divisor and of each entry in the second row. This makes the bottom row the sum rather than the difference of the two rows above it, and addition is easier to keep track of than subtraction. The division process performed above can thus be done as follows.

Step 1 Begin with a table that contains the coefficients of the divisor and dividend. (*Note:* Remember to supply zeros for missing coefficients in the dividend and to change the sign of the constant in the divisor.) Bring down the first coefficient of the dividend.

$$
\begin{array}{r|rrrr}
3 & 2 & -7 & 0 & 5 \\
& & & & \\
\hline
& 2 & & &
\end{array}
$$

Step 2 Multiply the 2 in the bottom row by the 3 in the upper left corner, and write the result in the second space of the second row. Then add this result to the number above it and write the sum in the bottom row.

$$
\begin{array}{r|rrrr}
3 & 2 & -7 & 0 & 5 \\
& & 6 & & \\
\hline
& 2 & -1 & &
\end{array}
$$

Step 3 Multiply the new element in the bottom row (the -1) by the 3 in the upper left corner, write the result in the next space of the second row, add it to the number above it, and write the sum in the bottom row.

$$
\begin{array}{r|rrrr}
3 & 2 & -7 & 0 & 5 \\
& & 6 & -3 & \\
\hline
& 2 & -1 & -3 &
\end{array}
$$

Step 4   Continue the process of multiplying the last element in the bottom row by the one in the upper left corner, putting the result in the second row, and then adding, until the table is complete.

$$
\begin{array}{r|rrrr}
3 & 2 & -7 & 0 & 5 \\
  &   & 6 & -3 & -9 \\
\hline
  & 2 & -1 & -3 & -4
\end{array}
$$

The last entry in the bottom row is the remainder and the rest of the entries are the coefficients of the quotient, so

$$2x^3 - 7x^2 + 5 = (x - 3)(2x^2 - x - 3) - 4$$

The remainder is also the value of the dividend when $x = 3$, by the Remainder Theorem, so if $P(x) = 2x^3 - 7x^2 + 5$, then $P(3) = -4$.

This process, though somewhat complicated to derive and describe, is really very simple to carry out in practice. The procedure is summarized in the following box.

SYNTHETIC DIVISION OF
$a_n x^n + a_{n-1} x^{n-1} + \cdots + a_1 x + a_0$
BY $x - c$

$$
\begin{array}{r|rrrrccrrr}
c & a_n & a_{n-1} & a_{n-2} & a_{n-3} & \cdots & a_2 & a_1 & a_0 \\
  &   & cb_{n-1} & cb_{n-2} & cb_{n-3} & \cdots & cb_2 & cb_1 & cb_0 \\
\hline
  & b_{n-1} & b_{n-2} & b_{n-3} & b_{n-4} & \cdots & b_1 & b_0 & r
\end{array}
$$

Each number in the bottom row is obtained by adding the numbers above it. In particular, $b_{n-1} = a_n$. The quotient is

$$b_{n-1} x^{n-1} + b_{n-2} x^{n-2} + \cdots + b_1 x + b_0$$

and the remainder is r.

EXAMPLE 5

Find the quotient and remainder when $3x^5 + 5x^4 - 4x^3 + 7x + 3$ is divided by $x + 2$.

SOLUTION

Since $x + 2 = x - (-2)$, the synthetic division table for this problem takes the following form.

$$
\begin{array}{r|rrrrrr}
-2 & 3 & 5 & -4 & 0 & 7 & 3 \\
   &   & -6 & 2 & 4 & -8 & 2 \\
\hline
   & 3 & -1 & -2 & 4 & -1 & 5
\end{array}
$$

The quotient is $3x^4 - x^3 - 2x^2 + 4x - 1$, and the remainder is 5. Thus

$$3x^5 + 5x^4 - 4x^3 + 7x + 3 = (x + 2)(3x^4 - x^3 - 2x^2 + 4x - 1) + 5$$

or $$\frac{3x^5 + 5x^4 - 4x^3 + 7x + 3}{x + 2} = 3x^4 - x^3 - 2x^2 + 4x - 1 + \frac{5}{x + 2}$$ ■

EXAMPLE 6

Let $P(x) = x^5 - 5x^4 + 20x^2 - 10x + 10$. Find $P(4)$.

SOLUTION

Since $P(4)$ is the remainder when $P(x)$ is divided by $x - 4$ (by the Remainder Theorem), we can use synthetic division.

$$
\begin{array}{r|rrrrr}
4 & 1 & -5 & 0 & 20 & -10 & 10 \\
  &   & 4 & -4 & -16 & 16 & 24 \\
\hline
  & 1 & -1 & -4 & 4 & 6 & 34
\end{array}
$$

The remainder is 34, so $P(4) = 34$.   ■

In Example 6 we could also have evaluated $P(4)$ by direct substitution:

$$P(4) = (4)^5 - 5(4)^4 + 20(4)^2 - 10(4) + 10$$
$$= 1024 - 1280 + 320 - 40 + 10 = 34$$

When evaluating polynomials at integers without the aid of a calculator, however, synthetic division usually involves somewhat less work.

EXERCISES 3.3

In Exercises 1–25 find the quotient and remainder.

1. $\dfrac{4x^2 - 3x + 2}{x - 2}$

2. $\dfrac{x^2 - 12x - 22}{x + 2}$

3. $\dfrac{x^3 - 3x^2 + 5x - 2}{x - 3}$

4. $\dfrac{2x^3 + 6x^2 - 40x + 1}{x - 7}$

5. $\dfrac{x^5 + x^3 + x}{x + 1}$

6. $\dfrac{x^6 - x^4 + x^2 - 1}{x - 1}$

7. $\dfrac{x^6 + 6x^5 + 15x^4 + 20x^3 + 15x^2 + 6x + 1}{x + 1}$

8. $\dfrac{x^3 - 9x^2 + 27x - 27}{x - 3}$

9. $\dfrac{x^3 + 6x + 3}{x^2 - 2x + 2}$

10. $\dfrac{3x^4 - 5x^3 - 20x - 5}{x^2 + x + 3}$

11. $\dfrac{6x^3 + 2x^2 + 22x}{2x^2 + 5}$

12. $\dfrac{9x^2 - x + 5}{3x^2 - 7x}$

13. $\dfrac{x^5 + x^4 + x^3 + x^2 + x + 1}{x^2 + x + 1}$

14. $\dfrac{x^6 + x^4 + x^2 + 1}{x^2 + x + 1}$

15. $\dfrac{4x^5 + 2x^4 - 8x^3 + x^2 + 5x + 1}{2x^3 + x^2 - 3x - 1}$

16. $\dfrac{x^6 - 27}{x^2 - 3}$

17. $\dfrac{6x - 10}{x^3 - 4x + 7}$

18. $\dfrac{2x^5 - 7x^4 - 13}{4x^2 - 6x + 8}$

19. $\dfrac{2x^3 + 3x^2 - 2x + 1}{x - \frac{1}{2}}$

20. $\dfrac{6x^4 + 10x^3 + 5x^2 + x + 1}{x + \frac{2}{3}}$

21. $\dfrac{x^{101} - 1}{x - 1}$

22. $\dfrac{x^{100} - 1}{x + 1}$

[*Hint:* In Exercises 21 and 22 you are obviously not being asked to carry out more than 100 steps of a division process. Think! Use the Remainder Theorem to help you, and look for a pattern when finding the quotient.]

23. $\dfrac{x^2 - x - 3}{2x - 4}$

$$\left[\textit{Hint: Note that } \frac{x^2 - x - 3}{2x - 4} = \frac{1}{2}\left(\frac{x^2 - x - 3}{x - 2}\right). \right]$$

24. $\dfrac{x^3 + 3x^2 + 4x + 3}{3x + 6}$

25. $\dfrac{x^4 - 2x^3 + x + 2}{2x - 1}$

In Exercises 26–39 use synthetic division and the Remainder Theorem to evaluate $P(c)$.

26. $P(x) = x^2 - 2x - 7, c = 3$

27. $P(x) = 3x^2 + 9x + 5, c = -4$

28. $P(x) = 2x^2 + 9x + 1, c = \frac{1}{2}$

29. $P(x) = 2x^2 + 9x + 1, c = 0.1$

30. $P(x) = x^3 + 3x^2 - 7x + 6, c = 2$

31. $P(x) = 2x^3 - 21x^2 + 9x - 200, c = 11$

32. $P(x) = 5x^4 + 30x^3 - 40x^2 + 36x + 14, c = -7$

33. $P(x) = 6x^5 + 10x^3 + x + 1, c = -2$

34. $P(x) = x^7 - 3x^2 - 1, c = 3$

35. $P(x) = -2x^6 + 7x^5 + 40x^4 - 7x^2 + 10x + 112,$
$c = -3$

36. $P(x) = 3x^3 + 4x^2 - 2x + 1, c = \frac{2}{3}$

37. $P(x) = x^3 - x + 1, c = \frac{1}{4}$

38. $P(x) = x^3 + 2x^2 - 3x - 8, c = \sqrt{3}$

39. $P(x) = -2x^3 + 3x^2 + 4x + 6, c = 1 + \sqrt{2}$

40. Let

$$P(x) = 6x^7 - 40x^6 + 16x^5 - 200x^4$$
$$- 60x^3 - 69x^2 + 13x - 139$$

Calculate $P(7)$ by using synthetic division, and by substituting $x = 7$ into the polynomial and evaluating directly.

In Exercises 41–44 use the Factor Theorem to show that $x - c$ is a factor of $P(x)$ for the given values of c.

41. $P(x) = x^3 - 3x^2 + 3x - 1, c = 1$

42. $P(x) = x^3 + 2x^2 - 3x - 10, c = 2$

43. $P(x) = 2x^3 + 7x^2 + 6x - 5, c = \frac{1}{2}$

44. $P(x) = x^4 + 3x^3 - 16x^2 - 27x + 63, c = 3, -3$

In Exercises 45 and 46 show that the given value(s) of c are zeros of $P(x)$, and find all other zeros of $P(x)$.

45. $P(x) = x^3 - x^2 - 11x + 15, c = 3$

46. $P(x) = 3x^4 - x^3 - 21x^2 - 11x + 6, c = \frac{1}{3}, -2$

47. Find a polynomial of degree 3 that has zeros 1, −2, and 3, and in which the coefficient of x^2 is 3.

48. Find a polynomial of degree 4 with integer coefficients that has zeros 1, −1, 2, and $\frac{1}{2}$.

49. Find a polynomial of degree 4 that has integer coefficients, constant coefficient 24, and zeros 3, 4, and −2.

50. Find the remainder when the polynomial $6x^{1000} - 17x^{562} + 12x + 26$ is divided by $x + 1$.

51. Is $x - 1$ a factor of $x^{567} - 3x^{400} + x^9 + 2$?

52. If we divide the polynomial $P(x) = x^4 + kx^2 - kx + 2$ by $x + 2$, the remainder is 36. What must the value of k be?

53. For what values of k will $x - 3$ be a factor of $P(x) = k^2x^2 + 2kx - 12$?

54. The quadratic equation $x^2 - 2kx + 12 = 0$ has two positive roots, one of which is three times the other. What is the value of k, and what are the two roots of the equation?

SECTION 3.4
RATIONAL ROOTS

In the preceding section we saw that if c is a root of a polynomial equation $P(x) = 0$ (or equivalently, a zero of the polynomial P), then $x - c$ is a factor of the polynomial. This means that factoring a polynomial into linear factors is really equivalent to finding its zeros. But we need some sort of criterion to tell us which numbers we should check when we are looking for zeros, since it is unlikely that we would find them all by trying numbers chosen at random. In this section we develop a systematic technique for finding all the *rational* zeros of a polynomial. [Remember that a **rational zero** of a polynomial $P(x)$ is a rational number r for which $P(r) = 0$.]

A *rational number* is one that can be expressed as a quotient (or ratio) of integers; for example, $\frac{2}{3}, \frac{-1}{5}, \frac{3}{1}$.

We begin with an illustration. By multiplying out the left side of the following equation, we see that

$$(x - 2)(x - 3)(x + 4) = x^3 - x^2 - 14x + 24$$

Thus the zeros of the polynomial $P(x) = x^3 - x^2 - 14x + 24$ are 2, 3, and -4, since those are the values that make each factor of P in turn equal 0. Notice that when multiplying out the left side of the displayed equation, the constant 24 on the right is obtained by multiplying $(-2) \times (-3) \times 4$. This means that the zeros of the polynomial all must divide its constant term evenly. The next theorem states that for polynomials like the one we are working with, this is always the case.

INTEGER ROOTS THEOREM

Any rational root of the polynomial equation with integer coefficients and leading coefficient 1

$$x^n + a_{n-1}x^{n-1} + \cdots + a_1 x + a_0 = 0$$

must be an *integer* that divides a_0 evenly.

EXAMPLE 1

What numbers could possibly be rational zeros of the polynomial $P(x) = x^3 + 3x^2 - 13x - 15$?

SOLUTION

By the Integer Roots Theorem, since the leading coefficient is 1, all the rational zeros of $P(x)$ must be integers that divide -15 evenly. Thus the only numbers that could be rational zeros of $P(x)$ are

$$1, \quad -1, \quad 3, \quad -3, \quad 5, \quad -5, \quad 15, \quad \text{and} \quad -15$$

In fact, by trying each one in turn, we see that $P(-1) = 0$, $P(3) = 0$, and

$P(-5) = 0$, so $P(x)$ factors as follows:

$$x^3 + 3x^2 - 13x - 15 = (x + 1)(x - 3)(x + 5)$$

(You should multiply out the right side of this equation to confirm this factorization.)

◼

In the remaining examples in this section, we will be using synthetic division (described in Section 3.3) for both evaluating and dividing polynomials. For those who prefer not to use synthetic division, simple substitution can be used to evaluate polynomials and long division can be used to divide them.

EXAMPLE 2

Find all rational roots of the following equation, and then solve it completely.

$$x^3 - 17x + 4 = 0$$

SOLUTION

$$
\begin{array}{r|rrrr}
1 & 1 & 0 & -17 & 4 \\
  &   & 1 & 1 & -16 \\
\hline
  & 1 & 1 & -16 & -12
\end{array}
$$

$$
\begin{array}{r|rrrr}
2 & 1 & 0 & -17 & 4 \\
  &   & 2 & 4 & -26 \\
\hline
  & 1 & 2 & -13 & -22
\end{array}
$$

$$
\begin{array}{r|rrrr}
4 & 1 & 0 & -17 & 4 \\
  &   & 4 & 16 & -4 \\
\hline
  & 1 & 4 & -1 & 0
\end{array}
$$

↑
Remainder is 0

Let $P(x) = x^3 - 17x + 4$. Since the leading coefficient of P is 1, all the rational roots are integer divisors of 4, so the possible candidates for rational solutions are ± 1, ± 2, and ± 4. Using synthetic division, we check these one at a time, starting with the positive candidates (see the margin). We find that 1 and 2 are not roots, but that 4 is a root because $P(4) = 0$. Thus

$$x^3 - 17x + 4 = (x - 4)(x^2 + 4x - 1)$$

The quadratic formula now gives us the zeros of the quotient $x^2 + 4x - 1$:

$$x = \frac{-4 \pm \sqrt{4^2 - 4 \cdot 1 \cdot (-1)}}{2} = -2 \pm \sqrt{5}$$

So the only rational root is 4, and the remaining roots are $-2 + \sqrt{5}$ and $-2 - \sqrt{5}$, which are irrational.

◼

If the leading coefficient of a polynomial is not 1, then its rational zeros are not necessarily all integers. For example, from the factorization

$$6x^2 - x - 12 = (2x - 3)(3x + 4)$$

we see that $\frac{3}{2}$ is a zero of the polynomial $6x^2 - x - 12$ (since $2x - 3 = 0$ implies $x = \frac{3}{2}$). But notice that 3, the numerator of this fraction, divides -12, the constant coefficient of the polynomial, and 2, the denominator of the fraction, divides 6, the leading coefficient of the polynomial. This is always the case for rational zeros of polynomials, as we see in the next theorem.

RATIONAL ROOTS THEOREM

If p/q is a rational root in lowest terms of the polynomial equation with integer coefficients

$$a_n x^n + a_{n-1} x^{n-1} + \cdots + a_1 x + a_0 = 0$$

(where $a_n \neq 0$ and $a_0 \neq 0$), then p divides a_0 evenly and q divides a_n evenly.

If $a_n = 1$ in the Rational Roots Theorem, then q must be 1 or -1, so p/q is an integer that divides a_0. Thus the Integer Roots Theorem is just a special case of the Rational Roots Theorem. A proof of the Rational Roots Theorem is given at the end of this section.

EXAMPLE 3

Find all solutions of the equation

$$2x^4 + x^3 - 17x^2 - 16x + 12 = 0$$

SOLUTION

By the Rational Roots Theorem, if p/q is a rational root of the equation, p must divide 12 and q must divide 2. The divisors of 12 are ± 1, ± 2, ± 3, ± 4, ± 6, and ± 12. The divisors of 2 are ± 1 and ± 2. By evaluating all possible quotients of the form

$$\frac{\text{divisor of } 12}{\text{divisor of } 2}$$

we see that the only possible rational roots are

$$\pm \tfrac{1}{1}, \quad \pm \tfrac{2}{1}, \quad \pm \tfrac{3}{1}, \quad \pm \tfrac{4}{1}, \quad \pm \tfrac{6}{1}, \quad \pm \tfrac{12}{1}, \quad \pm \tfrac{1}{2}, \quad \pm \tfrac{2}{2}, \quad \pm \tfrac{3}{2}, \quad \pm \tfrac{4}{2}, \quad \pm \tfrac{6}{2}, \quad \pm \tfrac{12}{2}$$

Simplifying the fractions and eliminating duplicates, we get

$$\pm 1, \quad \pm 2, \quad \pm 3, \quad \pm 4, \quad \pm 5, \quad \pm 6, \quad \pm 12, \quad \pm \tfrac{1}{2}, \quad \pm \tfrac{3}{2}$$

Using synthetic division to check through the list of possibilities (as in Example 2), we find that 1 and 2 are not roots, but 3 is a root. We divide by $x - 3$ to get

$$2x^4 + x^3 - 17x^2 - 16x + 12 = (x - 3)(2x^3 + 7x^2 + 4x - 4)$$

Rather than checking further through the list of possibilities, we will try instead to factor the quotient $2x^3 + 7x^2 + 4x - 4$. The possible rational zeros of this are ± 1, ± 2, ± 4, and $\pm \tfrac{1}{2}$. We have already found that 1 and 2 are not zeros of the original polynomial so they certainly cannot be zeros of the quotient either, and there is no need to check them again. Working through the list, we see that $\frac{1}{2}$ is a root, and dividing by $x - \frac{1}{2}$ gives

$$2x^3 + 7x^2 + 4x - 4 = \left(x - \tfrac{1}{2}\right)(2x^2 + 8x + 8)$$

$$= \left(x - \tfrac{1}{2}\right)2(x^2 + 4x + 4)$$

$$= 2\left(x - \tfrac{1}{2}\right)(x + 2)^2$$

So the original polynomial factors into

$$2(x + 2)^2\left(x - \tfrac{1}{2}\right)(x - 3) = (x + 2)^2(2x - 1)(x - 3)$$

Thus, by the Factor Theorem, the roots are -2, $\frac{1}{2}$, and 3. There are no irrational roots. ■

DESCARTES' RULE OF SIGNS

In some cases the following rule, discovered by the French philosopher and mathematician René Descartes around 1637 (see page 74), is helpful in eliminating candidates from lengthy lists of possible rational roots. Before we state the rule (which we do not prove), we must first explain what is meant by a *variation in sign*. If $P(x)$ is a polynomial with real coefficients, written with descending powers of x (and omitting powers with coefficient 0), then a **variation in sign** occurs whenever adjacent coefficients have opposite signs. For example,

$$P(x) = 5x^7 - 3x^5 - x^4 + 2x^2 + x - 3$$

has three variations in sign.

DESCARTES' RULE OF SIGNS

If $P(x)$ is a polynomial with real coefficients, then

1. The number of positive real zeros of $P(x)$ is either equal to the number of variations in sign in $P(x)$ or is less than that by an even whole number.
2. The number of negative real zeros of $P(x)$ is either equal to the number of variations in sign in $P(-x)$ or is less than that by an even whole number.

EXAMPLE 4

Use Descartes' Rule of Signs to determine the possible number of positive and negative real roots of the equation

$$3x^6 + 4x^5 + 3x^3 - x - 3 = 0$$

SOLUTION

The polynomial has one variation in sign and so there is one positive root. If $P(x)$ is the polynomial in the equation, then

$$P(-x) = 3(-x)^6 + 4(-x)^5 + 3(-x)^3 - (-x) - 3$$
$$= 3x^6 - 4x^5 - 3x^3 + x - 3$$

which has three variations in sign. There are either three or one negative root(s), making a total of either four or two real roots. The remaining roots are imaginary numbers, which we will study in Section 3.7. ■

In the next example we see how Descartes' Rule of Signs can be used to shorten the list of possible rational zeros of a polynomial provided by the Rational Roots Theorem. From now on, whenever an example involves using synthetic division several times with the same dividend, we will use an abbreviated version of the synthetic division table in which we omit the middle line and write the top line only once, with the result of each division written below this. For example, the three divisions we performed in Example 2 would be combined as follows:

$$
\begin{array}{r|rrrr}
 & 1 & 0 & -17 & 4 \\
\hline
1 & 1 & 1 & -16 & -12 \\
2 & 1 & 2 & -13 & -22 \\
4 & 1 & 4 & -1 & 0 \\
\end{array}
$$

EXAMPLE 5

Find all rational zeros of the polynomial

$$P(x) = 3x^4 + 22x^3 + 55x^2 + 52x + 12$$

SOLUTION

The possible rational zeros of $P(x)$ are ± 1, ± 2, ± 3, ± 4, ± 6, ± 12, $\pm \frac{1}{3}$, $\pm \frac{2}{3}$, and $\pm \frac{4}{3}$. But $P(x)$ has no variations in sign, and hence no positive real zeros, so we can eliminate all the positive numbers in our list. Now

$$P(-x) = 3x^4 - 22x^3 + 55x^2 - 52x + 12$$

has four variations in sign, and so the equation has either four, two, or zero negative real root(s).

We now use synthetic division to check the candidates for rational roots until we find one that gives a remainder of 0.

$$
\begin{array}{c|ccccc}
 & 3 & 22 & 55 & 52 & 12 \\
\hline
-1 & 3 & 19 & 36 & 16 & -4 \\
-2 & 3 & 16 & 23 & 6 & 0 \\
\end{array}
\qquad \leftarrow P(-2) = 0
$$

From the last line of the synthetic division table, we see that -2 is a zero, and that $P(x)$ factors as $(x + 2)(3x^3 + 16x^2 + 23x + 6)$. We continue by factoring the quotient $3x^3 + 16x^2 + 23x + 6$. Because -1 is not a zero, the list of possible rational zeros of the quotient at this point is -2, -3, -6, $-\frac{1}{3}$, and $-\frac{2}{3}$.

$$
\begin{array}{c|cccc}
 & 3 & 16 & 23 & 6 \\
\hline
-2 & 3 & 10 & 3 & 0 \\
\end{array}
$$

So $P(x) = (x + 2)(x + 2)(3x^2 + 10x + 3) = (x + 2)^2(x + 3)(3x + 1)$, and the zeros are -2, -3, and $-\frac{1}{3}$. ■

UPPER AND LOWER BOUNDS FOR ROOTS

We say that a is a **lower bound** and b is an **upper bound** for the roots of a polynomial equation if every real root c of the equation satisfies $a \leq c \leq b$. The next theorem helps us find such bounds for any polynomial equation.

THE UPPER AND LOWER BOUNDS THEOREM

Let $P(x)$ be a polynomial with real coefficients.

1. If we divide $P(x)$ by $x - b$ (where $b > 0$) using synthetic division, and if the row that contains the quotient and remainder has no negative entries, then b is an upper bound for the real roots of $P(x) = 0$.
2. If we divide $P(x)$ by $x - a$ (where $a < 0$) using synthetic division, and if the row that contains the quotient and remainder has entries that are alternately nonpositive and nonnegative, then a is a lower bound for the real roots of $P(x) = 0$.

A proof of this theorem is suggested in Exercises 61 and 62. Note that the phrase "alternately nonpositive and nonnegative" simply means that the signs of the numbers alternate, where 0 can be considered positive or negative as required.

EXAMPLE 6

Show that all the real roots of the equation $x^4 - 3x^2 + 2x - 5 = 0$ lie between -3 and 2.

SOLUTION

We divide the polynomial by $x - 2$ and $x + 3$ using synthetic division.

$$
\begin{array}{r|rrrrr}
 & 1 & 0 & -3 & 2 & -5 \\
\hline
2 & 1 & 2 & 1 & 4 & 3 \\
-3 & 1 & -3 & 6 & -16 & 43 \\
\end{array}
\quad
\begin{array}{l}
\leftarrow \text{all positive} \\
\leftarrow \text{alternate in sign}
\end{array}
$$

By the Upper and Lower Bounds Theorem, -3 is a lower bound and 2 is an upper bound for the roots. Since neither -3 nor 2 is a root, all the real roots lie between them. ∎

EXAMPLE 7

Factor completely the polynomial

$$P(x) = 2x^5 + 5x^4 - 8x^3 - 14x^2 + 6x + 9$$

SOLUTION

The possible rational zeros of $P(x)$ are $\pm\frac{1}{2}$, ± 1, $\pm\frac{3}{2}$, ± 3, $\pm\frac{9}{2}$, and ± 9. We check the positive candidates first, beginning with the smallest.

$$
\begin{array}{r|rrrrrr}
 & 2 & 5 & -8 & -14 & 6 & 9 \\
\hline
\frac{1}{2} & 2 & 6 & -5 & -\frac{33}{2} & -\frac{9}{4} & \frac{63}{8} \\
1 & 2 & 7 & -1 & -15 & -9 & 0 \\
\end{array}
\quad \leftarrow P(1) = 0
$$

So 1 is a zero, and $P(x)$ factors as $(x - 1)(2x^4 + 7x^3 - x^2 - 15x - 9)$. We continue by factoring the quotient.

$$
\begin{array}{r|rrrrr}
 & 2 & 7 & -1 & -15 & -9 \\
\hline
1 & 2 & 9 & 8 & -7 & -16 \\
\frac{3}{2} & 2 & 10 & 14 & 6 & 0 \\
\end{array}
\quad \leftarrow \text{all nonnegative}
$$

We see that $\frac{3}{2}$ is both a zero and an upper bound for the zeros of $P(x)$, so we do not need to check any further for positive zeros $\left(\text{since the remaining candidates are all} > \frac{3}{2}\right)$.

$$
\begin{aligned}
P(x) &= (x - 1)\left(x - \tfrac{3}{2}\right)(2x^3 + 10x^2 + 14x + 6) \\
&= (x - 1)(2x - 3)(x^3 + 5x^2 + 7x + 3)
\end{aligned}
$$

By Descartes' Rule of Signs, $x^3 + 5x^2 + 7x + 3$ has no positive zeros, so the only possible rational zeros are -1 and -3.

$$
\begin{array}{r}
\;\,1 \quad 5 \quad 7 \quad 3 \\
\hline
-1\,\big|\;1 \quad 4 \quad 3 \quad 0
\end{array}
$$

Therefore
$$
\begin{aligned}
P(x) &= (x - 1)(2x - 3)(x + 1)(x^2 + 4x + 3) \\
&= (x - 1)(2x - 3)(x + 1)^2(x + 3)
\end{aligned}
$$

This means that the zeros of P are 1, $\frac{3}{2}$, -1, and -3. ■

To summarize, the following procedure gives a systematic, efficient method for finding all rational roots of a polynomial equation.

Step 1   List all possible rational roots using the Rational Roots Theorem.

Step 2   Use Descartes' Rule of Signs to see how many positive and negative real roots the equation may have. (In some cases the positive or negative rational candidates may be eliminated completely by this step.)

Step 3   Check (from smallest to largest, in absolute value) the positive and the negative candidates for rational roots provided by Step 1. Stop when you find a root, reach an upper or lower bound, or have found the maximum number of positive or negative roots predicted by Descartes' Rule of Signs.

Step 4   If you find a root, repeat the process with the quotient. Remember that you do not need to check possible roots that have not worked at a previous stage. Once you reach a quotient that is quadratic or in some other way easily factorable, use the quadratic formula or other techniques you have learned to find the remaining roots.

■ PROOF OF THE RATIONAL ROOTS THEOREM

Let $P(x) = a_n x^n + a_{n-1} x^{n-1} + \cdots + a_1 x + a_0$ be a polynomial with integer coefficients (with $a_n \neq 0$ and $a_0 \neq 0$), and suppose that p/q is a rational number in lowest terms for which

$$
a_n\left(\frac{p}{q}\right)^n + a_{n-1}\left(\frac{p}{q}\right)^{n-1} + \cdots + a_1\left(\frac{p}{q}\right) + a_0 = 0
$$

so that p/q is a rational zero of $P(x)$. We are going to show that p divides a_0 evenly and that q divides a_n evenly.

First we multiply both sides of the equation by q^n, giving

$$a_n p^n + a_{n-1} p^{n-1} q + a_{n-2} p^{n-2} q^2 + \cdots + a_1 p q^{n-1} + a_0 q^n = 0 \qquad (1)$$

Subtracting $a_0 q^n$ from both sides of the equation and factoring p from the left, we get

$$p(a_n p^{n-1} + a_{n-1} p^{n-2} q + a_{n-2} p^{n-3} q^2 + \cdots + a_1 q^{n-1}) = -a_0 q^n$$

Now we can see that p divides the left side of the equation evenly, so it must divide the right side as well. But since p/q is in lowest terms, p and q have no integer factors in common, and so neither do p and q^n. Thus all the integer factors of p must go evenly into a_0, which means that p divides a_0.

If we take Equation 1 and subtract $a_n p^n$ from both sides, and then factor q from the left side, we get

$$q(a_{n-1} p^{n-1} + a_{n-2} p^{n-2} q + \cdots + a_1 p q^{n-2} + a_0 q^{n-1})' = -a_n p^n$$

Using the same reasoning as before, we can now show that q must divide a_n. This proves the Rational Roots Theorem. ∎

EXERCISES 3.4

In Exercises 1–4 list all possible rational roots given by the Rational Roots Theorem (but do not check to see which actually are roots).

1. $x^4 - 3x^2 + 100x - 7 = 0$

2. $x^3 - 4x^2 + 6x + 20 = 0$

3. $4x^5 - 3x^3 + x^2 - x + 15 = 0$

4. $6x^4 - x^3 + x^2 - 12 = 0$

In Exercises 5–12 use Descartes' Rule of Signs to determine how many positive and how many negative real zeros the given polynomial can have, and then determine the possible total number of real zeros.

5. $x^4 + 5x^3 - x^2 + 4x - 3$

6. $3x^5 - 4x^4 + 8x^3 - 5$

7. $2x^6 + 5x^4 - x^3 - 5x - 1$

8. $x^4 + x^3 + x^2 + x + 12$

9. $x^5 + 4x^3 - x^2 + 6x$

10. $125x^{12} + 76x^5 - 12x^2 - 1$

11. $4x^7 - 3x^5 + 5x^4 + x^3 - 3x^2 + 2x - 5$

12. $x^8 - x^5 + x^4 - x^3 + x^2 - x + 1$

In Exercises 13–16 show that the given values for a and b are lower and upper bounds, respectively, for the real roots of the equation.

13. $2x^3 + 5x^2 + x - 2 = 0$; $a = -3, b = 1$

14. $x^4 - 2x^3 - 9x^2 + 2x + 8 = 0$; $a = -3, b = 5$

15. $8x^3 + 10x^2 - 39x + 9 = 0$; $a = -3, b = 2$

16. $3x^4 - 17x^3 + 24x^2 - 9x + 1 = 0$; $a = 0, b = 6$

In Exercises 17–20 find integers that are upper and lower bounds for the real roots of the given equation.

17. $x^3 - 3x^2 + 4 = 0$

18. $2x^3 - 3x^2 - 8x + 12 = 0$

19. $x^4 - 2x^3 + x^2 - 9x + 2 = 0$

20. $x^5 - x^4 + 1 = 0$

In Exercises 21–56 find all rational roots of the equation, and then find the irrational roots, if any. Whenever appropriate, use the Rational Roots Theorem, the Upper and Lower Bounds Theorem, Descartes' Rule of Signs, the quadratic formula, or other factoring techniques.

21. $2x^2 + x - 6 = 0$

22. $3x^2 - 13x + 4 = 0$

23. $x^3 - x^2 + x - 1 = 0$

24. $x^3 - 2x^2 - x + 2 = 0$

25. $x^3 - 3x^2 - 4x + 12 = 0$

26. $x^3 - x^2 - 8x + 12 = 0$

27. $x^3 - 4x^2 + x + 6 = 0$

28. $x^3 + 3x^2 - 4 = 0$

29. $x^3 - 7x^2 + 14x - 8 = 0$

30. $x^3 - 4x^2 - 7x + 10 = 0$

31. $x^3 + 4x^2 - 11x + 6 = 0$

32. $x^3 + 7x^2 + 8x - 16 = 0$

33. $x^3 + 3x^2 + 6x + 4 = 0$

34. $x^3 - 2x^2 - 2x - 3 = 0$

35. $x^4 + 3x^2 - 4 = 0$

36. $x^4 - 5x^2 + 4 = 0$

37. $x^4 - 2x^3 - 3x^2 + 8x - 4 = 0$

38. $x^4 + 3x^3 - 7x^2 - 15x + 18 = 0$

39. $x^4 + 6x^3 + 7x^2 - 6x - 8 = 0$

40. $x^4 - x^3 - 23x^2 - 3x + 90 = 0$

41. $4x^4 - 25x^2 + 36 = 0$

42. $x^4 - x^3 - 5x^2 + 3x + 6 = 0$

43. $4x^3 - 7x + 3 = 0$

44. $2x^3 - 3x^2 - 2x + 3 = 0$

45. $4x^3 + 4x^2 - x - 1 = 0$

46. $8x^3 + 10x^2 - x - 3 = 0$

47. $2x^4 + 3x^3 - 4x^2 - 3x + 2 = 0$

48. $2x^4 + 15x^3 + 31x^2 + 20x + 4 = 0$

49. $4x^4 - 21x^2 + 5 = 0$

50. $6x^4 - 7x^3 - 8x^2 + 5x = 0$

51. $x^5 - 7x^4 + 9x^3 + 23x^2 - 50x + 24 = 0$

52. $8x^5 - 14x^4 - 22x^3 + 57x^2 - 35x + 6 = 0$

53. $x^4 - \frac{11}{4}x^3 - \frac{11}{2}x^2 + \frac{5}{4}x + 3 = 0$

54. $x^5 - \frac{1}{5}x^4 - 5x^3 + x^2 + 4x - \frac{4}{5} = 0$

55. $x^8 - 1 = 0$

56. $x^9 - 1 = 0$

Show that each polynomial in Exercises 57–60 does not have any rational zeros.

57. $x^3 - x - 2$ **58.** $2x^4 - x^3 + x + 2$

59. $3x^3 - x^2 - 6x + 12$ **60.** $x^{50} - 5x^{25} + x^2 - 1$

61. Let $P(x)$ be a polynomial with real coefficients, and let $b > 0$. Use the Division Algorithm to write

$$P(x) = (x - b) \cdot Q(x) + r$$

Suppose that $r \geq 0$ and that all the coefficients in $Q(x)$ are nonnegative. Let $z > b$.
   (a) Show that $P(z) > 0$.
   (b) Prove the first part of the Upper and Lower Bounds Theorem.

62. Use the first part of the Upper and Lower Bounds Theorem to prove the second part. [*Hint:* Show that if $P(x)$ satisfies the second part of the theorem, then $P(-x)$ satisfies the first part.]

63. Show that the equation

$$x^5 - x^4 - x^3 - 5x^2 - 12x - 6 = 0$$

has exactly one rational root, and then prove that it must have either two or four irrational roots.

64. Consider the equation $x^3 + kx + 4 = 0$, where k is an integer.
   (a) For which values of k does this equation have at least one rational root?
   (b) For which values of k does this equation have one rational root and two irrational roots?
   (c) What is the largest number of rational roots this equation can have?

65. For what integer values of k does the equation $x^3 - kx^2 + k^2x - 3 = 0$ have a rational root?

SECTION 3.5
IRRATIONAL ROOTS

For linear and quadratic equations we have formulas for the exact roots, including the irrational ones. In the preceding section we described a method for finding all the rational roots for equations of higher degree. Although there are formulas analogous to the quadratic formula for the roots of the general third- and fourth-degree equations, they are complicated, difficult to remember, and cumbersome to use. For example, the formula for one of the three roots of the general cubic equation $ax^3 + bx^2 + cx + d = 0$ is

$$x = \frac{-b}{3a} + \frac{1}{a}\sqrt[3]{\frac{-b^3}{27} + \frac{abc}{6} - \frac{a^2d}{2} + a\sqrt{\frac{a^2d^2}{4} - \frac{b^2c^2}{108} + \frac{b^3d}{27} + \frac{ac^3}{27} - \frac{abcd}{6}}}$$

$$+ \frac{1}{a}\sqrt[3]{\frac{-b^3}{27} + \frac{abc}{6} - \frac{a^2d}{2} - a\sqrt{\frac{a^2d^2}{4} - \frac{b^2c^2}{108} + \frac{b^3d}{27} + \frac{ac^3}{27} - \frac{abcd}{6}}}$$

For equations of degree 5 and higher, no such formulas exist. In fact, the French mathematician Evariste Galois proved in 1832 (shortly before his death at the age of 20) that it is *impossible* to construct formulas involving radicals and the usual algebraic operations for the roots of the general nth-degree polynomial equation for $n \geq 5$.

In some special cases the exact values of the irrational zeros of a higher-degree polynomial can be found by factoring and then using the quadratic formula, as we saw in Example 2 in Section 3.4. In this section we will demonstrate an effective numerical method for finding decimal approximations to irrational roots of polynomial equations to any desired accuracy. The technique depends on the following special case of the Intermediate Value Theorem, which is studied more fully in calculus courses.

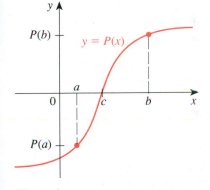

Figure 1

INTERMEDIATE VALUE THEOREM FOR POLYNOMIALS

If P is a polynomial with real coefficients and if $P(a)$ and $P(b)$ have opposite signs, then there is at least one value c between a and b for which $P(c) = 0$.

Although we will not prove this theorem, Figure 1 shows why it is at least intuitively plausible. A part of the graph of $y = P(x)$ is shown, with the case $P(a) < 0$ and $P(b) > 0$ illustrated.

EXAMPLE 1

(a) Show that the polynomial $P(x) = x^3 + 8x - 30$ has a zero between 2 and 3.
(b) Find an approximate value for this zero, correct to the nearest tenth.

SOLUTION

(a) Since $P(2) = -6$ and $P(3) = 21$, the Intermediate Value Theorem says that there is a number c between 2 and 3 for which $P(c) = 0$.
(b) We evaluate P at successive tenths between 2 and 3 until we find where P changes sign. This will "trap" the zero between successive tenths.

x	$P(x)$
2.1	-3.94
2.2	-1.75
2.3	0.57

opposite in sign

This means that the number c for which $P(c) = 0$ lies between 2.2 and 2.3. To see whether it is closer to 2.2 or 2.3, we check the value of P halfway between these two numbers:

$$P(2.25) \approx -0.61$$

Since the value of $P(x)$ is still negative when $x = 2.25$, the zero we are looking for is closer to 2.3 than to 2.2, as illustrated in Figure 2. Correct to the nearest tenth, the zero is 2.3.

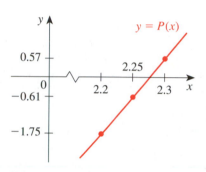

Figure 2

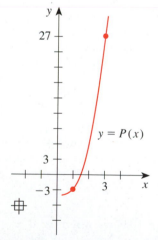

Figure 3
Part of the graph of
$P(x) = x^3 + x^2 - 2x - 3$

EXAMPLE 2

Show that the following equation has exactly one positive irrational root, and find the decimal value of this root correct to the nearest hundredth.

$$P(x) = x^3 + x^2 - 2x - 3 = 0$$

SOLUTION

By Descartes' Rule of Signs, the equation has exactly one positive real root. The only possible positive rational roots are 1 and 3, but $P(1) = -3$ and $P(3) = 27$, so neither actually turns out to be a root. So the positive real root must be irrational, and moreover it must lie between 1 and 3, since $P(1)$ and $P(3)$ are opposite in sign. We calculate $P(x)$ for enough values of x between 1 and 3 to locate this root between successive tenths. To help us decide where to start this process, we observe that $P(1)$ is much closer to 0 than is $P(3)$. This leads us to guess that the zero we are looking for is much closer to 1 than to 3 (see Figure 3), so we begin our search at $x = 1.3$.

x	P(x)
1.3	−1.713
1.4	−1.096
1.5	−0.375
1.6	0.456

⎱ opposite in sign

This means that the root lies somewhere between 1.5 and 1.6. We now attempt to locate it between successive hundredths.

x	P(x)
1.53	−1.138
1.54	−0.056
1.55	0.026

⎱ opposite in sign

So the root lies between 1.54 and 1.55. To see which of these is closer to the actual value, we calculate the value of P at 1.545, halfway between the two possibilities. We find that

$$P(1.545) \approx -0.015$$

Since $P(1.545)$ and $P(1.55)$ are opposite in sign, the root lies between 1.545 and 1.55 and thus is closer to 1.55 than to 1.54. So the positive irrational root is 1.55, correct to the nearest hundredth.   ■

The process described above can be continued to obtain the value of the root to any number of decimal places. The calculations involved are long and tedious, however. More efficient methods for approximating the roots of polynomial equations are studied in calculus courses.

EXAMPLE 3

A right triangle has an area of 8 m², and its hypotenuse is 1 m longer than one of the legs. Find the lengths of all sides of the triangle correct to two decimal places.

SOLUTION

Let the length of one leg be x. Then the hypotenuse has length $x + 1$, and the other leg (by the Pythagorean Theorem) has length $\sqrt{(x + 1)^2 - x^2} = \sqrt{2x + 1}$ (see Figure 4). Thus the area is

$$\tfrac{1}{2} \times \text{base} \times \text{height} = \tfrac{1}{2}x\sqrt{2x + 1} = 8$$

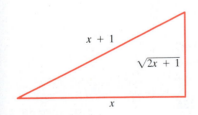

$x + 1$

$\sqrt{2x + 1}$

x

Figure 4

and so

$$x\sqrt{2x + 1} = 16$$
$$x^2(2x + 1) = 256$$
$$2x^3 + x^2 - 256 = 0$$

We are interested only in positive roots, and the equation has exactly one positive real root. Working through the rational possibilities, we get the following table.

x	$2x^3 + x^2 - 256$	
$\frac{1}{2}$	$-\frac{511}{2}$	
1	-253	
2	-236	
4	-112	
8	832	} opposite in sign

Because of the sign change, the positive real root must lie between 4 and 8, and it must be irrational. We locate the root between successive integers, then tenths, and then hundredths.

	$2x^3 + x^2 - 256$	
4	-112	
5	19	} sign change

The root lies between 4 and 5. Because 19 is closer to 0 than to -112, we guess that the root is closer to 5 than to 4, so we start our search at 4.7.

x	$2x^3 + x^2 - 256$	
4.7	-26.264	
4.8	-11.776	} sign change
4.9	3.308	

x	$2x^3 + x^2 - 256$	
4.87	-1.280	} sign change
4.88	0.243	

Evariste Galois (*1811–1832*) *is one of the very few mathematicians to have an entire theory named in his honor. Though he died while not yet 21, he completely settled the central problem in the theory of equations. He did so by describing a criterion that reveals whether any given equation can be solved by algebraic operations. Although Galois was one of the greatest mathematicians in the world at that time, no one knew it but him. He repeatedly sent his work to the eminent mathematicians Cauchy and Poisson, who either lost or did not understand his ideas. Galois wrote in a terse style and included few details, which probably played a role in his failure to pass the entrance exams at the Ecole Polytechnique in Paris. Galois was a political radical and spent several months in prison for his revolutionary activities. At the age of 20 his life came to a tragic end when he was killed in a duel over a love affair. The night before his duel, fearing that he would be killed, Galois wrote down the essence of his ideas and entrusted them to his friend Auguste Chevalier. He concluded by writing ". . . there will, I hope, be people who will find it to their advantage to decipher all this mess." That is just what the famous mathematician Camille Jordan did 14 years later.*

Since the value of the polynomial in our equation is -0.520 when $x = 4.875$, the root is closer to 4.88 than to 4.87. The legs of the triangle are 4.88 m and 3.28 m long, and the hypotenuse is 5.88 m long, all correct to two decimal places. ■

EXERCISES 3.5

In Exercises 1–6 show that the given polynomial has a zero between the given integers, and then find that zero correct to the given number of decimal places.

1. $x^3 - x - 1$; between 1 and 2; to the nearest tenth

2. $x^3 + 2x^2 - 10$; between 1 and 2; to one decimal place

3. $x^3 - 4x^2 + 2$; between 0 and 1; to two decimal places

4. $2x^4 - 4x^2 + 1$; between 1 and 2; to the nearest hundredth

5. $2x^4 - 4x^2 + 1$; between -1 and 0; to the nearest hundredth

6. $x^5 - x^3 + 1$; between -2 and -1; to two decimal places

In Exercises 7–12 find the indicated irrational root correct to two decimal places.

7. The positive root of $x^3 + 3x^2 + 3x - 1 = 0$

8. The real root of $x^3 + 4x^2 + 2 = 0$

9. The largest positive root of $x^4 - 8x^3 + 18x^2 - 8x - 2 = 0$

10. The negative root of $x^3 - 2x^2 - x + 3 = 0$

11. The positive root of $x^4 + 2x^3 + x^2 - 1 = 0$

12. The negative root of $x^4 + 2x^3 + x^2 - 1 = 0$

In Exercises 13–22 find all rational and irrational roots. Find the exact values of irrational roots using the quadratic formula whenever possible (as in Example 2, Section 3.4). Otherwise find approximate values correct to two decimal places.

13. $x^3 + 2x^2 - 6x - 4 = 0$

14. $3x^3 + 10x^2 + 6x + 1 = 0$

15. $2x^5 + 3x^4 + x^2 + x - 1 = 0$

16. $x^4 - 4x^3 + 2x^2 - 5x + 10 = 0$

17. $x^3 - 3x^2 + 3 = 0$

18. $x^3 + 2x^2 - 4x - 7 = 0$

19. $x^5 + x^4 - 4x^3 - x^2 + 5x - 2 = 0$

20. $x^5 + 11x^4 + 43x^3 + 73x^2 + 52x + 12 = 0$

21. $2x^4 - 7x^3 + 9x^2 + 5x - 4 = 0$

22. $5x^4 + 13x^3 + 9x^2 - 16x + 4 = 0$

23. A rectangle with an area of 10 ft^2 has a diagonal that is 2 ft longer than one of its sides. Find the dimensions of the rectangle (correct to the nearest thousandth of a foot).

24. An open box with a volume of 1500 cm^3 is to be constructed by taking a 20 cm by 40 cm piece of cardboard, cutting squares of side length x cm from each corner, and folding up the sides. Show that this can be done in two different ways, and find the exact dimensions of the box in each case.

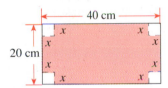

25. A rocket consists of a right circular cylinder of height 20 m surmounted by a cone whose height and diameter are equal and whose radius is the same as that of the cylindrical section. What should this radius be (correct to two decimals) if the total volume is to be $500\pi/3$ m^3?

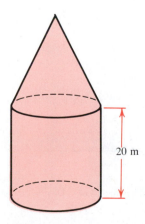

26. A rectangular box with a volume of $2\sqrt{2}$ ft³ has a square base. The diagonal of the box (between a pair of opposite corners) is 1 ft longer than each side of the base.

(a) If the base has sides of length x feet, show that

$$x^6 - 2x^5 - x^4 + 8 = 0$$

(b) Show that there are two different boxes that satisfy the given conditions. Find the dimensions in each case, to the nearest hundredth of a foot.

SECTION 3.6
USING GRAPHING DEVICES TO SOLVE POLYNOMIAL EQUATIONS

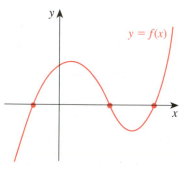

Figure 1

In Section 3.2 we used graphing calculators and computer graphics programs to help us graph polynomials. In this section we use the graphs that such devices produce, together with the theorems of Sections 3.4 and 3.5, to find the zeros of polynomials.

Recall that the x-intercepts of a function are the x-coordinates of the points at which the graph of the function intersects the x-axis (see Figure 1). Since the x-axis consists of all points in the coordinate plane whose y-coordinates are 0, this means that the x-intercepts of the function $y = f(x)$ are the solutions of the equation

$$0 = f(x)$$

So to solve equations of this form, we can use the x-intercepts of the graph of $y = f(x)$, as we see in the following examples.

EXAMPLE 1

Find all solutions of the equation

$$2x^3 - 15x^2 + 22x + 15 = 0$$

SOLUTION

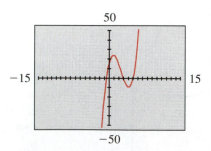

Figure 2
$P(x) = 2x^3 - 15x^2 + 22x + 15$

We begin by graphing the function $y = 2x^3 - 15x^2 + 22x + 15$ using a graphing device. Since we want to be sure to see all the y-intercepts of the curve on our graph, we initially choose a wide viewing rectangle. By the Rational Roots Theorem, 15 and -15 are the largest and smallest possible rational roots of the equation, so let's choose the viewing rectangle $[-15, 15]$ by $[-50, 50]$. In this rectangle, the graphing device gives the graph in Figure 2. (We have used a scale of 1 unit along the x-axis and 10 units along the y-axis.)

It appears that the graph has three x-intercepts, lying somewhere between -1 and 6. Since the x-intercepts are the only parts of the graph we are interested in here, we change the viewing rectangle to $[-1, 6]$ by $[-50, 50]$. This gives the

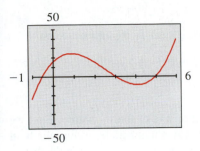

Figure 3
$P(x) = 2x^3 - 15x^2 + 22x + 15$

graph in Figure 3. The x-intercepts seem to be at $x = -\frac{1}{2}$, 3, and 5. To confirm this, we could use the zoom feature on the graphing device to get a closer look at each of the three x-intercepts. Or, to prove that these are in fact the solutions of the equation, we can substitute each of them into the equation:

$$2\left(-\tfrac{1}{2}\right)^3 - 15\left(-\tfrac{1}{2}\right)^2 + 22\left(-\tfrac{1}{2}\right) + 15 = -\tfrac{1}{4} - \tfrac{15}{4} - 11 + 15 = 0$$

$$2(3)^3 - 15(3)^2 + 22(3) + 15 = 54 - 135 + 66 + 15 = 0$$

$$2(5)^3 - 15(5)^2 + 22(5) + 15 = 250 - 375 + 110 + 15 = 0$$

The solutions of the equation are $-\frac{1}{2}$, 3, and 5.   ■

How can we be sure that the equation in Example 1 has no solutions other than the ones we found? Remember that the Factor Theorem says that if c is a solution of the polynomial equation $P(x) = 0$, then $x - c$ is a factor of $P(x)$. If the equation in Example 1 had *four* different solutions (call them c_1, c_2, c_3, and c_4), then the polynomial $2x^3 - 15x^2 + 22x + 15$ would have to be evenly divisible by $x - c_1$, $x - c_2$, $x - c_3$, and $x - c_4$, and hence by the product

$$(x - c_1)(x - c_2)(x - c_3)(x - c_4)$$

as well. But if this product is multiplied out, it contains the term x^4, so it is impossible for this to divide the third-degree polynomial $2x^3 - 15x^2 + 22x + 15$ evenly. This illustrates the general principle given in the following box.

THE NUMBER OF ROOTS OF A POLYNOMIAL EQUATION

If $P(x)$ is a polynomial of degree n, then the equation

$$P(x) = 0$$

can have at most n different real roots.

In Examples 2 and 3 we can see that the number of real roots of a polynomial equation of degree n can in fact be less than n. The preceding principle just says that there *cannot* be any *more* roots than the degree of the equation. We will consider this fact in more detail in Section 3.8, when we study the Complete Factorization Theorem.

Another way to ensure that we are finding all the real solutions of a polynomial equation is to use the Upper and Lower Bounds Theorem, as in the next example.

EXAMPLE 2

Find all real solutions of the following equation, correct to the nearest tenth:

$$3x^4 + 4x^3 - 7x^2 - 2x - 3 = 0$$

SOLUTION

This equation can have at most four real roots since it is of fourth degree. First we use the Upper and Lower Bounds Theorem to find two numbers between which all the solutions must lie. This will allow us to choose a viewing rectangle that is certain to contain all the x-intercepts of the polynomial function. We use synthetic division and proceed by trial and error:

$$
\begin{array}{r|rrrrr}
 & 3 & 4 & -7 & -2 & -3 \\
\hline
1 & 3 & 7 & 0 & -2 & -5 \\
2 & 3 & 10 & 13 & 24 & 45 \quad \leftarrow \text{all positive}
\end{array}
$$

Thus 2 is an upper bound for the roots of the equation. Now we look for a lower bound.

$$
\begin{array}{r|rrrrr}
 & 3 & 4 & -7 & -2 & -3 \\
\hline
-1 & 3 & 1 & -8 & 6 & -9 \\
-2 & 3 & -2 & -3 & 4 & -11 \\
-3 & 3 & -5 & 8 & -26 & 75 \quad \leftarrow \text{alternate in sign}
\end{array}
$$

Thus -3 is a lower bound for the solutions. This means that if we use the viewing rectangle $[-3, 2]$ by $[-20, 20]$ we are certain to see all the x-intercepts of the function $y = 3x^4 + 4x^3 - 7x^2 - 2x - 3$. With this rectangle we obtain the graph in Figure 4. There are two x-intercepts, one between -3 and -2, and the other between 1 and 2. To determine the values of these intercepts to the nearest tenth, we change the scale along the x-axis to 0.1, and then zoom in to examine each of them more closely. From the resulting graphs in Figures 5 and 6, we see that the negative x-intercept is close to -2.3, and the positive one is close to 1.3.

The solutions of the equation, to the nearest tenth, are -2.3 and 1.3.

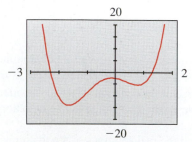

Figure 4
$y = 3x^4 + 4x^3 - 7x^2 - 2x - 3$

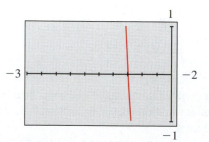

Figure 5
$y = 3x^4 + 4x^3 - 7x^2 - 2x - 3$

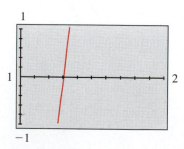

Figure 6
$y = 3x^4 + 4x^3 - 7x^2 - 2x - 3$ ∎

In the next example, we see that by zooming in closer and closer to an x-intercept, we can find the solution of an equation to any desired degree of accuracy (within the limits allowed by the graphing device we are using.)

EXAMPLE 3

Find all solutions, correct to three decimal places, of the equation

$$-3x^3 + 4x^2 + 5 = 0$$

SOLUTION

Since the polynomial in the equation has one sign change, it has one positive zero, by Descartes' Rule of Signs. Changing x to $-x$ in the polynomial gives

$$-3(-x)^3 + 4(-x)^2 + 5 = 3x^3 + 4x^2 + 5$$

which has no sign changes, so there are no negative solutions to the equation. Thus the equation has only one solution and it is positive. Graphing the function $y = -3x^3 + 4x^2 + 5$ in the viewing rectangle $[0, 3]$ by $[-10, 10]$ gives the graph in Figure 7. The solution of the equation is somewhat less than 2. To find its value correct to three decimal places, we change the scale on the x-axis to 0.001 and zoom in several times to get the graph in Figure 8. Using the cursor (or by inspection) we see that the x-intercept is 1.831 to the nearest thousandth. The only real solution of the equation is $x \approx 1.831$.

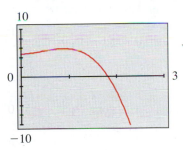

Figure 7
$y = -3x^3 + 4x^2 + 5$

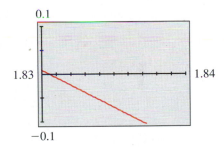

Figure 8
$y = -3x^3 + 4x^2 + 5$ ■

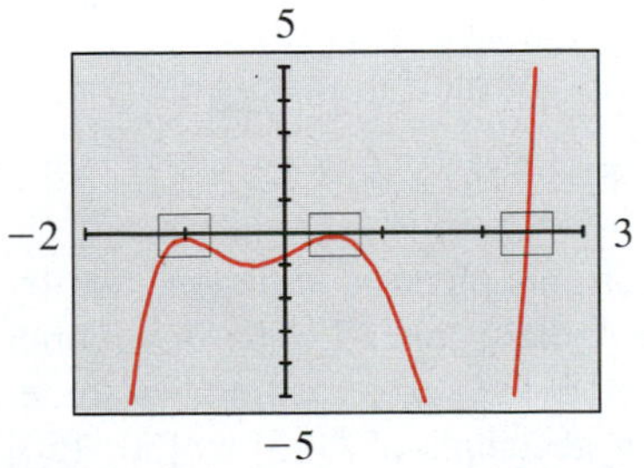

EXAMPLE 4

Find all real solutions of the following equation, correct to two decimals:

$$x^5 - 1.45x^4 - 3.19x^3 + 1.31x^2 + 1.50x - 0.62 = 0$$

SOLUTION

Using the viewing rectangle $[-2, 3]$ by $[-5, 5]$ we obtain the graph shown in Figure 9. The graph appears to touch the x-axis near -1 and $\frac{1}{2}$, and to cross the

Figure 9

x-axis between 2 and 3. However, if we zoom in to each of these three locations, we see that the graph does not actually touch the *x*-axis near $x = -1$, so there is no solution here (Figure 10). At $x = \frac{1}{2}$ the graph does touch the *x*-axis, so even though it does not *cross* this axis, this is still an *x*-intercept (Figure 11). You can verify by substitution that $x = 0.5$ is in fact an *exact* solution of the equation (although the calculations involved are a lot of work!). Zooming in several times on the remaining *x*-intercept shows that its value is 2.45, to the nearest hundredth (Figure 12). The solutions of the equation are

$$x = 0.50 \quad \text{and} \quad x \approx 2.45$$

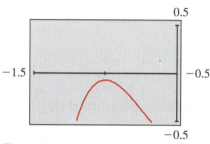

Figure 10

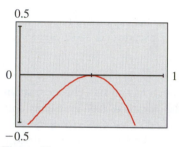

Figure 11

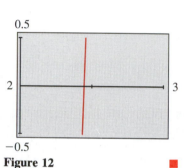

Figure 12 ■

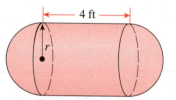

Figure 13

Volume of a cylinder: $V = \pi r^2 h$

Volume of a sphere: $V = \frac{4}{3}\pi r^3$

EXAMPLE 5

A fuel tank consists of a cylindrical center section that is 4 ft long and two hemispherical end sections, as shown in Figure 13. If the tank has a volume of 100 ft³, what is the radius *r* shown in the figure, to the nearest hundredth of a foot?

SOLUTION

Using the volume formula from the inside front cover of this book, we see that the volume of the cylindrical section of the tank is

$$\pi \cdot r^2 \cdot 4$$

The two hemispherical parts together form a complete sphere whose volume is

$$\frac{4}{3}\pi r^3$$

Because the total volume of the tank is 100 ft³, we get the following equation:

$$\frac{4}{3}\pi r^3 + 4\pi r^2 = 100$$

A negative solution for *r* would be meaningless in this physical situation, and by substitution we can verify that $r = 3$ leads to a tank that is over 226 ft³ in volume, much larger than the required 100 ft³. Thus we know the correct radius lies somewhere between 0 and 3 ft, and so we use a viewing rectangle of [0, 3] by [50, 150]

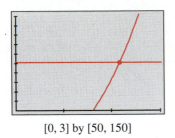

[0, 3] by [50, 150]

Figure 14
$y = \frac{4}{3}\pi x^3 + 4\pi x^2$ and $y = 100$

to graph the function $y = \frac{4}{3}\pi x^3 + 4\pi x^2$, as shown in Figure 14. Since we want the value of this function to be 100, we also graph the horizontal line $y = 100$ in the same viewing rectangle. The correct radius will be the x-coordinate of the point of intersection of the curve and the line. Using the cursor and zooming in we see that at the point of intersection $x \approx 2.15$, correct to two decimals. Thus the tank has a radius of about 2.15 ft. ■

Note that we also could have solved the equation in the preceding example by first writing it as

$$\frac{4}{3}\pi r^3 + 4\pi r^2 - 100 = 0$$

and then finding the x-intercept of the function $y = \frac{4}{3}\pi x^3 + 4\pi x^2 - 100$.

EXERCISES 3.6

In Exercises 1–8 use the given viewing rectangle to find the exact solutions of the equation. (You may assume that all the solutions of the equation can be found within the given viewing rectangle.)

1. $x^3 + 2x^2 - x - 2 = 0$, $[-3, 3]$ by $[-20, 20]$

2. $x^3 - 9x^2 + 23x - 15 = 0$, $[0, 6]$ by $[-20, 20]$

3. $2x^4 - 5x^3 - 14x^2 + 5x + 12 = 0$,
$[-2, 5]$ by $[-40, 40]$

4. $2x^4 - 5x^3 - 8x^2 + 17x - 6 = 0$,
$[-3, 5]$ by $[-60, 60]$

5. $2x^3 - x^2 = 8x + 5$, $[-2, 3]$ by $[-30, 30]$

6. $x^4 + 12x + 36 = 2x^3 + 11x^2$, $[-4, 4]$ by $[-50, 50]$

7. $3x^3 + 8x^2 + 5x + 2 = 0$, $[-3, 3]$ by $[-10, 10]$

8. $4x^4 + 4x^3 + 7x^2 = x + 2$, $[-2, 2]$ by $[-40, 40]$

In Exercises 9–20 find all real solutions of the equation, correct to two decimals.

9. $3x^3 + x^2 + x - 2 = 0$ **10.** $2x^3 - 8x^2 + 9x - 9 = 0$

11. $10x^4 - 9x^3 - 11x^2 + 5x - 3 = 0$

12. $3x^4 + 8x^3 + 2x^2 + 5x + 2 = 0$

13. $x^3 + 6 = 6x^2$ **14.** $x^4 + x^3 = 4$

15. $x^4 + 8x + 16 = 2x^3 + 8x^2$

16. $2x^5 + 9x^4 - 5x^2 = 21x^3 + 11x - 2$

17. $4.00x^4 + 4.00x^3 - 10.96x^2 - 5.88x + 9.09 = 0$

18. $x^5 + 2.00x^4 + 0.96x^3 + 5.00x^2 + 10.00x + 4.80 = 0$

19. $x^5 - 3x^4 - 4x^3 + 12x^2 + 4x - 12 = 0$

20. $3x^6 + 3x^5 = 7x^3 + x^2 - 2$

21. A grain silo consists of a cylindrical main section and a hemispherical roof. If the total volume of the silo (including the part inside the roof section) is 15,000 ft³, and the cylindrical part is 30 ft tall, what is the radius of the silo, to the nearest tenth of a foot?

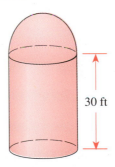

30 ft

22. Suppose a silo like the one described in Exercise 21 has a volume of 20,000 ft³, and the total height of the silo at the top of the roof is 40 ft. What is the approximate radius of the silo in this case?

23. A piece of sheet metal measuring 18 in. square is to be made into a box with an open top by cutting equal squares from each corner and then folding up and soldering the sides. The resulting box is to have a volume of 400 in^3. Show that this can be done in two different ways, and find the dimensions of the box correct to the nearest tenth of an inch for each of the two cases.

24. A rectangular building lot has an area of 5000 ft^2. A diagonal between opposite corners is measured to be 10 ft longer than one side of the lot. What are the dimensions of the lot, to the nearest foot?

SECTION 3.7
COMPLEX NUMBERS

$$2^2 = 4 \geqslant 0$$

$$\left(-\frac{1}{2}\right)^2 = \frac{1}{4} \geqslant 0$$

$$0^2 = 0 \geqslant 0$$

The square root of 25 is 5, since $5^2 = 25$, but -25 has no real square root, since the square of every real number is nonnegative. That is, $x^2 \geqslant 0$ for every real number x. But we need square roots of negative numbers to solve equations like $x^2 + 1 = 0$, whose solutions should be $x = \pm\sqrt{-1}$. The complex number system was invented to provide an environment in which such equations have solutions.

Complex numbers are expressions of the form $a + bi$, where a and b are real and the symbol i represents a number with the property $i^2 = -1$. The solutions of the equation $x^2 + 1 = 0$ are $\pm i$ in this number system.

Examples of Complex Numbers

$$i = \sqrt{-1}$$
$$5 + \tfrac{1}{3}i$$
$$0 - 4i = -4i$$
$$\sqrt{5} + 0i = \sqrt{5}$$

In the complex number $z = a + bi$, a is called the **real part** and b is called the **imaginary part** of z. For instance, the real part of $3 + 7i$ is 3 and the imaginary part is 7. Note that the real and imaginary parts are both real numbers. Two complex numbers are **equal** if their real and imaginary parts are the same. Any real number a can be thought of as a complex number with imaginary part 0; that is, $a = a + 0i$. Complex numbers with a nonzero imaginary part are called **imaginary numbers**; complex numbers of the form bi (with real part zero) are called **pure imaginary numbers**.

Although we use the term *imaginary* in this context, imaginary numbers should not be thought of as any less "real" (in the ordinary rather than the mathematical sense of that word) than negative or irrational numbers. All numbers (except possibly the positive integers) are creations of the human mind, the numbers -1 and $\sqrt{2}$ no less so than the number i. We study complex numbers because they complete, in a useful and elegant fashion, the study of the solution of polynomial equations. In fact, imaginary numbers are useful not just in algebra and mathematics, but in the other sciences as well. To give just one example, in electrical theory the *reactance* of a circuit is a quantity whose measure is an imaginary number.

ARITHMETIC OPERATIONS ON COMPLEX NUMBERS

Complex numbers are added, subtracted, multiplied, and divided just as we would any number of the form $a + b\sqrt{c}$. The only difference we must keep in mind is that $i^2 = -1$. Thus, in particular, the following calculation should be valid.

$$\begin{aligned}
(a + bi)(c + di) &= ac + (ad + bc)i + bdi^2 \\
&= ac + (ad + bc)i + bd(-1) \\
&= (ac - bd) + (ad + bc)i
\end{aligned}$$

We therefore define the sum, difference, and product of complex numbers as follows.

ADDING, SUBTRACTING, AND MULTIPLYING COMPLEX NUMBERS

$$(a + bi) + (c + di) = (a + c) + (b + d)i$$
$$(a + bi) - (c + di) = (a - c) + (b - d)i$$
$$(a + bi)(c + di) = (ac - bd) + (ad + bc)i$$

EXAMPLE 1

Express the following in the form $a + bi$:
(a) $(3 + 5i) + (4 - 2i)$ (b) $(3 + 5i) - (4 - 2i)$
(c) $(3 + 5i)(4 - 2i)$ (d) i^{23}

SOLUTION

(a) According to the definition, we add the real parts and we add the imaginary parts:

$$(3 + 5i) + (4 - 2i) = (3 + 4) + (5 - 2)i = 7 + 3i$$

(b) $(3 + 5i) - (4 - 2i) = (3 - 4) + [5 - (-2)]i = -1 + 7i$

(c) $(3 + 5i)(4 - 2i) = [3 \cdot 4 - 5(-2)] + [3(-2) + 5 \cdot 4]i = 22 + 14i$

(d) $i^{23} = i^{20+3} = (i^2)^{10}i^3 = (-1)^{10}i^2i = (1)(-1)i = -i$ ■

We now consider division of complex numbers. The process is much like rationalizing the denominator of a radical expression, which we considered in Section 1.2. We will use the following property of complex numbers.

COMPLEX CONJUGATES

Let $z = a + bi$. The **complex conjugate** of z is $\bar{z} = a - bi$. We have

$$z\bar{z} = (a + bi)(a - bi) = a^2 + b^2$$

This formula says that the product of a complex number and its complex conjugate is always a nonnegative real number. This fact is used to divide complex numbers, as in the following example.

EXAMPLE 2

Express the following in the form $a + bi$:   (a) $\dfrac{3 + 5i}{1 - 2i}$ (b) $\dfrac{7 + 3i}{4i}$

SOLUTION

(a) We multiply both numerator and denominator by the complex conjugate of the denominator, to make the new denominator a real number. The complex conjugate of $1 - 2i$ is $\overline{1 - 2i} = 1 + 2i$, and their product is $(1 - 2i)(1 + 2i) = 1^2 + 2^2 = 5$.

$$\frac{3 + 5i}{1 - 2i} = \left(\frac{3 + 5i}{1 - 2i}\right)\left(\frac{1 + 2i}{1 + 2i}\right) = \frac{-7 + 11i}{5} = -\frac{7}{5} + \frac{11}{5}i$$

(b) $$\frac{7 + 3i}{4i} = \left(\frac{7 + 3i}{4i}\right)\left(\frac{-4i}{-4i}\right) = \frac{12 - 28i}{16} = \frac{3}{4} - \frac{7}{4}i$$ ■

We summarize what we have learned about dividing complex numbers in the following box.

DIVIDING COMPLEX NUMBERS

To simplify the quotient $\frac{a + bi}{c + di}$, multiply the numerator and denominator by the complex conjugate of the denominator:

$$\frac{a + bi}{c + di} = \left(\frac{a + bi}{c + di}\right)\left(\frac{c - di}{c - di}\right) = \frac{(ac + bd) + (bc - ad)i}{c^2 + d^2}$$

Rather than memorize this entire formula, it is best just to remember the first step and then multiply out the numerator and denominator as usual.

Just as every positive real number r has two square roots $\left(\sqrt{r}$ and $-\sqrt{r}\right)$, every negative number has two square roots as well. Both are pure imaginary numbers, for if $r > 0$ is real, then

$$\left(i\sqrt{r}\right)^2 = i^2r = -r$$

and $$\left(-i\sqrt{r}\right)^2 = (-1)^2i^2r = -r$$

We call $i\sqrt{r}$ the **principal square root** of $-r$, and we will use the symbol $\sqrt{-r}$ to denote the principal square root. The other square root will then be $-\sqrt{-r} = -i\sqrt{r}$. Note that the two square roots of a negative real number are complex conjugates of each other.

SQUARE ROOTS OF NEGATIVE NUMBERS

If $-r < 0$, then the square roots of $-r$ are

$$i\sqrt{r} \quad \text{and} \quad -i\sqrt{r}$$

The **principal** square root of $-r$ is $i\sqrt{r}$.

We usually write $i\sqrt{b}$ instead of $\sqrt{b}i$ to avoid confusion with $\sqrt{bi}$.

EXAMPLE 3

Evaluate: (a) $\sqrt{-1}$ (b) $\sqrt{-16}$ (c) $\sqrt{-3}$

SOLUTION

(a) $\sqrt{-1} = i\sqrt{1} = i$ (b) $\sqrt{-16} = i\sqrt{16} = 4i$ (c) $\sqrt{-3} = i\sqrt{3}$ ■

Special care must be taken when performing calculations involving square roots of negative numbers. Although $\sqrt{a} \cdot \sqrt{b} = \sqrt{ab}$ when a and b are positive, this is *not* true when both are negative. For example,

$$\sqrt{-2} \cdot \sqrt{-3} = i\sqrt{2} \cdot i\sqrt{3} = i^2\sqrt{6} = -\sqrt{6}$$

but

$$\sqrt{(-2)(-3)} = \sqrt{6}$$

 so

$$\sqrt{-2} \cdot \sqrt{-3} \neq \sqrt{(-2)(-3)}$$

When multiplying radicals of negative numbers, express them first in the form $i\sqrt{r}$ (where $r > 0$) to avoid possible error.

EXAMPLE 4

Evaluate $\left(\sqrt{12} - \sqrt{-3}\right)\left(3 + \sqrt{-4}\right)$ and express in the standard form for complex numbers.

SOLUTION

$$
\begin{aligned}
\left(\sqrt{12} - \sqrt{-3}\right)\left(3 + \sqrt{-4}\right) &= \left(\sqrt{12} - i\sqrt{3}\right)\left(3 + i\sqrt{4}\right) \\
&= \left(2\sqrt{3} - i\sqrt{3}\right)\left(3 + 2i\right) \\
&= \left(6\sqrt{3} + 2\sqrt{3}\right) + i\left(2 \cdot 2\sqrt{3} - 3\sqrt{3}\right) \\
&= 8\sqrt{3} + i\sqrt{3}
\end{aligned}
$$

■

■ GRAPHING COMPLEX NUMBERS

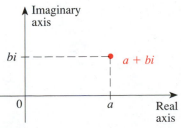

Figure 1

To graph real numbers or sets of real numbers, we have been using the number line, which has just one dimension. Complex numbers, however, have two components: the real part and the imaginary part. This suggests the following method for graphing complex numbers. We need two axes: one for the real part and one for the imaginary part. We call these the **real axis** and the **imaginary axis**, respectively. The plane determined by these two axes is called the **complex plane**. To graph the complex number $a + bi$, we plot the ordered pair of numbers (a, b) in this plane, as indicated in Figure 1.

Leonhard Euler (1707–1783) was
born in Basel, Switzerland, the son of
a pastor. At age 13 his father sent
him to the University at Basel to study
theology, but Euler soon decided to
devote himself to the sciences. Besides
theology he studied mathematics,
medicine, astronomy, physics, and
oriental languages. It is said that
Euler could calculate as effortlessly as
"men breathe or as eagles fly." One
hundred years before Euler, Fermat
(see page 52) had conjectured that
$2^{2^n} + 1$ is a prime number for all n.
The first five of these numbers are 5,
17, 257, 65537, and 4,294,967,297.
It is easy to show that the first four
are prime. The fifth was also thought
to be prime until Euler, with his phenomenal calculating ability, showed
that it is the product $641 \times 6,700,417$
and so is not prime. Euler published
more than any other mathematician in
history. His collected works comprise
75 large volumes. Although he was
blind for the last 17 years of his life,
he continued to work and publish. In
his writings he popularized the use of
the symbols π, e, and i, which you
will find in this textbook. One of
Euler's most lasting contributions is
the development of complex numbers.

EXAMPLE 5

Graph the complex numbers $z_1 = 2 + 3i$, $z_2 = 5 - 2i$, and $z_1 + z_2$.

SOLUTION

We have $z_1 + z_2 = (2 + 3i) + (5 - 2i) = 7 + i$. The graph is shown in
Figure 2.

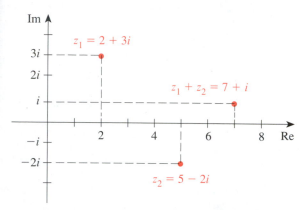

Figure 2

EXAMPLE 6

Graph the following sets of complex numbers:
(a) $S = \{a + bi \mid a \geq 0\}$ (b) $T = \{a + bi \mid a < 1, b \geq 0\}$

SOLUTION

(a) S is the set of complex numbers whose real part is nonnegative. Its graph is
shown in Figure 3(a).
(b) T is the set of complex numbers for which the real part is less than 1 and the
imaginary part is nonnegative. See Figure 3(b) for the graph.

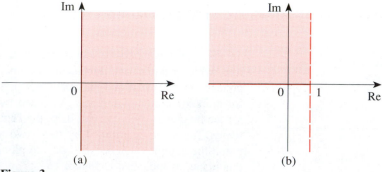

Figure 3

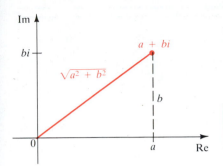

Figure 4

Recall from Section 1.1 that the absolute value of a real number can be thought of as its distance from the origin on the real number line. We define absolute value for complex numbers in a similar fashion. From Figure 4 we can see, using the Pythagorean Theorem, that the distance between $a + bi$ and the origin in the complex plane is $\sqrt{a^2 + b^2}$. This leads to the following definition:

> The **modulus** (or **absolute value**) of the complex number $z = a + bi$ is
> $$|z| = \sqrt{a^2 + b^2}$$

EXAMPLE 7

Find the moduli of the complex numbers $3 + 4i$ and $8 - 5i$.

SOLUTION

$$|3 + 4i| = \sqrt{3^2 + 4^2} = \sqrt{25} = 5$$
$$|8 - 5i| = \sqrt{8^2 + (-5)^2} = \sqrt{89}$$

EXAMPLE 8

Graph the following sets of complex numbers:
(a) $C = \{z \mid |z| = 1\}$ (b) $D = \{z \mid |z| \leq 1\}$

SOLUTION

(a) C is the set of complex numbers whose distance from the origin is 1. Thus C is a circle of radius 1 with center at the origin.
(b) D is the set of complex numbers whose distance from the origin is less than or equal to 1. Thus D is the disk that consists of all complex numbers on and inside the circle C of part (a).

The graphs of C and D are shown in Figure 5.

Figure 5

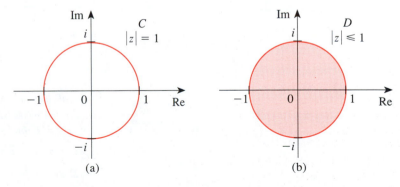

EXERCISES 3.7

In Exercises 1–6 find the real and imaginary parts of the complex number.

1. $5 - 6i$

2. $i\sqrt{7}$

3. $\dfrac{4 + i\sqrt{3}}{2}$

4. $\dfrac{3 - \sqrt{-6}}{\sqrt{3}}$

5. $\sqrt{2} + \sqrt{-3} - \sqrt{4} - \sqrt{-5}$

6. $\dfrac{2 + 4i}{\sqrt{-16}}$

In Exercises 7–40 evaluate the expression and write the result in the form $a + bi$.

7. $(3 + 2i) + (7 - 3i)$

8. $(6 - 4i) + (-3 + 7i)$

9. $\left(4 - \tfrac{1}{2}i\right) + \left(\tfrac{3}{2} + 5i\right)$

10. $(1 + i) - (2 - 3i)$

11. $(-12 + 8i) - (7 + 4i)$

12. $3i - (4 - i)$

13. $4 \cdot (2 + 6i)$

14. $2i\left(\tfrac{1}{2} - i\right)$

15. $(3 - i)(4 + i)$

16. $(4 - 7i)(1 + 3i)$

17. $(3 - 4i)(5 - 12i)$

18. $\left(\tfrac{2}{3} + 12i\right)\left(\tfrac{1}{6} + 24i\right)$

19. $\dfrac{1}{i}$

20. $\dfrac{1}{1 + i}$

21. $\dfrac{2 + 3i}{1 - 5i}$

22. $\dfrac{5 - i}{3 + 4i}$

23. $\dfrac{26 + 39i}{2 - 3i}$

24. $\dfrac{3}{4 - 3i}$

25. $\dfrac{4i}{3 - 2i}$

26. $(2 - 6i)^{-1}$

27. i^3

28. $(2i)^4$

29. i^{100}

30. i^{1002}

31. $\sqrt{-25}$

32. $\sqrt{\dfrac{-9}{4}}$

33. $\sqrt{-3}\sqrt{-12}$

34. $\sqrt{\tfrac{1}{3}}\sqrt{-27}$

35. $\left(3 - \sqrt{-5}\right)\left(1 + \sqrt{-15}\right)$

36. $\dfrac{1 - \sqrt{-1}}{1 + \sqrt{-1}}$

37. $\dfrac{2 + \sqrt{-7}}{1 + \sqrt{-14}}$

38. $\left(\sqrt{3} - \sqrt{-4}\right)\left(\sqrt{6} - \sqrt{-5}\right)$

39. $\dfrac{\sqrt{-\tfrac{1}{2}}}{\sqrt{-2}\sqrt{-9}}$

40. $\dfrac{\sqrt{-7}\sqrt{-49}}{\sqrt{28}}$

In Exercises 41–48 graph the complex number and find its modulus.

41. $3i$

42. -3

43. $5 + 2i$

44. $7 - 3i$

45. $\sqrt{3} + i$

46. $-1 - \dfrac{\sqrt{3}}{3}i$

47. $\dfrac{3 + 4i}{5}$

48. $\dfrac{-\sqrt{2} + i\sqrt{2}}{2}$

In Exercises 49 and 50 sketch the complex number z, and also sketch $2z$, $-z$, and $\tfrac{1}{2}z$ on the same complex plane.

49. $z = 1 + i$

50. $z = 2 - 3i$

In Exercises 51 and 52 sketch z_1, z_2, and $z_1 + z_2$.

51. $z_1 = 2 - i$, $z_2 = 2 + i$

52. $z_1 = -1 + i$, $z_2 = 2 - 3i$

In Exercises 53–60 sketch the given set of complex numbers.

53. $\{z = a + bi \mid a \le 0,\ b \ge 0\}$

54. $\{z = a + bi \mid a > 1,\ b > 1\}$

55. $\{z = a + bi \mid a + b < 2\}$

56. $\{z = a + bi \mid a \ge b\}$

57. $\{z \mid |z| = 3\}$

58. $\{z \mid |z| \ge 1\}$

59. $\{z \mid |z| < 2\}$

60. $\{z \mid 2 \le |z| \le 5\}$

Recall that the symbol $\bar{z}$ represents the complex conjugate of z. In Exercises 61–68 prove the given statement. Assume that $z = a + bi$ and $w = c + di$.

61. $\overline{z} + \overline{w} = \overline{z + w}$

62. $\overline{zw} = \overline{z} \cdot \overline{w}$

63. $(\bar{z})^2 = \overline{z^2}$

64. $\bar{\bar{z}} = z$

65. $z + \bar{z}$ is a real number

66. $z = \bar{z}$ if and only if z is real

67. $\dfrac{z}{\bar{z}} + \dfrac{\bar{z}}{z} = \dfrac{2(a^2 - b^2)}{a^2 + b^2}$

68. $\dfrac{z}{\bar{z}} - \dfrac{\bar{z}}{z} = \dfrac{4abi}{a^2 + b^2}$

69. Suppose that the equation

$$ax^2 + bx + c = 0$$

has two imaginary solutions (where a, b, and c are real numbers). Prove that the solutions are complex conjugates of each other. [*Hint:* Use the quadratic formula.]

SECTION 3.8
COMPLEX ROOTS AND THE FUNDAMENTAL THEOREM OF ALGEBRA

We have already seen that the quadratic equation $ax^2 + bx + c = 0$, with $a \neq 0$, has the solutions

$$x = \frac{-b \pm \sqrt{b^2 - 4ac}}{2a}$$

If $b^2 - 4ac < 0$, then the equation has no real solutions. But in the complex number system, this equation will always have solutions, because negative numbers have square roots in this expanded setting.

EXAMPLE 1

Solve the following equations:
(a) $x^2 + 9 = 0$ (b) $x^2 + 4x + 5 = 0$

SOLUTION

(a) $x^2 + 9 = 0$ means $x^2 = -9$, so $x = \pm\sqrt{-9} = \pm i\sqrt{9} = \pm 3i$. The solutions are $3i$ and $-3i$.

(b) By the quadratic formula,

$$x = \frac{-4 \pm \sqrt{4^2 - 4 \cdot 5}}{2}$$

$$= \frac{-4 \pm \sqrt{-4}}{2}$$

$$= \frac{-4 \pm 2i}{2}$$

so the solutions are $-2 + i$ and $-2 - i$.   ■

EXAMPLE 2

Find all the roots of the equation $x^6 - 1 = 0$.

SOLUTION

$$\begin{aligned}
x^6 - 1 &= (x^3)^2 - 1 \\
&= (x^3 - 1)(x^3 + 1) \\
&= (x - 1)(x^2 + x + 1)(x + 1)(x^2 - x + 1)
\end{aligned}$$

Setting each factor equal to zero, we see that 1 and -1 are real solutions of the equation. Using the quadratic formula on the remaining factors, we get

$$x = \frac{-1 \pm \sqrt{1 - 4}}{2} = \frac{-1 \pm \sqrt{-3}}{2}$$

and

$$x = \frac{1 \pm \sqrt{1 - 4}}{2} = \frac{1 \pm \sqrt{-3}}{2}$$

so the roots of the equation are

$$1, \quad -1, \quad -\tfrac{1}{2} + i\tfrac{\sqrt{3}}{2}, \quad -\tfrac{1}{2} - i\tfrac{\sqrt{3}}{2}, \quad \tfrac{1}{2} + i\tfrac{\sqrt{3}}{2}, \quad \tfrac{1}{2} - i\tfrac{\sqrt{3}}{2} \qquad \blacksquare$$

THE FUNDAMENTAL THEOREM OF ALGEBRA

It is a remarkable fact that adding just $\sqrt{-1}$ and its real multiples to the real number system, which is what we did when we created the complex numbers, is sufficient to provide a number system in which *every* polynomial equation has a root. Although we will not prove this fact (a proof requires mathematical expertise well beyond the scope of this book), it nevertheless forms the basis for much of our work in solving polynomial equations. This theorem was proved by the German mathematician C.F. Gauss in 1799.

FUNDAMENTAL THEOREM OF ALGEBRA

Every polynomial

$$P(x) = a_n x^n + a_{n-1} x^{n-1} + \cdots + a_1 x + a_0 \qquad (n \geq 1, \, a_n \neq 0)$$

with complex coefficients has at least one complex zero.

Since any real number is also a complex number, the theorem applies to polynomials with real coefficients as well.

Since every zero c of a polynomial corresponds to a factor of the form $x - c$ (the Factor Theorem), the Fundamental Theorem of Algebra ensures that we can factor any polynomial $P(x)$ of degree n as follows:

$$P(x) = (x - c_1) \cdot Q_1(x)$$

where $Q_1(x)$ is of degree $n - 1$ and c_1 is a zero of $P(x)$. But now applying the Fundamental Theorem to the quotient $Q_1(x)$ gives us the factorization

$$P(x) = (x - c_1) \cdot (x - c_2) \cdot Q_2(x)$$

where $Q_2(x)$ is of degree $n - 2$ and c_2 is a zero of $Q_1(x)$. Continuing this process for n steps, we will get a final quotient $Q_n(x)$ of degree 0, which is therefore a nonzero constant that we will call a. This proves the following corollary of the Fundamental Theorem of Algebra.

COMPLETE FACTORIZATION THEOREM

If $P(x)$ is a polynomial of degree $n > 0$, then there exist complex numbers $a, c_1, c_2, \ldots, c_n$ (with $a \neq 0$) such that

$$P(x) = a(x - c_1)(x - c_2) \cdots (x - c_n)$$

The number a is clearly the coefficient of x^n in $P(x)$. The numbers $c_1, c_2, \ldots, c_n$ are the zeros of $P(x)$ (by the Factor Theorem). These need not all be different. If the factor $x - c$ appears k times in the complete factorization of $P(x)$, we say that c is a zero of **multiplicity** k.

$P(x)$ can have no zeros other than $c_1, c_2, \ldots, c_n$, because if

$$P(c) = a(c - c_1)(c - c_2) \cdots (c - c_n) = 0$$

then at least one of the factors $c - c_i$ must be zero, so that $c = c_i$ for some $i \in \{1, 2, \ldots, n\}$. We have thus shown the following.

ZEROS THEOREM

Every polynomial of degree $n \geq 1$ has exactly n zeros, provided that a zero with multiplicity k is counted k times.

Note that in Example 2 the polynomial in the equation was of degree 6, and the equation had exactly 6 roots.

Carl Friedrich Gauss (1777–1855) is considered the greatest mathematician of modern times. He was referred to by his contemporaries as "The Prince of Mathematics." Gauss was born into a poor family; his father made a living as a mason. As a very small child he found a calculation error in his father's accounts. This was the first of many incidents that gave evidence of his mathematical precocity. At the age of 19 Gauss demonstrated that the regular 17-sided polygon can be constructed with straightedge and compass alone. This was remarkable because, since the time of Euclid, it was thought that the only regular polygons constructible in this way were the triangle and pentagon. Because of this discovery Gauss decided to pursue a career in mathematics instead of languages, his other passion. In his doctoral dissertation written at the age of 22, Gauss proved the Fundamental Theorem of Algebra: A polynomial of degree n with complex coefficients has n roots. His other accomplishments range over every branch of mathematics, as well as physics and astronomy.

EXAMPLE 3

Find the complete factorization and all five zeros of the polynomial

$$P(x) = 3x^5 + 24x^3 + 48x$$

SOLUTION

The terms of P have $3x$ as a common factor, so we get the following factorization:

$$P(x) = 3x(x^4 + 8x^2 + 16)$$
$$= 3x(x^2 + 4)^2$$

To factor $x^2 + 4$, note that $2i$ and $-2i$ are zeros of this polynomial. Thus $x^2 + 4 = (x - 2i)(x + 2i)$, and so

$$P(x) = 3x[(x - 2i)(x + 2i)]^2$$
$$= 3x(x - 2i)(x - 2i)(x + 2i)(x + 2i)$$

Setting each factor equal to zero in turn, we see that the zeros of P are 0, $2i$, and $-2i$. However, $2i$ and $-2i$ are each counted twice, since the factor of P that corresponds to each occurs twice in the factorization of P. Each of these is a zero of multiplicity 2 (or a *double* zero).   ■

EXAMPLE 4

Find a polynomial that satisfies the given description.
(a) A polynomial $P(x)$ of degree 3, with zeros 1, 2, and -4, and constant coefficient 16.
(b) A polynomial $Q(x)$ of degree 4, with zeros i, $-i$, 2, and -2, and value 25 when $x = 3$.
(c) A polynomial $R(x)$ of degree 4, with zeros -2 and 0, where -2 is a zero of multiplicity three.

SOLUTION

(a) From the description, we see that $P(x)$ has the complete factorization $a(x - 1)(x - 2)(x - (-4))$ for some a. Thus

$$P(x) = a(x^2 - 3x + 2)(x + 4)$$
$$= a(x^3 + x^2 - 10x + 8)$$
$$= ax^3 + ax^2 - 10ax + 8a$$

Since the constant coefficient is 16, we see that $a = 2$, so

$$P(x) = 2x^3 + 2x^2 - 20x + 16$$

(b)
$$Q(x) = a(x - i)(x - (-i))(x - 2)(x - (-2))$$
$$= a(x^2 + 1)(x^2 - 4)$$
$$= a(x^4 - 3x^2 - 4)$$

$$Q(3) = a(3^4 - 3 \cdot 3^2 - 4) = 50a = 25, \text{ so } a = \tfrac{1}{2}, \text{ and}$$

$$Q(x) = \tfrac{1}{2}x^4 - \tfrac{3}{2}x^2 - 2$$

(c)
$$R(x) = a(x - (-2))^3(x - 0)$$
$$= a(x + 2)^3 x$$
$$= a(x^2 + 4x + 4)(x + 2)x$$
$$= a(x^3 + 6x^2 + 12x + 8)x$$
$$= a(x^4 + 6x^3 + 12x^2 + 8x)$$

Since we are given no information about R other than its zeros and their multiplicity, we can choose any number we wish for a. If we pick $a = 1$ we get

$$R(x) = x^4 + 6x^3 + 12x^2 + 8x$$ ■

EXAMPLE 5

Find all five zeros of $P(x) = 3x^5 - 2x^4 - x^3 - 12x^2 - 4x$.

SOLUTION

By Descartes' Rule of Signs, $P(x)$ has one positive real zero and either three or one negative real zero(s). Since $x = 0$ is obviously a zero (but is neither positive nor negative), this proves that there are either five real zeros and no imaginary zero, or three real and two imaginary zeros. Checking through the list of possible rational zeros of $P(x)/x$, we see that $P(2) = 0$ and $P\left(-\tfrac{1}{3}\right) = 0$, so by the Factor Theorem, $x - 2$ and $x + \tfrac{1}{3}$ are factors. Thus

$$P(x) = x(x - 2)\left(x + \tfrac{1}{3}\right)(3x^2 + 3x + 6)$$

$$= x(x - 2)(3x + 1)(x^2 + x + 2)$$

The roots of the quadratic factor are

$$x = \frac{-1 \pm \sqrt{1 - 8}}{2}$$

$$= -\frac{1}{2} \pm i\frac{\sqrt{7}}{2}$$

so the zeros of $P(x)$ are 0, 2, $-\dfrac{1}{3}$, $-\dfrac{1}{2} + i\dfrac{\sqrt{7}}{2}$, and $-\dfrac{1}{2} - i\dfrac{\sqrt{7}}{2}$.   ■

EXAMPLE 6

Find all four zeros of $P(x) = x^4 + 2x^2 + 9$.

SOLUTION

$$
\begin{aligned}
P(x) &= x^4 + 2x^2 + 9 \\
&= x^4 + 6x^2 + 9 - 4x^2 && \text{(completing the square)} \\
&= (x^2 + 3)^2 - (2x)^2 \\
&= (x^2 - 2x + 3)(x^2 + 2x + 3) && \text{(difference of squares formula)}
\end{aligned}
$$

Using the quadratic formula on each factor, we get

$$
x = \frac{2 \pm \sqrt{4 - 12}}{2} = 1 \pm i\sqrt{2}
$$

and

$$
x = \frac{-2 \pm \sqrt{4 - 12}}{2} = -1 \pm i\sqrt{2}
$$

so the zeros of $P(x)$ are $1 + i\sqrt{2}$, $1 - i\sqrt{2}$, $-1 + i\sqrt{2}$, and $-1 - i\sqrt{2}$. ■

As you may have noticed from the examples so far, the imaginary roots of polynomial equations with real coefficients come in pairs. Whenever $a + bi$ is a root, so is its complex conjugate $a - bi$. This is always the case, as the following theorem states.

CONJUGATE ROOTS THEOREM

> If the polynomial $P(x)$ of degree $n > 0$ has real coefficients, and if the complex number z is a root of the equation $P(x) = 0$, then so is its complex conjugate $\bar{z}$.

The proof of this theorem is given at the end of this section.

EXAMPLE 7

Find a polynomial $P(x)$ of degree 5 that has integer coefficients, and zeros $\frac{1}{2}$, $3 - i$, and $2i$.

SOLUTION

Since $3 - i$ and $2i$ are zeros, so are $3 + i$ and $-2i$ by the Conjugate Roots Theorem. This means that $P(x)$ has all the factors in the following product:

$$
\left(x - \tfrac{1}{2}\right)[x - (3 - i)][x - (3 + i)](x - 2i)(x + 2i) = \left(x - \tfrac{1}{2}\right)(x^2 - 6x + 10)(x^2 + 4)
$$

$$
= x^5 - \tfrac{13}{2}x^4 + 17x^3 - 31x^2 + 52x - 20
$$

Multiplying by 2 to make all coefficients integers, we get

$$P(x) = 2x^5 - 13x^4 + 34x^3 - 62x^2 + 104x - 40$$

Any other solution must be an integer multiple of this one.   ■

EXAMPLE 8

Find all roots of the equation $x^4 - 12x^3 + 56x^2 - 120x + 96 = 0$, given that one root is $3 + i\sqrt{3}$.

SOLUTION

The complex conjugate of $3 + i\sqrt{3}$ is also a root, so both $x - \left(3 + i\sqrt{3}\right)$ and $x - \left(3 - i\sqrt{3}\right)$, and hence their product, must divide the polynomial in the equation.

$$\left[x - \left(3 + i\sqrt{3}\right)\right] \cdot \left[x - \left(3 - i\sqrt{3}\right)\right] = x^2 - 6x + 12$$

We divide $x^2 - 6x + 12$ into the original polynomial.

$$
\begin{array}{r}
x^2 - 6x + 8 \\
x^2 - 6x + 12\overline{)x^4 - 12x^3 + 56x^2 - 120x + 96} \\
\underline{x^4 - 6x^3 + 12x^2} \\
-6x^3 + 44x^2 - 120x \\
\underline{-6x^3 + 36x^2 - 72x} \\
8x^2 - 48x + 96 \\
\underline{8x^2 - 48x + 96} \\
0
\end{array}
$$

The quotient factors into $(x - 2)(x - 4)$, so the roots of the equation are 2, 4, $3 + i\sqrt{3}$ and $3 - i\sqrt{3}$.   ■

■ PROOF OF THE CONJUGATE ROOTS THEOREM

Let

$$P(x) = a_n x^n + a_{n-1}x^{n-1} + \cdots + a_1 x + a_0$$

where each coefficient is real. Suppose that $P(z) = 0$. To prove the Conjugate Roots Theorem, we must prove that $\bar{z}$ is also a zero of P. We use the facts that the complex conjugate of a sum of two complex numbers is the sum of the conjugates and that the conjugate of a product is the product of the conjugates. (See Exercises 61–66 of Section 3.7.)

$$P(\bar{z}) = a_n(\bar{z})^n + a_{n-1}(\bar{z})^{n-1} + \cdots + a_1\bar{z} + a_0$$
$$= \overline{a_n}\overline{z^n} + \overline{a_{n-1}}\overline{z^{n-1}} + \cdots + \overline{a_1}\,\overline{z} + \overline{a_0} \quad \text{(since the coefficients are real)}$$
$$= \overline{a_n z^n} + \overline{a_{n-1}z^{n-1}} + \cdots + \overline{a_1 z} + \overline{a_0}$$
$$= \overline{a_n z^n + a_{n-1}z^{n-1} + \cdots + a_1 z + a_0}$$
$$= \overline{P(z)} = \overline{0} = 0$$

This derivation shows that the conjugate $\bar{z}$ is also a root of $P(x) = 0$, and we have proved the theorem.   ∎

EXERCISES 3.8

In Exercises 1–14 find all the solutions of the equation.

1. $x^2 + 4 = 0$

2. $25x^2 + 9 = 0$

3. $x^2 - x + 1 = 0$

4. $x^2 + 2x + 2 = 0$

5. $x^2 + 4x + 8 = 0$

6. $2x^2 + 2x + 1 = 0$

7. $3x^2 - 5x + 4 = 0$

8. $2x^2 - 3x + 2 = 0$

9. $x^2 - 8x + 17 = 0$

10. $3x^2 - 4x + 2 = 0$

11. $t + 3 + \dfrac{3}{t} = 0$

12. $\theta^3 + \theta^2 + \theta = 0$

13. $z^2 - iz = 0$

14. $2z^3 + iz^2 = 0$

In Exercises 15–20 find a polynomial with integer coefficients that satisfies the given conditions.

15. $P(x)$ has degree 3, zeros 2 and i, and leading coefficient 1.

16. $Q(x)$ has degree 4, zeros $1 + i$ and 1, with 1 a zero of multiplicity 2, and $Q(0) = 4$.

17. $S(x)$ has degree 4 and zeros $1 + i$ and $3 - 4i$, and the coefficient of x^2 is 39.

18. $R(x)$ has degree 3, with 2 a zero of multiplicity 3.

19. $T(x)$ has constant coefficient 8 and is of the smallest possible degree consistent with having $1 - i$ as a zero of multiplicity 2.

20. $U(x)$ has degree 3, with $U(2i) = 0$, $U(2) = 0$, and $U(1) = -10$.

In Exercises 21–24 show that the indicated value of x is a solution of the given equation, and then find all solutions.

21. $x^3 - 2x^2 + 4x - 8 = 0$, $\quad x = 2i$

22. $x^3 + 5x^2 + 8x + 6 = 0$, $\quad x = -1 + i$

23. $2x^4 + 9x^2 + 4 = 0$, $\quad x = 2i$

24. $x^5 + 5x^3 + 4x = 0$, $\quad x = i$

In Exercises 25–42 find all solutions of the equation.

25. $x^4 - 1 = 0$

26. $x^3 - 64 = 0$

27. $x^3 + 8 = 0$

28. $x^4 + 4 = 0$

29. $x^4 - 16 = 0$

30. $16x^4 - 81 = 0$

31. $x^6 - 729 = 0$

32. $x^4 + 2x^2 + 1 = 0$

33. $x^4 + 10x^2 + 25 = 0$

34. $x^6 + 7x^3 - 8 = 0$

35. $x^3 + 2x^2 + 4x + 8 = 0$

36. $x^3 - 7x^2 + 17x - 15 = 0$

37. $x^3 - 2x^2 + 2x - 1 = 0$

38. $x^3 + 7x^2 + 18x + 18 = 0$

39. $x^3 - 3x^2 + 3x - 2 = 0$

40. $2x^3 - 8x^2 + 9x - 9 = 0$

41. $x^4 + x^3 + 7x^2 + 9x - 18 = 0$

42. $x^5 + x^3 + 8x^2 + 8 = 0$ [*Hint:* Factor by grouping.]

In Exercises 43–48 find the complete factorization of the polynomial.

43. $x^3 + 27$

44. $x^4 - 625$

45. $x^6 - 64$

46. $x^5 + 3x^3 + 2x$

47. $2x^3 + 7x^2 + 12x + 9$

48. $x^4 - x^3 + 7x^2 - 9x - 18$

49. Show that every polynomial with real coefficients and odd degree has at least one real root. [*Hint:* Use the Conjugate Roots Theorem.]

In Exercises 50 and 51 find the value of the polynomial at the given number.

50. $2x^3 + ix^2 - 3ix - (50 + 5i)$,   3

51. $3x^3 - x^2 + x - 4$, $2i$

52. (a) Show that $2i$ and $1 - i$ are both solutions of the equation

$$x^2 - (1 + i)x + (2 + 2i) = 0,$$

but that their complex conjugates $-2i$ and $1 + i$ are not.

(b) Explain why the result of part (a) does not violate the Conjugate Roots Theorem.

53. (a) Find the polynomial with *real* coefficients of the smallest possible degree that has i and $1 + i$ as zeros, and in which the coefficient of the highest power is 1.

(b) Find the polynomial with *complex* coefficients of the smallest possible degree that has i and $1 + i$ as zeros, and in which the coefficient of the highest power is 1.

54. The steps in this problem will provide a proof of the fact that it is impossible for the graphs of two different polynomials, each of degree less than or equal to n, to intersect at more than n points.

(a) Let

$$P(x) = a_nx^n + a_{n-1}x^{n-1} + \cdots + a_1x + a_0$$

and

$$Q(x) = b_nx^n + b_{n-1}x^{n-1} + \cdots + b_1x + b_0$$

Suppose that the graphs of P and Q intersect in the $n + 1$ points $(x_1, y_1), (x_2, y_2), \ldots, (x_n, y_n)$, and (x_{n+1}, y_{n+1}). Let $F(x) = P(x) - Q(x)$. Show that $F(x)$ has at least $n + 1$ zeros.

(b) Show that $F(x) = 0$ for all x. [*Hint:* Use the degree of F and the Zeros Theorem.]

(c) Conclude that P and Q are the same polynomial.

SECTION 3.9
RATIONAL FUNCTIONS

A **rational function** is a function of the form

$$y = r(x) = \frac{P(x)}{Q(x)}$$

where P and Q are polynomials. We assume that $P(x)$ and $Q(x)$ have no factors in common. Although polynomial functions are defined for all real values of x, rational functions are not defined for those values of x for which the denominator $Q(x)$ is 0. The x-intercepts (if any) of r are the zeros of the numerator $P(x)$, since a fraction is 0 only when its numerator is 0.

EXAMPLE 1

Find the domain, the x-intercepts, and the y-intercept of the function

$$r(x) = \frac{x^2 - 2x - 3}{2x^2 - x}$$

SOLUTION

We factor the numerator and denominator to write

$$r(x) = \frac{(x + 1)(x - 3)}{x(2x - 1)}$$

The function is defined for all x except those for which the denominator is 0, so the domain of $r(x)$ consists of all real numbers except 0 and $\frac{1}{2}$.

The x-intercepts are the zeros of the numerator, $x = -1$ and $x = 3$. The y-intercept is the value of the function when $x = 0$. Since the given function $r(x)$ is not defined for $x = 0$, the graph has no y-intercept. ■

The most important feature that distinguishes the graphs of rational functions is the presence of **asymptotes**. Before we formulate a precise definition of this word, we consider an example to illustrate the concept.

EXAMPLE 2

Sketch a graph of the function $y = \dfrac{x - 3}{x - 2}$.

SOLUTION

The function is not defined for $x = 2$, so we first examine the nature of the function for values of x near 2.

x	y
1	2
1.5	3
1.9	11
1.95	21
1.99	101
1.999	1001

x	y
3	0
2.5	−1
2.1	−9
2.05	−19
2.01	−99
2.001	−999

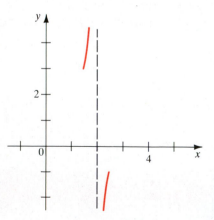

Figure 1

We see from the first table that as x approaches 2 from the left (that is, gets progressively closer to 2 while remaining smaller than 2), the values of y increase without bound. In fact, by taking x sufficiently close to 2 on the left, we can make y larger than any given number. We describe this situation by saying "y approaches infinity as x approaches 2 from the left," and we write this phrase using the notation

$$y \to \infty \quad \text{as} \quad x \to 2^-$$

Similarly, the other table shows that as x approaches 2 from the right, the values of y decrease without bound, and by taking x sufficiently close to 2 on the right, we can make y smaller than any given negative number. In this situation we say that y approaches negative infinity as x approaches 2 from the right, and we write

$$y \to -\infty \quad \text{as} \quad x \to 2^+$$

The graph of $y = r(x)$ therefore has the shape around $x = 2$ shown in Figure 1.

Now we examine the behavior of the function as x becomes progressively larger in absolute value (for both negative and positive x).

x	y
10	0.8750
100	0.9898
1000	0.9990
10,000	0.9999

x	y
-10	1.0833
-100	1.0098
-1000	1.0010
$-10,000$	1.0001

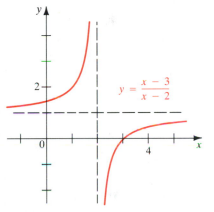

Figure 2

As $|x|$ becomes larger and larger, the values of y get progressively closer to 1. This means that the graph of $y = r(x)$ will approach the horizontal line $y = 1$ as x increases or decreases without bound. We express this by saying "y approaches 1 as x approaches infinity or negative infinity," and we write

$$y \to 1 \quad \text{as} \quad x \to \infty \qquad \text{and} \qquad y \to 1 \quad \text{as} \quad x \to -\infty$$

This means we can complete the graph of $y = r(x)$ as shown in Figure 2. ■

The line $x = 2$ is called a *vertical asymptote* of the graph in Example 2 and the line $y = 1$ is called a *horizontal asymptote*. Informally speaking, an asymptote of a function is a line that the graph of the function gets closer and closer to as one travels out along that line in either direction. More formally, we make the following definitions.

VERTICAL ASYMPTOTES

> The line $x = a$ is a **vertical asymptote** of the function $y = f(x)$ if $y \to \infty$ or $y \to -\infty$ as x approaches a from the right or the left.

If the function $y = f(x)$ is a rational function, then its vertical asymptotes are the lines $x = a$ where a is a zero of the denominator of the function, because only when the denominator is 0 does the function fail to have a real, finite value.

HORIZONTAL ASYMPTOTES

> The line $y = b$ is a **horizontal asymptote** of the function $y = f(x)$ if $y \to b$ as $x \to \infty$ or as $x \to -\infty$.

In the next example, we show the most efficient method for finding the horizontal asymptote of a rational function.

EXAMPLE 3

Find all vertical and horizontal asymptotes and the x- and y-intercepts of the following function. Use this information to graph the function.

$$y = r(x) = \frac{x^2 - x - 6}{2x^2 + 5x - 3}$$

SOLUTION

Factoring the numerator and denominator, we see that

$$r(x) = \frac{(x - 3)(x + 2)}{(2x - 1)(x + 3)}$$

The x-intercepts of the graph are the zeros of the numerator: $x = 3$ and $x = -2$. The y-intercept is $r(0) = (-6)/(-3) = 2$.

The horizontal asymptote (if it exists) will be the value that y approaches as $x \to \pm\infty$. To help us find this value, let us begin by dividing both the numerator and the denominator of $r(x)$ by x^2 (the highest power of x that appears in the function).

$$y = \frac{x^2 - x - 6}{2x^2 + 5x + 3} \cdot \frac{1/x^2}{1/x^2} = \frac{1 - \dfrac{1}{x} - \dfrac{6}{x^2}}{2 + \dfrac{5}{x} - \dfrac{3}{x^2}}$$

Any function of the form c/x^n approaches 0 as $x \to \pm\infty$ (if n is a positive integer).

For example, in the table we examine the values of $6/x^2$ (which appears in the above quotient) as x increases.

x	$6/x^2$	
10	0.06	
100	0.0006	
1000	0.000006	
10,000	0.00000006	← approaching 0

This means that as $x \to \pm\infty$,

$$y \to \frac{1 - 0 - 0}{2 + 0 - 0} = \frac{1}{2}$$

so that $y = \frac{1}{2}$ is the horizontal asymptote.

The vertical asymptotes occur where the function is undefined (or in other words, where the denominator is 0). This means that the vertical asymptotes here are $x = \frac{1}{2}$ and $x = -3$. To be able to graph the function, we need to know whether $y \to \infty$ or $y \to -\infty$ on each side of these vertical lines. Since there are only two choices, we need only determine the sign of y for values of x near the vertical asymptotes. As $x \to \frac{1}{2}^+$, the values of x are slightly larger than $\frac{1}{2}$, so

$$x - 3 < 0$$

$$x + 2 > 0$$

$$2x - 1 > 2 \cdot \frac{1}{2} - 1 = 0$$

and $$x + 3 > 0$$

This means that as $x \to \frac{1}{2}^+$, y is the quotient of one negative and three positive factors and therefore must be negative. So $y \to -\infty$ as $x \to \frac{1}{2}^+$. We can represent what happens here and at other sides of the vertical asymptotes schematically as in the following table:

as $x \to$		$\frac{1}{2}^+$	$\frac{1}{2}^-$	-3^+	-3^-
sign of $y =$	$\dfrac{(x - 3)(x + 2)}{(2x - 1)(x + 3)}$	$\dfrac{(-)(+)}{(+)(+)}$	$\dfrac{(-)(+)}{(-)(+)}$	$\dfrac{(-)(-)}{(-)(+)}$	$\dfrac{(-)(-)}{(-)(-)}$
$y \to$		$-\infty$	∞	$-\infty$	∞

Putting together this information about intercepts, asymptotes, and the behavior of the function near asymptotes, we obtain the partial graph in Figure 3, where the asymptotes have been plotted as broken lines.

Figure 3

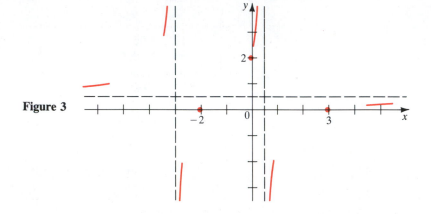

All we need to do now is to plot a few more points and fill in the rest of the graph, which is shown in Figure 4.

x	-4	-1	1	2	4	6
y	1.56	0.67	-1.5	-0.27	0.12	0.24

Figure 4

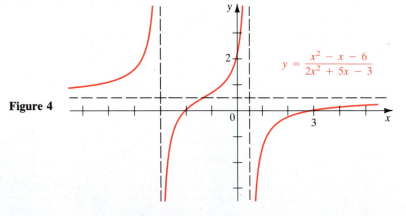

$$y = \frac{x^2 - x - 6}{2x^2 + 5x - 3}$$

We summarize the procedure to be followed in graphing rational functions by the following sequence of steps.

SKETCHING GRAPHS OF RATIONAL FUNCTIONS

1. **Factor.** Factor the numerator and denominator.
2. **Intercepts.** Find the x-intercepts by determining the zeros of the numerator, and the y-intercept from the value of the function at $x = 0$.
3. **Vertical asymptotes.** Find the vertical asymptotes by determining the zeros of the denominator, and then see if $y \to \infty$ or $y \to -\infty$ on each side of the vertical asymptote.
4. **Horizontal asymptote.** Find the horizontal asymptote (if any) by dividing both numerator and denominator by the highest power of x that appears in the denominator, and then letting $x \to \pm\infty$.
5. **Sketch the graph.** Sketch a partial graph using the information provided by the first four steps of this procedure. Then plot as many additional points as required to fill in the rest of the graph of the function.

Notice that we use three forms of the equation defining the rational function when we perform this analysis. In Example 3, the original form

$$r(x) = \frac{x^2 - x - 6}{2x^2 + 5x - 3}$$

easily gave us the y-intercept $r(0) = \frac{-6}{-3} = 2$. The factored form

$$r(x) = \frac{(x - 3)(x + 2)}{(2x - 1)(x + 3)}$$

gave us the x-intercepts and the vertical asymptotes. Finally, the form

$$r(x) = \frac{1 - \dfrac{1}{x} - \dfrac{6}{x^2}}{2 + \dfrac{5}{x} - \dfrac{3}{x^2}}$$

told us that the horizontal asymptote is $y = \frac{1}{2}$.

EXAMPLE 4

Graph the function $y = r(x) = \dfrac{x^2}{x^3 - 2x^2 - x + 2}$.

SOLUTION

Factoring the denominator, we get

Factor

$$r(x) = \frac{x^2}{(x - 2)(x - 1)(x + 1)}$$

Intercepts

Vertical asymptotes

The graph passes through the origin; this is simultaneously the y-intercept and the only x-intercept. The denominator is 0 when x is 2, 1, or -1, so the vertical asymptotes are $x = 2$, $x = 1$, and $x = -1$. The table shows the behavior of the function around the vertical asymptotes.

as $x \rightarrow$	2^+	2^-	1^+	1^-	-1^+	-1^-
sign of $y = \dfrac{x^2}{(x-2)(x-1)(x+1)}$	$\dfrac{(+)}{(+)(+)(+)}$	$\dfrac{(+)}{(-)(+)(+)}$	$\dfrac{(+)}{(-)(+)(+)}$	$\dfrac{(+)}{(-)(-)(+)}$	$\dfrac{(+)}{(-)(-)(+)}$	$\dfrac{(+)}{(-)(-)(-)}$
$y \rightarrow$	∞	$-\infty$	$-\infty$	∞	∞	$-\infty$

Dividing numerator and denominator by x^3 gives

Horizontal asymptote

$$r(x) = \frac{\dfrac{1}{x}}{1 - \dfrac{2}{x} - \dfrac{1}{x^2} + \dfrac{2}{x^3}}$$

so as $x \rightarrow \pm\infty$, $y \rightarrow \dfrac{0}{1 - 0 - 0 + 0} = 0$. The horizontal asymptote is $y = 0$ (the x-axis).

Calculating the coordinates of a few more points now provides us with enough information to sketch the graph of the function in Figure 5.

Sketch the graph

x	-2	$-\dfrac{1}{2}$	$\dfrac{1}{2}$	$\dfrac{3}{2}$	3
y	$-\dfrac{1}{3}$	$\dfrac{2}{15}$	$\dfrac{2}{9}$	$-3\dfrac{3}{5}$	$\dfrac{9}{8}$

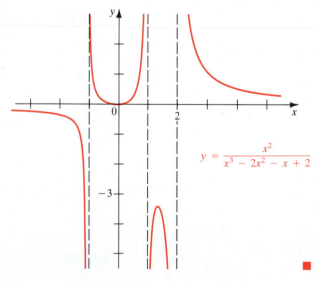

$$y = \frac{x^2}{x^3 - 2x^2 - x + 2}$$

Figure 5

Let
$$r(x) = \frac{P(x)}{Q(x)} = \frac{a_n x^n + a_{n-1} x^{n-1} + \cdots + a_1 x + a_0}{b_m x^m + b_{m-1} x^{m-1} + \cdots + b_1 x + b_0}$$

be a rational function. If the degrees of P and Q are the same (so that $n = m$), then we can see by dividing both numerator and denominator by x^n that $y = a_n/b_m$ is the horizontal asymptote of $y = r(x)$. This was the case in Example 3. If the degree of P is less than the degree of Q (that is, $n < m$), then dividing numerator and denominator by x^m shows that $y = 0$ is the horizontal asymptote, as in Example 4. Finally, if $n > m$ the same procedure shows that the function has no horizontal asymptotes, as the next example illustrates.

EXAMPLE 5

Graph the function $y = r(x) = \dfrac{x^3 + 1}{x - 2}$.

SOLUTION

Factoring the numerator, we see that

Factor

$$r(x) = \frac{(x + 1)(x^2 - x + 1)}{x - 2}$$

Intercepts

Vertical asymptote

so the x-intercept is -1, the y-intercept is $-\frac{1}{2}$, and the only vertical asymptote is $x = 2$. By performing the same kind of analysis as in the previous two examples, we can show that $y \to \infty$ as $x \to 2^+$, and $y \to -\infty$ as $x \to 2^-$.

Dividing numerator and denominator by x gives

$$y = \frac{x^2 + \dfrac{1}{x}}{1 - \dfrac{2}{x}}$$

Check for horizontal asymptote

As $x \to \infty$, the terms $1/x$ and $2/x$ approach 0, so since $x^2 \to \infty$ and the denominator approaches 1, $y \to \infty$. Similarly, as $x \to -\infty$, $y \to \infty$, so the function has no horizontal asymptote.

Plotting the points listed in the table allows us to complete the graph of the function in Figure 6.

Sketch the graph

x	-3	-2	1	3	4
y	5.2	1.75	-2	28	32.5

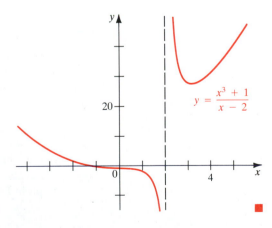

$$y = \frac{x^3 + 1}{x - 2}$$

Figure 6

As with the graphs of polynomial functions, we cannot tell exactly where the turning point of the right branch of the function in Example 5 lies without the use of calculus. But from the asymptotic behavior of the function we know that such a turning point must exist, and we could locate it more precisely by plotting more points in the vicinity of $x = 3$.

SLANT ASYMPTOTES

If $r(x) = P(x)/Q(x)$ is a rational function in which the degree of the numerator is one more than the degree of the denominator, we can use the Division Algorithm to express the function in the form

$$r(x) = ax + b + \frac{R(x)}{Q(x)}$$

where the degree of R is less than the degree of Q and $a \neq 0$. This means that as $x \to \pm\infty$, $R(x)/Q(x) \to 0$, so for large values of $|x|$, the graph of $y = r(x)$ approaches the graph of the line $y = ax + b$. In this situation we say that $y = ax + b$ is a **slant asymptote** or an **oblique asymptote**.

EXAMPLE 6

Graph the function

$$r(x) = \frac{x^2 - 4x - 5}{x - 3}$$

SOLUTION

Since the degree of the numerator is one more than the degree of the denominator, the function will have a slant asymptote. After dividing, we find

$$r(x) = x - 1 - \frac{8}{x - 3}$$

so the slant asymptote is the line $y = x - 1$. The line $x = 3$ is a vertical asymptote, and it is easy to see that $r(x) \to -\infty$ as $x \to 3^+$ and $r(x) \to \infty$ as $x \to 3^-$.

Factoring the numerator in the original formula for r gives

$$r(x) = \frac{(x + 1)(x - 5)}{x - 3}$$

so the x-intercepts are -1 and 5, and the y-intercept is $\frac{5}{3}$. Plotting the asymptotes, intercepts, and the additional points listed in the table allows us to complete the graph in Figure 7.

x	-2	1	2	4	6
y	-1.4	4	9	-5	2.33

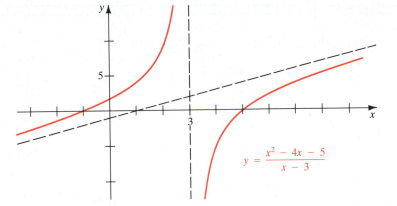

Figure 7

$$y = \frac{x^2 - 4x - 5}{x - 3}$$

When graphing a rational function in which the degree of the numerator is one plus the degree of the denominator, we must modify the fourth guideline given on page 239. Instead of finding the horizontal asymptote (which does not exist in this case), we find the slant asymptote using the method described in Example 6.

The following box summarizes what we have observed about horizontal and slant asymptotes.

HORIZONTAL AND SLANT ASYMPTOTES

Let

$$r(x) = \frac{P(x)}{Q(x)}$$

be a rational function with

$$P(x) = a_n x^n + a_{n-1} x^{n-1} + \cdots + a_1 x + a_0$$

of degree n and

$$Q(x) = b_m x^m + b_{m-1} x^{m-1} + \cdots + b_1 x + b_0$$

of degree m.

1. If $n < m$, then r has horizontal asymptote $y = 0$.

2. If $n = m$, then r has horizontal asymptote $y = \dfrac{a_n}{b_m}$.

3. If $n = m + 1$, then r has a slant asymptote.

4. If $n > m + 1$, then r has no horizontal or slant asymptote.

EXERCISES 3.9

In Exercises 1–6 find the x- and y-intercepts of the function.

1. $y = \dfrac{x - 6}{x + 1}$

2. $y = \dfrac{2}{x - 2}$

3. $y = \dfrac{x}{x^2 - 2x - 15}$

4. $y = \dfrac{x^2 - 2x - 15}{x}$

5. $y = \dfrac{x^2 + 10}{2x}$

6. $y = \dfrac{x^2 - 9}{x^3 - 1}$

In Exercises 7–16 find all asymptotes (including vertical, horizontal, and slant).

7. $y = \dfrac{5}{x + 3}$

8. $y = \dfrac{3x + 3}{x - 3}$

9. $y = \dfrac{x^2}{x^2 - x - 6}$

10. $y = \dfrac{2x - 4}{x^2 + 2x + 1}$

11. $y = \dfrac{6}{x^2 + 2}$

12. $y = \dfrac{(x - 1)(x - 2)}{(x - 3)(x - 4)}$

13. $y = \dfrac{x^2 + 2}{x - 1}$

14. $y = \dfrac{x^3 + 3x^2}{x^2 - 4}$

15. $y = \dfrac{2x^3 - x^2 - 8x + 4}{x + 3}$

16. $y = \dfrac{6x^4}{x^2 - 3}$

In Exercises 17–48 find intercepts and asymptotes, and then graph the given rational function.

17. $y = \dfrac{4}{x - 2}$

18. $y = \dfrac{9}{x + 3}$

19. $y = \dfrac{x - 1}{x - 2}$

20. $y = \dfrac{x + 9}{x - 3}$

21. $y = \dfrac{4x - 4}{x + 2}$

22. $y = \dfrac{2x + 6}{-6x + 3}$

23. $y = \dfrac{2x - 4}{x}$

24. $y = \dfrac{x}{2x - 4}$

25. $y = \dfrac{18}{(x - 3)^2}$

26. $y = \dfrac{x - 2}{(x + 1)^2}$

27. $y = \dfrac{4x + 8}{(x - 4)(x + 1)}$

28. $y = \dfrac{x - 9}{(x + 3)(x - 1)}$

29. $y = \dfrac{(x - 1)(x + 2)}{(x + 1)(x - 3)}$

30. $y = \dfrac{2x(x + 4)}{(x - 1)(x - 2)}$

31. $y = \dfrac{x^2 - 2x + 1}{x^2 + 2x + 1}$

32. $y = \dfrac{4x^2}{x^2 - 2x - 3}$

33. $y = \dfrac{2x^2 + 10x - 12}{x^2 + x - 6}$

34. $y = \dfrac{2x^2 + 2x - 4}{x^2 + x}$

35. $y = \dfrac{x^2 - x - 6}{x^2 + 3x}$

36. $y = \dfrac{x^2 + 3x}{x^2 - x - 6}$

37. $y = \dfrac{3x^2 + 6}{x^2 - 2x - 3}$

38. $y = \dfrac{5x^2 + 5}{x^2 + 4x + 4}$

39. $y = \dfrac{x^2}{x - 2}$

40. $y = \dfrac{x^2 + 2x}{x - 1}$

41. $y = \dfrac{x^2 - 2x - 8}{x}$

42. $y = \dfrac{3x - x^2}{2x - 2}$

43. $y = \dfrac{x^2 + 5x + 4}{x - 3}$

44. $y = \dfrac{x^2 + 4}{2x^2 + x - 1}$

45. $y = \dfrac{x^3 + x^2}{x^2 - 4}$

46. $y = \dfrac{2x^3 + 2x}{x^2 - 1}$

47. $y = \dfrac{x^3}{x - 2}$

48. $y = \dfrac{x^4 - 16}{x^2 - 1}$

In this chapter we adopted the convention that in rational functions, the numerator and denominator do not share common factors. In Exercises 49–53 we study the graphs of rational functions that do not satisfy this rule.

49. Show that the graph of

$$r(x) = \dfrac{3x^2 - 3x - 6}{x - 2}$$

is the line $y = 3x + 3$ with the point $(2, 9)$ removed. [*Hint:* Divide. What is the domain of r?]

In Exercises 50–53 graph the given rational function.

50. $y = \dfrac{x^2 + x - 20}{x + 5}$

51. $y = \dfrac{2x^2 - x - 1}{x - 1}$

52. $y = \dfrac{x^2 - 3x + 2}{x^2 - 4x + 4}$

53. $y = \dfrac{2x^2 - 5x - 3}{x^2 - 2x - 3}$

In Exercises 54–57 construct a rational function $y = P(x)/Q(x)$ that has the indicated properties, and in which the degrees of P and Q are as small as possible.

54. The function has vertical asymptote $x = 3$, horizontal asymptote $y = 0$, and y-intercept $-\frac{1}{3}$, and never crosses the x-axis.

55. The function has vertical asymptotes $x = 1$ and $x = -4$, horizontal asymptote $y = 1$, and x-intercepts 2 and 3.

56. The function has horizontal asymptote $y = 2$ but no vertical asymptote. The origin is the only x-intercept, and i is a zero of $Q(x)$.

57. The function has slant asymptote $y = 3x - 6$ and vertical asymptote $x = \frac{1}{2}$, and its graph passes through the origin.

58. Show that the function

$$y = \frac{x^6 + 10}{x^4 + 8x^2 + 15}$$

has no horizontal, vertical, or slant asymptotes, and no x-intercept.

SECTION 3.10
USING GRAPHING DEVICES
TO GRAPH RATIONAL FUNCTIONS

In the preceding section we learned how to find horizontal and vertical asymptotes for rational functions. Since the graph of a rational function is very steep near a vertical asymptote, some graphing calculators and computer graphing programs do not graph functions with vertical asymptotes properly (see Example 7 in Section 2.3). A sudden, nearly vertical jump in a graph produced by a graphing calculator often signals the presence of a vertical asymptote.

EXAMPLE 1

Graph the function

$$f(x) = \frac{2x + 1}{x - 3}$$

in the viewing rectangles $[-10, 10]$ by $[-10, 10]$ and $[-1000, 1000]$ by $[-10, 10]$. Interpret the result in each case.

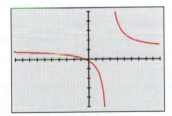

(a) $[-10, 10]$ by $[-10, 10]$

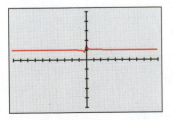

(b) $[-1000, 1000]$ by $[-10, 10]$

Figure 1
$f(x) = \dfrac{2x + 1}{x - 3}$

SOLUTION

The graph in the smaller viewing rectangle is shown in Figure 1(a). The graph consists of two parts because f is undefined when $x = 3$; the line $x = 3$ is a vertical asymptote. (On some graphing calculators, a vertical line may actually be shown on the screen. This is not part of the graph of the function and may be interpreted simply as an asymptote.)

In the very wide viewing rectangle shown in Figure 1(b), the graph of f looks almost like the horizontal line $y = 2$. So few points are plotted by the graphing device that the vertical asymptote disappears from the picture. This is because the calculator connects plotted points on the two sides of the vertical asymptote. We might make the same mistake ourselves if we were to graph the function simply by plotting points and joining them with a smooth curve, without analyzing the behavior of the function. However, this wide viewing rectangle shows the horizontal asymptote well, since a horizontal asymptote is a line that the graph approaches when $|x|$ is large. The horizontal asymptote $y = 2$ represents the end behavior of the function.   ■

In Section 3.9 we considered only horizontal and slant asymptotes as end behaviors for rational functions. In the next example we graph a function that behaves like a parabola for large values of $|x|$.

EXAMPLE 2

Graph the function

$$f(x) = \frac{x^3 - 2x^2 + 3}{x - 2}$$

in appropriate viewing rectangles to show the vertical asymptote and to determine its end behavior.

SOLUTION

First we graph the function in a narrow viewing rectangle to see the vertical asymptote. The function is undefined when $x = 2$, so we choose the viewing rectangle $[-4, 4]$ by $[-20, 20]$ and obtain the graph in Figure 2(a). The function has x-intercept -1, vertical asymptote $x = 2$, and a local minimum point on the right-hand branch of the graph, with approximate coordinates $(2.74, 11.56)$. To determine end behavior, we try a larger viewing rectangle—in this case $[-30, 30]$ by $[-200, 200]$. In the graph in Figure 2(b), the vertical asymptote has all but disappeared, and the graph looks like a parabola. To see why this is the case, we divide the denominator of f into the numerator and write the result in quotient-remainder form:

$$f(x) = x^2 + \frac{3}{x - 2}$$

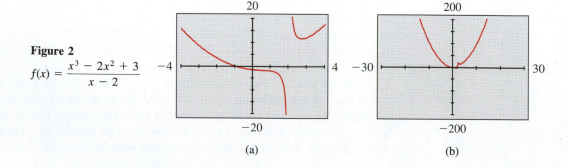

Figure 2
$$f(x) = \frac{x^3 - 2x^2 + 3}{x - 2}$$

(a)   (b)

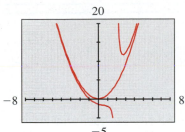

Figure 3

$$y = x^2 \text{ and } y = \frac{x^3 - 2x^2 + 3}{x - 2}$$

When $|x|$ is large, $3/(x - 2)$ is small; that is, as $x \to \pm\infty$, $3/(x - 2) \to 0$. This means that for large $|x|$, the graph of f will be close to the graph of $y = x^2$. Thus the end behavior of the function f is like that of the parabola $y = x^2$.

In Figure 3 the graphs of $y = (x^3 - 2x^2 + 3)/(x - 2)$ and $y = x^2$ are displayed in the viewing rectangle $[-8, 8]$ by $[-5, 20]$. From the figure we can see that the graphs of the two functions are very close to each other except in the vicinity of the vertical asymptote.   ■

Rational functions occur frequently in the applications of algebra to the sciences. In the next example we analyze the graph of a function from the theory of electricity.

EXAMPLE 3

When two resistors with resistances R_1 and R_2 are connected in parallel, their combined resistance R is given by the following formula:

$$R = \frac{R_1 R_2}{R_1 + R_2}$$

Suppose that a fixed 8-ohm resistor is connected in parallel with a variable resistor, as shown in Figure 4. If the resistance of the variable resistor is denoted by x, then the combined resistance R is a function of x. Graph R and give a physical interpretation of the graph.

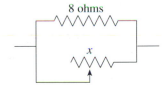

Figure 4

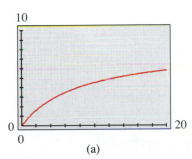

(a)

SOLUTION

Substituting $R_1 = 8$ and $R_2 = x$ into the formula gives the function

$$R(x) = \frac{8x}{8 + x}$$

Since resistance cannot be negative, this function has physical meaning only when $x > 0$. The function is graphed in Figure 5(a) using the viewing rectangle $[0, 20]$ by $[0, 10]$. The function has no vertical asymptotes when x is restricted to positive values. The combined resistance R increases as the variable resistance x increases. If we widen the viewing rectangle to $[0, 1000]$ by $[0, 10]$ we obtain the graph in Figure 5(b). For large x, the combined resistance R levels off, getting closer and closer to the horizontal asymptote $R = 8$. No matter how large the variable resistance x, the combined resistance is never greater than 8 ohms.   ■

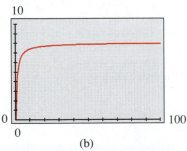

(b)

Figure 5

$$R(x) = \frac{8x}{8 + x}$$

⚑EXERCISES 3.10

In Exercises 1–4 graph the rational function in the given viewing rectangles. Determine all vertical and horizontal asymptotes from the graphs.

1. $y = \dfrac{3x + 7}{x + 2}$;   [−5, 5] by [−5, 10], [−100, 100] by [−5, 10]

2. $y = \dfrac{x}{5 - 2x}$;   [−2, 5] by [−5, 5], [−100, 100] by [−2, 5]

3. $y = \dfrac{3x^2 + 1}{x^2 - 9}$;   [−10, 10] by [−10, 10], [−100, 100] by [−10, 10]

4. $y = \dfrac{x^2 - 6}{x^2 - 3x}$;   [−5, 10] by [−5, 5], [−100, 100] by [−5, 5]

In Exercises 5–14 graph the rational function in an appropriate viewing rectangle and determine its vertical and horizontal asymptotes, its x- and y-intercepts, and all local extrema, correct to two decimals.

5. $y = \dfrac{7x - 14}{x}$

6. $y = \dfrac{2 - 12x}{4 + 3x}$

7. $y = \dfrac{4x}{x^2 - 4}$

8. $y = \dfrac{x^2 + 9}{x^2 - 9}$

9. $y = \dfrac{6x^2 - 6}{x^2 + 2}$

10. $y = \dfrac{x^2 + 1}{x^3 - 27}$

11. $y = \dfrac{4}{(x - 1)^2}$

12. $y = \dfrac{x^2 - 4x}{(x + 1)^2}$

13. $y = \dfrac{2x^2 + 3x - 2}{x^2}$

14. $y = \dfrac{1}{x^3 - 2x^2}$

In Exercises 15–18 graph the rational function f and determine all vertical asymptotes from your graph. Then graph f and g in a sufficiently large viewing rectangle to show that they have the same end behavior.

15. $f(x) = \dfrac{2x^2 + 6x + 6}{x + 3}$, $g(x) = 2x$

16. $f(x) = \dfrac{-x^3 + 6x^2 - 5}{x^2 - 2x}$, $g(x) = -x + 4$

17. $f(x) = \dfrac{x^3 - 2x^2 + 16}{x - 2}$, $g(x) = x^2$

18. $f(x) = \dfrac{-x^4 + 2x^3 - 2x}{(x - 1)^2}$, $g(x) = 1 - x^2$

In Exercises 19–22 graph the rational function and find all vertical asymptotes, x- and y-intercepts, and local extrema, correct to the nearest decimal. Then use long divison to find a polynomial that has the same end behavior as the rational function, and graph both in a sufficiently large viewing rectangle to verify that the end behaviors of the polynomial and the rational function are the same.

19. $y = \dfrac{2x^2 - 5x}{2x + 3}$

20. $y = \dfrac{x^4 - 3x^3 + x^2 - 3x + 3}{x^2 - 3x}$

21. $y = \dfrac{x^5}{x^3 - 1}$

22. $y = \dfrac{x^4}{x^2 - 2}$

23. Suppose a rocket is fired upward from the surface of the earth with an initial velocity v (measured in m/s). Then the maximum height h (in meters) reached by the rocket is given by the function

$$h(v) = \frac{Rv^2}{2gR - v^2}$$

where $R = 6.4 \times 10^6$ m is the radius of the earth and $g = 9.8$ m/s^2 is the acceleration due to gravity. Use a graphing device to sketch a graph of the function h. (Note that h and v must both be positive, so the viewing rectangle need not contain negative values.) What does the vertical asymptote represent physically?

24. As a train moves towards an observer, the pitch of its whistle sounds higher to the observer than it would if the train were at rest, because the crests of the sound waves are compressed closer together. This phenomenon is called the *Doppler effect*. The observed pitch P is a function of the speed v of the train, given by

$$P(v) = P_0 \left(\frac{s_0}{s_0 - v} \right)$$

where P_0 is the actual pitch of the whistle at the source and $s_0 = 332$ m/s is the speed of sound in air. Suppose that a train has a whistle pitched at $P_0 = 440$ Hz. Graph the function P using a graphing device. How can the vertical asymptote of this function be interpreted physically?

25. Give an example of a rational function that has horizontal asymptote $y = 2$ and vertical asymptotes $x = 1$ and $x = 4$, and whose graph passes through the origin. Graph your function using a graphing device to confirm that it has the required properties.

26. Give an example of a rational function that has vertical asymptotes $x = \pm 1$, whose end behavior is the same as the end behavior of $y = x^2$, and whose graph passes through the origin. Graph your function using a graphing device to confirm that it has the required properties.

SECTION 3.11
POLYNOMIAL AND RATIONAL INEQUALITIES

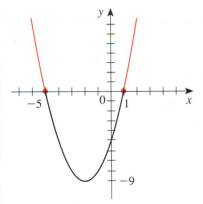

Figure 1

$f(x) > 0$ for $a < x < b$ and $c < x$

Functions with graphs that consist of a single unbroken curve are called **continuous**. We have seen from our work in Section 3.1 that polynomials are continuous functions. One important property of such functions, which we will use in this section to help us solve inequalities, is that if a and b are successive x-intercepts of the continuous function f, then either $f(x) > 0$ or $f(x) < 0$ for all values of x between a and b. For example, in Figure 1 $f(x) > 0$ if $a < x < b$ and $f(x) < 0$ if $b < x < c$. Moreover, assuming that the graph shows all the x-intercepts of the function, $f(x) < 0$ for $x < a$ and $f(x) > 0$ for $x > c$.

The x-intercepts of a polynomial are its zeros. We will use our knowledge of finding zeros of polynomials, together with the property of continuous functions we have just described, to solve polynomial inequalities. We have already encountered linear and quadratic inequalities in Section 1.7, but we are now able to solve more general types of inequalities as well.

EXAMPLE 1

Find all values of x for which $x^2 + 4x - 5 > 0$.

SOLUTION 1

Let $f(x) = x^2 + 4x - 5$. We first write

$$
\begin{aligned}
f(x) = x^2 + 4x - 5 &= (x + 5)(x - 1) &&\text{(factoring)} \\
&= (x^2 + 4x + 4) - 5 - 4 &&\text{(completing the square)} \\
&= (x + 2)^2 - 9
\end{aligned}
$$

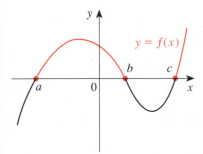

Figure 2

This is a parabola with vertex $(-2, -9)$ and with x-intercepts -5 and 1. Its graph is shown in Figure 2. We look for values of x for which $y = x^2 + 4x - 5 > 0$. From the figure, we see that this is the case where

$$x < -5 \quad \text{or} \quad x > 1$$

SOLUTION 2

Factoring, we get

$$(x + 5)(x - 1) > 0$$

The polynomial has zeros at -5 and 1. These points divide the real line into the intervals $(-\infty, -5)$, $(-5, 1)$, and $(1, \infty)$. We determine the sign of the factors and of the polynomial on each of these intervals in the following table. To determine the signs on an interval, it is enough to check the signs at a single, conveniently chosen point within the interval. For example, we used the test value $x = 0$ to determine the signs for the interval $-5 < x < 1$.

Interval	$x + 5$	$x - 1$	$(x + 5)(x - 1)$
$x < -5$	$-$	$-$	$+$
$-5 < x < 1$	$+$	$-$	$-$
$1 < x$	$+$	$+$	$+$

From the last column of the table, we see that $(x + 5)(x - 1)$ is positive if $x < -5$ or if $x > 1$, so the solution is

$$(-\infty, -5) \cup (1, \infty)$$

■

EXAMPLE 2

Solve the inequality $3x^3 + 1 \leq x^2 + 3x$.

SOLUTION

First we express the inequality in the form $P(x) \leq 0$ so that we can use the factoring technique of Example 1 (Solution 2) to solve the inequality:

$$3x^3 - x^2 - 3x + 1 \leq 0$$

We factor by grouping terms:

$$(3x^3 - x^2) - (3x - 1) \leq 0$$
$$(3x - 1)x^2 - (3x - 1) \leq 0$$
$$(3x - 1)(x^2 - 1) \leq 0$$
$$(3x - 1)(x - 1)(x + 1) \leq 0$$

The zeros of the polynomial are $\frac{1}{3}$, 1, and -1. These numbers divide the real line into four intervals: $(-\infty, -1)$, $\left(-1, \frac{1}{3}\right)$, $\left(\frac{1}{3}, 1\right)$, and $(1, \infty)$, so we check whether the polynomial is positive or negative at a test value in each of these intervals, as indicated in the following table.

Interval	$x + 1$	$3x - 1$	$x - 1$	$(x + 1)(3x - 1)(x - 1)$
$x < -1$	−	−	−	−
$-1 < x < \frac{1}{3}$	+	−	−	+
$\frac{1}{3} < x < 1$	+	+	−	−
$1 < x$	+	+	+	+

Note that when we substitute test values, we do not need to actually calculate the products in the last column of this table. We are interested only in the *sign* of the product, which can be determined by counting negative factors. The table shows us that $(-\infty, -1)$ and $\left(\frac{1}{3}, 1\right)$ are in the solution set. Finally, we note that the zeros of the polynomial also belong to the solution because the inequality is of the form $\leq$. The complete solution is therefore

$$(-\infty, -1] \cup \left[\tfrac{1}{3}, 1\right]$$ ■

RATIONAL INEQUALITIES

From the examples in Section 3.9 we can see that, in general, rational functions are not continuous. The vertical asymptotes of such functions break up their graphs into separate branches. Thus when we are solving inequalities that involve rational functions, we must take their vertical asymptotes (as well as their x-intercepts) into account when setting up the intervals on which the function does not change sign. This is the reason for the following definition:

CUT POINTS

> If $r(x) = P(x)/Q(x)$ is a rational function, then the **cut points** for r are the values of x at which either $P(x)$ or $Q(x)$ is 0.

If a and b are two successive cut points for the rational function r, then either $r(x) > 0$ or $r(x) < 0$ for all x between a and b. We use this fact to solve rational inequalities in the following examples.

EXAMPLE 3

Solve the inequality: $\dfrac{2x + 5}{x - 5} < 0.$

SOLUTION

The cut points of the given rational function are $-\frac{5}{2}$ and 5, so we must check the intervals $\left(-\infty, -\frac{5}{2}\right)$, $\left(-\frac{5}{2}, 5\right)$, and $(5, \infty)$. We will determine on which of these intervals $(2x + 5)/(x - 5) < 0$ by checking the sign of the function at test values chosen from each interval.

Interval	$2x + 5$	$x - 5$	$(2x + 5)/(x - 5)$
$x < -\frac{5}{2}$	$-$	$-$	$+$
$-\frac{5}{2} < x < 5$	$+$	$-$	$-$
$5 < x$	$+$	$+$	$+$

Only numbers from the middle interval satisfy the inequality. Since neither $-\frac{5}{2}$ nor 5 is itself a solution, the complete solution is $\left(-\frac{5}{2}, 5\right)$. ∎

EXAMPLE 4

Solve the inequality: $\quad \dfrac{2x + 1}{x - 3} - \dfrac{x}{x + 1} \geq 1$

SOLUTION

If this were an equation rather than an inequality, we could multiply both sides by $(x - 3)(x + 1)$ to clear the denominators, and then solve. But multiplying an inequality by a negative number reverses its direction (see Section 1.7). Since we do not know whether $(x - 3)(x + 1)$ will turn out to be positive or negative, we cannot simplify in this way. Instead we subtract 1 from both sides (to make the right side 0) and then combine the fractions.

$$\frac{2x + 1}{x - 3} - \frac{x}{x + 1} - 1 \geq 0$$

$$\frac{(2x + 1)(x + 1) - x(x - 3) - (x - 3)(x + 1)}{(x - 3)(x + 1)} \geq 0$$

$$\frac{(2x^2 + 3x + 1) - (x^2 - 3x) - (x^2 - 2x - 3)}{(x - 3)(x + 1)} \geq 0$$

$$\frac{8x + 4}{(x - 3)(x + 1)} \geq 0$$

$$\frac{4(2x + 1)}{(x - 3)(x + 1)} \geq 0$$

The cut points are -1, $-\frac{1}{2}$, and 3, so the intervals to be considered are $(-\infty, -1)$, $\left(-1, -\frac{1}{2}\right)$, $\left(-\frac{1}{2}, 3\right)$, and $(3, \infty)$. This leads to the following table.

Interval	$x + 1$	$2x + 1$	$x - 3$	$4(2x + 1)/[(x - 3)(x + 1)]$
$x < -1$	$-$	$-$	$-$	$-$
$-1 < x < -\frac{1}{2}$	$+$	$-$	$-$	$+$
$-\frac{1}{2} < x < 3$	$+$	$+$	$-$	$-$
$3 < x$	$+$	$+$	$+$	$+$

The numbers in the intervals $\left(-1, -\frac{1}{2}\right)$ and $(3, \infty)$ belong to the solution. Checking the cut points, we see that $-\frac{1}{2}$ is a solution, but that -1 and 3 are not, since the left-hand side of the inequality is undefined there. The complete solution is

$$\left(-1, -\tfrac{1}{2}\right] \cup (3, \infty) \qquad\blacksquare$$

To summarize, we solve inequalities involving polynomials or rational functions by performing the following sequence of steps:

SOLVING INEQUALITIES

1. If necessary, rewrite the inequality so that one side is 0. (Take care not to multiply by expressions that have an undetermined sign.)
2. Factor the polynomial (or the numerator and denominator of the rational function) into linear factors, and use these to find the cut points.
3. List the intervals determined by the cut points.
4. Check the sign of the polynomial or rational function on each interval by calculating the sign at some convenient test number chosen from the interval.
5. Check whether the inequality is satisfied by some or all of the cut points themselves. (This may happen if the inequality involves $\geq$ or $\leq$.)
6. Combine the information obtained in Steps 4 and 5 to get the complete solution set.

In the next example we introduce a short-hand form for the table we have been using in our work so far to do Step 4.

EXAMPLE 5

Solve the inequality: $\dfrac{x^2 + 2x - 5}{(x - 5)^2} \geq 0$

SOLUTION

Using the quadratic formula, we see that the zeros of the numerator are $1 + \sqrt{6}$ and

$1 - \sqrt{6}$ (or about 3.45 and -1.45). Thus the cut points are $1 - \sqrt{6}$, $1 + \sqrt{6}$, and 5, and the rational function factors into

$$\frac{\left[x - \left(1 - \sqrt{6}\right)\right]\left[x - \left(1 + \sqrt{6}\right)\right]}{(x - 5)^2}$$

Instead of making a table as before, we place the cut points on a number line and determine the sign of the function on each interval created by them (see Figure 3).

Figure 3

The numbers $1 - \sqrt{6}$ and $1 + \sqrt{6}$ do satisfy the inequality, but 5 does not. We have indicated this using solid and open circles in Figure 3. The solution set is thus

$$\left(-\infty, 1 - \sqrt{6}\right] \cup \left[1 + \sqrt{6}, 5\right) \cup (5, \infty) \qquad \blacksquare$$

USING GRAPHING DEVICES TO SOLVE INEQUALITIES

As we saw at the beginning of this section, any inequality of the form $f(x) > 0$ can be solved by analyzing the graph of $y = f(x)$. The solution is the set of all values of x for which $f(x)$ is positive; that is, the set of all x for which the point $(x, f(x))$ lies above the x-axis (see Figure 1). The same idea can also be used to solve inequalities of the form $f(x) \geq 0$, $f(x) < 0$, and $f(x) \leq 0$ graphically. This principle allows us to solve inequalities from the graphs produced by graphing devices, as in the next example.

EXAMPLE 6

Solve the inequality $2x^4 - 6x^3 - 5x^2 - 3x - 3 \leq 0$.

SOLUTION

First we graph the polynomial $P(x) = 2x^4 - 6x^3 - 5x^2 - 3x - 3$, as in Figure 4. To solve the inequality $P(x) \leq 0$ we must find all values of x for which the graph of the function lies on or below the x-axis. From the graph we see that the solution is the interval that lies between the two x-intercepts. By zooming in we find that the x-intercepts are -0.79 and 3.79 (to two decimal places), so the approximate solution of the inequality is the interval $[-0.79, 3.79]$. $\qquad \blacksquare$

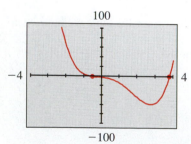

Figure 4
$P(x) = 2x^4 - 6x^3 - 5x^2 - 3x - 3$

EXERCISES 3.11

Solve the inequalities in Exercises 1–54.

1. $x^2 - x - 20 > 0$

2. $x^2 + 5x \leq 0$

3. $x^2 + 3x - 9 \geq 0$

4. $x^2 + 2x + 3 < 0$

5. $x^2 - 16 \leq 0$

6. $x^2 > 25$

7. $3x^2 + 11x - 4 \leq 0$

8. $6x^2 + 5x + 1 > 0$

9. $x(x - 2) < 8$

10. $3 + x^2 \leq 4x$

11. $x + 14 \geq 3x^2$

12. $2 > x(11 - 5x)$

13. $x^2 + 2 > x$

14. $2x^2 + 5x + 4 \leq 0$

15. $x^2 + 2x - 7 \leq 0$

16. $3x^2 - x + 1 > 0$

17. $9x < 2x^2 + 7$

18. $(x - 3)(x + 5)(2x + 11) < 0$

19. $(x - 1)(x + 2)(x - 3)(x + 4) \geq 0$

20. $(x + 5)^2(x + 3)(x - 1) > 0$

21. $(2x - 7)^4(x - 1)^3(x + 1) \leq 0$

22. $x^3 + 4x^2 - 4x - 16 \geq 0$

23. $2x^3 - x^2 - 18x + 9 < 0$

24. $2x^3 - x^2 + 18x - 9 < 0$

25. $x^4 + 3x^3 - x - 3 \geq 0$

26. $x(1 - x^2)^3 > 7(1 - x^2)^3$

27. $x^4 - 11x^2 - 18 < 0$

28. $4x^4 - 25x^2 + 36 \leq 0$

29. $x^3 + 6x > 5x^2$

30. $x^3 + x^2 - 17x + 15 \geq 0$

31. $x^2(7 - 6x) \leq 1$

32. $x^4 + 3x^3 - 3x^2 + 3x - 4 < 0$

33. $\dfrac{x - 1}{x - 10} < 0$

34. $\dfrac{3x - 7}{x + 2} \leq 0$

35. $\dfrac{2x + 5}{x^2 + 2x - 35} \geq 0$

36. $\dfrac{4x^2 - 25}{x^2 - 9} > 0$

37. $\dfrac{x}{x^2 + 2x - 2} \leq 0$

38. $\dfrac{x + 1}{2x^2 - 4x + 1} > 0$

39. $\dfrac{x^2 + 2x - 3}{3x^2 - 7x - 6} > 0$

40. $\dfrac{x - 1}{x^3 + 1} \geq 0$

41. $\dfrac{x^3 + 3x^2 - 9x - 27}{x + 4} \leq 0$

42. $\dfrac{x^2 - 16}{x^4 - 16} < 0$

43. $\dfrac{x - 3}{2x + 5} \geq 1$

44. $\dfrac{1}{x} + \dfrac{1}{x + 1} < \dfrac{2}{x + 2}$

45. $2 + \dfrac{1}{1 - x} \leq \dfrac{3}{x}$

46. $\dfrac{1}{x - 3} + \dfrac{1}{x + 2} \geq \dfrac{2x}{x^2 + x - 2}$

47. $\dfrac{(x - 1)^2}{(x + 1)(x + 2)} > 0$

48. $\dfrac{x^2 - 2x + 1}{x^3 + 3x^2 + 3x + 1} \leq 0$

49. $\dfrac{6}{x - 1} - \dfrac{6}{x} \geq 1$

50. $\dfrac{x}{2} \geq \dfrac{5}{x + 1} + 4$

51. $\dfrac{x + 2}{x + 3} < \dfrac{x - 1}{x - 2}$

52. $\dfrac{1}{x + 1} + \dfrac{1}{x + 2} \leq \dfrac{1}{x + 3}$

53. $\dfrac{(1 - x)^2}{\sqrt{x}} \geq 4\sqrt{x}(x - 1)$

54. $\frac{2}{3}x^{-1/3}(x + 2)^{1/2} + \frac{1}{2}x^{2/3}(x + 2)^{-1/2} < 0$

In Exercises 55–58 find all values of x for which the graph of f_1 lies above the graph of f_2.

55. $f_1(x) = x^2, \quad f_2(x) = 3x + 10$

56. $f_1(x) = \dfrac{1}{x}, \quad f_2(x) = \dfrac{1}{x - 1}$

57. $f_1(x) = 4x, \quad f_2(x) = \dfrac{1}{x}$

58. $f_1(x) = x^3 + x^2, \quad f_2(x) = \dfrac{1}{x}$

In Exercises 59–62 find the domain of the given function.

59. $f(x) = \sqrt{6 + x - x^2}$

60. $g(x) = \sqrt{\dfrac{5 + x}{5 - x}}$

61. $h(x) = \sqrt[4]{x^4 - 1}$

62. $k(x) = \dfrac{1}{\sqrt{x^4 - 5x^2 + 4}}$

63. Solve the inequality $(x - a)(x - b)(x - c)(x - d) \geq 0$, where $a < b < c < d$.

64. Solve

$$\frac{x^2 + (a - b)x - ab}{x + c} \leq 0$$

where $0 < a < b < c$.

 In Exercises 65–70, use a graphing device to solve the inequality, as in Example 6. Express your answer using

interval notation, with the endpoints of the intervals correct to two decimals.

65. $x^3 - 2x^2 - 5x + 6 \geq 0$

66. $2x^3 + x^2 - 8x - 4 \leq 0$

67. $2x^3 - 3x + 1 < 0$ **68.** $x^4 - 4x^3 + 8x > 0$

69. $5x^4 < 8x^3$ **70.** $x^5 + x^3 \geq x^2 + 6x$

CHAPTER 3 REVIEW

Define, state, or discuss the following.

1. Polynomial of degree n

2. The graph of $y = x^n$, n a positive integer

3. The graph of a polynomial

4. Turning points; local maxima and minima of polynomials

5. End behavior of a polynomial

6. Dividend, divisor, quotient, and remainder

7. Division Algorithm

8. Remainder Theorem

9. Synthetic division

10. Factor Theorem

11. Zero of a polynomial

12. Root of a polynomial equation

13. Rational Roots Theorem

14. Descartes' Rule of Signs

15. Upper and lower bounds for roots

16. Upper and Lower Bounds Theorem

17. Intermediate Value Theorem for Polynomials

18. Finding approximate values for irrational zeros of polynomials

19. Complex number

20. Modulus of a complex number

21. Fundamental Theorem of Algebra

22. Complete Factorization Theorem

23. Multiplicity of a zero

24. Zeros Theorem

25. Conjugate Roots Theorem

26. Rational function

27. Vertical, horizontal, and slant asymptotes

28. Polynomial and rational inequalities

REVIEW EXERCISES

In Exercises 1–6, graph the polynomial. Show clearly all x- and y-intercepts.

1. $y = (x - 2)^3 + 8$ **2.** $y = 32 - 2x^4$

3. $y = x^3 - 9x$ **4.** $y = x^3 - 5x^2 - 6x$

5. $y = x^3 - 5x^2 - 4x + 20$

6. $y = x^4 - 9x^2$

 In Exercises 7–10 use a graphing device to graph the polynomial. Find the x- and y-intercepts and the coordinates of all local extrema, correct to the nearest decimal. Describe the end behavior of the function.

7. $y = 2x^3 + x^2 - 18x - 9$

8. $y = x^4 - 8x^2 + 16$

9. $y = x^5 + x^2 - 5$ **10.** $y = 3x^5 + x^4 - 4x$

In Exercises 11–18 find the quotient and remainder.

11. $\dfrac{x^3 - x^2 + x - 11}{x - 3}$

12. $\dfrac{x^4 + 30x + 12}{2x + 6}$

13. $\dfrac{x^3 - x^2 - 11x + 6}{x^2 + 2x - 5}$

14. $\dfrac{x^5 - 3x^4 + 3x^3 + 20x - 6}{x^2 + 2x - 6}$

15. $\dfrac{x^4 - 25x^2 + 4x + 15}{x + 5}$

16. $\dfrac{2x^3 - x^2 - 5}{x - \frac{3}{2}}$

17. $\dfrac{x^4 + x^3 - 2x^2 - 3x - 1}{x - \sqrt{3}}$

18. $\dfrac{15x - 7}{5x + 12}$

In Exercises 19 and 20 find the indicated value of the given polynomial using the Remainder Theorem.

19. $P(x) = 2x^3 - 9x^2 - 7x + 13$;   find $P(5)$

20. $Q(x) = x^4 + 4x^3 + 7x^2 + 10x + 15$;   find $Q(-3)$

21. Show that $\frac{1}{2}$ is a zero of the polynomial

$$2x^4 + x^3 - 5x^2 + 10x - 4$$

22. Show using the Factor Theorem that $x + 4$ is a factor of the polynomial

$$x^5 + 4x^4 - 7x^3 - 23x^2 + 23x + 12$$

23. What is the remainder when the polynomial $x^{500} + 6x^{201} - x^2 - 2x + 4$ is divided by $x - 1$?

24. What is the remainder when $x^{101} - x^4 + 2$ is divided by $x + 1$?

In Exercises 25 and 26 list all possible rational roots (without testing to see if they actually are roots) and then determine the possible number of positive and negative real roots using Descartes' Rule of Signs.

25. $x^5 - 6x^3 - x^2 + 2x + 18 = 0$

26. $6x^4 + 3x^3 + x^2 + 3x + 4 = 0$

In Exercises 27–30 show that the equation has a root between the two given integers, and then find that root correct to the indicated number of decimal places.

27. $x^3 - 2x - 1 = 0$;
between 1 and 2, to the nearest tenth

28. $3x^3 + x^2 + 4 = 0$;
between -2 and -1, to one decimal place

29. $x^6 - 5x^4 + 10 = 0$;
between 1 and 2, to two decimal places

30. $x^6 - 5x^4 + 10 = 0$;
between 2 and 3, to the nearest hundredth

31. Find a polynomial of degree 3 with constant coefficient 12 and with zeros $-\frac{1}{2}$, 2, and 3.

32. Find a polynomial of degree 4 having integer coefficients and zeros $3i$ and 4, with 4 a double root.

33. Does there exist a polynomial of degree 4 with integer coefficients that has zeros i, $2i$, $3i$, and $4i$? If so, find it. If not, explain why not.

34. Prove that the equation $3x^4 + 5x^2 + 2 = 0$ has no real root.

In Exercises 35–44 find all rational, irrational, and imaginary roots (and state their multiplicities). Use Descartes' Rule of Signs, the Upper and Lower Bounds Theorem, the quadratic formula, or other factoring techniques to help you whenever possible.

35. $x^3 - 3x^2 - 13x + 15 = 0$

36. $2x^3 + 5x^2 - 6x - 9 = 0$

37. $x^4 + 6x^3 + 17x^2 + 28x + 20 = 0$

38. $x^4 + 7x^3 + 9x^2 - 17x - 20 = 0$

39. $x^5 - 3x^4 - x^3 + 11x^2 - 12x + 4 = 0$

40. $x^4 = 81$

41. $x^6 = 64$

42. $18x^3 + 3x^2 - 4x - 1 = 0$

43. $6x^4 - 18x^3 + 6x^2 - 30x + 36 = 0$

44. $x^4 + 15x^2 + 54 = 0$

45. Show that $2 - i\sqrt{2}$ is a root of

$$x^4 - 5x^3 + 8x^2 + 2x - 12 = 0$$

and find all other roots.

46. Show that i and $-1 + i\sqrt{6}$ are roots of

$$x^6 + x^5 + 7x^4 - 4x^3 + 13x^2 - 5x + 7 = 0$$

and find all other roots.

47. The polynomial $x^4 - x^2 - x - 2 = 0$ has one positive real root. Show that the root is irrational, and find a decimal approximation for it, correct to two decimal places.

48. (a) Show that the polynomial $x^4 + 2x^2 - x - 1$ has exactly two real zeros, and that neither is rational.
(b) Find decimal approximations of each of the two real roots, correct to the nearest hundredth.

 In Exercises 49–52 use a graphing device to find all real solutions of the equation.

49. $2x^2 = 5x + 3$

50. $x^3 + x^2 - 14x - 24 = 0$

51. $x^4 - 3x^3 - 3x^2 - 9x - 2 = 0$

52. $x^5 = x + 3$

In Exercises 53–58 graph the rational function. Show clearly all x- and y-intercepts and asymptotes.

53. $y = \dfrac{3x - 12}{x + 1}$

54. $y = \dfrac{1}{(x + 2)^2}$

55. $y = \dfrac{x - 2}{x^2 - 2x - 8}$

56. $y = \dfrac{2x^2 - 6x - 7}{x - 4}$

57. $y = \dfrac{x^2 - 9}{2x^2 + 1}$

58. $y = \dfrac{x^3 + 27}{x + 4}$

 In Exercises 59–62 use a graphing device to analyze the graph of the rational function. Find all x- and y-intercepts, vertical, horizontal, and slant asymptotes, and the coordinates of local extrema. If the function has no horizontal or slant asymptote, find a polynomial that has the same end behavior as the rational function.

59. $y = \dfrac{x - 3}{2x + 6}$

60. $y = \dfrac{2x - 7}{x^2 + 9}$

61. $y = \dfrac{x^3 + 8}{x^2 - x - 2}$

62. $y = \dfrac{2x^3 - x^2}{x + 1}$

In Exercises 63–68 solve the inequality.

63. $2x^2 \geq x + 3$

64. $x^3 - 3x^2 - 4x + 12 \leq 0$

65. $x^4 - 7x^2 - 18 < 0$

66. $\dfrac{5}{x^3 - x^2 - 4x + 4} < 0$

67. $\dfrac{3x + 1}{x + 2} \leq \dfrac{2}{3}$

68. $\dfrac{1}{x - 1} + \dfrac{2}{x + 1} \geq \dfrac{3}{x}$

In Exercises 69 and 70 find the domain of the function.

69. $f(x) = \sqrt{24 - x - 3x^2}$

70. $g(x) = \dfrac{1}{\sqrt[4]{x - x^4}}$

 In Exercises 71 and 72 use a graphing device to solve the inequality.

71. $x^4 + x^3 \leq 5x^2 + 4x - 5$

72. $x^5 - 4x^4 + 7x^3 - 12x + 2 > 0$

73. (a) Show that -1 is a root of the equation
$$2x^4 + 5x^3 + x + 4 = 0.$$
(b) Use the information from part (a) to show that $2x^3 + 3x^2 - 3x + 4 = 0$ has no positive real root. [*Hint:* Compare the coefficients of the latter polynomial to your synthetic division table from part (a).]

74. Find the coordinates of all points of intersection of the graphs of

$$y = x^4 + x^2 + 24x \quad \text{and} \quad y = 6x^3 + 20$$

CHAPTER 3 TEST

1. Graph the function $f(x) = x^3 - x^2 - 9x + 9$, showing clearly all x- and y-intercepts.

2. Let $P(x) = 2x^4 - 17x^3 + 53x^2 - 72x + 36$.
   (a) List all possible rational zeros of P given by the Rational Roots Theorem.
   (b) Use Descartes' Rule of Signs to determine how many positive and how many negative real roots the equation $P(x) = 0$ may have.
   (c) Show that 9 is an upper bound for the real roots of $P(x) = 0$ but is not itself a root.
   (d) Use the information from your answers to parts (a), (b), and (c) to construct a new, shorter list of possible rational zeros of P.
   (e) Determine which of the possible rational zeros actually are zeros of P.
   (f) Factor P in the form given in the Complete Factorization Theorem.

3. Find a fifth-degree polynomial with integer coefficients that has zeros $1 + 2i$ and -1, with -1 a zero of multiplicity 3.

4. Let
$$P(x) = x^{23} - 5x^{12} + 8x - 1$$
$$Q(x) = 3x^4 + x^2 - x - 15$$
$$R(x) = 4x^6 + x^4 + 2x^2 + 16$$

   (a) Explain why an even integer could not possibly be a zero of any of these three polynomials.
   (b) Does R have any real zeros? Why or why not?
   (c) How many real zeros does Q have? Why?
   (d) Show that P has no rational zeros.

5. Find a decimal approximation for the positive root of the equation $x^4 + 2x^2 - x - 4 = 0$ correct to the nearest tenth.

6. Find all roots of the equation
$$2x^4 - 17x^3 + 42x^2 - 25x + 4 = 0$$

7. Find all roots of the equation
$$x^4 + x^3 + 5x^2 + 4x + 4 = 0$$
   given that $-2i$ is one of its roots.

8. Let
$$r(x) = \frac{2x - 1}{x^2 - x - 2} \qquad s(x) = \frac{x^3 + 27}{x^2 + 4}$$
$$t(x) = \frac{x^3 - 9x}{x + 2} \qquad u(x) = \frac{x^2 + x - 6}{x^2 - 25}$$

   (a) Which of these four rational functions has a horizontal asymptote?
   (b) Which has a slant asymptote?
   (c) Which has no vertical asymptote?
   (d) Graph $y = u(x)$, showing clearly any asymptotes and x- and y-intercepts the function may have.

9. Solve: $x \leq \dfrac{6 - x}{2x - 5}$

10. Find the domain of f:
$$f(x) = \frac{1}{\sqrt{4 - 2x - x^2}}$$

11. (a) Choosing an appropriate viewing rectangle, graph the following function and find all its x-intercepts and local extrema, correct to two decimals.
$$P(x) = x^4 - 4x^3 + 8x$$

   (b) Use your graph from part (a) to solve the inequality
$$x^4 - 4x^3 + 8x \geq 0$$
   Express your answer in interval form, with the endpoints correct to two decimals.

FOCUS ON PROBLEM SOLVING

In solving a problem it is often necessary to **introduce something extra**—something that is not initially given in the problem. In the example here, we add something that helps make a connection between the given and the unknown, thereby making the solution to the problem clear.

The Impossible Museum Tour

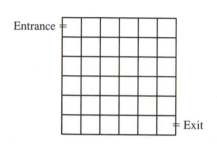

The floor plan of a museum is in the shape of a square with 6 square rooms to a side. Each room is connected to each adjacent room by a door. The museum entrance and exit are at diagonally opposite corners as shown in the figure.

A visitor making a tour of this museum would like to visit each room exactly once and then exit. Our task is to prove that this is not possible.

Trial and error will make it plausible to you that such a route is in fact not possible. Here are three examples of attempts:

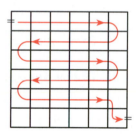 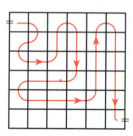 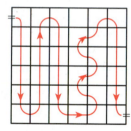

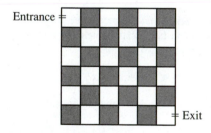

But these attempts do not *prove* that our proposed tour of the museum is impossible. How do we give a convincing argument for this? Here is a very clever idea. Let us imagine that the rooms are colored alternately black and white like a checkerboard. So there are 18 black rooms and 18 white rooms. We have introduced something extra that at first does not seem relevant to the problem. Indeed, the tour can or cannot be done regardless of the colors of the rooms. But notice how coloring the rooms in this fashion allows us to give a convincing argument that explains why this tour is impossible. We reason as follows:

If our visitor is in a white room, he must next go to a black room; and if he is in a black room, he must next go to a white room. The entrance is at a white room, so our visitor's path, in terms of colors of rooms, is

W B W B W B W B W B W B . . .

Since there is an even number of rooms and the visitor starts at a white room, he must end his tour at a black room. But the exit is at a white room, so the proposed tour is impossible!

260

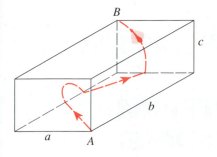

1. Is the museum tour possible if there are 5 rooms to a side instead of 6? Find a general result about when the tour is possible for an $n \times n$ museum.

2. A bug is at point A in one corner of a room and wants to go to point B in the diagonally opposite corner. Find the shortest path for the bug.

3. A Tibetan monk leaves the monastery at 7:00 A.M. and takes his usual path to the top of the mountain, arriving at 7:00 P.M. The following morning, he starts at 7:00 A.M. at the top and takes the same path back, arriving at 7:00 P.M. Show that there is exactly one point on the path where the monk will be at exactly the same time of day on both days.

4. A curve that starts and ends at the same point and does not intersect itself is called a *Jordan curve*. Such a curve always has an inside and an outside. Is the point A in the figure inside or outside the given Jordan curve? Find an easy method of determining whether a point is inside or outside a Jordan curve. [*Hint:* The point B is clearly outside the curve. Try adding a line segment connecting A and B.]

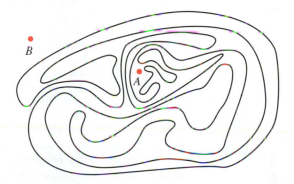

5. Find a polynomial of degree 3 with integer coefficients that has $2 - \sqrt[3]{2}$ as one of its zeros. How many other real zeros does the polynomial have?

6. If the equation $x^4 + ax^2 + bx + c = 0$ has roots 1, 2, and 3, find c.

7. A point P is located in the interior of a rectangle so that the distance from P to one corner is 5 cm, from P to the opposite corner is 14 cm, and from P to a third corner is 10 cm. What is the distance from P to the fourth corner?

8. Three tangent circles of radius 10 cm are drawn. All centers lie on the line AB. The tangent AC to the right-hand circle is drawn, intersecting the middle circle at D and E. Find the length $|DE|$.

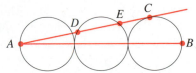

261

4 EXPONENTIAL AND LOGARITHMIC FUNCTIONS

Mathematics compares the most diverse phenomena and discovers the secret analogies which unite them.

JOSEPH FOURIER

The greatest mathematicians, as Archimedes, Newton, and Gauss, always united theory and applications in equal measure.

FELIX KLEIN

So far we have studied relatively simple functions such as polynomials and rational functions. We now turn our attention to two of the most important functions in mathematics, the exponential function and its inverse function, the logarithmic function. We use these functions to describe exponential growth in biology and economics and radioactive decay in physics and chemistry.

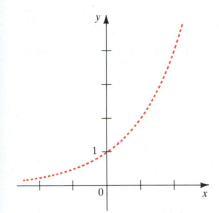

SECTION 4.1
EXPONENTIAL FUNCTIONS

In Section 1.2 we defined a^x if $a > 0$ and x is a rational number, but we have not yet defined irrational powers. For instance, what is meant by $2^{\sqrt{3}}$ or 5^{π}? To help us answer this question we first look at the graph of the function $y = 2^x$, where x is rational. A representation of this graph is shown in Figure 1.

We want to enlarge the domain of $y = 2^x$ to include both rational and irrational numbers. There are holes in the graph in Figure 1. We want to fill in the holes by defining $f(x) = 2^x$, where $x \in R$, so that f is an increasing function.

In particular, since

$$1.7 < \sqrt{3} < 1.8$$

we must have

$$2^{1.7} < 2^{\sqrt{3}} < 2^{1.8}$$

Figure 1
Representation of $y = 2^x$, x rational

and we know what $2^{1.7}$ and $2^{1.8}$ mean because 1.7 and 1.8 are rational numbers. Similarly, using better approximations for $\sqrt{3}$, we obtain better approximations for the value we need to give for $2^{\sqrt{3}}$:

$$1.73 < \sqrt{3} < 1.74 \quad \Rightarrow \quad 2^{1.73} < 2^{\sqrt{3}} < 2^{1.74}$$
$$1.732 < \sqrt{3} < 1.733 \quad \Rightarrow \quad 2^{1.732} < 2^{\sqrt{3}} < 2^{1.733}$$
$$1.7320 < \sqrt{3} < 1.7321 \quad \Rightarrow \quad 2^{1.7320} < 2^{\sqrt{3}} < 2^{1.7321}$$
$$1.73205 < \sqrt{3} < 1.73206 \quad \Rightarrow \quad 2^{1.73205} < 2^{\sqrt{3}} < 2^{1.73206}$$

Using advanced mathematics it can be shown that there is exactly one number that is greater than all of the numbers

$$2^{1.7}, \quad 2^{1.73}, \quad 2^{1.732}, \quad 2^{1.7320}, \quad 2^{1.73205}, \quad \ldots$$

and less than all of the numbers

$$2^{1.8}, \quad 2^{1.74}, \quad 2^{1.733}, \quad 2^{1.7321}, \quad 2^{1.73206}, \quad \ldots$$

We define $2^{\sqrt{3}}$ to be this number. Using the above approximation process we can compute it correct to 6 decimal places:

$$2^{\sqrt{3}} \approx 3.321997$$

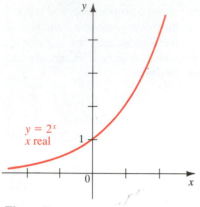

$y = 2^x$
x real

Figure 2

Similarly we can define 2^x (or a^x, if $a > 0$) where x is any irrational number. The graph of $f(x) = 2^x$, where $x \in R$, is shown in Figure 2. You can see that it is an increasing function, and in fact it increases very rapidly.

If $a > 0$, the **exponential function with base a** is defined by

$$f(x) = a^x$$

for every real number x.

It can be proved that the Laws of Exponents are still true when the exponents are real numbers.

To demonstrate just how quickly $f(x) = 2^x$ increases let us perform the following thought experiment. Suppose we start with a piece of paper a thousandth of an inch thick and we fold it in half 50 times. Each time we fold the paper in half, the thickness of the paper doubles, so the thickness of the resulting paper would be $2^{50}/1000$ inches. How thick do you think that is? It works out to be more than 17 million miles!

EXAMPLE 1

Draw the graphs of the following functions:

(a) $f(x) = 3^x$ (b) $g(x) = \left(\frac{1}{3}\right)^x$

SOLUTION

We calculate values of $f(x)$ and $g(x)$ and plot points to sketch the graphs in Figure 3.

x	$f(x) = 3^x$	$g(x) = \left(\frac{1}{3}\right)^x$
-3	$\frac{1}{27}$	27
-2	$\frac{1}{9}$	9
-1	$\frac{1}{3}$	3
0	1	1
1	3	$\frac{1}{3}$
2	9	$\frac{1}{9}$
3	27	$\frac{1}{27}$

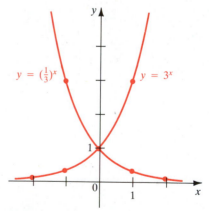

$y = \left(\frac{1}{3}\right)^x$ $y = 3^x$

Figure 3

Notice that

$$g(x) = \left(\frac{1}{3}\right)^x = \frac{1}{3^x} = 3^{-x} = f(-x)$$

and so the graph of g could have been obtained from the graph of f by reflecting in the y-axis. ■

Figure 4 shows the graphs of the exponential function $f(x) = a^x$ for various values of the base a. Notice that all of these graphs pass through the same point $(0, 1)$ because $a^0 = 1$ for $a \neq 0$.

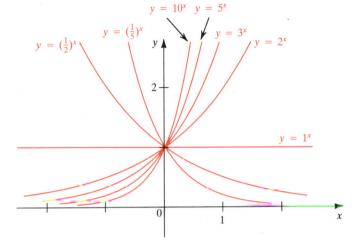

Figure 4

You can see from Figure 4 that there are basically three kinds of exponential functions $y = a^x$. If $0 < a < 1$, the exponential function decreases rapidly. If $a = 1$, it is constant. If $a > 1$, the function increases rapidly and the larger the base the more rapid the increase. These three cases are illustrated in Figure 5.

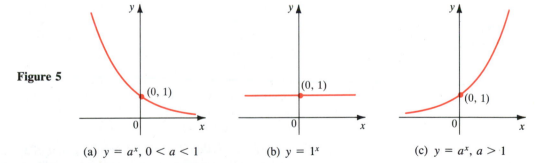

Figure 5

(a) $y = a^x$, $0 < a < 1$ (b) $y = 1^x$ (c) $y = a^x$, $a > 1$

Notice that, if $a > 1$, the graph of $y = a^x$ approaches zero as x decreases through negative values and so the x-axis is a horizontal asymptote. If $0 < a < 1$, the graph approaches zero as x increases indefinitely and again the x-axis is a horizontal asymptote. In both cases the graph never touches the x-axis because $a^x > 0$ for all x. Thus, for $a \neq 1$, the exponential function $f(x) = a^x$ has domain R and range $(0, \infty)$.

In the next two examples we show how to graph certain functions, not by plotting points but by taking the basic graphs of the exponential functions in Figures 4 and 5 and applying the shifting and reflecting transformations of Section 2.5.

EXAMPLE 2

Use the graph of $y = 2^x$ to sketch the graphs of the following functions:
(a) $y = 3 + 2^x$ (b) $y = -2^x$

SOLUTION

(a) The graph of $y = 3 + 2^x$ is obtained by starting with the graph of $y = 2^x$ in Figure 6(a) and shifting it three units upward. Notice from Figure 6(b) that the line $y = 3$ is a horizontal asymptote.

(b) Again we start with the graph of $y = 2^x$, but here we reflect in the x-axis to get the graph of $y = -2^x$ shown in Figure 6(c).

Figure 6

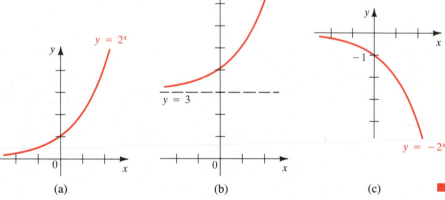

(a)   (b)   (c)   ■

Figure 7

EXAMPLE 3

(a) Use the graph of $y = 10^x$ to sketch the graph of $y = 10^{x-1} - 2$.
(b) State the asymptote, the domain, and the range of this function.

SOLUTION

(a) Recall from Section 2.5 that we get the graph of $y = f(x - 1)$ from the graph of $y = f(x)$ by shifting 1 unit to the right. Thus we get the graph of $y = 10^{x-1} - 2$ by shifting the graph of $y = 10^x$ one unit to the right and two units downward as in Figure 7.

(b) The horizontal asymptote is $y = -2$, the domain is R, and the range is

$$\{y \mid y > -2\} = (-2, \infty)$$ ■

THE EXPONENTIAL FUNCTION WITH BASE e

Any positive number can be used as the base for an exponential function, but some bases are used more frequently than others. We will see in the remaining sections of this chapter that the bases 2 and 10 are convenient for certain applications. But the most important base from the point of view of calculus is the number denoted by the letter e. This number is an irrational number, so we cannot write its exact value, but we can approximate it by using the following procedure.

The table in the margin shows the values, correct to five decimal places, of the expression $(1 + 1/n)^n$ for increasingly large values of n.

The number e is the number that the values of $(1 + 1/n)^n$ approach as n becomes large. (In calculus this idea is made more precise through the concept of a limit. See Exercise 39.) From the table it appears that, correct to 5 decimal places, we have

$$e \approx 2.71828$$

In fact it can be shown that the approximate value to 20 decimal places is

$$e \approx 2.71828182845904523536$$

The notation e for this number was chosen by the Swiss mathematician Leonhard Euler in 1727, probably because it is the first letter of the word *exponential*. We will see in the next section that the number e arises when we consider compound interest.

Since $2 < e < 3$, the graph of the exponential function $y = e^x$ lies between the graphs of $y = 2^x$ and $y = 3^x$ as in Figure 8. Since e is the most important base for exponential functions, scientific calculators have a special key for the function $y = e^x$.

n	$\left(1 + \frac{1}{n}\right)^n$
1	2.00000
5	2.48832
10	2.59374
100	2.70481
1000	2.71692
10,000	2.71815
100,000	2.71827
1,000,000	2.71828
10,000,000	2.71828

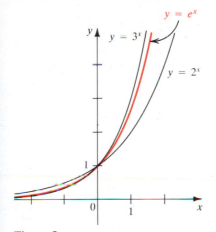

Figure 8

EXAMPLE 4

Sketch the graph of the function $f(x) = e^{-x}$.

SOLUTION

We start with the graph of $y = e^x$ and reflect in the y-axis to obtain $y = e^{-x}$ as in Figure 9. ∎

In the next example we use our knowledge of exponential functions to solve a type of equation that occurs frequently in calculus.

EXAMPLE 5

Solve the equation $3x^2e^x + x^3e^x = 0$.

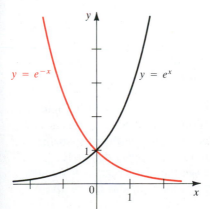

Figure 9

SOLUTION

First we factor the left side of the equation:

$$3x^2e^x + x^3e^x = 0$$
$$(3x^2 + x^3)e^x = 0$$
$$x^2(3 + x)e^x = 0$$

Remember that the graph of every exponential function lies above the x-axis and so $e^x > 0$ for all x. Therefore we can divide both sides of the equation by e^x:

$$x^2(3 + x) = 0$$
$$x = 0 \qquad \text{or} \qquad x = -3$$

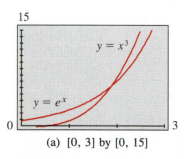

(a) [0, 3] by [0, 15]

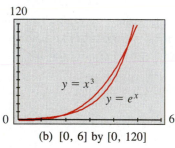

(b) [0, 6] by [0, 120]

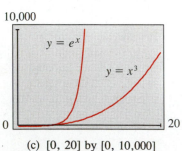

(c) [0, 20] by [0, 10,000]

Figure 10
Graphs of $y = e^x$ and $y = x^3$

 **EXAMPLE 6**

(a) Compare the rates of growth of the functions $f(x) = e^x$ and $g(x) = x^3$ by drawing the graphs of both functions in the following viewing rectangles:
(i) [0, 3] by [0, 15]   (ii) [0, 6] by [0, 120]
(iii) [0, 20] by [0, 10,000]

(b) Find the solutions of the equation $e^x = x^3$ correct to two decimal places.

SOLUTION

(a) (i) The graph in Figure 10(a) shows that the graph of $y = x^3$ catches up with, and becomes higher than, the graph of $y = e^x$ just to the left of $x = 2$.

(ii) However, Figure 10(b) shows that the graph of $y = e^x$ overtakes that of $y = x^3$ when x is between 4 and 5.

(iii) Figure 10(c) gives a more global view and shows that, when x is large, e^x is much larger than x^3.

(b) The solutions of the equation $e^x = x^3$ are the x-coordinates of the points of intersection of the curves $y = e^x$ and $y = x^3$. From the graphs in Figure 10 we see that there are two solutions of the equation, one between 1 and 2 and one between 4 and 5. Zooming in to the viewing rectangle [1, 2] by [4, 8], we see from Figure 11(a) that the smaller root lies between 1.8 and 1.9. So we zoom in further to the viewing rectangle [1.8, 1.9] by [6, 7] in Figure 11(b). By moving the cursor to the intersection point of the two curves, or by inspection and the fact that the x-scale is 0.01, we see that the smaller root is about 1.86. Similarly, by zooming in to the viewing rectangle [4.5, 4.6] by [90, 100] in Figure 11(c), we see that the larger root is about 4.54. Therefore, correct to two decimal places, the solutions of the given equation are 1.86 and 4.54.

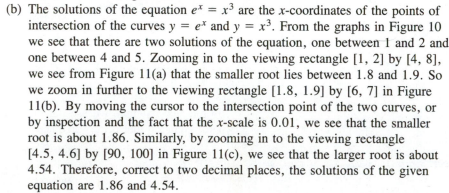

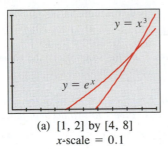

(a) $[1, 2]$ by $[4, 8]$
x-scale $= 0.1$

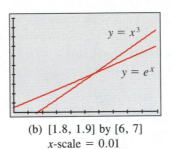

(b) $[1.8, 1.9]$ by $[6, 7]$
x-scale $= 0.01$

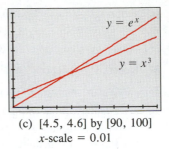

(c) $[4.5, 4.6]$ by $[90, 100]$
x-scale $= 0.01$

Figure 11
Locating the roots of $e^x = x^3$

EXERCISES 4.1

In Exercises 1–4 sketch the graph of the function by making a table of values. Use a calculator if necessary.

1. $f(x) = 6^x$ **2.** $f(x) = \left(\frac{3}{2}\right)^x$

3. $g(x) = \left(\frac{1}{4}\right)^x$ **4.** $h(x) = (1.1)^x$

5. On the same axes graph the functions $y = 4^x$ and $y = 7^x$.

6. On the same axes graph the functions

$$y = \left(\frac{2}{3}\right)^x \quad \text{and} \quad y = \left(\frac{4}{3}\right)^x$$

In Exercises 7–24 graph the given function, not by plotting points, but by starting from the graphs in Figures 4, 5, and 8. State the domain, range, and asymptote of each function.

7. $f(x) = -5^x$ **8.** $f(x) = 10^{-x}$

9. $g(x) = 2^x - 3$ **10.** $g(x) = 2^{x-3}$

11. $y = 4 + \left(\frac{1}{2}\right)^x$ **12.** $y = 6 - 3^x$

13. $y = 10^{x+3}$ **14.** $y = -\left(\frac{1}{5}\right)^x$

15. $y = -e^x$ **16.** $y = 1 - e^x$

17. $y = e^{-x} - 1$ **18.** $y = -e^{-x}$

19. $y = 5^{-2x}$ **20.** $y = 1 + 2^{x+1}$

21. $y = 5 - 2^{x-1}$ **22.** $y = 2 + 5(1 - 10^{-x})$

23. $y = 2^{|x|}$ **24.** $y = 2^{-|x|}$

25. Use a calculator to help graph the function $f(x) = e^{-x^2}$ for $x \geq 0$. Then use the fact that f is an even function to draw the rest of the graph.

26. Compare the functions $f(x) = x^2$ and $g(x) = 2^x$ by evaluating both of them for $x = 0, 1, 2, 3, 4, 5, 6, 7, 8, 9, 10, 15,$ and 20. Then draw the graphs of f and g on the same set of axes.

In Exercises 27–30 solve the equation.

27. $x^2 2^x - 2^x = 0$ **28.** $x^2 10^x - x10^x = 2(10^x)$

29. $4x^3 e^{-3x} - 3x^4 e^{-3x} = 0$ **30.** $x^2 e^x + xe^x - e^x = 0$

31. Solve the inequality $x^2 e^x - 2e^x < 0$.

32. If $f(x) = 10^x$, show that

$$\frac{f(x + h) - f(x)}{h} = 10^x\left(\frac{10^h - 1}{h}\right)$$

33. The hyperbolic cosine function is defined by

$$\cosh(x) = \frac{e^x + e^{-x}}{2}$$

Sketch the graphs of the functions $y = \frac{1}{2}e^x$ and $y = \frac{1}{2}e^{-x}$ on the same axes and use graphical addition (see Section 2.8) to draw the graph of $y = \cosh(x)$.

34. The hyperbolic sine function is defined by

$$\sinh(x) = \frac{e^x - e^{-x}}{2}$$

Graph this function using graphical addition as in Exercise 33.

In Exercises 35–38 use the definitions in Exercises 33 and 34 to prove the identity.

35. $\cosh(-x) = \cosh(x)$ **36.** $\sinh(-x) = -\sinh(x)$

37. $[\cosh(x)]^2 - [\sinh(x)]^2 = 1$

38. $\sinh(x + y) = \sinh(x)\cosh(y) + \cosh(x)\sinh(y)$

 39. (a) Draw the graph of the function $y = (1 + 1/x)^x$ in the viewing rectangle $[-10, 10]$ by $[-2, 15]$.
   (b) Illustrate the definition of the number e by graphing the curve $y = (1 + 1/x)^x$ and the line $y = e$ on the same screen using the viewing rectangle $[0, 40]$ by $[0, 4]$.

 40. Investigate the behavior of the function $f(x) = (1 - 1/x)^x$ as $x \to \infty$ by drawing its graph and the line $y = 1/e$ on the same screen using the viewing rectangle $[0, 20]$ by $[0, 1]$.

 41. (a) Compare the rates of growth of the functions $f(x) = 2^x$ and $g(x) = x^5$ by drawing the graphs of both functions in the following viewing rectangles:
   (i) $[0, 5]$ by $[0, 20]$ (ii) $[0, 25]$ by $[0, 10^7]$
   (iii) $[0, 50]$ by $[0, 10^8]$
   (b) Find the solutions of the equation $2^x = x^5$ correct to one decimal place.

 42. (a) Compare the rates of growth of the functions $f(x) = 3^x$ and $g(x) = x^4$ by drawing the graphs of

both functions in the following viewing rectangles:
   (i) $[-4, 4]$ by $[0, 20]$ (ii) $[0, 10]$ by $[0, 5000]$
   (iii) $[0, 20]$ by $[0, 10^5]$
   (b) Find the solutions of the equation $3^x = x^4$ correct to two decimal places.

 In Exercises 43–46 solve the given equation. Give the solutions correct to two decimal places.

43. $e^x = -x$ **44.** $2^{-x} = x - 1$

45. $4^{-x} = \sqrt{x}$ **46.** $e^{x^2} - 2 = x^3 - x$

 In Exercises 47–48 graph the function and comment on vertical and horizontal asymptotes.

47. $y = 2^{1/x}$ **48.** $y = \dfrac{e^x}{x}$

 In Exercises 49–52 find the local maximum and minimum values of the function and the values of x at which they occur. State your answers correct to two decimal places.

49. $f(x) = xe^{-x}$ **50.** $f(x) = e^x + e^{-3x}$

51. $g(x) = x^x$ **52.** $g(x) = \sqrt[x]{x}$

 In Exercises 53–54, find, approximately, (a) the intervals on which the function is increasing or decreasing and (b) the range of the function.

53. $y = 10^{x-x^2}$ **54.** $y = xe^x$

SECTION 4.2
APPLICATION: EXPONENTIAL GROWTH AND DECAY

The exponential function occurs very frequently in mathematical models of nature and society. In particular, we will see in this section that it occurs in the description of population growth, radioactive decay, and compound interest.

EXPONENTIAL GROWTH

EXAMPLE 1

Under ideal conditions a certain bacteria population is known to double every three hours. Suppose that there are initially 1000 bacteria.
(a) What is the size of the bacteria population after 15 h?
(b) What is the size after t hours?

SOLUTION

Let $n = n(t)$ be the number of bacteria after t hours. Then

$$n(0) = 1000$$
$$n(3) = 2 \cdot 1000$$
$$n(6) = 2(2 \cdot 1000) = 2^2 \cdot 1000$$
$$n(9) = 2(2^2 \cdot 1000) = 2^3 \cdot 1000$$
$$n(12) = 2(2^3 \cdot 1000) = 2^4 \cdot 1000$$
$$n(15) = 2(2^4 \cdot 1000) = 2^5 \cdot 1000$$

(a) After 15 h the number of bacteria is $n(15) = 2^5 \cdot 1000 = 32,000$.

(b) From the pattern above, it appears that the number of bacteria after t hours is

$$n(t) = 2^{t/3} \cdot 1000 \qquad\blacksquare$$

In general, suppose that the initial size of a population is n_0 and the **doubling period** is d; that is, the time required for the population to double is d. Then $n(t)$, the size of the population at time t, is

$$\boxed{n(t) = n_0 2^{t/d}} \tag{1}$$

where t and d are measured in the same unit of time.

EXAMPLE 2

A biologist makes a sample count of bacteria in a culture and finds that the population doubles every 20 min. The estimated bacteria count after 2 h is 32,000.
(a) What was the initial size of the culture at time $t = 0$?
(b) What is the estimated count after 10 h?
(c) What is the count after 2.5 h?
(d) Sketch the graph of the bacterial population function.

SOLUTION

(a) The doubling period is $d = 20$ min $= \frac{1}{3}$ h, so the population at time t is

$$n(t) = n_0 2^{t/(1/3)} = n_0 2^{3t}$$

where n_0 is the initial size. We are given that $n(2) = 32,000$, so

$$n_0 2^{3(2)} = 32,000$$

$$n_0 = \frac{32,000}{2^6} = \frac{32,000}{64} = 500$$

The initial count was 500 bacteria.

Standing Room Only

The population of the world was about five billion in 1986. The recently observed rate of increase is 2% per year. Using the exponential model for population growth and assuming that each person occupies an average of 4 ft^2 of the surface of the earth, we find that by the year 2556 there will be standing room only! (The total land surface area of the world is about 1.8×10^{15} ft^2.)

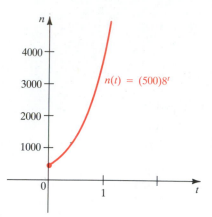

Figure 1

(b) From part (a) we have

$$n(t) = (500)2^{3t}$$

The population after 10 h is

$$n(10) = (500)2^{30} \approx 5.4 \times 10^{11}$$

(c) The population after 2.5 h is

$$n(2.5) = (500)2^{7.5} \approx 9.1 \times 10^4$$

(d) After noting that the domain is $t \geq 0$ and

$$n(t) = (500)2^{3t} = (500)(2^3)^t = (500)8^t$$

we draw the graph of $n(t)$ in Figure 1.

■

RADIOACTIVE DECAY

Radioactive substances decompose by spontaneously emitting radiation. The **half-life** of a radioactive material is the period of time during which any given amount decays until half of it remains.

EXAMPLE 3

An isotope of strontium, Sr^{90}, has a half-life of 25 years.
(a) Find the mass of Sr^{90} that remains from a sample of 24 mg after 125 years.
(b) Find the mass that remains after t years.
(c) Sketch the graph of the mass as a function of time.

SOLUTION

(a) Let $m(t)$ be the mass, in milligrams, that remains after t years. Then

$$m(0) = 24$$

$$m(25) = \frac{1}{2}(24)$$

$$m(50) = \frac{1}{2}\left(\frac{1}{2} \cdot 24\right) = \frac{1}{2^2}(24)$$

$$m(75) = \frac{1}{2}\left(\frac{1}{2^2} \cdot 24\right) = \frac{1}{2^3}(24)$$

$$m(100) = \frac{1}{2}\left(\frac{1}{2^3} \cdot 24\right) = \frac{1}{2^4}(24)$$

$$m(125) = \frac{1}{2}\left(\frac{1}{2^4} \cdot 24\right) = \frac{1}{2^5}(24)$$

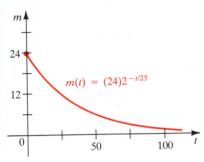

Figure 2

$m(t) = (24)2^{-t/25}$

The mass of Sr^{90} remaining after 125 yr is

$$m(125) = \frac{1}{2^5}(24) = \frac{24}{32} = \frac{3}{4} \text{ mg}$$

(b) From the pattern in part (a) it appears that the mass after t years is

$$m(t) = \frac{1}{2^{t/25}}(24) = (24)2^{-t/25}$$

(c) The graph of $m(t)$ is sketched in Figure 2. ■

In general we have

$$\boxed{m(t) = m_0 2^{-t/h}} \tag{2}$$

where $m(t)$ is the mass of a radioactive substance that remains after time t, m_0 is the initial mass, and h is the half-life (measured in the same unit of time as t). Some radioactive materials decay very slowly, having half-lives of thousands of years. Others decay very rapidly, with half-lives of less than a second.

Radiocarbon dating is a method archeologists use to determine the age of ancient objects. The carbon dioxide in the atmosphere always contains a fixed fraction of radioactive carbon, C^{14}, with a half-life of about 5730 years. Plants absorb carbon dioxide from the atmosphere, which then makes its way to animals through the food chain. Thus all living creatures contain the same fixed proportion of C^{14} to nonradioactive C^{12} as the atmosphere.

After an organism dies, it stops assimilating C^{14}, and the amount of C^{14} in it begins to decay exponentially. We can then determine the time elapsed since the death of an organism by measuring the amount of C^{14} left in it.

For example, if a donkey bone contains 73% as much C^{14} as a living donkey and it died t years ago, then by the formula for radioactive decay

$$0.73 = (1.00)2^{-t/5730}$$

We solve this to find $t \approx 2600$, so the bone is about 2600 years old.

EXAMPLE 4

Polonium-210 (Po^{210}) has a half-life of 140 days. If a sample has a mass of 300 mg, find the mass after 50 days.

SOLUTION

Since $m_0 = 300$ and $h = 140$, the mass that remains after t days is

$$m(t) = (300)2^{-t/140} \qquad \text{(from Equation 2)}$$

so

$$m(50) = (300)2^{-50/140} \approx 234$$

After 50 days the mass that remains is about 234 mg. ■

EXAMPLE 5

After 30 h a sample of Plutonium-243 (Pu^{243}) has decayed to $\frac{1}{64}$ of its original mass. Find the half-life of Pu^{243}.

SOLUTION

If h is the half-life and m_0 is the original amount, then

$$m(t) = m_0 2^{-t/h}$$

We are given that $m(30) = \frac{1}{64}m_0$, so

$$m_0 2^{-30/h} = \frac{1}{64}m_0$$

$$2^{-30/h} = \frac{1}{64} = 2^{-6} \qquad \text{(dividing by } m_0)$$

$$-\frac{30}{h} = -6 \qquad \text{(equating exponents)}$$

$$h = \frac{30}{6} = 5$$

The half-life of Pu^{243} is five hours.

COMPOUND INTEREST

If an amount of money P, called the **principal**, is invested at a simple interest rate r, then the interest after one time period is Pr and the amount of money is

$$A = P + Pr = P(1 + r)$$

For instance, if $P = \$1000$ and the interest rate is 12% per year, then $r = 0.12$ and the amount after one year is $\$1000(1 + 0.12) = \$1120$.

If the interest is reinvested, then the new principal is $P(1 + r)$ and the amount after another time period is

$$A = P(1 + r)(1 + r) = P(1 + r)^2$$

Similarly, after a third time period the amount is $P(1 + r)^3$, and in general after k periods it is

$$A = P(1 + r)^k$$

Notice that this is an exponential function with base $1 + r$.

For example, if the interest rate is 12% per year compounded semiannually, then the time period is 6 months and the interest rate per time period is 6%, or 0.06.

If interest is compounded n times per year, then in each time period the interest rate is r/n and there are nt time periods in t years, so the amount after t years is

$$A = P\left(1 + \frac{r}{n}\right)^{nt} \qquad (3)$$

EXAMPLE 6

A sum of $\$1000$ is invested at an interest rate of 12% per year. Find the amount in the account after 3 yr if interest is compounded:
(a) annually   (b) semiannually   (c) quarterly   (d) monthly   (e) daily

SOLUTION

We use Formula 3 with $P = \$1000$, $r = 0.12$, and $t = 3$.

(a) With annual compounding, $n = 1$:

$$A = 1000(1.12)^3 = \$1404.93$$

(b) With semiannual compounding, $n = 2$:

$$A = 1000\left(1 + \frac{0.12}{2}\right)^{2(3)} = 1000(1.06)^6 = \$1418.52$$

(c) With quarterly compounding, $n = 4$:

$$A = 1000\left(1 + \frac{0.12}{4}\right)^{4(3)} = 1000(1.03)^{12} = \$1425.76$$

(d) With monthly compounding, $n = 12$:

$$A = 1000\left(1 + \frac{0.12}{12}\right)^{12(3)} = 1000(1.01)^{36} = \$1430.77$$

(e) With daily compounding, $n = 365$:

$$A = 1000\left(1 + \frac{0.12}{365}\right)^{365(3)} = \$1433.24 \qquad \blacksquare$$

We see from Example 6 that the interest paid increases as the number of compounding periods (n) increases. In general, let us see what happens as n increases indefinitely. If we let $m = n/r$, then

$$A = P\left(1 + \frac{r}{n}\right)^{nt} = P\left[\left(1 + \frac{r}{n}\right)^{n/r}\right]^{rt} = P\left[\left(1 + \frac{1}{m}\right)^{m}\right]^{rt}$$

Recall that as m becomes large, the quantity $(1 + 1/m)^m$ approaches the number e. Thus the amount approaches

$$\boxed{A = Pe^{rt}} \qquad\qquad (4)$$

When interest is paid according to Formula 4, we refer to **continuous compounding of interest**.

EXAMPLE 7

Find the amount after three years if $1000 is invested at an interest rate of 12% per year compounded continuously.

SOLUTION

Using Formula 4 with $P = \$1000$, $r = 0.12$, and $t = 3$, we have

$$A = 1000e^{(0.12)3} = 1000e^{0.36} = \$1433.33$$

EXERCISES 4.2

1. A bacteria culture contains 1500 bacteria initially and doubles every half-hour. Find the size of the bacteria population
   (a) after t hours   **(b)** after 20 min   **(c)** after 24 h

2. The doubling period of a bacteria population is 15 min. At time $t = 1.25$ h an estimate of 80,000 was taken.
   (a) What was the initial population?
   (b) What is the size after t hours?
   (c) What is the size after 3 h?
   (d) What is the size after 40 min?
   (e) Sketch the graph of the population function.

3. A bacteria culture starts with 10,000 bacteria and the population triples every hour.
   (a) Find the number of bacteria after t hours.
   (b) Find the number of bacteria after 2.5 h.

4. A bacteria culture starts with 5000 bacteria. After 3 h the estimated count is 80,000.
   (a) Find the doubling period.
   (b) What is the population after 3.5 h?

5. The population of the world is doubling every 35 yr. In 1987 the total population was 5 billion. If the doubling period remains at 35 yr, find the projected world population
   (a) for the year 2001   **(b)** for the year 2100

6. The population of a certain city grows at a rate of 5% per year. The population in 1988 was 421,000. If the growth rate remains constant, find the projected population of the city for the years 2000 and 2030.

7. The half-life of Radium-226 (Ra^{226}) is 1590 yr. If a sample has a mass of 150 mg, find a formula for the mass that remains after t years. Then find the mass that remains after 1000 yr.

8. An isotope of sodium, Na^{24}, has a half-life of 15 h. Find the amount remaining from a 2-g sample after
   (a) 5 h   **(b)** 20 h   **(c)** 4 days

9. The half-life of Palladium-100 (Pd^{100}) is 4 days. After 20 days a sample of Pd^{100} has been reduced to a mass of 0.375 g.
   (a) What was the initial mass of the sample?
   (b) What is the mass after three days?
   (c) What is the mass after three weeks?

10. A certain amount of Vanadium-48 (V^{48}) decays to $\frac{1}{8}$ of its original mass in 48 days. What is the half-life of V^{48}?

11. If \$10,000 is invested at an interest rate of 10% per year, compounded semiannually, find the value of the investment after
   (a) 5 yr   **(b)** 10 yr   **(c)** 15 yr

12. If \$4000 is borrowed at a rate of 16% interest per year, compounded quarterly, find the amount due at the end of
   (a) 4 yr   **(b)** 6 yr   **(c)** 8 yr

13. If \$3000 is invested at a rate of 9% per year, find the amount of the investment at the end of 5 yr if the interest is compounded
   (a) annually   **(b)** semiannually   **(c)** monthly
   (d) weekly   **(e)** daily   **(f)** hourly
   (g) continuously

14. Which of the following would be the better investment?
   (a) An account paying $9\frac{1}{4}\%$ per year compounded semiannually.
   (b) An account paying 9% per year compounded continuously.

15. The **present value** of a sum of money is the amount that must be invested now, at a given rate of interest, to produce the desired sum at a later date. Find the present value of \$10,000 if interest is paid at a rate of 9% compounded semiannually for 3 yr.

16. A bacteria culture starts with 10,000 bacteria and the doubling period is 40 min. If the number of bacteria after t minutes is $n(t)$, draw the graph of $n(t)$ and use it to determine when the population of bacteria will reach 50,000.

17. The half-life of Polonium-210 is 140 days.
    (a) Find the mass $m(t)$ of a 300-mg sample after t days.
    (b) Use the graph of $m(t)$ to determine when the mass of this sample will decay to 200 mg.

 18. A sum of \$5000 is invested at an interest rate of 9% per year compounded semiannually.
    (a) Find the value $A(t)$ of the investment after t years.

    (b) Draw the graph of $A(t)$.
    (c) Use the graph of $A(t)$ to determine when this investment will amount to \$25,000.

 19. How long will it take for an investment to double in value if the interest rate is 8.5% per year compounded semiannually?

SECTION 4.3
LOGARITHMIC FUNCTIONS

If $a > 0$, where $a \neq 1$, then the exponential function $f(x) = a^x$ is a one-to-one function by the Horizontal Line Test (see Figure 1) and therefore has an inverse function.

Figure 1
$f(x) = a^x$ is one-to-one

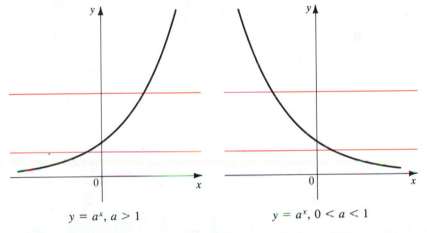

$y = a^x, a > 1$ $y = a^x, 0 < a < 1$

The inverse function f^{-1} of the exponential function $f(x) = a^x$ is called the **logarithmic function with base a** and is denoted by $\log_a$. Recall from Section 2.10 that the inverse function f^{-1} is defined by

$$f^{-1}(x) = y \quad \Leftrightarrow \quad f(y) = x$$

Thus the definition of the logarithmic function as the inverse of the exponential function means that

$$\log_a(x) = y \quad \Leftrightarrow \quad a^y = x$$

By tradition, the name of the logarithmic function is $\log_a$, not just a single letter.

When using the logarithmic function we usually omit the parentheses in the function notation and write

$$\log_a(x) = \log_a x$$

Therefore we can write the definition of the function $\log_a$ as follows:

We read $\log_a x = y$ as "log base a of x is y."

Let a be a positive number with $a \neq 1$. Then, for any positive number x, we define the function $\log_a$ by

$$\log_a x = y \quad \Leftrightarrow \quad a^y = x \qquad (1)$$

In words, this says that

$\log_a x$ is the exponent to which the base a must be raised to give x.

In using Equation 1 to switch back and forth between the logarithmic form $\log_a x = y$ and the exponential form $a^y = x$, it is helpful to notice that in both cases the base is the same:

$$\underset{\text{base}}{\overset{\text{exponent}}{\log_a x = y}} \quad \Leftrightarrow \quad \underset{\text{base}}{\overset{\text{exponent}}{a^y = x}}$$

EXAMPLE 1

Evaluate: (a) $\log_3 81$ (b) $\log_{16} 4$ (c) $\log_{10} 0.0001$

SOLUTION

(a) $\log_3 81 = 4$ because $3^4 = 81$

(b) $\log_{16} 4 = \frac{1}{2}$ because $16^{1/2} = 4$

(c) $\log_{10} 0.0001 = -4$ because $10^{-4} = 0.0001$ ∎

EXAMPLE 2

Express in exponential form:

(a) $\log_2\left(\frac{1}{2}\right) = -1$ (b) $\log_{10} 100{,}000 = 5$ (c) $\log_3 z = t$

SOLUTION

(a) $\log_2\left(\frac{1}{2}\right) = -1 \implies 2^{-1} = \frac{1}{2}$

(b) $\log_{10} 100{,}000 = 5 \implies 10^5 = 100{,}000$

(c) $\log_3 z = t \implies 3^t = z$ ∎

EXAMPLE 3

Express in logarithmic form:
(a) $1000 = 10^3$ (b) $2^{-3} = \frac{1}{8}$ (c) $s = 5^r$

SOLUTION

(a) $10^3 = 1000 \;\Rightarrow\; \log_{10} 1000 = 3$ (b) $2^{-3} = \frac{1}{8} \;\Rightarrow\; \log_2\!\left(\frac{1}{8}\right) = -3$

(c) $s = 5^r \;\Rightarrow\; \log_5 s = r$ ∎

EXAMPLE 4

Solve for x: $\log_2(25 - x) = 3$

SOLUTION

The first step is to rewrite the equation in exponential form.

$$\log_2(25 - x) = 3$$
$$2^3 = 25 - x$$
$$8 = 25 - x$$
$$x = 25 - 8 = 17$$
∎

EXAMPLE 5

Solve for x: $3^{x+2} = 7$

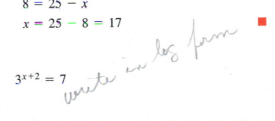
write in log form

SOLUTION

First we rewrite the equation in logarithmic form:

$$\log_3 7 = x + 2$$
$$x = (\log_3 7) - 2$$
∎

GRAPHS OF LOGARITHMIC FUNCTIONS

Recall that if a one-to-one function f has domain A and range B, then its inverse function f^{-1} has domain B and range A. Since the exponential function $f(x) = a^x$, where $a \neq 1$, has domain R and range $(0, \infty)$, we conclude that its inverse function, $f^{-1}(x) = \log_a x$ has domain $(0, \infty)$ and range R.

The graph of $f^{-1}(x) = \log_a x$ is obtained by reflecting the graph of $f(x) = a^x$ in the line $y = x$. Figure 2 shows the case where $a > 1$. (The most important logarithmic functions have base $a > 1$.) The fact that $y = a^x$ $(a > 1)$ is a very rapidly increasing function is reflected in the fact that $y = \log_a x$ is a very slowly increasing function (see Exercise 87).

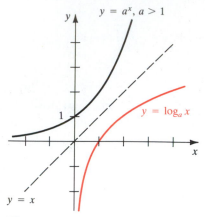

Figure 2

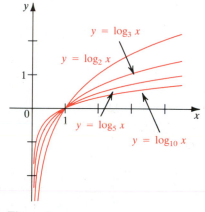

Figure 3

Notice that since $a^0 = 1$, we have

$$\boxed{\log_a 1 = 0}$$

and so the x-intercept of the function $y = \log_a x$ is 1. Notice also that because $y = a^x$ has the x-axis as a horizontal asymptote, the curve $y = \log_a x$ has the y-axis as a vertical asymptote. In fact, using the notation of Section 3.9, we can write

$$\log_a x \to -\infty \quad \text{as} \quad x \to 0^+$$

for the case $a > 1$.

Figure 3 shows the relationship among the graphs of the logarithmic functions with bases 2, 3, 5, and 10. These graphs were drawn by reflecting the graphs of $y = 2^x$, $y = 3^x$, $y = 5^x$, and $y = 10^x$ (see Figure 4 in Section 4.1) in the line $y = x$.

In the next two examples we graph logarithmic functions by starting with the basic graphs in Figure 3 and using the transformations of Section 2.5.

EXAMPLE 6

Sketch the graphs of the following functions:
(a) $y = -\log_2 x$ (b) $y = \log_2(-x)$

SOLUTION

(a) We start with the graph of $y = \log_2 x$ in Figure 4(a) and reflect about the x-axis to get the graph of $y = -\log_2 x$ in Figure 4(b).

(b) To obtain the graph of $y = \log_2(-x)$ we reflect the graph of $y = \log_2 x$ in the y-axis. See Figure 4(c).

Figure 4

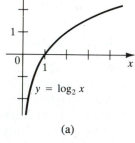

(a)

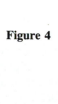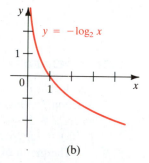

(b)

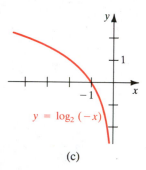

(c)

EXAMPLE 7

Find the domain of the function $f(x) = \log_{10}(x - 3)$ and sketch its graph.

SOLUTION

The domain of $y = \log_a x$ is the interval $(0, \infty)$, so $\log_{10} x$ is defined only when $x > 0$. Therefore the domain of $f(x) = \log_{10}(x - 3)$ is

$$\{x \mid x - 3 > 0\} = \{x \mid x > 3\} = (3, \infty)$$

The graph of f is obtained from the graph of $y = \log_{10} x$ by shifting three units to the right. Notice from Figure 5 that the line $x = 3$ is a vertical asymptote.

Figure 5

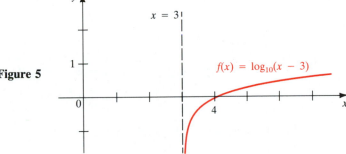

In Section 2.10 we saw that a function f and its inverse function f^{-1} satisfy the equations

dom(f) = the domain of f

$$f^{-1}(f(x)) = x \qquad x \in \text{dom}(f)$$
$$f(f^{-1}(x)) = x \qquad x \in \text{dom}(f^{-1})$$

When applied to $f(x) = a^x$ and $f^{-1}(x) = \log_a x$, these equations become

$$\boxed{\begin{aligned} \log_a(a^x) &= x & x \in R \\ a^{\log_a x} &= x & x > 0 \end{aligned}}$$

(2)

For instance, we have

$$\log_{10}(10^x) = x \qquad \text{and} \qquad 2^{\log_2 x} = x$$

NATURAL LOGARITHMS

Of all possible bases a for logarithms, it turns out that the most convenient choice of a base for the purposes of calculus is the number e, which was defined in Section 4.1. The logarithm with base e is called the **natural logarithm** and is given a special name and notation:

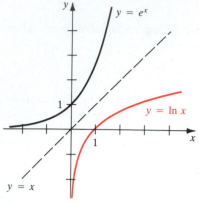

Figure 6

$$\ln x = \log_e x$$

(The abbreviation ln is short for *logarithmus naturalis*.)

Thus the natural logarithmic function $y = \ln x$ is the inverse function of the exponential function $y = e^x$; they are both graphed in Figure 6.

If we put $a = e$ and write ln for $\log_e$ in (1) and (2), then the defining properties of the natural logarithm become

$$\ln x = y \quad \Leftrightarrow \quad e^y = x \tag{3}$$

and

$$\begin{aligned} \ln(e^x) &= x & x \in R \\ e^{\ln x} &= x & x > 0 \end{aligned} \tag{4}$$

In particular, it is worth noting that

$$\ln e = 1 \qquad \text{and} \qquad \ln 1 = 0$$

EXAMPLE 8

Solve for x: $\qquad\qquad \ln x = 8$

SOLUTION 1

From (3) we see that

$$\ln x = 8 \qquad \text{means} \qquad e^8 = x$$

Therefore $x = e^8$.

SOLUTION 2

Start with the equation

$$\ln x = 8$$

To solve a logarithmic equation, exponentiate each side.

and apply the natural exponential function to both sides of the equation:

$$e^{\ln x} = e^8$$

The second equation in (4) says that $e^{\ln x} = x$. Therefore $x = e^8$.   ∎

EXAMPLE 9

Solve the equation $e^{3-2x} = 4$.

SOLUTION

To solve an exponential equation, take logarithms of each side.

We take natural logarithms of both sides of the equation and use (4):

$$\ln(e^{3-2x}) = \ln 4$$
$$3 - 2x = \ln 4$$
$$2x = 3 - \ln 4$$
$$x = \tfrac{1}{2}(3 - \ln 4)$$

∎

EXAMPLE 10

Find the domain of the function $f(x) = \ln(4 - x^2)$.

SOLUTION

As with any logarithmic function, $\ln x$ is defined when $x > 0$. Thus the domain of f is

$$\{x \mid 4 - x^2 > 0\} = \{x \mid x^2 < 4\} = \{x \mid |x| < 2\}$$
$$= \{x \mid -2 < x < 2\} = (-2, 2)$$

∎

EXAMPLE 11

Draw the graph of the function $y = x \ln(4 - x^2)$ and use it to find the asymptotes and local maximum and minimum values.

SOLUTION

As in Example 10, the domain of this function is the interval $(-2, 2)$, so we choose the viewing rectangle $[-3, 3]$ by $[-3, 3]$. The graph is shown in Figure 7 and we see that the lines $x = -2$ and $x = 2$ are vertical asymptotes.

There is a local maximum point to the right of $x = 1$ and a local minimum point to the left of $x = -1$. By zooming in and tracing along the graph with the cursor, we find that the local maximum value is approximately 1.13 and occurs when $x \approx 1.15$. Similarly (or by noticing that the function is odd) we find that the local minimum value is about -1.13 and occurs when $x \approx -1.15$. ∎

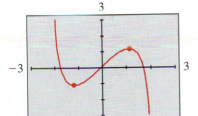

Figure 7
$y = x \ln(4 - x^2)$

odel 1 - 15

EXERCISES 4.3

In Exercises 1–8 express the equation in exponential form.

1. $\log_2 64 = 6$

2. $\log_5 1 = 0$

3. $\log_{10} 0.01 = -2$

4. $\log_8 4 = \tfrac{2}{3}$

5. $\log_8 512 = 3$

6. $\log_2\left(\tfrac{1}{16}\right) = -4$

7. $\log_a b = c$

8. $\log_r v = w$

In Exercises 9–16 express the equation in logarithmic form.

9. $2^3 = 8$

10. $10^5 = 100,000$

11. $10^{-4} = 0.0001$

12. $81^{1/2} = 9$

13. $4^{-3/2} = 0.125$

14. $6^{-1} = \tfrac{1}{6}$

15. $r^s = t$

16. $10^m = n$

In Exercises 17–36 evaluate the expression.

17. $\log_6 6^4$

18. $\log_2 32$

19. $\log_4 64$

20. $\log_8 8^{17}$

21. $\log_9 9$

22. $\log_6 1$

23. $\log_3(\frac{1}{27})$

24. $\log_5 125$

25. $\log_{10} \sqrt{10}$

26. $3^{\log_3 8}$

27. $\log_5 0.2$

28. $\log_4 8$

29. $\log_8 0.25$

30. $\log_9 \sqrt{3}$

31. $2^{\log_2 37}$

32. $\log_4 0.5$

33. $\ln e^4$

34. $e^{\ln \pi}$

35. $e^{\ln \sqrt{5}}$

36. $\ln(1/e)$

In Exercises 37–56 solve the given equation for x.

37. $\log_2 x = 10$

38. $\log_5 x = 4$

39. $\log_{10}(3x + 5) = 2$

40. $\log_3(2 - x) = 3$

41. $\log_x 16 = 4$

42. $\log_x 6 = \frac{1}{2}$

43. $5^x = 16$

44. $10^{-x} = 2$

45. $2^{1-x} = 3$

46. $3^{2x-1} = 5$

47. $\ln x = 10$

48. $\ln(2 + x) = 1$

49. $2 - \ln(3 - x) = 0$

50. $e^x = 10$

51. $e^{12x} = 17$

52. $e^{1-4x} = 2$

53. $\log_2(x^2 - x - 2) = 2$

54. $2^{2/\log_5 x} = \frac{1}{16}$

55. $\log_2(\log_3 x) = 4$

56. $10^{5^x} = 3$

57. Draw the graph of $y = 4^x$. Then use it to draw the graph of $y = \log_4 x$.

58. Most scientific calculators have keys for both LN and LOG (= $\log_{10}$). Use such a calculator to draw the graphs of $y = \ln x$ and $y = \log_{10} x$ on the same set of axes.

In Exercises 59–68 graph the given function, not by plotting points, but by starting from the graphs in Figures 2, 3, and 6. State the domain, range, and asymptote of each function.

59. $f(x) = \log_2(x - 4)$

60. $f(x) = -\log_{10} x$

61. $g(x) = \log_5(-x)$

62. $g(x) = \ln(x + 2)$

63. $y = 2 + \log_3 x$

64. $y = \log_3(x - 1) - 2$

65. $y = 1 - \log_{10} x$

66. $y = 1 + \ln(-x)$

67. $y = |\ln x|$

68. $y = \ln|x|$

In Exercises 69–74 find the domain of the function.

69. $f(x) = \log_{10}(2 + 5x)$

70. $f(x) = \log_2(10 - 3x)$

71. $g(x) = \log_3(x^2 - 1)$

72. $g(x) = \ln(x - x^2)$

73. $h(x) = \ln x + \ln(2 - x)$

74. $h(x) = \sqrt{x - 2} - \log_5(10 - x)$

 In Exercises 75–80 draw the graph of the function in a suitable viewing rectangle and use it to find the domain, the asymptotes, and the local maximum and minimum values.

75. $y = \log_{10}(1 - x^2)$

76. $y = \ln(x^2 - x)$

77. $y = x + \ln x$

78. $y = x(\ln x)^2$

79. $y = \dfrac{\ln x}{x}$

80. $y = x \log_{10}(x + 10)$

 81. Compare the rates of growth of the functions $f(x) = \ln x$ and $g(x) = \sqrt{x}$ by drawing their graphs on a common screen using the viewing rectangle $[-1, 30]$ by $[-1, 6]$.

82. **(a)** By drawing the graphs of the functions

$$f(x) = 1 + \ln(1 + x) \qquad \text{and} \qquad g(x) = \sqrt{x}$$

in a suitable viewing rectangle show that, even when a logarithmic function starts out higher than a root function, it is ultimately overtaken by the root function.

(b) Find, correct to two decimal places, the solutions of the equation $\sqrt{x} = 1 + \ln(1 + x)$.

 In Exercises 83–86 find the solutions of the equation correct to two decimal places.

83. $\ln x = 3 - x$

84. $\log_{10} x = x^2 - 2$

85. $x^3 - x = \log_{10}(x + 1)$

86. $x = \ln(4 - x^2)$

87. Suppose that the graph of $y = 2^x$ is drawn on a coordinate grid where the unit of measurement is an inch. Show that at a distance 2 ft to the right of the origin the height of the graph is about 265 mi. If the graph of $y = \log_2 x$ is drawn on the same set of axes, how far to the right of the origin do we have to go before the height of the curve reaches 2 ft?

88. Solve the inequality $3 \leq \log_2 x \leq 4$.

89. Solve the inequality $2 < 10^x < 5$.

90. Which is larger, $\log_4 17$ or $\log_5 24$?

91. (a) Find the domain of the function $f(x) = \log_2(\log_{10} x)$.
   (b) Find the inverse function of f.

92. (a) Find the domain of the function $f(x) = \ln(\ln(\ln x))$.
   (b) Find the inverse function of f.

93. Solve the equation $4^x - 2^{x+1} = 3$. [*Hint:* First write the equation as a quadratic equation in 2^x.]

94. (a) Find the inverse of the function

$$f(x) = \frac{2^x}{1 + 2^x}$$

(b) What is the domain of the inverse function?

SECTION 4.4
LAWS OF LOGARITHMS

Since logarithms are exponents, the Laws of Exponents give rise to the Laws of Logarithms. These properties give the logarithmic functions a wide range of application, as we will see in the next section.

LAWS OF LOGARITHMS

Suppose that $x > 0$, $y > 0$, and r is any real number. Then

1. $\log_a(xy) = \log_a x + \log_a y$

2. $\log_a\left(\dfrac{x}{y}\right) = \log_a x - \log_a y$

3. $\log_a(x^r) = r \log_a x$

Proof We make use of the equation $\log_a(a^x) = x$ that was given as Formula 2 in Section 4.3.

1. Let

$$\log_a x = b \qquad \text{and} \qquad \log_a y = c$$

When written in exponential form, these equations become

$$a^b = x \qquad \text{and} \qquad a^c = y$$

Thus
$$\log_a(xy) = \log_a(a^b a^c)$$
$$= \log_a(a^{b+c})$$
$$= b + c$$
$$= \log_a x + \log_a y$$

2. Using Law 1, we have

$$\log_a x = \log_a\left[\left(\frac{x}{y}\right)y\right] = \log_a\left(\frac{x}{y}\right) + \log_a y$$

so
$$\log_a\left(\frac{x}{y}\right) = \log_a x - \log_a y$$

John Napier (1550–1617) *was a Scottish landowner for whom mathematics was a hobby. Today, he is best known as the inventor of logarithms. He published his invention in 1614 under the title* A Description of the Marvelous Rule of Logarithms. *In Napier's time logarithms were used exclusively for simplifying complicated calculations. Since the advent of calculators and computers, logarithms are no longer used for this purpose. The logarithmic functions, however, have found many applications, some of which are described in this chapter. Napier wrote on many other topics. One of his more famous works is the book* A Plaine Discovery of the Whole Revelation of Saint John, *in which he predicted that the world would end in the year 1700.*

3. Let $\log_a x = b$. Then $a^b = x$, so

$$\log_a(x^r) = \log_a(a^b)^r = \log_a(a^{rb}) = rb = r \log_a x \qquad \blacksquare$$

We state the Laws of Logarithms in words as follows:

1. The logarithm of a product is the sum of the logarithms of the factors.
2. The logarithm of a quotient is the difference of the logarithms of the factors.
3. The logarithm of a power of a number is the exponent times the logarithm of the number.

As the following examples illustrate, these laws are used in both directions. Since the domain of any logarithmic function is the interval $(0, \infty)$, we assume that all quantities whose logarithms occur are positive.

EXAMPLE 1

Use the Laws of Logarithms to rewrite the following:

(a) $\log_2(6x)$ (b) $\log_{10} \sqrt{5}$ (c) $\log_5 x^3 y^6$ (d) $\ln \dfrac{ab}{\sqrt[3]{c}}$

SOLUTION

(a) $\log_2(6x) = \log_2 6 + \log_2 x$

(b) $\log_{10} \sqrt{5} = \log_{10} 5^{1/2} = \frac{1}{2} \log_{10} 5$

(c) $\log_5(x^3 y^6) = \log_5 x^3 + \log_5 y^6 = 3 \log_5 x + 6 \log_5 y$

(d) $\ln \dfrac{ab}{\sqrt[3]{c}} = \ln ab - \ln \sqrt[3]{c} = \ln a + \ln b - \ln c^{1/3}$
$\qquad\qquad = \ln a + \ln b - \frac{1}{3} \ln c$ $\qquad\qquad\qquad$ $\blacksquare$

EXAMPLE 2

Simplify the following:

(a) $\log_4 2 + \log_4 32$ (b) $\log_2 80 - \log_2 5$ (c) $-\frac{1}{3} \log_{10} 8$

SOLUTION

(a) $\log_4 2 + \log_4 32 = \log_4(2 \cdot 32) = \log_4 64 = 3$

(b) $\log_2 80 - \log_2 5 = \log_2\left(\frac{80}{5}\right) = \log_2 16 = 4$

(c) $-\frac{1}{3} \log_{10} 8 = \log_{10} 8^{-1/3} = \log_{10}\left(\frac{1}{2}\right)$ $\qquad\qquad$ $\blacksquare$

EXAMPLE 3

Express $3 \ln s + \frac{1}{2} \ln t - 4 \ln(t^2 + 1)$ as a single logarithm.

SOLUTION

$$3 \ln s + \tfrac{1}{2} \ln t - 4 \ln(t^2 + 1) = \ln s^3 + \ln t^{1/2} - \ln(t^2 + 1)^4$$

$$= \ln(s^3 t^{1/2}) - \ln(t^2 + 1)^4$$

$$= \ln\left(\frac{s^3 \sqrt{t}}{(t^2 + 1)^4}\right) \qquad \blacksquare$$

Warning Although the Laws of Logarithms tell us how to compute the logarithm of a product or a quotient, *there is no corresponding rule for the logarithm of a sum or a difference*. For instance,

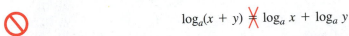

$$\log_a(x + y) \neq \log_a x + \log_a y$$

In fact we know that the right side is equal to $\log_a(xy)$.

Similarly we *cannot* write

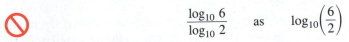

$$\frac{\log_{10} 6}{\log_{10} 2} \qquad \text{as} \qquad \log_{10}\left(\frac{6}{2}\right)$$

Nor can we write

$$(\log_2 x)^3 \qquad \text{as} \qquad 3 \log_2 x$$

EXAMPLE 4

Solve the equation $\log_{10}(x + 2) + \log_{10}(x - 1) = 1$.

SOLUTION

Using Law 1, we rewrite the equation as

$$\log_{10}[(x + 2)(x - 1)] = 1 \qquad (1)$$

or

$$\log_{10}[(x + 2)(x - 1)] = \log_{10} 10$$

Since the function $\log_{10}$ is one-to-one, we have

$$(x + 2)(x - 1) = 10 \qquad \text{[or raise 10 to both sides of (1)]}$$

$$x^2 + x - 12 = 0$$

$$(x + 4)(x - 3) = 0$$

$$x = -4 \qquad \text{or} \qquad x = 3$$

Let us check to see if these values satisfy the original equation. If $x = -4$, we have

$$\log_{10}(x + 2) = \log_{10}(-4 + 2) = \log_{10}(-2)$$

which is undefined, so $x = -4$ is an extraneous root. The only solution is $x = 3$, as you can verify. $\blacksquare$

■ **COMMON LOGARITHMS**

Logarithms with base 10 are called **common logarithms** and are often denoted by omitting the base:

$$\log x = \log_{10} x$$

It is standard practice to use common logarithms to solve exponential equations. For this reason they are found on scientific calculators.

EXAMPLE 5

Find the solution of the equation $3^{x+2} = 7$ correct to six decimal places.

SOLUTION

In Example 5 in Section 4.3 we obtained the solution in the form $x = \log_3 7 - 2$, but we cannot find the exact value of $\log_3 7$ and calculators do not have a $\log_3$ key. So instead we solve the given equation by taking common logarithms of both sides and using Law 3.

$$3^{x+2} = 7$$
$$\log(3^{x+2}) = \log 7$$
$$(x + 2) \log 3 = \log 7$$
$$x + 2 = \frac{\log 7}{\log 3}$$

Using a calculator, we get

$$x = \frac{\log 7}{\log 3} - 2 \approx -0.228756$$

■

Notice that in solving Example 5 we could have used natural logarithms instead of common logarithms. In fact, using the same steps, we get

$$x = \frac{\ln 7}{\ln 3} - 2 \approx -0.228756$$

■ **CHANGE OF BASE**

For some purposes it is useful to be able to change from logarithms in one base to logarithms in another base. Suppose that we are given $\log_a x$ and want to find $\log_b x$. Let

$$y = \log_b x$$

We write this in exponential form and take logarithms, with base a, of both sides.

$$b^y = x$$
$$\log_a(b^y) = \log_a x$$
$$y \log_a b = \log_a x$$
$$y = \frac{\log_a x}{\log_a b}$$

Thus we have proved the following formula.

CHANGE OF BASE FORMULA

$$\log_b x = \frac{\log_a x}{\log_a b}$$

In particular if we put $x = a$, then $\log_a a = 1$ and this formula becomes

$$\log_b a = \frac{1}{\log_a b}$$

EXAMPLE 6

Evaluate $\log_8 5$ correct to six decimal places.

SOLUTION

There is no $\log_8$ key on a calculator, but we use the change of base formula with $b = 8$ and $a = 10$ to convert to common logarithms:

$$\log_8 5 = \frac{\log_{10} 5}{\log_{10} 8} = \frac{\log 5}{\log 8} \approx 0.773976$$

EXERCISES 4.4

In Exercises 1–24 use the Laws of Logarithms to rewrite each expression in a form with no logarithms of products, quotients, or powers.

1. $\log_2 x(x - 1)$

2. $\log_5\left(\dfrac{x}{2}\right)$

3. $\log 7^{23}$

4. $\ln(\pi x)$

5. $\log_2(AB^2)$

6. $\log_6 \sqrt[4]{17}$

7. $\log_3(x\sqrt{y})$

8. $\log_2(xy)^{10}$

9. $\log_5 \sqrt[3]{x^2 + 1}$

10. $\log_a \dfrac{x^2}{yz^3}$

11. $\ln \sqrt{ab}$

12. $\ln \sqrt[3]{3r^2 s}$

13. $\log \dfrac{x^3 y^4}{z^6}$

14. $\log \dfrac{a^2}{b^4 \sqrt{c}}$

15. $\log_2 \dfrac{x(x^2 + 1)}{\sqrt{x^2 - 1}}$

16. $\log_5 \sqrt{\dfrac{x - 1}{x + 1}}$

17. $\ln x \sqrt{\dfrac{y}{z}}$

18. $\ln \dfrac{3x^2}{(x + 1)^{10}}$

✓ **19.** $\log \sqrt[4]{x^2 + y^2}$

20. $\log \dfrac{x}{\sqrt[3]{1 - x}}$

21. $\log \sqrt[3]{\dfrac{x^2 + 4}{(x^2 + 1)(x^3 - 7)^2}}$

22. $\log \sqrt{x\sqrt{y\sqrt{z}}}$

23. $\ln \dfrac{\sqrt{x}z^4}{\sqrt[3]{y^2 + 6y + 17}}$

24. $\log \dfrac{10^x}{x(x^2 + 1)(x^4 + 2)}$

In Exercises 25–34 evaluate the expression.

✓ **25.** $\log_5 \sqrt{125}$

26. $\log_2 112 - \log_2 7$

✓ **27.** $\log 2 + \log 5$

28. $\log \sqrt{0.1}$

✓ **29.** $\log_4 192 - \log_4 3$

30. $\log_{12} 9 + \log_{12} 16$

✓ **31.** $\ln 6 - \ln 15 + \ln 20$

32. $e^{3 \ln 5}$

✓ **33.** $10^{2 \log 4}$

34. $\log_2 8^{33}$

In Exercises 35–44 rewrite the expression as a single logarithm.

35. $\log_3 5 + 5 \log_3 2$

36. $\log 12 + \frac{1}{2} \log 7 - \log 2$

✓ **37.** $\log_2 A + \log_2 B - 2 \log_2 C$

38. $\log_5(x^2 - 1) - \log_5(x - 1)$

✓ **39.** $4 \log x - \frac{1}{3} \log(x^2 + 1) + 2 \log(x - 1)$

40. $\ln(a + b) + \ln(a - b) - 2 \ln c$

✓ **41.** $\ln 5 + 2 \ln x + 3 \ln(x^2 + 5)$

42. $\frac{1}{2}[\log_5 x + 2 \log_5 y - 3 \log_5 z]$

43. $\frac{1}{3} \log(2x + 1) + \frac{1}{2}[\log(x - 4) - \log(x^4 - x^2 - 1)]$

44. $\log_a b + c \log_a d - r \log_a s$

In Exercises 45–54 state whether the given equation is an identity.

✓ **45.** $\log_2(x - y) = \log_2 x - \log_2 y$

46. $\log_5 \dfrac{a}{b^2} = \log_5 a - 2 \log_5 b$

✓ **47.** $\log 2^z = z \log 2$

48. $(\log P)(\log Q) = \log P + \log Q$

✓ **49.** $\dfrac{\log a}{\log b} = \log a - \log b$

50. $(\log_2 7)^x = x \log_2 7$

✓ **51.** $\log_a a^a = a$

52. $\log(x - y) = \dfrac{\log x}{\log y}$

✓ **53.** $-\ln \dfrac{1}{A} = \ln A$

54. $r \ln s = \ln(s^r)$

55. For what value of x is it true that
$\log(x + 3) = \log x + \log 3$?

56. For what value of x is it true that $(\log x)^3 = 3 \log x$?

In Exercises 57–64 solve the equation for x.

✓ **57.** $\log_2 3 + \log_2 x = \log_2 5 + \log_2(x - 2)$

58. $2 \log x = \log 2 + \log(3x - 4)$

✓ **59.** $\log x + \log(x - 1) = \log 4x$

60. $\log_5 x + \log_5(x + 1) = \log_5 20$

61. $\log_5(x + 1) - \log_5(x - 1) = 2$

62. $\log x + \log(x - 3) = 1$

✓ **63.** $\log_9(x - 5) + \log_9(x + 3) = 1$

64. $\ln(x - 1) + \ln(x + 2) = 1$

In Exercises 65–76 find the solution of the equation correct to six decimal places.

65. $2^{x-5} = 3$

66. $8^{1-x} = 5$

✓ **67.** $10^{3-2x} = 13$

68. $3^{x/14} = 0.1$

69. $5^{-x/100} = 2$

70. $e^{3-5x} = 16$

✓ **71.** $e^{2x+1} = 200$

72. $\left(\frac{1}{4}\right)^x = 75$

73. $5^x = 4^{x+1}$

74. $10^{1-x} = 6^x$

75. $2^{3x+1} = 3^{x-2}$

76. $7^{x/2} = 5^{1-x}$

In Exercises 77–80 use the change of base formula and a calculator to evaluate the logarithm correct to six decimal places.

✓ **77.** $\log_2 7$

78. $\log_5 2$

✓ **79.** $\log_3 11$

80. $\log_6 92$

 81. Use the change of base formula to show that

$$\log_3 x = \dfrac{\ln x}{\ln 3}$$

Then use this fact to draw the graph of the function $f(x) = \log_3 x$.

82. Use the method of Exercise 81 to draw the graphs of the functions $y = \log_2 x$, $y = \ln x$, $y = \log_5 x$, and $y = \log_{10} x$ on the same screen using the viewing rectangle $[0, 5]$ by $[-3, 3]$.

83. Use the change of base formula to show that

$$\log e = \frac{1}{\ln 10}$$

84. Simplify $(\log_2 5)(\log_5 7)$.

85. Show that $-\ln\left(x - \sqrt{x^2 - 1}\right) = \ln\left(x + \sqrt{x^2 - 1}\right)$.

86. Solve the equation $(x - 1)^{\log(x-1)} = 100(x - 1)$.

87. Solve the inequality $\log(x - 2) + \log(9 - x) < 1$.

88. Solve the equation $\log_2 x + \log_4 x + \log_8 x = 11$.

89. Find the error:

$$\log 0.1 < 2 \log 0.1$$
$$= \log(0.1)^2$$
$$= \log 0.01$$
$$\log 0.1 < \log 0.01$$
$$0.1 < 0.01$$

SECTION 4.5
APPLICATIONS OF LOGARITHMS

Logarithms were invented by John Napier (1550–1617) to eliminate the tedious calculations involved in multiplying, dividing, and taking powers and roots of the large numbers that occur in astronomy and other sciences. With the advent of computers and calculators, logarithms are no longer important for such calculations. However, logarithms are useful for other reasons. They arise in problems of exponential growth and decay because logarithmic functions are inverses of exponential functions. Because of the Laws of Logarithms, they also turn out to be useful in the measurement of the loudness of sounds and the intensity of earthquakes, as well as many other phenomena.

EXPONENTIAL GROWTH

We saw in Section 4.2 that if the initial size of an animal or bacteria population is n_0 and the doubling period is d, then the size of the population at time t is

$$n(t) = n_0 2^{t/d} \tag{1}$$

Now that we are equipped with logarithms, we are in a position to answer questions concerning the time at which the population will reach a certain level.

EXAMPLE 1

A bacteria culture starts with 10,000 bacteria and the doubling period is 40 min. After how many minutes will there be 50,000 bacteria?

SOLUTION

Using Equation 1 with $n_0 = 10,000$ and $d = 40$, we have

$$n(t) = 10,000 \cdot 2^{t/40}$$

We want to find the value of t such that $n(t) = 50,000$. Thus we have to solve the exponential equation

$$10,000 \cdot 2^{t/40} = 50,000$$
$$2^{t/40} = 5$$

or

Taking common logarithms of both sides, we have

$$\log(2^{t/40}) = \log 5$$

$$\frac{t}{40} \log 2 = \log 5$$

$$t = 40 \frac{\log 5}{\log 2} \approx 93$$

The population will reach 50,000 in about 93 min.   ∎

Although we used common logarithms in Example 1, natural logarithms would have worked just as well.

RADIOACTIVE DECAY

In Section 4.2 we worked with the equation

$$m(t) = m_0 2^{-t/h} \tag{2}$$

where $m(t)$ is the mass of a radioactive substance after time t, m_0 is the initial mass, and h is the half-life. Now we can determine the time required for decay to a certain amount.

EXAMPLE 2

The half-life of Polonium-210 is 140 days. How long will it take a 300-mg sample to decay to a mass of 200 mg?

SOLUTION

Using Equation 2 with $m_0 = 300$ and $h = 140$, we see that the mass that remains after t days is

$$m(t) = (300)2^{-t/140}$$

The mass will be 200 mg when

$$(300)2^{-t/140} = 200$$

$$2^{-t/140} = \frac{200}{300} = \frac{2}{3}$$

$$-\frac{t}{140} \log 2 = \log\left(\frac{2}{3}\right)$$

$$t = -140 \frac{\log(2/3)}{\log 2} \approx 82$$

The time required is about 82 days.   ∎

NEWTON'S LAW OF COOLING

Newton's Law of Cooling states that the rate of cooling of an object is proportional to the temperature difference between the object and its surroundings, provided that the temperature difference is not too large. Using calculus, it can be deduced from this law that the temperature of the object at time t is

$$T(t) = T_s + D_0 e^{-kt} \qquad (3)$$

where T_s is the temperature of the surroundings, D_0 is the initial temperature difference, and k is a positive constant that is associated with the cooling object.

EXAMPLE 3

A cup of coffee has a temperature of 200 °F and is in a room that has a temperature of 70 °F. After 10 min the temperature of the coffee is 150 °F.
(a) What is the temperature of the coffee after 15 min?
(b) When will the coffee have cooled to 100 °F?
(c) Illustrate by sketching the graph of the temperature function.

SOLUTION

(a) The temperature of the room is $T_s = 70$ °F and the initial temperature difference is

$$D_0 = 200 - 70 = 130 \ °F$$

so, by Equation 3, the temperature after t minutes is

$$T(t) = 70 + 130e^{-kt}$$

When $t = 10$, the temperature is $T(10) = 150$, so we have

$$70 + 130e^{-10k} = 150$$
$$130e^{-10k} = 80$$
$$e^{-10k} = \frac{80}{130} = \frac{8}{13}$$
$$-10k = \ln(8/13)$$
$$k = -\tfrac{1}{10}\ln\!\left(\tfrac{8}{13}\right)$$

Putting this value of k into the expression for $T(t)$, we get

$$T(t) = 70 + 130e^{\ln(8/13)(t/10)}$$

So the temperature of the coffee after 15 min is

$$T(15) = 70 + 130e^{\ln(8/13)(15/10)} \approx 133 \ °F$$

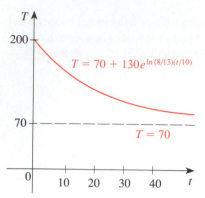

Figure 1
Temperature of coffee after t minutes

(b) The temperature will be 100 °F when

$$70 + 130e^{\ln(8/13)(t/10)} = 100$$
$$130e^{\ln(8/13)(t/10)} = 30$$
$$e^{\ln(8/13)(t/10)} = \frac{30}{130} = \frac{3}{13}$$

Take the natural logarithms of both sides:

$$\ln\left(\frac{8}{13}\right) \cdot \frac{t}{10} = \ln\left(\frac{3}{13}\right)$$
$$t = 10\,\frac{\ln(3/13)}{\ln(8/13)} \approx 30$$

The coffee will have cooled to 100 °F after about half an hour.

(c) The graph of the temperature function is sketched in Figure 1. Notice that the line $T = 70$ is a horizontal asymptote. (Why?)   ∎

■ COMPOUND INTEREST

In Section 4.2 we computed the amount of an investment or a loan using the formula

$$A = P\left(1 + \frac{r}{n}\right)^{nt} \tag{4}$$

where P is the principal, r is the interest rate, and interest is compounded n times per year for t years. Now we can use logarithms to determine the time it takes for the principal to increase to a given amount.

EXAMPLE 4

A sum of $5000 is invested at an interest rate of 9% per year compounded semi-annually. How long will it take for the money to double?

SOLUTION

In Formula 4 we have $P = \$5000$, $A = \$10,000$, $r = 0.09$, $n = 2$, and we solve for t:

$$5000(1 + .045)^{2t} = 10,000$$
$$(1.045)^{2t} = 2$$
$$2t \log 1.045 = \log 2$$
$$t = \frac{\log 2}{2 \log 1.045} \approx 7.9$$

Thus the money will double during the eighth year.   ∎

LOGARITHMIC SCALES

When physical quantities can vary over very large ranges it is often convenient to take their logarithms in order to have a more manageable set of numbers. We discuss three such situations: the pH scale in chemistry; the Richter scale, which measures the intensity of earthquakes; and the decibel scale, which measures the loudness of sounds. Other quantities that are measured on logarithmic scales are light intensity, information capacity, and radiation.

THE pH SCALE

Chemists measured the acidity of a solution by giving its hydrogen ion concentration until Sorensen, in 1909, proposed a more convenient measure. He defined

$$\text{pH} = -\log_{10}[\text{H}^+] = -\log[\text{H}^+] \tag{5}$$

where $[\text{H}^+]$ is the concentration of hydrogen ions measured in moles per liter (M). He did this to avoid very small numbers and negative exponents. For instance,

$$\text{if} \quad [\text{H}^+] = 10^{-4} \text{ M}, \qquad \text{then} \quad \text{pH} = -\log_{10}(10^{-4}) = -(-4) = 4$$

Solutions with a pH of 7 are called *neutral*; those with pH < 7 are called *acidic*; and those with pH > 7 are called *basic*. Notice that when the pH increases by one unit, $[\text{H}^+]$ decreases by a factor of 10.

EXAMPLE 5

(a) The hydrogen ion concentration of a sample of human blood was measured to be $[\text{H}^+] = 3.16 \times 10^{-8}$ M. Find the pH and classify as acidic or basic.
(b) The most acidic rainfall ever measured was in Scotland in 1974 and the pH was 2.4. Find the hydrogen ion concentration.

SOLUTION

(a) A calculator gives

$$\text{pH} = -\log[\text{H}^+] = -\log(3.16 \times 10^{-8}) \approx 7.5$$

Since this is greater than 7, the blood is basic.

(b) Writing Equation 5 in exponential form, we have

$$\log[\text{H}^+] = -\text{pH} \quad \Rightarrow \quad [\text{H}^+] = 10^{-\text{pH}}$$

and so

$$[\text{H}^+] = 10^{-2.4} \approx 4.0 \times 10^{-3} \text{ M}$$

THE RICHTER SCALE

In 1935 the American geologist Charles Richter (1900–1984) defined the magnitude of an earthquake to be

$$M = \log \frac{I}{S} \tag{6}$$

where I is the intensity of the earthquake (measured by the amplitude of a seismograph 100 km from the epicenter of the earthquake) and S is the intensity of a "standard" earthquake (where the amplitude is only 1 micron = 10^{-4} cm). Notice that the magnitude of the standard earthquake is

$$M = \log \frac{S}{S} = \log 1 = 0$$

Richter studied many earthquakes that occurred between 1900 and 1950. The largest had magnitude 8.9 on the Richter scale, the smallest had magnitude 0. This corresponds to a ratio of intensities of 800,000,000, so the Richter scale provides more manageable numbers to work with. For instance, an earthquake of magnitude 6 is ten times stronger than an earthquake of magnitude 5.

EXAMPLE 6

The 1906 earthquake in San Francisco had an estimated magnitude of 8.3 on the Richter scale. In the same year the strongest earthquake ever recorded occurred on the Colombia-Ecuador border and was four times as intense. What was the magnitude of the Columbia-Equador earthquake on the Richter scale?

SOLUTION

If I is the intensity of the San Francisco earthquake, then from Equation 6 we have

$$\log \frac{I}{S} = 8.3$$

The intensity of the Columbia-Ecuador earthquake was $4I$, so its magnitude was

$$\log \frac{4I}{S} = \log 4 + \log \frac{I}{S} = \log 4 + 8.3 \approx 8.9 \qquad \blacksquare$$

EXAMPLE 7

The 1989 Loma Prieta earthquake that shook San Francisco had a magnitude of 7.1 on the Richter scale. How many times more intense was the 1906 earthquake (see Example 6) than the 1989 earthquake?

SOLUTION

If I_1 and I_2 are the intensities of the 1906 and 1989 earthquakes, then we are required to find I_1/I_2. To relate this to the definition of magnitude in Equation 6 we divide numerator and denominator by S and we first find the common logarithm of I_1/I_2.

$$\log\left(\frac{I_1}{I_2}\right) = \log\frac{I_1/S}{I_2/S}$$

$$= \log\frac{I_1}{S} - \log\frac{I_2}{S}$$

$$= 8.3 - 7.1 = 1.2$$

Therefore

$$\frac{I_1}{I_2} = 10^{\log(I_1/I_2)} = 10^{1.2} \approx 16$$

The 1906 earthquake was about 16 times as intense as the 1989 earthquake. ■

THE DECIBEL SCALE

The ear is sensitive to an extremely wide range of sound intensities. We take as a reference intensity $I_0 = 10^{-12}$ watts/m^2 at a frequency of 1000 hertz, which measures a sound that is just barely audible (the threshold of hearing). The psychological sensation of loudness varies with the logarithm of the intensity (the Weber-Fechner Law) and so the **intensity level** β, measured in decibels, is defined as

$$\beta = 10 \log \frac{I}{I_0} \tag{7}$$

Notice that the intensity level of the barely audible reference sound is

$$\beta = 10 \log \frac{I_0}{I_0} = 10 \log 1 = 0 \text{ dB}$$

EXAMPLE 8

Find the intensity level of a jet plane taking off if the intensity was measured at 100 watts/m^2 at a distance of 40 m from the jet.

SOLUTION

From Equation 7 we see that the intensity level is

$$\beta = 10 \log \frac{I}{I_0} = 10 \log \frac{10^2}{10^{-12}}$$

$$= 10 \log 10^{14} = 140 \text{ dB}$$ ■

The following table lists decibel intensity levels for some common sounds ranging from the threshold of hearing to the jet takeoff of Example 8. The threshold of pain is about 120 dB.

Source of sound	β (dB)
Jet takeoff (40 m away)	140
Jackhammer	130
Rock concert (2 m from speakers)	120
Subway	100
Heavy traffic	80
Ordinary traffic	70
Normal conversation	50
Whisper	30
Rustling leaves	10–20
Threshold of hearing	0

EXERCISES 4.5

1. The hydrogen ion concentrations of samples of three substances are given. Calculate the pH.
   (a) lemon juice: $[H^+] = 5.0 \times 10^{-3}$ M
   (b) tomato juice: $[H^+] = 3.2 \times 10^{-4}$ M
   (c) seawater: $[H^+] = 5.0 \times 10^{-9}$ M

2. An unknown substance has a hydrogen ion concentration of $[H^+] = 3.1 \times 10^{-8}$ M. Find the pH and classify the substance as acidic or basic.

3. The pH readings of samples of the following substances are given. Calculate the hydrogen ion concentrations.
   (a) vinegar: pH = 3.0   (b) milk: pH = 6.5

4. The pH readings of glasses of beer and water are given. Find the hydrogen ion concentrations.
   (a) beer: pH = 4.6   (b) water: pH = 7.3

5. The hydrogen ion concentrations in cheese range from 4.0×10^{-7} M to 1.6×10^{-5} M. Find the corresponding range of pH readings.

6. The pH readings for wine vary from 2.8 to 3.8. Find the corresponding range of hydrogen ion concentrations.

7. A bacteria culture contains 1500 bacteria initially and doubles every half hour. After how many minutes will there be 4000 bacteria?

8. If a bacteria culture starts with 8000 bacteria and doubles every 20 min, when will the population reach a level of 30,000?

9. A bacteria culture starts with 20,000 bacteria. After an hour the count is 52,000. What is the doubling period?

10. The count in a bacteria culture was 400 after 2 h and 25,600 after 6 h.
    (a) What was the initial size of the culture?
    (b) Find the doubling period.
    (c) Find the size after 4.5 h.
    (d) When will the population be 50,000?

11. The population of the world is doubling about every 35 yr. In 1987 the total population was 5 billion. If the doubling period remains at 35 yr, in what year will the world population reach 50 billion?

12. The population of California was 10,586,223 in 1950 and 23,668,562 in 1980.
    (a) Assuming exponential growth during that period, find the doubling period.
    (b) Use the data to predict the population of California in the year 2000.

13. The half-life of Strontium-90 is 25 years. How long will it take a 50-mg sample to decay to a mass of 32 mg?

14. Radium-221 has a half-life of 30 s. How long will it take for 95% of a sample to decompose? Illustrate by sketching the graph of the mass function.

15. If 250 mg of a radioactive element decays to 200 mg in 48 h, find the half-life of the element.

16. After three days a sample of Radon-222 decayed to 58% of its original amount.
 (a) What is the half-life of Radon-222?
 (b) How long will it take the sample to decay to 20% of its original amount?

17. A roasted turkey is taken from an oven when its temperature has reached 185 °F and is placed on a table in a room where the temperature is 75 °F.
 (a) If the temperature of the turkey is 150 °F after half an hour, what is the temperature after 45 min?
 (b) When will the turkey cool to 100 °F?

18. A kettle full of water is brought to a boil in a room with temperature 20 °C. After 15 min the temperature of the water has decreased from 100 °C to 75 °C. Find the temperature after another 10 min. Illustrate by sketching the graph of the temperature function.

19. Find the length of time required for an investment of $5000 to grow to $8000 at an interest rate of 9.5% per year compounded quarterly.

20. Nancy wants to invest $4000 in saving certificates that bear an interest rate of 9.75% compounded semiannually. How long a time period should she choose in order to save an amount of $5000?

21. How long will it take for an investment to double in value if the interest rate is 8.5% per year compounded continuously?

22. A sum of $1000 was invested for 4 yr and the interest was compounded semiannually. If this sum amounted to $1435.77 after four years, what was the interest rate?

23. If one earthquake is 20 times as intense as another, how much larger is its magnitude on the Richter scale?

24. The 1906 earthquake in San Francisco had a magnitude of 8.3 on the Richter scale. At the same time in Japan there was an earthquake with magnitude 4.9 that caused only minor damage. How many times more intense was the San Francisco earthquake than the Japanese earthquake?

25. The Alaska earthquake of 1964 had a magnitude of 8.6 on the Richter scale. How many times more intense was this than the 1906 San Francisco earthquake?

26. The intensity of the sound of rush-hour traffic at a busy intersection was measured at 2.0×10^{-5} watts/m². Find the intensity level in decibels.

27. The intensity level of the sound of a subway train was measured at 98 dB. Find the intensity in watts/m².

28. The noise from a power mower was measured at 106 dB. The noise level at a rock concert was measured at 120 dB. Find the ratio of the intensity of the rock music to that of the power mower.

29. The figure shows an electric circuit containing a battery producing a voltage of 60 volts, a resistor with a resistance of 13 ohms, and an inductor with an inductance of 5 henries. Using calculus it can be shown that the current $I = I(t)$ (in amperes) t seconds after the switch is closed is

$$I = \tfrac{60}{13}(1 - e^{-13t/5})$$

Use this equation to express the time t as a function of the current I.

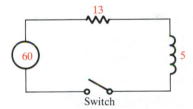

30. A **learning curve** is a graph of a function $P(t)$ that measures the performance of someone learning a skill as a function of the training time t. At first, the rate of learning is rapid. Then, as performance increases and approaches a maximal value M, the rate of learning decreases. It has been found that the function

$$P(t) = M - Ce^{-kt}$$

where k and C are positive constants and $C < M$, is a reasonable model for learning.
 (a) Sketch the graph of P.
 (b) Express the learning time t as a function of the performance level P.

31. It is a law of physics that the intensity of sound is inversely proportional to the square of the distance d from the source:

$$I = \frac{k}{d^2}$$

(a) Use this and Equation 7 to show that the decibel levels β_1 and β_2 at distances d_1 and d_2 from a sound source are related by the equation

$$\beta_2 = \beta_1 + 20 \log \frac{d_1}{d_2}$$

(b) The intensity level at a rock concert is 120 dB at a distance 2 m from the speakers. Find the intensity level at a distance of 10 m.

32. The table shows the mean distances d of the planets from the sun (taking the unit of measurement to be the distance from the earth to the sun) and their periods T (time of revolution in years). Try to discover a relationship between T and d. [*Hint:* Consider their logarithms.]

Planet	d	T
Mercury	0.387	0.241
Venus	0.723	0.615
Earth	1.000	1.000
Mars	1.523	1.881
Jupiter	5.203	11.861
Saturn	9.541	29.457
Uranus	19.190	84.008
Neptune	30.086	164.784
Pluto	39.507	248.35

CHAPTER 4 REVIEW

Define, state, or discuss each of the following.

1. The exponential function with base a

2. The number e

3. The exponential function with base e

4. Graphs of exponential functions

5. Doubling period

6. Population function

7. Half-life

8. Mass of a radioactive material

9. Compound interest

10. Continuous compounding of interest

11. The logarithmic function with base a

12. Natural logarithms

13. Common logarithms

14. Laws of Logarithms

15. Change of base formula

16. Newton's Law of Cooling

17. pH

18. The Richter scale

19. The decibel scale

REVIEW EXERCISES

In Exercises 1–12 sketch the graph of the function. State the domain, range, and asymptote.

1. $f(x) = \dfrac{1}{2^x}$

2. $g(x) = 3^{x-2}$

3. $y = 5 - 10^x$

4. $y = 1 + 5^{-x}$

5. $f(x) = \log_3(x - 1)$

6. $g(x) = \log(-x)$

7. $y = 2 - \log_2 x$

8. $y = 3 + \log_5(x + 4)$

9. $F(x) = e^x - 1$

10. $G(x) = \frac{1}{2}e^{x-1}$

11. $y = 2 \ln x$

12. $y = \ln(x^2)$

In Exercises 13–14 find the domain of the function.

13. $f(x) = 10^{x^2} + \log(1 - 2x)$

14. $g(x) = \ln(2 + x - x^2)$

In Exercises 15–18 write the equation in exponential form.

15. $\log_2 1024 = 10$

16. $\log_6 37 = x$

17. $\log x = y$

18. $\ln c = 17$

In Exercises 19–22 write the equation in logarithmic form.

19. $2^6 = 64$

20. $49^{-1/2} = \frac{1}{7}$

21. $10^x = 74$

22. $e^k = m$

In Exercises 23–38 evaluate the expression without using a calculator.

23. $\log_2 128$

24. $\log_8 1$

25. $10^{\log 45}$

26. $\log 0.000001$

27. $\ln(e^6)$

28. $\log_4 8$

29. $\log_3\left(\frac{1}{27}\right)$

30. $2^{\log_2 13}$

31. $\log_5 \sqrt{5}$

32. $e^{2 \ln 7}$

33. $\log 25 + \log 4$

34. $\log_3 \sqrt{243}$

35. $\log_2 16^{23}$

36. $\log_5 250 - \log_5 2$

37. $\log_8 6 - \log_8 3 + \log_8 2$

38. $\log \log 10^{100}$

In Exercises 39–44 rewrite the expression in a form with no logarithms of products, quotients, or powers.

39. $\log AB^2C^3$

40. $\log_2 x\sqrt{x^2 + 1}$

41. $\ln \sqrt{\dfrac{x^2 - 1}{x^2 + 1}}$

42. $\log \dfrac{4x^3}{y^2(x - 1)^5}$

43. $\log_5 \dfrac{x^2(1 - 5x)^{3/2}}{\sqrt{x^3 - x}}$

44. $\ln \dfrac{\sqrt[3]{x^4 + 12}}{(x + 16)\sqrt{x - 3}}$

In Exercises 45–50 rewrite the expression as a single logarithm.

45. $\log 6 + 4 \log 2$

46. $\log x + \log x^2y + 3 \log y$

47. $\frac{3}{2} \log_2(x - y) - 2 \log_2(x^2 + y^2)$

48. $\log_5 2 + \log_5(x + 1) - \frac{1}{3} \log_5(3x + 7)$

49. $\log(x - 2) + \log(x + 2) - \frac{1}{2} \log(x^2 + 4)$

50. $\frac{1}{2}[\ln(x - 4) + 5 \ln(x^2 + 4x)]$

In Exercises 51–60 solve the equation for x without using a calculator.

51. $\log_2(1 - x) = 4$

52. $2^{3x-5} = 7$

53. $5^{5-3x} = 26$

54. $\ln(2x - 3) = 14$

55. $e^{3x/4} = 10$

56. $2^{1-x} = 3^{2x+5}$

57. $\log x + \log(x + 1) = \log 12$

58. $\log_8(x + 5) - \log_8(x - 2) = 1$

59. $x^2e^{2x} + 2xe^{2x} = 8e^{2x}$

60. $2^{3^x} = 5$

In Exercises 61–64 use a calculator to find the solution of the equation correct to 6 decimal places.

61. $5^{-2x/3} = 0.63$

62. $2^{3x-5} = 7$

63. $5^{2x+1} = 3^{4x-1}$

64. $e^{-15k} = 10{,}000$

In Exercises 65–68 draw the graph of the function and use it to determine the asymptotes and local maximum and minimum values.

65. $y = e^{x/(x+2)}$

66. $y = 2x^2 - \ln x$

67. $y = \log(x^3 - x)$

68. $y = 10^x - 5^x$

In Exercises 69–70 find the solutions of the equation correct to two decimal places.

69. $3 \log x = 6 - 2x$

70. $4 - x^2 = e^{-2x}$

In Exercises 71–72 solve the inequality graphically.

71. $\ln x > x - 2$

72. $e^x < 4x^2$

73. Use a graph of $f(x) = e^x - 3e^{-x} - 4x$ to find, approximately, the intervals on which f is increasing or decreasing.

74. Find an equation of the line shown in the figure.

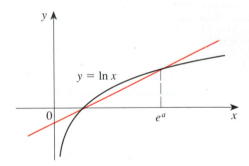

75. Evaluate $\log_4 15$ correct to 6 decimal places.

76. Solve the inequality $0.2 \leq \log x < 2$.

77. Which is larger, $\log_4 258$ or $\log_5 620$?

78. Find the inverse function of the function $f(x) = 2^{3^x}$ and state its domain and range.

79. The hydrogen ion concentration of fresh white eggs was measured as $[H^+] = 1.3 \times 10^{-8}$ M. Find the pH and classify as acidic or basic.

80. The pH of lime juice is 1.9. Find the hydrogen ion concentration.

81. A bacteria culture starts with 1000 bacteria and doubles every 25 min.
    (a) Find the population after an hour.
    (b) When will the population reach 5000?

82. A bacteria culture contains 10,000 bacteria initially. After an hour the bacteria count is 25,000.
    (a) Find the doubling period.
    (b) Find the population after 3 h.

83. Uranium-234 has a half-life of 2.7×10^5 years.
    (a) Find the amount remaining from a 10-mg sample after a thousand years.
    (b) How long will it take this sample to decompose until its mass is 7 mg?

84. A sample of Bismuth-210 decayed to 33% of its original mass after 8 days.

(a) Find the half-life of this element.
(b) Find the mass remaining after 12 days.

85. If $12,000 is invested at an interest rate of 10% per year, find the amount of the investment at the end of 3 years if the interest is compounded
    (a) semiannually   (b) monthly   (c) daily
    (d) continuously

86. A sum of $5000 is invested at an interest rate of $8\frac{1}{2}\%$ per year compounded semiannually.
    (a) Find the amount of the investment after a year and a half.
    (b) After what period of time will the investment amount to $7000?

87. If one earthquake has magnitude 6.5 on the Richter scale, what is the magnitude on the Richter scale of another that is 35 times as intense?

88. The noise from a jackhammer was measured at 132 dB. The sound of whispering was measured at 28 dB. Find the ratio of the intensity of the jackhammer to that of the whispering.

CHAPTER 4 TEST

1. Graph the functions $y = 2^x$ and $y = \log_2 x$ on the same axes.

2. Sketch the graph of the function $f(x) = \log(x - 3)$ and state the domain, range, and asymptote of f.

3. Evaluate:
   (a) $\log_8 4$ (b) $\log_6 4 + \log_6 9$

4. Use the Laws of Logarithms to rewrite the following expression without logarithms of products, quotients, powers, or roots.

   $$\log \sqrt{\frac{x^2 - 1}{x^3(y^2 + 1)^5}}$$

5. Write the following expression as a single logarithm.

   $$\ln x - 2 \ln(x^2 + 1) + \tfrac{1}{2} \ln(3 - x^4)$$

6. Solve for x without using a calculator:
   (a) $2^x = 10$ (b) $\ln(3 - x) = 4$

7. The initial size of a bacteria culture is 1000. After an hour the count is 8000.
   (a) Find the doubling period.
   (b) Find the population after t hours.
   (c) Find the population after 1.5 h.
   (d) When will the population reach 15,000?
   (e) Sketch the graph of the population function.

8. Find the domain of the function

   $$f(x) = \log(x + 4) + \log(8 - 5x)$$

 9. Let $f(x) = \dfrac{e^x}{x^3}$.

   (a) Graph f in an appropriate viewing rectangle.
   (b) What are the asymptotes of f?
   (c) Find, to two decimal places, the local minimum value of f and the value of x at which it occurs.
   (d) State the range of f.
   (e) Solve the equation $\dfrac{e^x}{x^3} = 2x + 1$. State the roots correct to two decimal places.

FOCUS ON PROBLEM SOLVING

One way to show that a statement is true is to show that its opposite leads to a contradiction. This type of **indirect reasoning** is an important problem-solving tool. The two examples we give here are problems that were solved over 2000 years ago but continue to have a profound influence in mathematics.

IRRATIONAL NUMBERS EXIST

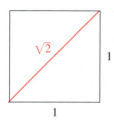

The Pythagoreans were fascinated by the beauty of the natural numbers: 1, 2, 3, ... They hoped that the length of every interval constructed in geometry would be a ratio of two natural numbers, and so the natural numbers would describe all of geometry. But what is the length of the diagonal of a square of side 1?

They knew (by the Pythagorean Theorem) that this length is $\sqrt{2}$. The question was: Is $\sqrt{2}$ the ratio of two natural numbers? In other words, is $\sqrt{2}$ a rational number? The Pythagoreans were able to prove that $\sqrt{2}$ is not rational. This was considered of such great importance (and rightly so, from our perspective 2000 years later) that, as legend has it, one hundred oxen were sacrificed to celebrate the discovery. The proof is a classic use of indirect reasoning.

Suppose $\sqrt{2}$ is rational, so that

$$\sqrt{2} = \frac{a}{b}$$

where a and b are natural numbers with no factors in common. Then $a = \sqrt{2}b$ and, squaring, we get

$$a^2 = 2b^2$$

This means that a^2 is an even number and so a is an even number. Then $a = 2m$, and from the fact that $a^2 = 2b^2$ we get

$$4m^2 = 2b^2 \qquad 2m^2 = b^2$$

Thus b is also even. So a and b have 2 as a common factor and this contradicts our assumption that a and b have no factors in common. Thus the assumption the $\sqrt{2}$ is rational leads to a contradiction and so $\sqrt{2}$ must be irrational.

THERE ARE INFINITELY MANY PRIMES

A prime number is one that has no factors other than 1 and itself. The first primes are

$$2, 3, 5, 7, 11, 13, 17, 19, 23, 29, 31, \ldots$$

How do we find the next prime? No formula is known that will produce the primes and no one has found a pattern for the location of the primes. One thing that is known is that there are infinitely many primes. Euclid gave a proof for this fact over 2000 years ago by a brilliant use of indirect reasoning. Here is his famous proof:

Suppose that there is only a finite number of primes. We list them as

$$p_1, p_2, p_3, \ldots, p_n$$

Then the number

$$N = p_1 \cdot p_2 \cdot p_3 \cdots \cdot p_n + 1$$

is not divisible by any prime (Why?) and so is itself prime. But N is not in our list of primes (Why?). This contradicts our assumption that the list contained all the primes. Thus there are infinitely many primes.

Eratosthenes (*circa 276–195* B.C.) *was a renowned Greek geographer, mathematician, and astronomer. He quite accurately calculated the circumference of the earth by an ingenious method (see Exercise 62, page 376). He is most famous, however, for his method for finding primes, now called the* sieve of Eratosthenes. *The method consists of listing the integers beginning with 2 (the first prime)* and then crossing out all multiples of 2, which are not prime. The next number remaining on the list is 3 (the second prime) so we again cross out all multiples of it. The next remaining number is 5 (the third prime) and we cross out all multiples of it, and so on. In this way all numbers that are not prime are crossed out and the remaining numbers are the primes.

Largest Known Prime

The search for large primes has a fascination for many people. At one point the record for the largest known prime was held by Laura Nickel and Curt Noll, two 18-year-old students at California State University, Hayward. In 1978 they showed using a computer that $2^{21,701} - 1$ is a prime number. At this writing the largest known prime is

$$2^{756,839} - 1$$

discovered in 1992 at the Harwell Laboratory in England using a Cray computer. In decimal notation this number contains 227,832 digits. If written out in full, it would occupy about 40 pages of this book. Numbers of the form $2^p - 1$, where p is prime, are called Mersenne numbers and are more readily checked for primality than others. That is why the largest known primes are of this form.

1. Prove that $\sqrt{3}$ is irrational.

2. Prove that $\log_2 5$ is irrrational.

3. How many positive integers less than 1000 are divisible by neither 5 nor 7?

4. (a) What is the smallest positive integer by which 12 can be multiplied to obtain a perfect cube?
   (b) What is the smallest positive integer by which 15 can be multiplied to obtain a perfect cube?
   (c) Can you find a rule for doing problems like (a) and (b) for any number n?

5. Evaluate $(\log_2 3)(\log_3 4)(\log_4 5) \cdots (\log_{31} 32)$.

6. Show that if $x > 0$ and $x \neq 1$, then

$$\frac{1}{\log_2 x} + \frac{1}{\log_3 x} + \frac{1}{\log_5 x} = \frac{1}{\log_{30} x}$$

7. Prove that at any party there are two people who know the same number of people. (Assume that if person A knows person B then B knows A. Assume also that everyone knows himself or herself.)

8. Suppose that each point in the plane is colored either red or blue. Show that there must always be two points of the same color that are exactly one unit apart.

9. Suppose that each point (x, y) in the plane, both of whose coordinates are rational numbers, represents a tree. If you are standing at the point $(0, 0)$, how far could you see in this forest?

5 TRIGONOMETRIC FUNCTIONS OF REAL NUMBERS

Trigonometry contains the science of continually undulating magnitude . . .

AUGUSTUS DE MORGAN

Trigonometry is one of the most versatile branches of mathematics. Ever since its invention in the ancient world, it has been important in both theoretical and practical applications. In modern times it has found applications in such diverse fields as signal processing in the telephone industry, coding of music on compact disc players, finding distances to stars, designing guidance systems in the space shuttle, producing CAT scans for medical use, and many others. It is an indispensable tool for electrical engineers, physicists, computer scientists, and for practically all of the sciences. You will see in studying trigonometry how some of these applications are possible.

The power and versatility of trigonometry stem from the fact that it can be viewed in two distinct ways. One of these sees trigonometry as the study of *functions of real numbers*, the other as the study of *functions of angles*. The trigonometric functions defined in these two different ways are identical—they assign the same value to a given real number (in the second case the real number is the measure of an angle). The difference is one of point of view. This difference is most clearly apparent when we consider the applications of trigonometry. One view lends itself to applications involving dynamic processes, such as harmonic motion, the study of sound waves, and the description of other periodic phenomena, whereas the other view lends itself to static applications, such as distance measurement, force, velocity, and, in general, applications involving measurements of length and direction.

To fully appreciate its uses, we must study both approaches to trigonometry. In this chapter we study trigonometric functions of real numbers, sketch their graphs, and describe their applications to harmonic motion. The angle approach is treated in Chapter 6. Either chapter may be studied first.

SECTION 5.1
THE UNIT CIRCLE

In this section we explore some properties of the circle of radius 1 centered at the origin. These properties are used in the next section to define the trigonometric functions.

THE UNIT CIRCLE

The set of points at a distance 1 from the origin is a circle of radius 1 (see Figure 1). By the distance formula, we see that a point (x, y) is on this circle if

$$\sqrt{(x - 0)^2 + (y - 0)^2} = 1$$
$$x^2 + y^2 = 1$$

307

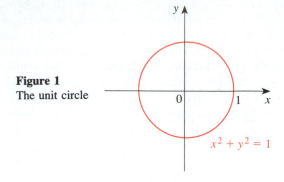

Figure 1
The unit circle

$x^2 + y^2 = 1$

THE UNIT CIRCLE

The **unit circle** is the circle of radius 1 centered at the origin in the xy-plane. Its equation is

$$x^2 + y^2 = 1$$

EXAMPLE 1

Show that the point $P\left(\dfrac{\sqrt{3}}{3}, \dfrac{\sqrt{2}}{\sqrt{3}}\right)$ is on the unit circle.

SOLUTION

We need to show that this point satisfies the equation of the unit circle, that is, $x^2 + y^2 = 1$. Since

$$\left(\frac{\sqrt{3}}{3}\right)^2 + \left(\frac{\sqrt{2}}{\sqrt{3}}\right)^2 = \frac{3}{9} + \frac{2}{3} = \frac{1}{3} + \frac{2}{3} = 1$$

P is on the unit circle. ∎

EXAMPLE 2

The point $P\left(\sqrt{3}/2, y\right)$ is on the unit circle in quadrant IV. Find its y-coordinate.

SOLUTION

Since the point is on the unit circle, we have

$$\left(\frac{\sqrt{3}}{2}\right)^2 + y^2 = 1$$

$$y^2 = 1 - \frac{3}{4} = \frac{1}{4}$$

$$y = \pm\frac{1}{2}$$

Since the point is in quadrant IV, its y-coordinate must be negative, so $y = -\frac{1}{2}$. ∎

TERMINAL POINTS ON THE UNIT CIRCLE

Suppose t is a real number. Let us mark off a distance t along the unit circle, starting at the point $(1, 0)$ and moving in a counterclockwise direction if t is positive and in a clockwise direction if t is negative (Figure 2). In this way we arrive at a point $P(x, y)$ on the unit circle. The point $P(x, y)$ obtained in this way is called the **terminal point** determined by the real number t.

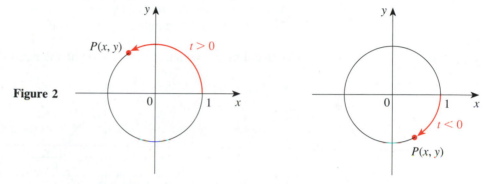

Figure 2

(a) Terminal point $P(x, y)$ determined by $t > 0$

(b) Terminal point $P(x, y)$ determined by $t < 0$

The circumference of the unit circle is $C = 2\pi(1) = 2\pi$. So, if a point starts at $(1, 0)$ and moves counterclockwise all the way around the unit circle and returns to $(1, 0)$, it travels a distance of 2π. To move halfway around the circle it travels a distance of $\frac{1}{2}(2\pi) = \pi$. To move a quarter of the distance around the unit circle, it travels a distance of $\frac{1}{4}(2\pi) = \pi/2$. Where does the point end up when it travels these distances along the circle? For example, when it travels a distance of π starting at $(1, 0)$, its terminal point is $(-1, 0)$. See Figure 3.

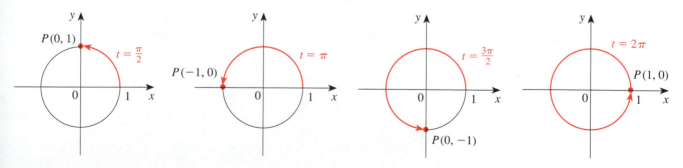

Figure 3
Terminal points determined by $t = \dfrac{\pi}{2}, \pi, \dfrac{3\pi}{2},$ and 2π.

EXAMPLE 3

Find the terminal point on the unit circle determined by the real number t.

(a) $t = 3\pi$ (b) $t = -\pi$ (c) $t = -\dfrac{\pi}{2}$

SOLUTION

From Figure 4 we get the following:

(a) The terminal point determined by 3π is $(-1, 0)$.
(b) The terminal point determined by $-\pi$ is $(-1, 0)$.
(c) The terminal point determined by $-\pi/2$ is $(0, -1)$.

Notice that different values of t can determine the same terminal point.

Figure 4

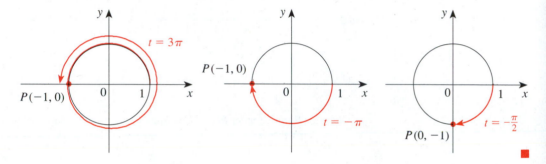

Finding the terminal point determined by $t = \pi/4$

The terminal point $P(x, y)$ determined by $t = \pi/4$ is the same distance from $(1, 0)$ as from $(0, 1)$ along the unit circle (see Figure 5). Since the unit circle is symmetric with respect to the line $y = x$, it follows that P lies on the line $y = x$. So P is the point of intersection of the circle $x^2 + y^2 = 1$ and the line $y = x$.

Figure 5

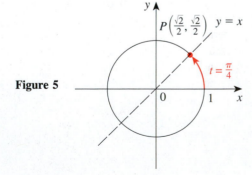

Substituting x for y in the equation of the circle, we get

$$x^2 + x^2 = 1$$

$$2x^2 = 1$$

$$x^2 = \frac{1}{2}$$

$$x = \pm\frac{1}{\sqrt{2}}$$

Since P is in the first quadrant, $x = 1/\sqrt{2}$ and since $y = x$, we have $y = 1/\sqrt{2}$ also. Thus the terminal point determined by $\pi/4$ is

$$P\left(\frac{1}{\sqrt{2}}, \frac{1}{\sqrt{2}}\right) = P\left(\frac{\sqrt{2}}{2}, \frac{\sqrt{2}}{2}\right)$$

EXAMPLE 4

Find the terminal point determined by t:

(a) $t = -\dfrac{\pi}{4}$ (b) $t = \dfrac{3\pi}{4}$

SOLUTION

(a) Let P be the terminal point determined by $-\pi/4$ and let Q be the terminal point determined by $\pi/4$. From Figure 6(a) we see that the point P has the same coordinates as Q except for sign. Since P is in quadrant IV, its x-coordinate is positive and its y-coordinate is negative. Thus the terminal point is

$$P\left(\frac{\sqrt{2}}{2}, -\frac{\sqrt{2}}{2}\right)$$

(b) Let P be the terminal point determined by $3\pi/4$ and let Q be the terminal point determined by $\pi/4$. From Figure 6(b) we see that the point P has the same coordinates as Q except for sign. Since P is in quadrant II, its x-coordinate is negative and its y-coordinate is positive. Thus the terminal point is

$$P\left(-\frac{\sqrt{2}}{2}, \frac{\sqrt{2}}{2}\right)$$

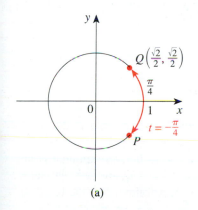

(a)

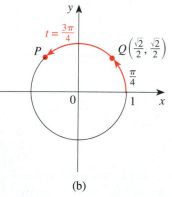

(b)

Figure 6

Finding the terminal point determined by $t = \pi/6$

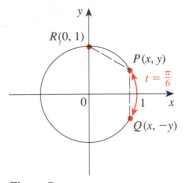

Figure 7

We now find the terminal point $P(x, y)$ determined by $t = \pi/6$. Let $Q(x, -y)$ and $R(0, 1)$ be as in Figure 7. The arcs PQ and PR are of equal length (they both have length $\pi/3$) and so they are subtended by equal chords, that is, $|PQ| = |PR|$. Thus

$$\sqrt{(x - x)^2 + (y + y)^2} = \sqrt{(x - 0)^2 + (y - 1)^2} \qquad \text{(by the Distance Formula)}$$
$$\sqrt{4y^2} = \sqrt{x^2 + y^2 - 2y + 1}$$
$$4y^2 = x^2 + y^2 - 2y + 1 \qquad \text{(squaring both sides)}$$
$$4y^2 = 1 - 2y + 1 \qquad \text{(because } x^2 + y^2 = 1\text{)}$$
$$2y^2 + y - 1 = 0 \qquad \text{(divide by 2)}$$
$$(2y - 1)(y + 1) = 0$$
$$y = \tfrac{1}{2} \qquad \text{or} \qquad y = -1$$

Since $P(x, y)$ is in quadrant I, it follows that $y = \tfrac{1}{2}$. To find x, we again use the fact that $P(x, y)$ is on the unit circle, that is, $x^2 + y^2 = 1$. From this we get

$$x^2 + \left(\tfrac{1}{2}\right)^2 = 1$$
$$x^2 = \tfrac{3}{4}$$
$$x = \frac{\sqrt{3}}{2} \qquad \text{(because } x > 0\text{)}$$

So we have found that the terminal point determined by $t = \pi/6$ is the point $P\left(\sqrt{3}/2, \tfrac{1}{2}\right)$.

Finding the terminal point determined by $t = \pi/3$

We can now use symmetry properties of the circle to find the terminal point for $t = \pi/3$. From Figure 8 we see that the points P and Q are symmetrically placed about the line $y = x$. Since $Q = Q\left(\sqrt{3}/2, \tfrac{1}{2}\right)$, it follows that the terminal point $P = P\left(\tfrac{1}{2}, \sqrt{3}/2\right)$.

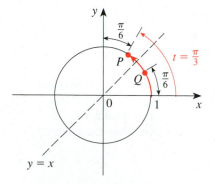

Figure 8

EXAMPLE 5

Find the terminal point on the unit circle determined by t:

(a) $t = -\dfrac{5\pi}{6}$ \qquad (b) $t = \dfrac{2\pi}{3}$

SOLUTION

(a) Let P be the terminal point determined by $-5\pi/6$ and let Q be the terminal point determined by $\pi/6$. From Figure 9(a) we see that the point P has the same coordinates as Q except for sign. Since P is in quadrant III, both its coordinates are negative. Thus the terminal point is $P\left(-\sqrt{3}/2, -\tfrac{1}{2}\right)$.

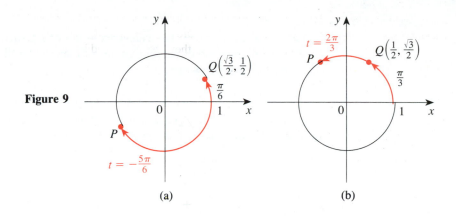

Figure 9

(a)　　　　　(b)

(b) Let P be the terminal point determined by $2\pi/3$ and let Q be the terminal point determined by $\pi/3$. From Figure 9(b) we see that the point P has the same coordinates as Q except for sign. Since P is in quadrant II, its x-coordinate is negative and its y-coordinate is positive. Thus the terminal point is $P\left(-\frac{1}{2}, \sqrt{3}/2\right)$. ■

We summarize the special terminal points we have found in Table 1 (see Figure 10).

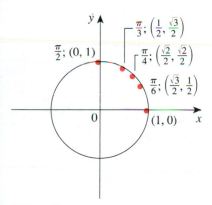

Figure 10

Table 1

t	Terminal point determined by t
0	$(1, 0)$
$\dfrac{\pi}{6}$	$\left(\dfrac{\sqrt{3}}{2}, \dfrac{1}{2}\right)$
$\dfrac{\pi}{4}$	$\left(\dfrac{\sqrt{2}}{2}, \dfrac{\sqrt{2}}{2}\right)$
$\dfrac{\pi}{3}$	$\left(\dfrac{1}{2}, \dfrac{\sqrt{3}}{2}\right)$
$\dfrac{\pi}{2}$	$(0, 1)$

THE REFERENCE NUMBER

From Examples 3, 4, and 5, we see that to find a terminal point in any quadrant we need only know the "corresponding" terminal point in the first quadrant. We give a procedure for finding such terminal points using the idea of the *reference number*.

REFERENCE NUMBER

Let t be a real number. The **reference number** $\bar{t}$ associated with t is the shortest distance along the unit circle between the terminal point determined by t and the x-axis.

Figure 11 shows that to find the reference number it is helpful to know the quadrant in which the terminal point determined by t lies.

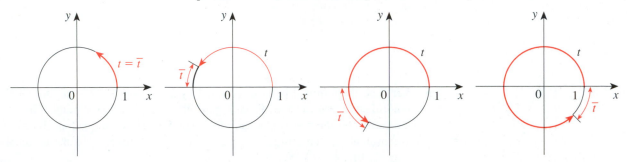

Figure 11
The reference number $\bar{t}$ for t

EXAMPLE 6

Find the reference numbers determined by the real number t:

(a) $t = \dfrac{5\pi}{6}$ (b) $t = \dfrac{7\pi}{4}$ (c) $t = -\dfrac{2\pi}{3}$ (d) $t = 5.80$

SOLUTION

From Figure 12 we find the reference numbers as follows:

(a) $\bar{t} = \pi - \dfrac{5\pi}{6} = \dfrac{\pi}{6}$ (b) $\bar{t} = 2\pi - \dfrac{7\pi}{4} = \dfrac{\pi}{4}$

(c) $\bar{t} = \pi - \dfrac{2\pi}{3} = \dfrac{\pi}{3}$ (d) $\bar{t} = 2\pi - 5.80 \approx 0.48$

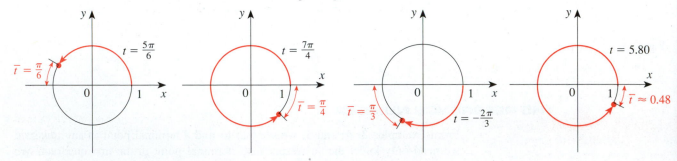

Figure 12

The value of π

The number π is the ratio of the circumference of a circle to its diameter. It was known since ancient times that this ratio is the same for all circles. The first systematic effort to find a numerical approximation for π was by Archimedes (ca. 240 B.C.), who proved that $\frac{22}{7} < \pi < \frac{223}{71}$ by finding the perimeters of regular polygons inscribed in and circumscribed about a circle. (See Problem 16, page 509.)

In about A.D. 480 the Chinese physicist Tsu Ch'ung-chih gave the approximation

$$\pi \approx \tfrac{355}{113} = 3.141592\ldots$$

which is correct to six decimals. This remained the most accurate estimation of π until the Dutch mathematician Adrianus Romanus (1593) used polygons having more than a billion sides to compute π correct to 15 decimals. In the seventeenth century, mathematicians began to use infinite series and trigonometric identities in the quest for π. (See Problem 15, page 509.) The Englishman William Shanks spent 15 years (1858–1873) using these methods to compute π to 707 decimals, but in 1946 it was found that his figures were wrong beginning with the 528th decimal. Today with the aid of computers it has become possible to routinely determine π correct to thousands and even millions of decimals. As of this writing, the record-holders are David and Gregory Chudnovsky, who in 1991 used their custom-built desktop computer to find the first 2.16 billion digits of π.

We use the reference number to find the terminal point $P(x, y)$ determined by any value of t using the following steps:

Step 1　Find the reference number $\bar{t}$.

Step 2　Find the terminal point $Q(x, y)$ determined by $\bar{t}$.

Step 3　The terminal point determined by t is $P(\pm x, \pm y)$, where the signs are chosen according to the quadrant in which this terminal point lies.

EXAMPLE 7

Find the terminal point determined by each value of t:

(a) $t = \dfrac{5\pi}{6}$　　(b) $t = \dfrac{7\pi}{4}$　　(c) $t = -\dfrac{2\pi}{3}$

SOLUTION

The reference numbers associated with these values of t were found in Example 6.

(a) The reference number is $\bar{t} = \pi/6$, which determines the terminal point $\left(\sqrt{3}/2, \frac{1}{2}\right)$ from Table 1. Since the terminal point determined by t is in quadrant II, its x-coordinate is negative and its y-coordinate is positive. Thus the desired terminal point is

$$\left(-\sqrt{3}/2, \tfrac{1}{2}\right)$$

(b) The reference number is $\bar{t} = \pi/4$, which determines the terminal point $\left(\sqrt{2}/2, \sqrt{2}/2\right)$ from Table 1. Since the terminal point is in quadrant IV, its x-coordinate is positive and its y-coordinate is negative. Thus the desired terminal point is

$$\left(\sqrt{2}/2, -\sqrt{2}/2\right)$$

(c) The reference number is $\bar{t} = \pi/3$, which determines the terminal point $\left(\frac{1}{2}, \sqrt{3}/2\right)$ from Table 1. Since the terminal point determined by t is in quadrant III, both its coordinates are negative. Thus the desired terminal point is

$$\left(-\tfrac{1}{2}, -\sqrt{3}/2\right) \qquad\blacksquare$$

Since the circumference of the unit circle is 2π, the terminal point determined by t is the same as that determined by $t + 2\pi$ or $t - 2\pi$. In general, we can add or subtract 2π any number of times without changing the terminal point determined by t. We use this observation in the next example to find terminal points for large t.

EXAMPLE 8

Find the terminal point determined by $t = \dfrac{29\pi}{6}$.

SOLUTION

Since

$$t = \frac{29\pi}{6} = 4\pi + \frac{5\pi}{6}$$

we see that the terminal point of t is the same as that of $5\pi/6$ (that is, we subtract 4π). So by Example 7(a) the terminal point is $\left(-\sqrt{3}/2, \frac{1}{2}\right)$. ∎

EXERCISES 5.1

In Exercises 1–6 show that the given point is on the unit circle.

1. $\left(\dfrac{3}{5}, \dfrac{4}{5}\right)$

2. $\left(\dfrac{40}{41}, \dfrac{9}{41}\right)$

3. $\left(\dfrac{6}{7}, -\dfrac{\sqrt{13}}{7}\right)$

4. $\left(-\dfrac{1}{3}, -\dfrac{2\sqrt{2}}{3}\right)$

5. $\left(-\dfrac{5}{13}, -\dfrac{12}{13}\right)$

6. $\left(\dfrac{\sqrt{5}}{5}, \dfrac{2\sqrt{5}}{5}\right)$

In Exercises 7–12 the point P is on the unit circle. Find P(x, y) from the given information.

7. The x-coordinate of P is $\frac{3}{5}$ and P is in quadrant I.

8. The y-coordinate of P is $-\frac{1}{3}$ and P is in quadrant III.

9. The x-coordinate of P is $\frac{2}{3}$ and the y-coordinate is negative.

10. The x-coordinate of P is positive and the y-coordinate of P is $-\sqrt{5}/5$.

11. The x-coordinate of P is $\sqrt{2}/3$ and P is in quadrant IV.

12. The x-coordinate of P is $-\frac{2}{5}$ and P is in quadrant II.

In Exercises 13–20 sketch on the unit circle the approximate location of the terminal point of the arc whose length is the given real number t [as always, the initial point of the arc is (1, 0)]. Determine the quadrant in which this terminal point lies.

13. $t = \dfrac{\pi}{8}$

14. $t = \dfrac{3\pi}{7}$

15. $t = -\dfrac{11\pi}{9}$

16. $t = 4$

17. $t = -1.7$

18. $t = 7$

19. $t = 10$

20. $t = 4.9$

In Exercises 21–30 find the terminal point P(x, y) on the unit circle determined by the given value of t.

21. $t = \dfrac{\pi}{4}$

22. $t = \dfrac{\pi}{2}$

23. $t = -\dfrac{\pi}{3}$

24. $t = \dfrac{7\pi}{6}$

25. $t = \pi$

26. $t = -\dfrac{5\pi}{6}$

27. $t = \dfrac{2\pi}{3}$

28. $t = -\dfrac{\pi}{2}$

29. $t = -\dfrac{3\pi}{4}$

30. $t = \dfrac{11\pi}{6}$

31. Suppose that the terminal point determined by t is the point $\left(\frac{3}{5}, \frac{4}{5}\right)$ on the unit circle. Find the terminal point determined by **(a)** $\pi - t$, **(b)** $-t$, **(c)** $\pi + t$, and **(d)** $2\pi + t$.

32. Suppose that the terminal point determined by t is the point $\left(\frac{1}{3}, 2\sqrt{2}/3\right)$ on the unit circle. Find the terminal point determined by **(a)** $-t$, **(b)** $4\pi + t$, **(c)** $\pi - t$, and **(d)** $t - \pi$.

In Exercises 33–36 find the reference number for the given value of t.

33. (a) $t = \dfrac{5\pi}{4}$ (b) $t = \dfrac{7\pi}{3}$

     (c) $t = -\dfrac{4\pi}{3}$ (d) $t = \dfrac{\pi}{6}$

34. (a) $t = \dfrac{5\pi}{7}$ (b) $t = -\dfrac{9\pi}{8}$

     (c) $t = 3.55$ (d) $t = -2.9$

35. (a) $t = -\dfrac{11\pi}{5}$ (b) $t = \dfrac{13\pi}{6}$

     (c) $t = \dfrac{7\pi}{3}$ (d) $t = -\dfrac{5\pi}{6}$

36. (a) $t = 3$ (b) $t = 6$
     (c) $t = -3$ (d) $t = -6$

In Exercises 37–48, find (a) the reference number for the given value of t and (b) the terminal point determined by t.

37. $t = \dfrac{3\pi}{4}$ **38.** $t = \dfrac{7\pi}{3}$

39. $t = -\dfrac{2\pi}{3}$ **40.** $t = -\dfrac{7\pi}{6}$

41. $t = \dfrac{13\pi}{4}$ **42.** $t = \dfrac{13\pi}{6}$

43. $t = \dfrac{7\pi}{6}$ **44.** $t = \dfrac{17\pi}{4}$

45. $t = -\dfrac{11\pi}{3}$ **46.** $t = \dfrac{31\pi}{6}$

47. $t = \dfrac{16\pi}{3}$ **48.** $t = -\dfrac{41\pi}{4}$

In Exercises 49–52 use the figure to find the terminal point determined by the given real number t, with coordinates correct to one decimal.

49. $t = 1$ **50.** $t = 2.5$

51. $t = -1.1$ **52.** $t = 4.2$

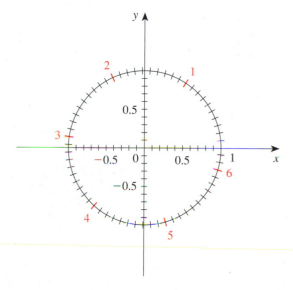

SECTION 5.2
TRIGONOMETRIC FUNCTIONS OF REAL NUMBERS

A function is a rule that assigns to each real number another real number. In this section we use properties of the unit circle from the preceding section to define certain functions of real numbers, the trigonometric functions.

THE TRIGONOMETRIC FUNCTIONS OF REAL NUMBERS

Recall that to find the terminal point $P(x, y)$ for a given real number t we move a distance t along the unit circle, starting at the point $(1, 0)$. We move in a counterclockwise direction if t is positive and in a clockwise direction if t is negative

(Figure 1). We now use the *x*- and *y*-coordinates of the point $P(x, y)$ to define several functions. For instance, we define the function called *sine* by assigning to each real number *t* the *y*-coordinate of the terminal point $P(x, y)$ determined by *t*. The functions *cosine*, *tangent*, *cosecant*, *secant*, and *cotangent* are also defined using the coordinates of $P(x, y)$.

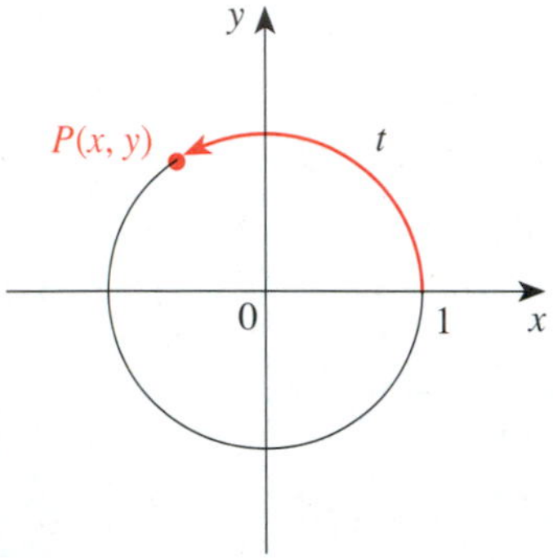

DEFINITION OF THE TRIGONOMETRIC FUNCTIONS

Let *t* be any real number and let $P(x, y)$ be the terminal point on the unit circle determined by *t*. We define

$$\sin t = y \qquad\qquad \cos t = x \qquad\qquad \tan t = \frac{y}{x} \quad (x \neq 0)$$

$$\csc t = \frac{1}{y} \quad (y \neq 0) \qquad \sec t = \frac{1}{x} \quad (x \neq 0) \qquad \cot t = \frac{x}{y} \quad (y \neq 0)$$

Figure 1

Because the trigonometric functions can be defined in terms of the unit circle, they are sometimes called the **circular functions**.

EXAMPLE 1

Find the six trigonometric functions of (a) $t = \dfrac{\pi}{3}$ and (b) $t = \dfrac{\pi}{4}$.

SOLUTION

(a) The terminal point determined by $t = \pi/3$ is $P\left(\frac{1}{2}, \sqrt{3}/2\right)$. Since the coordinates are $x = \frac{1}{2}$ and $y = \sqrt{3}/2$, we have

$$\sin\frac{\pi}{3} = \frac{\sqrt{3}}{2} \qquad \cos\frac{\pi}{3} = \frac{1}{2} \qquad \tan\frac{\pi}{3} = \frac{\sqrt{3}/2}{1/2} = \sqrt{3}$$

$$\csc\frac{\pi}{3} = \frac{2\sqrt{3}}{3} \qquad \sec\frac{\pi}{3} = 2 \qquad \cot\frac{\pi}{4} = \frac{1/2}{\sqrt{3}/2} = \frac{\sqrt{3}}{3}$$

(b) The terminal point determined by $t = \pi/4$ is $P\left(\sqrt{2}/2, \sqrt{2}/2\right)$. Since the coordinates are $x = \sqrt{2}/2$ and $y = \sqrt{2}/2$, we have

$$\sin\frac{\pi}{4} = \frac{\sqrt{2}}{2} \qquad \cos\frac{\pi}{4} = \frac{\sqrt{2}}{2} \qquad \tan\frac{\pi}{4} = \frac{\sqrt{2}/2}{\sqrt{2}/2} = 1$$

$$\csc\frac{\pi}{4} = \sqrt{2} \qquad \sec\frac{\pi}{4} = \sqrt{2} \qquad \cot\frac{\pi}{4} = \frac{\sqrt{2}/2}{\sqrt{2}/2} = 1 \qquad \blacksquare$$

Some special values of the trigonometric functions are listed in Table 1. This table is easily obtained from Table 1 of Section 5.1, together with the definitions of the trigonometric functions.

Table 1

t	$\sin t$	$\cos t$	$\tan t$	$\csc t$	$\sec t$	$\cot t$
$\dfrac{\pi}{6}$	$\dfrac{1}{2}$	$\dfrac{\sqrt{3}}{2}$	$\dfrac{\sqrt{3}}{3}$	2	$\dfrac{2\sqrt{3}}{3}$	$\sqrt{3}$
$\dfrac{\pi}{4}$	$\dfrac{\sqrt{2}}{2}$	$\dfrac{\sqrt{2}}{2}$	1	$\sqrt{2}$	$\sqrt{2}$	1
$\dfrac{\pi}{3}$	$\dfrac{\sqrt{3}}{2}$	$\dfrac{1}{2}$	$\sqrt{3}$	$\dfrac{2\sqrt{3}}{3}$	2	$\dfrac{\sqrt{3}}{3}$

EXAMPLE 2

Find the six trigonometric functions of (a) $t = 0$ and (b) $t = \dfrac{\pi}{2}$.

SOLUTION

(a) The terminal point determined by 0 is $P(1, 0)$. So,

$$\sin 0 = 0 \qquad \cos 0 = 1 \qquad \tan 0 = \frac{0}{1} = 0 \qquad \sec 0 = \frac{1}{1} = 1$$

The others, cot 0 and csc 0, are undefined because $y = 0$ occurs in the denominator in their definitions.

(b) The terminal point determined by $\pi/2$ is $P(0, 1)$. So,

$$\sin \frac{\pi}{2} = 1 \qquad \cos \frac{\pi}{2} = 0 \qquad \csc \frac{\pi}{2} = \frac{1}{1} = 1 \qquad \cot \frac{\pi}{2} = \frac{0}{1} = 0$$

But tan $\pi/2$ and sec $\pi/2$ are undefined because $x = 0$ appears in the denominator. ■

Example 2 shows that some of the trigonometric functions fail to be defined for certain real numbers. So we need to determine their domains. The functions sine and cosine are defined for all values of t. Since the functions cotangent and cosecant have y in the denominator of their definitions, they are not defined whenever the y-coordinate of the terminal point $P(x, y)$ determined by t is 0. This happens when $t = n\pi$ for any integer n, so their domains do not include these points. The functions tangent and secant have x in the denominator in their definitions, so they are not defined whenever $x = 0$. This happens when $t = (\pi/2) + n\pi$ for any integer n.

DOMAINS OF THE
TRIGONOMETRIC
FUNCTIONS

Function	Domain
sin, cos	all real numbers
tan, sec	all real numbers other than $\frac{\pi}{2} + n\pi$ for any integer n
cot, csc	all real numbers other than $n\pi$ for any integer n

■ VALUES OF THE TRIGONOMETRIC FUNCTIONS

To compute other values of the trigonometric functions we first determine their sign. The signs of the trigonometric functions depend on the quadrant in which the terminal point of t lies. For example, if the terminal point $P(x, y)$ determined by t lies in quadrant III, then both its coordinates are negative. So $\sin t$, $\cos t$, $\csc t$, and $\sec t$ are all negative, while $\tan t$ and $\cot t$ are positive. You can check the other entries in the following table.

SIGNS OF THE
TRIGONOMETRIC
FUNCTIONS

Quadrant	Positive functions	Negative functions
I	all	none
II	sin, csc	cos, sec, tan, cot
III	tan, cot	sin, csc, cos, sec
IV	cos, sec	sin, csc, tan, cot

EXAMPLE 3

(a) $\cos \frac{\pi}{3} > 0$, since the terminal point of $t = \frac{\pi}{3}$ is in quadrant I.

(b) $\tan 4 > 0$, since the terminal point of $t = 4$ is in quadrant I.

(c) If $\cos t < 0$ and $\sin t > 0$, then t must be in quadrant II. ■

In Section 5.1 we used the reference number to find the terminal point determined by a real number t. Since the trigonometric functions are defined in terms of the coordinates of terminal points, we can use the reference number to find values of the trigonometric functions. Suppose that $\bar{t}$ is the reference number for t. Then the terminal point of $\bar{t}$ has the same coordinates, except possibly for sign, as the terminal point of t. So the values of the trigonometric functions at t are the same, except possibly for sign, as their values at $\bar{t}$. We illustrate the procedure in the next example.

EXAMPLE 4

Find each value: (a) $\cos \dfrac{2\pi}{3}$ (b) $\tan\left(-\dfrac{\pi}{3}\right)$ (c) $\sin \dfrac{19\pi}{4}$

SOLUTION

(a) The reference number for $2\pi/3$ is $\pi/3$. Since $2\pi/3$ is in quadrant II, $\cos(2\pi/3)$ is negative. Thus

$$\cos \frac{2\pi}{3} = -\cos \frac{\pi}{3} = -\frac{1}{2}$$

$$\underset{\substack{\text{sign}}}{\uparrow} \quad \underset{\substack{\text{reference}\\\text{number}}}{\uparrow} \quad \underset{\substack{\text{from}\\\text{Table 1}}}{\uparrow}$$

(b) The reference number for $-\pi/3$ is $\pi/3$. Since $-\pi/3$ is in quadrant IV, $\tan(-\pi/3)$ is negative. Thus

$$\tan\left(-\frac{\pi}{3}\right) = -\tan \frac{\pi}{3} = -\sqrt{3}$$

$$\underset{\substack{\text{sign}}}{\uparrow} \quad \underset{\substack{\text{reference}\\\text{number}}}{\uparrow} \quad \underset{\substack{\text{from}\\\text{Table 1}}}{\uparrow}$$

(c) Since $(19\pi/4) - 4\pi = 3\pi/4$, the terminal points determined by $19\pi/4$ and $3\pi/4$ are the same. The reference number for $3\pi/4$ is $\pi/4$. Since $3\pi/4$ is in quadrant II, $\sin(3\pi/4)$ is positive. Thus

$$\sin \frac{19\pi}{4} = \sin \frac{3\pi}{4} = +\sin \frac{\pi}{4} = \frac{\sqrt{2}}{2}$$

$$\underset{\substack{\text{subtract } 4\pi}}{\uparrow} \qquad \underset{\substack{\text{sign}}}{\uparrow} \quad \underset{\substack{\text{reference}\\\text{number}}}{\uparrow} \quad \underset{\substack{\text{from}\\\text{Table 1}}}{\uparrow}$$

    So far we have been able to compute the values of the trigonometric functions only for certain values of t. In fact, we can compute the values of the trigonometric functions whenever t is a multiple of $\pi/6$, $\pi/4$, $\pi/3$, and $\pi/2$. How can we compute the trigonometric functions for other values of t? For example, how can we find $\sin 1.5$? One way is to carefully sketch a diagram and read the value (see Exercises 33–40). This method is not very accurate. Fortunately, programmed directly into scientific calculators are mathematical procedures (called *numerical methods*) that find the values of *sine*, *cosine*, and *tangent* correct to the number of digits in the display. The calculator must be put in *radian mode* to evaluate these

Numerical methods are ways of finding approximations for the values of functions. Since the values of polynomials are easy to compute, it is helpful to express functions that we wish to approximate in terms of polynomials. For example, using calculus, it can be shown that

$$\sin t = t - \frac{t^3}{3!} + \frac{t^5}{5!} - \frac{t^7}{7!} + \cdots$$

$$\cos t = 1 - \frac{t^2}{2!} + \frac{t^4}{4!} - \frac{t^6}{6!} + \cdots$$

where $n! = 1 \cdot 2 \cdot 3 \cdots n$ (for example, $4! = 1 \cdot 2 \cdot 3 \cdot 4 = 24$). These remarkable formulas were proved by the British mathematician Brook Taylor (1685–1731). The more terms we use from the series on the right-hand side of these formulas, the more accurate the values obtained for $\sin t$ and $\cos t$. For instance, if we use the first three terms of Taylor's series to find $\sin 0.9$, we get

$$\sin 0.9 \approx 0.9 - \frac{(0.9)^3}{3!} + \frac{(0.9)^5}{5!}$$

$$\approx 0.78342075$$

(Compare this with the value obtained using your calculator.) Formulas like these are programmed into your calculator, which uses them to compute functions like $\sin t$, $\cos t$, e^t, and $\log t$. The calculator uses enough terms to ensure that all displayed digits are accurate.

functions. To find values of cosecant, secant, and cotangent using a calculator, we need to use the *reciprocal relations*:

$$\csc t = \frac{1}{\sin t} \qquad \sec t = \frac{1}{\cos t} \qquad \cot t = \frac{1}{\tan t}$$

These identities follow from the definitions of the trigonometric functions. For instance, since $\sin t = y$ and $\csc t = 1/y$, we have $\csc t = 1/y = 1/(\sin t)$. The others follow similarly.

EXAMPLE 5

Making sure our calculator is set to radian mode and rounding the results to six decimal places, we get:

(a) $\sin 2.2 \approx 0.808496$ (b) $\cos 1.1 \approx 0.453596$

(c) $\cot 28 = \dfrac{1}{\tan 28} \approx -3.553286$ (d) $\csc 0.98 = \dfrac{1}{\sin 0.98} \approx 1.204098$ ■

Let us consider the relationship between the trigonometric functions of t and those of $-t$. From Figure 2 we see that

$$\sin(-t) = -y = -\sin t$$

$$\cos(-t) = x = \cos t$$

$$\tan(-t) = \frac{-y}{x} = -\frac{y}{x} = -\tan t$$

These equations show that sine and tangent are odd functions, whereas cosine is an even function. It is easy to see that the reciprocal of an even function is even, and the reciprocal of an odd function is odd (see Section 2.2). This fact, together with the reciprocal relations, completes our knowledge of the even-odd properties for all the trigonometric functions.

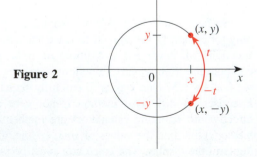

Figure 2

EVEN-ODD PROPERTIES

> Sine, cosecant, tangent, and cotangent are odd functions; cosine and secant are even functions.
>
> $$\sin(-t) = -\sin t \qquad \cos(-t) = \cot t \qquad \tan(-t) = -\tan t$$
> $$\csc(-t) = -\csc t \qquad \sec(-t) = \sec t \qquad \cot(-t) = -\cot t$$

EXAMPLE 6

Use the even-odd properties of the trigonometric functions to determine each value:

(a) $\sin\left(-\dfrac{\pi}{6}\right)$ (b) $\cos\left(-\dfrac{\pi}{4}\right)$ (c) $\tan\left(-\dfrac{\pi}{4}\right)$

SOLUTION

By the even-odd properties and Table 1, we have

(a) $\sin\left(-\dfrac{\pi}{6}\right) = -\sin\left(\dfrac{\pi}{6}\right) = -\dfrac{1}{2}$

(b) $\cos\left(-\dfrac{\pi}{4}\right) = \cos\left(\dfrac{\pi}{4}\right) = \dfrac{\sqrt{2}}{2}$

(c) $\tan\left(-\dfrac{\pi}{4}\right) = -\tan\left(\dfrac{\pi}{4}\right) = -1$ ∎

■ FUNDAMENTAL IDENTITIES

The trigonometric functions are related to each other through equations called **trigonometric identities**. The most important ones are the following.*

FUNDAMENTAL IDENTITIES

> **Reciprocal Identities**
>
> $$\csc t = \frac{1}{\sin t} \qquad \sec t = \frac{1}{\cos t} \qquad \cot t = \frac{1}{\tan t}$$
>
> $$\tan t = \frac{\sin t}{\cos t} \qquad \cot t = \frac{\cos t}{\sin t}$$
>
> **Pythagorean Identities**
>
> $$\sin^2 t + \cos^2 t = 1 \qquad \tan^2 t + 1 = \sec^2 t \qquad 1 + \cot^2 t = \csc^2 t$$

---

*We follow the usual convention of writing $\sin^2 t$ for $(\sin t)^2$. In general, we write $\sin^n t$ for $(\sin t)^n$ for all integers n except $n = -1$. The exponent $n = -1$ will be assigned another meaning in Section 7.6. Of course, the same convention applies to the other five trigonometric ratios.

**Relationship to Trigonometry
of Right Triangles**

If you have previously studied trigonometry of right triangles, you are probably wondering how the sine and cosine of an *angle* relate to those of this section. The following figures show how.

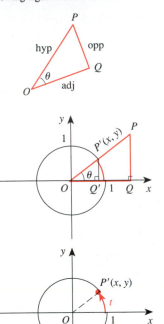

By the triangle definition,

$$\sin \theta = \frac{\text{opp}}{\text{hyp}} = \frac{PQ}{OP} = \frac{P'Q'}{OP'} = \frac{y}{1} = y$$

$$\cos \theta = \frac{\text{adj}}{\text{hyp}} = \frac{OQ}{OP} = \frac{OQ'}{OP'} = \frac{x}{1} = x$$

If θ is measured in radians, than $\theta = t$ and we can write

$$\sin t = y$$
$$\cos t = x$$

These are precisely the definitions of this section. So the two definitions give identical values.

Why then study trigonometry in two different ways? Because different applications require that we view the trigonometric functions differently (see Sections 5.6 and 6.2).

Proof   The reciprocal identities were proved on page 322, before Example 5. We now prove the Pythagorean identities. By definition, $\cos t = x$ and $\sin t = y$, where x and y are the coordinates of a point $P(x, y)$ on the unit circle. Since $P(x, y)$ is on the unit circle, we have $x^2 + y^2 = 1$. Thus

$$\sin^2 t + \cos^2 t = 1$$

Dividing both sides by $\cos^2 t$ (provided $\cos t \neq 0$), we get

$$\frac{\sin^2 t}{\cos^2 t} + \frac{\cos^2 t}{\cos^2 t} = \frac{1}{\cos^2 t}$$

$$\left(\frac{\sin t}{\cos t}\right)^2 + 1 = \left(\frac{1}{\cos t}\right)^2$$

$$\tan^2 t + 1 = \sec^2 t$$

We have used the reciprocal identities $\sin t/\cos t = \tan t$ and $1/\cos t = \sec t$. Similarly, dividing both sides of the first Pythagorean identity by $\sin^2 t$ (provided $\sin t \neq 0$) gives us $1 + \cot^2 t = \csc^2 t$.   ∎

As their name indicates, the fundamental identities play a central role in trigonometry. This is because they can be used to relate any trigonometric function to any other. So, if we know the value of any one of the trigonometric functions at t, then we can find the values of all the others at t.

EXAMPLE 7

If $\cos t = \frac{3}{5}$ and t is in quadrant IV, find the values of all the trigonometric functions at t.

SOLUTION

From the Pythagorean identities we have

$$\sin^2 t + \cos^2 t = 1$$

$$\sin^2 t + \left(\tfrac{3}{5}\right)^2 = 1$$

$$\sin^2 t = 1 - \tfrac{9}{25} = \tfrac{16}{25}$$

$$\sin t = \pm\tfrac{4}{5}$$

Since this point is in quadrant IV, $\sin t$ is negative, so $\sin t = -\frac{4}{5}$. Now that we know both $\sin t$ and $\cos t$, we can find the values of the other trigonometric functions using the reciprocal identities:

$$\sin t = -\frac{4}{5} \qquad \cos t = \frac{3}{5} \qquad \tan t = \frac{\sin t}{\cos t} = \frac{-\frac{4}{5}}{\frac{3}{5}} = -\frac{4}{3}$$

$$\csc t = \frac{1}{\sin t} = -\frac{5}{4} \qquad \sec t = \frac{1}{\cos t} = \frac{5}{3} \qquad \cot t = \frac{1}{\tan t} = -\frac{3}{4} \qquad \blacksquare$$

EXAMPLE 8

Write $\tan t$ in terms of $\cos t$, where t is in quadrant III.

SOLUTION

Since $\tan t = \sin t / \cos t$, we need to write $\sin t$ in terms of $\cos t$. By the Pythagorean identities we have

$$\sin^2 t + \cos^2 t = 1$$

$$\sin^2 t = 1 - \cos^2 t$$

$$\sin t = \pm\sqrt{1 - \cos^2 t}$$

Since $\sin t$ is negative in quadrant III, the negative sign applies here. Thus

$$\tan t = \frac{\sin t}{\cos t} = \frac{-\sqrt{1 - \cos^2 t}}{\cos t} \qquad \blacksquare$$

EXERCISES 5.2

In Exercises 1–20 find the exact value of the trigonometric function at the given real number.

1. (a) $\sin 0$
   (b) $\cos 0$

2. (a) $\sin \pi$
   (b) $\cos \pi$

3. (a) $\sin(-\pi)$
   (b) $\cos(-\pi)$

4. (a) $\cos \dfrac{\pi}{6}$
   (b) $\cos \dfrac{5\pi}{6}$

5. (a) $\sin \dfrac{\pi}{2}$
   (b) $\sin \dfrac{3\pi}{2}$

6. (a) $\sin \dfrac{7\pi}{6}$
   (b) $\cos \dfrac{7\pi}{6}$

7. (a) $\sin \dfrac{3\pi}{4}$
   (b) $\cos \dfrac{3\pi}{4}$

8. (a) $\cos \dfrac{\pi}{3}$
   (b) $\cos\left(-\dfrac{\pi}{3}\right)$

9. (a) $\sin \dfrac{\pi}{6}$
   (b) $\sin\left(-\dfrac{\pi}{6}\right)$

10. (a) $\cos \dfrac{\pi}{2}$
   (b) $\cos \dfrac{5\pi}{2}$

11. (a) $\sin \dfrac{5\pi}{6}$
   (b) $\sec \dfrac{5\pi}{6}$

12. (a) $\cos \dfrac{7\pi}{3}$
   (b) $\sec \dfrac{7\pi}{3}$

13. (a) $\tan \dfrac{\pi}{6}$
   (b) $\tan\left(-\dfrac{\pi}{6}\right)$

14. (a) $\tan \dfrac{\pi}{3}$
   (b) $\cot \dfrac{\pi}{3}$

15. (a) $\sec \dfrac{11\pi}{3}$
   (b) $\csc \dfrac{11\pi}{3}$

16. (a) $\sec \dfrac{13\pi}{6}$
   (b) $\sec\left(-\dfrac{13\pi}{6}\right)$

17. (a) $\sin \dfrac{9\pi}{4}$
   (b) $\csc \dfrac{9\pi}{4}$

18. (a) $\sec \pi$
   (b) $\csc \dfrac{\pi}{2}$

19. (a) $\tan\left(-\dfrac{\pi}{4}\right)$
   (b) $\cot\left(-\dfrac{\pi}{4}\right)$

20. (a) $\tan \dfrac{3\pi}{4}$
   (b) $\tan \dfrac{11\pi}{4}$

In Exercises 21–24 find the value of each of the six trigonometric functions (if it is defined) at the given real number t. Use your answers to complete the table.

21. $t = 0$

22. $t = \dfrac{\pi}{2}$

23. $t = \pi$

24. $t = \dfrac{3\pi}{2}$

[handwritten: homework 21 - 24   complete table]

t	$\sin t$	$\cos t$	$\tan t$	$\csc t$	$\sec t$	$\cot t$
0	0	1		undefined		
$\dfrac{\pi}{2}$						
π			0			undefined
$\dfrac{3\pi}{2}$						

In Exercises 25–32 the terminal point P(x, y) determined by t is given. Find sin t, cos t, and tan t.

25. $\left(\dfrac{3}{5}, \dfrac{4}{5}\right)$

26. $\left(-\dfrac{3}{5}, \dfrac{4}{5}\right)$

27. $\left(\dfrac{6}{7}, -\dfrac{\sqrt{13}}{7}\right)$

28. $\left(-\dfrac{1}{3}, -\dfrac{2\sqrt{2}}{3}\right)$

29. $\left(\dfrac{40}{41}, \dfrac{9}{41}\right)$

30. $\left(-\dfrac{3}{5}, -\dfrac{4}{5}\right)$

31. $\left(-\dfrac{5}{13}, -\dfrac{12}{13}\right)$

32. $\left(\dfrac{\sqrt{5}}{5}, \dfrac{2\sqrt{5}}{5}\right)$

In Exercises 33–40 find the approximate value of the given trigonometric function by (a) using the figure and (b) using a calculator. Compare the two values.

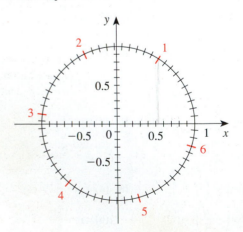

33. $\sin 1$

34. $\cos 0.8$

35. $\sin 1.2$

36. $\cos 5$

37. $\tan 0.8$

38. $\tan(-1.3)$

39. $\cos 4.1$

40. $\sin(-5.2)$

In Exercises 41–44 find the sign of the expression if the terminal point determined by t is in the given quadrant.

41. $\tan t \cos t$, t in quadrant II

42. $\sin^2 t \cos t$, t in quadrant IV   *[handwritten: (+)(+)= +]*

43. $\dfrac{\tan t \sin t}{\cot t}$, t in quadrant III

44. $\cos t \sec t$, t in any quadrant

In Exercises 45–48, from the information given, find the quadrant in which the terminal point determined by t lies.

45. $\sin t > 0$ and $\cos t < 0$

46. $\tan t > 0$ and $\sin t < 0$

47. $\csc t > 0$ and $\sec t < 0$

48. $\cos t < 0$ and $\cot t < 0$

In Exercises 49–60 write the first expression in terms of the second if the terminal point determined by t is in the given quadrant.

49. $\sin t$, $\cos t$; t in quadrant II

50. $\cos t$, $\sin t$; t in quadrant IV

51. $\tan t$, $\sin t$; t in quadrant IV

52. $\tan t$, $\cos t$; t in quadrant III   *[handwritten: $\tan t = \dfrac{\sin t}{\cos t} = \dfrac{\sqrt{1-\cos^2 t}}{\cos t}$]*

53. $\sec t$, $\tan t$; t in quadrant II

54. $\csc t$, $\cot t$; t in quadrant III

55. $\tan t$, $\sec t$; t in quadrant III

56. $\sin t$, $\sec t$; t in quadrant IV

57. $\tan^2 t$, $\sin t$; t in any quadrant

58. $\sec^2 t \sin^2 t$, $\cos t$; t in any quadrant

59. $\dfrac{1 - \sin^2 t}{\sec t}$, $\cos t$; t in any quadrant

60. $\dfrac{1 + \tan^2 t}{\cos t}$, $\sec t$; t in any quadrant

In Exercises 61–68 find the values of the trigonometric functions of t from the given information.

61. $\sin t = \dfrac{3}{5}$, t in quadrant II

62. $\cos t = -\frac{4}{5}$, t in quadrant III

63. $\tan t = -\frac{3}{4}$, $\cos t > 0$

64. $\sec t = 3$, t in quadrant IV

65. $\sec t = 2$, $\sin t < 0$

66. $\tan t = \frac{1}{4}$, t in quadrant III

67. $\sin t = -\frac{1}{4}$, $\sec t < 0$

68. $\tan t = -4$, t in quadrant II

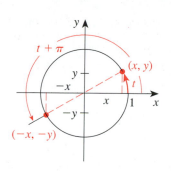

In Exercises 69–76 determine whether the given function is even, odd, or neither.

69. $f(x) = x^2 \sin x$

70. $f(x) = x^2 \cos 2x$

71. $f(x) = \sin x \cos x$

72. $f(x) = e^x \sin x$

73. $f(x) = |x| \cos x$

74. $f(x) = x \sin^3 x$

75. $f(x) = x^3 + \cos x$

76. $f(x) = \cos(\sin x)$

77. Prove the following "reduction formulas" from the diagram.

(a) $\sin(t + \pi) = -\sin t$

(b) $\cos(t + \pi) = -\cos t$

(c) $\tan(t + \pi) = \tan t$

78. Prove that triangle AOB is congruent to triangle CDO in the figure, and then deduce the following "reduction formulas."

(a) $\sin\left(t + \dfrac{\pi}{2}\right) = \cos t$

(b) $\cos\left(t + \dfrac{\pi}{2}\right) = -\sin t$

(c) $\tan\left(t + \dfrac{\pi}{2}\right) = -\cot t$

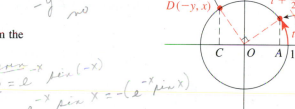

SECTION 5.3
TRIGONOMETRIC GRAPHS

The graph of a function helps us get a better idea of its behavior. So, in this section we sketch graphs of the sine and cosine functions and certain transformations of these functions. The other trigonometric functions are sketched in the next section.

GRAPHS OF THE SINE AND COSINE FUNCTIONS

To help us sketch the graphs of the sine and cosine functions, we first observe that these functions repeat their values in a regular fashion. To see exactly how this happens, we recall that the circumference of the unit circle is 2π. It follows that the terminal point $P(x, y)$ determined by the real number t is the same as that determined by $t + 2\pi$. Since the sine and cosine functions are defined in terms of the coordinates of $P(x, y)$, it follows that their values are unchanged by the addition of any integer multiple of 2π. In other words,

$$\sin(t + 2n\pi) = \sin t \quad \text{for any integer } n$$

$$\cos(t + 2n\pi) = \cos t \quad \text{for any integer } n$$

Thus the sine and cosine functions are *periodic* according to the following definition: A function f is **periodic** if there is a positive number p such that $f(t + p) = f(t)$ for every t. The least such positive number (if it exists) is the **period** of f. If f has period p, then the graph of f on any interval of length p is called **one complete period** of f.

PERIODIC PROPERTIES OF SINE AND COSINE

The function sine has period 2π: $\sin(t + 2\pi) = \sin t$

The function cosine has period 2π: $\cos(t + 2\pi) = \cos t$

So the sine and cosine functions repeat their values in any interval of length 2π. To sketch their graphs, we first sketch the graph of one period. To sketch the graphs on the interval $0 \le t \le 2\pi$, we could try to make a table of values and use those points to draw the graph. Since no such table can be complete, let us instead look more closely at the definitions of these functions.

Recall that $\sin t$ is the y-coordinate of the terminal point $P(x, y)$ on the unit circle determined by the real number t (Figure 1). How does the y-coordinate of this point vary as t increases? It is easy to see that the y-coordinate of $P(x, y)$ increases to 1, then decreases to -1 repeatedly as the point $P(x, y)$ travels around the unit circle. In fact, as t increases from 0 to $\pi/2$, $y = \sin t$ increases from 0 to 1. As t increases from $\pi/2$ to π, the value of $y = \sin t$ decreases from 1 to 0. Table 1 shows the variation of the sine and cosine functions for t between 0 and 2π.

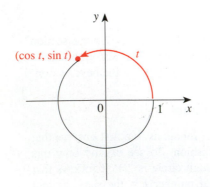

Figure 1

Table 1

t	$\sin t$	$\cos t$
$0 \to \dfrac{\pi}{2}$	$0 \to 1$	$1 \to 0$
$\dfrac{\pi}{2} \to \pi$	$1 \to 0$	$0 \to -1$
$\pi \to \dfrac{3\pi}{2}$	$0 \to -1$	$-1 \to 0$
$\dfrac{3\pi}{2} \to 2\pi$	$-1 \to 0$	$0 \to 1$

To draw the graphs more accurately, we find a few other values of sin t and cos t in Table 2. We could find still other values with the aid of a calculator.

Table 2	t	0	$\dfrac{\pi}{6}$	$\dfrac{\pi}{3}$	$\dfrac{\pi}{2}$	$\dfrac{2\pi}{3}$	$\dfrac{5\pi}{6}$	π	$\dfrac{7\pi}{6}$	$\dfrac{4\pi}{3}$	$\dfrac{3\pi}{2}$	$\dfrac{5\pi}{3}$	$\dfrac{11\pi}{6}$	2π
	$\sin t$	0	$\dfrac{1}{2}$	$\dfrac{\sqrt{3}}{2}$	1	$\dfrac{\sqrt{3}}{2}$	$\dfrac{1}{2}$	0	$-\dfrac{1}{2}$	$-\dfrac{\sqrt{3}}{2}$	-1	$-\dfrac{\sqrt{3}}{2}$	$-\dfrac{1}{2}$	0
	$\cos t$	1	$\dfrac{\sqrt{3}}{2}$	$\dfrac{1}{2}$	0	$-\dfrac{1}{2}$	$-\dfrac{\sqrt{3}}{2}$	-1	$-\dfrac{\sqrt{3}}{2}$	$-\dfrac{1}{2}$	0	$\dfrac{1}{2}$	$\dfrac{\sqrt{3}}{2}$	1

Now we use this information to sketch the graphs of the functions sin t and cos t for t between 0 and 2π in Figures 2 and 3. These are the graphs of one period. Using the fact that these functions are periodic with period 2π, we get their complete graphs by continuing the same pattern to the left and to the right in every successive interval of length 2π.

The graph of the sine function is symmetric with respect to the origin. This is as expected, since sine is an odd function. Since the cosine function is an even function, its graph is symmetric with respect to the y-axis.

Figure 2
Graph of sin t

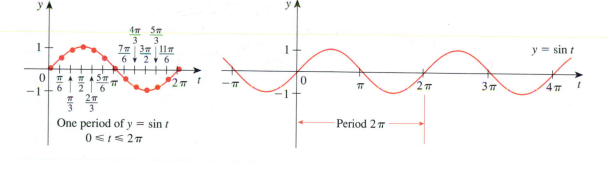

One period of $y = \sin t$
$0 \le t \le 2\pi$

Period 2π

Figure 3
Graph of cos t

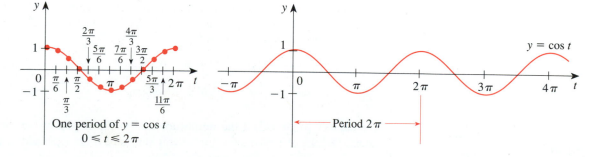

One period of $y = \cos t$
$0 \le t \le 2\pi$

Period 2π

GRAPHS OF TRANSFORMATIONS OF SINE AND COSINE

We now consider graphs of functions that are transformations of the sine and cosine functions. Thus the graphing techniques of Section 2.5 are very useful here. The graphs we obtain are important for understanding applications to physical situations such as harmonic motion (Section 5.6). But some of them are beautiful graphs that are interesting in their own right.

It is traditional to use the letter x to denote the variable in the domain of a function. So, from here on we use the letter x and write $y = \sin x$, $y = \cos x$, $y = \tan x$, and so on to denote these functions.

EXAMPLE 1

Sketch the graphs of (a) $f(x) = 2 + \cos x$ and (b) $g(x) = -\cos x$.

SOLUTION

(a) The graph of $y = f(x)$ is the same as the graph of $y = \cos x$, but shifted up two units [see Figure 4(a)].
(b) The graph of $y = g(x)$ in Figure 4(b) is the reflection of the graph of $y = \cos x$ in the x-axis.

Figure 4

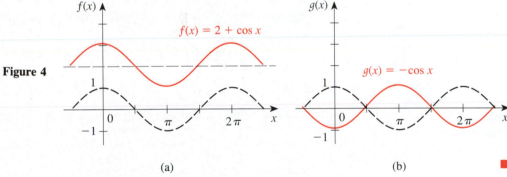

(a)   (b)

Let us sketch the graph of $y = 2 \sin x$. We start with the graph of $y = \sin x$ and multiply the y-coordinate of each point by 2. This has the effect of stretching the graph vertically by a factor of 2 [see Figure 5(a)]. To graph $y = \frac{1}{2} \sin x$ we start with the graph of $y = \sin x$ and multiply the y-coordinate of each point by $\frac{1}{2}$. This has the effect of shrinking the graph vertically by a factor of 2 [see Figure 5(b)].

In general, for the functions

$$y = a \sin x$$

and

$$y = a \cos x$$

the number $|a|$ is called the **amplitude** and is the largest value these functions attain. Graphs of $y = a \sin x$ for several values of a are sketched in Figure 6.

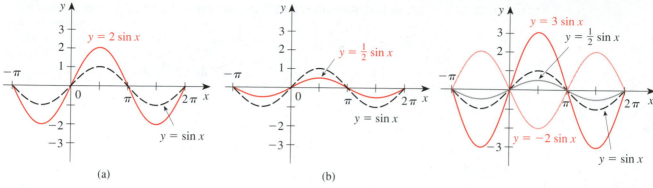

(a)

(b)

Figure 5

Figure 6

EXAMPLE 2

Find the amplitude of $y = -3 \cos x$ and sketch the graph.

SOLUTION

The amplitude is $|-3| = 3$. So the largest value the graph attains is 3 and the smallest value is -3. To sketch the graph, we begin with the graph of $y = \cos x$, stretch the graph vertically by a factor of 3, and reflect in the x-axis, arriving at the graph in Figure 7.

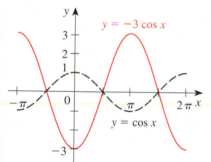

Figure 7

Since the sine and cosine functions have period 2π, the functions

$$y = a \sin kx \qquad \text{and} \qquad y = a \cos kx \qquad (k > 0)$$

complete one period as kx varies from 0 to 2π, that is, for $0 \leq kx \leq 2\pi$ or for $0 \leq x \leq 2\pi/k$. So these functions complete one period as x varies between 0 and $2\pi/k$ and thus have period $2\pi/k$. The graphs of these functions are called **sine curves** and **cosine curves**, respectively.

SINE AND COSINE CURVES

The sine and cosine curves

$$y = a \sin kx \qquad \text{and} \qquad y = a \cos kx \qquad (k > 0)$$

have amplitude $|a|$ and period $2\pi/k$.

An appropriate interval on which to sketch one complete period is $[0, 2\pi/k]$.

To see how the value of k affects the graph of $y = \sin kx$, let us sketch the sine curve $y = \sin 2x$. Since the period is $2\pi/2 = \pi$, the graph completes one period in the interval $0 \le x \le \pi$ [see Figure 8(a)]. For the sine curve $y = \sin \frac{1}{2}x$ the period is $2\pi \div \frac{1}{2} = 4\pi$, and so the graph completes one period in the interval $0 \le x \le 4\pi$ [see Figure 8(b)]. We see that the effect is to compress the graph if $k > 1$ and to stretch the graph if $k < 1$.

Figure 8

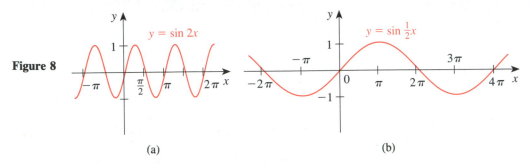

(a)   (b)

For comparison we sketch in Figure 9 the graphs of one period of the sine curve $y = a \sin kx$ for several values of k.

Figure 9

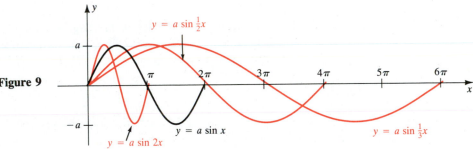

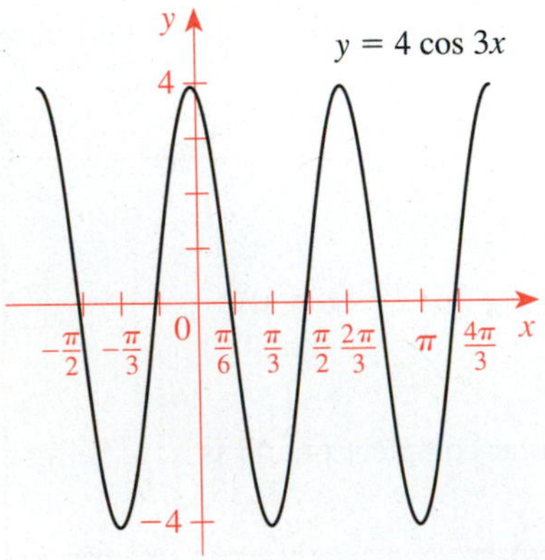

EXAMPLE 3

Find the amplitude and period and sketch the graph:

(a) $y = 4 \cos 3x$ (b) $y = -2 \sin \frac{1}{2}x$

SOLUTION

(a) For $y = 4 \cos 3x$,

$$\text{amplitude} = |a| = 4$$

$$\text{period} = \frac{2\pi}{3}$$

The graph is sketched in Figure 10.

Figure 10

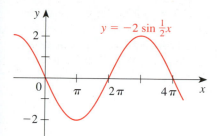

Figure 11

(b) For $y = -2 \sin \frac{1}{2}x$,

$$\text{amplitude} = |a| = |-2| = 2$$

$$\text{period} = \frac{2\pi}{1/2} = 4\pi$$

The graph is sketched in Figure 11.

The graphs of functions of the form

$$y = a \sin k(x - b)$$

and

$$y = a \cos k(x - b)$$

are simply sine and cosine curves that are shifted horizontally by an amount $|b|$. They are shifted to the right if $b > 0$ or to the left if $b < 0$. The number b is the **phase shift**. We summarize the properties of these functions.

SHIFTED SINE AND COSINE CURVES

The sine and cosine curves

$$y = a \sin k(x - b) \qquad \text{and} \qquad y = a \cos k(x - b) \qquad (k > 0)$$

have amplitude $|a|$, period $2\pi/k$, and phase shift b.

An appropriate interval on which to sketch one complete period is $[b, b + (2\pi/k)]$.

The graphs of $y = \sin\left(x - \dfrac{\pi}{3}\right)$ and $y = \sin\left(x + \dfrac{\pi}{6}\right)$ are sketched in Figure 12.

Figure 12

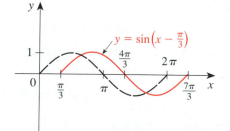

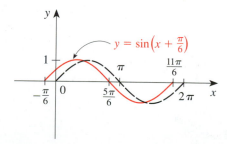

EXAMPLE 4

Find the amplitude, period, and phase shift of $y = 3 \sin 2\left(x - \dfrac{\pi}{4}\right)$, and sketch the graph of one complete period.

SOLUTION

We have

$$\text{amplitude} = |a| = 3$$

$$\text{period} = \frac{2\pi}{2} = \pi$$

$$\text{phase shift} = \frac{\pi}{4}$$

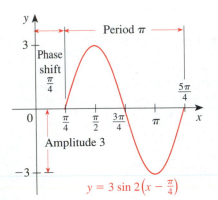

Since the phase shift is $\pi/4$ and the period is π, one complete period occurs on the interval

$$\left[\frac{\pi}{4}, \frac{\pi}{4} + \pi\right] = \left[\frac{\pi}{4}, \frac{5\pi}{4}\right]$$

As an aid in sketching the graph we divide this interval into four equal parts, then sketch a sine curve with amplitude 3 as in Figure 13. ■

Figure 13

EXAMPLE 5

Find the amplitude, period, and phase shift of the function

$$y = \frac{3}{4} \cos\left(2x - \frac{2\pi}{3}\right)$$

Sketch the graph of one period.

SOLUTION

We first write this function in the form $y = a \cos k(x - b)$. To do this we factor 2 from the expression $\left(2x - \dfrac{2\pi}{3}\right)$ to get

$$y = \frac{3}{4} \cos 2\left(x - \frac{\pi}{3}\right)$$

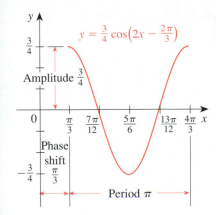

Figure 14

Thus we have

$$\text{amplitude} = |a| = \frac{3}{4}$$

$$\text{period} = \frac{2\pi}{k} = \frac{2\pi}{2} = \pi$$

$$\text{phase shift} = b = \frac{\pi}{3}$$

From this information, it follows that one period of this cosine curve begins at $\pi/3$ and ends at $\pi + (\pi/3) = 4\pi/3$. To sketch the graph over the interval $[\pi/3, 4\pi/3]$ we divide this interval into four equal parts and sketch a cosine curve with amplitude $\frac{3}{4}$ as shown in Figure 14. ■

EXERCISES 5.3

In Exercises 1–10 sketch the graph of the given function.

1. $y = 2 + \sin x$

2. $y = -\sin x$

3. $y = -1 + \cos x$

4. $y = 4 - \cos x$

5. $y = 3 \sin x$

6. $y = -2 \cos x$

7. $y = 3 + 3 \sin x$

8. $y = 4 - 2 \cos x$

9. $y = |\sin x|$

10. $y = |\cos x|$

In Exercises 11–20 find the amplitude and period, and sketch the graph.

11. $y = 3 \sin 3x$

12. $y = -2 \sin 2\pi x$ *don't graph*

13. $y = 10 \sin \frac{1}{2}x$

14. $y = \cos 10\pi x$

15. $y = -\cos \frac{1}{3}x$

16. $y = \sin(-2x)$

17. $y = 3 \cos 3\pi x$

18. $y = 5 - 2 \sin 2x$

19. $y = \sin \frac{2}{3}x$

20. $y = -\cos(-x)$

In Exercises 21–36 find the amplitude, period, and phase shift, and sketch the graph.

21. $y = \cos\left(x - \frac{\pi}{2}\right)$

22. $y = \sin(x + 1)$

23. $y = 2 \sin\left(\pi x - \frac{\pi}{3}\right)$

24. $y = -2 \cos\left(x - \frac{\pi}{12}\right)$

25. $y = 2 \sin\left(\frac{\pi}{2} - x\right)$

26. $y = -4 \sin 2\left(x + \frac{\pi}{2}\right)$

27. $y = 5 \cos\left(3x - \frac{\pi}{4}\right)$

28. $y = \frac{1}{2} \cos\left(\frac{\pi}{2} - 2\pi x\right)$

29. $y = 2 + 2 \sin\left(\frac{2}{3}x - \frac{\pi}{6}\right)$

30. $y = \sin \frac{1}{2}\left(x + \frac{\pi}{4}\right)$

31. $y = 3 \cos \pi\left(x + \frac{1}{2}\right)$

32. $y = 1 + \cos\left(3x + \frac{\pi}{2}\right)$

33. $y = -2 \cos\left(2x - \frac{\pi}{3}\right)$

34. $y = 3 + 2 \sin 3(x + 1)$

35. $y = \sin(3x + \pi)$

36. $y = \cos\left(\frac{\pi}{2} - x\right)$

In Exercises 37–42 the graph of a sine or cosine curve is given.

(a) Find the amplitude, period, and phase shift.

(b) Write an equation that represents the curve in the form $y = a \sin k(x - b)$ or $y = a \cos k(x - b)$.

37.

38.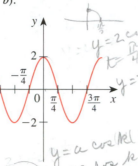

39.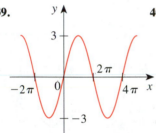

40.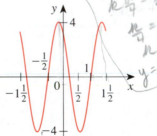

$3 \sin \frac{1}{2} x$

41.

42.

In Exercises 43–46 determine whether the function whose graph is shown is periodic. If so, determine its period from the graph.

43.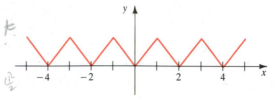

44.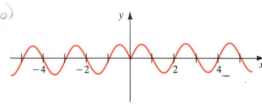

45.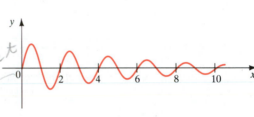

46.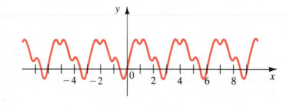

In Exercises 47–54 determine the period of the function.

47. $f(x) = 1 + \sin x$

48. $f(x) = 2 \sin x + 5 \cos x$

49. $f(x) = \sin^2 x$

50. $f(x) = \cos 3x$

51. $f(x) = e^{\sin x}$

52. $f(x) = 2^{\sin 2x}$

53. $f(x) = \tan x + \cot x$

54. $f(x) = \sin(\cos x)$

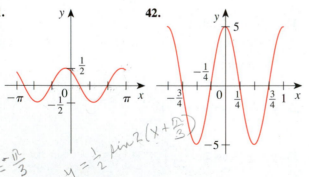

SECTION 5.4
MORE TRIGONOMETRIC GRAPHS

In this section we sketch the graphs of the tangent, cotangent, secant, and cosecant functions, and transformations of these functions.

GRAPHS OF THE TANGENT, COTANGENT, SECANT, AND COSECANT FUNCTIONS

We begin by stating the periodic properties of these functions. Recall that sine and cosine have period 2π. Since cosecant and secant are the reciprocals of sine and cosine, respectively, they also have period 2π (see Exercise 51). Tangent and cotangent, however, have period π (see Exercise 77(c) of Section 5.2).

PERIODIC PROPERTIES

The functions tangent and cotangent have period π:

$$\tan(x + \pi) = \tan x \quad \text{and} \quad \cot(x + \pi) = \cot x$$

The functions cosecant and secant have period 2π:

$$\csc(x + 2\pi) = \csc x \quad \text{and} \quad \sec(x + 2\pi) = \sec x$$

x	$\tan x$
0	0
$\dfrac{\pi}{6}$	0.58
$\dfrac{\pi}{4}$	1
$\dfrac{\pi}{3}$	1.73
1.4	5.80
1.5	14.10
1.55	48.08
1.57	1255.77
1.5707	10,381.33

We first sketch the graph of tangent. Since it has period π, we need only sketch the graph on any interval of length π, and then repeat the pattern to the left and to the right. We sketch the graph on the interval $(-\pi/2, \pi/2)$. Since $\tan \pi/2$ and $\tan(-\pi/2)$ are not defined, we need to be careful in sketching the graph at points near $\pi/2$ and $-\pi/2$. As x gets near $\pi/2$ through values less than $\pi/2$, the value of $\tan x$ becomes large. To see this, notice that as x gets close to $\pi/2$, $\cos x$ approaches 0 and $\sin x$ approaches 1 and so $\tan x = \sin x/\cos x$ is large. A table of values of $\tan x$ for x close to $\pi/2$ (≈ 1.570796) is shown in the margin.

Thus, by choosing x close enough to $\pi/2$ through values less than $\pi/2$, we can make the value of $\tan x$ larger than any given positive number. In a similar way, by choosing x close to $-\pi/2$ through values greater than $-\pi/2$, we can make $\tan x$ smaller than any given negative number. In the notation of Section 3.9 we have

$$\tan x \to \infty \quad \text{as} \quad x \to \frac{\pi}{2}^{+}$$

$$\tan x \to -\infty \quad \text{as} \quad x \to \frac{\pi}{2}^{-}$$

Thus $x = \pi/2$ and $x = -\pi/2$ are vertical asymptotes (see Section 3.9). With the information we have so far we sketch the graph of $\tan x$ for $-\pi/2 < x < \pi/2$ in

Figure 1. The complete graph of tangent (see Figure 5) is now obtained using the fact that tangent is periodic with period π.

The graph of $y = \cot x$ is sketched on the interval $[0, \pi]$ by a similar analysis (see Figure 2). Since $\cot x$ is undefined for $x = n\pi$, with n an integer, its complete graph (in Figure 5) has vertical asymptotes at these values.

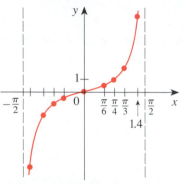

Figure 1
One period of $y = \tan x$

Figure 2
One period of $y = \cot x$

To sketch the graphs of the cosecant and secant functions we use the reciprocal identities

$$\csc x = \frac{1}{\sin x} \qquad \text{and} \qquad \sec x = \frac{1}{\cos x}$$

So, to graph $y = \csc x$ we take the reciprocals of the y-coordinates of the points of the graph of $y = \sin x$ (see Figure 3). Similarly, to graph $y = \sec x$ we take the reciprocals of the y-coordinates of the points of the graph of $y = \cos x$ (see Figure 4).

Let us consider more closely the graph of the function $y = \csc x$ on the interval $0 < x < \pi$. We need to examine the values of the function near 0 and π, since at

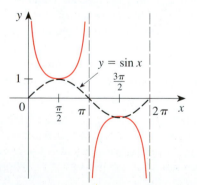

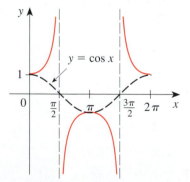

Figure 3
One period of $y = \csc x$

Figure 4
One period of $y = \sec x$

these values $\sin x = 0$ and thus $\csc x$ is undefined. We see that

$$\csc x \to \infty \quad \text{as} \quad x \to 0^+$$

$$\csc x \to \infty \quad \text{as} \quad x \to \pi^-$$

Thus the lines $x = 0$ and $x = \pi$ are vertical asymptotes. The graph in the interval $\pi < x < 2\pi$ is sketched in the same way. The values of $\csc x$ in that interval are the same as those in the interval $0 < x < \pi$ except for sign (see Figure 3). The complete graph in Figure 5 is now obtained from the fact that the function cosecant is periodic with period 2π. Note that the graph has vertical asymptotes at the points where $\sin x = 0$, that is, at $x = n\pi$, for n an integer.

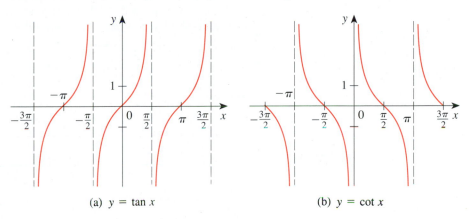

(a) $y = \tan x$ (b) $y = \cot x$

Figure 5

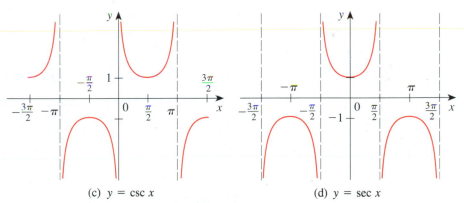

(c) $y = \csc x$ (d) $y = \sec x$

    The graph of $y = \sec x$ is sketched in a similar manner. Observe that the domain of $\sec x$ is the set of all real numbers other than $x = (\pi/2) + n\pi$, for n an integer, so the graph has vertical asymptotes at those points.

    It is apparent that the graphs of $y = \tan x$, $y = \cot x$, and $y = \csc x$ are symmetric about the origin, whereas that of $y = \sec x$ is symmetric about the y-axis. This is because tangent, cotangent, and cosecant are odd functions, whereas secant is an even function.

■ GRAPHS OF FUNCTIONS INVOLVING TANGENT AND COTANGENT

We now consider graphs of transformations of the tangent and cotangent functions.

EXAMPLE 1

Sketch the graph of (a) $y = 2 \tan x$ and (b) $y = -\tan x$.

SOLUTION

We first sketch the graph of $y = \tan x$.

(a) To graph $y = 2 \tan x$ we multiply the y-coordinate of each point on the graph of $y = \tan x$ by 2. The resulting graph is sketched in Figure 6(a).

(b) The graph of $y = -\tan x$ in Figure 6(b) is obtained from that of $y = \tan x$ by reflecting in the x-axis.

Figure 6

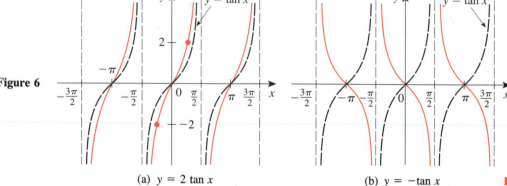

(a) $y = 2 \tan x$ (b) $y = -\tan x$ ■

Since the tangent and cotangent functions have period π, the functions

$$y = a \tan kx \qquad \text{and} \qquad y = a \cot kx \qquad (k > 0)$$

complete one period as kx varies from 0 to π, that is, for $0 \le kx \le \pi$ or for $0 \le x \le \pi/k$. So they have period π/k.

TANGENT AND COTANGENT CURVES

> The functions
>
> $$y = a \tan kx \qquad \text{and} \qquad y = a \cot kx \qquad (k > 0)$$
>
> have period π/k.

Thus one complete period of the graphs of these functions occurs on any interval of length π/k. To sketch a complete period of these graphs it is convenient to select an interval between vertical asymptotes. Thus:

To graph one period of $y = a \tan kx$, an appropriate interval is $(-\pi/(2k),\ \pi/(2k))$.

To graph one period of $y = a \cot kx$, an appropriate interval is $(0,\ \pi/k)$.

EXAMPLE 2

Sketch the graphs of (a) $y = \tan 2x$ and (b) $y = \tan 2\left(x - \dfrac{\pi}{4}\right)$.

SOLUTION

(a) The period is $\pi/2$ and an appropriate interval is $(-\pi/4,\ \pi/4)$. The end-points $x = -\pi/4$ and $x = \pi/4$ are vertical asymptotes. Thus we sketch on $(-\pi/4,\ \pi/4)$ one complete period of the function. The graph has the same shape as that of the tangent function, but is compressed horizontally by a factor of 2. We then repeat that portion of the graph to the left and to the right. See Figure 7(a).

(b) The graph is the same as that in part (a), but is shifted $\pi/4$ unit to the right, as shown in Figure 7(b).

Figure 7

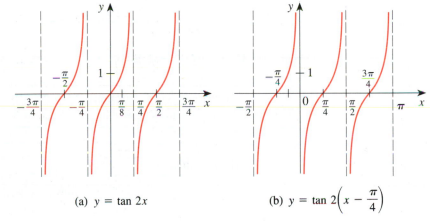

(a) $y = \tan 2x$ (b) $y = \tan 2\left(x - \dfrac{\pi}{4}\right)$ ■

EXAMPLE 3

Sketch the graph of $y = 2 \cot\left(3x - \dfrac{\pi}{2}\right)$.

SOLUTION

We first put this in the form $y = a \cot k(x - b)$ by factoring 3 from the expression $3x - \dfrac{\pi}{2}$. Thus

$$y = 2 \cot\left(3x - \frac{\pi}{2}\right) = 2 \cot 3\left(x - \frac{\pi}{6}\right)$$

Thus the graph is the same as that of $y = 2 \cot 3x$, but is shifted $\pi/6$ unit to the right. The period of $y = 2 \cot 3x$ is $\pi/3$ and an appropriate interval is $(0, \pi/3)$. To get the corresponding interval for the desired graph we shift this interval to the right by $\pi/6$. This gives

$$\left(0 + \frac{\pi}{6}, \frac{\pi}{3} + \frac{\pi}{6}\right) = \left(\frac{\pi}{6}, \frac{\pi}{2}\right)$$

Finally we graph one period in the shape of cotangent on the interval $(\pi/6, \pi/2)$ and repeat that portion of the graph to the left and to the right. See Figure 8.

Figure 8

$y = 2 \cot\left(3x - \dfrac{\pi}{2}\right)$

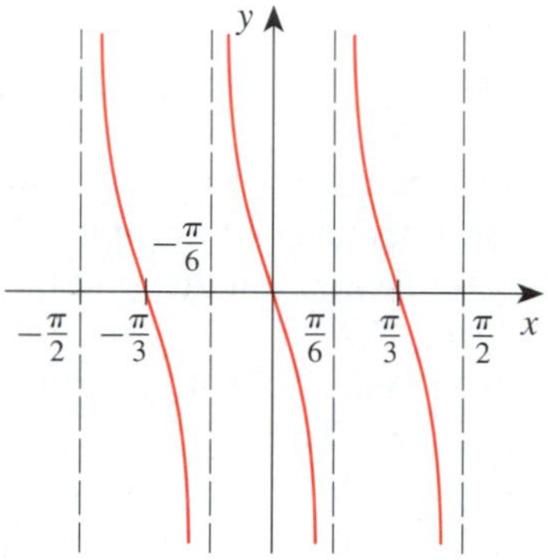

GRAPHS INVOLVING THE SECANT AND COSECANT FUNCTIONS

We have already observed that the cosecant and secant functions are the reciprocals of the sine and cosine functions. Thus the following result is the counterpart of the result for sine and cosine curves in Section 5.3.

COSECANT AND SECANT CURVES

The functions

$$y = a \csc kx \qquad \text{and} \qquad y = a \sec kx \qquad (k > 0)$$

have period $2\pi/k$.

An appropriate interval on which to sketch one complete period is $[0, 2\pi/k]$.

EXAMPLE 4

Sketch the graph of (a) $y = \frac{1}{2} \csc 2x$ and (b) $y = \frac{1}{2} \csc\left(2x + \dfrac{\pi}{2}\right)$.

SOLUTION

(a) The period is $2\pi/2 = \pi$. An appropriate interval is $[0, \pi]$ and the asymptotes occur in this interval whenever $\sin 2x = 0$. So the asymptotes in this interval are $x = 0$, $x = \pi/2$, and $x = \pi$. With this information we sketch on the interval $[0, \pi]$ a graph with the same general shape as that of one period of the cosecant function. The complete graph in Figure 9(a) is obtained by repeating this portion of the graph to the left and to the right.

Figure 9

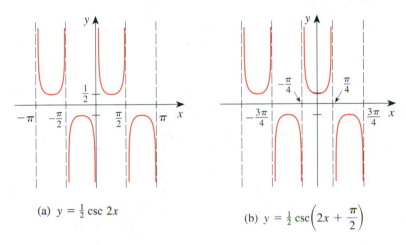

(a) $y = \frac{1}{2} \csc 2x$

(b) $y = \frac{1}{2} \csc\left(2x + \dfrac{\pi}{2}\right)$

(b) We first write

$$y = \frac{1}{2} \csc\left(2x + \frac{\pi}{2}\right)$$

$$= \frac{1}{2} \csc 2\left(x + \frac{\pi}{4}\right)$$

From this we see that the graph is the same as that in part (a), shifted $\pi/4$ unit to the left. The graph is sketched in Figure 9(b). ∎

EXAMPLE 5

Sketch the graph of $y = 3 \sec \frac{1}{2}x$.

SOLUTION

The period is $2\pi \div \frac{1}{2} = 4\pi$. An appropriate interval is $[0, 4\pi]$ and the asymptotes occur in this interval wherever $\cos \frac{1}{2}x = 0$. Thus the asymptotes in this interval are $x = \pi$, $x = 3\pi$. With this information we sketch on the interval $[0, 4\pi]$ a graph with the same general shape as that of one period of the secant function. The complete graph in Figure 10 is obtained by repeating this portion of the graph to the left and to the right. ∎

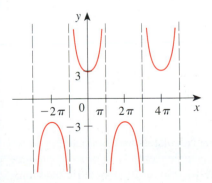

Figure 10
$y = 3 \sec \frac{1}{2}x$

EXERCISES 5.4

In Exercises 1–50 find the period and sketch the graph of the given function.

1. $y = 3 \tan x$

2. $y = -3 \tan x$

3. $y = \frac{1}{2} \tan x$

4. $y = -\frac{1}{2} \tan x$

5. $y = 4 \cot x$

6. $y = \frac{1}{4} \cot x$

7. $y = 2 \csc x$

8. $y = \frac{1}{2} \csc x$

9. $y = 4 \sec x$

10. $y = \frac{1}{4} \sec x$

11. $y = \tan\left(x + \frac{\pi}{2}\right)$

12. $y = \tan\left(x - \frac{\pi}{4}\right)$

13. $y = \csc\left(x + \frac{\pi}{2}\right)$

14. $y = \sec\left(x - \frac{\pi}{4}\right)$

15. $y = \cot\left(x + \frac{\pi}{4}\right)$

16. $y = 2 \csc\left(x - \frac{\pi}{3}\right)$

17. $y = \frac{1}{2} \sec\left(x - \frac{\pi}{6}\right)$

18. $y = 3 \csc\left(x - \frac{\pi}{2}\right)$

19. $y = \tan 2x$

20. $y = \tan \frac{1}{2}x$

21. $y = \tan \pi x$

22. $y = \cot \frac{\pi}{2}x$

23. $y = \sec 2x$

24. $y = 5 \csc 3x$

25. $y = \csc 2x$

26. $y = \csc \frac{1}{2}x$

27. $y = 2 \tan 3x$

28. $y = 2 \tan \frac{\pi}{2}x$

29. $y = 5 \csc 3x$

30. $y = 5 \sec 2\pi x$

31. $y = \tan 2\left(x + \frac{\pi}{2}\right)$

32. $y = \csc 2\left(x + \frac{\pi}{2}\right)$

33. $y = \tan 2(x - \pi)$

34. $y = \sec 2\left(x - \frac{\pi}{2}\right)$

35. $y = \cot\left(2x - \frac{\pi}{2}\right)$

36. $y = \frac{1}{2} \tan(\pi x - \pi)$

37. $y = 2 \csc\left(\pi x - \frac{\pi}{3}\right)$

38. $y = 2 \sec\left(\frac{1}{2}x - \frac{\pi}{3}\right)$

39. $y = 2 \csc\left(x - \frac{\pi}{2}\right)$

40. $y = 4 \csc \frac{1}{3}\left(x + \frac{\pi}{2}\right)$

41. $y = 5 \sec\left(3x - \frac{\pi}{2}\right)$

42. $y = \frac{1}{2} \sec(2\pi x - \pi)$

43. $y = \tan\left(\frac{2}{3}x - \frac{\pi}{6}\right)$

44. $y = \tan \frac{1}{2}\left(x + \frac{\pi}{4}\right)$

45. $y = 3 \sec \pi\left(x + \frac{1}{2}\right)$

46. $y = \sec\left(3x + \frac{\pi}{2}\right)$

47. $y = -2 \tan\left(2x - \frac{\pi}{3}\right)$

48. $y = 2 \csc(3x + 3)$

49. $y = \sec(3x + \pi)$

50. $y = \tan\left(\frac{1}{2}x - \frac{\pi}{2}\right)$

51. (a) Prove that if f is periodic with period p, then $1/f$ is also periodic with period p.
   (b) Prove that cosecant and secant have period 2π.

SECTION 5.5
USING GRAPHING DEVICES IN TRIGONOMETRY

In Section 2.3 we discussed the use of graphing calculators and computers, and we saw that it is important to choose a viewing rectangle carefully in order to produce a reasonable graph of a function. This is especially true for trigonometric functions; Example 1 shows that, if care is not taken, it is easy to produce a very misleading graph of a trigonometric function.

EXAMPLE 1

Graph the function $f(x) = \sin 50x$ in an appropriate viewing rectangle.

SOLUTION

Figure 1(a) shows the graph of f produced by a graphing calculator using the viewing rectangle $[-12, 12]$ by $[-1.5, 1.5]$. At first glance the graph appears to be reasonable. But if we change the viewing rectangle to the ones shown in the following parts of Figure 1, the graphs look very different. Something strange is happening.

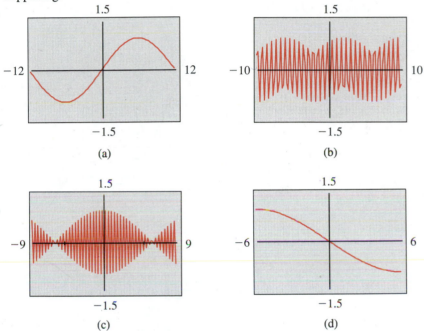

Figure 1

Graphs of $f(x) = \sin 50x$ in different viewing rectangles

In order to explain the big differences in appearance of these graphs and to find an appropriate viewing rectangle, we need to find the period of the function $y = \sin 50x$. From Section 5.3 we know that the period is

$$\frac{2\pi}{50} = \frac{\pi}{25} \approx 0.126$$

This suggests that we should deal only with small values of x in order to show just a few oscillations of the graph. If we choose the viewing rectangle $[-0.25, 0.25]$ by $[-1.5, 1.5]$, we get the graph shown in Figure 2.

Now we see what went wrong in Figure 1. The oscillations of $y = \sin 50x$ are so rapid that when the calculator plots points and joins them, it misses most of the maximum and minimum points and therefore gives a very misleading impression of the graph. ∎

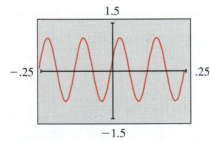

Figure 2

$f(x) = \sin 50x$

*The function in Example 2 is **periodic** with period 2π. In general, functions that are sums of functions from the following list*

$$1, \quad \cos kx, \quad \cos 2kx, \quad \cos 3kx, \ldots$$
$$\sin kx, \quad \sin 2kx, \quad \sin 3kx, \ldots$$

are periodic. Although these functions appear to be special, they are actually fundamental to describing all periodic functions that arise in practice. It is a remarkable fact, discovered by the French mathematician J. B. J. Fourier, that nearly every periodic function can be written as a sum (usually an infinite sum) of these functions (see Section 10.5). This is remarkable because it means that any situation where periodic variation occurs can be described mathematically using the functions sine and cosine. A modern application of Fourier's discovery is the digital encoding of sound on compact discs.

EXAMPLE 2

Draw the graphs of $f(x) = 2 \cos x$, $g(x) = \sin 2x$, and $h(x) = 2 \cos x + \sin 2x$ on a common screen to illustrate the method of graphical addition.

SOLUTION

Notice that $h = f + g$, so its graph is obtained by adding the corresponding y-coordinates of the graphs of f and g. The graphs of f, g, and h are shown in Figure 3.

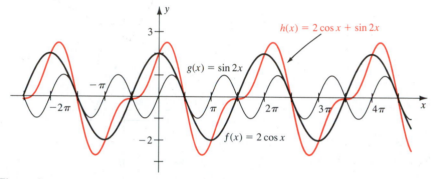

Figure 3

EXAMPLE 3

Find the maximum and minimum values of the function $h(x) = 2 \cos x + \sin 2x$ that was graphed in Example 2.

SOLUTION

Since h has period 2π, we first graph h in the viewing rectangle $[0, 6.28]$ by $[-3, 3]$ in Figure 4(a). We see from this graph that h attains its maximum value when $x \approx 0.5$ and its minimum value when $x \approx 2.6$. So we zoom in to the viewing rectangles shown in parts (b) and (c) of Figure 4 and we find that h has a maximum value of about 2.598 when $x \approx 0.52$ and a minimum value of about -2.598 when

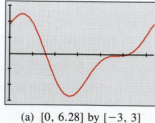
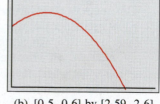
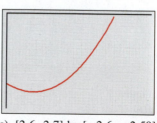

Figure 4
$y = 2 \cos x + \sin 2x$ (a) $[0, 6.28]$ by $[-3, 3]$ (b) $[0.5, 0.6]$ by $[2.59, 2.6]$ (c) $[2.6, 2.7]$ by $[-2.6, -2.59]$

$x \approx 2.62$. Because h is periodic, these maximum and minimum values are attained infinitely many times. ∎

EXAMPLE 4

Find the values of x in the interval $[-\pi/2, \pi/2]$ that satisfy the equation $\sin x = \frac{1}{3}$.

SOLUTION

We solve the equation graphically using the methods of Section 3.6. In Figure 5(a) we graph the curve $y = \sin x$ and the line $y = \frac{1}{3}$ in the viewing rectangle $[-1.57, 1.57]$ by $[-1.5, 1.5]$. The curves intersect when $x \approx 0.3$, so we zoom in to the viewing rectangle shown in Figure 5(b). From this diagram we see that, correct to two decimal places, the solution of the equation is $x \approx 0.34$.

Figure 5

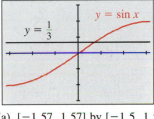

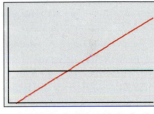

(a) $[-1.57, 1.57]$ by $[-1.5, 1.5]$ (b) $[0.3, 0.4]$ by $[0.3, 0.4]$ ∎

EXAMPLE 5

Draw the graphs of the functions $y = e^{-x}$, $y = -e^{-x}$, and $y = e^{-x} \cos 6\pi x$ on a common screen. Comment on and explain the relationships among the graphs.

SOLUTION

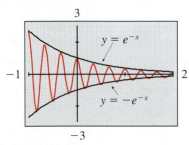

Figure 6
$y = e^{-x} \cos 6\pi x$

Figure 6 shows all three graphs in the viewing rectangle $[-1, 2]$ by $[-3, 3]$. It appears that the graph of $y = e^{-x} \cos 6\pi x$ lies between the graphs of the exponential functions $y = e^{-x}$ and $y = -e^{-x}$.

To explain this we recall that the values of $\cos 6\pi x$ are between -1 and 1, that is,

$$-1 \leq \cos 6\pi x \leq 1$$

for all values of x. Multiplying the inequalities by e^{-x}, and noting that $e^{-x} \geq 0$, we get

$$-e^{-x} \leq e^{-x} \cos 6\pi x \leq e^{-x}$$

This explains why the exponential functions form a boundary for the graph of $y = e^{-x} \cos 6\pi x$. (Note that the graphs touch when $\cos 6\pi x = \pm 1$.) ∎

Jean Baptiste Joseph Fourier (1768–1830) is responsible for the most powerful application of the trigonometric functions (see marginal note on page 346). He used sums of these functions to describe such physical phenomena as the transmission of sound and the flow of heat.

Fourier was orphaned as a young boy and was educated in a military school, where he became a mathematics teacher at the age of 20. He later was appointed professor at the École Polytechnique but resigned this position to accompany Napoleon on his expedition to Egypt, where Fourier served as governor. After his return to France he began conducting experiments on heat. The French Academy refused to publish his early papers on this subject due to his lack of rigor. Fourier was later to become Secretary of the Academy and in this capacity had his papers published in their original form. Probably because of his study of heat and his years in the deserts of Egypt, Fourier became obsessed with keeping himself warm. He wore several layers of clothes, even in the summer, and kept his rooms at unbearably high temperatures. Evidently, this lifestyle overburdened his heart and contributed to his death at the age of 63.

Example 5 shows that the function $y = e^{-x}$ controls the amplitude of the graph of $y = e^{-x} \cos 6\pi x$. In general, if $f(x) = a(x) \sin kx$ or $a(x) \cos kx$, the function a determines how the amplitude of f varies, and the graph of f lies between the graphs of $y = -a(x)$ and $y = a(x)$. Here is another example.

EXAMPLE 6

Graph the function $f(x) = \cos 2\pi x \cos 16\pi x$.

SOLUTION

The graph is shown in Figure 7. Although it was drawn by a computer, we could have drawn it by hand, by first sketching the boundary curves $y = \cos 2\pi x$ and $y = -\cos 2\pi x$. The graph of f is a cosine curve that lies between the graphs of these two functions.

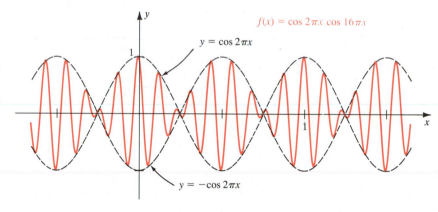

Figure 7

EXAMPLE 7

The function $f(t) = \dfrac{\sin t}{t}$ is very important in calculus. Graph this function and comment on its behavior when t is close to 0.

SOLUTION

The viewing rectangle $[-15, 15]$ by $[-0.5, 1.5]$ shown in Figure 8(a) gives a good global view of the graph of f. The viewing rectangle $[-1, 1]$ by $[-0.5, 1.5]$ in Figure 8(b) focuses on the behavior of f when $t \approx 0$. Notice that although $f(t)$ is not defined when $t = 0$ (in other words, 0 is not in the domain of f), the values of f seem to approach 1 when t gets close to 0. This fact is crucial in calculus.

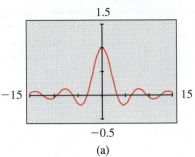

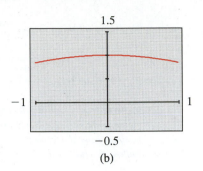

Figure 8

$$f(t) = \frac{\sin t}{t}$$

(a)   (b)

The function in Example 7 can be written as

$$f(t) = \left(\frac{1}{t}\right) \cdot \sin t$$

and may thus be viewed as a sine function whose amplitude is controlled by the function $a(t) = 1/t$.

EXERCISES 5.5

In Exercises 1–8 determine an appropriate viewing rectangle for the given function, and use it to draw the graph.

1. $f(x) = \cos 100x$ **2.** $f(x) = 3 \sin 120x$

3. $f(x) = \sin(x/40)$ **4.** $f(x) = \cos(x/80)$

5. $y = \tan 25x$ **6.** $y = \csc 40x$

7. $y = e^{\sin 20x}$ **8.** $y = \sqrt{\tan 10\pi x}$

In Exercises 9–12 draw the graphs of f, g, and f + g on a common screen to illustrate graphical addition.

9. $f(x) = x,\quad g(x) = \sin x$

10. $f(x) = x,\quad g(x) = \cos x$

11. $f(x) = \sin x,\quad g(x) = \sin 2x$

12. $f(x) = x^2,\quad g(x) = \sin 2x$

In Exercises 13–18 find the maximum and minimum values of the function.

13. $y = \sin x + \sin 2x$ **14.** $y = x^2 + \sin 2x$

15. $y = x - 2 \sin x,\ 0 \leqslant x \leqslant 2\pi$

16. $y = \sin x + \cos x$

17. $y = 2 \sin x + \sin^2 x$

18. $y = \dfrac{\cos x}{2 + \sin x}$

In Exercises 19–24 find all solutions of the given equation that lie in the interval $[0, \pi]$. State each answer correct to two decimal places.

19. $\cos x = 0.4$ **20.** $\tan x = 2$

21. $\csc x = 3$ **22.** $\cos x = x$

23. $2 \sin x = x$ **24.** $\cot x = \ln x$

In Exercises 25–30 graph the three functions on a common screen. How are the graphs related?

25. $y = x^2,\quad y = -x^2,\quad y = x^2 \sin x$

26. $y = x,\quad y = -x,\quad y = x \cos x$

27. $y = e^x,\quad y = -e^x,\quad y = e^x \sin 5\pi x$

28. $y = \dfrac{1}{1 + x^2}$, $y = -\dfrac{1}{1 + x^2}$, $y = \dfrac{\cos 2\pi x}{1 + x^2}$

29. $y = \cos 3\pi x$, $y = -\cos 3\pi x$, $y = \cos 3\pi x \cos 21\pi x$

30. $y = \sin 2\pi x$, $y = -\sin 2\pi x$, $y = \sin 2\pi x \sin 10\pi x$

In Exercises 31–36, (a) use a graphing device to graph the function; (b) determine from the graph whether the function is periodic and, if so, determine the period; and (c) determine from the graph whether the function is odd, even, or neither.

31. $y = |\sin x|$

32. $y = \sin |x|$

33. $y = e^{\sin x}$

34. $y = 2^{\cos x}$

35. $y = \sin^2 x$

36. $y = \sin(x^2)$

37. Let $f(t) = \dfrac{1 - \cos t}{t}$.

  (a) Is the function f even, odd, or neither?

  (b) Find the t-intercepts of the graph of f.

  (c) Draw the graph of f in an appropriate viewing rectangle.

  (d) Describe the behavior of the function as $t \to \infty$ and as $t \to -\infty$.

  (e) Notice that $f(t)$ is not defined when $t = 0$. What happens as $t \to 0$?

38. This exercise explores the effect of the inner function g on a composite function $y = f(g(x))$.

  (a) Graph the function $y = \sin(\sqrt{x})$ using the viewing rectangle [0, 400] by [−1.5, 1.5]. How does this graph differ from the graph of the sine function?

  (b) Graph the function $y = \sin(x^2)$ using the viewing rectangle [−5, 5] by [−1.5, 1.5]. How does this graph differ from the graph of the sine function?

SECTION 5.6
HARMONIC MOTION

Motion that is caused by vibration or oscillation is common in nature. A weight suspended from a spring that has been compressed and then allowed to vibrate vertically is a simple example. This same "back and forth" motion also occurs in such diverse phenomena as sound waves, light waves, alternating electrical currents, pulsating stars, and nuclear magnetic resonance, to name a few. All these phenomena exhibit a kind of motion called *harmonic motion*.

The trigonometric functions are ideal for mathematically describing oscillatory motion. It is not hard to see why this is so. A glance at the graphs of the sine and cosine functions shows that these functions themselves exhibit oscillatory behavior. Figure 1 is the graph of $y = \sin t$. If we think of t as time, we see that as time

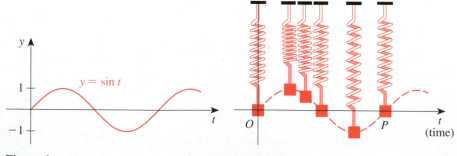

Figure 1
$y = \sin t$

Figure 2

increases, $y = \sin t$ moves up and down through values between -1 and 1. For instance, the variation of the sine function is mimicked in the motion of a vibrating mass on a spring (Figure 2). In the context of physical applications, the term **cycle** refers to one complete vibration of an object. In Figure 2 the mass completes the cycle of its motion between points O and P.

SIMPLE HARMONIC MOTION

If the equation describing the displacement y of an object at time t is

$$y = a \sin \omega t \qquad \text{or} \qquad y = a \cos \omega t$$

then the object is in **simple harmonic motion**. In this case

    amplitude $= |a|$ (the maximum displacement of the object)
    period $= 2\pi/\omega$ (time required to complete one cycle)
    frequency $= \omega/(2\pi)$ (number of cycles per unit of time)

Notice that the functions

$$y = a \sin 2\pi\nu t \qquad \text{and} \qquad y = a \cos 2\pi\nu t$$

have frequency ν, because $2\pi\nu/(2\pi) = \nu$. Since we can immediately read the frequency from these equations, we often write equations of simple harmonic motion in this form.

EXAMPLE 1

The displacement of the mass in Figure 3 is given by

$$y = 10 \sin 4\pi t$$

where y is measured in inches and t in seconds. Describe the motion of the mass. (We assume here that there is no friction, so the spring and weight will oscillate in the same way indefinitely.)

SOLUTION

The mass is in simple harmonic motion. The amplitude, period, and frequency of the motion are

$$\text{amplitude} = |a| = 10$$

$$\text{period} = \frac{2\pi}{\omega} = \frac{2\pi}{4\pi} = \frac{1}{2}$$

$$\text{frequency} = \frac{4\pi}{2\pi} = 2$$

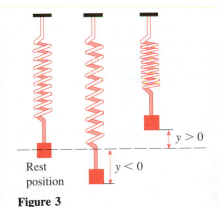

$y > 0$

$y < 0$

Rest position

Figure 3

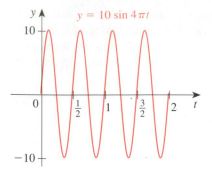

Figure 4

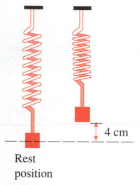

Rest position

Figure 5

Thus the maximum displacement of the mass is 10 cm, it takes $\frac{1}{2}$ s to complete one cycle, and it goes through two complete cycles every second. At $t = 0$, the displacement is $y = 10 \sin 4\pi 0 = 0$, so the mass is initially at its rest position. A nice description of the motion is provided by the graph of its displacement at time t in Figure 4. Notice that y is the displacement of the mass above its rest position. So when $y > 0$, the mass is above its rest position and when $y < 0$, it is below its rest position.   ■

The main difference between the two equations describing simple harmonic motion is the starting point. At $t = 0$ we get

$$y = a \sin \omega \cdot 0 = 0 \qquad \text{or} \qquad y = a \cos \omega \cdot 0 = a$$

In other words, in the first case the motion "starts" with zero displacement, while in the second case the motion "starts" with the displacement at maximum (at the amplitude a). Of course, we may think of the motion as "starting" at any point in its cycle. In this case the equations for simple harmonic motion take the form

$$y = a \sin \omega(t + \theta) \qquad \text{and} \qquad y = a \cos \omega(t + \theta)$$

where θ determines the starting point.

EXAMPLE 2

A mass is suspended from a spring as shown in Figure 5. The spring is compressed a distance of 4 cm and then released. It is observed that the mass returns to the compressed position after $\frac{1}{3}$ s. Describe the motion of the mass.

SOLUTION

The motion of the mass is given by one of the equations for simple harmonic motion. The amplitude of the motion is 4 cm. Since this amplitude is reached at time $t = 0$ when the mass is released, the appropriate equation to use is

$$y = a \cos \omega t$$

The period $p = \frac{1}{3}$, so the frequency is $1/p = 3$. The motion of the mass is described by the equation

$$y = 4 \cos(2\pi)3t$$
$$y = 4 \cos 6\pi t$$

where y is the displacement from the rest position at time t. Notice that when $t = 0$, the displacement is $y = 4$, as we expect.   ■

An important situation where simple harmonic motion occurs is in the production of sound. Sound is produced by a regular variation in air pressure from the normal pressure. If the pressure varies in simple harmonic motion, then a pure sound is produced. The tone of the sound depends on the frequency and the loudness on the amplitude.

EXAMPLE 3

A tuba player plays the note E and sustains the sound for some time. Assuming that the sound is a pure E, its equation is given by

$$V(t) = 0.2 \sin 80\pi t$$

where $V(t)$ is the variation in pressure from normal pressure at time t, measured in lb/in². Find the frequency, amplitude, and period of the sound. If the tuba player increases the loudness of the note, how does the equation for $V(t)$ change? If the player is playing the note incorrectly and it is a little flat, how does this affect $V(t)$?

$y = 0.2 \sin 80\pi t$

Figure 6

SOLUTION

The frequency is $80\pi/(2\pi) = 40$, the period is $\frac{1}{40}$, and the amplitude is 0.2. If the player increases the loudness, the amplitude of the wave increases. In other words, the number 0.2 is replaced by a larger number. If the note is flat, then the frequency is less than 40. In this case the coefficient of t is less than 80π. ■

EXAMPLE 4

A variable star is one whose brightness alternately increases and decreases. For the most visible variable star, Delta Cephei, the time between periods of maximum brightness is 5.4 days. The average brightness (or magnitude) of the star is 4.0 and its brightness varies by ± 0.35 magnitude. Express the brightness as a function of time.

SOLUTION

The brightness increases and decreases from the average brightness in simple harmonic fashion. Let us measure the time in days, taking $t = 0$ to be a time when the star is at its average brightness. Since the period of the star is 5.4 days,

$$\frac{2\pi}{\omega} = 5.4 \qquad \text{or} \qquad \omega = \frac{2\pi}{5.4}$$

Thus the variation $V(t)$ from average brightness is

$$V(t) = a \sin \omega t = 0.35 \sin \frac{2\pi}{5.4} t$$

The graph of $V(t)$ in Figure 7 shows that at time $t = 1.35$ days, the star is at maximum brightness (0.35 magnitude above its average brightness) and at $t = 4.05$

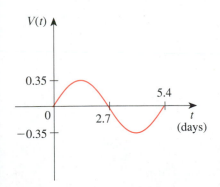

Figure 7
Variation from average brightness

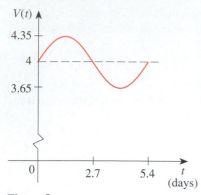

Figure 8
Brightness

days, it is at its minimum brightness (0.35 magnitude below its average brightness). The brightness $B(t)$ at time t is given by

$$B(t) = 4.0 + 0.35 \sin \frac{2\pi}{5.4}t$$

The graph of $B(t)$ is sketched in Figure 8.   ■

Another situation where simple harmonic motion occurs is in alternating current (AC) generators. Alternating current is produced when an armature rotates about its axis in a magnetic field.

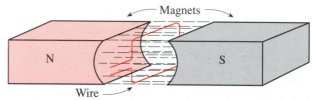

Figure 9

Figure 9 represents a simple version of such a generator. As the wire passes through the magnetic field, a voltage E is generated in the wire. It can be shown that the voltage generated is given by

$$E(t) = E_0 \cos \omega t$$

where E_0 is the maximum voltage produced (which depends on the strength of the magnetic field) and $\omega/(2\pi)$ is the number of revolutions per second of the armature (the frequency).

EXAMPLE 5

Ordinary 110-V household alternating current varies from $+155$ V to -155 V with a frequency of 60 Hz (cycles per second). Find an equation that describes this variation in voltage.

SOLUTION

The variation in voltage is simple harmonic. Since the frequency is 60 cycles per second, we have

$$\frac{\omega}{2\pi} = 60 \qquad \text{or} \qquad \omega = 120\pi$$

Let us take $t = 0$ to be a time when the voltage is $+155$ V. Then

$$E(t) = a \cos \omega t = 155 \cos 120\pi t$$ ■

There are numerous other applications of simple harmonic motion. Some of these are described in Exercises 5.6.

Why do we say that household current is 110 V when the maximum voltage produced is 155 V? From the symmetry of the cosine function, we see that the average voltage produced is zero. This average value would be the same for all AC generators and so gives no information about the voltage generated. To obtain a more informative measure of voltage, engineers use the **root-mean-square** *(**RMS**) method. It can be shown that the RMS voltage is $1/\sqrt{2}$ times the maximum voltage. So, for household current the RMS voltage is*

$$155 \times \frac{1}{\sqrt{2}} \approx 110 \text{ V}$$

■ DAMPED HARMONIC MOTION

The spring in Figure 2 is assumed to oscillate in a frictionless environment. In this hypothetical case the amplitude of the oscillation will not change. In the presence of friction, however, the motion of the spring will eventually "die down," that is, the amplitude of the motion will decrease with time. Motion of this type is called *damped harmonic motion*.

DAMPED HARMONIC MOTION

> If the equation describing the displacement y of an object at time t is
>
> $$y = ke^{-ct} \sin \omega t \qquad \text{or} \qquad y = ke^{-ct} \cos \omega t$$
>
> then the object is in **damped harmonic motion**. The constant c is the **damping constant**.

Damped harmonic motion is simply harmonic motion for which the amplitude is governed by the function $a(t) = ke^{-ct}$. Figure 10 shows the difference between harmonic motion and damped harmonic motion.

Figure 10

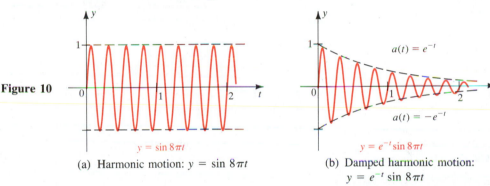

(a) Harmonic motion: $y = \sin 8\pi t$

(b) Damped harmonic motion: $y = e^{-t} \sin 8\pi t$

The larger the damping constant c, the quicker the oscillation dies down. When a guitar string is plucked and then allowed to vibrate freely, a point on that string undergoes damped harmonic motion. We can hear the damping of the motion as the sound produced by the vibration of the string fades. How fast the damping of the string occurs (as measured by the size of the constant c) is a property of the size of the string and the material it is made of. Another example of damped harmonic motion is the motion that a shock absorber on a car undergoes when the car hits a bump in the road. In this case the shock absorber is engineered to damp the motion as quickly as possible (large c) and to have the frequency as small as possible (small ω). On the other hand, the sound produced by a tuba player playing a note, say C, is undamped as long as the player can maintain the loudness of the note. The electromagnetic waves that produce light move in simple harmonic motion that is not damped.

AM and FM Radio

Radio transmissions consist of sound waves superimposed on a harmonic electromagnetic wave form called the carrier signal.

Sound wave

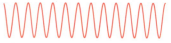

Carrier signal

*There are two types of radio transmission, called **amplitude modulation** (**AM**) and **frequency modulation** (**FM**). In AM broadcasting the sound wave changes, or **modulates**, the amplitude of the carrier, but the frequency remains unchanged.*

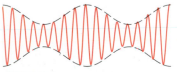

AM signal

In FM broadcasting the sound wave modulates the frequency, but the amplitude remains the same.

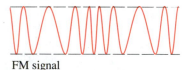

FM signal

EXAMPLE 6

The G-string on a violin is pulled a distance of 0.5 cm above its rest position, then released and allowed to vibrate. The damping constant c for this string is determined to be 1.4. Suppose that the note produced is a pure G (frequency = 200 cycles per second). Find the equation that describes the motion of the point at which the string was plucked.

SOLUTION

Let P be the point at which the string was plucked. We will find a function $f(t)$ that gives the distance at time t of the point P from its original rest position. Since the maximum displacement occurs at $t = 0$, we find an equation in the form

$$y = ke^{-ct} \cos \omega t$$

From this equation, we see that $f(0) = k$. But we know that the original displacement of the string is 0.5 cm. Thus $k = 0.5$. Since the frequency of the vibration is 200, we have $\omega/(2\pi) = 200$ or $\omega = (200)(2\pi)$. Finally, since we know that the damping constant is 1.4, we get

$$f(t) = 0.5e^{-1.4t} \cos 400\pi t \qquad \blacksquare$$

EXAMPLE 7

A stone is dropped in a calm lake, causing waves to form. The up-and-down motion of a point on the surface of the water is damped harmonic motion. At some time the amplitude of the wave is measured, and 20 s later it is found that the amplitude had dropped to $\frac{1}{10}$ of its value. Find the damping constant c.

SOLUTION

The amplitude is governed by the coefficient ke^{-ct} in the equations for damped harmonic motion. Thus the amplitude at time t is ke^{-ct}, and 20 s later, it is $ke^{-c(t+20)}$. So, from the given information

$$\frac{ke^{-ct}}{ke^{-c(t+20)}} = 10$$

We now solve this equation for c. Canceling k and using the Laws of Exponents, we get

$$e^{-ct+c(t+20)} = 10$$

$$e^{-ct+ct+20c} = e^{20c} = 10$$

Taking the natural logarithm of each side gives

$$20c = \ln(10)$$
$$c = \tfrac{1}{20} \ln(10) \approx \tfrac{1}{20}(2.30) \approx 0.12$$

Thus the damping constant is $c \approx 0.12$. ■

EXERCISES 5.6

In Exercises 1–10 the given function specifies the position of an object that is moving along an axis in simple harmonic motion, where t is given in seconds and f(t) is in centimeters.
(a) Find the amplitude, period, and frequency of the motion.
(b) What is the position of the object at time t = 0?
(c) Sketch a graph of one complete period of the function.

1. $f(t) = 4 \sin 8\pi t$

2. $f(t) = 10 \sin \tfrac{1}{2}t$

3. $f(t) = 0.3 \cos 2t$

4. $f(t) = -2 \cos\left(\dfrac{3\pi}{2}t\right)$

5. $f(t) = 1000 \sin 2\pi t$

6. $f(t) = \tfrac{1}{20} \cos\left(\dfrac{\pi}{3}t\right)$

7. $f(t) = -\cos \pi\left(t + \tfrac{1}{2}\right)$

8. $f(t) = 2 \sin\left(\dfrac{\pi}{12}t\right)$

9. $f(t) = 8 \sin\left(3\pi t + \dfrac{\pi}{6}\right)$

10. $f(t) = 5 \cos\left(2t + \dfrac{\pi}{3}\right)$

11. A point P in simple harmonic motion completes two cycles every minute, and the amplitude of the motion is 6 ft. Find a function that describes the motion of P.

12. A point P is in simple harmonic motion with amplitude 10 cm and frequency 3 cycles per second. Find a function that describes the motion of the point P.

13. A point P is in simple harmonic motion with amplitude 2 m and period $\tfrac{1}{3}$ s. Find a function that describes the motion.

14. The graph gives the displacement $d(t)$ at time t of an object in simple harmonic motion. Find a formula for the function $d(t)$.

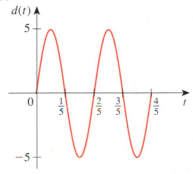

15. The graph shows the variation of the water level relative to mean sea level in the Tacoma harbor for a particular 24-hour period. Assuming that this variation is simple harmonic, find an equation of the form $y = a \sin \omega t$ that describes the variation in water level as a function of the number of hours after midnight.

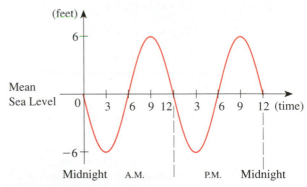

16. A mass suspended from a spring is oscillating with a frequency of 5 cycles per second. If the maximum displacement from rest position is 2 cm, find an equation that describes the motion.

17. A mass oscillating on a spring is observed to complete one cycle of its motion in $\frac{1}{10}$ s. It is also observed that the amplitude of the motion is 8 cm. Find an equation that describes the motion of this mass.

18. A mass suspended from a spring is pulled down a distance of 2 ft from its rest position. The mass is released at time $t = 0$ and allowed to oscillate. If the mass returns to this position after 1 s, find an equation that describes its motion.

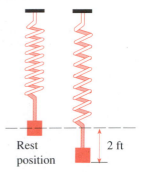

Rest position   2 ft

19. A mass is suspended on a spring. The spring is compressed so that the mass is 5 cm above its rest position. The mass is released at time $t = 0$ and allowed to oscillate. It is observed that the mass reaches its lowest point $\frac{1}{2}$ s after it is released. Find an equation that describes the motion of the mass.

20. The frequency of oscillation of an object suspended on a spring depends on the stiffness k of the spring (called the *spring constant*) and the mass m of the object. If the spring is compressed a distance a and then allowed to oscillate, its displacement is given by

$$f(t) = a \cos \sqrt{k/m}\, t$$

  (a) A 10-g mass is suspended from a spring with stiffness $k = 3$. If the spring is compressed a distance 5 cm and then released, find the equation that describes the oscillation of the spring.

  (b) Find a general formula for the frequency (in terms of k and m).

  (c) How is the frequency affected if the mass is increased? Is the oscillation faster or slower?

  (d) How is the frequency affected if a stiffer spring is used (larger k)? Is the oscillation faster or slower?

21. A ferris wheel has a radius of 10 m, and the bottom of the wheel is 1 m above the ground. If the ferris wheel makes one complete revolution every 20 s, find an equa-

tion that gives the height above the ground of a person on the ferris wheel as a function of time.

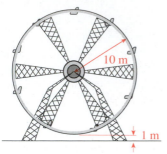

22. The Bay of Fundy in Nova Scotia has the highest tides in the world. In one section of the bay the water level rises to 21 ft above mean sea level and then drops 21 ft below mean sea level every 12-hour period. Assuming that the motion of the tides is simple harmonic, find an equation that describes the height of the tide in the Bay of Fundy above mean sea level. Sketch a graph that shows the level of the tides over a 24-hour period.

23. Normal 110 V household alternating current actually varies from -155 V to $+155$ V with a frequency of 60 Hz (60 cycles per second). Suppose that at time $t = 0$ the voltage produced is zero and is increasing. Write an equation that describes the variation in the voltage of household current with respect to time.

24. The graph shows an oscilloscope reading of the variation in voltage of an AC current produced by a simple generator.

  (a) Find the maximum voltage produced.

  (b) Find the frequency (cycles per second) of the generator.

  (c) How many revolutions per second does the armature in the generator make?

  (d) Find a formula that describes the variation in voltage as a function of time.

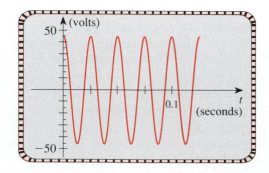

25. The armature in an electric generator is rotating at the rate of 100 revolutions per second (RPS). If the maximum voltage produced is 310 V (see the marginal note on page 354), find an equation that describe this variation in voltage. What is the RMS voltage?

26. The variable star Zeta Gemini has a period of 10 days. The average brightness of the star is 3.8 magnitudes and the maximum variation from the average is 0.2 magnitude. Assuming that the variation in brightness is simple harmonic, find an equation that gives the brightness of the star as a function of time.

27. Astronomers believe that the radius of a variable star increases and decreases with the brightness of the star. The variable star Delta Cephei (Example 4) has an average radius of 20 million miles and changes by a maximum of 1.5 million miles from this average during a single pulsation. Find an equation that describes the radius of this star as a function of time.

28. The pendulum in a grandfather clock makes one complete swing every 2 s. The maximum angle that the pendulum makes with respect to its rest position is 10°. We know from physical principles that the angle θ between the pendulum and its rest position changes in simple harmonic fashion. Find an equation that describes the size of the angle θ as a function of time. (Take $t = 0$ to be a time when the pendulum is vertical.)

29. A train whistle produces a sharp, pure tone. If the train is moving toward us, the pitch of the whistle increases and then decreases as the train moves away from us. This phenomenon is known as the *Doppler effect*. Suppose that the maximum change in pressure caused by the whistle is 10 N/m² and the frequency is 20 cycles per second.
    (a) Find an equation that describes the variation in pressure as a function of time, and sketch the graph.
    (b) How does the graph in part (a) change as the train moves toward us? Away from us?

30. The number of hours of daylight varies through the course of a year, the longest day being June 21 and the shortest December 21. The average length of daylight is 12 h and the variation from this average is simple harmonic. In Philadelphia, daylight lasts 14 h 50 min on June 21 and 9 h 10 min on December 21. Find an equation that gives the number of hours of daylight in Philadelphia as a function of the number of days after **(a)** March 21, **(b)** June 21, and **(c)** January 1.

   The graph in the figure shows the number of hours of daylight for different latitudes. Use the graph and the latitude of your city to find an equation that gives the number of hours of daylight in your city as a function of the number of days after March 21. (Graph taken from *Daylight, Twilight, Darkness and Time*, by Lucia Carolyn Harrison, New York: Silver, Burdett and Company, 1935, p. 40.)

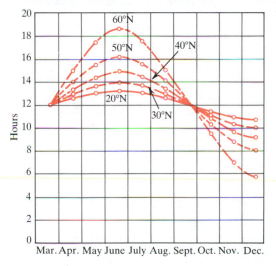

Graph of the relative length of day from March 21 to December 21, inclusive, at the latitudes indicated.

31. A spring-mass system is compressed a distance of 10 cm from its rest position, then released. The mass is in damped harmonic motion. The damping constant for this spring is $c = 2.1$ and the spring oscillates with a frequency of 5 cycles per second. Find a function that describes the motion of the mass if $t = 0$ is taken to be the instant the spring is released.

32. A strong gust of wind strikes a tall building, causing it to sway back and forth in damped harmonic motion. The frequency of the oscillation is 0.5 cycle per second and the damping constant is $c = 0.9$. Find an equation that

describes the motion of the building. (Assume $k = 1$ and take $t = 0$ to be the instant when the gust of wind strikes the building.)

33. When a car hits a certain bump on the road, a shock absorber on the car is compressed a distance of 6 in., then released (see figure). The shock absorber vibrates in damped harmonic motion with a frequency of 2 cycles per

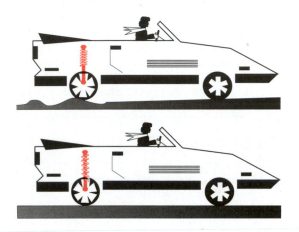

second. The damping constant for this particular shock absorber is 2.8.
   (a) Find an equation that describes the displacement of the shock absorber from its rest position as a function of time. Take $t = 0$ to be the instant that the shock absorber is released.
   (b) How long does it take for the amplitude of the vibration to decrease to 0.5 in.?

34. A tuning fork is struck and oscillates in damped harmonic motion. The amplitude of the motion is measured, and 3 s later it is found that the amplitude has dropped to $\frac{1}{4}$ of this value. Find the damping constant c for this tuning fork.

35. A guitar string is pulled at the point P a distance of 3 cm above its rest position. It is then released and vibrates in damped harmonic motion with a frequency of 165 cycles per second. After 2 s it is observed that the amplitude of the vibration at the point P is 0.6 cm.
   (a) Find the damping constant c.
   (b) Find an equation that describes the position of the point P above its rest position as a function of time. Take $t = 0$ to be the instant that the string is released.

CHAPTER 5 REVIEW

Define, state, or discuss each of the following.

1. The unit circle
2. Terminal point determined by a real number t
3. Reference number
4. The trigonometric functions: sin, cos, tan, csc, sec, cot
5. Domains of the trigonometric functions
6. Signs of the trigonometric functions
7. Even-odd properties
8. Fundamental identities
9. Reciprocal identities
10. Pythagorean identities
11. Periodic function

12. Periods of the trigonometric functions
13. Graphs of the trigonometric functions
14. Trigonometric graphs
15. Sine curve, cosine curve
16. Amplitude
17. Period
18. Phase shift
19. Simple harmonic motion
20. Frequency
21. Damped harmonic motion
22. Damping constant

REVIEW EXERCISES

In Exercises 1 and 2, a point P(x, y) is given.
(a) Show that the given point P is on the unit circle.
(b) Suppose that the given point is the terminal point determined by t. Find sin t, cos t, and tan t.

1. $\left(\dfrac{\sqrt{3}}{2}, \dfrac{1}{2}\right)$

2. $\left(-\dfrac{5}{13}, \dfrac{12}{13}\right)$

In Exercises 3–6 a real number t is given.
(a) Find the reference number for t.
(b) Find the terminal point P(x, y) on the unit circle determined by t.
(c) Find the six trigonometric functions of t.

3. $t = \dfrac{\pi}{6}$

4. $t = \dfrac{11\pi}{6}$

5. $t = \dfrac{9\pi}{4}$

6. $t = -\dfrac{5\pi}{3}$

In Exercises 7–16 find the value of the trigonometric function. If possible, give the exact value; otherwise, use a calculator to find an approximate value correct to five decimal places.

7. (a) $\sin \dfrac{3\pi}{4}$ **(b)** $\cos \dfrac{3\pi}{4}$

8. (a) $\tan \dfrac{\pi}{3}$ **(b)** $\tan\left(-\dfrac{\pi}{3}\right)$

9. (a) $\sin 1.1$ **(b)** $\cos 1.1$

10. (a) $\cos \dfrac{\pi}{5}$ **(b)** $\cos\left(-\dfrac{\pi}{5}\right)$

11. (a) $\cos \dfrac{9\pi}{2}$ **(b)** $\sec \dfrac{9\pi}{2}$

12. (a) $\sin \dfrac{\pi}{7}$ **(b)** $\csc \dfrac{\pi}{7}$

13. (a) $\tan \dfrac{5\pi}{2}$ **(b)** $\cot \dfrac{5\pi}{2}$

14. (a) $\sin 2\pi$ **(b)** $\csc 2\pi$

15. (a) $\tan \dfrac{\pi}{8}$ **(b)** $\cot \dfrac{\pi}{8}$

16. (a) $\cos \dfrac{\pi}{3}$ **(b)** $\sin \dfrac{\pi}{6}$

In Exercises 17–20 use the fundamental identities to write the first expression in terms of the second.

17. $\dfrac{\tan t}{\cos t}$, $\sin t$

18. $\tan^2 t \sec t$, $\cos t$

19. $\tan t$, $\sin t$; t in quadrant IV

20. $\sec t$, $\sin t$; t in quadrant II

In Exercises 21–24 find the values of the remaining trigonometric functions at t from the given information.

21. $\sin t = \frac{5}{13}$, $\cos t = -\frac{12}{13}$

22. $\sin t = -\frac{1}{2}$, $\cos t > 0$

23. $\cot t = -\frac{1}{2}$, $\csc t = \sqrt{5}/2$

24. $\cos t = -\frac{3}{5}$, $\tan t < 0$

25. If $\tan t = \frac{1}{4}$ and t is in quadrant III, find $\sec t + \cot t$.

26. If $\sin t = -\frac{8}{17}$ and t is in quadrant IV, find $\csc t + \sec t$.

27. If $\cos t = \frac{3}{5}$ and t is in quadrant I, find $\tan t + \sec t$.

28. If $\sec t = -5$ and t is in quadrant II, find $\sin^2 t + \cos^2 t$.

In Exercises 29–36 (a) find the amplitude, period, and phase shift and (b) sketch the graph.

29. $y = 10 \cos \frac{1}{2}x$

30. $y = 4 \sin 2\pi x$

31. $y = -\sin \frac{1}{2}x$

32. $y = 2 \sin\left(x - \dfrac{\pi}{4}\right)$

33. $y = 3 \sin(2x - 2)$

34. $y = \cos 2\left(x - \dfrac{\pi}{2}\right)$

35. $y = -\cos\left(\dfrac{\pi}{2}x + \dfrac{\pi}{6}\right)$

36. $y = 10 \sin\left(2x - \dfrac{\pi}{2}\right)$

In Exercises 37–40 the graph of one period of a function of the form y = a sin k(x − b) is shown. Determine the function.

37.

38.

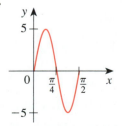

39.

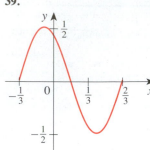

40.

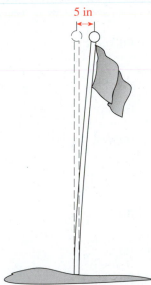

$\left(-\frac{2\pi}{3}, -4\right)$

In Exercises 41–48 find the period, and sketch the graph.

41. $y = 3 \tan x$

42. $y = \tan \pi x$

43. $y = 2 \cot\left(x - \frac{\pi}{2}\right)$

44. $y = \sec\left(\frac{1}{2}x - \frac{\pi}{2}\right)$

45. $y = 4 \csc(2x - \pi)$

46. $y = \tan\left(x + \frac{\pi}{6}\right)$

47. $y = \tan\left(\frac{1}{2}x - \frac{\pi}{8}\right)$

48. $y = -4 \sec 4\pi x$

 In Exercises 49–54, (a) use a graphing device to graph the given function; (b) determine from the graph whether the function is periodic and, if so, determine the period; and (c) determine from the graph whether the function is odd, even, or neither.

49. $y = |\cos x|$

50. $y = \sin(\cos x)$

51. $y = \cos(2^{0.1x})$

52. $y = 1 + 2^{\cos x}$

53. $y = e^{-|x|} \cos 3x$

54. $y = \ln x \sin 3x \quad (x > 0)$

 In Exercises 55–58 graph the three functions on a common screen. How are the graphs related?

55. $y = x, \quad y = -x, \quad y = x \sin x$

56. $y = 2^x, \quad y = 2^{-x}, \quad y = 2^{-x} \cos 4\pi x$

57. $y = x, \quad y = \sin 4x, \quad y = x + \sin 4x$

58. $y = \sin^2 x, \quad y = \cos^2 x, \quad y = \sin^2 x + \cos^2 x$

 In Exercises 59 and 60 find the maximum and minimum values of the function.

59. $y = \cos x + \sin 2x$

60. $y = \cos x + \sin^2 x$

61. Find the solutions of $\sin x = 0.3$ in the interval $[0, 2\pi]$.

62. Find the solutions of $\cos 3x = x$ in the interval $[0, \pi]$.

63. Let $f(t) = \dfrac{\sin^2 t}{t}$.

   (a) Is the function f even, odd, or neither?

   (b) Find the t-intercepts of the graph of f.

   (c) Draw the graph of f in an appropriate viewing rectangle.

   (d) Describe the behavior of the function as $t \to \infty$ and as $t \to -\infty$.

   (e) Notice that $f(t)$ is not defined when $t = 0$. What happens as $t \to 0$?

64. A point P moving in simple harmonic motion completes 8 cycles every second. If the amplitude of the motion is 50 cm, find an equation that describes the motion of P as a function of time. Assume the point P is at its maximum displacement when $t = 0$.

65. The top of a flagpole is displaced a distance of 5 in. from its rest position by a sharp gust of wind. It then vibrates in damped harmonic motion with frequency 2 cycles per second. If the damping constant for this flagpole is $c = 0.3$, write an equation in the form $y = ke^{-ct} \cos \omega t$ that describes the displacement of the top of the flagpole as a function of time. Take $t = 0$ to be the instant the top of the flagpole is at its maximum displacement.

5 in

66. To avoid excessive vibration, the flagpole in Exercise 65 is replaced by another pole. To test this new flagpole, its top is pulled 5 in. from its rest position and then released. It is observed that the pole vibrates at a frequency of

2 cycles per second. Two seconds later it is found that the amplitude of the vibration has dropped to 1 in.
(a) Find the damping constant for this flagpole.
(b) Write an equation in the form $y = ke^{-ct} \cos \omega t$ that gives the displacement of the top of the flagpole as a function of time. Take $t = 0$ to be the instant the top of the flagpole is at its maximum displacement.

67. A mass suspended from a spring oscillates in simple harmonic motion at a frequency of 4 cycles per second. The distance from the highest point to the lowest point of the oscillation is 100 cm. Find an equation that describes the distance of the mass from its rest position as a function of time. Assume the mass is at its lowest point when $t = 0$.

68. The graph in the figure is an oscilloscope reading of the displacement of a shock absorber that is vibrating in damped harmonic motion. Find the equation of the motion of the shock absorber in the form $y = ke^{-ct} \cos \omega t$.

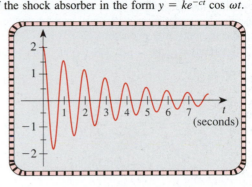

CHAPTER 5 TEST

1. The point $P(x, y)$ is on the unit circle in quadrant IV. If $x = \frac{5}{13}$, find y.

2. The point P in the figure has y-coordinate $\frac{3}{5}$. Find
   (a) $\sin t$ (b) $\cos(-t)$
   (c) $\tan(\pi - t)$ (d) $\sec(\pi + t)$

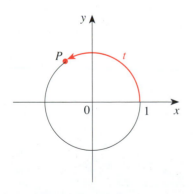

3. Express $\tan t$ in terms of $\sin t$ for t in quadrant II.

4. If $\cos t = -\frac{8}{17}$ and t is in quadrant III, find $\tan t \cot t + \csc t$.

In Problems 5 and 6, (a) find the amplitude, period, and phase shift of the given function and (b) sketch the graph.

5. $y = 3 \sin 2x$

6. $y = \cos\left(\dfrac{1}{2}x - \dfrac{\pi}{6}\right)$

In Problems 7 and 8 find the period and sketch the graph.

7. $y = -\csc 2x$

8. $y = \tan\left(2x - \dfrac{\pi}{2}\right)$

9. The graph shown is one period of a function of the form $y = a \sin k(x - b)$. Determine the function.

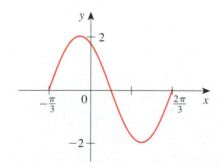

10. The graph shows the variation of the water level relative to mean sea level in the Long Beach harbor for a particular 24-hour period. Assuming that this variation is simple harmonic, find an equation of the form $y = a \cos \omega t$ that describes the variation in water level as a function of the number of hours after midnight.

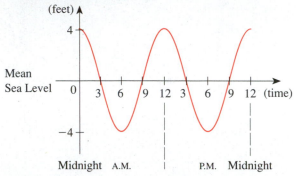

11. A mass suspended from a spring oscillates in simple harmonic motion. The mass completes 2 cycles every second and the distance between the highest point and the lowest point of the oscillation is 10 cm. Find an equation of the form $y = a \sin \omega t$ that gives the distance of the mass from its rest position as a function of time.

 In Problems 12 and 13, (a) use a graphing device to graph the function in an appropriate viewing rectangle; (b) determine from the graph if the function is even, odd, or neither; (c) determine from the graph if the function is periodic and, if so, find the period; and (d) find the minimum and maximum values of the function.

12. $y = \dfrac{\cos x}{1 + x^2}$

13. $y = e^{-\sin 3x}$

14. Let $y_1 = \cos(\sin x)$ and $y_2 = \sin(\cos x)$.
   (a) Sketch the graphs of y_1 and y_2 in the same viewing rectangle.
   (b) Determine the period of each of these functions from its graph.
   (c) Find an inequality between $\sin(\cos x)$ and $\cos(\sin x)$ that is valid for all x.

FOCUS ON PROBLEM SOLVING

For many problems the solution involves **drawing a diagram**, even when no diagram is immediately suggested by the problem itself. To illustrate this we present two very different situations where drawing a diagram is the key to solving the problem.

EXAMPLE 1

A Trigonometric Equation

Find the number of solutions of the equation

$$\sin x = \frac{x}{10}$$

SOLUTION

There is no obvious algebraic technique for solving this equation, so let us sketch $y = \sin x$ and $y = x/10$ on the same axes to see if we can count the solutions using the graph (see Figure 1). The solutions of the equation occur at those values of x where the two graphs intersect. From the graph we see that there are seven solutions.

Figure 1

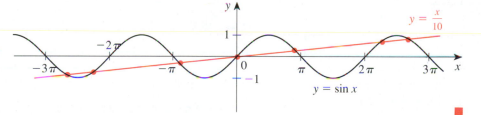

EXAMPLE 2

A Counting Problem

In a survey of 60 students it is found that 35 study French, 30 study German, 11 Russian, 12 study French and German, 4 French and Russian, 5 German and Russian, and 2 study all three languages. How many of these students study none of these languages?

SOLUTION

At first glance it may seem that all we need to do is add up the number of students studying French, German, and Russian and then subtract from 60. But, since

365

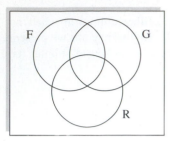

F = French G = German
R = Russian

Figure 2

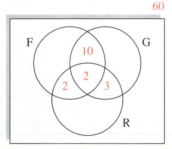

Figure 3

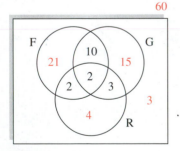

Figure 4

$35 + 30 + 11 = 76$, which is more than 60, this will clearly not work. To solve this problem we must keep track of how many students study each possible combination of these languages. This is difficult to do without organizing the information in some pictorial form. One way is to draw a **Venn diagram**, which consists of circles drawn within a rectangle, with the circles representing the various sets that occur in the problem. (This idea is attributed to the English logician John Venn). In the Venn diagram in Figure 2, the circles represent the students that study the three given languages and the rectangle represents all the students that were interviewed.

Since 2 students study all three languages, we put the number 2 in the intersection of all three circles in Figure 3. Of the 12 students that study French and German, 2 study all three languages, so just 10 study *only* French and German. Thus we put the number 10 in the portion of the diagram that belongs to F and G but not to R. Similarly, 2 students study only French and Russian and 3 study only German and Russian. This gives the distribution shown in Figure 3.

Now we use the Venn diagram in Figure 3 to determine how many students study just one language. Since 35 students study French, and since we have already accounted for $10 + 2 + 2 = 14$ of them in the diagram of Figure 3, the number of students who study only French must be $35 - 14 = 21$. We put this number in the appropriate section in circle F in Figure 4. By similar reasoning, we see that 15 study only German and 4 study only Russian.

To find the total number of students who study at least one of the four languages, we add all the numbers inside the circles in Figure 4:

$$21 + 15 + 4 + 10 + 3 + 2 + 2 = 57$$

Thus the number of students who study none of these languages is $60 - 57 = 3$. ∎

PROBLEMS

1. Find the number of solutions of the equation $\sin x = \dfrac{x}{100}$.

2. Find the number of solutions of the equation $\cos x = \dfrac{x^2}{400}$.

3. Among 100 business executives, 68 have an American Express card, 52 have a Visa card, and 52 have a MasterCard. Some executives have more than one card: 33 have both an American Express card and a MasterCard, 35 have both an American Express and a Visa card, 37 have both a MasterCard and a Visa card, and 30 have all three cards. How many of the business executives have none of these three cards?

4. A city has two newspapers, the *Morning Telegraph* and the *Evening Standard*. In a survey of 100 people, 54 say they read the *Morning Telegraph*, 42 read the *Evening Standard*, and 14 read neither. How many read both newspapers?

5. In a group of 95 pet owners, 65 own a dog, 43 own a cat, and 30 own a bird. We know that 15 own a dog and a bird, 8 own a cat and a bird, and 23 own a dog and a cat, and that 10 own a bird but do not own a cat or a dog.
   (a) Of these owners, how many own none of these types of pets?
   (b) How many own a cat and a dog, but not a bird?

6. In a group of 400 college professors, it is found that 120 earn less than $25,000 and 260 earn $50,000 or less annually. In this group, 63 own a boat, and of these, 5 earn less than $25,000 a year. Of those who earn more than $50,000, it is known that 100 do not own a boat. How many of those professors who earn between $25,000 and $50,000 (inclusive) do not own a boat?

7. Find the exact value of

$$\sin \frac{\pi}{100} + \sin \frac{2\pi}{100} + \sin \frac{3\pi}{100} + \sin \frac{4\pi}{100} + \cdots + \sin \frac{199\pi}{100} + \sin \frac{200\pi}{100}$$

where the sum contains the sines of all numbers of the form $k\pi/100$, with k varying from 1 to 200.

8. A person starts at a point P on the earth's surface and walks 1 mi south, then 1 mi east, then 1 mi north, and finds himself back at P, where he started. Describe all points P for which this is possible. (There are infinitely many.)

9. The Indian mathematician Bhaskara sketched the two figures shown here and wrote below them "Behold!" Explain how his sketches prove the Pythagorean Theorem.

6 TRIGONOMETRIC FUNCTIONS OF ANGLES

The greatest mathematicians, as Archimedes, Newton, and Gauss, always united theory and applications in equal measure.

FELIX KLEIN

In this chapter we present a different approach to the trigonometric functions. Here we view them as functions of angles, that is, functions that associate to each *angle* a real number. For example, the function sine is defined here as one that assigns to each angle θ a real number $\sin \theta$. Of course, the measure of the angle θ is a real number, so we may again view these trigonometric functions of angles as functions of real numbers. Viewed in this way, they are identical with the trigonometric functions defined in Chapter 5. Thus the difference is one of point of view, but the resulting trigonometric functions are the same. The two approaches are independent of each other, so either Chapter 5 or Chapter 6 may be studied first.

Why do we study trigonometry in two different ways? Because different applications require that we view these functions differently: In some applications, viewing the trigonometric functions as functions of a real number is appropriate. For example, when we apply the trigonometric functions to the study of harmonic motion, we must view them as functions of time, a real number. On the other hand, for many applications it is imperative to view them as functions of angles. For example, to find the distance to the sun, we can first accurately measure the angle formed by the sun, earth, and moon. From this information and a knowledge of the trigonometric functions of *angles*, we can easily solve this remarkable problem (see Exercise 57 of Section 6.2). Put simply, these functions are so useful and important because they can be viewed in these different ways.

We begin this chapter by describing an elegant and natural way of measuring angles, called *radian measure*. In Section 6.2 we define the trigonometric functions of acute angles as certain ratios of sides of right triangles and give several applications. In Section 6.3 we extend the definition of the trigonometric functions to include all angles and describe the relationship of these trigonometric functions of angles to those of real numbers. The remaining sections contain applications of the trigonometric functions of angles in situations where the angles are not necessarily acute.

SECTION 6.1
ANGLE MEASURE

An **angle** AOB consists of two rays R_1 and R_2 with a common vertex O (see Figure 1). We often interpret an angle as a rotation of the ray R_1 onto R_2. In this case R_1 is called the **initial side** and R_2 is called the **terminal side** of the angle. If the

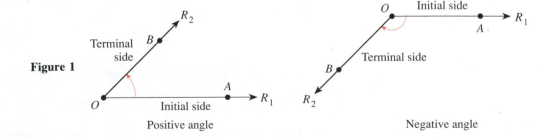

Figure 1

Positive angle   Negative angle

rotation is counterclockwise the angle is considered **positive**, and if the rotation is clockwise the angle is considered **negative**.

■ ANGLE MEASURE

The **measure** of an angle is the amount of rotation about the vertex required to move R_1 onto R_2. Intuitively, we see that this is how much the angle "opens." One unit of measurement for angles is the **degree**. An angle of measure one degree is formed by rotating the initial side 1/360 of a complete revolution. In calculus and other branches of mathematics a more natural method of measuring angles is used— *radian measure*. The amount an angle opens is measured along the arc of a circle of radius 1 with center at the vertex of the angle.

DEFINITION OF RADIAN MEASURE

If a circle of radius 1 is drawn with the vertex of an angle at its center, then the measure of this angle in **radians** is the length of the arc that subtends the angle (see Figure 2).

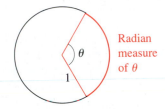

Figure 2

The circumference of the circle of radius 1 is 2π, and so a complete revolution has measure 2π radians, a straight angle has measure π radians, and a right angle has measure $\pi/2$ radians. An angle that is subtended by an arc of length 2 along the unit circle has radian measure 2 (see Figure 3).

Since a complete revolution measured in degrees is 360° and measured in radians is 2π radians, we get the following simple relationship between these two methods of angle measurement.

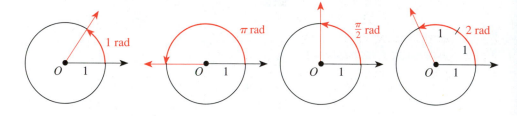

Figure 3
Radian measure

RELATIONSHIP BETWEEN DEGREES AND RADIANS

$$180° = \pi \text{ rad} \qquad 1 \text{ rad} = \left(\frac{180}{\pi}\right)° \qquad 1° = \left(\frac{\pi}{180}\right) \text{ rad}$$

I. To convert degrees to radians, multiply by $\dfrac{\pi}{180}$.

II. To convert radians to degrees, multiply by $\dfrac{180}{\pi}$.

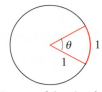

Measure of $\theta = 1$ rad
Measure of $\theta \approx 57.296°$

Figure 4

To get some idea of the size of a radian, notice that

$$1 \text{ rad} \approx 57.296° \quad \text{and} \quad 1° \approx 0.01745 \text{ rad}$$

An angle θ of measure 1 rad is sketched in Figure 4.

EXAMPLE 1

(a) Find the radian measure of 60°.

(b) Express $\dfrac{\pi}{6}$ rad in degrees.

SOLUTION

The relationship between degrees and radians gives:

(a) $60° = 60\left(\dfrac{\pi}{180}\right) \text{ rad} = \dfrac{\pi}{3} \text{ rad}$

(b) $\dfrac{\pi}{6} \text{ rad} = \left(\dfrac{\pi}{6}\right)\left(\dfrac{180}{\pi}\right) = 30°$ ∎

A note on terminology: We often use a phrase such as "a 30° angle" to mean *an angle whose measure is* 30°. Also, for an angle θ we write $\theta = 30°$ or $\theta = \pi/6$ to mean *the measure of θ is* 30° *or* $\pi/6$ *radian*. When no unit is given, the angle is assumed to be measured in radians.

■ ANGLES IN STANDARD POSITION

An angle is in **standard position** if it is drawn in the xy-plane with its vertex at the origin and its initial side on the positive x-axis. Figure 5 gives examples of angles in standard position.

Figure 5
Angles in standard position

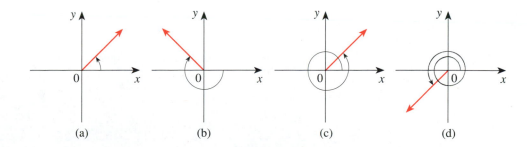

(a)   (b)   (c)   (d)

Two angles in standard position are **coterminal** if their sides coincide. In Figure 5, the angles in (a) and (c) are coterminal.

EXAMPLE 2

(a) Find angles that are coterminal with the angle $\theta = 30°$ in standard position.

(b) Find angles that are coterminal with the angle $\theta = \dfrac{\pi}{3}$ in standard position.

SOLUTION

(a) To find positive angles that are coterminal with θ we add any multiple of 360°. Thus

$$30° + 360° = 390° \qquad \text{and} \qquad 30° + 720° = 750°$$

are coterminal with $\theta = 30°$. To find negative angles that are coterminal with θ we subtract any multiple of 360°. Thus

$$30° - 360° = -330° \qquad \text{and} \qquad 30° - 720° = -690°$$

are coterminal with θ. See Figure 6.

Figure 6

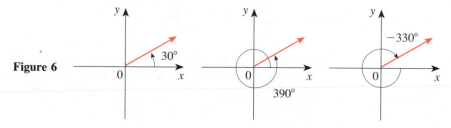

(b) To find positive angles that are coterminal with θ we add any multiple of 2π. Thus

$$\frac{\pi}{3} + 2\pi = \frac{7\pi}{3} \qquad \text{and} \qquad \frac{\pi}{3} + 4\pi = \frac{13\pi}{3}$$

are coterminal with $\theta = \pi/3$. To find negative angles that are coterminal with θ we subtract any multiple of 2π. Thus

$$\frac{\pi}{3} - 2\pi = -\frac{5\pi}{3} \qquad \text{and} \qquad \frac{\pi}{3} - 4\pi = -\frac{11\pi}{3}$$

are coterminal with θ. See Figure 7.

Figure 7

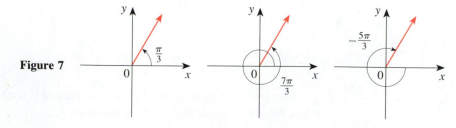

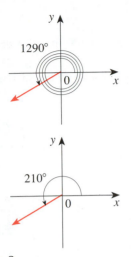

Figure 8

EXAMPLE 3

Find an angle with measure between 0° and 360° that is coterminal with the angle of measure 1290° in standard position.

SOLUTION

We can subtract 360° as many times as we wish from 1290° and the resulting angle will be coterminal with 1290°. Thus 1290° − 360° = 930° is coterminal with 1290° and so is the angle 1290° − 2(360)° = 1290° − 720° = 570°.

To find the angle that we want between 0° and 360°, we need to subtract 360° from 1290° as many times as necessary. An efficient way to do this is to determine how many times 360° goes into 1290°, that is, to divide 1290 by 360, and the remainder will be the angle we are looking for. We see that 360 goes into 1290 three times with a remainder of 210. Thus 210° is the desired angle (see Figure 8). ∎

■ **LENGTH OF A CIRCULAR ARC**

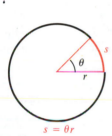

$s = \theta r$

Figure 9

An angle whose radian measure is θ is subtended by an arc that is the fraction $\theta/(2\pi)$ of the circumference of a circle. Thus in a circle of radius r, the length s of an arc that subtends the angle θ (see Figure 9) is

$$s = \frac{\theta}{2\pi} \times (\text{circumference of circle})$$

$$= \frac{\theta}{2\pi}(2\pi r) = \theta r$$

LENGTH OF A CIRCULAR ARC

In a circle of radius r the length s of an arc that subtends a central angle of θ *radians* is

$$s = r\theta$$

Solving for θ, we get the important formula

$$\theta = \frac{s}{r}$$

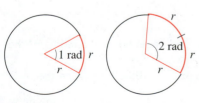

Figure 10

This formula allows us to define the radian measure using a circle of any radius r: The radian measure of an angle θ is s/r, where s is the length of the circular arc that subtends θ in a circle of radius r. Thus the radian measure of θ is the number of "radiuses" that can fit in the arc that subtends θ; hence the term "radians" (see Figure 10). The circle of radius 1 is used for convenience.

EXAMPLE 4

(a) Find the length of an arc of a circle with radius 10 m that subtends a central angle of 30°.

(b) A central angle θ in a circle of radius 4 m is subtended by an arc of length 6 m. Find the measure of θ in radians.

SOLUTION

(a) From Example 1(b) we see that $30° = \pi/6$ rad. So the length of the arc is

The formula $s = r\theta$ is true only when θ is measured in radians.

$$s = r\theta = (10)\frac{\pi}{6} = \frac{5\pi}{3} \text{ m}$$

(b) By the formula $\theta = s/r$, we have

$$\theta = \frac{s}{r} = \frac{6}{4} = \frac{3}{2} \text{ rad}$$

■

■ AREA OF A CIRCULAR SECTOR

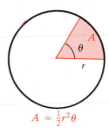

$A = \frac{1}{2}r^2\theta$

Figure 11

The area of a circle of radius r is $A = \pi r^2$. A sector of this circle with central angle θ has an area that is the fraction $\theta/(2\pi)$ of the area of the entire circle. So the area of this sector is

$$A = \frac{\theta}{2\pi} \times \text{(area of circle)}$$

$$= \frac{\theta}{2\pi}(\pi r^2) = \frac{1}{2}r^2\theta$$

AREA OF A CIRCULAR SECTOR

In a circle of radius r the area A of a sector with central angle of θ *radians* is

$$A = \tfrac{1}{2}r^2\theta$$

EXAMPLE 5

(a) Find the area of a sector of a circle with central angle 60° if the radius of the circle is 3 m.

(b) A sector of a circle of radius 100 m has an area of 10,000 m². Find the central angle of the sector.

SOLUTION

(a) To use the formula for the area of a circular sector we must find the central angle of the sector in radians: $60° = 60(\pi/180)$ rad $= \pi/3$ rad. Thus the area of the sector is

The formula $A = \frac{1}{2}r^2\theta$ is true only when θ is measured in radians.

$$A = \frac{1}{2}r^2\theta = \frac{1}{2}(3)^2\left(\frac{\pi}{3}\right) = \frac{3\pi}{2}\ \text{m}^2$$

(b) We need to find θ. Solving for θ in the formula, we get

$$\theta = \frac{2A}{r^2} = \frac{2(10,000)}{100^2} = 2\ \text{rad}$$

Thus the central angle of the sector has measure 2 rad. ∎

odd 1–41

EXERCISES 6.1

In Exercises 1–9 find the radian measure of the angle with the given degree measure. $\times \frac{\pi}{180}$

1. 40°   **2.** 330°   **3.** 72°

4. −30°   **5.** 45°   **6.** −80°

7. 765°   **8.** −150°   **9.** 36°

In Exercises 10–18 find the degree measure of the angle with the given radian measure. $\times \frac{180}{\pi}$

10. $\dfrac{3\pi}{4}$ **11.** $-\dfrac{7\pi}{2}$ **12.** $\dfrac{5\pi}{6}$

13. 2   **14.** 1.5   **15.** $\dfrac{2\pi}{9}$

16. $-\dfrac{\pi}{12}$ **17.** $\dfrac{\pi}{5}$ **18.** $\dfrac{\pi}{18}$

In Exercises 19–27 the measure of an angle in standard position is given. Find two positive angles and two negative angles that are coterminal with the given angle.

19. 300°   **20.** 135°   **21.** $\dfrac{3\pi}{4}$

22. $\dfrac{11\pi}{6}$ **23.** −388°   **24.** −50°

25. 3.5   **26.** $-\dfrac{\pi}{5}$ **27.** 497°

In Exercises 28–36 the measures of two angles in standard position are given. Determine whether the angles are coterminal.

28. 50°,   340°   **29.** −30°,   330°

30. 70°,   430°   **31.** $\dfrac{32\pi}{3}$, $\dfrac{11\pi}{3}$

32. 22π, 2π **33.** $\dfrac{5\pi}{6}$, $\dfrac{10\pi}{3}$

34. 155°,   875°   **35.** 260°,   −460°

36. 57°,   777°

In Exercises 37–42 find an angle between 0° and 360° that is coterminal with the given angle.

37. 733°   **38.** 361°

39. 2223°   **40.** −100°

41. −800°   **42.** 1270°

In Exercises 43–48 find an angle between 0 and 2π that is coterminal with the given angle.

43. $\dfrac{12\pi}{5}$ **44.** $-\dfrac{7\pi}{3}$ **45.** 87π

46. 10 **47.** $\dfrac{17\pi}{4}$ **48.** $\dfrac{51\pi}{2}$

49. Find the length of the arc *s* in the figure.

$\theta =$

$\dfrac{11}{9}\pi = \dfrac{s}{5}$

$\dfrac{55}{9}\pi$

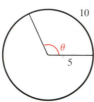

220

$\begin{array}{r}360\\-140\\\hline 220\end{array} \cdot \dfrac{\pi}{180}$

$\dfrac{11}{9}\pi$

50. Find the angle θ in the figure.

$\theta = \dfrac{10}{5} = 2$

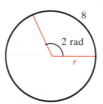

51. Find the radius *r* of the circle in the figure.

$2 = \dfrac{8}{r}$

$r = 4$

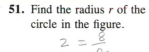

52. Find the length of an arc that subtends a central angle of 45° in a circle of radius 10 m. $\dfrac{\pi}{4} = \dfrac{s}{10}$ $s = \dfrac{5}{2}\pi$

53. Find the length of an arc that subtends a central angle of 2 rad in a circle of radius 2 mi.

54. A central angle θ in a circle of radius 5 m is subtended by an arc of length 6 m. Find the measure of θ in degrees and in radians.

55. An arc of length 100 m subtends a central angle of radius 50 m. Find the measure of θ in degrees and in radians.

56. A circular arc of length 3 ft subtends a central angle of 25°. Find the radius of the circle.

57. Find the radius of the circle if an arc of length 6 m on the circle subtends a central angle of $\pi/6$ rad.

58. How many revolutions will a car wheel of diameter 30 in. make as the car travels a distance of one mile?

59. Pittsburgh, Pennsylvania, and Miami, Florida, are approximately on the same meridian. Pittsburgh has a latitude of 40.5° N and Miami, 25.5° N. Find the distance between these two cities. (The radius of the earth is 3960 mi.)

$\theta = 15 \cdot \dfrac{\pi}{180}$

$= \dfrac{\pi}{12}$

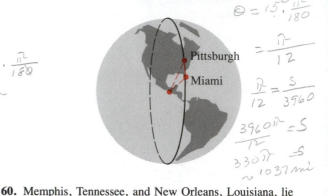

$\dfrac{\pi}{12} = \dfrac{s}{3960}$

$\dfrac{3960\pi}{12} = s$

$330\pi \quad s$

$\sim 1037\,mi$

60. Memphis, Tennessee, and New Orleans, Louisiana, lie approximately on the same meridian. Memphis has latitude 30° N and New Orleans, 35° N. Find the distance between these two cities. (The radius of the earth is 3960 mi.)

61. Find the distance that the earth travels in one day in its path around the sun. Assume that a year has 365 days and that the path of the earth around the sun is a circle of radius 93 million miles. (The path of the earth around the sun is actually an *ellipse* with the sun at one focus. This ellipse, however, has very small eccentricity, so it is nearly circular.)

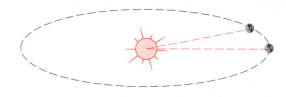

62. The Greek mathematician Eratosthenes (ca. 276–195 B.C.) measured the circumference of the earth from the following observations. He noticed that on a certain day the sun shone directly down a deep well in Syene (modern Aswan). At the same time in Alexandria, 500 miles north (on the same meridian), the rays of the sun shone at an angle of 7.2° to the zenith. Use this information and the figure to find the radius and circumference of the earth.

(The data used in this problem are more accurate than those available to Eratosthenes.)

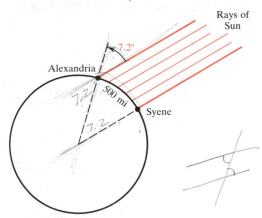

63. Find the distance along an arc on the surface of the earth that subtends a central angle of 1 minute $\left(1 \; minute = \frac{1}{60} \; degree\right)$. This distance is called a *nautical mile*. (The radius of the earth is 3960 mi.)

64. Find the area of the sector shown in the figure.

(a)   **(b)**

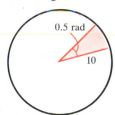

65. Find the area of a sector with central angle 1 rad in a circle of radius 10 m.

66. A sector of a circle has a central angle of 60°. Find the area of the sector if the radius of the circle is 3 mi.

67. The area of a sector of a circle with a central angle of 2 rad is 16 m². Find the radius of the circle.

68. Find the radius of a circle if the area of a quarter of the circle is 9π ft².

69. Find the central angle of a sector of a circle of radius 8 m if the area of the sector is 5 m².

70. A sector of a circle of radius 24 mi has an area of 288 mi². Find the central angle of the sector.

71. The area of a circle is 72 cm². Find the area of a sector of this circle that subtends a central angle of $\pi/6$ rad.

72. Three circles with radii 1, 2, and 3 ft are externally tangent to one another, as shown in the figure. Find the area of the sector of the circle of radius 1 that is cut off by the line segments joining the center of that circle to the centers of the other two circles.

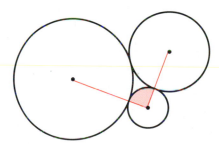

SECTION 6.2
TRIGONOMETRY OF RIGHT TRIANGLES

In this section we study certain ratios of the sides of right triangles, called trigonometric ratios, and give several applications. The trigonometric ratios are consistent with the trigonometric functions of real numbers. The relationship between them is explained in Section 6.3.

■ TRIGONOMETRIC RATIOS

Consider a right triangle with θ as one of its acute angles. We define the following ratios, called trigonometric ratios (see Figure 1):

THE TRIGONOMETRIC RATIOS

$$\sin\,\theta = \frac{\text{opposite}}{\text{hypotenuse}} \qquad \cos\,\theta = \frac{\text{adjacent}}{\text{hypotenuse}} \qquad \tan\,\theta = \frac{\text{opposite}}{\text{adjacent}}$$

$$\csc\,\theta = \frac{\text{hypotenuse}}{\text{opposite}} \qquad \sec\,\theta = \frac{\text{hypotenuse}}{\text{adjacent}} \qquad \cot\,\theta = \frac{\text{adjacent}}{\text{opposite}}$$

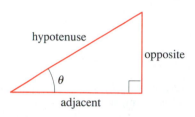

Figure 1

The symbols we use for these ratios are abbreviations for their full names: **sine, cosine, tangent, cosecant, secant, cotangent.** Since any two right triangles with angle θ are similar, these ratios are the same, regardless of the size of the triangle; they depend only on the angle θ (see Figure 2).

Figure 2

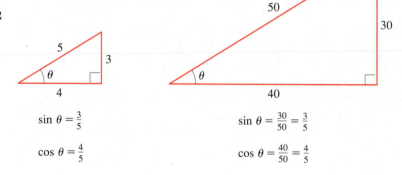

$$\sin \theta = \frac{3}{5} \qquad\qquad \sin \theta = \frac{30}{50} = \frac{3}{5}$$

$$\cos \theta = \frac{4}{5} \qquad\qquad \cos \theta = \frac{40}{50} = \frac{4}{5}$$

EXAMPLE 1

Find the six trigonometric ratios of the angle θ in Figure 3.

SOLUTION

$$\sin \theta = \frac{2}{3} \qquad\qquad \cos \theta = \frac{\sqrt{5}}{3} \qquad\qquad \tan \theta = \frac{2}{\sqrt{5}}$$

$$\csc \theta = \frac{3}{2} \qquad\qquad \sec \theta = \frac{3}{\sqrt{5}} \qquad\qquad \cot \theta = \frac{\sqrt{5}}{2}$$

Figure 3

EXAMPLE 2

If $\cos \alpha = \frac{3}{4}$, find the other five trigonometric ratios of α.

SOLUTION

Since $\cos \alpha$ is defined as the ratio of the adjacent side to the hypotenuse, we draw a triangle with hypotenuse of length 4 and a side of length 3 adjacent to α. If the oposite side is x, then by the Pythagorean Theorem, $3^2 + x^2 = 4^2$ or $x^2 = 7$, so $x = \sqrt{7}$. We now use the triangle in Figure 4 to write the ratios:

$$\sin \alpha = \frac{\sqrt{7}}{4} \qquad \cos \alpha = \frac{3}{4} \qquad \tan \alpha = \frac{\sqrt{7}}{3}$$

$$\csc \alpha = \frac{4}{\sqrt{7}} \qquad \sec \alpha = \frac{4}{3} \qquad \cot \alpha = \frac{3}{\sqrt{7}}$$ ■

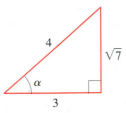

Figure 4

SPECIAL TRIANGLES

The following right triangles have ratios that can be calculated easily from the Pythagorean Theorem. Since they will appear often in the following chapters, we mention them here.

The first triangle is obtained by drawing a diagonal in a square of side 1 (see Figure 5). By the Pythagorean Theorem this diagonal has length $\sqrt{2}$. The resulting triangle has angles 45°, 45°, and 90° (or $\pi/4$, $\pi/4$, and $\pi/2$). To get the second triangle we start with an equilateral triangle ABC of side 2 and draw the perpendicular bisector DB of the base, as in Figure 6. By the Pythagorean Theorem the length of DB is $\sqrt{3}$. Since DB also bisects angle ABC, we obtain a triangle with angles 30°, 60°, and 90° (or $\pi/6$, $\pi/3$, and $\pi/2$).

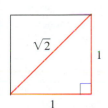

Figure 5

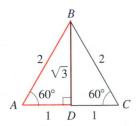

Figure 6

EXAMPLE 3

Find: (a) $\sin 30°$ (b) $\cos \dfrac{\pi}{4}$ (c) $\tan \dfrac{\pi}{6}$

SOLUTION

From Figures 5 and 6 we get

(a) $\sin 30° = \dfrac{1}{2}$ (b) $\cos 45° = \dfrac{1}{\sqrt{2}} = \dfrac{\sqrt{2}}{2}$

(c) $\tan \dfrac{\pi}{6} = \dfrac{1}{\sqrt{3}} = \dfrac{\sqrt{3}}{3}$ ■

As we have seen in Example 3, the values of the trigonometric ratios for the angles with measure 30°, 45°, and 60° (or $\pi/6$, $\pi/4$, and $\pi/3$) can be calculated from the two special right triangles constructed in Figures 5 and 6. We list these in Table 1.

Table 1

θ in degrees	θ in radians	sin θ	cos θ	tan θ	csc θ	sec θ	cot θ
30°	$\dfrac{\pi}{6}$	$\dfrac{1}{2}$	$\dfrac{\sqrt{3}}{2}$	$\dfrac{\sqrt{3}}{3}$	2	$\dfrac{2\sqrt{3}}{3}$	$\sqrt{3}$
45°	$\dfrac{\pi}{4}$	$\dfrac{\sqrt{2}}{2}$	$\dfrac{\sqrt{2}}{2}$	1	$\sqrt{2}$	$\sqrt{2}$	1
60°	$\dfrac{\pi}{3}$	$\dfrac{\sqrt{3}}{2}$	$\dfrac{1}{2}$	$\sqrt{3}$	$\dfrac{2\sqrt{3}}{3}$	2	$\dfrac{\sqrt{3}}{3}$

Hipparchus (*circa 140* B.C.) *is considered the founder of trigonometry. He constructed tables for a function closely related to the modern sine function and evaluated for angles at half-degree intervals. These are considered the first trigonometric tables. The main use to which he put his tables was to calculate the paths of the planets through the heavens.*

It is useful to remember these special trigonometric ratios, since they occur often. Of course they can be recalled easily if we remember the triangles from which they are obtained.

To find the values of the trigonometric ratios for other angles we use a calculator. Mathematical methods (called *numerical methods*) used in finding the trigonometric ratios are programmed directly into scientific calculators. For instance, when the [SIN] key is pressed, the calculator computes an approximation to the value of the sine of the given angle. In this book we will always use a calculator to find the values of the trigonometric ratios for angles other than the three special angles given in Table 1. Calculators give the values of sine, cosine, and tangent. The other ratios can be easily calculated from these using the following **reciprocal relations**:

$$\csc t = \frac{1}{\sin t} \qquad \sec t = \frac{1}{\cos t} \qquad \cot t = \frac{1}{\tan t}$$

You should check that these relations follow immediately from the definitions of the trigonometric ratios.

We follow the convention that when we write sin *t, where t is a real number, we mean the sine of the angle whose radian measure is t.* For instance, sin 1 means the sine of the angle whose radian measure is 1. When using a calculator to find an approximate value for this number, we set our calculator to radian mode and find that

$$\sin 1 \approx 0.841471$$

If we want to find the sine of the angle whose measure is 1° we write sin 1° and, with our calculator in degree mode, we get

$$\sin 1° \approx 0.0174524$$

EXAMPLE 4

With our calculator in degree mode, and writing the results correct to five decimal places, we find:

(a) $\sin 17° \approx 0.29237$ (b) $\sec 88° = \dfrac{1}{\cos 88°} \approx 28.65371$

With our claculator in radian mode, and writing the results correct to five decimal places, we find:

(c) $\cos 1.2 \approx 0.36236$ (d) $\cot 1.54 = \dfrac{1}{\tan 1.54} \approx 0.03081$ ■

APPLICATIONS OF TRIGONOMETRY OF RIGHT TRIANGLES

A triangle has six parts: three angles and three sides. To **solve a triangle** means to determine all of its parts from the information known about the triangle, that is, to determine the lengths of the three sides and the measures of the three angles.

EXAMPLE 5

Solve triangle ABC, shown in Figure 7.

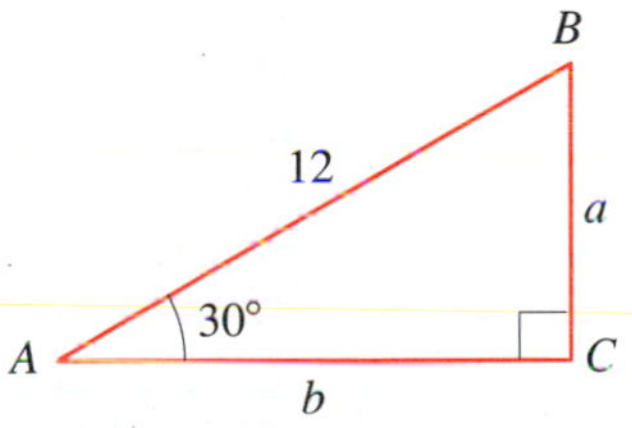

SOLUTION

It is clear that $\angle B = 60°$. To find a, we look for an equation that relates a to the lengths and angles we already know. In this case we have $\sin 30° = \dfrac{a}{12}$, so

$$a = 12 \sin 30° = 12\left(\tfrac{1}{2}\right) = 6$$

Similarly, $\cos 30° = \dfrac{b}{12}$, so

$$b = 12 \cos 30° = 12\left(\frac{\sqrt{3}}{2}\right) = 6\sqrt{3}$$ ■

Figure 7

It is very useful to know that, using the information given in Figure 8, the lengths of the legs of a right triangle are

$$a = r \sin \theta \quad \text{and} \quad b = r \cos \theta$$

The ability to solve right triangles using the trigonometric ratios is fundamental to many problems in navigation, surveying, astronomy, and the measurement of distances. The applications we consider in this section always involve right triangles, but as we will see in the next three sections, trigonometry is also useful in solving triangles that are not right triangles.

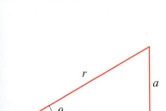

Figure 8

To discuss the next examples we need some terminology: If an observer is looking at an object, then the line from the observer's eye to the object is called the **line of sight** (Figure 9). If the object being observed is above the horizontal, then the angle between the line of sight and the horizontal is called the **angle of elevation**. If the object is below the horizontal, then the angle between the line of sight and the horizontal is called the **angle of depression**. In many of the examples and exercises, angles of elevation and depression will be given for a hypothetical observer at ground level. If the line of sight follows a physical object, such as an inclined plane or a hillside, we use the term **angle of inclination**.

Figure 9

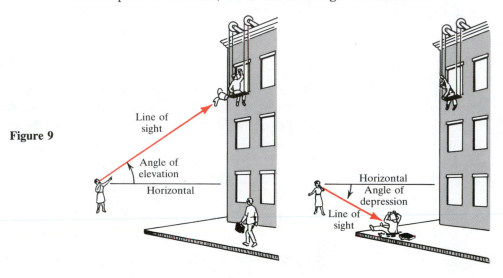

The next example gives a remarkable application of trigonometry to the problem of measurement. Although the example is simple, the result is fundamental to the method of applying the trigonometric ratios to such problems. In the example, we measure the height of a tall tree without having to climb it!

EXAMPLE 6

A giant redwood tree casts a shadow that is 532 ft long. Find the height of the tree if the angle of elevation of the sun is 25.7°.

SOLUTION

Let the height of the tree be h. From Figure 10 we see that

$$\frac{h}{532} = \tan 25.7°$$

so

$$h = 532 \tan 25.7°$$
$$\approx 532(0.48127) \approx 256$$

Therefore the height of the tree is about 256 ft.   ■

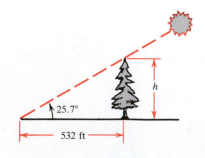

Figure 10

EXAMPLE 7

From a point on the ground 500 ft from the base of a building, it is observed that the angle of elevation to the top of the building is 24° and the angle of elevation to the top of a flagpole atop the building is 27°. Find the height of the building and the length of the flagpole.

SOLUTION

Figure 11 illustrates the situation. The height of the building is found in the same way that we found the height of the tree in Example 6.

$$\frac{h}{500} = \tan 24°$$

$$h = 500 \tan 24°$$

$$\approx 500(0.4452) \approx 223$$

The height of the building is approximately 223 ft. To find the length of the flagpole, let us first find the height from the ground to the top of the flagpole:

$$\frac{k}{500} = \tan 27°$$

$$k = 500 \tan 27° \approx 500(0.5095) \approx 255$$

To find the length of the flagpole, we subtract h from k. So the length of the flagpole is approximately $255 - 223 = 32$ ft. ∎

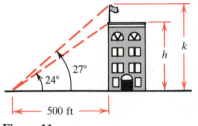

Figure 11

EXAMPLE 8

A pilot is flying over Washington, D.C. He does not know his elevation but he sights the Washington monument, which he knows to be 555 ft high. He measures the angle of depression to the top of the monument to be 17° and the angle of depression to the bottom of the monument to be 19°. Find the elevation of the pilot.

SOLUTION

From Figure 12 we see that

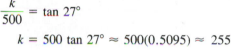

$$\tan 17° = \frac{h - 555}{a} \qquad \text{and} \qquad \tan 19° = \frac{h}{a}$$

Figure 12

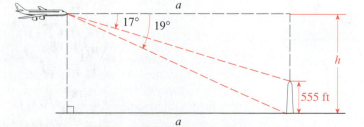

We have two unknowns a and h and neither of the two right triangles in Figure 12 can be solved independently. To find h we eliminate a from the two equations. We do this by solving for a in each equation to get

$$a = \frac{h - 555}{\tan 17°} \quad \text{and} \quad a = \frac{h}{\tan 19°}$$

Equating these two expressions gives

$$\frac{h - 555}{\tan 17°} = \frac{h}{\tan 19°}$$

$$\tan 19°(h - 555) = h \tan 17°$$

$$h \tan 19° - h \tan 17° = 555 \tan 19°$$

$$h(\tan 19° - \tan 17°) = 555 \tan 19°$$

$$h = \frac{555 \tan 19°}{\tan 19° - \tan 17°}$$

$$\approx \frac{555(0.3443)}{0.3443 - 0.3057} \approx 4951.2$$

Thus the elevation of the pilot is approximately 4950 ft.   ■

The trigonometric ratios can also be used to find angles, as the following example shows.

EXAMPLE 9

A 40-ft ladder leans against a building. If the base of the ladder is 6 ft from the base of the building, what is the angle formed by the ladder and the building?

SOLUTION

We sketch a diagram as in Figure 13. If θ is the angle between the ladder and the building, then

$$\sin \theta = \tfrac{6}{40} = 0.15$$

So θ is the angle whose sine is 0.15. To find the angle θ we use a calculator and use the $\boxed{\text{INV}}$ $\boxed{\text{SIN}}$ keys. Making sure our calculator is set to degree mode, we get

$$\theta \approx 8.6°$$ ■

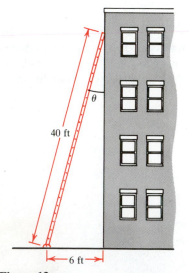

Figure 13

On some calculators the $\boxed{\text{INV}}$ $\boxed{\text{SIN}}$ is called $\boxed{\text{ARCSIN}}$ or $\boxed{\text{SIN}^{-1}}$. When this key is pressed, the calculator works "backward" to compute the angle that corresponds to the given value of sine. The same is true for the $\boxed{\text{INV}}$ $\boxed{\text{COS}}$ and $\boxed{\text{INV}}$ $\boxed{\text{TAN}}$ keys. These inverse trigonometric functions will be studied in more detail in Section 7.6.

EXERCISES 6.2

In Exercises 1–6 find the values of the six trigonometric ratios of the angle θ in the triangle shown.

1.
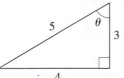

2.
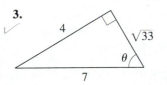

3.
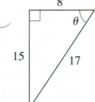

4.
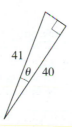

5.
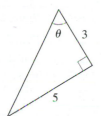

6.
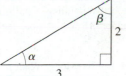

In Exercises 7–10 find (a) sin α and cos β, (b) tan α and cot β, and (c) sec α and csc β.

7.
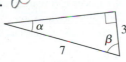

8.
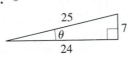

9.
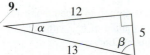

10.
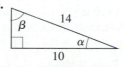

In Exercises 11–16 find the side labeled x. In Exercises 15 and 16, state your answer correct to five decimal places.

11.
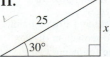

12.
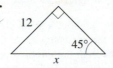

13.

14.
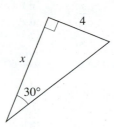

15. tan 36 = $\frac{12}{x}$

  $x = \frac{12}{\tan 36}$
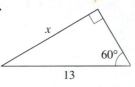

16.
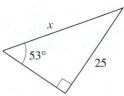

In Exercises 17–20 find x and y in terms of θ.   also use csc + cot.

  sin θ = $\frac{10}{y}$

  y = $\frac{10}{\sin θ}$

  tan θ = $\frac{10}{x}$

  x = $\frac{10}{\tan θ}$

17.
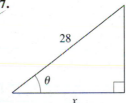

18.
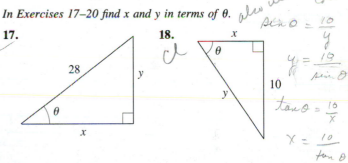

19.
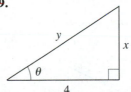

20.
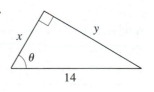

In Exercises 21–26 find the other five trigonometric ratios of θ.

21. sin θ = $\frac{3}{5}$

22. cos θ = $\frac{2}{7}$

23. cot θ = 1

24. tan θ = $\sqrt{3}$

25. sec θ = 7

26. csc θ = $\frac{13}{12}$

In Exercises 27–34 evaluate the given expression.

27. $\sin \dfrac{\pi}{6} + \cos \dfrac{\pi}{6}$

28. $\sin 30° \csc 30°$

29. $\sin 30° \cos 60° + \sin 60° \cos 30°$

30. $(\sin 60°)^2 + (\cos 60°)^2$

31. $\log_3\!\left(\tan \dfrac{\pi}{3} \right)$

32. $\log_2\!\left(\sin \dfrac{\pi}{4} \right)$

33. $(\cos 30°)^2 - (\sin 30°)^2$

34. $\left(\sin \dfrac{\pi}{3} \cos \dfrac{\pi}{4} - \sin \dfrac{\pi}{4} \cos \dfrac{\pi}{3} \right)^2$

In Exercises 35 and 36 use the figure to estimate the given trigonometric ratios.

35. $\sin 15°, \quad \cos 15°, \quad \tan 15°$

36. $\sin 40°, \quad \cos 40°, \quad \tan 40°$

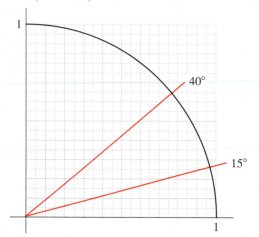

In Exercises 37–40 solve the given right triangle.

37.

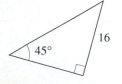

38.

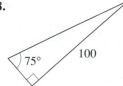

39.

40.

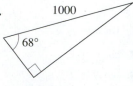

41. The angle of elevation to the top of the Empire State Building in New York is found to be 11° from the ground at a distance of 1 mi from the base of the building. Using this information, find the height of the Empire State Building.

42. A plane is flying within sight of the Gateway Arch in St. Louis, Missouri, at an elevation of 35,000 ft. The pilot would like to estimate her distance from the Gateway Arch. She finds that the angle of depression to a point on the ground below the arch is 22°.
   (a) What is the distance between the plane and the arch?
   (b) What is the distance between a point on the ground directly below the plane and the Gateway Arch?

43. A laser beam is to be directed toward the center of the moon, but the beam strays 0.5° from its intended path.
   (a) How far has the beam diverged from its assigned target when it reaches the moon? (The distance from the earth to the moon is 240,000 mi.)
   (b) The radius of the moon is about 1000 mi. Will the beam strike the moon?

44. From the top of a 200-ft lighthouse, the angle of depression to a ship in the ocean is 23°. How far is the ship from the base of the lighthouse?

45. A 20-ft ladder leans against a building so that the angle between the ground and the ladder is 72°. How high does the ladder reach on the building?

46. A 20-ft ladder is leaning against a building. If the base of the ladder is 6 ft from the base of the building, what is the angle of elevation of the ladder? How high does the ladder reach on the building?

47. A 96-ft tree casts a shadow that is 120 ft long. What is the angle of elevation of the sun?

48. A 600-ft guy wire is attached to the top of a communication tower. If the wire makes an angle of 65° with the ground, how tall is the communication tower?

49. A man is lying on the beach, flying a kite. He holds the end of the kite string at ground level, and estimates the angle of elevation of the kite to be 50°. If the string is 450 ft long, how high is the kite above the ground?

50. The height of a steep cliff is to be measured from a point on the opposite side of the river. Find the height of the cliff from the information given in the following figure.

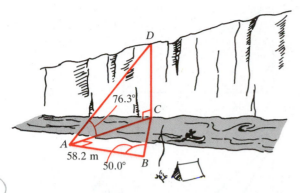

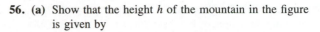

51. A water tower is located 325 ft from a building (see the figure). From a window in the building it is observed that the angle of elevation to the top of the tower is 39° and the angle of depression to the bottom of the tower is 25°. How tall is the tower? How high is the window?

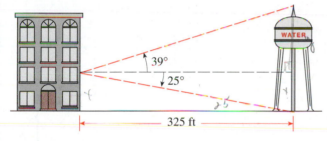

52. An airplane is flying at an elevation of 5150 ft, directly above a straight highway. Two motorists are driving cars on the highway on opposite sides of the plane, and the angle of depression to one car is 35° and to the other is 52°. How far apart are the cars?

53. If both cars in Exercise 52 are on one side of the plane and if the angle of depression to one car is 38° and to the other car is 52°, how far apart are the cars?

54. A hot-air balloon is floating above a straight road. To estimate their height above the ground, the balloonists simultaneously measure the angle of depression to two consecutive mileposts on the road on the same side of the balloon. The angles of depression are found to be 20° and 22°. How high is the balloon?

55. To estimate the height of a mountain above a level plain, the angle of elevation to the top of the mountain is measured to be 32°. One thousand feet closer to the mountain along the plain, it is found that the angle of elevation is 35°. Estimate the height of the mountain.

56. (a) Show that the height h of the mountain in the figure is given by

$$h = d \, \frac{\tan \beta \, \tan \alpha}{\tan \beta - \tan \alpha}$$

(b) Use the formula in part (a) to find the height h of the mountain if $\alpha \approx 25°$, $\beta \approx 29°$, and $d \approx 800$ ft.

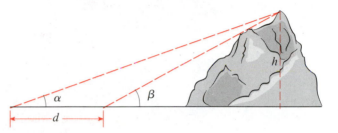

57. When the moon is exactly half full, the angle formed by the earth, moon, and sun is a right angle (see the figure). At that time the angle formed by the sun, earth, and moon is measured to be 89.85°. If the distance from the earth to the moon is 240,000 mi, estimate the distance from the earth to the sun.

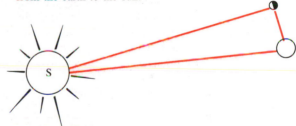

58. To find the distance to the sun as in Exercise 57, we needed to know the distance to the moon. Here is a way of estimating that distance: When the moon is seen at the zenith at a point A on the earth, it is observed to be at the horizon from point B (see the figure). Points A and B are 6155 mi apart and the radius of the earth is 3960 mi.
(a) Find the angle θ in degrees.
(b) Estimate the distance from point A to the moon.

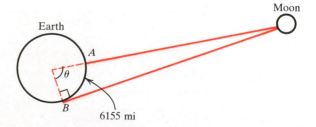

59. In Exercise 62 of Section 6.1 a method was given for finding the radius of the earth. Here is a more modern method of doing this. From a satellite 600 mi above the earth it is observed that the angle formed by the vertical and the line of sight to the horizon is 60.276°. Use this information to find the radius of the earth.

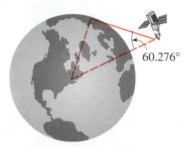

60. To find the distance to nearby stars, the method of parallax is used. The idea is to find a triangle with the star at one vertex and with a base as large as possible. To do this, the star is observed at two different times exactly 6 months apart and its apparent change in position is recorded. From these two observations $\angle E_1 S E_2$ can be calculated. (The times are chosen so that $\angle E_1 S E_2$ is as large as possible, which guarantees that $\angle E_1 O S$ is 90°.) The angle $E_2 S O$ is called the *parallax* of the star. Alpha Centauri, the star nearest to earth, has a parallax of 0.000211°. Estimate the distance to this star. (Take the distance from the earth to the sun to be 9.3×10^7 mi.)

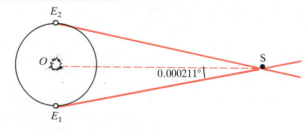

In Exercises 61–64 find x correct to one decimal place.

61.

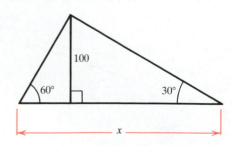

62.

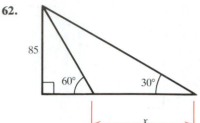

63.   **64.**

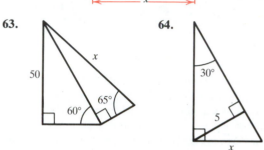

In Exercises 65 and 66 find x in terms of θ.

65.   **66.**

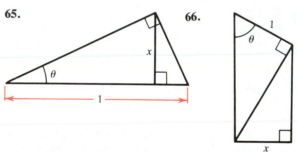

67. Express the lengths a, b, c, and d in the figure in terms of the trigonometric ratios of θ.

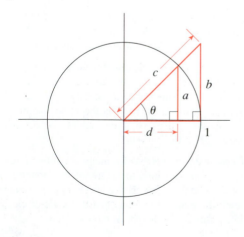

In Exercises 68–70 the following two facts from geometry are needed.
(1) *An angle inscribed in a semicircle is a right angle.*
(2) *The central angle subtended by a chord of a circle is twice the angle subtended by that chord on the circle.*

68. Express the length d in the diagram in terms of θ.

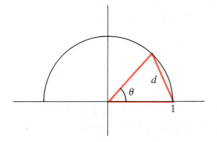

69. From the following figure show that

$$\sin 2\theta = 2 \sin \theta \cos \theta$$

[*Hint:* Find the area of triangle ABC in two ways.]

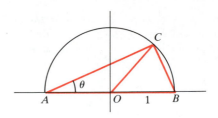

70. (a) Show from the figure that

$$\tan \theta = \frac{\sin 2\theta}{1 + \cos 2\theta}$$

(b) Use part (a) to find the exact values of tan 22.5° and tan 15°.

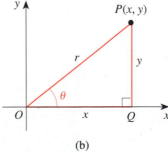

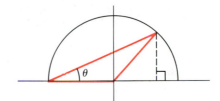

■ **SECTION 6.3**
TRIGONOMETRIC FUNCTIONS OF ANGLES

In the preceding section we defined the trigonometric ratios for acute angles. Here we extend the trigonometric ratios to all angles by defining the trigonometric functions of angles. With these functions we can solve practical problems which involve angles that are not necessarily acute. We study one such application in this section and many others in Sections 6.4 and 6.5.

■ **TRIGONOMETRIC FUNCTIONS OF ANGLES**

Let POQ be a right triangle with acute angle θ as shown in Figure 1(a). Place θ in standard position as shown in Figure 1(b). Then $P = P(x, y)$ is a point on the

Figure 1

(a)   (b)

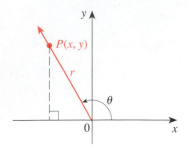

Figure 2

terminal side of θ. In triangle POQ the opposite side has length y and the adjacent side has length x. Using the Pythagorean Theorem, we see that the hypotenuse has length $r = \sqrt{x^2 + y^2}$. So

$$\sin \theta = \frac{y}{r} \qquad \cos \theta = \frac{x}{r} \qquad \tan \theta = \frac{y}{x}$$

The other trigonometric ratios can be found in the same way.

These observations allow us to extend the definition of the trigonometric ratios to any angle. We begin by defining the trigonometric functions of angles as follows (see Figure 2).

DEFINITION OF THE TRIGONOMETRIC FUNCTIONS OF ANGLES

Let θ be an angle in standard position and let $P(x, y)$ be a point on the terminal side. If $r = \sqrt{x^2 + y^2}$ is the distance from the origin to the point $P(x, y)$, then

$$\sin \theta = \frac{y}{r} \qquad\qquad \cos \theta = \frac{x}{r} \qquad\qquad \tan \theta = \frac{y}{x} \quad (x \neq 0)$$

$$\csc \theta = \frac{r}{y} \quad (y \neq 0) \qquad \sec \theta = \frac{r}{x} \quad (x \neq 0) \qquad \cot \theta = \frac{x}{y} \quad (y \neq 0)$$

It is a crucial fact that the values of the trigonometric functions do *not* depend on the choice of the point $P(x, y)$. This is because if $P'(x', y')$ is any other point on the terminal side, as in Figure 3, then triangles POQ and $P'OQ'$ are similar.

Figure 3

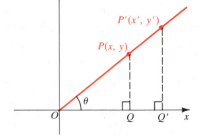

If we replace the angle θ by its radian measure (which is a real number), we may consider the trigonometric functions of angles as functions of real numbers. Viewed in this way the trigonometric functions of angles are identical with those of real numbers defined in Chapter 5 in the sense that they assign identical values to a given real number. (See the marginal comment on page 391.) Although the values are identical, the definitions are not. In this chapter we *choose* to think of $\sin \theta$ and $\cos \theta$ as functions of an *angle* whose measure is (the real number) θ. This point of view leads to applications involving angles.

◼ FINDING VALUES OF THE TRIGONOMETRIC FUNCTIONS FOR ANY ANGLE

Relationship to the Trigonometric Functions of Real Numbers

Let θ be an angle in standard position and choose $P(x, y)$ on the terminal side so that its distance r to the origin is 1, as in the figure.

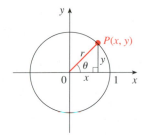

By the definition of the trigonometric functions of the *angle* θ, we have

$$\sin \theta = \frac{y}{1} = y \qquad \cos \theta = \frac{x}{1} = x$$

On the other hand, $P(x, y)$ is also the terminal point of an arc of length t.

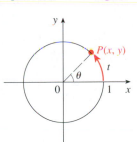

By the definition of the trigonometric functions of the *real number* t, we have

$$\sin t = y \qquad \text{and} \qquad \cos t = x$$

If θ is measured in radians, then $\theta = t$. Comparing the two ways of defining the trigonometric functions, we see that they give identical values. In other words, viewed as functions, they assign identical values to a given real number (the real number is the radian measure of θ in one case or the length t of a circular arc in the other).

From the definition we see that the values of the trigonometric functions are all positive if the angle θ has its terminal side in quadrant I. This is because x and y are positive in this quadrant. [Of course, r is always positive, since it is simply the distance from the origin to the point $P(x, y)$.] If the terminal side of θ is in quadrant II, however, then x is negative and y is positive. So in that quadrant sin θ and csc θ are positive and all the other trigonometric functions have negative values. Table 1 indicates which trigonometric values are positive in each quadrant. A mnemonic device, shown in Figure 4, can be used to remember the signs of the trigonometric functions. For example, the letter C in quadrant IV reminds us that the values of cos θ are positive in that quadrant (so sec θ will also be positive in that quadrant, because sec $\theta = 1/\cos \theta$) and the rest are negative.

Table 1

Quadrant containing θ	Positive trigonometric functions
I	All
II	sin θ, csc θ
III	tan θ, cot θ
IV	cos θ, sec θ

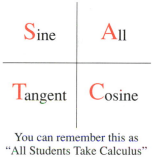

You can remember this as "All Students Take Calculus"

Figure 4

Since division by zero is an undefined operation, certain trigonometric functions are not defined for certain angles. For example, tan 90° = y/x is undefined because $x = 0$. The angles for which the trigonometric functions may be undefined are the angles for which either the x- or y-coordinate of a point on the terminal side of the angle is zero. These are **quadrantal angles**—angles that are coterminal with the coordinate axes.

We now turn our attention to finding the values of the trigonometric functions for angles that are not acute.

EXAMPLE 1

Find (a) cos 135° and (b) tan 390°.

SOLUTION

(a) From Figure 5 we see that cos 135° = $-x/r$. But cos 45° = x/r, and since

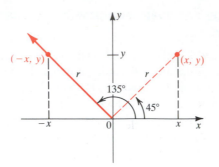

Figure 5

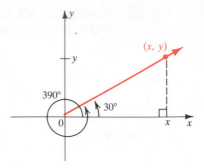

Figure 6

$\cos 45° = \sqrt{2}/2$, we have

$$\cos 135° = -\frac{\sqrt{2}}{2}$$

(b) The angles 390° and 30° are coterminal. From Figure 6 it is clear that $\tan 390° = \tan 30°$ and, since $\tan 30° = \sqrt{3}/3$, we have

$$\tan 390° = \frac{\sqrt{3}}{3}$$

■

From Example 1 we see that it is not difficult to find trigonometric functions of angles that are not acute: The trigonometric functions for angles that are not acute have the same value, except possibly for sign, as the corresponding trigonometric functions of an acute angle. That acute angle will be called the *reference angle*.

REFERENCE ANGLE

Let θ be an angle in standard position. The **reference angle $\bar{\theta}$** associated with θ is the acute angle formed by the terminal side of θ and the x-axis.

Figure 7 shows that to find a reference angle it is useful to know the quadrant in which the terminal side of the angle lies.

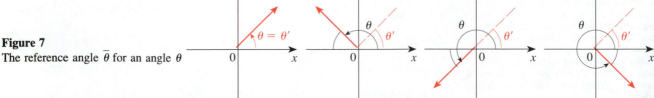

Figure 7
The reference angle $\bar{\theta}$ for an angle θ

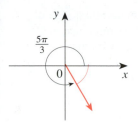

Figure 8

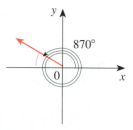

Figure 9

EXAMPLE 2

Find the reference angle for each angle:

(a) $\theta = \dfrac{5\pi}{3}$ (b) $\theta = 870°$

SOLUTION

(a) The reference angle is the acute angle formed by the terminal side of the angle $5\pi/3$ and the x-axis (see Figure 8). Since the terminal side of this angle is in quadrant IV, the reference angle is

$$2\pi - \frac{5\pi}{3} = \frac{\pi}{3}$$

(b) The angles $870°$ and $150°$ are coterminal [because $870 - 2(360) = 150$]. Thus the terminal side of this angle is in quadrant II (see Figure 9). So the reference angle is

$$180° - 150° = 30° \qquad\blacksquare$$

The method used in Example 1 to find the trigonometric functions for any angle θ can now be summarized in the following steps:

Step 1   Find the reference angle $\bar{\theta}$ associated with the angle θ.

Step 2   Determine the sign of the trigonometric function of θ.

Step 3   The value of the trigonometric function of θ is the same, except possibly for sign, as the value of the trigonometric function of $\bar{\theta}$.

EXAMPLE 3

Find (a) $\sin 240°$ and (b) $\cot 495°$.

SOLUTION

(a) This angle has its terminal side in quadrant III, as shown in Figure 10. Thus the reference angle is $240° - 180° = 60°$ and the value of $\sin 240°$ is negative. Thus

$$\sin 240° = -\sin 60° = -\frac{\sqrt{3}}{2}$$

$\qquad\qquad\qquad\qquad\quad \uparrow \qquad \uparrow$
$\qquad\qquad\qquad\qquad\; \text{sign}\;\; \text{reference angle}$

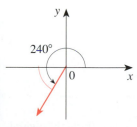

$\dfrac{S\,|\,A}{T\,|\,C}$ *sin 240° is negative.*

Figure 10

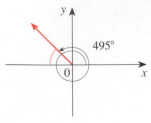

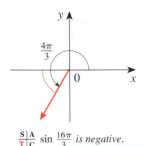

$\dfrac{S \mid A}{T \mid C}$ tan 495° *is negative,*
so cot 495° *is negative.*

Figure 11

(b) The angle 495° is coterminal with the angle 135° and the terminal side of this angle is in quadrant II, as shown in Figure 11. So the reference angle is 180° − 135° = 45° and the value of cot 495° is negative. We have

$$\cot 495° = \cot 135° = -\cot 45° = -1$$

↑   ↑   ↑
coterminal angles   sign   reference angle

■

EXAMPLE 4

Find (a) $\sin \dfrac{16\pi}{3}$ and (b) $\sec\left(-\dfrac{\pi}{4}\right)$.

SOLUTION

(a) The angle $16\pi/3$ is coterminal with $4\pi/3$ and these angles are in quadrant III (see Figure 12). Thus the reference angle is $(4\pi/3) - \pi = \pi/3$. Since the value of sine is negative in quadrant III, we have

$$\sin \dfrac{16\pi}{3} = \sin \dfrac{4\pi}{3} = -\sin \dfrac{\pi}{3} = -\dfrac{\sqrt{3}}{2}$$

↑   ↑   ↑
coterminal angles   sign   reference angle

$\dfrac{S \mid A}{T \mid C}$ $\sin \dfrac{16\pi}{3}$ *is negative.*

Figure 12

(b) The angle $-\pi/4$ is in quadrant IV, and its reference angle is $\pi/4$ (see Figure 13). Since secant is positive in this quadrant, we get

$$\sec\left(-\dfrac{\pi}{4}\right) = +\sec \dfrac{\pi}{4} = \dfrac{\sqrt{2}}{2}$$

↑   ↑
sign   reference angle

Figure 13

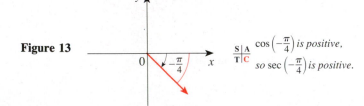

$\dfrac{S \mid A}{T \mid C}$ $\cos\left(-\dfrac{\pi}{4}\right)$ *is positive,*
so $\sec\left(-\dfrac{\pi}{4}\right)$ *is positive.*

■

■   **TRIGONOMETRIC IDENTITIES**

The trigonometric functions of angles are related to each other through several important equations called **trigonometric identities**. We have already encountered the reciprocal identities. These identities continue to hold for any angle θ, provided both sides of the equations are defined. The Pythagorean identities are a consequence of the Pythagorean Theorem.*

FUNDAMENTAL IDENTITIES

Reciprocal Identities

$$\csc \theta = \frac{1}{\sin \theta} \qquad \sec \theta = \frac{1}{\cos \theta} \qquad \cot \theta = \frac{1}{\tan \theta}$$

$$\tan \theta = \frac{\sin \theta}{\cos \theta} \qquad \cot \theta = \frac{\cos \theta}{\sin \theta}$$

Pythagorean Identities

$$\sin^2\theta + \cos^2\theta = 1 \qquad \tan^2\theta + 1 = \sec^2\theta \qquad 1 + \cot^2\theta = \csc^2\theta$$

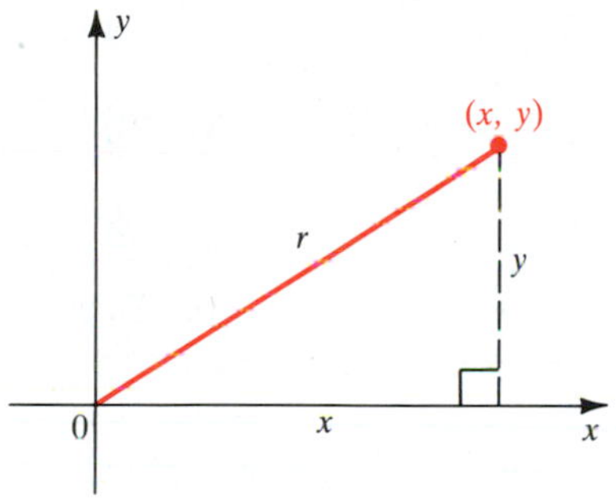

Figure 14

Proof We prove the Pythagorean identities. Applying the Pythagorean Theorem to the triangle in Figure 14 gives

$$\sin^2\theta + \cos^2\theta = \left(\frac{y}{r}\right)^2 + \left(\frac{x}{r}\right)^2 = \frac{x^2 + y^2}{r^2} = \frac{r^2}{r^2} = 1$$

Thus

$$\sin^2\theta + \cos^2\theta = 1 \qquad (1)$$

(Although the proof has been indicated for acute angles, you should check that a similar argument shows that the identity holds for all angles θ.) Dividing both sides of (1) by $\cos^2\theta$ (provided $\cos \theta \neq 0$), we get

$$\frac{\sin^2\theta}{\cos^2\theta} + \frac{\cos^2\theta}{\cos^2\theta} = \frac{1}{\cos^2\theta}$$

$$\left(\frac{\sin \theta}{\cos \theta}\right)^2 + 1 = \left(\frac{1}{\cos \theta}\right)^2$$

Since $\sin \theta/\cos \theta = \tan \theta$ and $1/\cos \theta = \sec \theta$, this becomes

$$\tan^2\theta + 1 = \sec^2\theta$$

Similarly, dividing both sides of (1) by $\sin^2\theta$ (provided $\sin \theta \neq 0$) gives us $1 + \cot^2\theta = \csc^2\theta$. ∎

---

*We follow the usual convention of writing $\sin^2\theta$ for $(\sin \theta)^2$. In general, we write $\sin^n\theta$ for $(\sin \theta)^n$ for all integers n except $n = -1$. The exponent $n = -1$ will be assigned another meaning in Section 7.6. Of course, the same convention applies to the other five trigonometric functions.

These identities can be used to write any trigonometric functions in terms of any other. For example, from (1) we get

$$\sin \theta = \pm\sqrt{1 - \cos^2\theta} \qquad (2)$$

where the sign depends on the quadrant. For instance, if θ is in quadrant II, then $\sin \theta$ is positive and hence

$$\sin \theta = \sqrt{1 - \cos^2\theta}$$

whereas if θ is in quadrant III, $\sin \theta$ is negative and so

$$\sin \theta = -\sqrt{1 - \cos^2\theta}$$

EXAMPLE 5

Write $\tan \theta$ in terms of $\sin \theta$, where θ is in quadrant II.

SOLUTION

Since $\tan \theta = \sin \theta/\cos \theta$, we need to write $\cos \theta$ in terms of $\sin \theta$. By (2),

$$\cos \theta = \pm\sqrt{1 - \sin^2\theta}$$

and since $\cos \theta$ is negative in quadrant II, the negative sign applies here. Thus

$$\tan \theta = \frac{\sin \theta}{\cos \theta} = \frac{\sin \theta}{-\sqrt{1 - \sin^2\theta}} \qquad \blacksquare$$

EXAMPLE 6

If $\tan \theta = \frac{2}{3}$ and θ is in quadrant III, find $\cos \theta$.

SOLUTION 1

We need to write $\cos \theta$ in terms of $\tan \theta$. From $\tan^2\theta + 1 = \sec^2\theta$ we get $\sec \theta = \pm\sqrt{\tan^2\theta + 1}$. In quadrant III, $\sec \theta$ is negative, so

$$\sec \theta = -\sqrt{\tan^2\theta + 1}$$

Thus,

$$\cos \theta = \frac{1}{\sec \theta} = \frac{1}{-\sqrt{\tan^2\theta + 1}}$$

$$= \frac{1}{-\sqrt{\left(\frac{2}{3}\right)^2 + 1}} = \frac{1}{-\sqrt{\frac{13}{9}}} = -\frac{3}{\sqrt{13}}$$

SOLUTION 2

This problem can be solved more easily using the method of Example 2 of Section 6.2. Recall that, except for sign, the values of the trigonometric functions of

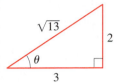

Figure 15

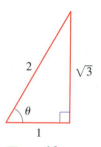

Figure 16

any angle are the same as those of an acute angle (the reference angle). So, ignoring the sign for the moment, let us draw a right triangle with an acute angle α satisfying $\tan \alpha = \frac{2}{3}$ (see Figure 15). By the Pythagorean Theorem the hypotenuse of this triangle has length $\sqrt{13}$. From the triangle in Figure 15 we immediately see that $\cos \alpha = 3/\sqrt{13}$. Since θ is in quadrant III, $\cos \theta$ is negative, and so

$$\cos \theta = -\frac{3}{\sqrt{13}}$$ ■

EXAMPLE 7

If $\sec \theta = 2$ and θ is in quadrant IV, find the other five trigonometric functions of θ.

SOLUTION

We draw a triangle as in Figure 16 so that $\sec \theta = 2$. Taking into account the fact that θ is in quadrant IV, we get

$$\sin \theta = -\frac{\sqrt{3}}{2} \qquad \cos \theta = \frac{1}{2} \qquad \tan \theta = -\sqrt{3}$$

$$\csc \theta = -\frac{2}{\sqrt{3}} \qquad \sec \theta = 2 \qquad \cot \theta = -\frac{1}{\sqrt{3}}$$ ■

■ AREAS OF TRIANGLES

We end this section by giving an application of the trigonometric functions which involves angles that are not necessarily acute. More extensive applications appear in the next two sections.

The area of a triangle is $\mathcal{A} = \frac{1}{2}(\text{base}) \times (\text{height})$. If we know two sides and the included angle of a triangle, then we can find the height using the trigonometric functions, and from this we can find the area.

If θ is an acute angle then the height of the triangle in Figure 17(a) is given by $h = b \sin \theta$. Thus the area is

$$\mathcal{A} = \tfrac{1}{2} \times \text{base} \times \text{height} = \tfrac{1}{2}ab \sin \theta$$

If the angle θ is not acute, then from Figure 17(b) we see that the height of the triangle is

$$h = b \sin(180° - \theta) = b \sin \theta$$

This is so because the reference angle of $180° - \theta$ is the angle θ. Thus, in this case also, the area of the triangle is

$$\mathcal{A} = \tfrac{1}{2} \times \text{base} \times \text{height} = \tfrac{1}{2}ab \sin \theta$$

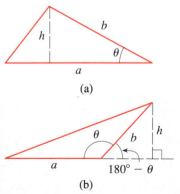

Figure 17

AREA OF A TRIANGLE

The area $\mathcal{A}$ of a triangle with sides of lengths a and b and with included angle θ is

$$\mathcal{A} = \tfrac{1}{2}ab \sin \theta$$

EXAMPLE 8

Find the area of triangle ABC if $|AB| = 10$, $|BC| = 3$, and if the angle included by these two sides is $120°$.

SOLUTION

By the formula for the area of a triangle we have

$$\mathcal{A} = \tfrac{1}{2}ab \sin \theta$$
$$= \tfrac{1}{2}(10)(3) \sin(120°)$$
$$= 15 \sin(60°) \quad \text{(reference angle)}$$
$$= 15 \frac{\sqrt{3}}{2}$$

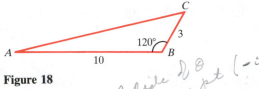

Figure 18

terminal side of θ
passes thru pt $(-5, -12)$
$(-7, -24)$
find 6 trig funct'ns

EXERCISES 6.3

In Exercises 1–8 find the reference angle for the given angle.

1. (a) $225°$ **(b)** $-35°$ **(c)** $181°$

2. (a) $290°$ **(b)** $750°$ **(c)** $570°$

3. (a) $-230°$ **(b)** $-310°$ **(c)** $-1234°$

4. (a) $335°$ **(b)** $-95°$ **(c)** $165°$

5. (a) $\dfrac{3\pi}{5}$ **(b)** $\dfrac{7\pi}{6}$ **(c)** $-\dfrac{2\pi}{3}$

6. (a) $\dfrac{17\pi}{3}$ **(b)** $-\dfrac{\pi}{4}$ **(c)** 3

7. (a) 1.7 **(b)** -7 **(c)** -0.7

8. (a) $\dfrac{23\pi}{11}$ **(b)** $\dfrac{23}{11}$ **(c)** $\dfrac{17\pi}{7}$

In Exercises 9–38 find the exact value of each trigonometric function.

9. $\sin 150°$ **10.** $\cos 225°$ **11.** $\sin 135°$

12. $\tan 330°$ **13.** $\sin(-60°)$ **14.** $\sec(-60°)$

15. $\csc(-630°)$ **16.** $\cot 210°$ **17.** $\cos 570°$

18. $\sec 120°$ **19.** $\tan 750°$ **20.** $\cos 660°$

21. $\sin 450°$ **22.** $\tan 450°$ **23.** $\csc 330°$

24. $\sin \dfrac{2\pi}{3}$ **25.** $\sin \dfrac{5\pi}{3}$ **26.** $\sin \dfrac{3\pi}{2}$

27. $\cos \dfrac{7\pi}{3}$ **28.** $\cos\left(-\dfrac{7\pi}{3}\right)$ **29.** $\tan \dfrac{5\pi}{6}$

30. $\sec \dfrac{17\pi}{3}$ **31.** $\csc \dfrac{5\pi}{4}$ **32.** $\cot\left(-\dfrac{\pi}{4}\right)$

33. $\cos \dfrac{7\pi}{4}$ **34.** $\tan \dfrac{5\pi}{2}$ **35.** $\csc 17\pi$

36. $\tan \dfrac{7\pi}{2}$ **37.** $\sin \dfrac{11\pi}{6}$ **38.** $\cos \dfrac{7\pi}{6}$

In Exercises 39–42 find the sign of the expression if θ is in the given quadrant.

39. $\sin\theta\cos\theta$,   θ in quadrant II

40. $\tan^2\theta\sec\theta$,   θ in quadrant IV

41. $\dfrac{\sin\theta\sec\theta}{\cot\theta}$,   θ in quadrant III

42. $\sin\theta\csc\theta$.   θ in any quadrant

In Exercises 43–46 find the quadrant in which θ lies from the information given.

43. $\sin\theta < 0$ and $\cos\theta < 0$ **44.** $\tan\theta < 0$ and $\sin\theta < 0$

45. $\sec\theta > 0$ and $\tan\theta < 0$ **46.** $\csc\theta > 0$ and $\cos\theta < 0$

In Exercises 47–54 write the first trigonometric function in terms of the second for θ in the given quadrant.

47. $\tan\theta$, $\cos\theta$;   θ in quadrant III

48. $\cot\theta$, $\sin\theta$;   θ in quadrant II

49. $\cos\theta$, $\sin\theta$;   θ in quadrant IV

50. $\sec\theta$, $\sin\theta$;   θ in quadrant I

51. $\sec\theta$, $\tan\theta$;   θ in quadrant II

52. $\csc\theta$, $\cot\theta$;   θ in quadrant III

53. $\sin\theta$, $\sec\theta$;   θ in quadrant IV

54. $\cos\theta$, $\tan\theta$;   θ in quadrant IV

In Exercises 55–62 find the values of the trigonometric functions of θ from the information given.

55. $\sin\theta = \frac{3}{5}$,   θ in quadrant II

56. $\cos\theta = -\frac{7}{12}$,   θ in quadrant III

57. $\tan\theta = -\frac{3}{4}$, $\cos\theta > 0$

58. $\sec\theta = 5$, $\sin\theta < 0$

59. $\csc\theta = 2$,   θ in quadrant I

60. $\cot\theta = \frac{1}{4}$, $\sin\theta < 0$

61. $\cos\theta = -\frac{2}{7}$, $\tan\theta < 0$

62. $\tan\theta = -4$, $\sin\theta > 0$

63. If $\theta = \pi/3$, find the value of each expression:
(a) $\sin 2\theta$, $2\sin\theta$ (b) $\sin\frac{1}{2}\theta$, $\dfrac{\sin\theta}{2}$
(c) $\sin^2\theta$, $\sin(\theta^2)$

64. Find the area of a triangle with sides of length 7 and 9 and included angle 72°.

65. Find the area of a triangle with sides of length 10 and 22 and included angle 10°.

66. Find the area of the triangle in the figure.

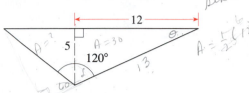

67. Find the area of an equilateral triangle with side of length 10.

68. A triangle has an area of 16 in² and two of the sides of the triangle have lengths 5 in. and 7 in. Find the angle included by these two sides.

69. Find the area of the region shaded in the figure.

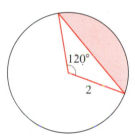

70. Deduce the following identities from the figure.
(a) $\sin^2\theta = \dfrac{1 - \cos 2\theta}{2}$

[*Hint:* Find the length of *DB* in two different ways.]

(b) $\cos^2\theta = \dfrac{1 + \cos 2\theta}{2}$

[*Hint:* Find the length *AD* in two different ways.]

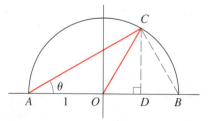

71. Use the identities given in Exercise 70 to find the exact values.
(a) $\sin 15°$, $\cos 15°$ (b) $\sin 22.5°$, $\cos 22.5°$
(c) $\sin 7.5°$, $\cos 7.5°$

SECTION 6.4
THE LAW OF SINES

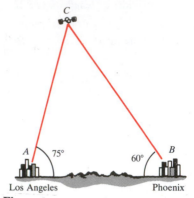

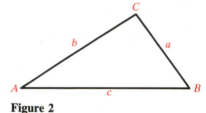

Figure 1

In Section 6.3 we used the trigonometric ratios to solve right triangles. Here we see how the trigonometric functions can be used to solve *oblique triangles*—that is, triangles with no right angles. This is important because many applications involve oblique triangles. Here is an example.

EXAMPLE 1

A satellite orbiting the earth passes directly overhead at observation stations in Phoenix and Los Angeles, 340 miles apart. At an instant when the satellite is between these two stations, its angle of elevation is simultaneously observed to be 60° at Phoenix and 75° at Los Angeles. How far is the satellite from Los Angeles? In other words, find the distance AC in Figure 1.

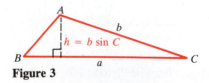

Figure 2

To solve such problems we need to know a relationship between the sides and angles of a triangle. Such relationships exist; two of these are the Law of Sines and the Law of Cosines. In this section we study the Law of Sines and in the next section, the Law of Cosines. To state these laws (or formulas) more easily we follow the convention of labeling the angles of a triangle as A, B, C, and the lengths of the corresponding opposite sides as a, b, c, as in Figure 2.

THE LAW OF SINES

The **Law of Sines** says that in any triangle the lengths of the sides are proportional to the sines of the corresponding opposite angles.

THE LAW OF SINES

In triangle ABC we have

$$\frac{\sin A}{a} = \frac{\sin B}{b} = \frac{\sin C}{c}$$

Proof   To see why the Law of Sines is true, refer to Figure 3. By the formula in Section 6.3, the area of triangle ABC is $\frac{1}{2}ab \sin C$. By the same formula, the area of this triangle is also $\frac{1}{2}ac \sin B$ and $\frac{1}{2}bc \sin A$. Thus

$$\tfrac{1}{2}bc \sin A = \tfrac{1}{2}ac \sin B = \tfrac{1}{2}ab \sin C$$

Multiplying by $2/(abc)$ gives the Law of Sines.   ∎

Figure 3

We can now solve the problem in Example 1.

SOLUTION TO EXAMPLE 1

Whenever two angles in a triangle are known, the third angle can be determined immediately because the sum of the angles of a triangle is 180°. In this case $\angle C = 180° - (75° + 60°) = 45°$ (see Figure 4). We have

$$\frac{\sin B}{b} = \frac{\sin C}{c}$$

or

$$\frac{\sin 60°}{b} = \frac{\sin 45°}{340}$$

Solving for b, we get

$$b = \frac{340 \sin 60°}{\sin 45°} \approx 416$$

Thus the distance of the satellite from Los Angeles is approximately 416 miles.

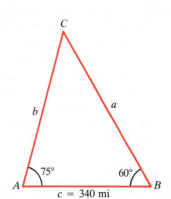

Figure 4

EXAMPLE 2

Solve the triangle in Figure 5.

SOLUTION

First, $\angle B = 180° - (20° + 25°) = 135°$. Since side c is known, we use the relation

$$\frac{\sin A}{a} = \frac{\sin C}{c}$$

to find side a. Thus

$$a = \frac{c \sin A}{\sin C} = \frac{80.4 \sin 20°}{\sin 25°} \approx 65.1$$

Similarly, to find b we use

$$\frac{\sin B}{b} = \frac{\sin C}{c}$$

and solve to get

$$b = \frac{c \sin B}{\sin C} = \frac{80.4 \sin 135°}{\sin 25°} \approx 134.5$$

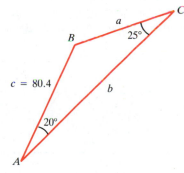

Figure 5

■ THE AMBIGUOUS CASE

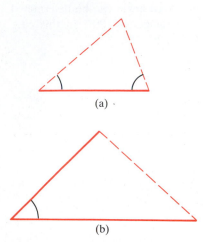
(a)

(b)

Figure 6

To solve a triangle it is necessary to have enough information about the sides and angles to completely determine the triangle. To decide whether this is the case, it is often helpful to sketch the triangle using the information provided. For instance, if we are given two angles and the included side of a triangle, then it is clear that one and only one triangle is formed [see Figure 6(a)]. Similarly, if two sides and the included angle are known, then a unique triangle is determined [Figure 6(b)]. But if, for example, we know all three angles of a triangle, we cannot solve such a triangle, since there are many triangles with the same three angles. (All these triangles would be similar, of course.) So we will not consider this case.

In general, a triangle is determined by three of its parts (angles and sides) as long as at least one of these parts is a side. So, the possibilities are as follows:

Case 1   *SAA*   A side and two angles

Case 2   *SSA*   Two sides and the angle opposite one of those sides

Case 3   *SAS*   Two sides and the included angle

Case 4   *SSS*   Three sides

In each of these cases, except Case 2, a unique triangle is determined by the information given. Case 2 presents a situation where there may be two triangles, one triangle, or no triangle with the given properties. For this reason Case 2 is sometimes called the **ambiguous case**. To see how this can be, we sketch in Figure 7 the possibilities if angle A and sides a and b are given. In part (a) no solution is possible, since side a is too short to complete the triangle. In part (b) the solution is a right triangle. In part (c) two solutions are possible, and in part (d) there is a unique triangle with the given properties. We illustrate these possibilities in the following examples.

Figure 7
The ambiguous case

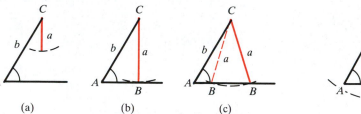
(a)   (b)   (c)   (d)

EXAMPLE 3

In triangle ABC, $\angle A = 45°$, $a = 7\sqrt{2}$, and $b = 7$. Solve the triangle.

SOLUTION

We first sketch the triangle with the information we have. Our sketch in Figure 8 is necessarily tentative, since we do not yet know what the other angles of the

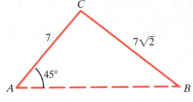

Figure 8

We consider only angles smaller than 180°, since no triangle can contain an angle of 180° or larger.

triangle are. Nevertheless, sketching the triangle will help us see what the possibilities can be.

We first find $\angle B$. By the Law of Sines,

$$\frac{\sin A}{a} = \frac{\sin B}{b}$$

so

$$\sin B = \frac{b \sin A}{a} = \frac{7}{7\sqrt{2}} \sin 45° = \frac{1}{\sqrt{2}} \frac{\sqrt{2}}{2} = \frac{1}{2}$$

Which angles B have $\sin B = \frac{1}{2}$? From the preceding section we know that there are two such angles smaller than 180°. They are 30° and 150°. Which of these possibilities is compatible with what we know about triangle ABC? Since $\angle A = 45°$, we cannot have $\angle B = 150°$, because $45° + 150° > 180°$. So $\angle B = 30°$ and the remaining angle is $\angle C = 180° - (30° + 45°) = 105°$.

It remains to find side c. Again using the Law of Sines, we have

$$\frac{\sin B}{b} = \frac{\sin C}{c}$$

so

$$c = \frac{b \sin C}{\sin B} = \frac{7 \sin 105°}{\sin 30°} = \frac{7 \sin 105°}{\frac{1}{2}} = 14 \sin 105°$$

An approximation for this is $c \approx 13.5$.   ■

In Example 3 there were two possibilities for angle B, and one of these possibilities was not compatible with the rest of the information we knew about the triangle. Notice that if $\sin A < 1$ we must check the angle and its supplement as possibilities, because any angle smaller than 180° can be in the triangle. To decide whether either possibility works, we see whether the resulting sum of the angles of the triangle exceeds 180°. It can happen, as in Figure 7(c), that both possibilities are compatible with the given information. In that case two different triangles are solutions to the problem.

EXAMPLE 4

Solve triangle ABC if $\angle A = 43.1°$, $a = 186.2$, and $b = 248.6$.

SOLUTION

From the given information we sketch the triangle in Figure 9. We see that side a may be drawn in two possible positions to complete the triangle. From the Law of Sines, we get

$$\sin B = \frac{b \sin A}{a} = \frac{248.6 \sin 43.1°}{186.2} \approx 0.91225$$

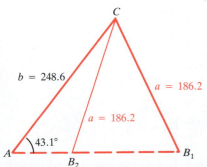

Figure 9

There are two possible angles B between $0°$ and $180°$ such that $\sin B = 0.91225$. Using a calculator, we find that one of these angles is approximately $65.8°$. The other is approximately $180° - 65.8° = 114.2°$. We denote these two angles by B_1 and B_2 so that

$$\angle B_1 \approx 65.8° \quad \text{and} \quad \angle B_2 \approx 114.2°$$

Thus there are two possible triangles that satisfy the given conditions: triangle $A_1 B_1 C_1$ and triangle $A_2 B_2 C_2$. We first solve triangle $A_1 B_1 C_1$. Clearly

$$\angle C_1 \approx 180° - (43.1° + 65.8°) = 71.1°$$

By the Law of Sines, we get

$$c_1 = \frac{a_1 \sin C_1}{\sin A_1} \approx \frac{186.2 \sin 71.1°}{\sin 43.1°} \approx 257.8$$

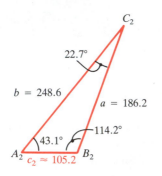

To solve triangle $A_2 B_2 C_2$, notice that

$$\angle C_2 \approx 180° - (43.1° + 114.2°) = 22.7°$$

By the Law of Sines,

$$c_2 = \frac{a_2 \sin C_2}{\sin A_2} \approx \frac{186.2 \sin 22.7°}{\sin 43.1°} \approx 105.2$$

Triangles $A_1 B_1 C_1$ and $A_2 B_2 C_2$ are sketched in Figure 10. ∎

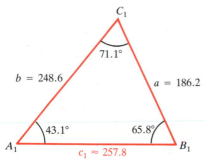

The next example presents a situation for which no triangle is compatible with the given data.

Figure 10

EXAMPLE 5

In triangle ABC, $\angle A = 42°$, $a = 70$, and $b = 122$. Solve the triangle.

SOLUTION

Let us try to find $\angle B$. By the Law of Sines, we have

$$\frac{\sin A}{a} = \frac{\sin B}{b}$$

so

$$\sin B = \frac{b \sin A}{a} = \frac{122 \sin 42°}{70} \approx 1.17$$

Since the sine of an angle is never greater than 1, we conclude that no triangle satisfies the conditions in this problem. ∎

EXERCISES 6.4

In Exercises 1–6 use the Law of Sines to find the indicated side or angle.

1.

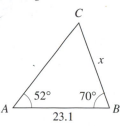

2.

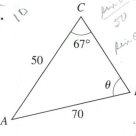

3.

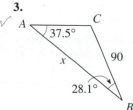

4.

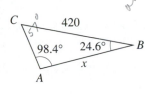

5.

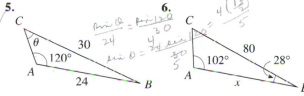

6.

In Exercises 7–8 solve the triangle using the Law of Sines.

7.

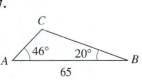

8.

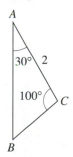

In Exercises 9–14 sketch each triangle and then solve the triangle using the Law of Sines.

9. $\angle A = 50°$, $\angle B = 68°$, $c = 230$

10. $\angle A = 23°$, $\angle B = 110°$, $c = 50$

11. $\angle A = 30°$, $\angle C = 65°$, $b = 10$

12. $\angle A = 22°$, $\angle B = 95°$, $a = 420$

13. $\angle B = 29°$, $\angle C = 51°$, $b = 44$

14. $\angle B = 10°$, $\angle C = 100°$, $c = 115$

In Exercises 15–22 use the Law of Sines to solve for all possible triangles that satisfy the given conditions.

15. $a = 28$, $b = 15$, $\angle A = 110°$

16. $a = 30$, $c = 40$, $\angle A = 37°$

17. $a = 20$, $c = 45$, $\angle A = 125°$

18. $b = 45$, $c = 42$, $\angle C = 38°$

19. $b = 25$, $c = 30$, $\angle B = 25°$

20. $a = 75$, $b = 100$, $\angle A = 30°$

21. $a = 50$, $b = 100$, $\angle A = 50°$

22. $a = 100$, $b = 80$, $\angle A = 135°$

23. To find the distance across a river, a surveyor chooses points A and B that are 200 ft apart on one side of the river (see the figure). She then chooses a reference point C on the opposite side of the river and finds that $\angle BAC \approx 82°$ and $\angle ABC \approx 52°$. Approximate the distance from A to C.

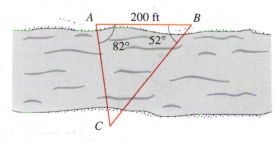

24. A pilot is flying over a straight highway. He determines the angles of depression to two mileposts, 5 mi apart, to be 32° and 48°, as shown in the figure.
   (a) Find the distance of the plane from point A.
   (b) Find the elevation of the plane.

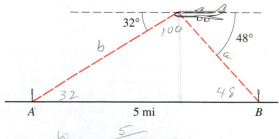

25. The path of a satellite orbiting the earth causes it to pass directly over two tracking stations A and B, which are 50 mi apart. When the satellite is on one side of the two stations, the angles of elevation at A and B are measured to be 87.0° and 84.2°, respectively.

(a) How far is the satellite from station A?

(b) How high is the satellite above the ground?

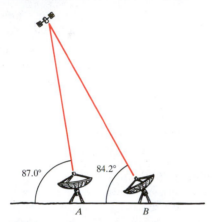

26. A tree on a hillside casts a shadow 215 ft down the hill (see the figure). If the angle of inclination of the hillside is 22° to the horizontal and if the angle of elevation of the sun is 52°, find the height of the tree.

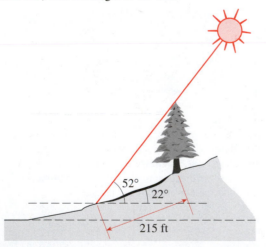

27. A communication tower is located at the top of a steep hill, as shown in the figure. The angle of inclination of the hill is 58°. A guy wire is to be attached to the top of the tower and to the ground, 100 m downhill from the base of the tower. The angle α in the figure is determined to be 12°. Find the length of cable required for the guy wire.

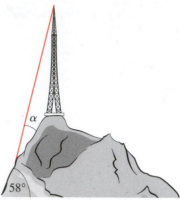

28. Points A and B are separated by a lake. To find the distance between them, a surveyor locates a point C on land such that $\angle CAB = 48.6°$. He also measures CA as 312 ft and CB as 527 ft. Find the distance between A and B.

29. To calculate the height of a mountain, angles α, β, and distance d are determined as in the figure.

(a) Find the length of BC in terms of α, β, and d.

(b) Show that the height h of the mountain is given by the formula

$$h = d \frac{\sin \alpha \sin \beta}{\sin(\beta - \alpha)}$$

(c) Use the formula in part (b) to find the height of a mountain if $\alpha \approx 25°$, $\beta \approx 29°$, and $d \approx 800$ ft.

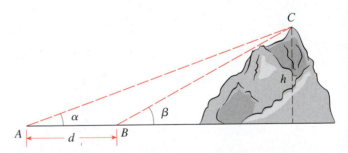

30. Observers at P and Q are located on the side of a hill that is inclined 32° to the horizontal, as shown in the figure. The observer at P determines the angle of elevation to a hot-air balloon to be 62°. At the same instant the observer at Q measures the angle of elevation to the balloon to be 71°. If P is 60 m down the hill from Q, find the distance from Q to the balloon.

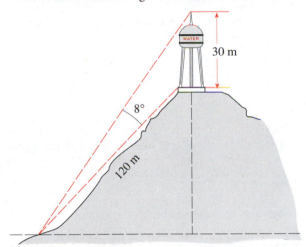

31. A water tower 30 m tall is located at the top of a hill. From a distance of 120 m down the hill, it is observed that the angle formed between the top and base of the tower is 8°. Find the angle of inclination of the hill.

32. For the triangle in the figure, find **(a)** ∠*BCD* and **(b)** ∠*DCA*.

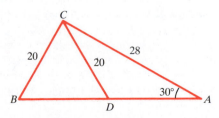

33. In this problem we develop criteria for determining whether there are two solutions, one solution, or no solution to a triangle in the ambiguous case. Suppose we are given ∠*A* and sides *a* and *b*. Sketch triangles like the ones in Figure 7 to show the following:
   (a) If $a \geq b$, then there is exactly one solution.
   (b) If $a < b$, then there are three possibilities:

   If $a < b \sin A$, then there is no solution.

   If $a = b \sin A$, then there is exactly one solution.

   If $a > b \sin A$, then there are two solutions.

34. Refer to Exercise 33. If ∠*A* = 30° and *b* = 100, find the range of values of *a* for which the triangle has **(a)** two solutions, **(b)** one solution, and **(c)** no solution.

35. In triangle *ABC*, ∠*A* = 40°, *a* = 15, and *b* = 20.
   (a) Show that there are two triangles, *ABC* and *A′B′C′*, that satisfy these conditions.
   (b) Show that the areas of the triangles in part (a) are proportional to the sines of the angles *C* and *C′*, that is,

$$\frac{\text{area of } \triangle ABC}{\text{area of } \triangle A'B'C'} = \frac{\sin C}{\sin C'}$$

36. Show that given the three angles *A*, *B*, *C* of a triangle and one side, say *a*, then the area of the triangle is

$$\text{area} = \frac{a^2 \sin B \sin C}{2 \sin A}$$

37. (a) Let *r* be the radius of the circle in which triangle *ABC* is inscribed (see the figure). Prove that

$$2r = \frac{a}{\sin A} = \frac{b}{\sin B} = \frac{c}{\sin C}$$

[*Hint:* ∠*AOD* = ∠*ACB*.]
   (b) Show that the area of triangle *ABC* is $abc/(4r)$.

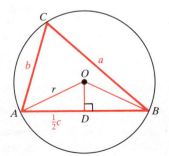

SECTION 6.5
THE LAW OF COSINES

The Law of Sines cannot be used directly to solve triangles if we know two sides and the angle between them or if we know all three sides (these are Cases 3 and 4 of the preceding section). In these two cases the **Law of Cosines** applies.

THE LAW OF COSINES

In any triangle ABC we have

$$a^2 = b^2 + c^2 - 2bc \cos A$$

$$b^2 = a^2 + c^2 - 2ac \cos B$$

$$c^2 = a^2 + b^2 - 2ab \cos C$$

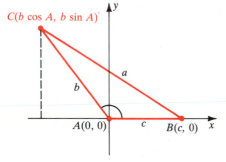

Figure 1

Proof   To prove the Law of Cosines, place triangle ABC so that angle A is at the origin, as shown in Figure 1. The coordinates of the vertices B and C are $(c, 0)$ and $(b \cos A, b \sin A)$, respectively. (You should check that the coordinates of these points would be the same if we had drawn angle A to be an acute angle.) Using the Distance Formula, we get

$$a^2 = (b \cos A - c)^2 + (b \sin A - 0)^2$$
$$= b^2 \cos^2 A - 2bc \cos A + c^2 + b^2 \sin^2 A$$
$$= b^2(\cos^2 A + \sin^2 A) - 2bc \cos A + c^2$$
$$= b^2 + c^2 - 2bc \cos A \qquad \text{(because } \sin^2 A + \cos^2 A = 1\text{)}$$

This proves the first formula. The other two formulas are obtained in the same way by placing each of the other vertices of the triangle at the origin and repeating the preceding argument.   ∎

In words, the Law of Cosines says that the square of any side of a triangle is equal to the sum of the squares of the other two sides minus twice the product of those two sides times the cosine of the included angle.

If one of the angles of a triangle, say $\angle C$, is a right angle, then $\cos C = 0$ and the Law of Cosines reduces to the Pythagorean Theorem, $c^2 = a^2 + b^2$. Thus the Pythagorean Theorem is a special case of the Law of Cosines.

EXAMPLE 1

In triangle ABC it is known that $\angle A = 46.53°$, $b = 10.5$, and $c = 18.0$. Solve the triangle.

SOLUTION

We can find a using the Law of Cosines:

$$a^2 = b^2 + c^2 - 2bc \cos A$$
$$= (10.5)^2 + (18.0)^2 - 2(10.5)(18.0)(\cos 46.53°) \approx 174.20$$

Thus $a \approx \sqrt{174.20} \approx 13.19832$. The two remaining angles can now be found using the Law of Sines. We have

$$\sin B = \frac{b \sin A}{a} \approx \frac{10.5 \sin 46.53°}{13.19832} \approx 0.57736$$

So B is the angle whose sine is 0.57736. For this we use our calculator to get $\angle B \approx 35.3°$. Since angle B can have measure between $0°$ and $180°$, another possibility for angle B is $\angle B = 180° - 35.3° = 144.7°$. It is a simple matter to choose between these two possibilities, since the largest angle in a triangle must be opposite the longest side. So the correct choice is $\angle B \approx 35.3°$. In this case $\angle C \approx 180° - (46.5° + 35.3°) = 98.2°$ and indeed the largest angle $\angle C$ is opposite the longest side $c = 18$.

To summarize:

$$\angle B \approx 35.3°, \angle C \approx 98.2°, \text{ and } a \approx 13.2. \qquad \blacksquare$$

EXAMPLE 2

The sides of a triangle are $a = 5$, $b = 8$, and $c = 12$ (see Figure 2). Find the angles of the triangle.

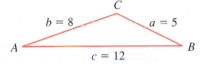

Figure 2

SOLUTION

We first find $\angle A$. From the Law of Cosines, $a^2 = b^2 + c^2 - 2bc \cos A$. Solving for $\cos A$, we get

$$\cos A = \frac{b^2 + c^2 - a^2}{2bc} = \frac{8^2 + 12^2 - 5^2}{2(8)(12)} = \frac{183}{192} = 0.953125$$

Using a calculator, we find that $\angle A \approx 18°$. In the same way the equations

$$\cos B = \frac{a^2 + c^2 - b^2}{2ac} = \frac{5^2 + 12^2 - 8^2}{2(5)(12)} = 0.875$$

$$\cos C = \frac{a^2 + b^2 - c^2}{2ab} = \frac{5^2 + 8^2 - 12^2}{2(5)(8)} = -0.6875$$

give $\angle B \approx 29°$ and $\angle C \approx 133°$. Of course, once two angles are calculated the third can more easily be found from the fact that the sum of the angles of a triangle is $180°$. However, it is a good idea to calculate all three angles using the Law of Cosines and add the three angles as a check on our computations. $\qquad \blacksquare$

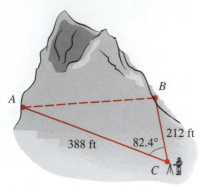

Figure 3

EXAMPLE 3

A tunnel is to be built through a mountain. To estimate the length of the tunnel, a surveyor makes the measurements shown in Figure 3. Use the surveyor's data to approximate the length of the tunnel.

SOLUTION

To approximate the length c of the tunnel, we use the Law of Cosines:

$$c^2 = a^2 + b^2 - 2ab \cos C$$
$$= 388^2 + 212^2 - 2(388)(212) \cos 82.4°$$
$$\approx 173730.2367$$
$$c \approx \sqrt{173730.2367} \approx 416.8$$

Thus the tunnel will be approximately 417 ft long. ∎

■ AREA OF TRIANGLES

An interesting application of the Law of Cosines is a formula for finding the area of a triangle from the lengths of its three sides.

HERON'S FORMULA

> The area $\mathcal{A}$ of triangle ABC is given by
>
> $$\mathcal{A} = \sqrt{s(s - a)(s - b)(s - c)}$$
>
> where $s = \frac{1}{2}(a + b + c)$ is the **semiperimeter** of the triangle, that is, s is half the perimeter.

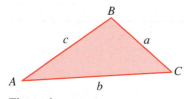

Figure 4

Proof We start with the formula $\mathcal{A} = \frac{1}{2}ab \sin C$ from Section 6.3. Thus

$$\mathcal{A}^2 = \tfrac{1}{4}a^2b^2 \sin^2 C$$
$$= \tfrac{1}{4}a^2b^2(1 - \cos^2 C)$$
$$= \tfrac{1}{4}a^2b^2(1 - \cos C)(1 + \cos C)$$

Next, we write the expressions $1 - \cos C$ and $1 + \cos C$ in terms of a, b, and c. By the Law of Cosines, we have

$$\cos C = \frac{a^2 + b^2 - c^2}{2ab}$$

Thus

$$1 + \cos C = 1 + \frac{a^2 + b^2 - c^2}{2ab}$$

$$= \frac{2ab + a^2 + b^2 - c^2}{2ab}$$

$$= \frac{(a + b)^2 - c^2}{2ab}$$

$$= \frac{(a + b + c)(a + b - c)}{2ab}$$

Similarly,

$$1 - \cos C = \frac{(c + a - b)(c - a + b)}{2ab}$$

Substituting these expressions in the formula that we obtained for $\mathcal{A}^2$ gives

$$\mathcal{A}^2 = \tfrac{1}{4}a^2b^2 \frac{(a + b + c)(a + b - c)}{2ab} \frac{(c + a - b)(c - a + b)}{2ab}$$

$$= \frac{(a + b + c)}{2} \frac{(a + b - c)}{2} \frac{(c + a - b)}{2} \frac{(c - a + b)}{2}$$

$$= s(s - c)(s - b)(s - a)$$

We leave it as an exercise to show that each factor in the last expression equals the corresponding factor in the preceding expression. Heron's Formula now follows by taking the square root of each side. ∎

EXAMPLE 4

A businessman wishes to buy a triangular lot in a busy downtown location (see Figure 5). The lot frontages on the three adjacent streets are 125 ft, 280 ft, and 315 ft. Find the area of the lot.

SOLUTION

The semiperimeter of the lot is

$$s = \frac{125 + 280 + 315}{2} = 360$$

By Heron's Formula the area is

$$\mathcal{A} = \sqrt{360(360 - 125)(360 - 280)(360 - 315)} \approx 17{,}451.6$$

Thus the area is approximately 17,452 ft².

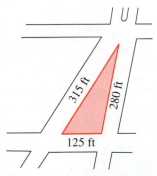

315 ft

280 ft

125 ft

Figure 5

EXERCISES 6.5

In Exercises 1–6 use the Law of Cosines to determine the indicated side or angle.

1.

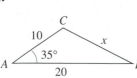

2.

$x^2 = 10^2 + 16^2 - 2(10)(16) \cos 98$

20.01

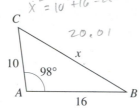

3. ✓

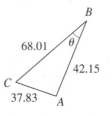

4.

$15 - 4.6^2 = 60.1^2 + 122.5^2 - 2(60.1)(122.5) \cos \theta$

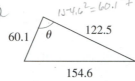

5. ✓

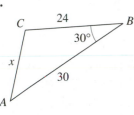

6.

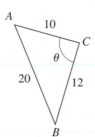

In Exercises 7–16 solve triangle ABC.

7.

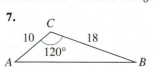

8.

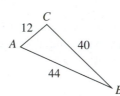

9. ✓ $a = 3.0$, $b = 4.0$, $\angle C = 53°$

10. $b = 60$, $c = 30$, $\angle A = 70°$

11. ✓ $a = 20$, $b = 25$, $c = 22$

12. $a = 10$, $b = 12$, $c = 16$

13. $b = 125$, $c = 162$, $\angle B = 40°$

14. $a = 65$, $c = 50$, $\angle C = 52°$

15. $a = 50$, $b = 65$, $\angle A = 55°$

16. $a = 73.5$, $\angle B = 61°$, $\angle C = 83°$

In Exercises 17–24 find the indicated side or angle. (Use either the Law of Sines or the Law of Cosines, as appropriate.)

17.

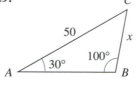

18.

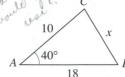

19.

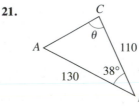

20.

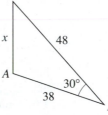

21.

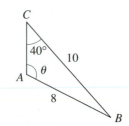

22.

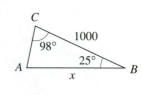

23. ✓

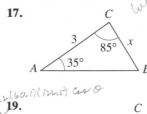

24.

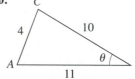

25. In order to find the distance across a small lake, the measurements in the figure were made by a surveyor. Find the distance across the lake using this information.

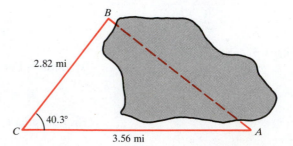

26. A parallelogram has sides of lengths 3 and 5 and one of its angles is 50°. Find the length of the diagonals.

27. Two straight roads diverge at an angle of 65°. Two cars leave the intersection at 2:00 P.M., one traveling at 50 mi/h and the other at 30 mi/h. How far apart are the cars at 2:30 P.M.?

28. A car travels along a straight road, heading east for one hour, then traveling for 30 min on another road that leads northeast. If the car has maintained a constant speed of 40 mi/h, how far is it from its starting position?

29. A pilot flies in a straight path for 1 h 30 min. She then makes a course correction, heading 10° to the right of her original course, and flies 2 h in the new direction. If she maintains a constant speed of 625 mi/h, how far is she from her starting position?

30. Two boats leave the same port at the same time. One travels at a speed of 30 mi/h in the direction N50°E and the other travels at a speed of 26 mi/h in a direction S70°E (see the figure). How far apart are the two boats after one hour?

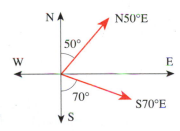

31. A triangular field has sides of length 22, 36, and 44 yd. Find the largest angle.

32. Two tugboats that are 120 ft apart pull a barge, as shown in the figure. If the length of one cable is 212 ft and the length of the other is 230 ft, find the angle formed by the two cables.

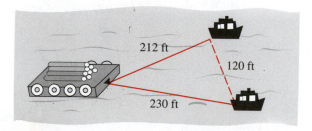

33. A boy is flying two kites at the same time. He has 380 ft of line out to one kite and 420 ft to the other. He estimates the angle between the two lines to be 30°. Approximate the distance between the kites.

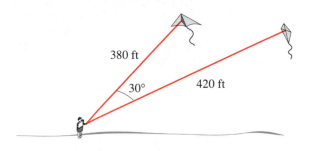

34. A 125-ft tower is located on the side of a mountain that is inclined 32° to the horizontal. A guy wire is to be attached to the top of the tower and anchored at a point 55 ft downhill from the base of the tower. Find the shortest length of wire needed.

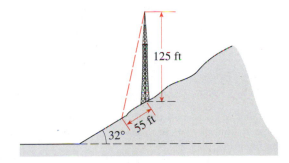

35. A steep mountain is inclined 74° to the horizontal and rises 3400 ft above the surrounding plain. A cable car is to be installed from a point 800 ft from the base to the top of the mountain (see the figure). Find the shortest length of cable needed.

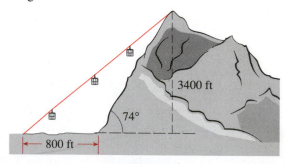

36. The CN Tower in Toronto is the highest tower in the world. A woman on the observation deck, 1150 ft above the ground, wants to determine the distance between two landmarks on the ground below. She observes that the angle formed by the lines of sight to these two landmarks is 43°. She also observes that the angle between the vertical and the line of sight to one of the landmarks is 62° and to the other landmark is 54° (see the figure). Find the distance between the two landmarks.

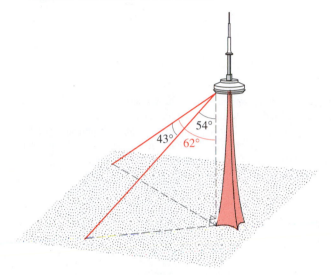

37. Three circles of radii 4 cm, 5 cm, and 6 cm are mutually tangent. Find the area enclosed between the circles (see the figure).

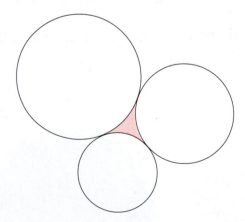

38. Prove that in triangle ABC,

$$a = b \cos C + c \cos B$$

$$b = c \cos A + a \cos C$$

$$c = a \cos B + b \cos A$$

These are called the *Projection Laws*. [*Hint:* To get the first equation, add together the second and third equations in the Law of Cosines and solve for a.]

39. A surveyor wishes to find the distance between two points A and B on the opposite side of a river. On her side of the river, she chooses two points C and D that are 20 m apart and measures the angles shown in the figure. Find the distance between A and B.

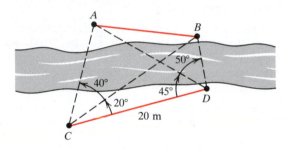

40. Find the area of a triangle with sides of length 12 m, 18 m, and 24 m.

41. Land in downtown Columbia is valued at $20 a square foot. What is the value of a triangular lot with sides of length 112 ft, 148 ft, and 190 ft?

42. Find the area of the quadrilateral in the figure, correct to two decimals.

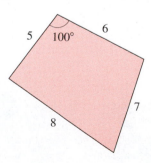

CHAPTER 6 REVIEW

Define, state, or discuss each of the following.

1. Angle

2. Positive angle, negative angle

3. Angle measure

4. Degree

5. Radian

6. Angles in standard position

7. Coterminal angles

8. Length of a circular arc

9. Area of a circular sector

10. The trigonometric ratios

11. Solving right triangles

12. Trigonometric functions of angles

13. Reference angle

14. Fundamental identities

15. Reciprocal identities

16. Pythagorean identities

17. Area of a triangle

18. The Law of Sines

19. The ambiguous case

20. The Law of Cosines

21. Heron's Formula

REVIEW EXERCISES

In Exercises 1 and 2 find the radian measure that corresponds to the given degree measure.

1. **(a)** 70°   **(b)** 420°   **(c)** −240°   **(d)** −40°

2. **(a)** 24°   **(b)** −330°   **(c)** 750°   **(d)** 5°

In Exercises 3 and 4 find the degree measure that corresponds to the given radian measure.

3. **(a)** $\dfrac{7\pi}{2}$ **(b)** $-\dfrac{\pi}{3}$ **(c)** $\dfrac{7\pi}{4}$ **(d)** 2.1

4. **(a)** 8   **(b)** $-\dfrac{5}{2}$ **(c)** $\dfrac{11\pi}{6}$ **(d)** $\dfrac{3\pi}{5}$

5. Find the length of an arc of a circle of radius 8 m if the arc subtends a central angle of 1 rad.

6. Find the measure of a central angle θ in a circle of radius 5 ft if the angle is subtended by an arc of length 7 ft.

7. A circular arc of length 100 ft subtends a central angle of 70°. Find the radius of the circle.

8. How many revolutions will a car wheel of diameter 28 in. make over a period of half an hour if the car is traveling at 60 mi/h?

9. New York and Los Angeles are 2450 miles apart. Find the angle that the arc between these two cities subtends at the center of the earth. (The radius of the earth is 3960 mi.)

10. Find the area of a sector with central angle 2 rad in a circle of radius 5 m.

11. Find the area of a sector with central angle 52° in a circle of radius 200 ft.

12. A sector in a circle of radius 25 ft has an area of 125 ft². Find the central angle of the sector.

In Exercises 13 and 14 find the values of the six trigonometric ratios of θ.

13.

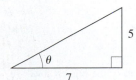

14.

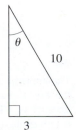

In Exercises 15–18 find the sides labeled x and y, correct to two decimal places.

15.

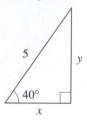

16.

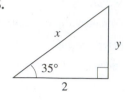

17.

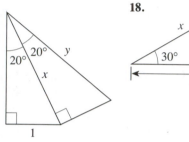

18.

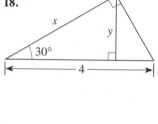

In Exercises 19 and 20 solve the given triangle.

19.

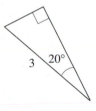

20.

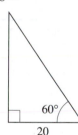

21. Express the lengths a and b in the figure in terms of the trigonometric ratios of θ.

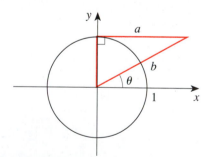

22. The angle of elevation to the top of the highest tower in the world, the CN Tower in Toronto, is 28.81° from a distance of 1 km from the base of the tower. Find the height of the tower.

23. Find the perimeter of a regular hexagon that is inscribed in a circle of radius 8 m.

24. The pistons in a car engine move up and down repeatedly to turn the crankshaft, as shown in the figure. Find the height of the point P above the center O of the crankshaft in terms of the angle θ.

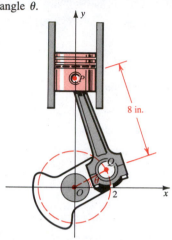

25. As viewed from the earth, the angle subtended by the full moon is 0.518°. Use this information and the fact that the distance AB from the earth to the moon is 236,900 mi to find the radius of the moon.

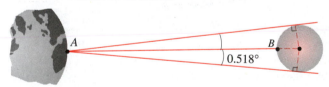

26. A pilot measures the angles of depression to two ships to be 40° and 52° (see the figure). If the pilot is flying at an elevation of 35,000 ft, find the distance between the two ships.

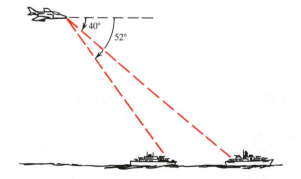

In Exercises 27–38 find the exact value.

27. $\sin 315°$

28. $\csc \dfrac{9\pi}{4}$

29. $\tan(-135°)$

30. $\cos \dfrac{5\pi}{6}$

31. $\cot\left(-\dfrac{22\pi}{3}\right)$

32. $\sin 405°$

33. $\cos 585°$

34. $\sec \dfrac{22\pi}{3}$

35. $\csc \dfrac{8\pi}{3}$

36. $\sec \dfrac{13\pi}{6}$

37. $\cot(-390°)$

38. $\tan \dfrac{23\pi}{4}$

39. Find the values of the six trigonometric ratios of the angle θ in standard position if the point $(-5, 12)$ is on the terminal side of θ.

40. Find $\sin \theta$ if θ is in standard position and its terminal side intersects the circle of radius 1 centered at the origin at the point $\left(-\sqrt{3}/2, \frac{1}{2}\right)$.

41. Find the acute angle that is formed by the line $y - \sqrt{3}\,x + 1 = 0$ and the x-axis.

42. Find the six trigonometric ratios of the angle θ in standard position if its terminal side is in quadrant III and is parallel to the line $4y - 2x - 1 = 0$.

In Exercises 43–46 write the first expression in terms of the second for θ in the given quadrant.

43. $\tan \theta$, $\cos \theta$; θ in quadrant II

44. $\sec \theta$, $\sin \theta$; θ in quadrant III

45. $\tan^2\theta$, $\sin \theta$; θ in any quadrant

46. $\csc^2\theta \cos^2\theta$, $\sin \theta$; θ in any quadrant

In Exercises 47–50 find the values of the six trigonometric functions of θ from the information given.

47. $\tan \theta = \sqrt{7}/3$, $\sec \theta = \frac{4}{3}$

48. $\sec \theta = \frac{41}{40}$, $\csc \theta = -\frac{41}{9}$

49. $\sin \theta = \frac{3}{5}$, $\cos \theta < 0$

50. $\sec \theta = -\frac{13}{5}$, $\tan \theta > 0$

51. If $\tan \theta = -\frac{1}{2}$ for θ in quadrant II, find $\sin \theta + \cos \theta$.

52. If $\sin \theta = \frac{1}{2}$ for θ in quadrant I, find $\tan \theta + \sec \theta$.

53. If $\tan \theta = -1$, find $\sin^2\theta + \cos^2\theta$.

54. If $\cos \theta = -\sqrt{3}/2$ and $\pi/2 < \theta < \pi$, find $\sin 2\theta$.

In Exercises 55–60 find the side labeled x.

55.

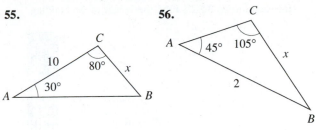

56.

57.

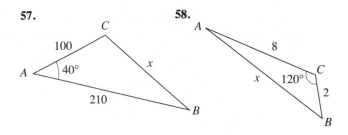

58.

59.

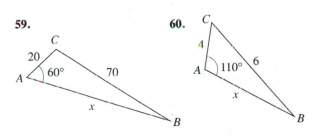

60.

61. Two ships leave a port at the same time. One travels at 20 mi/h in a direction N32°E, and the other travels at 28 mi/h in a direction S42°E. How far apart are the two ships after 2 h?

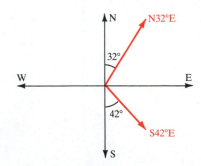

62. From a point A on the ground, the angle of elevation to the top of a tall building is 24.1°. From a point B, which

is 600 ft closer to the building, the angle of elevation is measured to be 30.2°. Find the height of the building.

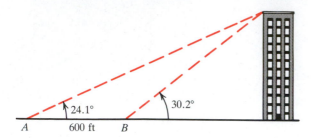

63. Find the distance between the points A and B on the opposite side of a lake from the information given in the figure.

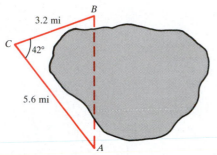

64. A boat is cruising the ocean off a straight shoreline. Points A and B are 120 miles apart on the shore, as shown in the figure. It is found that $\angle A = 42.3°$ and

$\angle B = 68.9°$. Find the shortest distance from the boat to the shore.

65. In order to measure the height of an inaccessible cliff on the opposite side of a river, a surveyor makes the measurements shown in the figure. Find the height of the cliff.

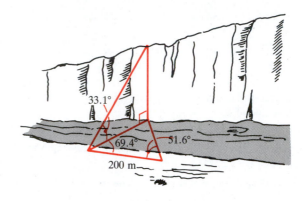

66. Find the area of a triangle with sides of lengths 8 and 14 and included angle 35°.

67. Find the area of a triangle with sides of lengths 5, 6, and 8.

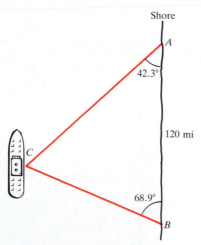

CHAPTER 6 TEST

1. Find the radian measures that correspond to the following degree measures: 300°, −18°, 225°.

2. Find the degree measures that correspond to the following radian measures: $\dfrac{5\pi}{6}$, $-\dfrac{11\pi}{4}$, 2.4.

3. Find the exact values of the following:
   (a) sin 405°
   (b) tan(−150°)
   (c) sec $\dfrac{5\pi}{3}$
   (d) csc $\dfrac{5\pi}{2}$

4. Find tan θ + sin θ for the angle θ shown in the figure.

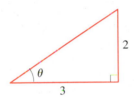

5. Find the lengths a and b shown in the figure in terms of θ.

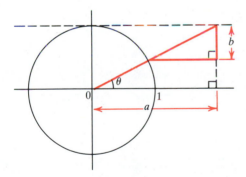

6. If cos θ = $-\frac{1}{3}$ and θ is in quadrant III, find tan θ cot θ + csc θ.

7. If sin θ = $\frac{5}{13}$ and tan θ = $-\frac{5}{12}$, find sec θ.

8. Express tan θ in terms of sec θ for θ in quadrant II.

9. The base of the ladder in the figure is 6 ft from the building, and the angle formed by the ladder and the ground is 73°. How high up the building does the ladder touch?

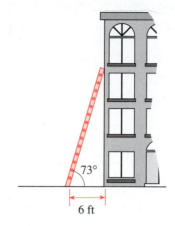

73°

6 ft

In Problems 10–13 find the side labeled x.

10.

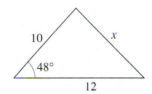

11.

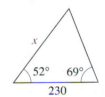

12.

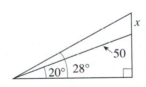

13.

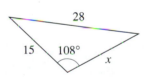

Problems 14 and 15 refer to the figure.

14. Find the area of the shaded region.

15. Find the perimeter of the shaded region.

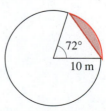

72°

10 m

Problems 16 and 17 refer to the triangle in the figure.

16. Find the angle opposite the longest side.

17. Find the area of the triangle.

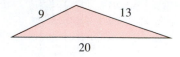

9   13

20

18. Two wires hold down a balloon as shown in the figure. The distance between the anchor points is 100 ft and the shorter wire is 550 ft. How high is the balloon above the ground?

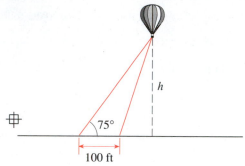

h

75°

100 ft

FOCUS ON PROBLEM SOLVING

It is often necessary to combine different problem-solving principles. The principles of **drawing a diagram** and **adding something extra** are used in Euler's solution of the "Königsberg bridge problem." The problem itself is interesting, but not of great practical importance. However, Euler's solution is so brilliant that it is the basis for a whole field of mathematics called *network theory*, which has practical applications to electric circuits and economics.

THE KÖNIGSBERG BRIDGE PROBLEM

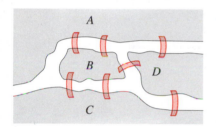

Figure 1

The old city of Königsberg is situated on both banks of the Pregel River and on two islands in the river. Seven bridges connect the parts of the city, as shown in Figure 1. According to local folklore it was impossible to cross all seven bridges in a continuous walk without recrossing any. When Euler heard of the problem he realized that an interesting mathematical principle must be at work here. To discover the principle Euler first stripped away the nonessential parts of the problem by drawing a simpler diagram of the city where the land masses are represented by points and the bridges by lines connecting the points (Figure 2). The problem now is to draw the diagram without lifting the pencil and without retracing any lines. To show that this is impossible, Euler added something extra by labeling each point with the number of line segments that end at the point (Figure 3). He then argued that every pass through a point accounts for two line ends (one entering and one leaving). Thus each point, except possibly the starting point and the ending point, must be labeled by an even number in order for the walk to be possible. Since there are more than two points labeled by odd numbers, the walk is impossible!

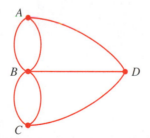

Figure 2

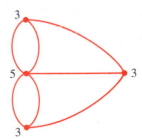

Figure 3

The beauty of Euler's solution is that it is completely general. In modern language a diagram consisting of points and connecting lines is called a *network*, each point is a *vertex*, each line an *edge*, and the number of edges ending at a vertex is the *order* of the vertex. A network is *traversable* if it can be drawn in one stroke without retracing any edge. Euler's reasoning about the Königsberg bridge problem shows the following: *If a network has more than two vertices with odd order, then it is not traversable.*

PROBLEMS

In Exercises 1–6, determine if each network is traversable. If it is, find a path that traverses it.

1.

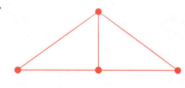

2.

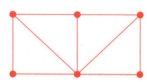

3.

4.

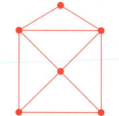

5.

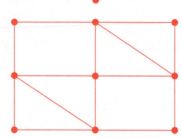

6.

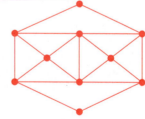

7. When two bubbles cling together in midair, their common surface is part of a sphere whose center D lies on the line passing through the centers of the bubbles. Also, the angles ACB and ACD both have measure 60° (see the figure).

(a) Show that the radius r of the common face is given by

$$r = \frac{ab}{a - b}$$

(b) If the radii of the two bubbles are 4 cm and 3 cm, respectively, find the radius of the common face.

(c) What shape does the common face take if the two bubbles have radii of the same length?

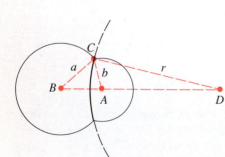

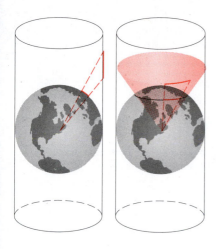

8. In order to draw a map of the spherical earth on a flat sheet of paper, several different ingenious methods have been developed. One of these is the Mercator projection. In this method each point on the spherical earth is projected onto a circumscribed cylinder tangent to the sphere at every point on the equator by a line through the center of the earth (see the figure). By what factor are the following distances distorted using this method of projection? That is, what is the ratio of the projected distance on the cylinder to the actual distance on the sphere?
   (a) The distance between 20° and 21° North latitude along a meridian
   (b) The distance between 40° and 41° North latitude along a meridian
   (c) The distance between 80° and 81° North latitude along a meridian
   (d) The distance between two points that are 1° apart on the 20th parallel
   (e) The distance between two points that are 1° apart on the 40th parallel
   (f) The distance between two points that are 1° apart on the 80th parallel

9. A ball of radius a is inside a cone so that the surface of the cone is tangent to the ball. A larger ball of radius b fits inside the cone in such a way that it is tangent to both the ball of radius a and the sides of the cone. Express b in terms of a and the angle θ shown in the figure.

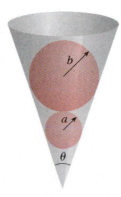

10. Two circles of radius 1 are placed so that their centers are 1 unit apart. Find the area of the region common to both circles.

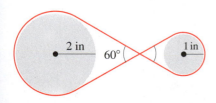

11. Two pulleys of radius 1 in. and 2 in. are connected by a belt, as shown in the figure. The belt crosses itself at an acute angle of 60°. Find the length of the belt.

12. Suppose that the lengths of the sides of a triangle are rational numbers. Prove that the cosine of each angle of the triangle is a rational number.

7 ANALYTIC TRIGONOMETRY

The function of mathematics in providing answers is often less important than its function in providing understanding.

BEN NOBLE

There are many relations among the trigonometric functions. The identities we have encountered in Chapters 5 and 6 are basic examples of these. In this chapter we study some of the deeper properties that relate the trigonometric functions to one another. We find identities for trigonometric functions of sums and differences of real numbers, multiple-angle formulas, and other related identities.

Because of these properties, the trigonometric functions are very useful for certain mathematical applications. In Sections 7.7 and 7.8 we see how they are used in the study of complex numbers and vectors. In Chapter 9 the trigonometric functions will be applied to the study of rotations of graphs in the plane, polar coordinates, and parametric equations.

SECTION 7.1
TRIGONOMETRIC IDENTITIES

We have been working with equations and identities throughout this book. Let us take a closer look at what these concepts mean.

EQUATIONS AND IDENTITIES

An **equation** is a statement that two mathematical expressions are equal. Thus

$$1 + 1 = 2$$

is an equation. An equation may contain a variable; for example,

$$3x + 2 = 14$$

is an equation. This kind of equation is valid only for certain values of x and so it is sometimes called a **conditional equation**. Such an equation implies a question: For what values of the variable x is the statement $3x + 2 = 14$ true? The value $x = 4$ makes this equation true, and for all other values of x it is false. In general, each value of x that makes an equation true is called a **solution** of the equation.

Here are more examples of equations:

(A) $x + 2 = 5$	(B) $x + 1 = 1 + x$
$2x^2 + 3x = x - 3$	$(x + 1)^2 = x^2 + 2x + 1$
$y + 2x + 1 = y + x$	$(x + y)^2 = x^2 + 2xy + y^2$
$e^{2x} + 5e^x + 6 = 0$	$e^{x+y} = e^x e^y$
$\ln x = 0$	$\ln(xy) = \ln x + \ln y$
$\sin^2 x - 1 = 0$	$\sin^2 x + \cos^2 x = 1$

What is the difference between the equations in list (A) and those in (B)? The important difference is that the equations in (B) are true for *every* value of the variables for which both sides are defined. Equations with this property are called **identities**. For some equations, it is easy to see that they are identities. For example, the equation

$$x + 1 = 1 + x$$

is clearly an identity, whereas it takes more work to show that

$$\sin^2 x + \cos^2 x = 1$$

is also an identity.

In this section we are interested in equations and identities that involve the trigonometric functions. We first review some of the trigonometric identities from Chapters 5 and 6.

FUNDAMENTAL TRIGONOMETRIC IDENTITIES

Reciprocal Identities

$$\csc x = \frac{1}{\sin x} \qquad \sec x = \frac{1}{\cos x} \qquad \cot x = \frac{1}{\tan x}$$

$$\tan x = \frac{\sin x}{\cos x} \qquad \cot x = \frac{\cos x}{\sin x}$$

Pythagorean Identities

$$\sin^2 x + \cos^2 x = 1$$

$$\tan^2 x + 1 = \sec^2 x$$

$$1 + \cot^2 x = \csc^2 x$$

Even–odd Identities

$$\sin(-x) = -\sin x$$

$$\cos(-x) = \cos x$$

$$\tan(-x) = -\tan x$$

Identities help us to write the same expression in different ways. It is often possible to rewrite a complicated-looking expression as a much simpler one. In Example 1 a trigonometric expression involving the sine, secant, and tangent functions turns out to be equal to the cosine function.

EXAMPLE 1

Simplify the expression $(1 + \sin x)(\sec x - \tan x)$.

SOLUTION

We use the fundamental identities.

$$(1 + \sin x)(\sec x - \tan x) = (1 + \sin x)\left(\frac{1}{\cos x} - \frac{\sin x}{\cos x}\right) \qquad \text{(reciprocal identities)}$$

$$= (1 + \sin x)\left(\frac{1 - \sin x}{\cos x}\right) \qquad \text{(common denominator)}$$

$$= \frac{1 - \sin^2 x}{\cos x} \qquad \text{(difference of squares)}$$

$$= \frac{\cos^2 x}{\cos x} \qquad \text{(Pythagorean identity)}$$

$$= \cos x \qquad \blacksquare$$

■ **PROVING TRIGONOMETRIC IDENTITIES**

Many identities follow from the fundamental identities. In the following examples we learn how to prove that a given trigonometric equation is an identity, and in the process we see how to discover new identities.

First, it is easy to decide when a given equation is *not* an identity. All that needs to be done is to show that the equation does not hold for some value of the variable (or variables). Thus, the equation

$$\sin x + \cos x = 1$$

is not an identity, because when $x = \pi/4$ we have

$$\sin \frac{\pi}{4} + \cos \frac{\pi}{4} = \frac{\sqrt{2}}{2} + \frac{\sqrt{2}}{2} = \sqrt{2} \neq 1$$

To verify that a trigonometric equation is an identity, we transform one side of the equation into the other side by a series of steps, each of which is itself an identity. The next three examples illustrate this procedure.

EXAMPLE 2

Verify the identity $\tan x \sin x + \cos x = \sec x$.

SOLUTION

We start with the left-hand side of this equation and transform it into the right-hand side. In each step a trigonometric identity or an algebraic identity is used. Supply the reasons for each step.

$$\tan x \sin x + \cos x = \left(\frac{\sin x}{\cos x}\right)\sin x + \cos x$$

$$= \frac{\sin^2 x}{\cos x} + \cos x$$

$$= \frac{\sin^2 x + \cos^2 x}{\cos x}$$

$$= \frac{1}{\cos x} = \sec x \qquad \blacksquare$$

In Example 2 it is not easy to see how to change the right-hand side into the left side, but it is possible to do so. Just notice that each of the steps is reversible. In other words, if we start with the last expression in the proof (this is the right-hand side of the equation) and read backward through the steps, we see that each step is an identity and so the right-hand side is transformed into the left-hand side. You will probably agree, however, that it is more difficult to see the steps necessary to arrive at the left-hand side from the right-hand side. *In general it is better to change the more complicated side of the equation into the simpler side.*

EXAMPLE 3

Prove that the following equation is an identity:

$$\frac{\sin x \,(\sin x - \csc x) + \cot^2 x}{\cos^2 x} = \cot^2 x$$

SOLUTION

Since the left-hand side appears more complicated, we transform it into the other side. Supply the reasons for each step:

$$\frac{\sin x \,(\sin x - \csc x) + \cot^2 x}{\cos^2 x} = \frac{\sin^2 x - \sin x \csc x + \cot^2 x}{\cos^2 x}$$

$$= \frac{\sin^2 x - \sin x \left(\dfrac{1}{\sin x}\right) + \cot^2 x}{\cos^2 x}$$

$$= \frac{\sin^2 x - 1 + \cot^2 x}{\cos^2 x}$$

$$= \frac{\sin^2 x - 1}{\cos^2 x} + \frac{\cot^2 x}{\cos^2 x}$$

$$= \frac{-\cos^2 x}{\cos^2 x} + \left(\frac{\cos^2 x}{\sin^2 x}\right)\frac{1}{\cos^2 x}$$

$$= -1 + \frac{1}{\sin^2 x}$$

$$= \csc^2 x - 1 = \cot^2 x \qquad \blacksquare$$

EXAMPLE 4

Verify the identity

$$2 \tan x \sec x = \frac{1}{1 - \sin x} - \frac{1}{1 + \sin x}$$

SOLUTION

Finding a common denominator and combining the fractions on the right-hand side of this equation, we get

$$\frac{1}{1 - \sin x} - \frac{1}{1 + \sin x} = \frac{(1 + \sin x) - (1 - \sin x)}{(1 - \sin x)(1 + \sin x)} = \frac{2 \sin x}{1 - \sin^2 x}$$

$$= \frac{2 \sin x}{\cos^2 x} = 2 \frac{\sin x}{\cos x} \left(\frac{1}{\cos x} \right) = 2 \tan x \sec x \quad ∎$$

In Example 5 we introduce something extra to the problem by multiplying the numerator and the denominator by a trigonometric expression, chosen so that we can simplify the result.

EXAMPLE 5

Verify the identity

$$\frac{\cot x}{1 - \sin x} = \sec x \, (\csc x + 1)$$

SOLUTION

We multiply the numerator and denominator of the left-hand side of this equation by $1 + \sin x$.

$$\frac{\cot x}{1 - \sin x} = \frac{\cot x}{1 - \sin x} \frac{1 + \sin x}{1 + \sin x} = \frac{\cot x \, (1 + \sin x)}{1 - \sin^2 x}$$

$$= \frac{\cot x \, (1 + \sin x)}{\cos^2 x} = \frac{\cos x}{\sin x} \frac{(1 + \sin x)}{\cos^2 x}$$

$$= \frac{1 + \sin x}{\sin x \cos x} = \frac{1}{\sin x \cos x} + \frac{\sin x}{\sin x \cos x}$$

$$= \csc x \sec x + \sec x = \sec x \, (\csc x + 1) \quad ∎$$

Here is another method for proving that an equation is an identity. If we can transform each side of the equation separately, by way of identities, to arrive at the same result, then the equation is an identity. Example 6 illustrates this procedure.

EXAMPLE 6

Prove that the following equation is an identity:

$$\frac{1}{1 - \sin z} = \sec^2 z + \tan z \sec z$$

SOLUTION

We transform the left-hand side to get

$$\frac{1}{1 - \sin z} = \frac{1}{1 - \sin z} \frac{1 + \sin z}{1 + \sin z} = \frac{1 + \sin z}{1 - \sin^2 z} = \frac{1 + \sin z}{\cos^2 z}$$

From the right-hand side we get

$$\sec^2 z + \tan z \sec z = \frac{1}{\cos^2 z} + \frac{\sin z}{\cos z} \frac{1}{\cos z} = \frac{1}{\cos^2 z} + \frac{\sin z}{\cos^2 z} = \frac{1 + \sin z}{\cos^2 z}$$

It follows that

$$\frac{1}{1 - \sin z} = \sec^2 z + \tan z \sec z \qquad \blacksquare$$

There are often several ways to prove that a given identity is true. For instance, in Example 6 we could have continued to manipulate the left-hand side to arrive at the other side. In all the preceding examples we could have started with either side and transformed it to the other side by way of identities. Indeed, it is not possible to give rules for proving identities. You will find, however, that with the suggestions made in this section and some practice, it will in most cases be easy to see what to do to prove a particular identity.

 As a final warning, notice that to prove an identity we do *not* perform the same operation on both sides of the equation. Remember that we do not know at first that the equation we are given really is an identity. It is our task to prove that it is. Indeed, if we start with an equation that is not an identity, such as

$$\sin x = -\sin x \qquad (1)$$

and square both sides, we get the equation

$$\sin^2 x = \sin^2 x \qquad (2)$$

which is clearly an identity. Does this mean that the original equation is an identity? Of course not. The problem here is that the operation of squaring is not **reversible** in the sense that we cannot arrive back at (1) from (2) by taking square roots (reversing the procedure). Only operations that are reversible will necessarily transform an identity into an identity.

We end this section by describing the technique of *trigonometric substitution*, which we use to convert algebraic expressions to trigonometric ones. This is often useful in calculus, for instance, in finding the area of a circle or an ellipse.

EXAMPLE 7

Substitute $\sin\theta$ for x in the expression $\sqrt{1 - x^2}$ and simplify. Assume that $-\pi/2 \leq \theta \leq \pi/2$.

SOLUTION

Setting $x = \sin\theta$, we have

$$\sqrt{1 - x^2} = \sqrt{1 - \sin^2\theta}$$
$$= \sqrt{\cos^2\theta}$$
$$= \cos\theta$$

The last equality is true, because $\cos\theta \geq 0$ for the values of θ in question.   ■

EXERCISES 7.1

In Exercises 1–8 write each trigonometric expression in terms of sine and cosine, and then simplify.

1. $\cos x \tan x$

2. $\sec\alpha + \tan\alpha$

3. $\sec^2 x - \tan^2 x$

4. $(\tan w - \csc w)(\sin w \cos w)$

5. $\tan A + \cot A$

6. $\dfrac{\tan x + \cot x}{\sec x \csc x}$

7. $\cos u + \tan u \sin u$

8. $\cos^2 x (1 + \tan^2 x)$

In Exercises 9–24 simplify the trigonometric expression.

9. $\dfrac{\cos x \sec x}{\cot x}$

10. $\cos^3 x + \sin^2 x \cos x$

11. $\dfrac{1 + \sin y}{1 + \csc y}$

12. $\dfrac{\tan x}{\sec(-x)}$

13. $\dfrac{\sec^2 x - 1}{\sec^2 x}$

14. $\dfrac{\sec x - \cos x}{\tan x}$

15. $\dfrac{1 + \csc x}{\cos x + \cot x}$

16. $\dfrac{\sin x}{\csc x} + \dfrac{\cos x}{\sec x}$

17. $\dfrac{1 + \sin u}{\cos u} + \dfrac{\cos u}{1 + \sin u}$

18. $\tan x \cos x \csc x$

19. $\dfrac{2 + \tan^2 x}{\sec^2 x} - 1$

20. $\dfrac{1 + \cot A}{\csc A}$

21. $\tan\theta + \cos(-\theta) + \tan(-\theta)$

22. $\dfrac{\cos x}{\sec x + \tan x}$

23. $\dfrac{\sin t}{1 - \cos t} - \csc t$

24. $(\sec x - \tan x)^2 (1 + \sin x)$

In Exercises 25–110 verify the identity.

25. $\tan x \cot x = 1$

26. $\sin x \csc x = 1$

27. $\sin\theta \cot\theta = \cos\theta$

28. $\dfrac{\tan x}{\sec x} = \sin x$

29. $\dfrac{\cos u \sec u}{\tan u} = \cot u$

30. $\dfrac{\cot x \sec x}{\csc x} = 1$

31. $\dfrac{\tan y}{\csc y} = \sec y - \cos y$

32. $\dfrac{\cos v}{\sec v \sin v} = \csc v - \sin v$

33. $\sin B + \cos B \cot B = \csc B$

34. $\cos x + \sin x \tan x = \sec x$

35. $\sin(-x)\tan(-x) = \sin^2 x \sec x$

36. $\cos(-x) - \sin(-x) = \cos x + \sin x$

37. $\cot(-\alpha)\cos(-\alpha) + \sin(-\alpha) = -\csc \alpha$

38. $\csc x[\csc x + \sin(-x)] = \cot^2 x$

39. $(1 - \sin x)(1 + \sin x) = \cos^2 x$

40. $(\sin x + \cos x)^2 = 1 + 2\sin x \cos x$

41. $(1 - \cos \beta)(1 + \cos \beta) = \dfrac{1}{\csc^2 \beta}$

42. $\dfrac{\cos x}{\sec x} + \dfrac{\sin x}{\csc x} = 1$

43. $\dfrac{(\sin x + \cos x)^2}{\sin^2 x - \cos^2 x} = \dfrac{\sin^2 x - \cos^2 x}{(\sin x - \cos x)^2}$

44. $(\sin x + \cos x)^4 = (1 + 2\sin x \cos x)^2$

45. $\dfrac{\sec t - \cos t}{\sec t} = \sin^2 t$

46. $\dfrac{1 - \sin x}{1 + \sin x} = (\sec x - \tan x)^2$ $zwoop$

47. $\dfrac{1}{1 - \sin^2 y} = 1 + \tan^2 y$

48. $\csc x - \sin x = \cos x \cot x$

49. $(\cot x - \csc x)(\cos x + 1) = -\sin x$

50. $\sin^4 \theta - \cos^4 \theta = \sin^2 \theta - \cos^2 \theta$

51. $(1 - \cos^2 x)(1 + \cot^2 x) = 1$

52. $\cos^2 x - \sin^2 x = 2\cos^2 x - 1$

53. $2\cos^2 x - 1 = 1 - 2\sin^2 x$

54. $\tan y + \cot y = \sec y \csc y$

55. $\dfrac{1 - \cos \alpha}{\sin \alpha} = \dfrac{\sin \alpha}{1 + \cos \alpha}$

56. $\csc x \cos^2 x + \sin x = \csc x$

57. $\dfrac{1}{1 - \sin x} = \sec x (\sec x + \tan x)$

58. $\sin^2 \alpha + \cos^2 \alpha + \tan^2 \alpha = \sec^2 \alpha$

59. $\dfrac{\sin x - 1}{\sin x + 1} = \dfrac{-\cos^2 x}{(\sin x + 1)^2}$

60. $\dfrac{\sin w}{\sin w + \cos w} = \dfrac{\tan w}{1 + \tan w}$

61. $\dfrac{(\sin t + \cos t)^2}{\sin t \cos t} = 2 + \sec t \csc t$

62. $\sec t \csc t (\tan t + \cot t) = \sec^2 t + \csc^2 t$

63. $\dfrac{1 + \tan^2 u}{1 - \tan^2 u} = \dfrac{1}{\cos^2 u - \sin^2 u}$

64. $\dfrac{1 + \sec^2 x}{1 + \tan^2 x} = 1 + \cos^2 x$

65. $\dfrac{\sec x}{\sec x - \tan x} = \sec x (\sec x + \tan x)$

66. $\dfrac{\sec x + \csc x}{\tan x + \cot x} = \sin x + \cos x$

67. $\sec v - \tan v = \dfrac{1}{\sec v + \tan v}$

68. $\csc x - \cot x = \dfrac{1}{\csc x + \cot x}$

69. $\dfrac{1 + \sin x}{1 - \sin x} = 2\sec x (\sec x + \tan x) - 1$

70. $\dfrac{\sin A}{1 - \cos A} - \cot A = \csc A$

71. $\dfrac{\sin x + \cos x}{\sec x + \csc x} = \sin x \cos x$

72. $\dfrac{1 - \cos x}{\sin x} + \dfrac{\sin x}{1 - \cos x} = 2\csc x$

73. $\dfrac{\csc x - \cot x}{\sec x - 1} = \cot x$ **74.** $\dfrac{\csc^2 x - \cot^2 x}{\sec^2 x} = \cos^2 x$

75. $\tan^2 u - \sin^2 u = \tan^2 u \sin^2 u$

76. $\cot^2 u - \cos^2 u = \cot^2 u \cos^2 u$

77. $\dfrac{\tan v \sin v}{\tan v + \sin v} = \dfrac{\tan v - \sin v}{\tan v \sin v}$

78. $\dfrac{\cot v \cos v}{\cot v + \cos v} = \dfrac{\cot v - \cos v}{\cot v \cos v}$

79. $\sec^4 x - \tan^4 x = \sec^2 x + \tan^2 x$

80. $\dfrac{\cos \theta}{1 - \sin \theta} = \sec \theta + \tan \theta$

81. $\dfrac{\cos \theta}{1 - \sin \theta} = \dfrac{\sin \theta - \csc \theta}{\cos \theta - \cot \theta}$

82. $\dfrac{1 + \tan x}{1 - \tan x} = \dfrac{\cos x + \sin x}{\cos x - \sin x}$

83. $\dfrac{\cos^2 t + \tan^2 t - 1}{\sin^2 t} = \tan^2 t$

84. $\dfrac{1}{1 - \sin x} - \dfrac{1}{1 + \sin x} = 2 \sec x \tan x$

85. $\dfrac{1}{\sec x + \tan x} + \dfrac{1}{\sec x - \tan x} = 2 \sec x$

86. $\dfrac{1 + \sin x}{1 - \sin x} - \dfrac{1 - \sin x}{1 + \sin x} = 4 \tan x \sec x$

87. $(\tan x + \cot x)^2 = \sec^2 x + \csc^2 x$

88. $\tan^2 x - \cot^2 x = \sec^2 x - \csc^2 x$

89. $\dfrac{\sec u - 1}{\sec u + 1} = \dfrac{1 - \cos u}{1 + \cos u}$

90. $\dfrac{\sec u - 1}{\sec u + 1} = \dfrac{\tan u - \sin u}{\tan u + \sin u}$

91. $\dfrac{\cot x + 1}{\cot x - 1} = \dfrac{1 + \tan x}{1 - \tan x}$

92. $\dfrac{\sin^3 x + \cos^3 x}{\sin x + \cos x} = 1 - \sin x \cos x$

93. $\dfrac{\sin^3 x - \csc^3 x}{\sin x - \csc x} = \sin^2 x + \csc^2 x + 1$

94. $\dfrac{\tan v - \cot v}{\tan^2 v - \cot^2 v} = \sin v \cos v$

95. $\dfrac{1 + \sin x}{1 - \sin x} = (\tan x + \sec x)^2$

96. $\dfrac{1 + \cos x}{1 - \cos x} = (\cot x + \csc x)^2$

97. $\dfrac{(\sin x + \cos x)^2}{(\sin x - \cos x)^2} = \dfrac{\sec x + 2 \sin x}{\sec x - 2 \sin x}$

98. $\dfrac{\tan x + \tan y}{\cot x + \cot y} = \tan x \tan y$

99. $(\tan x + \cot x)^4 = \csc^4 x \sec^4 x$

100. $\dfrac{\sin^2 x - \tan^2 x}{\cos^2 x - \cot^2 x} = \tan^6 x$

101. $(\sin \alpha - \tan \alpha)(\cos \alpha - \cot \alpha) = (\cos \alpha - 1)(\sin \alpha - 1)$

102. $\sin^6 \beta + \cos^6 \beta + 3 \sin^2 \beta \cos^2 \beta = 1$

103. $\dfrac{1 + \cos x + \sin x}{1 + \cos x - \sin x} = \dfrac{1 + \sin x}{\cos x}$

104. $\ln|\tan x \sin x| = 2 \ln|\sin x| + \ln|\sec x|$

105. $\ln|\tan x + \cot x| = 2 \ln|\sec x| - \ln|\tan x|$

106. $\ln|\sec x + \tan x| = -\ln|\sec x - \tan x|$

107. $\ln|\csc x - \cot x| = -\ln|\csc x + \cot x|$

108. $\ln|\tan x| + \ln|\cot x| = 0$

109. $e^{\sin^2 x} e^{\tan^2 x} = e^{\sec^2 x} e^{-\cos^2 x}$

110. $e^{x + 2 \ln|\sin x|} = e^x \sin^2 x$

In Exercises 111–116 make the indicated trigonometric substitution in the given algebraic expression and simplify (see Example 7). In Exercises 114–116, assume $a > 0$.

111. $\dfrac{x}{\sqrt{1 - x^2}}; \quad x = \sin \theta \quad \left(-\dfrac{\pi}{2} < \theta < \dfrac{\pi}{2}\right)$

112. $\sqrt{1 + x^2}; \quad x = \tan \theta \quad \left(-\dfrac{\pi}{2} \le \theta \le \dfrac{\pi}{2}\right)$

113. $\sqrt{x^2 - 1}; \quad x = \sec \theta \quad (0 \le \theta < \pi/2)$

114. $\dfrac{1}{x^2 \sqrt{a^2 + x^2}}; \quad x = a \tan \theta \quad \left(-\dfrac{\pi}{2} \le \theta \le \dfrac{\pi}{2}\right)$

115. $\sqrt{a^2 - x^2}; \quad x = a \sin \theta \quad (-\pi/2 \le \theta \le \pi/2)$

116. $\dfrac{\sqrt{x^2 - a^2}}{x}; \quad x = a \sec \theta \quad \left(0 \le \theta < \dfrac{\pi}{2}\right)$

In Exercises 117–122 show that the equation is not an identity.

117. $\sin 2x = 2 \sin x$

118. $\sin(x + y) = \sin x + \sin y$

119. $\sec^2 x + \csc^2 x = 1$ **120.** $\tan x + \cot x = 1$

121. $\sin(x - \pi) = \sin x$

122. $\dfrac{1}{\sin x + \cos x} = \csc x + \sec x$

In Exercises 123–128 determine whether the equation is an identity.

123. $\dfrac{1}{1 + \sin x} + \dfrac{1}{1 - \sin x} = 2 \sec^2 x$

124. $\dfrac{1}{\sec t - \tan t} - \dfrac{1}{\sec t + \tan t} = 2 \tan t$

125. $\dfrac{1}{\sec^2 y} - \dfrac{1}{\csc^2 y} = 1$

126. $\dfrac{\cot x - 1}{1 - \tan x} = \cot x$

127. $\dfrac{\sin x - \tan x}{\tan x} = 1 + \cos x$

128. $\sin^4 t - \cos^4 t + \cos^2 t = 1 + \sin^2 t$

 In Exercises 129–132, graph f and g. Do the graphs suggest that the equation $f(x) = g(x)$ is an identity? Prove your answer.

129. $f(x) = \cos^2 x - \sin^2 x,\quad g(x) = 1 - 2 \sin^2 x$

130. $f(x) = \tan x\,(1 + \sin x),\quad g(x) = \dfrac{\sin x \cos x}{1 + \sin x}$

131. $f(x) = (\sin x + \cos x)^2,\quad g(x) = 1$

132. $f(x) = \cos^4 x - \sin^4 x,\quad g(x) = 2 \cos^2 x - 1$

SECTION 7.2
TRIGONOMETRIC EQUATIONS

A trigonometric equation is an equation that contains trigonometric functions. For example,

$$2 \sin x - 1 = 0 \qquad 4 \cos^2 x - 2 \cos x + 3 = 0 \qquad (\sin x - 1)(\tan x + 2) = 0$$

are trigonometric equations. We are interested in solving such equations; that is, we would like to find all the values of the variable x that make the equation true. The first of these equations

$$2 \sin x - 1 = 0$$
or
$$\sin x = \tfrac{1}{2}$$

is the simplest kind of trigonometric equation. The solutions to this equation in the interval $[0, 2\pi)$ are $x = \pi/6$ and $x = 5\pi/6$. But since the sine function is periodic with period 2π, any integer multiple of 2π added to one of these solutions gives another solution. Thus all the solutions are

$$x = \frac{\pi}{6} + 2k\pi, \qquad x = \frac{5\pi}{6} + 2k\pi$$

for any integer k. Figure 1 shows a graphical representation of the solutions of this equation.

In general, as in this example, if a trigonometric equation has one solution, then it has infinitely many solutions. To find all the solutions of such an equation we need only find the solutions in an appropriate interval, and then use the fact that the trigonometric functions are periodic.

To solve more complicated trigonometric equations we use algebraic techniques in order to reduce the equation to simple equations like the one we just solved. The following examples illustrate this technique.

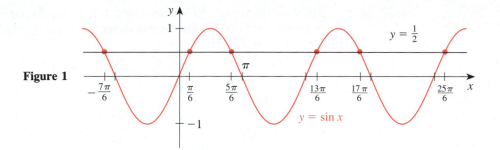

Figure 1

EXAMPLE 1

Solve the equation $4 \cos^2 x - 8 \cos x + 3 = 0$.

SOLUTION

Factoring the left-hand side of this equation, we get

$$(2 \cos x - 1)(2 \cos x - 3) = 0$$

$$2 \cos x - 1 = 0 \qquad \text{or} \qquad 2 \cos x - 3 = 0$$

$$\cos x = \tfrac{1}{2} \qquad\qquad\qquad \cos x = \tfrac{3}{2}$$

Since $\cos x$ is never greater than 1, we see that $\cos x = \frac{3}{2}$ has no solution. In the interval $[0, 2\pi)$, the equation $\cos x = \frac{1}{2}$ has solutions $x = \pi/3$ and $x = 5\pi/3$. Because the cosine function is periodic with period 2π, all the solutions are of the form

$$x = \frac{\pi}{3} + 2k\pi, \qquad x = \frac{5\pi}{3} + 2k\pi$$

for any integer k. ■

EXAMPLE 2

Find all the solutions of the equation $\tan x \sin x - \tan x - \sin x + 1 = 0$.

SOLUTION

Factoring $\tan x$ from the first two terms of the left-hand side of the equation, we get

$$\tan x (\sin x - 1) - \sin x + 1 = 0$$

$$\tan x (\sin x - 1) - (\sin x - 1) = 0$$

$$(\sin x - 1)(\tan x - 1) = 0$$

$$\sin x - 1 = 0 \qquad \text{or} \qquad \tan x - 1 = 0$$

$$\sin x = 1 \qquad\qquad\qquad \tan x = 1$$

The equation $\sin x = 1$ has the solution $x = \pi/2$ in the interval $[0, 2\pi)$. Since sine is periodic with period 2π, we get all the solutions by adding integer multiples of 2π to $x = \pi/2$:

$$x = \frac{\pi}{2} + 2k\pi$$

for any integer k. Since the function tan is periodic with period π, we need to find the solutions to $\tan x = 1$ in any interval of length π. A convenient interval is $(-\pi/2, \pi/2)$, in which $\tan x = 1$ has the solution $x = \pi/4$. This gives

$$x = \frac{\pi}{4} + k\pi$$

for any integer k. It follows that the solutions to the original equation are

$$x = \frac{\pi}{2} + 2k\pi, \qquad x = \frac{\pi}{4} + k\pi$$

for any integer k.

■

EXAMPLE 3

Solve the trigonometric equation $\tan^4 2x - 9 = 0$.

SOLUTION

Adding 9 to both sides of this equation, we get

$$\tan^4 2x = 9$$

Taking the fourth root gives

$$\tan 2x = \sqrt{3} \qquad \text{or} \qquad \tan 2x = -\sqrt{3}$$

Thus, in the interval $(-\pi/2, \pi/2)$, we have

$$2x = \frac{\pi}{3} \qquad \text{or} \qquad 2x = -\frac{\pi}{3}$$

Since the tangent function is periodic with period π, all the solutions are given by

$$2x = \frac{\pi}{3} + k\pi \qquad \text{or} \qquad 2x = -\frac{\pi}{3} + k\pi$$

for any integer k. Dividing both sides of these equations by 2 gives us the solution to the equation of this example:

$$x = \frac{\pi}{6} + \frac{k}{2}\pi, \qquad x = -\frac{\pi}{6} + \frac{k}{2}\pi$$

for any integer k.   ■

Trigonometric identities are useful tools for solving trigonometric equations. They can be used to transform an equation into an equivalent equation that is simpler to solve. The next example illustrates this.

EXAMPLE 4

Solve the equation $3 \sin x = 2 \cos^2 x$ in the interval $[0, 2\pi]$.

SOLUTION

Using the identity $\cos^2 x = 1 - \sin^2 x$, we get an equivalent equation that involves only the sine function:

$$3 \sin x = 2(1 - \sin^2 x)$$
$$2 \sin^2 x + 3 \sin x - 2 = 0$$
$$(2 \sin x - 1)(\sin x + 2) = 0$$

$$2 \sin x - 1 = 0 \qquad \text{or} \qquad \sin x + 2 = 0$$
$$\sin x = \tfrac{1}{2} \qquad\qquad\qquad \sin x = -2$$

Since $-1 \le \sin x \le 1$, the equation $\sin x = -2$ has no solution. The solutions of the given equation are thus the solutions of $\sin x = \tfrac{1}{2}$. These are $x = \pi/6, 5\pi/6$.   ■

EXAMPLE 5

Find the points of intersection of the graphs of $f(x) = \sin x$ and $g(x) = \cos x$.

SOLUTION

The graphs of f and g are shown in Figure 2. The graphs intersect at the points

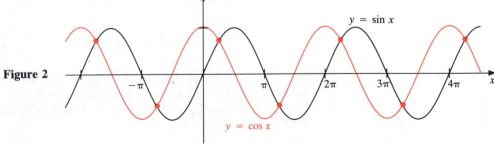

Figure 2

where $f(x) = g(x)$. So we need to find the solutions of the equation

$$\sin x = \cos x$$

Notice that the numbers x for which $\cos x = 0$ are not solutions of this equation. For $\cos x \neq 0$, we can divide both sides of the equation by $\cos x$ to get

$$\frac{\sin x}{\cos x} = 1 \qquad \text{or} \qquad \tan x = 1$$

The solution of this last equation on the interval $(-\pi/2, \pi/2)$ is $x = \pi/4$. Since the tangent function is periodic with period π, the solutions are

$$x = \frac{\pi}{4} + k\pi$$

for any integer k.   ■

EXAMPLE 6

Find the solutions of $\cos 2x \sec x = 2 \cos 2x$ in the interval $[0, 2\pi]$.

SOLUTION

We subtract $2 \cos 2x$ from both sides of the equation and factor out $\cos 2x$ to get

$$\cos 2x \, (\sec x - 2) = 0$$

$$\cos 2x = 0 \qquad \text{or} \qquad \sec x = 2$$

We begin by solving $\cos 2x = 0$. Since we are seeking solutions in the interval $0 \leqslant x \leqslant 2\pi$, we have $0 \leqslant 2x \leqslant 4\pi$. In this interval, $\cos 2x = 0$ has the solutions

$$2x = \frac{\pi}{2}, \frac{3\pi}{2}, \frac{5\pi}{2}, \frac{7\pi}{2}$$

so

$$x = \frac{\pi}{4}, \frac{3\pi}{4}, \frac{5\pi}{4}, \frac{7\pi}{4}$$

Now we solve $\sec x = 2$. In the interval $[0, 2\pi]$, the solutions of $\sec x = 2$ are $x = \pi/3$ and $x = 5\pi/3$. Thus all the solutions of the given equation are

$$x = \frac{\pi}{4}, \frac{\pi}{3}, \frac{3\pi}{4}, \frac{5\pi}{4}, \frac{5\pi}{3}, \frac{7\pi}{4}$$ ■

Notice that in Example 6 we did not divide both sides by $\cos 2x$. It would have been wrong to do so, since we could be dividing by zero. Indeed, if we divide both sides by $\cos 2x$, then we lose all solutions of the given equation that are solutions of $\cos 2x = 0$.

When solving an equation, any operation we perform on it must produce an *equivalent* equation in the sense that the new equation has the same roots as the original equation. An example of an operation that may *not* give an equivalent equation is "squaring both sides." Consider the equation $\sin x = -\sin x$. It is easy to see that the solutions to this equation are $x = k\pi$ for any integer k. But if we square both sides of this equation, we introduce many new solutions. In fact, the equation $\sin^2 x = \sin^2 x$ is obviously an identity and so has every real number x as a solution. Another operation that may introduce new roots to an equation is multiplying both sides by an expression that itself may be zero. For instance,

$$\sin x = 0$$

has the solutions $x = k\pi$ for any integer k. Multiplying both sides by $\cos x$ gives the equation

$$\cos x \sin x = 0$$

which has as additional solutions the roots of $\cos x = 0$.

 If we perform an operation that may introduce new roots of an equation, then we must check that the solutions obtained are not extraneous; that is, we must verify that they satisfy the original equation.

EXAMPLE 7

Solve the equation $\cos x + 1 = \sin x$ in the interval $[0, 2\pi)$.

SOLUTION

To get an equation that involves either sine only or cosine only, we square both sides and use the Pythagorean identities.

$$(\cos x + 1)^2 = \sin^2 x$$
$$\cos^2 x + 2 \cos x + 1 = \sin^2 x$$
$$\cos^2 x + 2 \cos x + 1 = 1 - \cos^2 x$$
$$2 \cos^2 x + 2 \cos x = 0$$
$$2 \cos x (\cos x + 1) = 0$$
$$\cos x = 0 \quad \text{or} \quad \cos x + 1 = 0$$

From these we get the possible solutions

$$x = \frac{\pi}{2}, \ \frac{3\pi}{2}, \ \pi$$

Since we may have introduced extraneous roots by squaring both sides of the equation, we must check to see whether each of these values for x satisfies the original equation:

$$x = \frac{\pi}{2} \qquad\qquad\qquad x = \frac{3\pi}{2} \qquad\qquad\qquad x = \pi$$

$$\cos \frac{\pi}{2} + 1 = \sin \frac{\pi}{2} \qquad \cos \frac{3\pi}{2} + 1 = \sin \frac{3\pi}{2} \qquad \cos \pi + 1 = \sin \pi$$

$$0 + 1 = 1 \qquad\qquad\qquad 0 + 1 = -1 \qquad\qquad\qquad -1 + 1 = 0$$

True   False   True

Thus the solutions of the given equation in the interval $[0, 2\pi)$ are $\pi/2$ and π.

EXAMPLE 8

Solve the equation $2 \cos x + \tan x - \sec x = 0$ in the interval $[0, 2\pi)$.

SOLUTION

Writing the equation in terms of sine and cosine gives

$$2 \cos x + \frac{\sin x}{\cos x} - \frac{1}{\cos x} = 0$$

Multiplying both sides by $\cos x$, we have

$$2 \cos^2 x + \sin x - 1 = 0$$

To get an equation that involves the sine function only, we use the identity $\cos^2 x = 1 - \sin^2 x$ to get

$$2(1 - \sin^2 x) + \sin x - 1 = 0$$
$$2 - 2 \sin^2 x + \sin x - 1 = 0$$
$$2 \sin^2 x - \sin x - 1 = 0$$
$$(2 \sin x + 1)(\sin x - 1) = 0$$

$$2 \sin x + 1 = 0 \qquad \text{or} \qquad \sin x - 1 = 0$$
$$\sin x = -\tfrac{1}{2} \qquad\qquad\qquad \sin x = 1$$

In the interval $[0, 2\pi)$ the first of these equations has solutions $x = 7\pi/6$, $11\pi/6$ and the other has the solution $x = \pi/2$. Since we have multiplied both sides of the equation by $\cos x$, we must check to see whether any of these solutions are extraneous. Notice that neither $\tan x$ nor $\sec x$ is defined for $x = \pi/2$, and so this value is not a solution of the given equation. It is easy to check that the other two values do satisfy the equation. Thus the solutions of the original equation are

$$x = \frac{7\pi}{6}, \frac{11\pi}{6}$$

The equation in the next example involves both algebraic and trigonometric expressions. In most cases we cannot find exact solutions to such equations. With the aid of a graphing device we are able to approximate the solutions.

EXAMPLE 9

Solve the equation $\sin x = x^2$.

SOLUTION

We solve the equation graphically (see Section 5.5). In Figure 3(a) we use a graphing device to graph $y = \sin x$ and $y = x^2$ in the viewing rectangle $[-3.14, 3.14]$ by $[-1.5, 1.5]$. We see that the curves intersect at two points. One of these is the origin and a little thought reveals that $x = 0$ actually is a solution. To find the other point we zoom in to the viewing rectangle shown in Figure 3(b). From the graph we see that, correct to three decimal places, the other solution of the equation is $x \approx 0.877$. Thus the solutions of the given equation are $x = 0$ and $x \approx 0.877$.

Figure 3
$y = \sin x$ and $y = x^2$

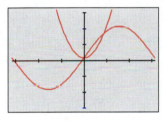

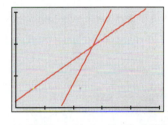

(a) $[-3.14, 3.14]$ by $[-1.5, 1.5]$ (b) $[0.85, 0.90]$ by $[0.75, 0.78]$ ■

We will return to the topic of trigonometric equations in later sections of this chapter.

EXERCISES 7.2

In Exercises 1–16 find all the solutions of the equation.

1. $2 \sin x - \sqrt{3} = 0$ **2.** $\tan x + 1 = 0$

3. $(\tan x - 1)(2 \cos x - \sqrt{3}) = 0$

4. $\sqrt{3} \sin 2x = \cos 2x$

5. $4 \cos^2 x - 3 = 0$ **6.** $\cos \dfrac{x}{2} - 1 = 0$

7. $\cos x \sin x - 2 \cos x = 0$ **8.** $\sin x + \cos x = 3$

9. $\sin(-x) + 3 \sin x = 1$ **10.** $\sin(-x) = \sin x + 1$

11. $\sin^2 x = 4 - 2 \cos^2 x$ **12.** $\dfrac{1 - \sin x}{1 + \sin x} = -3$

13. $\csc 3x = \sin 3x$ **14.** $\sin^2 x + \tfrac{1}{2} = \sqrt{2} \sin x$

15. $4 \cos^2 x - 4 \cos x + 1 = 0$

16. $2 \cos^2 x + \sin x = 1$

In Exercises 17–48 solve the equation in the interval $[0, 2\pi)$.

17. $2 \sin x \tan x - \tan x = 1 - 2 \sin x$

18. $\tan^2 3x + \cot^2 3x = 2$

19. $\sec x \tan x - \cos x \cot x = \sin x$

20. $\tan^5 x - 9 \tan x = 0$

good for grouping, factoring
sin x = √3 cos x change to tan.

21. $3 \tan^3 x - 3 \tan^2 x - \tan x + 1 = 0$

22. $4 \sin x \cos x + 2 \sin x - 2 \cos x - 1 = 0$

23. $\tan x - 3 \cot x = 0$ *change to all sines*

24. $2 \sin x - 2\sqrt{3} \cos x - \sqrt{3} \tan x + 3 = 0$
    $2(\sin x - \sqrt{3} \cos x) - \frac{\sqrt{3}}{\cos x}(\sin x - \sqrt{3}\cos x) = 0$

25. $\sec x - \csc x - 2 \tan x + 2 = 0$

26. $2 \sin^2 x - \cos x = 1$ 27. $4 \cos^2 x + 2 \sin^2 x = 3$

28. $\cos^2 x - \frac{7}{2} \sin x = -1$ 29. $\cos^2 \pi x - \sin^2 \pi x = 0$

30. $8 \sin^4 x - 10 \sin^2 x + 3 = 0$

31. $\sec x - \tan x = \cos x$ 32. $\tan^2 x + \sec^2 x = 3$

33. $\cos x - 2 = \sin x$ 34. $\tan 3x + 1 = \sec 3x$

35. $\cot x - \csc x = 1$

36. $\sin x + 2 \cot x = 2 \csc x$

37. $\tan^3 x - \tan^2 x - 3 \tan x + 3 = 0$

38. $x \sin x + 1 = x + \sin x$ 39. $\sin 2x = 2 \tan 2x$

40. $3 \sec^2 x + 4 \cos^2 x = 7$ 41. $\csc(x + 1) = \sin(x + 1)$

42. $2 \sin\left(\frac{\pi}{2} + x\right) = 1$

43. $2 \cos x + \tan x - \sec x = 0$

44. $\sin(\cos x) = 0$ 45. $\ln(2 - \sin^2 x) = 0$

46. $\ln|1 - \tan^2 x| = 2 \ln|1 + \tan x|$

47. $\log_2(2 \cos x) = 0$ 48. $\log_3(2 \sin x) = \frac{1}{2}$

In Exercises 49–52 sketch the graphs of f and g on the same axes and find their points of intersection.

49. $f(x) = 3 \cos x + 1$; $g(x) = \cos x - 1$

50. $f(x) = \sin 2x$; $g(x) = 2 \sin 2x + 1$

51. $f(x) = \tan x$; $g(x) = \sqrt{3}$

52. $f(x) = \sin x - 1$; $g(x) = \cos x$

In Exercises 53 and 54 find the points of intersection of the graphs of f and g.

53. $f(x) = \tan x \sin x$; $g(x) = 2 - \cos x$

54. $f(x) = \ln(e + \sin^2 x)$; $g(x) = 1$

55. Find $\dfrac{1}{\tan x} + \dfrac{1}{\cot x}$ if $\tan x + \cot x = \dfrac{9}{2}$.

56. Find $\dfrac{1}{\sin x} - \dfrac{1}{\csc x}$ if $\sin x - \csc x = 3$.

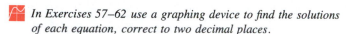 *In Exercises 57–62 use a graphing device to find the solutions of each equation, correct to two decimal places.*

57. $\sin 2x = x$ 58. $\cos x = \dfrac{x}{3}$

59. $2^{\sin x} = x$ 60. $\sin x = x^3$

61. $\dfrac{\cos x}{1 + x^2} = x^2$ 62. $\cos x = \frac{1}{2}(e^x + e^{-x})$

SECTION 7.3
ADDITION AND SUBTRACTION FORMULAS

In this section we derive identities for trigonometric functions of sums and differences. The main identities are the **addition and subtraction formulas** for sine and cosine.

ADDITION AND SUBTRACTION FORMULAS FOR SINE AND COSINE

$$\sin(s + t) = \sin s \cos t + \cos s \sin t$$

$$\sin(s - t) = \sin s \cos t - \cos s \sin t$$

$$\cos(s + t) = \cos s \cos t - \sin s \sin t$$

$$\cos(s - t) = \cos s \cos t + \sin s \sin t$$

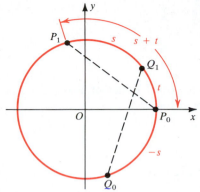

Figure 1

We first prove the addition formula for cosine (the third formula in the box). We will then use it to prove the others.

To prove the formula $\cos(s + t) = \cos s \cos t - \sin s \sin t$, we use Figure 1. In the figure, the distances t, $s + t$, and $-s$ have been marked on the unit circle, starting at $P_0(1, 0)$ and terminating at Q_1, P_1, and Q_0, respectively. The coordinates of these points are

$$P_0(1, 0) \qquad\qquad Q_0(\cos(-s), \sin(-s))$$
$$P_1(\cos(s + t), \sin(s + t)) \qquad Q_1(\cos t, \sin t)$$

Since $\cos(-s) = \cos s$ and $\sin(-s) = -\sin s$, it follows that the point Q_0 has the coordinates $Q_0(\cos s, -\sin s)$. Notice that the distances between P_0 and P_1 and between Q_0 and Q_1 measured along the arc of the circle are equal. Since equal arcs are subtended by equal chords, it follows that $|P_0P_1| = |Q_0Q_1|$. Using the Distance Formula, we get

$$\sqrt{[\cos(s + t) - 1]^2 + [\sin(s + t) - 0]^2} = \sqrt{[\cos t - \cos s]^2 + [\sin t + \sin s]^2}$$

Squaring both sides and expanding, we have

these add to 1

$$\cos^2(s + t) - 2\cos(s + t) + 1 + \sin^2(s + t)$$
$$= \cos^2 t - 2\cos s \cos t + \cos^2 s + \sin^2 t + 2\sin s \sin t + \sin^2 s$$

these add to 1

these add to 1

Using the Pythagorean identity $\sin^2 z + \cos^2 z = 1$ three times gives

$$2 - 2\cos(s + t) = 2 - 2\cos s \cos t + 2\sin s \sin t$$

Finally, subtracting 2 from each side and dividing both sides by -2, we get

$$\cos(s + t) = \cos s \cos t - \sin s \sin t$$

which proves the addition formula for cosine.

The subtraction formula for cosine is obtained by replacing t by $-t$ in the addition formula as follows:

$$\cos(s - t) = \cos(s + (-t))$$
$$= \cos s \cos(-t) - \sin s \sin(-t)$$
$$= \cos s \cos t + \sin s \sin t$$

This proves the subtraction formula for cosine. Before proving the other formulas, we give some applications of the ones we have already proved.

EXAMPLE 1

Find the exact value of cos 75°.

SOLUTION

Notice that 75° = 45° + 30°. Since we know the exact values of sine and cosine at 45° and 30°, we use the addition formula for cosine to get

$$\cos 75° = \cos(45° + 30°)$$
$$= \cos 45° \cos 30° - \sin 45° \sin 30°$$
$$= \frac{\sqrt{2}}{2}\frac{\sqrt{3}}{2} - \frac{\sqrt{2}}{2}\frac{1}{2} = \frac{\sqrt{2}\sqrt{3} - \sqrt{2}}{4} = \frac{\sqrt{6} - \sqrt{2}}{4} \qquad ■$$

EXAMPLE 2

Prove each identity: (a) $\cos\left(\dfrac{\pi}{2} - u\right) = \sin u$ (b) $\sin\left(\dfrac{\pi}{2} - v\right) = \cos v$

SOLUTION

(a) By the subtraction formula for cosine,

$$\cos\left(\frac{\pi}{2} - u\right) = \cos\frac{\pi}{2}\cos u + \sin\frac{\pi}{2}\sin u$$
$$= 0 \cdot \cos u + 1 \cdot \sin u = \sin u$$

(b) We replace u by $\dfrac{\pi}{2} - v$ in part (a) to get

$$\cos\left[\frac{\pi}{2} - \left(\frac{\pi}{2} - v\right)\right] = \sin\left(\frac{\pi}{2} - v\right)$$
$$\cos v = \sin\left(\frac{\pi}{2} - v\right) \qquad ■$$

Sine and cosine are called *cofunctions* of each other. Similarly, tangent and cotangent are cofunctions of each other, as are secant and cosecant. The identities in Example 2 are called **cofunction identities**. The other cofunction identities in the following box are a consequence of these two and the reciprocal identities (see Exercises 15–18).

COFUNCTION IDENTITIES

$$\sin\left(\frac{\pi}{2} - u\right) = \cos u \qquad \tan\left(\frac{\pi}{2} - u\right) = \cot u \qquad \sec\left(\frac{\pi}{2} - u\right) = \csc u$$

$$\cos\left(\frac{\pi}{2} - u\right) = \sin u \qquad \cot\left(\frac{\pi}{2} - u\right) = \tan u \qquad \csc\left(\frac{\pi}{2} - u\right) = \sec u$$

For acute angles, the cofunction identities have an interesting interpretation. Since u and $(\pi/2) - u$ are complementary angles, these identities say that the value of a trigonometric function of an angle is the same as that of the *co*function of its *co*mplementary angle. (This explains the prefix *co-* in the names of the trigonometric functions; it stands for "complementary.") Figure 2 shows, for instance, that for an acute angle u, we have

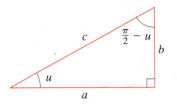

Figure 2

$$\sin\left(\frac{\pi}{2} - u\right) = \frac{a}{c} = \cos u$$

$$\tan\left(\frac{\pi}{2} - u\right) = \frac{a}{b} = \cot u$$

$$\sec\left(\frac{\pi}{2} - u\right) = \frac{c}{a} = \csc u$$

We now return to proving the remaining addition and subtraction formulas. The addition formula for sine can be derived with the aid of the cofunction identities. We have

$$\sin(s + t) = \cos\left[\frac{\pi}{2} - (s + t)\right] \qquad \text{(cofunction identity)}$$

$$= \cos\left[\left(\frac{\pi}{2} - s\right) - t\right]$$

$$= \cos\left(\frac{\pi}{2} - s\right)\cos t + \sin\left(\frac{\pi}{2} - s\right)\sin t \qquad \text{(subtraction formula for cosine)}$$

$$= \sin s \cos t + \cos s \sin t \qquad \text{(cofunction identity)}$$

This proves the addition formula for sine. The subtraction formula for sine is obtained from this formula in the same way that the subtraction formula for cosine is obtained from the addition formula for cosine.

EXAMPLE 3

Find the exact values of (a) $\sin 105°$ and (b) $\cos \dfrac{\pi}{12}$.

SOLUTION

(a) Since $105° = 60° + 45°$, the addition formula for sine gives

$$\sin 105° = \sin(60° + 45°)$$

$$= \sin 60° \cos 45° + \cos 60° \sin 45°$$

$$= \frac{\sqrt{3}}{2}\frac{\sqrt{2}}{2} + \frac{1}{2}\frac{\sqrt{2}}{2}$$

$$= \frac{\sqrt{6} + \sqrt{2}}{4}$$

(b) Since $\dfrac{\pi}{12} = \dfrac{\pi}{4} - \dfrac{\pi}{6}$, the subtraction formula for cosine gives

$$\cos\frac{\pi}{12} = \cos\!\left(\frac{\pi}{4} - \frac{\pi}{6}\right)$$

$$= \cos\frac{\pi}{4}\cos\frac{\pi}{6} + \sin\frac{\pi}{4}\sin\frac{\pi}{6}$$

$$= \frac{\sqrt{2}}{2}\frac{\sqrt{3}}{2} + \frac{\sqrt{2}}{2}\frac{1}{2} = \frac{\sqrt{6}+\sqrt{2}}{4} \quad \blacksquare$$

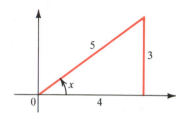

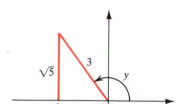

Figure 3

EXAMPLE 4

If $\sin x = \frac{3}{5}$, x in quadrant I, and $\cos y = -\frac{2}{3}$, y in quadrant II, find $\sin(x + y)$.

SOLUTION

To apply the addition formula for sine we need to find $\cos x$ and $\sin y$. A simple way to do this is to use the definitions of the trigonometric functions and sketch a graph to represent the angles x and y as shown in Figure 3. Thus $\cos x = \frac{4}{5}$ and $\sin y = \sqrt{5}/3$. The addition formula for sine now gives

$$\sin(x + y) = \sin x \cos y + \cos x \sin y$$

$$= \frac{3}{5}\!\left(-\frac{2}{3}\right) + \frac{4}{5}\frac{\sqrt{5}}{3} = \frac{-6 + 4\sqrt{5}}{15} = \frac{2(2\sqrt{5} - 3)}{15} \quad \blacksquare$$

We now obtain addition and subtraction formulas for tangent.

ADDITION AND SUBTRACTION FORMULAS FOR TANGENT

$$\tan(s + t) = \frac{\tan s + \tan t}{1 - \tan s \tan t} \qquad \tan(s - t) = \frac{\tan s - \tan t}{1 + \tan s \tan t}$$

Proof   To prove the addition formula we use the reciprocal relations and the addition formulas for sine and cosine. We have

$$\tan(s + t) = \frac{\sin(s + t)}{\cos(s + t)} \qquad \text{(reciprocal identity)}$$

$$= \frac{\sin s \cos t + \cos s \sin t}{\cos s \cos t - \sin s \sin t} \qquad \text{(addition formulas for sine and cosine)}$$

$$= \frac{\dfrac{\sin s \cos t}{\cos s \cos t} + \dfrac{\cos s \sin t}{\cos s \cos t}}{\dfrac{\cos s \cos t}{\cos s \cos t} - \dfrac{\sin s \sin t}{\cos s \cos t}} \qquad \begin{array}{l}\text{(divide numerator and denominator by}\\ \cos s \cos t)\end{array}$$

$$= \frac{\tan s + \tan t}{1 - \tan s \tan t} \qquad \text{(reciprocal identity)}$$

As before, to obtain the subtraction formula we replace t by $-t$ in the addition formula. ∎

EXAMPLE 5

Verify the identity

$$\frac{1 + \tan x}{1 - \tan x} = \tan\left(\frac{\pi}{4} + x\right)$$

SOLUTION

Starting with the right-hand side and using the addition formula for tangent, we get

$$\tan\left(\frac{\pi}{4} + x\right) = \frac{\tan \dfrac{\pi}{4} + \tan x}{1 - \tan \dfrac{\pi}{4} \tan x}$$

$$= \frac{1 + \tan x}{1 - \tan x} \qquad\qquad ∎$$

We end this section with an example containing a typical use of the addition and subtraction identities in calculus.

EXAMPLE 6

If $f(x) = \sin x$, show that

$$\frac{f(x + h) - f(x)}{h} = \sin x \left(\frac{\cos h - 1}{h}\right) + \cos x \left(\frac{\sin h}{h}\right)$$

SOLUTION

$$\frac{f(x + h) - f(x)}{h} = \frac{\sin(x + h) - \sin x}{h}$$

$$= \frac{\sin x \cos h + \cos x \sin h - \sin x}{h}$$

$$= \frac{\sin x (\cos h - 1) + \cos x \sin h}{h}$$

$$= \sin x \left(\frac{\cos h - 1}{h}\right) + \cos x \left(\frac{\sin h}{h}\right) \qquad ∎$$

EXERCISES 7.3

In Exercises 1–6 find the exact value of each expression.

1. $\sin 15°$

2. $\cos 135°$

3. $\tan 105°$

4. $\cot \dfrac{\pi}{12}$

5. $\sin\left(-\dfrac{11\pi}{12}\right)$

6. $\dfrac{\sqrt{2}}{2}\cos\dfrac{\pi}{12} - \dfrac{\sqrt{2}}{2}\sin\dfrac{\pi}{12}$

In Exercises 7–10 write each expression as a trigonometric function of one number, and find its exact value.

7. $\sin 18° \cos 27° + \cos 18° \sin 27°$

8. $\cos\dfrac{3\pi}{7}\cos\dfrac{2\pi}{21} + \sin\dfrac{3\pi}{7}\sin\dfrac{2\pi}{21}$

9. $\dfrac{\tan 73° - \tan 13°}{1 + \tan 73° \tan 13°}$

10. $\cos\dfrac{13\pi}{15}\cos\left(-\dfrac{\pi}{5}\right) - \sin\dfrac{13\pi}{15}\sin\left(-\dfrac{\pi}{5}\right)$

11. If α and β are two angles such that α is in quadrant I, β is in quadrant II, $\cos \alpha = \frac{5}{13}$, and $\cos \beta = -\frac{3}{5}$, find $\sin(\alpha - \beta)$ and $\sin(\alpha + \beta)$.

12. If α and β are two angles such that α is in quadrant II, $\sin \alpha = \frac{3}{5}$, and $\tan \beta = \frac{4}{3}$, find $\tan(\alpha + \beta)$ and $\tan(\alpha - \beta)$.

13. If α and β are two angles such that α and β are in quadrants III and II, respectively, $\cos \alpha = -\frac{3}{5}$, and $\csc \beta = \frac{5}{4}$, find $\sin(\alpha + \beta)$ and $\cos(\alpha + \beta)$.

14. If α and β are two acute angles such that $\cos \beta = \frac{24}{25}$ and $\sin(\alpha + \beta) = \frac{4}{5}$, find $\sin \alpha$ and $\cos \alpha$.

In Exercises 15–18 prove each cofunction identity using the cofunction identities for sine and cosine.

15. $\tan\left(\dfrac{\pi}{2} - u\right) = \cot u$

16. $\cot\left(\dfrac{\pi}{2} - u\right) = \tan u$

17. $\sec\left(\dfrac{\pi}{2} - u\right) = \csc u$

18. $\csc\left(\dfrac{\pi}{2} - u\right) = \sec u$

In Exercises 19–45 prove each identity.

19. $\sin\left(x - \dfrac{\pi}{2}\right) = -\cos x$

20. $\cos\left(x - \dfrac{\pi}{2}\right) = \sin x$

21. $\sin(x - \pi) = -\sin x$

22. $\cos(x - \pi) = -\cos x$

23. $\tan(x - \pi) = \tan x$

24. $\sin\left(\dfrac{\pi}{2} - x\right) = \sin\left(\dfrac{\pi}{2} + x\right)$

25. $\cos\left(x - \dfrac{\pi}{2}\right) + \cos\left(x + \dfrac{\pi}{2}\right) = 0$

26. $\cos\left(x + \dfrac{\pi}{6}\right) + \sin\left(x - \dfrac{\pi}{3}\right) = 0$

27. $\cos\left(x - \dfrac{\pi}{3}\right) = \frac{1}{2}\left(\cos x + \sqrt{3} \sin x\right)$

28. $\cos\left(x + \dfrac{\pi}{4}\right) = -\sin\left(x - \dfrac{\pi}{4}\right)$

29. $\tan\left(x + \dfrac{\pi}{4}\right) = \dfrac{1 + \tan x}{1 - \tan x}$

30. $\tan\left(x - \dfrac{\pi}{4}\right) = \dfrac{\tan x - 1}{\tan x + 1}$

31. $\sin(x + y) + \sin(x - y) = 2 \sin x \cos y$

32. $\sin(x + y) - \sin(x - y) = 2 \cos x \sin y$

33. $\cos(x + y) + \cos(x - y) = 2 \cos x \cos y$

34. $\cos(x + y) - \cos(x - y) = -2 \sin x \sin y$

35. $\cot(x - y) = \dfrac{\cot x \cot y + 1}{\cot y - \cot x}$

36. $\cot(x + y) = \dfrac{\cot x \cot y - 1}{\cot x + \cot y}$

37. $\tan x - \tan y = \dfrac{\sin(x - y)}{\cos x \cos y}$

38. $1 - \tan x \tan y = \dfrac{\cos(x + y)}{\cos x \cos y}$

39. $\dfrac{\sin(x - y)}{\cos(x + y)} = \dfrac{\tan x - \tan y}{1 - \tan x \tan y}$

40. $\dfrac{\sin(x + y) - \sin(x - y)}{\cos(x + y) + \cos(x - y)} = \tan y$

41. $\dfrac{\sin \pi x}{\sin x} - \dfrac{\cos \pi x}{\cos x} = \dfrac{\sin(\pi - 1)x}{\sin x \cos x}$

42. $\cos(x + y) \cos(x - y) = \cos^2 x - \sin^2 y$

43. $\sin(x - y) \cos y + \cos(x - y) \sin y = \sin x$

44. $\cos(x + y) \cos y + \sin(x + y) \sin y = \cos x$

45. $\cot x - \tan y = \dfrac{\cos(x + y)}{\sin x \cos y}$

In Exercises 46–49 solve each trigonometric equation in the interval $[0, 2\pi)$.

46. $\cos x \cos 3x = \sin x \sin 3x$

47. $\cos x \cos 2x + \sin x \sin 2x = \frac{1}{2}$

48. $\sin 2x \cos x + \cos 2x \sin x = \sqrt{3}/2$

49. $\sin 3x \cos x = \cos 3x \sin x$

In Exercises 50–52 prove each identity.

50. $\sin(x + y + z) = \sin x \cos y \cos z + \cos x \sin y \cos z$
$\qquad + \cos x \cos y \sin z - \sin x \sin y \sin z$

51. $\tan(x - y) + \tan(y - z) + \tan(z - x)$
$\qquad = \tan(x - y) \tan(y - z) \tan(z - x)$

52. $\cot(y - z) \cot(z - x) + \cot(x - y) \cot(z - x)$
$\qquad + \cot(x - y) \cot(y - z) = 1$

53. Show that if $\beta - \alpha = \pi/2$, then

$$\sin(x + \alpha) + \cos(x + \beta) = 0$$

54. Let $g(x) = \cos x$. Show that

$$\frac{g(x + h) - g(x)}{h} = -\cos x \left(\frac{1 - \cos h}{h} \right) - \sin x \left(\frac{\sin h}{h} \right)$$

55. Refer to the figure. Show that $\alpha + \beta = \gamma$ and find $\tan \gamma$.

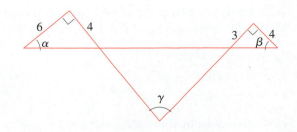

56. (a) If L is a line in the plane and θ is the angle formed by the line and the x-axis as shown in the figure, show that the slope m of the line is given by

$$m = \tan \theta$$

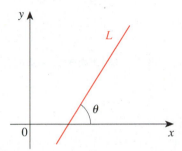

(b) Let L_1 and L_2 be two nonparallel lines in the plane with slopes m_1 and m_2, respectively. Let ψ be the acute angle formed by the two lines as shown in the figure. Show that

$$\tan \psi = \frac{m_2 - m_1}{1 + m_1 m_2}$$

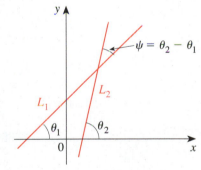

(c) Find the acute angle formed by the two lines $y = \frac{1}{3}x + 1$ and $y = -\frac{1}{2}x - 3$.

(d) Show that if two lines are perpendicular, then the slope of one is the negative reciprocal of the slope of the other.   [*Hint:* First find an expression for $\cot \psi$.]

In Exercises 57 and 58, (a) graph the function and make a conjecture, and (b) prove that your conjecture is true.

57. $y = \sin^2\left(x + \dfrac{\pi}{4}\right) + \sin^2\left(x - \dfrac{\pi}{4}\right)$

58. $y = -\frac{1}{2}[\cos(x + \pi) + \cos(x - \pi)]$

SECTION 7.4
DOUBLE–ANGLE AND HALF–ANGLE FORMULAS

The identities we consider in this section are consequences of the addition formulas. The **double-angle formulas** allow us to find the values of the trigonometric functions at $2x$ from their values at x. The **half-angle formulas** relate the values of the trigonometric functions at $\frac{1}{2}x$ to their values at x.

DOUBLE–ANGLE FORMULAS

The formulas in the following box are consequences of the addition formulas, which we proved in the preceding section.

DOUBLE-ANGLE FORMULAS

$$\sin 2x = 2 \sin x \cos x$$
$$\cos 2x = \cos^2 x - \sin^2 x$$
$$\cos 2x = 1 - 2 \sin^2 x$$
$$\cos 2x = 2 \cos^2 x - 1$$
$$\tan 2x = \frac{2 \tan x}{1 - \tan^2 x}$$

Proof   Setting $x = y$ in the addition formula for sine, we get

$$\sin 2x = \sin(x + x)$$
$$= \sin x \cos x + \sin x \cos x$$
$$= 2 \sin x \cos x$$

Setting $x = y$ in the addition formula for cosine gives

$$\cos 2x = \cos(x + x)$$
$$= \cos x \cos x - \sin x \sin x$$
$$= \cos^2 x - \sin^2 x$$

The other two formulas for $\cos 2x$ are obtained from the formula we just proved and the Pythagorean identity. Making the substitution $\cos^2 x = 1 - \sin^2 x$ gives

$$\cos 2x = \cos^2 x - \sin^2 x$$
$$= (1 - \sin^2 x) - \sin^2 x$$
$$= 1 - 2 \sin^2 x$$

The other formula is obtained in the same way, by substituting $\sin^2 x = 1 - \cos^2 x$.

Finally, we obtain the double-angle formula for tangent by setting $x = y$ in the addition formula for tangent. ∎

EXAMPLE 1

If $\cos x = -\frac{2}{3}$ and x is in quadrant II, find $\sin 2x$ and $\cos 2x$.

SOLUTION

Using one of the double-angle formulas for cosine, we get

$$\cos 2x = 2 \cos^2 x - 1$$

$$= 2\left(-\frac{2}{3}\right)^2 - 1 = \frac{8}{9} - 1 = -\frac{1}{9}$$

To use the formula $\sin 2x = 2 \sin x \cos x$, we need to find $\sin x$ first. We have

$$\sin x = \sqrt{1 - \cos^2 x} = \sqrt{1 - \left(\frac{2}{3}\right)^2} = \frac{\sqrt{5}}{3}$$

where we have used the positive square root because $\sin x$ is positive in quadrant II. Thus

$$\sin 2x = 2 \sin x \cos x$$

$$= 2\left(\frac{\sqrt{5}}{3}\right)\left(-\frac{2}{3}\right) = -\frac{4\sqrt{5}}{9} \qquad ∎$$

EXAMPLE 2

Write $\cos 3x$ in terms of $\cos x$.

SOLUTION

Supply the reasons for each step:

$$\begin{aligned}
\cos 3x &= \cos(2x + x)\\
&= \cos 2x \cos x - \sin 2x \sin x\\
&= (2 \cos^2 x - 1)\cos x - (2 \sin x \cos x)\sin x\\
&= 2 \cos^3 x - \cos x - 2 \sin^2 x \cos x\\
&= 2 \cos^3 x - \cos x - 2 \cos x (1 - \cos^2 x)\\
&= 2 \cos^3 x - \cos x - 2 \cos x + 2 \cos^3 x\\
&= 4 \cos^3 x - 3 \cos x \qquad ∎
\end{aligned}$$

Example 2 shows that $\cos 3x$ can be written as a polynomial of degree 3 in $\cos x$. The identity $\cos 2x = 2 \cos^2 x - 1$ shows that $\cos 2x$ is a polynomial of degree 2 in cos x. In fact, for any natural number n, we can write cos nx as a polynomial in cos x of degree n (see Exercise 63). The analogous result for sin nx is not true in general.

EXAMPLE 3

Prove the identity: $\dfrac{\sin 3x}{\sin x \cos x} = 4 \cos x - \sec x$

SOLUTION

Supply reasons for each step:

$$\frac{\sin 3x}{\sin x \cos x} = \frac{\sin(x + 2x)}{\sin x \cos x}$$

$$= \frac{\sin x \cos 2x + \cos x \sin 2x}{\sin x \cos x}$$

$$= \frac{\sin x \,(2 \cos^2 x - 1) + \cos x \,(2 \sin x \cos x)}{\sin x \cos x}$$

$$= \frac{\sin x \,(2 \cos^2 x - 1)}{\sin x \cos x} + \frac{\cos x \,(2 \sin x \cos x)}{\sin x \cos x}$$

$$= \frac{2 \cos^2 x - 1}{\cos x} + 2 \cos x$$

$$= 2 \cos x - \frac{1}{\cos x} + 2 \cos x$$

$$= 4 \cos x - \sec x \qquad \blacksquare$$

HALF-ANGLE FORMULAS

If we know the value of a trigonometric function at $2x$, then the following identities allow us to find the value at x (half of $2x$). For this reason they are called half-angle formulas.

HALF-ANGLE FORMULAS

$$\sin^2 x = \frac{1 - \cos 2x}{2} \qquad \cos^2 x = \frac{1 + \cos 2x}{2}$$

$$\tan^2 x = \frac{1 - \cos 2x}{1 + \cos 2x}$$

Proof   The first formula is obtained by solving for $\sin^2 x$ in the double-angle formula $\cos 2x = 1 - 2 \sin^2 x$. Similarly, the second identity is obtained by solving for $\cos^2 x$ in the double-angle identity $\cos 2x = 2 \cos^2 x - 1$. The last identity follows from the first two and the reciprocal identities as follows:

$$\tan^2 x = \frac{\sin^2 x}{\cos^2 x} = \frac{\dfrac{1 - \cos 2x}{2}}{\dfrac{1 + \cos 2x}{2}} = \frac{1 - \cos 2x}{1 + \cos 2x} \qquad \blacksquare$$

The half-angle identities allow us to write any trigonometric expression involving even powers of sine and cosine in terms of the first power of cosine only. This technique is important in calculus.

EXAMPLE 4

Express $\sin^2 x \cos^2 x$ in terms of the first power of cosine.

SOLUTION

Using the half-angle formulas repeatedly gives

$$\sin^2 x \cos^2 x = \left(\frac{1 - \cos 2x}{2}\right)\left(\frac{1 + \cos 2x}{2}\right)$$

$$= \frac{1 - \cos^2 2x}{4} = \frac{1}{4} - \frac{1}{4}\cos^2 2x$$

$$= \frac{1}{4} - \frac{1}{4}\left(\frac{1 + \cos 4x}{2}\right) = \frac{1}{4} - \frac{1}{8} - \frac{\cos 4x}{8}$$

$$= \frac{1}{8} - \frac{1}{8}\cos 4x = \frac{1}{8}(1 - \cos 4x)$$

Another way of obtaining this identity is to use the double-angle formula for sine in the form $\sin x \cos x = \frac{1}{2}\sin 2x$. Thus

$$\sin^2 x \cos^2 x = \frac{1}{4}\sin^2 2x = \frac{1}{4}\left(\frac{1 - \cos 4x}{2}\right) = \frac{1}{8}(1 - \cos 4x) \qquad \blacksquare$$

EXAMPLE 5

Find the exact value of $\cos 15°$.

SOLUTION

Using the half-angle formula for cosine with $x = 15°$, we get

$$\cos^2 15° = \frac{1 + \cos 2(15°)}{2} = \frac{1 + \cos 30°}{2}$$

$$= \frac{1 + \dfrac{\sqrt{3}}{2}}{2} = \frac{2 + \sqrt{3}}{4}$$

Thus

$$\cos 15° = \sqrt{\frac{2 + \sqrt{3}}{4}} = \frac{1}{2}\sqrt{2 + \sqrt{3}}$$

where we have chosen the positive square root because 15° is in quadrant I. ■

In Example 5 we were able to find the exact value of cos 15° because the right-hand side of the half-angle formula for cosine required us to find cos 2(15°), that is, cos 30°, a quantity whose exact value we know. This shows how the half-angle formulas are useful for finding the value of the trigonometric functions of half an angle when we know the value at that angle. To see this more clearly, let us substitute $x = u/2$ in the half-angle formulas and take the square root of each side. This gives another way of writing the half-angle formulas.

HALF-ANGLE FORMULAS

$$\sin \frac{u}{2} = \pm \sqrt{\frac{1 - \cos u}{2}} \qquad \cos \frac{u}{2} = \pm \sqrt{\frac{1 + \cos u}{2}}$$

$$\tan \frac{u}{2} = \pm \sqrt{\frac{1 - \cos u}{1 + \cos u}}$$

In these equations the choice of the + or − sign depends on the quadrant in which $u/2$ lies.

EXAMPLE 6

Find the exact value of sin 7.5°.

SOLUTION

In Example 5 we found the exact value of cos 15°. Since 7.5° is half of 15°, we can use the half-angle formula for sine with $u = 15°$ to get

$$\sin 7.5° = \sin\left(\frac{15}{2}\right)° = \sqrt{\frac{1 - \cos 15°}{2}}$$

$$= \sqrt{\frac{1 - \frac{1}{2}\sqrt{2 + \sqrt{3}}}{2}}$$

$$= \sqrt{\frac{2 - \sqrt{2 + \sqrt{3}}}{4}}$$

$$= \frac{1}{2}\sqrt{2 - \sqrt{2 + \sqrt{3}}}$$

We have chosen the positive sign because 7.5° is in quadrant I. ■

We now find two more half-angle formulas for tangent function. Multiplying the numerator and denominator in the half-angle formula for tangent by $1 - \cos u$, we get

$$\tan \frac{u}{2} = \pm \sqrt{\left(\frac{1 - \cos u}{1 + \cos u}\right)\left(\frac{1 - \cos u}{1 - \cos u}\right)}$$

$$= \pm \sqrt{\frac{(1 - \cos u)^2}{1 - \cos^2 u}}$$

$$= \pm \frac{|1 - \cos u|}{|\sin u|}$$

Now, $1 - \cos u$ is nonnegative for all values of u. It is also true that $\sin u$ and $\tan(u/2)$ always have the same sign. (Verify this.) It follows that

$$\tan \frac{u}{2} = \frac{1 - \cos u}{\sin u}$$

Another half-angle identity for tangent can be derived from this by multiplying numerator and denominator by $1 + \cos u$.

HALF-ANGLE FORMULAS FOR TANGENT

$$\tan \frac{u}{2} = \frac{1 - \cos u}{\sin u} \qquad \tan \frac{u}{2} = \frac{\sin u}{1 + \cos u}$$

EXAMPLE 7

Find $\tan(u/2)$ if $\sin u = \frac{2}{5}$ and u is in quadrant II.

SOLUTION

To use the half-angle formulas for tangent, we first need to find $\cos u$. Since cosine is negative in quadrant II, we have

$$\cos u = -\sqrt{1 - \sin^2 u}$$

$$= -\sqrt{1 - \left(\tfrac{2}{5}\right)^2} = -\frac{\sqrt{21}}{5}$$

Thus,

$$\tan \frac{u}{2} = \frac{1 - \cos u}{\sin u}$$

$$= \frac{1 + \left(\sqrt{21}/5\right)}{2/5} = \frac{5 + \sqrt{21}}{2} \qquad \blacksquare$$

EXAMPLE 8

Let $f(x) = \cos 3x + \cos x$.

(a) Sketch the graph of f on the interval $[0, 2\pi]$.
(b) Find the approximate values of the x-intercepts.
(c) Find the exact values of the x-intercepts by solving the equation $f(x) = 0$.

SOLUTION

(a) The graph is sketched in Figure 1 with the aid of a graphing calculator.
(b) The x-intercepts can be found approximately from the graph (see Section 5.5). Zooming in and using the cursor, we see that the x-intercepts are

$$0.79, \ 1.57, \ 2.36, \ 3.93, \ 4.71, \ 5.50$$

correct to two decimal places.

(c) To find the x-intercepts exactly, we need to solve the equation

$$\cos 3x + \cos x = 0$$

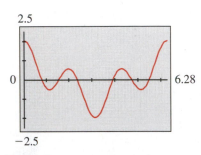

Figure 1
$f(x) = \cos 3x + \cos x$

on the interval $[0, 2\pi]$. First we use identities to simplify the equation. In Example 2 we found that $\cos 3x = 4\cos^3 x - 3\cos x$. Thus

$$\cos 3x + \cos x = 0$$
$$4\cos^3 x - 3\cos x + \cos x = 0$$
$$4\cos^3 x - 2\cos x = 0$$
$$2\cos^3 x - \cos x = 0$$
$$\cos x\,(2\cos^2 x - 1) = 0$$
$$\cos x = 0 \quad \text{or} \quad 2\cos^2 x - 1 = 0$$

The equation $\cos x = 0$ has the solutions $x = \pi/2, \ 3\pi/2$. The other equation gives

$$\cos^2 x = \frac{1}{2}$$

$$\cos x = \pm\frac{1}{\sqrt{2}} = \pm\frac{\sqrt{2}}{2}$$

with solutions $x = \pi/4, \ 3\pi/4, \ 5\pi/4, \ 7\pi/4$. Thus the x-intercepts of the graph on the interval $[0, 2\pi]$ are

$$\frac{\pi}{4}, \ \frac{\pi}{2}, \ \frac{3\pi}{4}, \ \frac{5\pi}{4}, \ \frac{3\pi}{2}, \ \frac{7\pi}{4}$$

EXERCISES 7.4

In Exercises 1 and 2 use a double-angle formula to write the given expression as a trigonometric expression involving half the angle.

1. (a) $\sin 88°$ (b) $\cos \dfrac{\pi}{4}$ (c) $\sin 4\theta$

2. (a) $\cos 48°$ (b) $\tan \dfrac{\pi}{6}$ (c) $\tan 8\theta$

In Exercises 3 and 4 use a half-angle formula to write the given expression as a trigonometric expression involving twice the angle.

3. (a) $\tan 13°$ (b) $\sin 28°$ (c) $\tan \dfrac{\theta}{8}$

4. (a) $\cos \dfrac{\pi}{12}$ (b) $\tan \dfrac{\pi}{30}$ (c) $\sin \dfrac{\theta}{4}$

In Exercises 5–10 simplify the expression by using a double-angle formula or a half-angle formula.

5. (a) $2 \sin 18° \cos 18°$ (b) $2 \sin 3\theta \cos 3\theta$

6. (a) $\dfrac{2 \tan 7°}{1 - \tan^2 7°}$ (b) $\dfrac{2 \tan 7\theta}{1 - \tan^2 7\theta}$

7. (a) $\cos^2 34° - \sin^2 34°$ (b) $\cos^2 5\theta - \sin^2 5\theta$

8. (a) $\cos^2 \dfrac{\theta}{2} - \sin^2 \dfrac{\theta}{2}$ (b) $2 \sin \dfrac{\theta}{2} \cos \dfrac{\theta}{2}$

9. (a) $\dfrac{\sin 8°}{1 + \cos 8°}$ (b) $\dfrac{1 - \cos 4\theta}{\sin 4\theta}$

10. (a) $\sqrt{\dfrac{1 - \cos 30°}{2}}$ (b) $\sqrt{\dfrac{1 - \cos 8\theta}{2}}$

In Exercises 11–18 use an appropriate half-angle formula to find the exact value of the expression.

11. $\sin 15°$ **12.** $\tan 15°$ **13.** $\sin 7.5°$

14. $\cos 22.5°$ **15.** $\cos 11.25°$ **16.** $\tan \dfrac{\pi}{8}$

17. $\sin \dfrac{\pi}{12}$ **18.** $\cos \dfrac{5\pi}{12}$

In Exercises 19–24 find $\sin 2x$, $\cos 2x$, and $\tan 2x$ from the given information.

19. $\sin x = \dfrac{5}{13}$, x in quadrant I

20. $\cos x = \dfrac{4}{5}$, $\csc x < 0$

21. $\tan x = -\dfrac{4}{3}$, x in quadrant II

22. $\csc x = 4$, $\tan x < 0$

23. $\sin x = -\dfrac{3}{5}$, x in quadrant III

24. $\cot x = \dfrac{2}{3}$, $\sin x > 0$

In Exercises 25–30 find $\sin \dfrac{x}{2}$, $\cos \dfrac{x}{2}$, and $\tan \dfrac{x}{2}$ from the given information.

25. $\sin x = \dfrac{3}{5}$, $0° < x < 90°$

26. $\cos x = -\dfrac{4}{5}$, $180° < x < 270°$

27. $\csc x = 3$, $90° < x < 180°$

28. $\tan x = 1$, $0° < x < 90°$

29. $\sec x = \dfrac{3}{2}$, $270° < x < 360°$

30. $\cot x = 5$, $\csc x < 0$

In Exercises 31–47 prove that the equation is an identity.

31. $\cos^2 5x - \sin^2 5x = \cos 10x$

32. $\sin 8x = 2 \sin 4x \cos 4x$

33. $2 \sin \dfrac{x}{2} \cos \dfrac{x}{2} = \sin x$ **34.** $\dfrac{1 - \cos 2x}{\sin 2x} = \tan x$

35. $(\sin x + \cos x)^2 = 1 + \sin 2x$

36. $\dfrac{2 \tan \dfrac{x}{2}}{1 + \tan^2 \dfrac{x}{2}} = \sin x$ **37.** $\dfrac{\sin 4x}{\sin x} = 4 \cos x \cos 2x$

38. $\dfrac{1 + \sin 2x}{\sin 2x} = 1 + \dfrac{1}{2} \sec x \csc x$

39. $\dfrac{2(\tan x - \cot x)}{\tan^2 x - \cot^2 x} = \sin 2x$

40. $\cot 2x = \dfrac{1 - \tan^2 x}{2 \tan x}$　　**41.** $\tan x = \dfrac{\sin 2x}{1 + \cos 2x}$

42. $\tan 3x = \dfrac{3 \tan x - \tan^3 x}{1 - 3 \tan^2 x}$

43. $4(\sin^6 x + \cos^6 x) = 4 - 3 \sin^2 2x$

44. $(1 - \cos 4x)(2 + \tan^2 x + \cot^2 x) = 8$

45. $\dfrac{\sin 3x + \cos 3x}{\cos x - \sin x} = 1 + 4 \sin x \cos x$

46. $\cos^4 x - \sin^4 x = \cos 2x$

47. $\tan^2\left(\dfrac{x}{2} + \dfrac{\pi}{4}\right) = \dfrac{1 + \sin x}{1 - \sin x}$

48. Use the identity $\sin 2x = 2 \sin x \cos x$, n times to show that

$$\sin(2^n x) = 2^n \sin x \cos x \cos 2x \cos 4x \cdots \cos 2^{n-1} x$$

 49. **(a)** Sketch the graph of $f(x) = \dfrac{\sin 3x}{\sin x} - \dfrac{\cos 3x}{\cos x}$ and make a conjecture.
　(b) Prove the conjecture you made in part (a).

 50. **(a)** Sketch the graph of $f(x) = \cos 2x + 2 \sin^2 x$ and make a conjecture.
　(b) Prove the conjecture you made in part (a).

In Exercises 51–56 solve each equation in the interval $[0, 2\pi)$.

✓ **51.** $\sin 2x - \cos x = 0$　　**52.** $\sin 2\pi x + \sin \pi x = 0$

✓ **53.** $\cos 2x + \sin x = 0$　　**54.** $\tan \dfrac{x}{2} - \sin x = 0$

55. $\cos 2x + \cos x = 2$　　**56.** $\tan x + \cot x = 4 \sin 2x$

 In Exercises 57–60, (a) sketch the graph of $f(x)$ *over the interval* $[0, 2\pi)$; *(b) find approximate values of the x-intercepts from the graph, and (c) find the exact values of the x-intercepts by solving the equation* $f(x) = 0$.

57. $f(x) = \cos 3x - \cos x$

58. $f(x) = \sin x + \sin 2x$

59. $f(x) = 1 - \sin x \cos x$

60. $f(x) = \sin x \cos 2x + 2 \sin^2 x - 1$

61. The lower right-hand corner of a long piece of paper that is 6 in. wide is folded over to the left-hand edge as shown in the figure. The length L of the fold depends on the angle θ. Show that

$$L = \dfrac{3}{\sin \theta \cos^2 \theta}$$

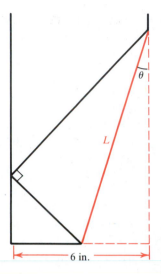

6 in.

62. Let $3x = \frac{1}{3}\pi$ and let $y = \cos x$. Use the result of Example 2 to show that y satisfies the equation

$$8y^3 - 6y - 1 = 0$$

[*Note:* This equation has roots of a certain kind that are used in showing that the angle $\frac{1}{3}\pi$ cannot be trisected using a ruler and compass only.]

63. **(a)** Show that there is a polynomial $P(t)$ of degree 4 such that $\cos 4x = P(\cos x)$ (see Example 2).
　(b) Show that there is a polynomial $P(t)$ of degree 5 such that $\cos 5x = P(\cos x)$.
　[*Note:* In general, there is a polynomial $P(t)$ of degree n such that $\cos nx = P(\cos x)$. These polynomials are named after the Russian mathematician Tchebycheff.]

64. If $\sin x + \cos x = \alpha$, show the following:
　(a) $\sin 2x = \alpha^2 - 1$
　(b) $\sin^3 x + \cos^3 x = \frac{1}{2}(3\alpha - \alpha^3)$
　[*Hint:* Find $(\sin x + \cos x)^3$ and use part (a).]

65. In triangle ABC (shown in the figure) the line segment s bisects angle C. Show that the length of s is given by

$$s = \frac{2ab \cos x}{a + b}$$

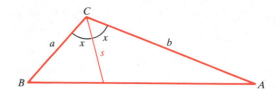

SECTION 7.5
FURTHER IDENTITIES

In this section we derive several trigonometric identities that are useful in calculus. We use some of these to write products of sines and cosines as sums, others give sums of sines and cosines as products.

■ PRODUCT–TO–SUM IDENTITIES

It is possible to write the product $\sin u \cos v$ as a sum of trigonometric functions. To see this, consider the addition and subtraction formulas for the sine function:

$$\sin(u + v) = \sin u \cos v + \cos u \sin v$$

$$\sin(u - v) = \sin u \cos v - \cos u \sin v$$

Adding the left- and right-hand sides of these identities gives

$$\sin(u + v) + \sin(u - v) = 2 \sin u \cos v$$

or

$$\sin u \cos v = \tfrac{1}{2}[\sin(u + v) + \sin(u - v)]$$

The other three **product-to-sum identities** follow from the addition formulas in a similar way.

PRODUCT-TO-SUM IDENTITIES

$$\sin u \cos v = \tfrac{1}{2}[\sin(u + v) + \sin(u - v)] \tag{1}$$

$$\cos u \sin v = \tfrac{1}{2}[\sin(u + v) - \sin(u - v)] \tag{2}$$

$$\cos u \cos v = \tfrac{1}{2}[\cos(u + v) + \cos(u - v)] \tag{3}$$

$$\sin u \sin v = \tfrac{1}{2}[\cos(u - v) - \cos(u + v)] \tag{4}$$

EXAMPLE 1

Express $\sin 3x \sin 5x$ as a sum of trigonometric functions.

SOLUTION

Using Formula 4 with $u = 3x$ and $v = 5x$ and the fact that cosine is an even function, we get

$$\sin 3x \sin 5x = \tfrac{1}{2}\big[\cos(3x - 5x) - \cos(3x + 5x)\big]$$
$$= \tfrac{1}{2}\cos(-2x) - \tfrac{1}{2}\cos 8x$$
$$= \tfrac{1}{2}\cos 2x - \tfrac{1}{2}\cos 8x \qquad \blacksquare$$

EXAMPLE 2

Find the value of the product $\sin 37.5° \cos 7.5°$.

SOLUTION

Using Formula 1 with $u = 37.5°$ and $v = 7.5°$ gives

$$\sin 37.5° \cos 7.5° = \tfrac{1}{2}\big[\sin(37.5° + 7.5°) + \sin(37.5° - 7.5°)\big]$$
$$= \tfrac{1}{2}\big[\sin 45° + \sin 30°\big]$$
$$= \frac{1}{2}\left(\frac{\sqrt{2}}{2} + \frac{1}{2}\right) = \frac{\sqrt{2} + 1}{4}$$

We also could have obtained this value by using Formula 2. $\blacksquare$

■ SUM–TO–PRODUCT IDENTITIES

The product-to-sum identities can also be used as sum-to-product identities. This is because the right-hand side of each of the product-to-sum identities is a sum and the left side is a product. For example, in order to write

$$\sin x + \sin y$$

as a product, we use Formula 1 with the appropriate choice for u and v. To find out what u and v should be, notice that we want

$$\begin{cases} x = u + v \\ y = u - v \end{cases}$$

Solving these two equations simultaneously for u and v gives

$$u = \frac{x + y}{2} \qquad \text{and} \qquad v = \frac{x - y}{2}$$

Substituting for u, v, $u + v$, and $u - v$ in Formula 1 gives

$$\sin \frac{x + y}{2} \cos \frac{x - y}{2} = \tfrac{1}{2}\big[\sin x + \sin y\big]$$

or
$$\sin x + \sin y = 2 \sin \frac{x+y}{2} \cos \frac{x-y}{2}$$

The remaining three of the following **sum-to-product identities** are obtained in a similar manner.

SUM-TO-PRODUCT IDENTITIES

$$\sin x + \sin y = 2 \sin \frac{x+y}{2} \cos \frac{x-y}{2} \qquad (5)$$

$$\sin x - \sin y = 2 \cos \frac{x+y}{2} \sin \frac{x-y}{2} \qquad (6)$$

$$\cos x + \cos y = 2 \cos \frac{x+y}{2} \cos \frac{x-y}{2} \qquad (7)$$

$$\cos x - \cos y = -2 \sin \frac{x+y}{2} \sin \frac{x-y}{2} \qquad (8)$$

EXAMPLE 3

Write $\sin 7x + \sin 3x$ as a product.

SOLUTION

Formula 5 gives

$$\sin 7x + \sin 3x = 2 \sin \frac{7x+3x}{2} \cos \frac{7x-3x}{2}$$
$$= 2 \sin 5x \cos 2x \qquad \blacksquare$$

EXAMPLE 4

Verify the identity:
$$\frac{\sin 3x - \sin x}{\cos 3x + \cos x} = \tan x$$

SOLUTION

We apply Formula 6 to the numerator and Formula 7 to the denominator.

$$\frac{\sin 3x - \sin x}{\cos 3x + \cos x} = \frac{2 \cos \dfrac{3x+x}{2} \sin \dfrac{3x-x}{2}}{2 \cos \dfrac{3x+x}{2} \cos \dfrac{3x-x}{2}}$$

$$= \frac{2 \cos 2x \sin x}{2 \cos 2x \cos x}$$

$$= \frac{\sin x}{\cos x} = \tan x \qquad \blacksquare$$

EXAMPLE 5

Solve the equation $\sin 3x - \sin x = 0$.

SOLUTION

We first use Formula 6 to write

$$\sin 3x - \sin x = 2 \cos \frac{3x + x}{2} \sin \frac{3x - x}{2}$$

$$= 2 \cos 2x \sin x$$

Thus the equation is equivalent to

$$2 \cos 2x \sin x = 0$$

So $\qquad \cos 2x = 0 \qquad \text{or} \qquad \sin x = 0$

The solutions of these two equations are

$$2x = \frac{\pi}{2} + k\pi \qquad \text{and} \qquad x = k\pi$$

Thus the solutions of the given equation are

$$x = \frac{(2k + 1)\pi}{4}, \qquad x = k\pi$$

for any integer k.   ■

EXPRESSIONS OF THE FORM $A \sin x + B \cos x$

We can write expressions of the form $A \sin x + B \cos x$ in terms of a single trigonometric function. For example, consider

$$\frac{1}{2} \sin x + \frac{\sqrt{3}}{2} \cos x$$

If we set $\phi = \pi/3$, then $\cos \phi = \frac{1}{2}$ and $\sin \phi = \sqrt{3}/2$ and we can write

$$\frac{1}{2} \sin x + \frac{\sqrt{3}}{2} \cos x = \cos \phi \sin x + \sin \phi \cos x$$

$$= \sin(x + \phi) = \sin\left(x + \frac{\pi}{3}\right)$$

We are able to do this because the coefficients $\frac{1}{2}$ and $\sqrt{3}/2$ are precisely the cosine and sine of a particular number—in this case, $\pi/3$. We can use this same idea in general to write $A \sin x + B \cos x$ in the form $k \sin(x + \phi)$. We start by multiplying numerator and denominator by $\sqrt{A^2 + B^2}$ to get

$$A \sin x + B \cos x = \sqrt{A^2 + B^2} \left(\frac{A}{\sqrt{A^2 + B^2}} \sin x + \frac{B}{\sqrt{A^2 + B^2}} \cos x \right)$$

We need a number ϕ with the property that

$$\cos \phi = \frac{A}{\sqrt{A^2 + B^2}} \qquad \text{and} \qquad \sin \phi = \frac{B}{\sqrt{A^2 + B^2}}$$

Figure 1 shows that the point (A, B) in the plane determines a number ϕ with precisely this property. With this ϕ we have

$$A \sin x + B \cos x = \sqrt{A^2 + B^2} \, (\cos \phi \sin x + \sin \phi \cos x)$$
$$= \sqrt{A^2 + B^2} \, \sin(x + \phi)$$

We have proved the following theorem.

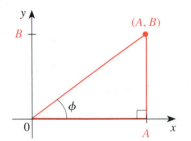

Figure 1

THEOREM

> If A and B are real numbers, then
>
> $$A \sin x + B \cos x = k \sin(x + \phi)$$
>
> where $k = \sqrt{A^2 + B^2}$ and ϕ satisfies
>
> $$\cos \phi = \frac{A}{\sqrt{A^2 + B^2}} \qquad \text{and} \qquad \sin \phi = \frac{B}{\sqrt{A^2 + B^2}}$$

EXAMPLE 6

Express $3 \sin x + 4 \cos x$ in the form $k \sin(x + \phi)$.

SOLUTION

By the theorem, $k = \sqrt{A^2 + B^2} = \sqrt{3^2 + 4^2} = 5$. The angle ϕ has the property that $\sin \phi = \frac{4}{5}$ and $\cos \phi = \frac{3}{5}$. Using a calculator we find $\phi \approx 53.1°$. Thus

$$3 \sin x + 4 \cos x \approx 5 \sin(x + 53.1°) \qquad \blacksquare$$

EXAMPLE 7

(a) Graph $f(x) = -\sqrt{3} \sin 2x + \cos 2x$.
(b) Find the x-intercepts of the graph.

SOLUTION

We first write the expression in the form $k \sin(x + \phi)$. Since $A = -\sqrt{3}$ and $B = 1$, we have $k = \sqrt{A^2 + B^2} = \sqrt{3 + 1} = 2$. The angle ϕ satisfies $\sin \phi = \frac{1}{2}$

and $\cos \phi = -\sqrt{3}/2$. From the signs of these quantities we conclude that ϕ is in quadrant II. Thus $\phi = 2\pi/3$. By the theorem we can write

$$f(x) = -\sqrt{3} \sin 2x + \cos 2x = 2 \sin\left(2x + \frac{2\pi}{3}\right)$$

(a) Using the form

$$f(x) = 2 \sin 2\left(x + \frac{\pi}{3}\right)$$

we see that the graph is a sine curve with amplitude 2, period $2\pi/2 = \pi$, and phase shift $-\pi/3$. The graph is sketched in Figure 2.

(b) To find the x-intercepts we need to solve the equation

$$f(x) = -\sqrt{3} \sin 2x + \cos 2x = 0$$

By part (a), this is equivalent to solving

$$2 \sin 2\left(x + \frac{\pi}{3}\right) = 0$$

Since $\sin u = 0$ when $u = k\pi$, the solutions satisfy

$$2\left(x + \frac{\pi}{3}\right) = k\pi$$

Thus the x-intercepts occur at

$$x = \frac{k\pi}{2} - \frac{\pi}{3}$$

for any integer k.   ■

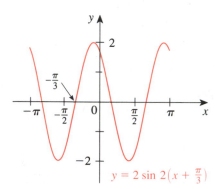

Figure 2

$y = 2 \sin 2\left(x + \frac{\pi}{3}\right)$

EXERCISES 7.5

In Exercises 1–4 write the product as a sum.

1. $\sin 2x \cos 3x$

2. $\sin x \sin 5x$

3. $3 \cos 4x \cos 7x$

4. $11 \sin \dfrac{x}{2} \cos \dfrac{x}{4}$

In Exercises 5–8 find the value of the product.

5. $2 \sin 52.5° \sin 97.5°$

6. $3 \cos 37.5° \cos 7.5°$

7. $\tan 52.5° \tan 97.5°$

8. $\cos 37.5° \sin 7.5°$

In Exercises 9–16 write the sum as a product.

9. $\sin 5x + \sin 3x$

10. $\sin x - \sin 4x$

11. $\cos 4x - \cos 6x$

12. $\cos 9x + \cos 2x$

13. $\sin 2x - \sin 7x$

14. $\sin 3x + \sin 4x$

15. $\cos 11\pi x + \cos 9\pi x$

16. $\cos \dfrac{x}{2} - \cos \dfrac{5x}{2}$

In Exercises 17–20 find the value of the sum.

17. $\sin 75° + \sin 15°$

18. $\sin 105° - \sin 15°$

19. $\cos 255° - \cos 195°$

20. $\cos \dfrac{\pi}{12} + \cos \dfrac{5\pi}{12}$

In Exercises 21–28 verify each identity.

21. $\dfrac{\sin x + \sin 5x}{\cos x + \cos 5x} = \tan 3x$

22. $\dfrac{\sin 3x + \sin 7x}{\cos 3x - \cos 7x} = \cot 2x$

23. $\dfrac{\sin 10x}{\sin 9x + \sin x} = \dfrac{\cos 5x}{\cos 4x}$

24. $\dfrac{\sin x + \sin 3x + \sin 5x}{\cos x + \cos 3x + \cos 5x} = \tan 3x$

25. $\dfrac{\sin x + \sin y}{\cos x + \cos y} = \tan\left(\dfrac{x + y}{2}\right)$

26. $\dfrac{\cos x - \cos y}{\sin x + \sin y} = -\tan\left(\dfrac{x - y}{2}\right)$

27. $\tan y = \dfrac{\sin(x + y) - \sin(x - y)}{\cos(x + y) + \cos(x - y)}$

28. $\cot x = \dfrac{\sin(x + y) - \sin(x - y)}{\cos(x - y) - \cos(x + y)}$

In Exercises 29–36 solve each equation (see Example 5).

29. $\sin x + \sin 3x = 0$ **30.** $\cos 5x - \cos 7x = 0$

31. $\cos 4x + \cos 2x = \cos x$

32. $\sin x + \sin 3x + \sin 5x = 0$

33. $\sin 5x - \sin 3x = \cos 4x$

34. $\cos 3x - \cos x - \sin 2x = 0$

35. $\sin 3x - \sin x - \cos 2x = 0$

36. $\sin 4x - \sin 2x - \sqrt{2}\cos 3x = 0$

37. Show that $\sin 45° + \sin 15° = \sin 75°$.

38. Show that $\cos 87° + \cos 33° = \sin 63°$.

39. If $A + B + C = \pi/2$, show that

$$\sin 2A + \sin 2B + \sin 2C = 4\cos A \cos B \cos C$$

40. If $A + B + C = \pi$, show that

$$\tan A + \tan B + \tan C = \tan A \tan B \tan C$$

41. Prove the identity

$$\dfrac{\sin x + \sin 2x + \sin 3x + \sin 4x + \sin 5x}{\cos x + \cos 2x + \cos 3x + \cos 4x + \cos 5x} = \tan 3x$$

In Exercises 42–45 write the expression in terms of sine only.

42. $-\sqrt{3}\sin x + \cos x$ **43.** $\sin x + \cos x$

44. $5(\sin 2x - \cos 2x)$ **45.** $3\sin \pi x + 3\sqrt{3}\cos \pi x$

In Exercises 46 and 47 graph each function. [Hint: First express the function in terms of sine only.]

46. $f(x) = \sin x + \cos x$

47. $g(x) = \cos 2x + \sqrt{3}\sin 2x$

In Exercises 48–51 solve the equation in the interval $[0, 2\pi)$ by the method of Example 7.

48. $\sin x - \cos x = \sqrt{2}$ **49.** $\sin 3x + \cos 3x = 1$

50. $\sin 2x - \sqrt{3}\cos 2x = 2$

51. $3\sqrt{3}\cos \pi x + 3\sin \pi x = 0$

 52. A digital delay-device echoes an input signal by repeating it a fixed length of time after it is received. If such a device receives the pure note $f_1(t) = 5\sin t$ and echoes the pure note $f_2(t) = 5\cos t$, then the combined sound is $f(t) = f_1(t) + f_2(t)$.
   (a) Graph $y = f(t)$ and observe that the graph has the form of a sine curve $y = k\sin(t + \phi)$.
   (b) Find k and ϕ.

 53. Let $f(x) = \sin 6x + \sin 7x$.
   (a) Graph $y = f(x)$.
   (b) Verify that $f(x) = 2\cos\frac{1}{2}x \sin\frac{13}{2}x$.
   (c) Graph $y = 2\cos\frac{1}{2}x$ and $y = -2\cos\frac{1}{2}x$, together with the graph in part (a), in the same viewing rectangle. How are these graphs related to that in part (a)?

 54. When two pure notes that are close in frequency are played together, their sounds interfere to produce *beats*; that is, the loudness (or amplitude) of the sound alternately increases and decreases. If the two notes are given by

$$f_1(t) = \cos 11t \quad \text{and} \quad f_2(t) = \cos 13t$$

the resulting sound is $f(t) = f_1(t) + f_2(t)$.
   (a) Graph the function $y = f(t)$.
   (b) Verify that $f(t) = 2\cos t \cos 12t$.
   (c) Graph $y = 2\cos t$ and $y = -2\cos t$, together with the graph in part (a), in the same viewing rectangle. How do these graphs describe the variation in the loudness of the sound?

SECTION 7.6
INVERSE TRIGONOMETRIC FUNCTIONS

If f is a one-to-one function with domain A and range B, then its inverse f^{-1} is the function with domain B and range A defined by

$$f^{-1}(x) = y \iff f(y) = x$$

(See Section 2.10.) In other words, f^{-1} is the rule that reverses the action of f. Figure 1 gives a graphical representation of the actions of f and f^{-1}.

In order for a function to have an inverse, it must be one-to-one. Since the trigonometric functions are not one-to-one, they do not have inverses. It is possible, however, to restrict the domains of the trigonometric functions in such a way that the resulting functions are one-to-one.

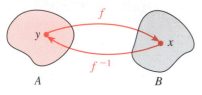

Figure 1
$f^{-1}(x) = y \iff f(y) = x$

THE INVERSE SINE FUNCTION

Let us first consider the sine function. There are many ways to restrict the domain of sine so that the new function is one-to-one. A natural way to do this is to restrict the domain to the interval $[-\pi/2, \pi/2]$. The reason for this choice is that sine attains each of its values exactly once on this interval. We write $\mathrm{Sin}\, x$ (with a capital S) for the new function, which has the domain $[-\pi/2, \pi/2]$ and has the same values as $\sin x$ on this interval. The graphs of $\sin x$ and $\mathrm{Sin}\, x$ are shown in Figure 2. The function $\mathrm{Sin}\, x$ is one-to-one (by the Horizontal Line Test), and so has an inverse.

Figure 2

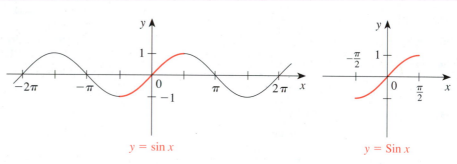

$y = \sin x$ $y = \mathrm{Sin}\, x$

The inverse of the function $\mathrm{Sin}\, x$ is the function $\mathrm{Sin}^{-1}x$ defined by

$$\mathrm{Sin}^{-1}x = y \iff \mathrm{Sin}\, y = x$$

for $-1 \le x \le 1$ and $-\pi/2 \le y \le \pi/2$. The graph of $y = \mathrm{Sin}^{-1}x$ is shown in Figure 3; it is obtained by reflecting the graph of $y = \mathrm{Sin}\, x$ in the line $y = x$. It is customary to write $\mathrm{Sin}^{-1}x$ simply as $\sin^{-1}x$.

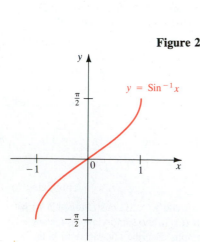

Figure 3

**DEFINITION OF THE
INVERSE SINE FUNCTION**

The **inverse sine function** is the function $\sin^{-1}$ with domain $[-1, 1]$ and range $[-\pi/2, \pi/2]$ defined by

$$\sin^{-1}x = y \iff \sin y = x$$

The inverse sine function is also called **arcsine** and is denoted by **arcsin.**

Thus $\sin^{-1}x$ *is the number in the interval* $[-\pi/2, \pi/2]$ *whose sine is x.* In other words, $\sin(\sin^{-1}x) = x$. In fact, from the general properties of inverse functions studied in Section 2.10, we have the following relations:

$$\sin(\sin^{-1}x) = x \qquad \text{for } -1 \leq x \leq 1$$

$$\sin^{-1}(\sin x) = x \qquad \text{for } -\frac{\pi}{2} \leq x \leq \frac{\pi}{2}$$

EXAMPLE 1

Find (a) $\sin^{-1}\frac{1}{2}$, (b) $\sin^{-1}\left(-\frac{1}{2}\right)$, and (c) $\sin^{-1}\frac{3}{2}$.

SOLUTION

(a) The number in the interval $[-\pi/2, \pi/2]$ whose sine is $\frac{1}{2}$ is $\pi/6$. Thus $\sin^{-1}\frac{1}{2} = \pi/6$.

(b) Again, $\sin^{-1}\left(-\frac{1}{2}\right)$ is the number in the interval $[-\pi/2, \pi/2]$ whose sine is $-\frac{1}{2}$. Since $\sin(-\pi/6) = -\frac{1}{2}$, we have $\sin^{-1}\left(-\frac{1}{2}\right) = -\pi/6$.

(c) Since $\frac{3}{2} > 1$, it is not in the domain of $\sin^{-1}x$, and so $\sin^{-1}\frac{3}{2}$ is not defined.

EXAMPLE 2

Find approximate values for (a) $\sin^{-1}(0.82)$ and (b) $\sin^{-1}\frac{1}{3}$.

SOLUTION

Since no rational multiple of π has a sine of 0.82 or $\frac{1}{3}$, we use a calculator to approximate these values. Using the $\boxed{\text{INV}}$ $\boxed{\text{SIN}}$, or $\boxed{\text{SIN}^{-1}}$, or $\boxed{\text{ARCSIN}}$ key on the calculator (making sure the calculator is in radian mode), we get

(a) $\sin^{-1}(0.82) \approx 0.96141$ $\qquad$ (b) $\sin^{-1}\frac{1}{3} \approx 0.33984$

EXAMPLE 3

Find $\cos\left(\sin^{-1}\frac{3}{5}\right)$.

SOLUTION 1

It is easy to find $\sin\left(\sin^{-1}\frac{3}{5}\right)$. In fact, by the properties of inverse functions, this value is exactly $\frac{3}{5}$. To find $\cos\left(\sin^{-1}\frac{3}{5}\right)$, we reduce this to the easier problem, by writing the cosine function in terms of the sine function. Let $u = \sin^{-1}\frac{3}{5}$. Since $-\pi/2 \leq u \leq \pi/2$, $\cos u$ is positive and we can write

$$\cos u = +\sqrt{1 - \sin^2 u}$$

Thus

$$\cos\left(\sin^{-1}\tfrac{3}{5}\right) = \sqrt{1 - \sin^2\left(\sin^{-1}\tfrac{3}{5}\right)}$$

$$= \sqrt{1 - \left(\tfrac{3}{5}\right)^2} = \sqrt{1 - \tfrac{9}{25}} = \sqrt{\tfrac{16}{25}} = \tfrac{4}{5}$$

SOLUTION 2

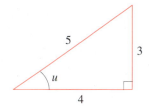

Again let $u = \sin^{-1}\frac{3}{5}$. Then u is the number in the interval $[-\pi/2,\ \pi/2]$ whose sine is $\frac{3}{5}$, Let us interpret u as an angle and draw a right triangle with u as one of its acute angles, with opposite side 3 and hypotenuse 5 (see Figure 4). The remaining leg of the triangle is found by the Pythagorean Theorem to be 4. From the diagram we get

Figure 4

$$\cos\left(\sin^{-1}\tfrac{3}{5}\right) = \cos u = \tfrac{4}{5}$$ ■

From Solution 2 of Example 3 we can immediately find the values of the other trigonometric functions of $u = \sin^{-1}\frac{3}{5}$ from the triangle. Thus $\tan\left(\sin^{-1}\frac{3}{5}\right) = \frac{3}{4}$, $\sec\left(\sin^{-1}\frac{3}{5}\right) = \frac{5}{4}$, and $\csc\left(\sin^{-1}\frac{3}{5}\right) = \frac{5}{3}$.

■ THE INVERSE COSINE FUNCTION

If the domain of the cosine function is restricted to the interval $[0,\ \pi]$, the resulting function is one-to-one and so has an inverse. We choose this interval because on it, cosine attains each of its values exactly once (see Figure 5).

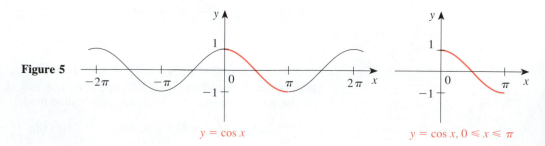

Figure 5

$y = \cos x$

$y = \cos x, 0 \leq x \leq \pi$

DEFINITION OF THE INVERSE COSINE FUNCTION

The **inverse cosine function** is the function **cos⁻¹** with domain $[-1, 1]$ and range $[0, \pi]$ defined by

$$\cos^{-1}x = y \iff \cos y = x$$

The inverse cosine function is also called **arccosine** and is denoted by **arccos.**

Thus, *y = cos⁻¹x is the number in the interval* $[0, \pi]$ *whose cosine is x.* The following relations hold:

$$\cos(\cos^{-1}x) = x \qquad \text{for } -1 \leqslant x \leqslant 1$$

$$\cos^{-1}(\cos x) = x \qquad \text{for } 0 \leqslant x \leqslant \pi$$

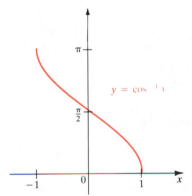

Figure 6

The graph of $y = \cos^{-1}x$ is sketched in Figure 6; it is obtained by reflecting the graph of $y = \cos x$, $0 \leqslant x \leqslant \pi$, in the line $y = x$.

EXAMPLE 4

Find (a) $\cos^{-1}\left(\sqrt{3}/2\right)$, (b) $\cos^{-1}0$, and (c) $\cos^{-1}\frac{5}{7}$.

SOLUTION

(a) The number in the interval $[0, \pi]$ whose cosine is $\sqrt{3}/2$ is $\pi/6$. Thus $\cos^{-1}\left(\sqrt{3}/2\right) = \pi/6$.

(b) Since $\cos(\pi/2) = 0$, it follows that $\cos^{-1}0 = \pi/2$.

(c) Since no rational multiple of π has cosine $\frac{5}{7}$, we use a calculator to find this value approximately: $\cos^{-1}\frac{5}{7} \approx 0.77519$. ∎

EXAMPLE 5

Write $\sin(\cos^{-1}x)$ and $\tan(\cos^{-1}x)$ as algebraic expressions in x for $-1 \leqslant x \leqslant 1$.

SOLUTION 1

Let $u = \cos^{-1}x$. We need to find $\sin u$ and $\tan u$ in terms of x. As in Example 3, the idea here is to write sine and tangent in terms of cosine. We have

$$\sin u = \pm\sqrt{1 - \cos^2u} \qquad \text{and} \qquad \tan u = \frac{\sin u}{\cos u} = \frac{\pm\sqrt{1 - \cos^2u}}{\cos u}$$

To choose the proper signs, note that since $u = \cos^{-1}x$, u lies in the interval $[0, \pi]$. Since $\sin u$ is positive on this interval, the $+$ sign is the correct choice. Now substituting $u = \cos^{-1}x$ in the displayed equations and using the relation $\cos(\cos^{-1}x) = x$ gives

$$\sin(\cos^{-1}x) = \sqrt{1 - x^2} \qquad \text{and} \qquad \tan(\cos^{-1}x) = \frac{\sqrt{1 - x^2}}{x}$$

SOLUTION 2

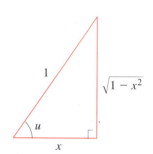

Figure 7

Let $u = \cos^{-1}x$ and let us interpret u as radian measure. Thus u is an angle with $\cos u = x$. We sketch in Figure 7 a right triangle which has the property that $\cos u = x$. From the triangle in the figure we have $\sin u = \sqrt{1 - x^2}$ and $\tan u = \sqrt{1 - x^2}/x$. Thus, when $-1 \le x \le 1$, we

$$\sin(\cos^{-1}x) = \sqrt{1 - x^2} \qquad \text{and} \qquad \tan(\cos^{-1}x) = \frac{\sqrt{1 - x^2}}{x} \qquad ■$$

Note   Some remarks about Solution 2 are in order. Strictly speaking, we can sketch a right triangle only if the angle u is acute. It turns out, however, that the method works for any u. In fact, except for sign, the triangles give the correct relations

$$\sin(\cos^{-1}x) = \pm\sqrt{1 - x^2} \qquad \text{and} \qquad \tan(\cos^{-1}x) = \frac{\pm\sqrt{1 - x^2}}{x}$$

Which sign should we choose? Fortunately, the domains of the inverse trigonometric functions have been chosen specifically so that the positive sign is always the correct choice. You can easily check this as follows. Let $u = \cos^{-1}x$. If we let $0 \le x \le 1$, then $0 \le u \le \pi/2$, hence $\sin u$ and $\tan u$ are positive. If $-1 \le x \le 0$, then $\pi/2 \le u \le \pi$, hence $\sin u$ is positive and $\tan u$ is negative. Thus, in each case the positive sign for the square root is the correct choice. In general, it is a valid operation to sketch a triangle in order to find $S(T^{-1}(x))$, where S and T are any trigonometric functions. This technique is a device for quickly solving problems like the ones in Example 5 and is often used in calculus.

EXAMPLE 6

(a) Write $\sin(2 \cos^{-1}x)$ as an algebraic expression in x for $-1 \le x \le 1$.

(b) Find $\sin\left(2 \cos^{-1}\frac{3}{5}\right)$.

SOLUTION

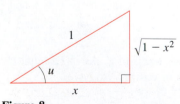

Figure 8

(a) Let $u = \cos^{-1}x$ and sketch a triangle as in Figure 8. We need to find $\sin 2u$, but from the triangle we can find the trigonometric function only of u, not of

$2u$. The double-angle identity for sine is useful here. We have

$$\sin 2u = 2 \sin u \cos u$$

From the triangle we have

$$\sin 2u = 2 \sin u \cos u = 2\sqrt{1 - x^2}\, x$$

Thus,

$$\sin(2 \cos^{-1}x) = 2x\sqrt{1 - x^2}$$

(b) Setting $x = \frac{3}{5}$ in part (a) gives

$$\sin\left(2 \cos^{-1}\tfrac{3}{5}\right) = 2\left(\tfrac{3}{5}\right)\sqrt{1 - \left(\tfrac{3}{5}\right)^2} = \tfrac{6}{5} \cdot \tfrac{4}{5} = \tfrac{24}{25} \qquad ■$$

■ THE INVERSE TANGENT FUNCTION

We restrict the domain of the tangent function to the interval $(-\pi/2,\ \pi/2)$ in order to obtain a one-to-one function.

DEFINITION OF THE INVERSE TANGENT FUNCTION

The **inverse tangent function** is the function **tan^{-1}** with domain R and range $(-\pi/2,\ \pi/2)$ defined by

$$\tan^{-1}x = y \quad \Leftrightarrow \quad \tan y = x$$

The inverse tangent function is also called **arctangent** and is denoted by **arctan.**

Thus $\tan^{-1}x$ *is the number in the interval $(-\pi/2,\ \pi/2)$ whose tangent is x.* The inverse function relations give

$$\tan(\tan^{-1}x) = x \qquad \text{for } x \in R$$

$$\tan^{-1}(\tan x) = x \qquad \text{for } -\frac{\pi}{2} < x < \frac{\pi}{2}$$

Figure 9 shows the graph of $y = \tan x$ on the interval $(-\pi/2,\ \pi/2)$ and the graph of its inverse function, $y = \tan^{-1}x$.

Figure 9

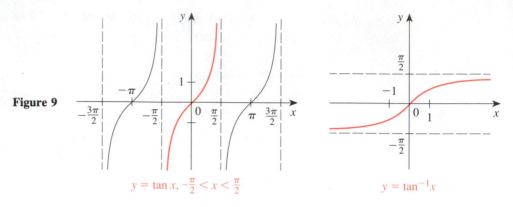

$$y = \tan x, \ -\frac{\pi}{2} < x < \frac{\pi}{2}$$

$$y = \tan^{-1}x$$

EXAMPLE 7

Find (a) $\tan^{-1}1$, (b) $\tan^{-1}\sqrt{3}$, and (c) $\tan^{-1}(-20)$.

SOLUTION

(a) The number in the interval $(-\pi/2, \pi/2)$ with tangent 1 is $\pi/4$. Thus $\tan^{-1}1 = \pi/4$.

(b) Since $\tan(\pi/3) = \sqrt{3}$, we have $\tan^{-1}\sqrt{3} = \pi/3$.

(c) We use a calculator to find that $\tan^{-1}(-20) \approx -1.52084$. ■

EXAMPLE 8

Write $\sin(\cos^{-1}x + \tan^{-1}y)$ as an algebraic expression in x and y for $-1 \leqslant x \leqslant 1$ and y any real number.

SOLUTION

Let $u = \cos^{-1}x$ and $v = \tan^{-1}y$. In Figure 10 we sketch right triangles with acute angles u and v such that $\cos u = x$ and $\tan v = y$. By the addition formula for sine and the triangles in the figure, we get

$$\sin(\cos^{-1}x + \tan^{-1}y) = \sin(u + v)$$

$$= \sin u \cos v + \cos u \sin v$$

$$= \sqrt{1 - x^2}\ \frac{1}{\sqrt{1 + y^2}} + x\ \frac{y}{\sqrt{1 + y^2}}$$

$$= \frac{1}{\sqrt{1 + y^2}}\left(\sqrt{1 - x^2} + xy\right)$$

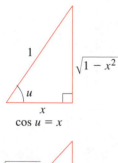

$$\cos u = x$$

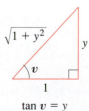

$$\tan v = y$$

Figure 10

Note that the relations between the trigonometric functions of u and v expressed in these triangles hold for all vaules of u and v in question. (See the remarks following Example 5.) ■

■ THE INVERSE SECANT, COSECANT, AND COTANGENT FUNCTIONS

To define the inverse functions of the secant, cosecant, and cotangent functions we restrict the domain of each function to a set on which it is one-to-one and on which it attains all its values. Although any interval satisfying these criteria is appropriate, we choose to restrict the domains in a way that simplifies the choice of sign in computations involving inverse trigonometric functions. The choices we make are also appropriate for calculus. This explains the seemingly strange restriction for the domain of the secant and cosecant functions. We end this section by displaying the graphs of the secant, cosecant, and cotangent functions with their restricted domains and the graphs of their inverse functions.

Figure 11
The inverse secant function

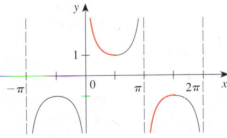

$$y = \sec x,\, 0 \le x < \tfrac{\pi}{2},\, \pi \le x < \tfrac{3\pi}{2}$$

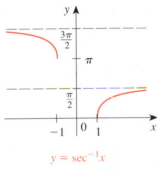

$$y = \sec^{-1}x$$

Figure 12
The inverse cosecant function

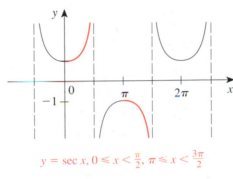

$$y = \csc x,\, 0 < x \le \tfrac{\pi}{2},\, \pi < x \le \tfrac{3\pi}{2}$$

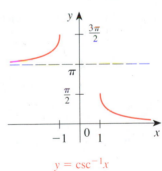

$$y = \csc^{-1}x$$

Figure 13
The inverse cotangant function

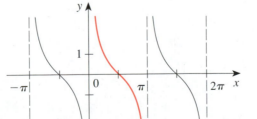

$$y = \cot x,\, 0 < x < \pi$$

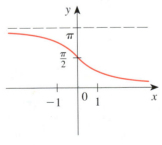

$$y = \cot^{-1}x$$

EXERCISES 7.6

In Exercises 1–8 find the exact value of each expression, if it is defined.

1. (a) $\sin^{-1}\frac{1}{2}$ (b) $\cos^{-1}\frac{1}{2}$ (c) $\cos^{-1}2$

2. (a) $\sin^{-1}\frac{\sqrt{3}}{2}$ (b) $\cos^{-1}\frac{\sqrt{3}}{2}$ (c) $\cos^{-1}\left(-\frac{\sqrt{3}}{2}\right)$

3. (a) $\sin^{-1}\frac{\sqrt{2}}{2}$ (b) $\cos^{-1}\frac{\sqrt{2}}{2}$ (c) $\sin^{-1}\left(-\frac{\sqrt{2}}{2}\right)$

4. (a) $\tan^{-1}\sqrt{3}$ (b) $\tan^{-1}\left(-\sqrt{3}\right)$ (c) $\sin^{-1}\sqrt{3}$

5. (a) $\sin^{-1}1$ (b) $\cos^{-1}1$ (c) $\cos^{-1}(-1)$

6. (a) $\tan^{-1}1$ (b) $\tan^{-1}(-1)$ (c) $\tan^{-1}0$

7. (a) $\tan^{-1}\frac{\sqrt{3}}{3}$ (b) $\tan^{-1}\left(-\frac{\sqrt{3}}{3}\right)$ (c) $\sin^{-1}(-2)$

8. (a) $\sin^{-1}0$ (b) $\cos^{-1}0$ (c) $\cos^{-1}\left(-\frac{1}{2}\right)$

In Exercises 9–10 use a calculator to find an approximate value of each expression correct to five decimal places.

9. (a) $\sin^{-1}(0.7688)$ (b) $\cos^{-1}(-0.5014)$
(c) $\tan^{-1}(15.2000)$

10. (a) $\cos^{-1}(0.3388)$ (b) $\tan^{-1}(1.0000)$
(c) $\cos^{-1}(0.9800)$

In Exercises 11–24 find the exact value of the expression.

11. $\sin\left(\sin^{-1}\frac{1}{2}\right)$

12. $\cos\left(\sin^{-1}\frac{1}{2}\right)$

13. $\tan\left(\sin^{-1}\frac{1}{2}\right)$

14. $\sin(\sin^{-1}0)$

15. $\cos\left(\sin^{-1}\frac{\sqrt{3}}{2}\right)$

16. $\tan\left(\sin^{-1}\frac{\sqrt{2}}{2}\right)$

17. $\sin\left(\tan^{-1}\sqrt{3}\right)$

18. $\tan^{-1}\left(\tan\frac{3\pi}{4}\right)$

19. $\cos^{-1}\left(\cos\frac{\pi}{3}\right)$

20. $\cos^{-1}\left(\cos\frac{4\pi}{3}\right)$

21. $\tan^{-1}\left(2\sin\frac{\pi}{3}\right)$

22. $\cos^{-1}\left(\sqrt{3}\sin\frac{\pi}{6}\right)$

23. $\tan^{-1}\left(\sin\frac{\pi}{3}+\cos\frac{\pi}{6}\right)$

24. $\cos^{-1}\left(\frac{1}{2}\tan\frac{\pi}{3}-\sin\frac{\pi}{3}\right)$

In Exercises 25–38 evaluate the expression. Do not use a calculator.

25. $\sin\left(\cos^{-1}\frac{3}{5}\right)$

26. $\tan\left(\sin^{-1}\frac{4}{5}\right)$

27. $\sin\left(\tan^{-1}\frac{12}{5}\right)$

28. $\cos(\tan^{-1}5)$

29. $\sin\left(2\cos^{-1}\frac{3}{5}\right)$

30. $\tan\left(2\tan^{-1}\frac{5}{13}\right)$

31. $\cos\left(\frac{1}{2}\sin^{-1}\frac{\sqrt{3}}{2}\right)$

32. $\tan\left(\frac{1}{2}\sin^{-1}\frac{15}{17}\right)$

33. $\sin\left(\sin^{-1}\frac{1}{2}+\cos^{-1}\frac{1}{2}\right)$

34. $\cos\left(\sin^{-1}\frac{3}{5}-\cos^{-1}\frac{3}{5}\right)$

35. $\sin\left(\sin^{-1}\frac{5}{13}-\cos^{-1}\frac{12}{13}\right)$

36. $\tan\left(\sin^{-1}\frac{1}{2}+\cos^{-1}\frac{1}{3}\right)$

37. $\cos\left(\tan^{-1}3+\cos^{-1}\frac{1}{2}\right)$

38. $\cos\left(2\sin^{-1}\frac{3}{5}+\tan^{-1}\frac{3}{4}\right)$

In Exercises 39–48 rewrite each expression as an algebraic expression in x.

39. $\cos(\sin^{-1}x)$

40. $\sin(\tan^{-1}x)$

41. $\tan(\sin^{-1}x)$

42. $\sin(2\cos^{-1}x)$

43. $\cos(2\tan^{-1}x)$

44. $\sin(2\sin^{-1}x)$

45. $\tan\left(\frac{1}{2}\sin^{-1}x\right)$

46. $\cos(\cos^{-1}x+\sin^{-1}x)$

47. $\sin(\tan^{-1}x-\sin^{-1}x)$

48. $\sin(2\sin^{-1}x+\cos^{-1}x)$

In Exercises 49 and 50 write each expression as an algebraic expression in x and y.

49. $\tan(\tan^{-1}x+\sin^{-1}y)$ **50.** $\tan(\tan^{-1}x-\tan^{-1}2y)$

51. A 50-ft pole casts a shadow of length s as in the figure. Express the angle θ of elevation of the sun in terms of the length of the shadow.

52. A 680-ft rope anchors a hot-air balloon as in the figure. Express the height h of the balloon as a function of the angle θ the rope forms with the horizontal.

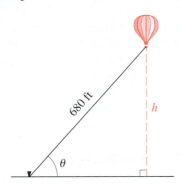

53. A lighthouse is on an island that is 2 miles off a straight shoreline (see the figure). Express the angle θ formed by the beam of light and the shoreline in terms of the distance d in the figure.

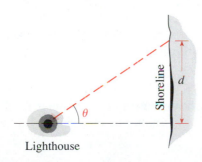

54. A painting that is 2 m high hangs in a museum with its bottom edge 3 m above the floor. A person whose eye level is h meters above the floor stands at a distance of x meters directly in front of the painting. The size that the painting appears to the viewer is determined by the size of the angle θ that the painting subtends at the viewer's eyes (see the figure). The larger θ is, the larger the painting appears to the viewer. The angle θ depends on the distance x; in other words, the angle θ is a function of x. Show that

$$\theta = \tan^{-1}\left(\frac{2x}{x^2 + (3-h)(5-h)}\right)$$

[*Hint:* Use the subtraction formula for tangent and the fact that $\theta = \alpha - \beta$.]

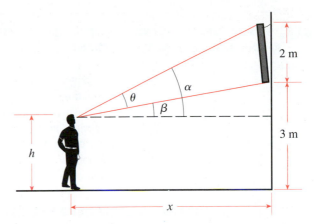

In Exercises 55–60, (a) solve the trigonometric equation on the given interval, and express the solutions in terms of inverse trigonometric functions; and (b) use a calculator to approximate the solutions you found in part (a) to five decimal places.

55. $\tan^2 x - \tan x - 2 = 0$ on $(-\pi/2,\ \pi/2)$

56. $2 \sin 2x - \cos x = 0$ on $[0,\ \pi/2)$

57. $\sin 2x\ (\sec^2 x - 2) = 0$ on $[0,\ \pi)$

58. $\cos 2x - 5 \cos x + 4 = 0$ on $[0,\ \pi)$

59. $3 \sin^2 x - 7 \sin x + 2 = 0$ on $[0,\ 2\pi)$

60. $\tan^4 x - 13 \tan^2 x + 36 = 0$ on $[0,\ \pi/2]$

In Exercises 61–64 prove each identity.

61. $\sin^{-1}(-x) = -\sin^{-1}x,\quad |x| \le 1$

62. $\cos^{-1}x = \sin^{-1}\sqrt{1 - x^2},\quad |x| \le 1$

63. $\tan^{-1}x = \sin^{-1}\dfrac{x}{\sqrt{1 + x^2}}$

64. $\cos^{-1}(-x) = \pi - \cos^{-1}x,\quad |x| \le 1$

In Exercises 65 and 66, (a) graph the given function and make a conjecture, and (b) prove that your conjecture is true.

65. $y = \tan^{-1}x + \tan^{-1}\dfrac{1}{x}$ **66.** $y = \sin^{-1}x + \cos^{-1}x$

In Exercises 67 and 68 (a) use a graphing device to find all solutions to the given equation, correct to two decimal places, and (b) find the exact solution.

67. $\sin^{-1}x - \cos^{-1}x = 0$ **68.** $\tan^{-1}x + \tan^{-1}2x = \dfrac{\pi}{4}$

TRIGONOMETRIC FORM OF COMPLEX NUMBERS; DeMOIVRE'S THEOREM

Complex numbers were introduced in Chapter 3 in order to solve certain algebraic equations. The applications of complex numbers go far beyond this initial use, however. Complex numbers are now used routinely in physics, electrical engineering, aerospace engineering, and many other fields. In Chapter 3 we learned to perform arithmetic operations on complex numbers and to represent these numbers graphically. In this section we represent complex numbers using the functions sine and cosine. This will enable us to find the nth roots of complex numbers.

TRIGONOMETRIC FORM OF COMPLEX NUMBERS

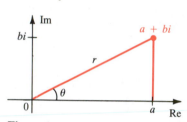

Figure 1

Let $z = a + bi$ be a complex number and let us draw in the complex plane the line segment joining the origin to the point $a + bi$ (see Figure 1). Let the length of this line segment be denoted by $r = |z| = \sqrt{a^2 + b^2}$. (Recall that $|z|$ is the **modulus** of z.) If θ is an angle in standard position whose terminal side coincides with this line segment, then by the definitions of sine and cosine (see Section 6.3),

$$a = r \cos \theta \quad \text{and} \quad b = r \sin \theta$$

so $z = r \cos \theta + ir \sin \theta = r(\cos \theta + i \sin \theta)$. It follows that any complex number z can be represented in **trigonometric form** as

$$z = r(\cos \theta + i \sin \theta)$$

where $r = |z|$. The angle θ is called the **argument** of the complex number z. The argument of z is not unique, but any two arguments of z differ by a multiple of 2π. We have shown the following.

TRIGONOMETRIC FORM OF COMPLEX NUMBERS

If $z = a + bi$ is a complex number, then z has the trigonometric form

$$z = r(\cos \theta + i \sin \theta)$$

where $r = |z| = \sqrt{a^2 + b^2}$ and θ is an argument of z.

EXAMPLE 1

Write each complex number in trigonometric form.

(a) $1 + i$ (b) $-1 + \sqrt{3}i$ (c) $-4\sqrt{3} - 4i$ (d) $3 + 4i$

SOLUTION

These complex numbers are graphed in Figure 2. The graphs help us find their arguments.

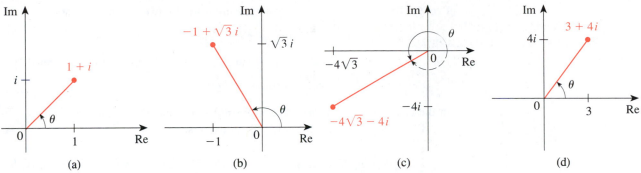

Figure 2

(a) An argument is $\theta = \pi/4$ and $r = \sqrt{1 + 1} = \sqrt{2}$. Thus

$$1 + i = \sqrt{2}\left(\cos \frac{\pi}{4} + i \sin \frac{\pi}{4}\right)$$

(b) An argument is $\theta = 2\pi/3$ and $r = \sqrt{1 + 3} = 2$. Thus

$$-1 + \sqrt{3}\,i = 2\left(\cos \frac{2\pi}{3} + i \sin \frac{2\pi}{3}\right)$$

(c) An argument is $\theta = 7\pi/6$ (or we could use $\theta = -5\pi/6$), and $r = \sqrt{48 + 16} = 8$. Thus

$$-4\sqrt{3} - 4i = 8\left(\cos \frac{7\pi}{6} + i \sin \frac{7\pi}{6}\right)$$

(d) An argument is $\theta = \tan^{-1}\frac{4}{3}$ and $r = \sqrt{3^2 + 4^2} = 5$. So

$$3 + 4i = 5\left[\cos\left(\tan^{-1}\tfrac{4}{3}\right) + i \sin\left(\tan^{-1}\tfrac{4}{3}\right)\right]$$
■

The addition formulas for sine and cosine discussed in Section 7.3 greatly simplify the multiplication and division of complex numbers in trigonometric form. The following theorem shows how.

MULTIPLICATION AND DIVISION OF COMPLEX NUMBERS

If the two complex numbers z_1 and z_2 have the trigonometric forms

$$z_1 = r_1(\cos \theta_1 + i \sin \theta_1) \qquad \text{and} \qquad z_2 = r_2(\cos \theta_2 + i \sin \theta_2)$$

then

$$z_1 z_2 = r_1 r_2[\cos(\theta_1 + \theta_2) + i \sin(\theta_1 + \theta_2)] \tag{1}$$

$$\frac{z_1}{z_2} = \frac{r_1}{r_2}[\cos(\theta_1 - \theta_2) + i \sin(\theta_1 - \theta_2)], \qquad z_2 \neq 0 \tag{2}$$

Proof   To prove Formula 1 we simply multiply the two complex numbers.

$$z_1 z_2 = r_1 r_2 (\cos \theta_1 + i \sin \theta_1)(\cos \theta_2 + i \sin \theta_2)$$
$$= r_1 r_2 [\cos \theta_1 \cos \theta_2 - \sin \theta_1 \sin \theta_2) + i (\sin \theta_1 \cos \theta_2 + \cos \theta_1 \sin \theta_2)]$$
$$= r_1 r_2 [\cos(\theta_1 + \theta_2) + i \sin(\theta_1 + \theta_2)]$$

In the last step we used the addition formulas for sine and cosine.
    The proof of Formula 2 is left as an exercise.   ∎

The preceding theorem says: *To multiply two complex numbers we multiply the moduli and add the arguments*, and *to divide two complex numbers we divide the moduli and subtract the arguments*.

EXAMPLE 2

Let

$$z_1 = 2\left(\cos \frac{\pi}{4} + i \sin \frac{\pi}{4}\right) \qquad \text{and} \qquad z_2 = 5\left(\cos \frac{\pi}{3} + i \sin \frac{\pi}{3}\right)$$

Find (a) $z_1 z_2$ and (b) z_1/z_2.

SOLUTION

(a) By Formula 1,

$$z_1 z_2 = (2)(5)\left[\cos\left(\frac{\pi}{4} + \frac{\pi}{3}\right) + i \sin\left(\frac{\pi}{4} + \frac{\pi}{3}\right)\right] = 10\left(\cos \frac{7\pi}{12} + i \sin \frac{7\pi}{12}\right)$$

To approximate the answer we use a calculator in radian mode and get

$$z_1 z_2 \approx 10(-0.2588 + 0.9659i) = -2.588 + 9.659i$$

(b) By Formula 2,

$$\frac{z_1}{z_2} = \frac{2}{5}\left[\cos\left(\frac{\pi}{4} - \frac{\pi}{3}\right) + i \sin\left(\frac{\pi}{4} - \frac{\pi}{3}\right)\right]$$
$$= \frac{2}{5}\left[\cos\left(-\frac{\pi}{12}\right) + i \sin\left(-\frac{\pi}{12}\right)\right]$$
$$= \frac{2}{5}\left(\cos \frac{\pi}{12} - i \sin \frac{\pi}{12}\right)$$

Using a calculator in radian mode gives the approximate answer

$$\frac{z_1}{z_2} \approx \frac{2}{5}(0.9659 - 0.2588i) = 0.3864 - 0.1035i$$ ∎

EXAMPLE 3

Let $z_1 = -1 + \sqrt{3}\, i$ and $z_2 = -4\sqrt{3} - 4i$. Use the trigonometric form to find $z_1 z_2$ and z_1/z_2.

SOLUTION

From Example 1 we have

$$z_1 = -1 + \sqrt{3}\, i = 2\left(\cos \frac{2\pi}{3} + i \sin \frac{2\pi}{3}\right)$$

and

$$z_2 = -4\sqrt{3} - 4i = 8\left(\cos \frac{7\pi}{6} + i \sin \frac{7\pi}{6}\right)$$

By Formulas 1 and 2 we have

$$z_1 z_2 = (2)(8)\left[\cos\left(\frac{2\pi}{3} + \frac{7\pi}{6}\right) + i \sin\left(\frac{2\pi}{3} + \frac{7\pi}{6}\right)\right]$$

$$= 16\left(\cos \frac{11\pi}{6} + i \sin \frac{11\pi}{6}\right)$$

$$= 16\left(\frac{\sqrt{3}}{2} - \frac{1}{2}i\right) = 8\left(\sqrt{3} - i\right)$$

$$\frac{z_1}{z_2} = \frac{2}{8}\left[\cos\left(\frac{2\pi}{3} - \frac{7\pi}{6}\right) + i \sin\left(\frac{2\pi}{3} - \frac{7\pi}{6}\right)\right]$$

$$= \frac{1}{4}\left[\cos\left(-\frac{\pi}{2}\right) + i \sin\left(-\frac{\pi}{2}\right)\right] = -\frac{1}{4}i$$

You should check that these same answers are obtained if we do the operations algebraically as in Chapter 3. ■

DeMOIVRE'S THEOREM

Repeated use of the multiplication formula gives a useful formula for raising a complex number to a power n for any positive integer n. Let z be a complex number written in trigonometric form

$$z = r(\cos \theta + i \sin \theta)$$

Then by Formula 1,

$$z^2 = zz = r^2[\cos(\theta + \theta) + i \sin(\theta + \theta)]$$
$$= r^2(\cos 2\theta + i \sin 2\theta)$$

By Formula 1 again, this time applied to z and z^2, we get

$$z^3 = z^2 z = r^3[\cos(2\theta + \theta) + i \sin(2\theta + \theta)]$$
$$= r^3(\cos 3\theta + i \sin 3\theta)$$

Repeating this argument, we see that for any positive integer n,

$$z^n = r^n(\cos n\theta + i \sin n\theta)$$

A similar argument using the division formula (2) shows that the preceding formula holds for negative integers also. We have proved the following theorem.

DeMOIVRE'S THEOREM

> If $z = r(\cos \theta + i \sin \theta)$, then for any integer n,
>
> $$z^n = r^n(\cos n\theta + i \sin n\theta)$$

This says that *to take the nth power of a complex number we take the nth power of the modulus and multiply the argument by n.*

EXAMPLE 4

Find $\left(\frac{1}{2} + \frac{1}{2}i\right)^{10}$.

SOLUTION

Since $\frac{1}{2} + \frac{1}{2}i = \frac{1}{2}(1 + i)$, it follows from Example 1(a) that

$$\frac{1}{2} + \frac{1}{2}i = \frac{\sqrt{2}}{2}\left(\cos \frac{\pi}{4} + i \sin \frac{\pi}{4}\right)$$

So by DeMoivre's Theorem,

$$\left(\frac{1}{2} + \frac{1}{2}i\right)^{10} = \left(\frac{\sqrt{2}}{2}\right)^{10}\left(\cos \frac{10\pi}{4} + i \sin \frac{10\pi}{4}\right)$$

$$= \frac{2^5}{2^{10}}\left(\cos \frac{5\pi}{2} + i \sin \frac{5\pi}{2}\right) = \frac{1}{32}i \qquad \blacksquare$$

■ *n*TH ROOTS OF COMPLEX NUMBERS

To find the nth roots of the complex number z we need to find a complex number w such that

$$w^n = z$$

Let us write z in trigonometric form:

$$z = r(\cos\theta + i\sin\theta)$$

One nth root of z is

$$w = r^{1/n}\left(\cos\frac{\theta}{n} + i\sin\frac{\theta}{n}\right)$$

since by DeMoivre's Theorem, $w^n = z$. But the argument θ of z can be replaced by $\theta + 2k\pi$ for any integer k. Thus other nth roots of z are

$$w = r^{1/n}\left[\cos\left(\frac{\theta + 2k\pi}{n}\right) + i\sin\left(\frac{\theta + 2k\pi}{n}\right)\right]$$

Since this expression gives a different value of w for $k = 0, 1, 2, \ldots, n - 1$, we have proved the following theorem.

nTH ROOTS OF COMPLEX NUMBERS

> If $z = r(\cos\theta + i\sin\theta)$ and n is a positive integer, then z has the n distinct nth roots
>
> $$w_k = r^{1/n}\left[\cos\left(\frac{\theta + 2k\pi}{n}\right) + i\sin\left(\frac{\theta + 2k\pi}{n}\right)\right]$$
>
> for $k = 0, 1, 2, \ldots, n - 1$.

The following observations help us use the formula.

1. The modulus of each nth root is $r^{1/n}$.
2. The argument of the first root is θ/n.
3. We repeatedly add $\dfrac{2\pi}{n}$ to get the argument of each successive root.

These observations show that, when graphed, the nth roots of z are spaced equally on the circle of radius $r^{1/n}$.

EXAMPLE 5

Find the six sixth roots of $z = -8$ and graph these roots in the complex plane.

SOLUTION

In trigonometric form, $z = 8(\cos\pi + i\sin\pi)$. Applying the formula for nth roots with $n = 6$, we get

$$w_k = 8^{1/6}\left[\cos\left(\frac{\pi + 2k\pi}{6}\right) + i\sin\left(\frac{\pi + 2k\pi}{6}\right)\right]$$

for $k = 0, 1, 2, 3, 4, 5$. Thus the six sixth roots of -8 are

We add $2\pi/6 = \pi/3$ to each argument to get the argument of the next root.

$$w_0 = 8^{1/6}\left(\cos\frac{\pi}{6} + i\sin\frac{\pi}{6}\right) = \sqrt{2}\left(\frac{\sqrt{3}}{2} + \frac{1}{2}i\right)$$

$$w_1 = 8^{1/6}\left(\cos\frac{\pi}{2} + i\sin\frac{\pi}{2}\right) = \sqrt{2}\,i$$

$$w_2 = 8^{1/6}\left(\cos\frac{5\pi}{6} + i\sin\frac{5\pi}{6}\right) = \sqrt{2}\left(-\frac{\sqrt{3}}{2} + \frac{1}{2}i\right)$$

$$w_3 = 8^{1/6}\left(\cos\frac{7\pi}{6} + i\sin\frac{7\pi}{6}\right) = \sqrt{2}\left(-\frac{\sqrt{3}}{2} - \frac{1}{2}i\right)$$

$$w_4 = 8^{1/6}\left(\cos\frac{3\pi}{2} + i\sin\frac{3\pi}{2}\right) = -\sqrt{2}\,i$$

$$w_5 = 8^{1/6}\left(\cos\frac{11\pi}{6} + i\sin\frac{11\pi}{6}\right) = \sqrt{2}\left(\frac{\sqrt{3}}{2} - \frac{1}{2}i\right)$$

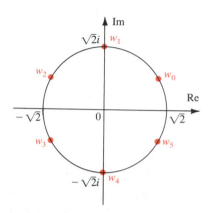

Figure 3
The six sixth roots of $z = -8$

All these points lie on the circle of radius $\sqrt{2}$, as shown in Figure 3. ■

When finding roots of complex numbers we sometimes write the argument θ of the complex number in degrees. In this case the nth roots are obtained from the formula

$$w_k = r^{1/n}\left[\cos\left(\frac{\theta + 360°k}{n}\right) + i\sin\left(\frac{\theta + 360°k}{n}\right)\right]$$

for $k = 0, 1, 2, \ldots, n - 1$.

EXAMPLE 6

Find the three cube roots of $z = 4\sqrt{2} + 4\sqrt{2}\,i$ and graph these roots in the complex plane.

SOLUTION

First we write z in trigonometric form using degrees. Thus

$$z = 8(\cos 45° + i\sin 45°)$$

Applying the formula for nth roots (in degrees) with $n = 3$, we find the cube roots of z are of the form

$$w_k = 8^{1/3}\left[\cos\left(\frac{45° + 360°k}{3}\right) + i\sin\left(\frac{45° + 360°k}{3}\right)\right]$$

where $k = 0, 1, 2$. Thus the three cube roots are

$$w_0 = 2(\cos 15° + i \sin 15°)$$

$$w_1 = 2(\cos 135° + i \sin 135°)$$

$$w_2 = 2(\cos 255° + i \sin 255°)$$

We add $360°/3 = 120°$ to each argument to get the argument of the next root.

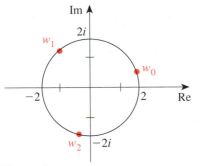

The three cube roots of z are graphed in Figure 4. These roots are spaced equally on the circle of radius 2. ∎

EXAMPLE 7

Solve the equation $z^6 + 8 = 0$.

SOLUTION

This equation can be written $z^6 = -8$. Thus the solutions are the sixth roots of -8, which we found in Example 5. ∎

Figure 4
The three cube roots of
$z = 4\sqrt{2} + 4\sqrt{2}\,i$

EXERCISES 7.7

In Exercises 1–24 write each complex number in trigonometric form with argument θ between 0 and 2π.

1. $1 + i$
2. $1 - \sqrt{3}\,i$
3. $\sqrt{2} - \sqrt{2}\,i$
4. $1 - i$
5. $2\sqrt{3} - 2i$
6. $-1 + i$
7. $-\sqrt{2}\,i$
8. $-3 - 3\sqrt{3}\,i$
9. $5 + 5i$
10. 4
11. $4\sqrt{3} - 4i$
12. $8i$
13. -20
14. $\sqrt{3} + i$
15. $3 + 4i$
16. $i(2 - 2i)$
17. $3i(1 + i)$
18. $2(1 - i)$
19. $4(\sqrt{3} + i)$
20. $-3 - 3i$
21. $2 + i$
22. $3 + \sqrt{3}\,i$
23. $\sqrt{2} + \sqrt{2}\,i$
24. $-\pi i$

In Exercises 25–34 write z_1 and z_2 in trigonometric form and then use Formulas 1 and 2 to find the product $z_1 z_2$ and the quotients z_1/z_2 and $1/z_1$.

25. $z_1 = \sqrt{3} + i, \quad z_2 = 1 + \sqrt{3}\,i$
26. $z_1 = \sqrt{2} - \sqrt{2}\,i, \quad z_2 = 1 - i$
27. $z_1 = 2\sqrt{3} - 2i, \quad z_2 = -1 + i$
28. $z_1 = -\sqrt{2}\,i, \quad z_2 = -3 - 3\sqrt{3}\,i$
29. $z_1 = 5 + 5i, \quad z_2 = 4$
30. $z_1 = 4\sqrt{3} - 4i, \quad z_2 = 8i$
31. $z_1 = -20, \quad z_2 = \sqrt{3} + i$
32. $z_1 = 3 + 4i, \quad z_2 = 2 - 2i$
33. $z_1 = 3i(1 + i), \quad z_2 = 2(1 - i)$
34. $z_1 = 4(\sqrt{3} + i), \quad z_2 = -3 - 3i$

In Exercises 35–46 find the indicated power using DeMoivre's Theorem.

35. $(1 + i)^{20}$
36. $\left(1 - \sqrt{3}\,i\right)^5$
37. $\left(2\sqrt{3} + 2i\right)^5$
38. $(1 - i)^8$
39. $\left(\dfrac{\sqrt{2}}{2} + \dfrac{\sqrt{2}}{2}i\right)^{12}$
40. $\left(\sqrt{3} - i\right)^{-10}$
41. $(2 - 2i)^8$
42. $\left(-\dfrac{1}{2} - \dfrac{\sqrt{3}}{2}i\right)^{15}$
43. $(-1 - i)^7$
44. $\left(3 + \sqrt{3}\,i\right)^4$
45. $\left(2\sqrt{3} + 2i\right)^{-5}$
46. $(1 - i)^{-8}$

In Exercises 47–56 find the indicated roots, and sketch the roots in the complex plane.

47. The square roots of $4\sqrt{3} + 4i$

48. The cube roots of $4\sqrt{3} + 4i$

49. The fourth roots of $-81i$

50. The fifth roots of 32   **51.** The eighth roots of 1

52. The cube roots of $1 + i$ **53.** The cube roots of i

54. The fifth roots of i **55.** The fourth roots of -1

56. The fifth roots of $-16 - 16\sqrt{3}\,i$

In Exercises 57–64 solve each equation.

57. $z^4 + 1 = 0$ **58.** $z^8 - i = 0$

59. $z^3 - 4\sqrt{3} - 4i = 0$ **60.** $z^2 + z + 1 = 0$

61. $z^6 - 1 = 0$ **62.** $z^3 + 1 = -i$

63. $iz^2 - 4z - 3i = 0$ **64.** $z^3 - 1 = 0$

65. Show that the sum of the four fourth roots of 1 is zero. (It is true in general that the sum of the n nth roots of z is zero.)

66. Find the product of the four fourth roots of 1.

67. (a) Let $w = \cos\dfrac{2\pi}{n} + i\sin\dfrac{2\pi}{n}$ where n is a positive integer. Show that $1, w, w^2, w^3, \ldots, w^{n-1}$ are the n nth roots of 1.

(b) If $z \neq 0$ is any complex number and $s^n = z$, show that the n nth roots of z are $s, sw, sw^2, sw^3, \ldots, sw^{n-1}$.

SECTION 7.8
VECTORS

In applications of mathematics certain quantities are determined completely by their magnitude—for example, length, mass, area, temperature, and energy. We speak of a length of 5 m or a mass of 3 kg; only one number is needed to describe each of these quantities. Such a quantity is called a **scalar**.

On the other hand, to describe the displacement of an object, two numbers are required: the *magnitude* and the *direction* of the displacement. Thus to describe an airplane's flight, we must give the distance traveled as well as the direction of travel. To describe the velocity of a moving object we must specify both the *speed* and the *direction* of travel. Quantities such as displacement, velocity, acceleration, and force that involve magnitude as well as direction are called *directed quantities*. One way to represent such quantities mathematically is through the use of *vectors*.

GEOMETRIC DESCRIPTION OF VECTORS

A **vector** in the plane is a line segment with an assigned direction. We sketch a vector as in Figure 1 with an arrowhead to specify the direction. We denote this vector by $\overrightarrow{AB}$. The point A is the **initial point** and B is the **terminal point** of the vector $\overrightarrow{AB}$. The length of the line segment AB is called the **magnitude** or **length** of the vector and is denoted by $|AB|$. We use boldface letters to denote vectors. Thus we write $\mathbf{u} = \overrightarrow{AB}$.

Two vectors are considered **equal** if they have equal magnitudes and the same direction. Thus all the vectors in Figure 2 are equal. This definition of equality makes sense if we think of a vector as representing a displacement. Two such displacements are the same if they have equal magnitudes and the same direction. The initial and terminal points of a vector depend on the initial position of the

Figure 1

object to which this displacement is applied. So the vectors in Figure 2 can be thought of as the *same* displacement applied to objects in different locations in the plane.

Figure 2

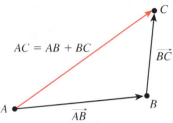

$$AC = AB + BC$$

Figure 3

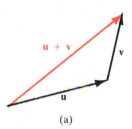

(a)

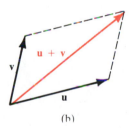

(b)

Figure 4

If the displacement $\mathbf{u} = \overrightarrow{AB}$ is followed by the displacement $\mathbf{v} = \overrightarrow{BC}$, then the resulting displacement is $\overrightarrow{AC}$ as shown in Figure 3. In other words, the single displacement represented by the vector $\overrightarrow{AC}$ has the same effect as the other two displacements together. We call the vector $\overrightarrow{AC}$ the **sum** of the vectors $\overrightarrow{AB}$ and $\overrightarrow{BC}$ and we write $\overrightarrow{AC} = \overrightarrow{AB} + \overrightarrow{BC}$. (The **zero vector**, denoted by $\mathbf{0}$, represents no displacement). Thus, to find the sum of any two vectors $\mathbf{u}$ and $\mathbf{v}$, we sketch vectors equal to $\mathbf{u}$ and $\mathbf{v}$ with the initial point of one at the terminal point of the other [see Figure 4(a)]. If we draw $\mathbf{u}$ and $\mathbf{v}$ starting at the same point, then $\mathbf{u} + \mathbf{v}$ is the vector that is the diagonal of the parallelogram formed by $\mathbf{u}$ and $\mathbf{v}$, as shown in Figure 4(b).

If a is a real number and $\mathbf{v}$ is a vector, we define a new vector $a\mathbf{v}$ as follows: The vector $a\mathbf{v}$ has magnitude $|a||\mathbf{v}|$ and has the same direction as $\mathbf{v}$ if $a > 0$, or has direction opposite to $\mathbf{v}$ if $a < 0$. If $a = 0$, then $a\mathbf{v} = \mathbf{0}$, the zero vector. This process is called **multiplication of a vector by a scalar.** Multiplication of a vector by a positive scalar has the effect of stretching or shrinking a vector, while (at the same time) preserving its direction. Multiplication by a negative scalar has a similar effect on the length of a vector but reverses the direction. Figure 5 shows graphs of the vector $a\mathbf{v}$ for different values of a. We write the vector $(-1)\mathbf{v}$ as $-\mathbf{v}$. Thus $-\mathbf{v}$ is the vector with the same length as $\mathbf{v}$ but with the opposite direction. The **difference** of two vectors $\mathbf{u}$ and $\mathbf{v}$ is defined by $\mathbf{u} - \mathbf{v} = \mathbf{u} + (-\mathbf{v})$. Figure 6 shows that the vector $\mathbf{u} - \mathbf{v}$ is the other diagonal of the parallelogram formed by $\mathbf{u}$ and $\mathbf{v}$.

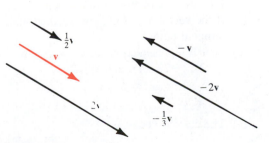

Figure 5
Multiplication of a vector by a scalar

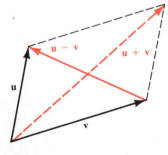

Figure 6
Subtraction of vectors

■ ANALYTIC DESCRIPTION OF VECTORS

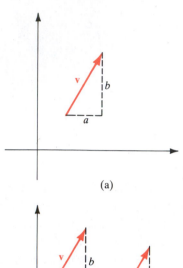

(a)

(b)

Figure 7

So far we have been discussing vectors geometrically. We now give a method for describing vectors using coordinates (that is, analytically). Consider a vector **v** in the coordinate plane as in Figure 7(a). To go from the initial point of **v** to the terminal point, we move a units to the right and b units upward. We represent the vector **v** as an ordered pair of real numbers

$$\mathbf{v} = \langle a, b \rangle$$

where a is the **horizontal component** of **v** and b is the **vertical component** of **v**. We must remember that a vector represents a magnitude and a direction, not a particular arrow in the plane. Thus, again, the vector $\langle a, b \rangle$ has many different representations, depending on its initial point [see Figure 7(b)].

The relationship between a geometric representation of a vector and the analytic one can be seen from Figure 8 as follows:

> If a vector **v** is represented in the plane with initial point $P(x_1, y_1)$ and terminal point $Q(x_2, y_2)$, then
>
> $$\mathbf{v} = \langle x_2 - x_1, y_2 - y_1 \rangle$$

Figure 8

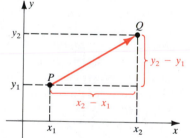

EXAMPLE 1

(a) Find the vector **v** with initial point $(-2, 5)$ and terminal point $(3, 7)$.

(b) If the vector $\mathbf{v} = \langle 3, 7 \rangle$ is sketched with initial point $(2, 4)$, what is its terminal point?

(c) Sketch representations of the vector $\mathbf{u} = \langle 2, 3 \rangle$ with initial points at $(0, 0)$, $(2, 2)$, $(-2, -1)$, and $(1, 4)$.

SOLUTION

(a) The desired vector is

$$\mathbf{v} = \langle 3 - (-2), 7 - 5 \rangle = \langle 5, 2 \rangle$$

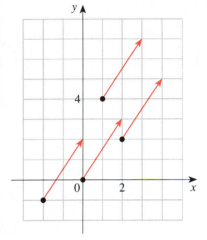

Figure 9

(b) Let the terminal point of **v** be (x, y). Then

$$\langle x - 2, y - 4 \rangle = \langle 3, 7 \rangle$$

So $x - 2 = 3$ and $y - 4 = 7$, or $x = 5$ and $y = 11$. The terminal point is $(5, 11)$.

(c) The vectors are sketched in Figure 9. ∎

We now give analytic definitions of the various operations on vectors that we have described geometrically. We start with equality of vectors. We have said that two vectors are equal if they have equal magnitudes and the same direction. For the vectors $\mathbf{u} = \langle a_1, b_1 \rangle$ and $\mathbf{v} = \langle a_2, b_2 \rangle$, this means that $a_1 = a_2$ and $b_1 = b_2$. In other words, two vectors are **equal** if and only if their corresponding components are equal. Thus all the arrows in Figure 7(b) represent the same vector, as do all the arrows in Figure 9.

The length of a vector has the following meaning in terms of components:

MAGNITUDE OF A VECTOR

The **magnitude** or **length** of a vector $\mathbf{v} = \langle a, b \rangle$ is

$$|\mathbf{v}| = \sqrt{a^2 + b^2}$$

EXAMPLE 2

Find the magnitude of each vector:

(a) $\mathbf{v} = \langle 2, -3 \rangle$ (b) $\mathbf{u} = \langle 5, 0 \rangle$ (c) $\mathbf{w} = \left\langle \frac{3}{5}, \frac{4}{5} \right\rangle$

SOLUTION

(a) $|\mathbf{v}| = \sqrt{2^2 + (-3)^2} = \sqrt{13}$ (b) $|\mathbf{u}| = \sqrt{5^2 + 0^2} = \sqrt{25} = 5$

(c) $|\mathbf{w}| = \sqrt{\left(\frac{3}{5}\right)^2 + \left(\frac{4}{5}\right)^2} = \sqrt{\frac{9}{25} + \frac{16}{25}} = 1$ ∎

The following definitions of addition, subtraction, and scalar multiplication of vectors correspond to the geometric descriptions given earlier. Figure 10 shows how the analytic definition of addition corresponds to the geometric one.

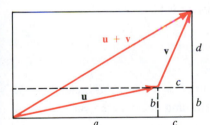

Figure 10

ALGEBRAIC OPERATIONS ON VECTORS

If $\mathbf{u} = \langle a_1, b_1 \rangle$ and $\mathbf{v} = \langle a_2, b_2 \rangle$, then

$$\mathbf{u} + \mathbf{v} = \langle a_1 + a_2, b_1 + b_2 \rangle$$
$$\mathbf{u} - \mathbf{v} = \langle a_1 - a_2, b_1 - b_2 \rangle$$
$$c\mathbf{u} = \langle ca_1, cb_1 \rangle, \qquad c \in R$$

EXAMPLE 3

If $\mathbf{u} = \langle 2, -3 \rangle$ and $\mathbf{v} = \langle -1, 2 \rangle$, find $\mathbf{u} + \mathbf{v}$, $\mathbf{u} - \mathbf{v}$, $2\mathbf{u}$, $-3\mathbf{v}$, and $2\mathbf{u} + 3\mathbf{v}$.

SOLUTION

By the definitions of the vector operations, we have

$$\mathbf{u} + \mathbf{v} = \langle 2, -3 \rangle + \langle -1, 2 \rangle = \langle 1, -1 \rangle$$
$$\mathbf{u} - \mathbf{v} = \langle 2, -3 \rangle - \langle -1, 2 \rangle = \langle 3, -5 \rangle$$
$$2\mathbf{u} = 2\langle 2, -3 \rangle = \langle 4, -6 \rangle$$
$$-3\mathbf{v} = -3\langle -1, 2 \rangle = \langle 3, -6 \rangle$$
$$2\mathbf{u} + 3\mathbf{v} = 2\langle 2, -3 \rangle + 3\langle -1, 2 \rangle = \langle 4, -6 \rangle + \langle -3, 6 \rangle = \langle 1, 0 \rangle \quad \blacksquare$$

The following rules for vector operations follow easily from the definitions. We state them here and leave the proofs as exercises. The **zero vector** is the vector $\mathbf{0} = \langle 0, 0 \rangle$. It plays the same role for addition of vectors as the number 0 does for addition of real numbers.

PROPERTIES OF VECTORS

Vector addition	**Mutiplication by a scalar**
$\mathbf{u} + \mathbf{v} = \mathbf{v} + \mathbf{u}$	$c(\mathbf{u} + \mathbf{v}) = c\mathbf{u} + c\mathbf{v}$
$\mathbf{u} + (\mathbf{v} + \mathbf{w}) = (\mathbf{u} + \mathbf{v}) + \mathbf{w}$	$(c + d)\mathbf{u} = c\mathbf{u} + d\mathbf{u}$
$\mathbf{u} + \mathbf{0} = \mathbf{u}$	$(cd)\mathbf{u} = c(d\mathbf{u}) = d(c\mathbf{u})$
$\mathbf{u} + (-\mathbf{u}) = \mathbf{0}$	$1\mathbf{u} = \mathbf{u}$
	$0\mathbf{u} = \mathbf{0}$
Length of a vector	$c\mathbf{0} = \mathbf{0}$
$|c\mathbf{u}| = |c||\mathbf{u}|$	

A vector of length 1 is called a **unit vector.** For instance, in Example 2(c), the vector $\mathbf{w} = \langle \frac{3}{5}, \frac{4}{5} \rangle$ is a unit vector. Two useful unit vectors are $\mathbf{i}$ and $\mathbf{j}$, defined by

$$\mathbf{i} = \langle 1, 0 \rangle, \qquad \mathbf{j} = \langle 0, 1 \rangle$$

These vectors are special because any vector can be expressed in terms of them.

VECTORS IN TERMS OF i AND j

The vector $\mathbf{v} = \langle a, b \rangle$ can be expressed in terms of $\mathbf{i}$ and $\mathbf{j}$ by

$$\mathbf{v} = \langle a, b \rangle = a\mathbf{i} + b\mathbf{j}$$

EXAMPLE 4

(a) Write the vector $\mathbf{u} = \langle 5, -8 \rangle$ in terms of $\mathbf{i}$ and $\mathbf{j}$.
(b) If $\mathbf{u} = 3\mathbf{i} + 2\mathbf{j}$ and $\mathbf{v} = -\mathbf{i} + 6\mathbf{j}$, write $2\mathbf{u} + 5\mathbf{v}$ in terms of $\mathbf{i}$ and $\mathbf{j}$.

SOLUTION

(a) $\mathbf{u} = 5\mathbf{i} + (-8)\mathbf{j} = 5\mathbf{i} - 8\mathbf{j}$.
(b) The properties of addition and scalar multiplication of vectors show that we can manipulate vectors in the same way as algebraic expressions. Thus

$$
\begin{aligned}
2\mathbf{u} + 5\mathbf{v} &= 2(3\mathbf{i} + 2\mathbf{j}) + 5(-\mathbf{i} + 6\mathbf{j}) \\
&= (6\mathbf{i} + 4\mathbf{j}) + (-5\mathbf{i} + 30\mathbf{j}) \\
&= \mathbf{i} + 34\mathbf{j}
\end{aligned}
$$
■

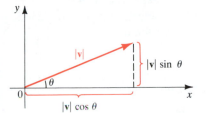

Figure 11

Let $\mathbf{v}$ be a vector in the plane with its initial point at the origin. The **direction** of $\mathbf{v}$ is θ, the smallest positive angle in standard position formed by the positive x-axis and $\mathbf{v}$ (see Figure 11). If we know the magnitude and direction of a vector, then Figure 11 shows that we can find the horizontal and vertical components of the vector.

HORIZONTAL AND VERTICAL COMPONENTS OF A VECTOR

Let $\mathbf{v}$ be a vector with magnitude $|\mathbf{v}|$ and direction θ. Then $\mathbf{v} = \langle a, b \rangle = a\mathbf{i} + b\mathbf{j}$, where

$$
a = |\mathbf{v}| \cos \theta \qquad \text{and} \qquad b = |\mathbf{v}| \sin \theta
$$

EXAMPLE 5

(a) A vector $\mathbf{v}$ has length 8 and direction $\pi/3$. Find the horizontal and vertical components and write $\mathbf{v}$ in terms of $\mathbf{i}$ and $\mathbf{j}$.
(b) Find the direction of the vector $\mathbf{u} = -\sqrt{3}\,\mathbf{i} + \mathbf{j}$.

SOLUTION

(a) We have $\mathbf{v} = \langle a, b \rangle$, where the components are given by

$$
a = 8 \cos \frac{\pi}{3} = 4 \qquad \text{and} \qquad b = 8 \sin \frac{\pi}{3} = 4\sqrt{3}
$$

Thus $\mathbf{v} = \langle 4, 4\sqrt{3} \rangle = 4\mathbf{i} + 4\sqrt{3}\,\mathbf{j}$.
(b) From Figure 12 we see that the direction θ has the property that

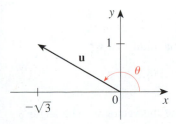

Figure 12

$$
\tan \theta = \frac{1}{-\sqrt{3}} = -\frac{\sqrt{3}}{3}
$$

Thus the reference angle for θ is $\pi/6$. Since the terminal point of the vector **u** is in quadrant II, it follows that $\theta = 5\pi/6$. ∎

APPLICATIONS TO VELOCITY AND FORCE

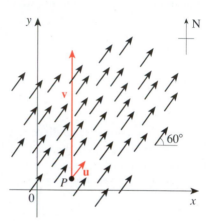

Figure 13

The **velocity** of a moving object is described by a vector whose direction is the direction of motion and whose magnitude is the speed. Figure 13 shows some vectors **u**, representing the velocity of wind flowing at 40 mi/h in the direction N30°E, and a vector **v**, representing the velocity of an airplane flying through this wind at the point P. The speed of the airplane (in still air) is 300 mi/h and the pilot heads his plane straight north. It is obvious from experience that wind affects the speed and the direction of an airplane. Figure 14 indicates that the true course of the plane (relative to the ground) is given by the vector **w** = **u** + **v**.

EXAMPLE 6

Assume the airplane and the wind are as described in Figure 13.

(a) Find the true velocity of the airplane as a vector.
(b) Find the true speed and direction of the airplane.

SOLUTION

The velocity of the airplane in still air is **v** = 0**i** + 300**j** = 300**j**. We must also write the velocity **u** of the wind as a vector. By the formulas for the component of a vector, we have

$$\begin{aligned}
\mathbf{u} &= (40 \cos 60°)\mathbf{i} + (40 \sin 60°)\mathbf{j} \\
&= 20\mathbf{i} + 20\sqrt{3}\,\mathbf{j} \\
&\approx 20\mathbf{i} + 34.64\mathbf{j}
\end{aligned}$$

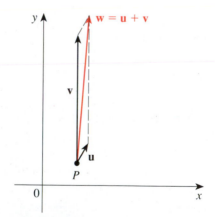

Figure 14

(a) The true velocity of the airplane is given by the vector **w** = **u** + **v**.

$$\begin{aligned}
\mathbf{w} = \mathbf{u} + \mathbf{v} &= \left(20\mathbf{i} + 20\sqrt{3}\,\mathbf{j}\right) + (300\mathbf{j}) \\
&= 20\mathbf{i} + \left(20\sqrt{3} + 300\right)\mathbf{j} \\
&\approx 20\mathbf{i} + 334.64\mathbf{j}
\end{aligned}$$

(b) The true speed of the airplane is given by the magnitude of **w**:

$$|\mathbf{w}| \approx \sqrt{(20)^2 + (334.64)^2} \approx 335.2 \text{ mi/h}$$

The direction of the airplane is the direction θ of the vector **w**. The angle θ

has the property that tan $\theta \approx 334.64/20 = 16.732$ and so $\theta \approx 86.6°$. Thus the airplane is heading in the direction N3.4°E. ∎

EXAMPLE 7

A woman launches a boat from one shore of a straight river and wants to land at the point directly on the opposite shore. If the speed of the boat (in still water) is 10 mi/h and the river is flowing east at the rate of 5 mi/h, in what direction should she head the boat in order to arrive at the desired landing point?

SOLUTION

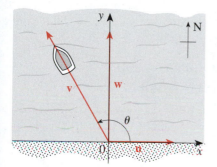

Figure 15

We choose a coordinate system with the origin at the initial position of the boat as shown in Figure 15. Let **u** and **v** represent the velocities of the river and the boat, respectively. Clearly, **u** = 5**i** and, since the speed of the boat is 10 mi/h, we have $|\mathbf{v}| = 10$, so

$$\mathbf{v} = (10 \cos \theta)\mathbf{i} + (10 \sin \theta)\mathbf{j}$$

where the angle θ is as shown in Figure 15. The true course of the boat is given by the vector **w** = **u** + **v**. We have

$$\mathbf{w} = \mathbf{u} + \mathbf{v} = 5\mathbf{i} + (10 \cos \theta)\mathbf{i} + (10 \sin \theta)\mathbf{j}$$
$$= (5 + 10 \cos \theta)\mathbf{i} + (10 \sin \theta)\mathbf{j}$$

Since the woman wants to land at a point directly across the river, her direction should have horizontal component 0. In other words, she should choose θ in such a way that

$$5 + 10 \cos \theta = 0$$
$$\cos \theta = -\frac{1}{2}$$
$$\theta = 120°$$

Thus she should head her boat in the direction $\theta = 120°$ (or N30°W), as shown in Figure 15. ∎

Force is also represented by a vector. Intuitively, we can think of force as describing a push or a pull on an object—for example, a horizontal push of a book across a table or the downward pull of the earth's gravity on a ball. Force is measured in pounds (or in newtons, in the metric system). For instance, a man weighing 200 lb exerts a force of 200 lb downward on the ground. If several forces are acting on an object, the **resultant force** experienced by the object is the vector sum of these forces.

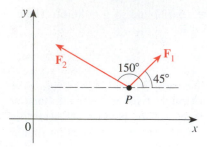

Figure 16

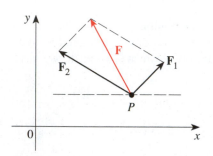

Figure 17

EXAMPLE 8

Two forces $\mathbf{F}_1$ and $\mathbf{F}_2$ with magnitudes 10 lb and 20 lb, respectively, act on an object at a point P as shown in Figure 16. Find the resultant force acting at P.

SOLUTION

We write $\mathbf{F}_1$ and $\mathbf{F}_2$ in terms of their components:

$$\mathbf{F}_1 = (10 \cos 45°)\mathbf{i} + (10 \sin 45°)\mathbf{j} = 10\frac{\sqrt{2}}{2}\mathbf{i} + 10\frac{\sqrt{2}}{2}\mathbf{j} = 5\sqrt{2}\,\mathbf{i} + 5\sqrt{2}\,\mathbf{j}$$

$$\mathbf{F}_2 = (20 \cos 150°)\mathbf{i} + (20 \sin 150°)\mathbf{j} = -20\frac{\sqrt{3}}{2}\mathbf{i} + 20\left(\frac{1}{2}\right)\mathbf{j} = -10\sqrt{3}\,\mathbf{i} + 10\mathbf{j}$$

So the resultant force $\mathbf{F}$ is

$$
\begin{aligned}
\mathbf{F} &= \mathbf{F}_1 + \mathbf{F}_2 \\
&= \left(5\sqrt{2}\,\mathbf{i} + 5\sqrt{2}\,\mathbf{j}\right) + \left(-10\sqrt{3}\,\mathbf{i} + 10\mathbf{j}\right) \\
&= \left(5\sqrt{2} - 10\sqrt{3}\right)\mathbf{i} + \left(5\sqrt{2} + 10\right)\mathbf{j} \\
&\approx -10\mathbf{i} + 17\mathbf{j}
\end{aligned}
$$

The resultant force $\mathbf{F}$ is shown in Figure 17. ■

EXERCISES 7.8

In Exercises 1 and 2 sketch vectors that represent $2\mathbf{v}$, $3\mathbf{v}$, $\frac{1}{2}\mathbf{v}$, $-\mathbf{v}$, $-2\mathbf{v}$, *and* $-\frac{1}{4}\mathbf{v}$ *for the vector* $\mathbf{v}$ *shown.*

1.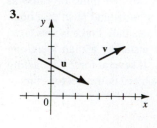

2.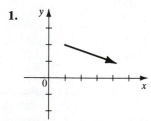

In Exercises 3 and 4 two vectors $\mathbf{u}$ *and* $\mathbf{v}$ *are shown. Sketch* $\mathbf{u} + \mathbf{v}$, $\mathbf{u} - \mathbf{v}$, $\mathbf{v} - \mathbf{u}$, $2\mathbf{v} + \mathbf{u}$, *and* $\mathbf{v} - 2\mathbf{u}$.

3.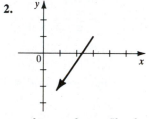

4.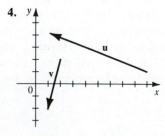

In Exercises 5–10 find the vector with initial point P and terminal point Q.

5. $P(3, 2)$, $Q(8, 9)$ **6.** $P(1, 1)$, $Q(9, 9)$

7. $P(5, 3)$, $Q(1, 0)$ **8.** $P(-1, 3)$, $Q(-6, -1)$

9. $P(-1, -1)$, $Q(-1, 1)$

10. $P(-8, -6)$, $Q(-1, -1)$

In Exercises 11–16 find $2\mathbf{u}$, $-3\mathbf{v}$, $\mathbf{u} + \mathbf{v}$, *and* $3\mathbf{u} - 4\mathbf{v}$ *for the given vectors* $\mathbf{u}$ *and* $\mathbf{v}$.

11. $\mathbf{u} = \langle 2, 7 \rangle$, $\mathbf{v} = \langle 3, 1 \rangle$

12. $\mathbf{u} = \langle -2, 5 \rangle$, $\mathbf{v} = \langle 2, -8 \rangle$

13. $\mathbf{u} = \langle 0, -1 \rangle$, $\mathbf{v} = \langle -2, 0 \rangle$

14. $\mathbf{u} = \mathbf{i}$, $\mathbf{v} = -2\mathbf{j}$

15. $\mathbf{u} = 2\mathbf{i}$, $\mathbf{v} = 3\mathbf{i} - 2\mathbf{j}$

16. $\mathbf{u} = \mathbf{i} + \mathbf{j}$, $\mathbf{v} = \mathbf{i} - \mathbf{j}$

In Exercises 17–20 find $|\mathbf{u}|$, $|\mathbf{v}|$, $|2\mathbf{u}|$, $\left|\frac{1}{2}\mathbf{v}\right|$, $|\mathbf{u} + \mathbf{v}|$, $|\mathbf{u} - \mathbf{v}|$, *and* $|\mathbf{u}| - |\mathbf{v}|$.

17. $\mathbf{u} = 2\mathbf{i} + \mathbf{j}$, $\quad \mathbf{v} = 3\mathbf{i} - 2\mathbf{j}$

18. $\mathbf{u} = -2\mathbf{i} + 3\mathbf{j}$, $\mathbf{v} = \mathbf{i} - 2\mathbf{j}$

19. $\mathbf{u} = \langle 10, -1 \rangle$, $\quad \mathbf{v} = \langle -2, -2 \rangle$

20. $\mathbf{u} = \langle -6, 6 \rangle$, $\quad \mathbf{v} = \langle -2, -1 \rangle$

In Exercises 21–26 find the horizontal and vertical components of each vector with given length and direction, and write the vector as a linear combination of the vectors $\mathbf{i}$ *and* $\mathbf{j}$.

21. $|\mathbf{v}| = 40$, $\theta = 30°$ **22.** $|\mathbf{v}| = 50$, $\theta = 120°$

23. $|\mathbf{v}| = 1$, $\theta = 225°$ **24.** $|\mathbf{v}| = 800$, $\theta = 125°$

25. $|\mathbf{v}| = 4$, $\theta = 10°$ **26.** $|\mathbf{v}| = \sqrt{3}$, $\theta = 300°$

27. A man pushes a lawn mower with a force of 30 lb exerted at an angle of 30° to the ground. Find the horizontal and vertical components of the force.

28. A jet is flying in a direction N20°E with a speed of 500 mi/h. Find the north and east components of the velocity.

In Exercises 29–34, find the magnitude and direction (in degrees) of the vector.

29. $\mathbf{v} = \langle 3, 4 \rangle$ **30.** $\mathbf{v} = \left\langle -\dfrac{\sqrt{2}}{2}, -\dfrac{\sqrt{2}}{2} \right\rangle$

31. $\mathbf{v} = \langle -12, 5 \rangle$ **32.** $\mathbf{v} = \langle 40, 9 \rangle$

33. $\mathbf{v} = \mathbf{i} + \sqrt{3}\,\mathbf{j}$ **34.** $\mathbf{v} = \mathbf{i} + \mathbf{j}$

35. A pilot heads his jet due east. The jet has a speed of 425 mi/h in still air. The wind is blowing due north with a speed of 40 mi/h.
(a) Find the true velocity of the jet as a vector.
(b) Find the true speed and direction of the jet.

36. A jet is flying through a wind that is blowing with a speed of 55 mi/h in the direction N30°E. The jet has a speed of 765 mi/h in still air and the pilot heads the jet in the direction N45°E.
(a) Find the true velocity of the jet as a vector.
(b) Find the true speed and direction of the jet.

37. Find the true speed and direction of the jet in Exercise 36 if the pilot heads the plane in the direction N30°W.

38. In what direction should the pilot in Exercise 36 head the plane in order for the true course to be due north?

39. A straight river flows east at a speed of 10 mi/h. A boater starts at the south shore of the river and heads in a direction 60° from the shore (see the figure). The motorboat has a speed of 20 mi/h in still water.
(a) Find the true velocity of the motorboat as a vector.
(b) Find the true speed and direction of the motorboat.

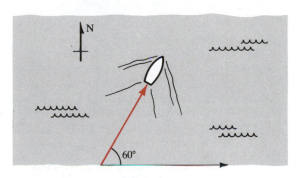

40. The boater in Exercise 39 wants to arrive at a point on the north shore of the river directly opposite the starting point. In what direction should she head her boat?

41. A boat heads in the direction N72°E. The speed of the boat in still water is 24 mi/h. The water is flowing directly south. It is observed that the true direction of the boat is directly east. Find the speed of the water and the true speed of the boat.

42. A woman walks due west on the deck of an ocean liner at 2 mi/h. The ocean liner is moving due north at a speed of 25 mi/h. Find the speed and direction of the woman relative to the surface of the water.

The forces $\mathbf{F}_1$, $\mathbf{F}_2$, . . . , $\mathbf{F}_n$ *acting at the same point P are said to be in equilibrium if the resultant force is zero, that is, if* $\mathbf{F}_1 + \mathbf{F}_2 + \cdots + \mathbf{F}_n = \mathbf{0}$. *In Exercises 43–48 find* (a) *the resultant forces acting at P, and* (b) *the additional force required (if any) in order for the forces to be in equilibrium.*

43. $\mathbf{F}_1 = \langle 2, 5 \rangle$, $\quad \mathbf{F}_2 = \langle 3, -8 \rangle$

44. $\mathbf{F}_1 = \langle 3, -7 \rangle$, $\quad \mathbf{F}_2 = \langle 4, -2 \rangle$, $\quad \mathbf{F}_3 = \langle -7, 9 \rangle$

45. $\mathbf{F}_1 = 4\mathbf{i} - \mathbf{j}$, $\quad \mathbf{F}_2 = 3\mathbf{i} - 7\mathbf{j}$,
$\mathbf{F}_3 = -8\mathbf{i} + 3\mathbf{j}$, $\quad \mathbf{F}_4 = \mathbf{i} + \mathbf{j}$

46. $\mathbf{F}_1 = \mathbf{i} - \mathbf{j}$, $\quad \mathbf{F}_2 = \mathbf{i} + \mathbf{j}$, $\quad \mathbf{F}_3 = -2\mathbf{i} + \mathbf{j}$

47.

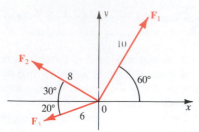

48.

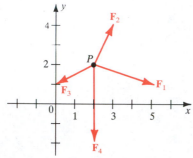

49. A 100-lb weight hangs from a string as shown in the figure. Find the tensions T_1 and T_2 in the string.

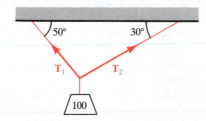

50. Two cranes are lifting an object that weighs 18,278 lb, as shown in the figure. Find the tensions T_1 and T_2.

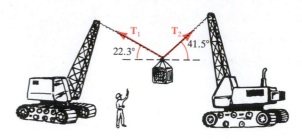

In Exercises 51–61 **u**, **v**, *and* **w** *are vectors and a and b are scalars. Prove the given property.*

51. $u + v = v + u$

52. $v + (v + w) = (u + v) + w$

53. $u + 0 = u$ **54.** $u + (-u) = 0$

55. $a(u + v) = au + av$ **56.** $(a + b)u = au + bu$

57. $(ab)u = a(bu) = b(au)$ **58.** $1u = u$

59. $0u = 0$ **60.** $a0 = 0$

61. $|au| = |a|\,|u|$

62. Let $m = \langle 1, 1 \rangle$ and $n = \langle 0, 1 \rangle$. Show that every vector **v** in the plane can be written in the form $v = am + bn$ where a and b are scalars.

SECTION 7.9
THE DOT PRODUCT

In this section we define an operation on vectors called the dot product. This concept is very useful in calculus and in applications of vectors to physics and engineering.

■ THE DOT PRODUCT OF VECTORS

We begin by defining the dot product of two vectors.

DEFINITION OF THE DOT PRODUCT

If $u = \langle a_1, b_1 \rangle$ and $v = \langle a_2, b_2 \rangle$ are vectors, then their **dot product**, denoted by $u \cdot v$, is defined by

$$u \cdot v = a_1 a_2 + b_1 b_2$$

Thus, to find the dot product of **u** and **v** we multiply corresponding components and add. The result is *not* a vector; it is a real number, or scalar.

EXAMPLE 1

(a) If $\mathbf{u} = \langle 3, -2 \rangle$ and $\mathbf{v} = \langle 4, 5 \rangle$, then

$$\mathbf{u} \cdot \mathbf{v} = (3)(4) + (-2)(5) = 2$$

(b) If $\mathbf{u} = 2\mathbf{i} + \mathbf{j}$ and $\mathbf{v} = 5\mathbf{i} - 6\mathbf{j}$, then

$$\mathbf{u} \cdot \mathbf{v} = (2)(5) + (1)(-6) = 4 \qquad \blacksquare$$

The proofs of the following properties of the dot product follow easily from the definition.

PROPERTIES OF THE DOT PRODUCT

$$\mathbf{u} \cdot \mathbf{v} = \mathbf{v} \cdot \mathbf{u}$$

$$(a\mathbf{u}) \cdot \mathbf{v} = a(\mathbf{u} \cdot \mathbf{v}) = \mathbf{u} \cdot (a\mathbf{v})$$

$$(\mathbf{u} + \mathbf{v}) \cdot \mathbf{w} = \mathbf{u} \cdot \mathbf{w} + \mathbf{v} \cdot \mathbf{w}$$

$$|\mathbf{u}|^2 = \mathbf{u} \cdot \mathbf{u}$$

Proof   We prove only the last property. The proofs of the others are left as exercises. Let $\mathbf{u} = \langle a, b \rangle$. Then

$$\mathbf{u} \cdot \mathbf{u} = \langle a, b \rangle \cdot \langle a, b \rangle = a^2 + b^2 = |\mathbf{u}|^2 \qquad \blacksquare$$

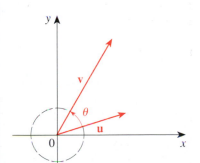

Figure 1

Let **u** and **v** be vectors and sketch them with initial points at the origin. We define the **angle θ between u and v** to be the smaller of the angles formed by these representations of **u** and **v** (see Figure 1). Thus $0 \le \theta \le \pi$. The next theorem relates the angle between two vectors to their dot product.

THEOREM: THE DOT PRODUCT

If θ is the angle between two nonzero vectors **u** and **v**, then

$$\mathbf{u} \cdot \mathbf{v} = |\mathbf{u}| \, |\mathbf{v}| \cos \theta$$

Proof   The proof is a nice application of the Law of Cosines. Applying the Law of Cosines to triangle AOB in Figure 2 gives

$$|\mathbf{u} - \mathbf{v}|^2 = |\mathbf{u}|^2 + |\mathbf{v}|^2 - 2|\mathbf{u}| \, |\mathbf{v}| \cos \theta$$

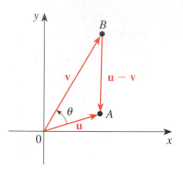

Figure 2

Using the properties of the dot product, we can write the left-hand side as follows:

$$|\mathbf{u} - \mathbf{v}|^2 = (\mathbf{u} - \mathbf{v}) \cdot (\mathbf{u} - \mathbf{v})$$
$$= \mathbf{u} \cdot \mathbf{u} - \mathbf{u} \cdot \mathbf{v} - \mathbf{v} \cdot \mathbf{u} + \mathbf{v} \cdot \mathbf{v}$$
$$= |\mathbf{u}|^2 - 2(\mathbf{u} \cdot \mathbf{v}) + |\mathbf{v}|^2$$

Equating the right-hand sides of the displayed equations, we get

$$|\mathbf{u}|^2 - 2(\mathbf{u} \cdot \mathbf{v}) + |\mathbf{v}|^2 = |\mathbf{u}|^2 + |\mathbf{v}|^2 - 2|\mathbf{u}|\,|\mathbf{v}| \cos \theta$$
$$-2(\mathbf{u} \cdot \mathbf{v}) = -2|\mathbf{u}|\,|\mathbf{v}| \cos \theta$$
$$\mathbf{u} \cdot \mathbf{v} = |\mathbf{u}|\,|\mathbf{v}| \cos \theta$$

This proves the theorem.   ∎

The Dot Product Theorem is useful because it allows us to find the angle between two vectors if we know the components of the vectors. The angle is obtained simply by solving the equation in the Dot Product Theorem for $\cos \theta$. We state this important result explicitly.

COROLLARY: ANGLE BETWEEN TWO VECTORS

If θ is the angle between two nonzero vectors $\mathbf{u}$ and $\mathbf{v}$, then

$$\cos \theta = \frac{\mathbf{u} \cdot \mathbf{v}}{|\mathbf{u}|\,|\mathbf{v}|}$$

EXAMPLE 2

Find the angle between the vectors $\mathbf{u} = \langle 2, 5 \rangle$ and $\mathbf{v} = \langle 4, -3 \rangle$.

SOLUTION

By the corollary we have

$$\cos \theta = \frac{\mathbf{u} \cdot \mathbf{v}}{|\mathbf{u}|\,|\mathbf{v}|} = \frac{(2)(4) + (5)(-3)}{\sqrt{4 + 25}\,\sqrt{16 + 9}} = \frac{-7}{5\sqrt{29}}$$

Thus the angle between $\mathbf{u}$ and $\mathbf{v}$ is

$$\theta = \cos^{-1}\!\left(\frac{-7}{5\sqrt{29}}\right) \approx 105.1°$$

■

Two nonzero vectors $\mathbf{u}$ and $\mathbf{v}$ are called **perpendicular**, or **orthogonal**, if the angle between them is $\pi/2$. The following theorem shows that we can determine if two vectors are perpendicular by finding their dot product.

ORTHOGONAL VECTORS

> Two nonzero vectors **u** and **v** are perpendicular if and only if **u** · **v** = 0.

Proof   If **u** and **v** are perpendicular, then the angle between them is $\pi/2$ and so

$$\mathbf{u} \cdot \mathbf{v} = |\mathbf{u}|\,|\mathbf{v}|\cos\frac{\pi}{2} = 0$$

Conversely, if **u** · **v** = 0, then

$$|\mathbf{u}|\,|\mathbf{v}|\cos\theta = 0$$

Since **u** and **v** are nonzero vectors, we conclude that $\cos\theta = 0$, and so $\theta = \pi/2$. Thus **u** and **v** are orthogonal.   ■

EXAMPLE 3

Determine whether the vectors in each pair are perpendicular.

(a) $\mathbf{u} = \langle 3, 5 \rangle$ and $\mathbf{v} = \langle 2, -8 \rangle$ (b) $\mathbf{u} = \langle 2, 1 \rangle$ and $\mathbf{v} = \langle -1, 2 \rangle$

SOLUTION

(a) $\mathbf{u} \cdot \mathbf{v} = (3)(2) + (5)(-8) = -34$, so **u** and **v** are not perpendicular.
(b) $\mathbf{u} \cdot \mathbf{v} = (2)(-1) + (1)(2) = 0$, so **u** and **v** are perpendicular.   ■

■ THE COMPONENT OF u ALONG v

The **component of u along v** (or the **component of u in the direction of v**) is defined to be

$$|\mathbf{u}|\cos\theta$$

where θ is the angle between **u** and **v**. Figure 3 gives a geometric interpretation of this concept. Intuitively, the component of **u** along **v** is the magnitude of the portion of **u** that points in the direction of **v**. Notice that the component of **u** along **v** is negative if $\pi/2 < \theta \leq \pi$.

Figure 3

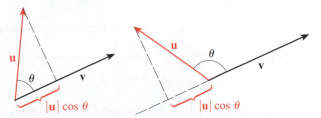

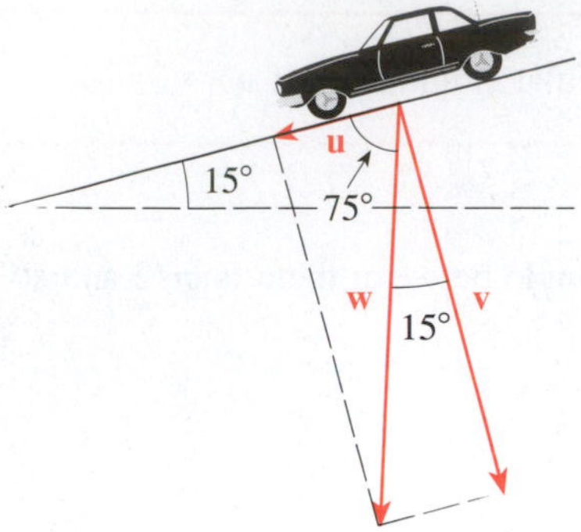

Figure 4

Components of vectors are useful in analyzing forces. For example, suppose that a car weighing 3000 lb is parked on a driveway that is inclined 15° to the horizontal. The car exerts a force **w** of 3000 lb directly downward. What is the actual force caused by the weight of the car on the driveway? The answer is not 3000 lb, because part of this force causes the car to roll down the driveway and the other part presses the car against the driveway. Let **u** be a vector parallel to the driveway and let **v** be a vector perpendicular to the driveway so that **w** = **u** + **v** as shown in Figure 4. (We say that **w** is **resolved** into the vectors **u** and **v**.)

EXAMPLE 4

Refer to the situation illustrated in Figure 4.

(a) Find the magnitude of the force required to prevent the car from rolling down the driveway.
(b) Find the magnitude of the force experienced by the driveway due to the weight of the car.

SOLUTION

(a) The magnitude of the part of the force **w** that causes the car to roll down the driveway is

$$|\mathbf{u}| = \text{the component of } \mathbf{w} \text{ along } \mathbf{u} = 3000 \cos 75° \approx 776$$

Thus the force needed to prevent the car from rolling down the driveway is about 776 lb.

(b) The magnitude of the force exerted by the car on the driveway is

$$|\mathbf{v}| = \text{the component of } \mathbf{w} \text{ along } \mathbf{v} = 3000 \cos 15° \approx 2898$$

Thus the force experienced by the driveway is about 2898 lb.   ■

The component of **u** along **v** can be computed using dot products:

$$|\mathbf{u}| \cos \theta = \frac{|\mathbf{v}| \, |\mathbf{u}| \cos \theta}{|\mathbf{v}|} = \frac{\mathbf{u} \cdot \mathbf{v}}{|\mathbf{v}|}$$

We have shown the following.

> The component of **u** along **v** is $\dfrac{\mathbf{u} \cdot \mathbf{v}}{|\mathbf{v}|}$.

EXAMPLE 5

Let **u** = $\langle 1, 4 \rangle$ and **v** = $\langle -2, 1 \rangle$. Find the component of **u** along **v**.

SOLUTION

We have

$$\text{component of } \mathbf{u} \text{ along } \mathbf{v} = \frac{\mathbf{u} \cdot \mathbf{v}}{|\mathbf{v}|} = \frac{(1)(-2) + (4)(1)}{\sqrt{4 + 1}} = \frac{2}{\sqrt{5}}$$ ∎

■ WORK

One use of the dot product occurs in calculating work. The term *work* is used in everyday language to mean the total amount of effort required to perform a task. In physics *work* has a technical meaning that conforms to this intuitive meaning. If a constant force of magnitude F moves an object through a distance d along a straight line, then the **work** done is

$$W = Fd \quad \text{or} \quad \text{Work} = \text{force} \times \text{distance}$$

If F is measured in pounds and d in feet, then the unit of work is a foot-pound (ft-lb). For example, how much work is done in lifting a 20-lb weight 6 ft off the ground. Since a force of 20 lb is required to lift this weight and since the weight moves through a distance of 6 ft, the amount of work done is

$$W = Fd = (20)(6) = 120 \text{ ft-lb}$$

This formula applies only when the force is directed along the direction of motion. In the general case, if the force $\mathbf{F}$ moves an object from P to Q, as in Figure 5, then only the component of the force in the direction of $\mathbf{D} = \overrightarrow{PQ}$ affects the object. Thus the effective magnitude of the force on the object is

$$\text{component of } \mathbf{F} \text{ along } \mathbf{D} = |\mathbf{F}| \cos \theta$$

Thus, the work done is

$$W = \text{force} \times \text{distance} = (|\mathbf{F}| \cos \theta) |\mathbf{D}| = |\mathbf{F}| |\mathbf{D}| \cos \theta = \mathbf{F} \cdot \mathbf{D}$$

We have derived the following simple formula for calculating work.

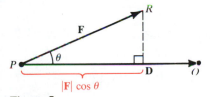

Figure 5

WORK

The **work** W done by a force $\mathbf{F}$ in moving along a vector $\mathbf{D}$ is

$$W = \mathbf{F} \cdot \mathbf{D}$$

EXAMPLE 6

A force is given by the vector $\mathbf{F} = \langle 2, 3 \rangle$ and moves an object from the point $(1, 3)$ to the point $(5, 9)$. Find the work done.

SOLUTION

The displacement vector is

$$\mathbf{D} = \langle 5 - 1,\, 9 - 3 \rangle = \langle 4,\, 6 \rangle$$

So the work done is

$$W = \mathbf{F} \cdot \mathbf{D} = \langle 2,\, 3 \rangle \cdot \langle 4,\, 6 \rangle = 26$$

If the unit of force is pounds and the distance is measured in feet, then the work done is 26 ft-lb. ■

EXAMPLE 7

A man pulls a wagon horizontally by exerting a force of 20 lb on the handle. If the handle makes an angle of 60° with the horizontal, find the work done in moving the wagon 100 ft.

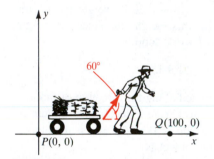

Figure 6

SOLUTION

We choose a coordinate system with the origin at the initial position of the wagon (see Figure 6). Thus the wagon moves from the point $P(0,\, 0)$ to the point $Q(100,\, 0)$. The vector that represents this displacement is

$$\mathbf{D} = 100\mathbf{i}$$

The force on the handle can be written in terms of components (see Section 7.8) as

$$\mathbf{F} = (20 \cos 60°)\mathbf{i} + (20 \sin 60°)\mathbf{j} = 10\mathbf{i} + 10\sqrt{3}\,\mathbf{j}$$

Thus the work done is

$$W = \mathbf{F} \cdot \mathbf{D} = \left(10\mathbf{i} + 10\sqrt{3}\,\mathbf{j}\right) \cdot (100\mathbf{i}) = 1000 \text{ ft-lb}$$ ■

EXERCISES 7.9

In Exercises 1–8 find (a) $\mathbf{u} \cdot \mathbf{v}$ *and (b) the angle between* $\mathbf{u}$ *and* $\mathbf{v}$ *to the nearest degree.*

1. $\mathbf{u} = \langle 2,\, 0 \rangle$, $\mathbf{v} = \langle 1,\, 1 \rangle$

2. $\mathbf{u} = \mathbf{i} + \sqrt{3}\,\mathbf{j}$, $\mathbf{v} = -\sqrt{3}\,\mathbf{i} + \mathbf{j}$

3. $\mathbf{u} = \langle 2,\, 7 \rangle$, $\mathbf{v} = \langle 3,\, 1 \rangle$

4. $\mathbf{u} = \langle -6,\, 6 \rangle$, $\mathbf{v} = \langle 1,\, -1 \rangle$

5. $\mathbf{u} = \langle 3,\, -2 \rangle$, $\mathbf{v} = \langle 1,\, 2 \rangle$

6. $\mathbf{u} = 2\mathbf{i} + \mathbf{j}$, $\mathbf{v} = 3\mathbf{i} - 2\mathbf{j}$

7. $\mathbf{u} = -5\mathbf{j}$, $\mathbf{v} = -\mathbf{i} - \sqrt{3}\,\mathbf{j}$

8. $\mathbf{u} = \mathbf{i} + \mathbf{j}$, $\mathbf{v} = \mathbf{i} - \mathbf{j}$

In Exercises 9–14 determine whether the given vectors are orthogonal.

9. $\mathbf{u} = \langle 6,\, 4 \rangle$, $\mathbf{v} = \langle -2,\, 3 \rangle$

10. $\mathbf{u} = \langle 0,\, -5 \rangle$, $\mathbf{v} = \langle 4,\, 0 \rangle$

11. $\mathbf{u} = \langle -2,\, 6 \rangle$, $\mathbf{v} = \langle 4,\, 2 \rangle$

12. $\mathbf{u} = 2\mathbf{i}$, $\mathbf{v} = -7\mathbf{j}$

13. $\mathbf{u} = 2\mathbf{i} - 8\mathbf{j}$, $\mathbf{v} = -12\mathbf{i} - 3\mathbf{j}$

14. $\mathbf{u} = 4\mathbf{i}$, $\mathbf{v} = -\mathbf{i} + 3\mathbf{j}$

In Exercises 15–18, $\mathbf{u} = 2\mathbf{i} + \mathbf{j}$, $\mathbf{v} = \mathbf{i} - 3\mathbf{j}$, and $\mathbf{w} = 3\mathbf{i} + 4\mathbf{j}$. Find the indicated quantity.

15. $\mathbf{u} \cdot \mathbf{v} + \mathbf{u} \cdot \mathbf{w}$ **16.** $\mathbf{u} \cdot (\mathbf{v} + \mathbf{w})$

17. $(\mathbf{u} + \mathbf{v}) \cdot (\mathbf{u} - \mathbf{v})$ **18.** $(\mathbf{u} \cdot \mathbf{v})(\mathbf{u} \cdot \mathbf{w})$

In Exercises 19–22 find the component of $\mathbf{u}$ along $\mathbf{v}$.

19. $\mathbf{u} = \langle 4, 6 \rangle$, $\mathbf{v} = \langle 3, -4 \rangle$

20. $\mathbf{u} = \langle -3, 5 \rangle$, $\mathbf{v} = \langle 1/\sqrt{2}, 1/\sqrt{2} \rangle$

21. $\mathbf{u} = 7\mathbf{i} - 24\mathbf{j}$, $\mathbf{v} = \mathbf{j}$

22. $\mathbf{u} = 7\mathbf{i}$, $\mathbf{v} = 8\mathbf{i} + 6\mathbf{j}$

In Exercises 23–26 find the work done by the force $\mathbf{F}$ in moving an object from P to Q.

23. $\mathbf{F} = 4\mathbf{i} - 5\mathbf{j}$; $P(0, 0)$, $Q(3, 8)$

24. $\mathbf{F} = 400\mathbf{i} + 50\mathbf{j}$; $P(-1, 1)$, $Q(200, 1)$

25. $\mathbf{F} = 10\mathbf{i} + 3\mathbf{j}$; $P(2, 3)$, $Q(6, -2)$

26. $\mathbf{F} = -4\mathbf{i} + 20\mathbf{j}$; $P(0, 10)$, $Q(5, 25)$

27. The force $\mathbf{F} = 4\mathbf{i} - 7\mathbf{j}$ moves an object 4 ft along the *x*-axis in the positive direction. Find the work done if the unit of force is the pound.

28. A constant force $\mathbf{F} = \langle 2, 8 \rangle$ moves an object along a straight line from the point $(2, 5)$ to the point $(11, 13)$. Find the work done if the distance is measured in feet and the force is measured in pounds.

29. A lawn mower is pushed a distance of 200 ft along a horizontal path by a constant force of 50 lb. The handle of the lawn mower is held at an angle of 30° from the horizontal (see the figure). Find the work done.

30. A car drives 500 ft on a road that is inclined 12° to the horizontal, as shown in the figure. The car weighs 2500 lb. Thus gravity acts straight down on the car with a constant force $\mathbf{F} = -2500\mathbf{j}$. Find the work done by the car in overcoming gravity.

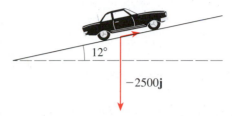

31. A car is on a driveway that is inclined 25° to the horizontal. If the car weighs 2755 lb, find the force required to keep it from rolling down the driveway.

32. A car is on a driveway that is inclined 10° to the horizontal. A force of 490 lb is required to keep the car from rolling down the driveway.
(a) Find the weight of the car.
(b) Find the force the car exerts against the driveway.

33. A package that weighs 200 lb is placed on an inclined plane. If a force of 80 lb is just sufficient to keep the package from sliding, find the angle of inclination of the plane. (Ignore the effects of friction.)

34. A sailboat has its sail inclined in the direction N20°E. The wind is blowing into the sail in the direction S45°W with a force of 220 lb (see the figure).

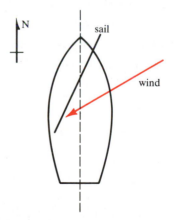

(a) Find the effective force of the wind on the sail. [*Hint:* Find the components of the wind parallel to the sail and perpendicular to the sail. The component of the wind parallel to the sail slips by and does not push on the sail.]

(b) If the keel of the ship is aligned due north, find the effective force of the wind that drives the boat forward. [*Hint:* Only the component of the force found in part (a) that is parallel to the keel drives the boat forward.]

In Exercises 35–38, **u**, **v**, *and* **w** *are vectors and a is a scalar. Prove the given property.*

35. $\mathbf{u} \cdot \mathbf{v} = \mathbf{v} \cdot \mathbf{u}$ **36.** $(a\mathbf{u}) \cdot \mathbf{v} = a(\mathbf{u} \cdot \mathbf{v}) = \mathbf{u} \cdot (a\mathbf{v})$

37. $(\mathbf{u} + \mathbf{v}) \cdot \mathbf{w} = \mathbf{u} \cdot \mathbf{w} + \mathbf{v} \cdot \mathbf{w}$

38. $(\mathbf{u} - \mathbf{v}) \cdot (\mathbf{u} + \mathbf{v}) = |\mathbf{u}|^2 - |\mathbf{v}|^2$

CHAPTER 7 REVIEW

Define, state, or discuss each of the following.

1. Equation

2. Identity

3. Proving identities

4. Trigonometric equation

5. Trigonometric identity

6. Solving trigonometric equations

7. Addition formulas for sine, cosine, and tangent

8. Subtraction formulas for sine, cosine, and tangent

9. Double-angle formulas for sine, cosine, and tangent

10. Half-angle formulas for sine, cosine, and tangent

11. Product-to-sum identities

12. Sum-to-product identities

13. The inverse trigonometric functions

14. Trigonometric form of complex numbers

15. Multiplying complex numbers in trigonometric form

16. DeMoivre's Theorem

17. *n*th roots of complex numbers

18. Scalar and vector quantities

19. Vectors

20. Components of a vector

21. Addition of vectors

22. Scalar multiplication of vectors

23. Magnitude or length of a vector

24. The vectors **i** and **j**

25. Direction of a vector

26. Dot product of vectors

27. Angle between two vectors

28. Orthogonal vectors

29. The component of **u** along **v**

30. Work

REVIEW EXERCISES

In Exercises 1–22 verify the identity.

1. $\cos^2 x \csc x - \csc x = -\sin x$

2. $\dfrac{1}{1 - \sin^2 x} = 1 + \tan^2 x$

3. $\dfrac{\cos^2 x - \tan^2 x}{\sin^2 x} = \cot^2 x - \sec^2 x$

4. $\dfrac{1 + \sec x}{\sec x} = \dfrac{\sin^2 x}{1 - \cos x}$

5. $\dfrac{\cos^2 x}{1 - \sin x} = \dfrac{\cos x}{\sec x - \tan x}$

6. $(1 - \tan x)(1 - \cot x) = 2 - \sec x \csc x$

7. $\sin^2 x \cot^2 x + \cos^2 x \tan^2 x = 1$

8. $(\tan x + \cot x)^2 = \csc^2 x \sec^2 x$

9. $\dfrac{\sin 2x}{1 + \cos 2x} = \tan x$

10. $\dfrac{\cos(x + y)}{\cos x \sin y} = \cot y - \tan x$

11. $\tan\left(\dfrac{x}{2}\right) = \csc x - \cot x$

12. $\dfrac{\sin(x + y) + \sin(x - y)}{\cos(x + y) + \cos(x - y)} = \tan x$

13. $\sin(x + y)\sin(x - y) = \sin^2 x - \sin^2 y$

14. $\csc x - \tan \dfrac{x}{2} = \cot x$

15. $1 + \tan x \tan \dfrac{x}{2} = \sec x$

16. $\dfrac{\sin 3x + \cos 3x}{\cos x - \sin x} = 1 + 2 \sin 2x$

17. $\left(\cos \dfrac{x}{2} - \sin \dfrac{x}{2}\right)^2 = 1 - \sin x$

18. $\dfrac{\cos 3x - \cos 7x}{\sin 3x + \sin 7x} = \tan 2x$

19. $\dfrac{\sin 2x}{\sin x} - \dfrac{\cos 2x}{\cos x} = \sec x$

20. $(\cos x + \cos y)^2 + (\sin x - \sin y)^2 = 2 + 2 \cos(x + y)$

21. $\tan\left(x + \dfrac{\pi}{4}\right) = \dfrac{1 + \tan x}{1 - \tan x}$

22. $\dfrac{\sec x - 1}{\sin x \sec x} = \tan \dfrac{x}{2}$

In Exercises 23–26, (a) graph f and g. (b) Do the graphs suggest that the equation f(x) = g(x) is an identity? Prove your answer.

23. $f(x) = 1 - \left(\cos \dfrac{x}{2} - \sin \dfrac{x}{2}\right)^2$, $g(x) = \sin x$

24. $f(x) = \sin x + \cos x$, $g(x) = \sqrt{\sin^2 x + \cos^2 x}$

25. $f(x) = \tan x \tan \dfrac{x}{2}$, $g(x) = \dfrac{1}{\cos x}$

26. $f(x) = 1 - 8 \sin^2 x + 8 \sin^4 x$, $g(x) = \cos 4x$

In Exercises 27–28, (a) graph the given function(s) and make a conjecture, and (b) prove your conjecture.

27. $f(x) = 2 \sin^2 3x + \cos 6x$

28. $f(x) = \sin x \cot \dfrac{x}{2}$, $g(x) = \cos x$

In Exercises 29–44 solve the equation in the interval $[0, 2\pi)$.

29. $\cos x \sin x - \sin x = 0$

30. $\sin x - 2 \sin^2 x = 0$

31. $2 \sin^2 x - 5 \sin x + 2 = 0$

32. $\sin x - \cos x - \tan x = -1$

33. $2 \cos^2 x - 7 \cos x + 3 = 0$

34. $4 \sin^2 x + 2 \cos^2 x = 3$

35. $\dfrac{1 - \cos x}{1 + \cos x} = 3$

36. $\sin x = \cos 2x$

37. $\tan^3 x + \tan^2 x - 3 \tan x - 3 = 0$

38. $\cos 2x \csc^2 x = 2 \cos 2x$

39. $\tan \tfrac{1}{2}x + 2 \sin 2x = \csc x$

40. $\cos 3x + \cos 2x + \cos x = 0$

41. $\tan x + \sec x = \sqrt{3}$

42. $2 \cos x - 3 \tan x = 0$

43. $\cos x = x^2 - 1$

44. $e^{\sin x} = x$

In Exercises 45–54 find the exact value of each expression.

45. $\cos 15°$

46. $\sin \dfrac{5\pi}{12}$

47. $\tan \dfrac{\pi}{8}$

48. $2 \sin \dfrac{\pi}{12} \cos \dfrac{\pi}{12}$

49. $\sin 5° \cos 40° + \cos 5° \sin 40°$

50. $\dfrac{\tan 66° - \tan 6°}{1 + \tan 66° \tan 6°}$

51. $\cos^2 \dfrac{\pi}{8} - \sin^2 \dfrac{\pi}{8}$

52. $\dfrac{1}{2} \cos \dfrac{\pi}{12} + \dfrac{\sqrt{3}}{2} \sin \dfrac{\pi}{12}$

53. $\cos 37.5° \cos 7.5°$

54. $\cos 67.5° + \cos 22.5°$

In Exercises 55–60 find the exact value of the expression given that sec x = $\frac{3}{2}$, csc y = 3, and x and y are in quadrant I.

55. sin(x + y)

56. cos(x − y)

57. tan(x + y)

58. sin 2x

59. cos $\frac{y}{2}$

60. tan $\frac{y}{2}$

In Exercises 61–68 find the exact value of each expression.

61. $\sin^{-1}(\sqrt{3}/2)$

62. $\tan^{-1}(\sqrt{3}/3)$

63. $\cos(\tan^{-1}\sqrt{3})$

64. $\sin[\cos^{-1}(\sqrt{3}/2)]$

65. $\tan(\sin^{-1}\frac{2}{5})$

66. $\sin(\cos^{-1}\frac{3}{8})$

67. $\cos(2\sin^{-1}\frac{1}{3})$

68. $\cos(\sin^{-1}\frac{5}{13} - \cos^{-1}\frac{4}{5})$

In Exercises 69 and 70 express θ in terms of x.

69.

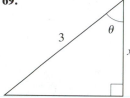

70.

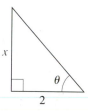

71. The figure shows a satellite in stationary orbit a distance h above the earth. The distance between two points P and Q farthest apart that the satellite can "see" is labeled s in the figure. Express s as a function of h. The radius of the earth is 3960 mi. [*Hint:* First express s in terms of θ.]

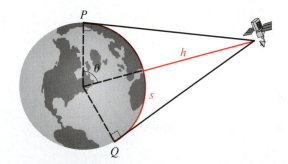

72. How high does the satellite in Exercise 71 have to be in order to see both Los Angeles and New York, 2450 miles apart?

In Exercises 73–78 write each complex number in trigonometric form with argument between 0 and 2π.

73. 4 + 4i

74. −10i

75. 5 + 3i

76. 1 + $\sqrt{3}$ i

77. i(1 + i)

78. −20

In Exercises 79–82 find the indicated power.

79. $(1 - \sqrt{3}\,i)^4$

80. $(1 + i)^8$

81. $(\sqrt{3} + i)^{-4}$

82. $(\frac{1}{2} + \frac{\sqrt{3}}{2}i)^{20}$

In Exercises 83–86 find the indicated roots.

83. The square roots of −16i

84. The cube roots of 4 + 4$\sqrt{3}$ i

85. The sixth roots of 1

86. The eighth roots of i

*In Exercises 87 and 88 find the vectors **u** + **v**, **u** − **v**, 2**u**, and 3**u** − 2**v** for the given vectors **u** and **v**.*

87. u = ⟨−2, 3⟩,   **v** = ⟨8, 1⟩

88. u = 2**i** + **j**,   **v** = **i** − 2**j**

*In Exercises 89–94, **u** = **i** + 2**j** and **v** = 2**i** − 2**j**. Find the indicated vector or scalar.*

89. |**u**|

90. |**u**|**v**

91. |**u** + **v**|

92. u · **v**

93. u · (**u** + **v**)

94. |**u** · **v**|**u**

95. Find the vector **u** with initial point $P(0, 3)$ and terminal point $Q(3, -1)$.

96. Find the vector **u** having length |**u**| = 20 and direction $\theta = 60°$.

97. If the vector 5**i** − 8**j** is placed in the plane with its initial point at $P(5, 6)$, find its terminal point.

98. Find the direction of the vector 2**i** − 5**j**.

99. Two tugboats are pulling a barge, as shown in the figure. One pulls with a force of 2.0×10^4 lb in the direction N30°E and the other with a force of 3.4×10^4 lb in the direction N10°W.
   (a) Find the resultant force on the barge as a vector.
   (b) Find the magnitude and direction of the resultant force.

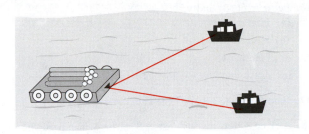

*In Exercises 101–104, are **u** and **v** orthogonal? If not, find the angle between them.*

101. $\mathbf{u} = \langle -4, 2 \rangle$, $\mathbf{v} = \langle 3, 6 \rangle$

102. $\mathbf{u} = \langle 5, 3 \rangle$, $\mathbf{v} = \langle -2, 6 \rangle$

103. $\mathbf{u} = 2\mathbf{i} + \mathbf{j}$, $\mathbf{v} = \mathbf{i} + 3\mathbf{j}$

104. $\mathbf{u} = \mathbf{i} - \mathbf{j}$, $\mathbf{v} = \mathbf{i} + \mathbf{j}$

105. Find the work done by the force $\mathbf{F} = 2\mathbf{i} + 9\mathbf{j}$ in moving an object from the point $(1, 1)$ to the point $(7, -1)$.

100. An airplane heads N60°E at 600 mi/h in still air. A tail wind begins to blow in the direction N30°E at 50 mi/h.
(a) Find the velocity of the airplane as a vector.
(b) Find the true speed and direction of the airplane.

106. A force $\mathbf{F}$ with magnitude 250 lb moves an object in the direction of the vector $\mathbf{D}$ a distance of 20 ft. If the work done is 3800 ft-lb, find the angle between $\mathbf{F}$ and $\mathbf{D}$.

CHAPTER 7 TEST

1. Verify each identity.
(a) $\dfrac{\tan x}{1 - \cos x} = \csc x \, (1 + \sec x)$

(b) $\dfrac{2 \tan x}{1 + \tan^2 x} = \sin 2x$

2. Solve each trigonometric equation on the interval $[0, 2\pi)$.
(a) $2 \cos^2 x + 5 \cos x + 2 = 0$

(b) $\sin 2x + 2 \sin^2 \dfrac{x}{2} = 1$

3. Let $x = 2 \sin \theta$, $-\pi/2 < \theta < \pi/2$. Simplify the expression

$$\frac{x}{\sqrt{4 - x^2}}$$

4. Find the exact value of each expression.
(a) $\sin 8° \cos 22° + \cos 8° \sin 22°$
(b) $\sin 15°$

5. For the angles α and β in the figure, find (a) $\cos(\alpha + \beta)$ and (b) $\tan \dfrac{\alpha}{2}$.

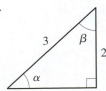

6. If $\sin x = \frac{3}{5}$, x in quadrant I, and $\cos y = \frac{5}{13}$, y in quadrant IV, find $\sin(x + y)$.

7. (a) Write $\sin 3x \cos 5x$ as a sum of trigonometric functions.
(b) Write $\sin 2x - \sin 5x$ as a product of trigonometric functions.

8. If $\sin \theta = -\frac{4}{5}$ and θ is in quadrant III, find $\tan \dfrac{\theta}{2}$.

9. Sketch the graphs of $y = \sin x$ and $y = \sin^{-1} x$, and specify the domain of each function.

10. Express θ in each figure in terms of x.
(a) (b)

11. Find the exact value of each expression.
(a) $\cos\left(\tan^{-1} \frac{9}{40}\right)$ (b) $\tan^{-1}\left(\sin \dfrac{\pi}{3} + \cos \dfrac{\pi}{6}\right)$

12. Let $z = 1 + \sqrt{3}\,i$.
(a) Write z in trigonometric form.
(b) Find the complex number z^9.

13. Let

$$z_1 = 4\left(\cos \frac{7\pi}{12} + i \sin \frac{7\pi}{12}\right)$$

and

$$z_2 = 2\left(\cos \frac{5\pi}{12} + i \sin \frac{5\pi}{12}\right)$$

Find $z_1 z_2$ and $\dfrac{z_1}{z_2}$.

14. Find the cube roots of $27i$ and sketch these roots in the complex plane.

15. Let **u** be the vector with initial point $P(3, -1)$ and terminal point $Q(-3, 9)$.
 (a) Express **u** in terms of **i** and **j**.
 (b) Find the length of **u**.
 (c) Find a vector of length 1 that has the same direction as **u**.

16. Let $\mathbf{u} = \langle 1, 3 \rangle$ and $\mathbf{v} = \langle -6, 2 \rangle$.
 (a) Find $\mathbf{u} - 3\mathbf{v}$.
 (b) Find $|\mathbf{u} + \mathbf{v}|$.
 (c) Find $\mathbf{u} \cdot \mathbf{v}$.
 (d) Are **u** and **v** perpendicular?

17. Find the angle between the vectors $3\mathbf{i} + 2\mathbf{j}$ and $5\mathbf{i} - \mathbf{j}$.

18. A river is flowing due east at 8 mi/h. A man heads his motorboat in a direction N30°E in the river. The speed of the motorboat in still water is 12 mi/h.
 (a) Express the true velocity of the motorboat as a vector.
 (b) Find the true speed and direction of the motorboat.

19. Find the work done by the force $\mathbf{F} = 3\mathbf{i} - 5\mathbf{j}$ in moving an object from the point $(2, 2)$ to the point $(7, -13)$.

FOCUS ON PROBLEM SOLVING

It is sometimes necessary to solve a problem by **taking cases**, or in other words, by accounting for all possibilities. We split the problem into several cases and give different arguments for each of the cases. For instance, we used this strategy in Section 1.7 to solve inequalities (see the second solution to Example 6). We also used it in dealing with absolute value (see Examples 3 and 5 in Section 1.8). In fact, because of the nature of absolute value, it is often essential to use the strategy of taking cases in solving problems involving absolute value.

EXAMPLE

Solve the inequality $|x - 3| + |x + 2| < 11$.

SOLUTION

Recall the definition of absolute value:

$$|x| = \begin{cases} x & \text{if } x \geq 0 \\ -x & \text{if } x < 0 \end{cases}$$

It follows that

$$|x - 3| = \begin{cases} x - 3 & \text{if } x - 3 \geq 0 \\ -(x - 3) & \text{if } x - 3 < 0 \end{cases}$$

$$= \begin{cases} x - 3 & \text{if } x \geq 3 \\ -x + 3 & \text{if } x < 3 \end{cases}$$

Similarly

$$|x + 2| = \begin{cases} x + 2 & \text{if } x + 2 \geq 0 \\ -(x + 2) & \text{if } x + 2 < 0 \end{cases}$$

$$= \begin{cases} x + 2 & \text{if } x \geq -2 \\ -x - 2 & \text{if } x < -2 \end{cases}$$

These expressions show that there are three cases to be considered:

$$x < -2 \qquad -2 \leq x < 3 \qquad x \geq 3$$

Case I If $x < -2$, we have

$$|x - 3| + |x + 2| < 11$$
$$-x + 3 - x - 2 < 11$$
$$-2x < 10$$
$$x > -5$$

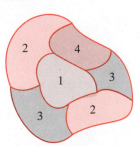

Case II   If $-2 \leq x < 3$, the given inequality becomes

$$-x + 3 + x + 2 < 11$$
$$5 < 11 \quad \text{(always true)}$$

Case III   If $x \geq 3$, the inequality becomes

$$x - 3 + x + 2 < 11$$
$$2x < 12$$
$$x < 6$$

Combining Cases I, II, and III, we see that the inequality is satisfied when $-5 < x < 6$. So the solution is the interval $(-5, 6)$. ■

PROBLEMS

In Problems 1 and 2 solve the inequality.

1. $|x + 1| + |x + 4| \leq 5$ **2.** $|x - 1| - |x - 3| \geq 5$

In Problems 3 and 4 solve the equation.

3. $|2x - 1| - |x + 5| = 3$ **4.** $|x - 1| + |x - 2| + |x - 3| = 1$

5. Solve the inequality $|\tan x| \leq 1$.

6. Solve the equation $|\sin x| = \sin x + 2 \cos x$, $0 \leq x \leq 2\pi$.

7. Each letter in the following multiplication represents a different digit. Find the value of each letter.

$$\begin{array}{r} \text{ABCDE} \\ \times \quad 4 \\ \hline \text{EDCBA} \end{array}$$

8. Find every positive integer that gives a perfect square if 132 is added to it and another perfect square if 200 is added to it.

9. Solve the following problem, which was first posed in the nineteenth century: "Every day at noon a ship leaves New York for Le Havre and another ship leaves Le Havre for New York. The trip takes seven full days. How many New York-to-Le Havre ships will the ship leaving Le Havre today meet during its trip to New York?"

10. A farmer has a fox, a goose, and a sack of grain. He wishes to cross a river in a rowboat that can hold only himself and one of these three. The farmer's problem is that foxes eat geese and geese eat grain. How can he safely ferry all three of these possessions to the other side?

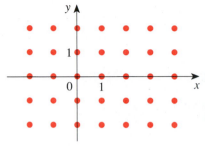

Lattice points in the plane

11. Points (m, n) in the coordinate plane, both of whose coordinates are integers, are called *lattice points*. Show that it is impossible for any equilateral triangle to have each of its vertices at a lattice point. [*Hint:* See Section 7.3, Exercise 56(b).]

12. Sketch the graph of each function:
   (a) $f(x) = \sin(\sin^{-1}x)$ **(b)** $f(x) = \sin^{-1}(\sin x)$

13. Find $\angle A + \angle B + \angle C$ in the figure.

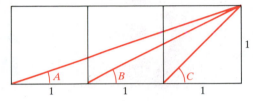
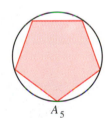

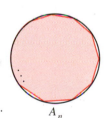

14. If $0 < \theta < \pi/2$ and $\sin 2\theta = a$, find $\sin \theta + \cos \theta$.

15. Show that $\dfrac{\pi}{4} = \tan^{-1}\frac{1}{2} + \tan^{-1}\frac{1}{5} + \tan^{-1}\frac{1}{8}$.   [*Note:* This identity was used by Zacharias Dase in 1844 to calculate the decimal expansion of π to 200 places.]

16. **(a)** Find a formula for the area A_n of the regular polygon with n sides inscribed in a circle of radius 1 (see the figure). Express the answer in terms of $\sin(\pi/n)$ and $\cos(\pi/n)$.

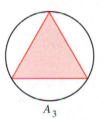

A_3 A_4 A_5 $\cdots$ A_n

   (b) Find A_3, A_4, A_{100}, and A_{1000}, correct to six decimals. Notice that the values get closer and closer to π. Why?

17. One of nine eggs is lighter than the rest. The other eight all have exactly the same weight. How can you determine which is the lighter egg with exactly two weighings on a balance?

18. Among 12 similar coins there is one counterfeit. It is not known whether the counterfeit coin is lighter or heavier than a genuine coin. Using a balance three times, how can the counterfeit be identified and in the process determined to be lighter or heavier than a genuine coin?

SYSTEMS OF EQUATIONS AND INEQUALITIES

As the sun eclipses the stars by his brilliancy, so the man of knowledge will eclipse the fame of others in assemblies of the people if he proposes algebraic problems, and still more if he solves them.

BRAHMAGUPTA, 628 A.D.

Many of the problems to which we can apply the techniques of algebra give rise to sets of equations with several unknowns, rather than to just a single equation in a single variable. A set of equations with common variables is called a **system** of equations, and in this chapter we develop techniques for finding simultaneous solutions of systems. We first consider pairs of linear equations with two unknowns, the simplest case of this situation. To help us solve linear equations in an arbitrary number of variables, we study the algebra of matrices and determinants. We also study systems of inequalities and linear programming, which is an optimization technique widely used in business and the social sciences.

SECTION 8.1
PAIRS OF LINES

In Section 1.10, we saw that the graph of any equation of the form

$$Ax + By = C$$

is a line. Let us consider a **system** of two such equations:

$$\begin{cases} ax + by = c \\ dx + ey = f \end{cases}$$

A **solution** of this system is an ordered pair of numbers (x_0, y_0) that simultaneously makes each equation a true statement when x is replaced by x_0 and y by y_0. This means that the point (x_0, y_0) lies on both of the lines in the system, and so it must be a point at which they intersect. For example, $(2, 6)$ is a solution of the system

$$\begin{cases} 3x - y = 0 \\ 5x + 2y = 22 \end{cases}$$

because

$$3(2) - (6) = 0$$

and

$$5(2) + 2(6) = 22$$

Graphing the lines given by these equations, we see in Figure 1 that $(2, 6)$ is their point of intersection. The graph also shows that there can be no other solutions of the system because the lines do not intersect anywhere else.

In general, there are three situations that can occur when we graph two linear equations. The graphs may intersect at a single point (Figure 2), they may be parallel with no intersection points (Figure 3), or the two equations may just be different equations for the same line (Figure 4). This means that the system can have one solution, no solution, or infinitely many solutions, since each solution corresponds to an intersection point of the lines.

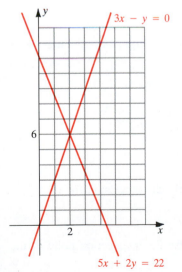

Figure 1

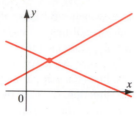

Figure 2

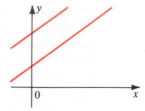

Figure 3

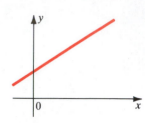

Figure 4

There are two basic methods for solving systems of two linear equations. The first, called the **method of substitution**, is perhaps the more obvious, and we use it in Example 1. The second method is easier to extend to situations where we have more equations and more variables. We use it in the remaining examples.

Substitution method

EXAMPLE 1

Solve the system

$$\begin{cases} 4x - 3y = 11 \\ 6x + 2y = -3 \end{cases}$$

and graph the lines.

SOLUTION

Solving the second equation for y in terms of x, we get

$$y = -3x - \tfrac{3}{2}$$

Now we can substitute this expression for y into the first equation, which gives us an equation that involves only the variable x:

$$4x - 3\left(-3x - \tfrac{3}{2}\right) = 11$$
$$13x + \tfrac{9}{2} = 11$$
$$13x = \tfrac{13}{2}$$
$$x = \tfrac{1}{2}$$

We now substitute this value for x back into the original expression for y:

$$y = -3\left(\tfrac{1}{2}\right) - \tfrac{3}{2} = -3$$

The solution of the system is $\left(\tfrac{1}{2}, -3\right)$, which is also the intersection point of the lines in the system (see Figure 5). ∎

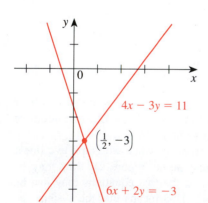

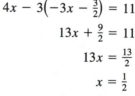

Figure 5

Another way of solving pairs of linear equations is to eliminate either x or y from the equations by adding a suitable multiple of one to the other. This is called the **elimination method** and we illustrate it in the next example.

Elimination method

EXAMPLE 2

Solve the system

$$\begin{cases} x - 3y = 6 \\ -2x + 5y = -5 \end{cases}$$

SOLUTION

If we multiply both sides of the first equation by 2, the coefficients of x in the two equations are negatives of each other:

$$\begin{cases} 2x - 6y = 12 \\ -2x + 5y = -5 \end{cases}$$

Adding corresponding sides of the two equations eliminates the variable x, and we can solve for y:

$$-y = 7 \qquad \text{or} \qquad y = -7$$

At this point we could substitute this value for y into either of the original equations and solve for x. We use the first one, because it looks a little easier.

$$x - 3(-7) = 6$$
$$x + 21 = 6$$
$$x = -15$$

The solution of the system is $(-15, -7)$. As a check on our answer, we make sure that the point satisfies the second equation as well:

$$-2(-15) + 5(-7) = -5$$
$$30 - 35 = -5$$

This is true, so our answer satisfies both equations. ■

EXAMPLE 3

Solve the system

$$\begin{cases} 8x - 2y = 5 \\ -12x + 3y = 7 \end{cases}$$

and graph the lines.

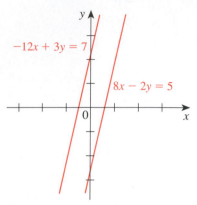

Figure 6

SOLUTION

This time we try to find a suitable combination of the two equations to eliminate the variable y. Multiplying the first equation by 3 and the second by 2 gives

$$\begin{cases} 24x - 6y = 15 \\ -24x + 6y = 14 \end{cases}$$

Adding the two equations eliminates *both* x and y in this case, and we end up with $0 = 29$, which is obviously false. This means that the given system is **inconsistent**; there is no solution, since no matter what values we assign to x and y, we end up with the false statement $0 = 29$. In slope-intercept form the equations in the system are

$$y = 4x - \tfrac{5}{2} \qquad \text{and} \qquad y = 4x + \tfrac{7}{3}$$

These lines are parallel, with different y-intercepts (see Figure 6), so there is no intersection point.   ■

EXAMPLE 4

Solve the system

$$\begin{cases} 3x - 6y = 12 \\ 4x - 8y = 16 \end{cases}$$

SOLUTION

Multiplying the first equation by 4 and the second by 3, in preparation for subtracting the equations to eliminate x, gives

$$\begin{cases} 12x - 24y = 48 \\ 12x - 24y = 48 \end{cases}$$

We see that the two equations in the original system are just different ways of expressing the equation of one single line. The coordinates of any point on this line give a solution of the system. Writing the equation in slope-intercept form, we have $y = \tfrac{1}{2}x - 2$, so any pair of the form

$$\left(x, \tfrac{1}{2}x - 2\right)$$

where x can be any real number, is a solution of the system. There are infinitely many solutions.   ■

APPLIED LINEAR SYSTEMS

Frequently when we use equations to solve problems in the sciences or in other areas we obtain systems like the ones we have been considering. The next two examples illustrate such situations.

EXAMPLE 5

A woman rows a boat upstream from one point on a river to another point 4 mi away in $1\frac{1}{2}$ h. The return trip, traveling with the current, takes only 45 min. How fast does she row relative to the water, and at what speed is the current flowing?

SOLUTION

In this and any other problem that involves distance, time, and speed, we make use of the fundamental relationship between these quantities,

$$\text{speed} = \frac{\text{distance}}{\text{time}}$$

or the equivalent formulations

$$\text{distance} = \text{speed} \times \text{time} \qquad \text{and} \qquad \text{time} = \frac{\text{distance}}{\text{speed}}$$

We use these equations to put the English sentences of the problem into mathematical form. Since we are asked to find the rowing speed and the speed of the current, we give names to these quantities. Let

$$x = \text{rowing speed in miles per hour}$$

and

$$y = \text{speed of the current in miles per hour}$$

When she is traveling with the current (downstream), she will be moving at a total of $x + y$ miles per hour, but upstream she moves at $x - y$ miles per hour, since the current decreases her net speed. The distance both upstream and downstream is 4 mi, so using the fact that distance = speed × time for both parts of the trip, we get the equations

$$4 = (x - y) \cdot \tfrac{3}{2} \qquad \text{and} \qquad 4 = (x + y) \cdot \tfrac{3}{4}$$

[*Note*: All times have been converted to hours, since we are expressing the speeds in miles per *hour*.] If we multiply the equations by 2 and 4, respectively, to clear the denominators, we get the system

$$\begin{cases} 3x - 3y = 8 \\ 3x + 3y = 16 \end{cases}$$

Adding the equations eliminates the variable y:

$$6x = 24$$
$$x = 4$$

Substituting this into the first equation in the system (although the second works just as well) and solving for y gives

$$3(4) - 3y = 8$$
$$-3y = 8 - 12$$
$$y = \tfrac{4}{3}$$

The woman rows at 4 mi/h and the current flows at $1\tfrac{1}{3}$ mi/h. ■

Many students find that the hardest part of solving a word problem is getting started. Remember the following two key steps:

> **1.** Assign letters to denote the variable quantities in the problem. Usually the last sentence of the problem tells you what is being asked for, so this is what the variable names will represent.
>
> **2.** Translate the given information from English sentences into equivalent mathematical equations involving the variables. Note that an equation is just a sentence written using mathematical notation.

EXAMPLE 6

A vintner wishes to fortify wine that contains 10% alcohol by adding to it some 70% alcohol solution. The resulting mixture is to have an alcoholic strength of 16% and is to fill 1000 one-liter bottles. How many liters of the wine and of the alcohol solution should he use?

SOLUTION

Let

$$x = \text{number of liters of wine to be used}$$
$$y = \text{number of liters of alcohol solution to be used}$$

To help us translate the information in the problem into equations, we organize the given data in a table:

	Wine	Alcohol solution	Resulting mixture
Volume	x	y	1000
Percent alcohol	10%	70%	16%
Amount of alcohol	$(0.10)x$	$(0.70)y$	$(0.16)1000$

The volume of the mixture must be the total of the two volumes the vintner is adding together, so

$$x + y = 1000$$

Similarly, the amount of alcohol in the mixture must be the total of the alcohol contributed by the wine and by the alcohol solution, which means that

$$(0.10)x + (0.70)y = (0.16)1000$$
$$(0.10)x + (0.70)y = 160$$
$$x + 7y = 1600$$

Thus we must solve the system

$$\begin{cases} x + y = 1000 \\ x + 7y = 1600 \end{cases}$$

Subtracting the first equation from the second eliminates the x, and we get

$$6y = 600$$
$$y = 100$$

We now substitute this into the first equation and solve:

$$x + 100 = 1000$$
$$x = 900$$

The vintner should use 900 L of wine and 100 L of the alcohol solution.   ■

EXERCISES 8.1

In Exercises 1–6 graph each pair of lines on a single set of axes. Determine whether the lines are parallel, and if they are not parallel, estimate the coordinates of their point of intersection from your graph.

1. $\begin{cases} 2x + y = 8 \\ 3x - 2y = 12 \end{cases}$

2. $\begin{cases} 3x + 2y = 3 \\ -x + 5y = 16 \end{cases}$

3. $\begin{cases} 6x - 3y = 6 \\ -10x + 5y = 0 \end{cases}$

4. $\begin{cases} 3x + 5y = 15 \\ x + \frac{5}{3}y = 5 \end{cases}$

5. $\begin{cases} 2x + 5y = 15 \\ 4x + 8y = 22 \end{cases}$

6. $\begin{cases} -4x + 14y = 28 \\ 10x - 35y = 70 \end{cases}$

In Exercises 7–12 solve the system using the substitution method.

7. $\begin{cases} 3x - 5y = 1 \\ -x + 4y = 2 \end{cases}$

8. $\begin{cases} 4x - 3y = 28 \\ 9x - y = -6 \end{cases}$

9. $\begin{cases} 5x + 2y = 11 \\ 3x + 6y = 9 \end{cases}$

10. $\begin{cases} -4x + 12y = 0 \\ 12x + 4y = 160 \end{cases}$

11. $\begin{cases} \frac{1}{2}x + \frac{1}{3}y = 2 \\ \frac{1}{5}x - \frac{2}{3}y = 8 \end{cases}$

12. $\begin{cases} 0.2x - 0.2y = -1.8 \\ -0.3x + 0.5y = 3.3 \end{cases}$

In Exercises 13–26 solve the system using the elimination method. If a system has infinitely many solutions, express them in the form given in Example 4.

13. $\begin{cases} x - 3y = 6 \\ -2x + 5y = -10 \end{cases}$

14. $\begin{cases} 4x + 2y = 16 \\ x - 5y = 70 \end{cases}$

15. $\begin{cases} 2x - 30y = -12 \\ -3x + 45y = 18 \end{cases}$

16. $\begin{cases} 18x + y = 30 \\ 12x - 3y = -24 \end{cases}$

17. $\begin{cases} 3x + 5y = 17 \\ 7x + 9y = 29 \end{cases}$

18. $\begin{cases} 2x - 3y = -8 \\ 14x - 21y = 3 \end{cases}$

19. $\begin{cases} 8s - 3t = -3 \\ 5s - 2t = -1 \end{cases}$

20. $\begin{cases} u - 30v = -5 \\ -3u + 80v = 5 \end{cases}$

21. $\begin{cases} \frac{1}{2}x + \frac{3}{5}y = 3 \\ \frac{1}{3}x + 2y = -6 \end{cases}$

22. $\begin{cases} \frac{3}{2}x - \frac{1}{3}y = \frac{1}{2} \\ 2x - \frac{1}{2}y = -\frac{1}{2} \end{cases}$

23. $\begin{cases} 0.2r + 0.3s = 0.16 \\ -1.2r + 4s = 1.36 \end{cases}$

24. $\begin{cases} 4.8x - 1.6y = 8 \\ 3.6x - 1.2y = 6 \end{cases}$

25. $\begin{cases} \sqrt{3}x + \sqrt{2}y = 5 \\ 2\sqrt{6}x + 4y = \sqrt{5} \end{cases}$

26. $\begin{cases} \sqrt{10}w - \sqrt{2}z = -2 + 5\sqrt{2} \\ \sqrt{2}w + 2\sqrt{5}z = 3\sqrt{10} \end{cases}$

In Exercises 27–30 solve the system by first making a substitution that will turn the equations into equations of lines and then using the methods discussed in the text. The appropriate substitutions are given in Exercises 27 and 28, but you must determine them for yourself in Exercises 29 and 30.

27. $\begin{cases} \dfrac{2}{u} + \dfrac{1}{v} = 1 \\ \dfrac{3}{u} - \dfrac{2}{v} = 1 \end{cases}$

$$\left[\text{Let } x = \frac{1}{u},\ y = \frac{1}{v}.\right]$$

28. $\begin{cases} 2r^2 + 3s^2 = 11 \\ 6r^2 - s^2 = 23 \end{cases}$

[Let $x = r^2$, $y = s^2$.]

29. $\begin{cases} 2z^3 + \frac{1}{2}w^3 = 2 \\ -3z^3 + \frac{3}{2}w^3 = 15 \end{cases}$

30. $\begin{cases} \dfrac{2}{x} - \dfrac{4}{y^2} = 8 \\ \dfrac{1}{x} - \dfrac{3}{y^2} = 6 \end{cases}$

Solve the systems in Exercises 31–34.

31. $\begin{cases} \dfrac{2x - 5}{3} + \dfrac{y - 1}{6} = \dfrac{1}{2} \\ \dfrac{x}{5} + \dfrac{3y - 6}{12} = 1 \end{cases}$

32. $\begin{cases} x - 3y = 4x - 6y - 10 \\ 2x = 12y + 10 \end{cases}$

33. $x - 2y = 2x + 2y = 1$ **34.** $x = 2x + y = 2y + 1$

In Exercises 35–38 find x and y in terms of a and b.

35. $\begin{cases} x + y = 0 \\ x + ay = 1 \end{cases}$ $(a \neq 1)$

36. $\begin{cases} ax + by = 0 \\ x + y = 1 \end{cases}$ $(a \neq b)$

37. $\begin{cases} ax + by = 1 \\ bx + ay = 1 \end{cases}$ $(a^2 - b^2 \neq 0)$

38. $\begin{cases} ax + by = 0 \\ a^2x + b^2y = 1 \end{cases}$ $(a \neq 0,\ b \neq 0,\ a \neq b)$

39. Find two numbers whose sum is 34 and whose difference is 10.

40. The sum of two numbers is twice their difference. The larger number is 6 more than twice the smaller. Find the numbers.

41. A man has 14 coins in his pocket, all of which are dimes and quarters. If the total value of his change is $2.75, how many dimes and how many quarters does he have?

42. The admission fee at an amusement park is $1.50 for children and $4.00 for adults. On a certain day, 2200 people entered the park and the admission fees collected totaled $5050. How many children and how many adults were admitted?

43. A man flies a small airplane from Fargo to Bismarck, North Dakota—a distance of 180 mi. Because he is flying into a head wind, the trip takes him 2 h. On the way back, the wind is still blowing at the same speed, so the return trip takes only 1 h 12 min. What is his speed in still air, and how fast is the wind blowing?

44. A boat on a river travels downstream between two points 20 mi apart in 1 h. The return trip against the current takes $2\frac{1}{2}$ h. What is the boat's speed, and how fast does the current in the river flow?

45. A woman keeps fit by bicycling and running every day. On Monday she spends $\frac{1}{2}$ h at each activity, covering a total of $12\frac{1}{2}$ mi. On Tuesday, she runs for 12 min and cycles for 45 min, and covers a total of 16 mi. Assuming her running and cycling speeds do not change from day to day, find these speeds.

46. A biologist has two brine solutions, one containing 5% salt and another containing 20% salt. How many milliliters of each solution should he mix to obtain 1 L of a solution that contains 14% salt?

47. A researcher performs an experiment to test a hypothesis that involves the nutrients niacin and retinol. She wishes to feed one of her groups of laboratory rats a diet that contains precisely 32 units of niacin and 22,000 units of retinol per day. She has two types of commercial pellet foods available. Food A contains 0.12 unit of niacin and 100 units of retinol per gram. Food B contains 0.20 unit of niacin and 50 units of retinol per gram. How many grams of each food should she feed this group of rats each day?

48. A customer in a coffee shop wishes to purchase a blend of two coffees: Kenyan, costing $3.50 a pound, and Sri Lankan, costing $5.60 a pound. He ends up buying 3 lb of such a blend, which costs him $11.55. How many pounds of each kind went into the mixture?

49. A chemist has two large containers of sulfuric acid solution, with different concentrations of acid in each container. Blending 300 mL of the first solution and 600 mL of the second gives a mixture that is 15% acid, whereas 100 mL of the first mixed with 500 mL of the second gives a $12\frac{1}{2}$% acid mixture. What is the concentration of sulfuric acid in each of the original containers?

50. John and Mary leave their house at the same time and drive off in opposite directions. John drives at 60 mi/h and travels 35 mi farther than Mary, who drives at 40 mi/h. Mary's trip takes 15 min longer than John's. For what length of time does each of them drive?

51. A business executive normally leaves her office at 5:00 P.M. to drive home. On Monday, she is able to leave the office at 4:45 P.M. She is able to drive at 30 mi/h and arrives home 19 min earlier than usual. On Tuesday she leaves the office at 5:20 P.M. Because the traffic is so heavy, she drives at an average speed of only 15 mi/h and arrives home 36 min later than usual. What is her usual speed on the way home, and at what time does she usually get there?

52. The sum of the digits of a two-digit number is 7. When the digits are reversed, the number is increased by 27. Find the number.

53. The sum of the digits of a two-digit number is 9. When the digits are reversed, the value of the number is decreased to $\frac{3}{8}$ of its original value. What is the number?

54. Find the area of the triangle that lies in the first quadrant (with its base on the x-axis) and that is bounded by the lines $y = 2x - 4$ and $y = -4x + 20$.

55. Find the equation of the parabola that passes through the origin and through the points $(1, 12)$ and $(3, 6)$. [*Hint:* Recall that the general equation of a parabola of this type is $y = ax^2 + bx + c$.]

56. Find the equation of the parabola that passes through the point $(-1, 8)$, that has y-intercept 14, and whose vertex has x-coordinate -2. [*Hint:* Recall that the equation of a parabola with vertex (h, k) is of the form $y = a(x - h)^2 + k$.]

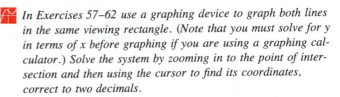 *In Exercises 57–62 use a graphing device to graph both lines in the same viewing rectangle. (Note that you must solve for y in terms of x before graphing if you are using a graphing calculator.) Solve the system by zooming in to the point of intersection and then using the cursor to find its coordinates, correct to two decimals.*

57. $\begin{cases} 0.21x + 3.17y = 9.51 \\ 2.35x - 1.17y = 5.89 \end{cases}$

58. $\begin{cases} 18.72x - 14.91y = 12.33 \\ 6.21x - 12.92y = 17.82 \end{cases}$

59. $\begin{cases} 2371x - 6552y = 13{,}591 \\ 9815x + 992y = 618{,}555 \end{cases}$

60. $\begin{cases} -435x + 912y = 0 \\ 132x + 455y = 994 \end{cases}$

61. $\begin{cases} \sqrt{3}x + \sqrt{5}y = \sqrt{7} \\ -\sqrt{2}x + \sqrt{11}y = -\sqrt{13} \end{cases}$

62. $\begin{cases} \frac{1}{23}x - \frac{1}{42}y = 3 \\ \pi x + \pi^2 y = 3.14 \end{cases}$

SECTION 8.2
SYSTEMS OF LINEAR EQUATIONS

A **linear equation in n variables** is an equation which can be put in the form

$$a_1 x_1 + a_2 x_2 + \cdots + a_n x_n = c$$

where $a_1, a_2, \ldots, a_n$ and c are real numbers, and $x_1, x_2, \ldots, x_n$ are the variables. If the number of variables is no more than three or four, we generally use x, y, z, and w instead of x_1, x_2, x_3, and x_4. Such equations are called linear because if we have just two variables, the equation is

$$a_1 x + a_2 y = c$$

which is the equation of a line. The equations

$$6x_1 - 3x_2 + \sqrt{5}x_3 = 1000$$

and

$$x + y + z = 2w - \tfrac{1}{2}$$

are thus linear, but the equations

$$x^2 + 3y - \sqrt{z} = 5$$

and

$$x_1x_2 + 6x_3 = -6$$

are not. Each term of a linear equation is either a constant or a constant multiple of one of the variables.

We are going to adapt the elimination method introduced in Section 8.1 to solve systems of linear equations in any number of variables. We begin with an example to show how the method works before formally describing the technique.

EXAMPLE 1

Solve the system

$$\begin{cases} x + 2y + 4z = 7 \\ -x + y + 2z = 5 \\ 2x + 3y + 3z = 7 \end{cases}$$

SOLUTION

We first eliminate the x from the second and third equations. If we add the first equation to the second, we get

$$\begin{aligned} x + 2y + 4z &= 7 \\ -x + y + 2z &= 5 \\ \hline 3y + 6z &= 12 \end{aligned}$$

which does not contain the variable x. Similarly, if we add -2 times the first equation to the third, we get

$$\begin{aligned} -2x - 4y - 8z &= -14 \\ 2x + 3y + 3z &= 7 \\ \hline -y - 5z &= -7 \end{aligned}$$

which also does not contain x. This means that the original system is equivalent to the simpler system

$$\begin{cases} x + 2y + 4z = 7 \\ 3y + 6z = 12 \\ -y - 5z = -7 \end{cases}$$

Since each term in the second equation has a common factor of 3, we multiply both sides by $\frac{1}{3}$.

$$\begin{cases} x + 2y + 4z = 7 \\ y + 2z = 4 \\ -y - 5z = -7 \end{cases}$$

We now eliminate the y from the last equation by adding the second equation to it. This gives

$$\begin{cases} x + 2y + 4z = 7 \\ y + 2z = 4 \\ -3z = -3 \end{cases}$$

Finally, multiplying the last equation by $-\frac{1}{3}$, we get

$$\begin{cases} x + 2y + 4z = 7 \\ y + 2z = 4 \\ z = 1 \end{cases}$$

We now know from the third equation that $z = 1$. Substituting this into the second equation allows us to solve for y:

$$y + 2(1) = 4$$
$$y = 4 - 2 = 2$$

Putting these values for y and z into the first equation now gives us x:

$$x + 2(2) + 4(1) = 7$$
$$x = 7 - 4 - 4 = -1$$

Thus the simultaneous solution to the system is the ordered triple $(-1, 2, 1)$; that is, $x = -1$, $y = 2$, and $z = 1$. ■

The technique we have been using here is called **Gaussian elimination** in honor of the German mathematician C. F. Gauss (see page 228). The method consists of using algebraic operations to change the linear system we are solving into an equivalent system in triangular form. A system of three equations in the three variables x, y, and z is in **triangular form** if the second equation does not have x in it and the third has neither x nor y. The last system in the solution to Example 1 is in triangular form. We solve a system in this form by using **back substitution.** That is, we substitute the value of z obtained from the last equation back into the equation that involves just y and z, and solve for y. Then we substitute y and z back into the first equation and solve for z.

The algebraic operations we are permitted to use to change the system to triangular form are called the **elementary row operations.** They are listed in the following box.

<table>
<tr><td>ELEMENTARY ROW
OPERATIONS</td><td>1. Add a multiple of one equation to another.
2. Multiply an equation by a nonzero constant.
3. Rearrange the order of the equations.</td></tr>
</table>

None of these operations changes the solutions of an equation, so after performing them we always end up with an equivalent system, that is, one that has the same solution.

If we examine the solution to Example 1, we see that the variables x, y, and z act simply as place-holders in our computations. It is only the coefficients of the variables and the constants that actually enter into the calculations. We will use this fact to simplify our notation for the Gaussian elimination process. Instead of writing the equations in a system out in full, we write only the coefficients and constants in a rectangular array, called the **matrix form** of the system. The matrix form of the system of Example 1 is as follows.

System of linear equations   **Matrix form**

$$\begin{cases} x + 2y + 4z = 7 \\ -x + y + 2z = 5 \\ 2x + 3y + 3z = 7 \end{cases} \qquad \begin{bmatrix} 1 & 2 & 4 & 7 \\ -1 & 1 & 2 & 5 \\ 2 & 3 & 3 & 7 \end{bmatrix}$$

The **rows** of the matrix are the horizontal lists of numbers in the array. For example, the first row of the matrix in the preceding display is [1 2 4 7]. Each row of the matrix form of a system represents an equation. To further simplify the notation, we use the following symbols to represent the elementary row operations:

<table>
<tr><td rowspan="4">NOTATION FOR THE
ELEMENTARY ROW
OPERATIONS</td><td>**Symbol**</td><td>**Meaning**</td></tr>
<tr><td>$R_i \rightarrow R_i + kR_j$</td><td>Change the ith row by adding k times row j to it</td></tr>
<tr><td>kR_i</td><td>Multiply the ith row by k</td></tr>
<tr><td>$R_i \leftrightarrow R_j$</td><td>Interchange the ith and jth rows</td></tr>
</table>

In the next example, we compare the two ways of writing systems of linear equations.

EXAMPLE 2

Solve the following system, using Gaussian elimination.

System	**Matrix form**

$$\begin{cases} x - y + 3z = 4 \\ x + 2y - 2z = 10 \\ 3x - y + 5z = 14 \end{cases} \qquad \begin{bmatrix} 1 & -1 & 3 & 4 \\ 1 & 2 & -2 & 10 \\ 3 & -1 & 5 & 14 \end{bmatrix}$$

SOLUTION

In each step we write both the full form of the system and the shorthand matrix form. First we eliminate x from the second and third equations.

Subtract the first equation from the second.
Subtract 3 times the first from the third.

$$\begin{cases} x - y + 3z = 4 \\ \quad 3y - 5z = 6 \\ \quad 2y - 4z = 2 \end{cases} \xrightarrow[\substack{R_2 \rightarrow R_2 - R_1 \\ R_3 \rightarrow R_3 - 3R_1}]{} \begin{bmatrix} 1 & -1 & 3 & 4 \\ 0 & 3 & -5 & 6 \\ 0 & 2 & -4 & 2 \end{bmatrix}$$

Multiply the third equation by $\frac{1}{2}$.

$$\begin{cases} x - y + 3z = 4 \\ \quad 3y - 5z = 6 \\ \quad y - 2z = 1 \end{cases} \xrightarrow[\quad]{\frac{1}{2}R_3} \begin{bmatrix} 1 & -1 & 3 & 4 \\ 0 & 3 & -5 & 6 \\ 0 & 1 & -2 & 1 \end{bmatrix}$$

Subtract 3 times the third equation from the second (to eliminate y from the second equation).

$$\begin{cases} x - y + 3z = 4 \\ \qquad\qquad z = 3 \\ \quad y - 2z = 1 \end{cases} \xrightarrow[\quad]{R_2 \rightarrow R_2 - 3R_3} \begin{bmatrix} 1 & -1 & 3 & 4 \\ 0 & 0 & 1 & 3 \\ 0 & 1 & -2 & 1 \end{bmatrix}$$

Interchange the second and third equations (to put the system into triangular form).

$$\begin{cases} x - y + 3z = 4 \\ \quad y - 2z = 1 \\ \qquad\qquad z = 3 \end{cases} \xrightarrow[\quad]{R_2 \leftrightarrow R_3} \begin{bmatrix} 1 & -1 & 3 & 4 \\ 0 & 1 & -2 & 1 \\ 0 & 0 & 1 & 3 \end{bmatrix}$$

Using $z = 3$, we now back-substitute in the second equation to get

$$y - 2(3) = 1$$
$$y = 7$$

We then substitute $y = 7$ and $z = 3$ back into the first equation to get x:

$$x - (7) + 3(3) = 4$$
$$x = 2$$

Thus the solution of the system is $(2, 7, 3)$. ■

A rectangular array of numbers like the ones we have been using is called a **matrix**. We will study the algebra of matrices in Sections 8.4 and 8.5, but for now we just use them as a shorthand device when solving systems of equations.

Note that the goal in Gaussian elimination is to end up with a system in triangular form, one that can be easily solved. Each elementary row operation we perform should take us one step closer to that goal. At each stage in the process there will be several possible operations to choose from, so there is no single "right" way to solve such a problem. Of course, no matter what route we decide to take to arrive at the triangular form, the final answer will be the same.

EXAMPLE 3

Solve the system

$$\begin{cases} 3x + y + 4z + w = 6 \\ 2x \quad\quad + 3z + 4w = 13 \\ \quad\quad y - 2z - w = 0 \\ x - y + z + w = 3 \end{cases}$$

SOLUTION

In matrix form the system is

$$\begin{bmatrix} 3 & 1 & 4 & 1 & 6 \\ 2 & 0 & 3 & 4 & 13 \\ 0 & 1 & -2 & -1 & 0 \\ 1 & -1 & 1 & 1 & 3 \end{bmatrix}$$

Since it is advantageous to have a 1 in the upper left corner, we first rearrange the order of the rows and then create as many zeros as possible in the first column.

$$\begin{bmatrix} 1 & -1 & 1 & 1 & 3 \\ 0 & 1 & -2 & -1 & 0 \\ 3 & 1 & 4 & 1 & 6 \\ 2 & 0 & 3 & 4 & 13 \end{bmatrix} \xrightarrow[\begin{subarray}{c} R_3 \to R_3 - 3R_1 \\ R_4 \to R_4 - 2R_1 \end{subarray}]{} \begin{bmatrix} 1 & -1 & 1 & 1 & 3 \\ 0 & 1 & -2 & -1 & 0 \\ 0 & 4 & 1 & -2 & -3 \\ 0 & 2 & 1 & 2 & 7 \end{bmatrix}$$

To get this into triangular form, we must change the 4 and the 2 in the second column to zeros, so we continue as follows:

$$\xrightarrow[\begin{subarray}{c} R_3 \to R_3 - 4R_2 \\ R_4 \to R_4 - 2R_2 \end{subarray}]{} \begin{bmatrix} 1 & -1 & 1 & 1 & 3 \\ 0 & 1 & -2 & -1 & 0 \\ 0 & 0 & 9 & 2 & -3 \\ 0 & 0 & 5 & 4 & 7 \end{bmatrix}$$

We would now like to change the 5 to a 0. Although we might be tempted to do this by subtracting five times row 1 from row 4, this will not work because that would eliminate the first two zeros in that row, which we worked so hard to get. So instead we perform the following operations on the last two rows:

$$\xrightarrow{R_3 \to R_3 - 2R_4} \begin{bmatrix} 1 & -1 & 1 & 1 & 3 \\ 0 & 1 & -2 & -1 & 0 \\ 0 & 0 & -1 & -6 & -17 \\ 0 & 0 & 5 & 4 & 7 \end{bmatrix} \xrightarrow{R_4 \to R_4 + 5R_3}$$

$$\begin{bmatrix} 1 & -1 & 1 & 1 & 3 \\ 0 & 1 & -2 & -1 & 0 \\ 0 & 0 & -1 & -6 & -17 \\ 0 & 0 & 0 & -26 & -78 \end{bmatrix} \xrightarrow[-\frac{1}{26}R_4]{-R_3} \begin{bmatrix} 1 & -1 & 1 & 1 & 3 \\ 0 & 1 & -2 & -1 & 0 \\ 0 & 0 & 1 & 6 & 17 \\ 0 & 0 & 0 & 1 & 3 \end{bmatrix}$$

The last equation tells us that $w = 3$, so working backward through the equations as before, we get

$$z + 6w = 17 \qquad\qquad y - 2z - w = 0 \qquad\qquad x - y + z + w = 3$$
$$z + 6(3) = 17 \qquad\quad y - 2(-1) - 3 = 0 \qquad\quad x - 1 + (-1) + 3 = 3$$
$$z = -1 \qquad\qquad\qquad\quad y = 1 \qquad\qquad\qquad\qquad\quad x = 2$$

The solution is $(2, 1, -1, 3)$.   ■

Linear equations, often containing hundreds or even thousands of variables, occur frequently in the applications of algebra to the sciences and to other fields. For now, we consider an example that involves only three variables.

EXAMPLE 4

A nutritionist is performing an experiment on student volunteers. He wishes to feed one of his subjects a daily diet that consists of a combination of three commercial diet foods: MiniCal, SloStarve, and SlimQuick. For the experiment it is important that the subject consume exactly 500 mg of potassium, 75 g of protein, and 1150 units of vitamin D every day. The amounts of these nutrients in one ounce of each food are given in the following table.

	MiniCal	**SloStarve**	**SlimQuick**
Potassium (mg)	50	75	10
Protein (g)	5	10	3
Vitamin D (units)	90	100	50

How many ounces of each food should the subject eat every day to satisfy the nutrient requirements exactly?

SOLUTION

Let x, y, and z represent the number of ounces of MiniCal, SloStarve, and SlimQuick, respectively, that the subject should eat every day. This means that he will get $50x$ mg of potassium from MiniCal, $75y$ mg from SloStarve, and $10z$ mg from SlimQuick, for a total of $50x + 75y + 10z$ mg potassium in all. Since the potassium requirement is 500 mg, we get the equation

$$50x + 75y + 10z = 500$$

Similar reasoning for the protein and vitamin D requirements leads to

$$5x + 10y + 3z = 75$$

and

$$90x + 100y + 50z = 1150$$

Dividing the first equation by 5 and the third by 10 gives the system

$$\begin{cases} 10x + 15y + 2z = 100 \\ 5x + 10y + 3z = 75 \\ 9x + 10y + 5z = 115 \end{cases}$$

We solve this using Gaussian elimination.

$$\begin{bmatrix} 10 & 15 & 2 & 100 \\ 5 & 10 & 3 & 75 \\ 9 & 10 & 5 & 115 \end{bmatrix} \xrightarrow{R_1 \to R_1 - R_3} \begin{bmatrix} 1 & 5 & -3 & -15 \\ 5 & 10 & 3 & 75 \\ 9 & 10 & 5 & 115 \end{bmatrix}$$

$$\xrightarrow[R_3 \to R_3 - 9R_1]{R_2 \to R_2 - 5R_1} \begin{bmatrix} 1 & 5 & -3 & -15 \\ 0 & -15 & 18 & 150 \\ 0 & -35 & 32 & 250 \end{bmatrix} \xrightarrow{-\frac{1}{3}R_2} \begin{bmatrix} 1 & 5 & -3 & -15 \\ 0 & 5 & -6 & -50 \\ 0 & -35 & 32 & 250 \end{bmatrix}$$

$$\xrightarrow{R_3 \to R_3 + 7R_2} \begin{bmatrix} 1 & 5 & -3 & -15 \\ 0 & 5 & -6 & -50 \\ 0 & 0 & -10 & 100 \end{bmatrix} \xrightarrow{-\frac{1}{10}R_3} \begin{bmatrix} 1 & 5 & -3 & -15 \\ 0 & 5 & -6 & -50 \\ 0 & 0 & 1 & 10 \end{bmatrix}$$

Now we work backward through the equations to get $z = 10$, $y = 2$, and $x = 5$. The subject should be fed 5 oz of MiniCal, 2 oz of SloStarve, and 10 oz of SlimQuick every day. ∎

A more practical application might involve dozens of foods and nutrients, rather than just three. As you may imagine, such a problem would be almost impossible to solve without the assistance of a computer.

EXERCISES 8.2

In Exercises 1–6 state whether the equation or system of equations is linear.

1. $6x - 3y + 1000z - w = \sqrt{13}$

2. $6xy - 3yz + 15zx = 0$

3. $x_1^2 + x_2^2 + x_3^2 = 36$

4. $e^2 x_1 + \pi x_2 - \sqrt{5} = x_3 - \frac{1}{2}x_4$

5. $\begin{cases} x - 3xy + 5y = 0 \\ 12x + 321y = 123 \end{cases}$

6. $\begin{cases} x - 3y = 15z + \dfrac{1}{\sqrt{3}} \\ \quad x = 3z \\ y - z = \dfrac{x}{\sqrt{47}} \end{cases}$

In Exercises 7–10 write a system of equations that corresponds to the given matrix.

7. $\begin{bmatrix} 2 & 3 & 1 \\ 4 & 2 & 3 \end{bmatrix}$

8. $\begin{bmatrix} 1 & 2 & 4 & 6 \\ 3 & -1 & 2 & 4 \\ -1 & -1 & 0 & 7 \end{bmatrix}$

9. $\begin{bmatrix} 0 & 1 & 0 & 0 \\ 1 & 0 & 1 & 0 \\ 0 & -2 & 2 & 7 \end{bmatrix}$

10. $\begin{bmatrix} 1 & 2 & 3 & 4 & 5 \\ -1 & 0 & 1 & 0 & 6 \\ 2 & 3 & 5 & 0 & 0 \\ 0 & 1 & 1 & 0 & -2 \end{bmatrix}$

Use Gaussian elimination to solve each system in Exercises 11–30.

11. $\begin{cases} x + y - z = 2 \\ 2y + z = 8 \\ 2x + 3y - 5z = 1 \end{cases}$

12. $\begin{cases} x + y + 6z = 3 \\ x + y + 3z = 3 \\ x + 2y + 4z = 7 \end{cases}$

13. $\begin{cases} x + y + z = 2 \\ 2x - 3y + 2z = 4 \\ 4x + y - 3z = 1 \end{cases}$

14. $\begin{cases} x + y + z = 4 \\ -x + 2y + 3z = 17 \\ 2x - y = -7 \end{cases}$

15. $\begin{cases} x_1 + 2x_2 - x_3 = 9 \\ 2x_1 \quad - x_3 = -2 \\ 3x_1 + 5x_2 + 2x_3 = 22 \end{cases}$

16. $\begin{cases} 2x_1 + x_2 = 7 \\ 2x_1 - x_2 + x_3 = 6 \\ 3x_1 - 2x_2 + 4x_3 = 11 \end{cases}$

17. $\begin{cases} 2x - 3y - z = 13 \\ -x + 2y - 5z = 6 \\ 5x - y - z = 49 \end{cases}$

18. $\begin{cases} 10x + 10y - 20z = 60 \\ 15x + 20y + 30z = -25 \\ -5x + 30y - 10z = 45 \end{cases}$

19. $\begin{cases} 0.1x + y - 0.2z = 0.6 \\ -0.2x + 1.1y + 0.6z = -1.6 \\ 0.3x + 0.2y + z = -1.4 \end{cases}$

20. $\begin{cases} \frac{1}{2}x + \frac{1}{3}y - \frac{1}{6}z = 8 \\ x - \frac{2}{3}y + \frac{1}{3}z = -4 \\ -\frac{1}{3}x + \frac{1}{2}y - z = 10 \end{cases}$

21. $\begin{cases} 3x + y + z = \frac{3}{2} \\ 3x + 12z = -5 \\ 2y - 4z = 4 \end{cases}$

22. $\begin{cases} x - y + 2z = 1.9 \\ 5x - 6y + z = 8 \\ 7x + y - 2z = -1.1 \end{cases}$

23. $\begin{cases} x + y - z - w = 6 \\ 2x + z - 3w = 8 \\ x - y + 4w = -10 \\ 3x + 5y - z - w = 20 \end{cases}$

24. $\begin{cases} -x + 2y + z - 3w = 3 \\ 3x - 4y + z + w = 9 \\ -x - y + z + w = 0 \\ 2x + y + 4z - 2w = 3 \end{cases}$

25. $\begin{cases} x_1 - x_2 + x_3 + 2x_4 + 3x_5 = 0 \\ -x_1 - 2x_2 + x_3 - 2x_4 + x_5 = 7 \\ -x_1 + x_2 + x_4 - x_5 = -4 \\ 2x_1 - 2x_2 + 3x_3 - x_4 = 12 \\ x_1 + x_3 - x_4 - 5x_5 = 5 \end{cases}$

26.
$$\begin{cases} x + y + z + w + u + v = 12 \\ y - z \quad + u - v = -1 \\ 2x \quad - 2z + 4w \quad - 4v = -6 \\ 3y - z \quad + \quad v = 4 \\ x - y + z - w + u - v = 0 \\ -x - y + z + w \quad = 2 \end{cases}$$

27.
$$\begin{cases} x + y = -2 \\ 2y + z = 1 \\ x - 3z = -20 \end{cases}$$

28.
$$\begin{cases} 3x + 5z = 56 \\ 4y - 2z = 14 \\ 7x + 4y = 77 \end{cases}$$

29.
$$\begin{cases} x_1 + 7x_3 = -20 \\ 2x_1 - 5x_2 = 7 \\ -3x_2 + x_3 = 0 \end{cases}$$

30.
$$\begin{cases} x_1 + x_2 - x_3 = 0 \\ x_1 + 3x_4 = 13 \\ 3x_2 - 2x_3 = 0 \\ 2x_1 + 5x_3 = 17 \end{cases}$$

31. A doctor recommends that one of her patients takes 50 mg each of niacin, riboflavin, and thiamin daily to help alleviate a deficiency. Looking into his medicine chest at home, the patient finds three brands of vitamin pills. The amounts of the relevant vitamins per pill are given in the table.

	VitaMax	**Vitron**	**VitaPlus**
Niacin (mg)	5	10	15
Riboflavin (mg)	15	20	0
Thiamin (mg)	10	10	10

How many pills of each type should he take every day to fulfill the doctor's prescription?

32. A chemist has three containers of acid solution at various concentrations. The first is 10% acid, the second is 20%, and the third is 40%. How many milliliters of each should he mix together to make 100 mL of acid at 18% concentration, if he has to use four times as much of the 10% solution as the 40% solution?

33. The drawer of a cash register contains 30 coins (pennies, nickels, dimes, and quarters). The total value of the coins is $3.31. The total number of pennies and nickels combined is the same as the total number of dimes and quarters combined. The total value of the quarters is five times the total value of the dimes. How many coins of each type are there?

34. A small school has 100 students who occupy three classrooms: rooms A, B, and C. After the first period of the school day, half the students in room A move to room B, one fifth of the students in room B move to room C, and one third of the students in room C move to room A. Nevertheless, the total number of students in each room remains the same after this shift. How many students are there in each room?

35. A hotel offers three classes of accommodation: standard, deluxe, and first-class rooms. A group of ten employees of a manufacturing company attend a trade convention and stay in this hotel. If six of them take standard rooms, two take deluxe, and two take first-class, the total hotel bill will be $530 per day. If five stay in standard rooms, four in deluxe, and only one in a first-class room, the bill will decrease to $510 per day. If they splurge and have three stay in standard rooms, three in deluxe, and four in first-class rooms, the bill will be $645 per day. How much is the daily rate for each type of room?

36. Amanda, Bryce, and Corey enter a race in which they have to run, swim, and cycle over a marked course. Amanda runs at an average speed of 10 mi/h, swims at 4 mi/h, and cycles at 20 mi/h during this race. Bryce runs at $7\frac{1}{2}$ mi/h, swims at 6 mi/h, and cycles at 15 mi/h. Corey runs at 15 mi/h, swims at 3 mi/h, and cycles at 40 mi/h. Corey finishes first with a total time of 1 h 45 min. Amanda comes in second with a time of 2 h 30 min. Bryce finishes last with a time of 3 h. How many miles long is each part of the race?

37. Determine a, b, and c so that the graph of the parabola $y = ax^2 + bx + c$ passes through the points $(-2, 24)$, $(1, 3)$, and $(3, 9)$.

38. Determine a, b, c, and d so that the points $(1, 1)$, $(2, 45)$, $(-1, -3)$, and $(-3, 225)$ all lie on the graph of the function $f(x) = ax^4 + bx^2 + cx + d$.

In Exercises 39 and 40 solve the system.

39. $5x + 2y = 4x - z = 4y + 3z = 1$

40. $6x + 2y = 2z - 2y = 3w - 7y = x + z = 2$

SECTION 8.3
INCONSISTENT AND DEPENDENT SYSTEMS

All the systems of linear equations that we considered in the last section had one unique solution for each of the unknowns. But as we saw in Section 8.1, a system of two linear equations in two variables can have one solution, no solution, or infinitely many solutions. The same cases arise when we study linear systems with more equations and more variables. To solve a general system of linear equations, we use Gaussian elimination to reduce the matrix that represents the system to a special form, which we now describe.

ECHELON FORM OF A MATRIX

A matrix is in **echelon form** if it has the following properties:

1. The first nonzero number in each row (reading from left to right) is a 1. This is called the **leading entry** of the row.
2. The leading entry in each row is to the right of the leading entry in the row immediately above it.
3. Rows that consist entirely of 0's are at the bottom of the matrix.

For example, the matrix on the left below is in echelon form. However, the matrix on the right is not, since the leading entries of successive rows do not step down to the right.

$$\begin{bmatrix} 1 & 3 & -6 & 10 & 0 \\ 0 & 0 & 1 & 4 & -3 \\ 0 & 0 & 0 & 1 & \frac{1}{2} \\ 0 & 0 & 0 & 0 & 0 \end{bmatrix} \qquad \begin{bmatrix} 0 & 1 & -\frac{1}{2} & 0 & 7 \\ 1 & 0 & 3 & 4 & -5 \\ 0 & 0 & 0 & 1 & 0.4 \\ 0 & 1 & 1 & 0 & 0 \end{bmatrix}$$

If the matrix form of a system of equations is in echelon form, we call the variables that correspond to the leading entries the **leading variables** of the system. For example, the matrix in echelon form on the left above corresponds to the system

$$\begin{cases} x + 3y - 6z + 10w = 0 \\ \qquad\qquad z + 4w = -3 \\ \qquad\qquad\qquad w = \frac{1}{2} \\ \qquad\qquad\qquad 0 = 0 \end{cases}$$

In this system, x, z, and w are leading variables, but y is not a leading variable. In Example 1 we see how to determine when a system has no solution.

EXAMPLE 1

Solve the system

$$\begin{cases} x - 3y + 2z = 12 \\ 2x - 5y + 5z = 14 \\ x - 2y + 3z = 20 \end{cases}$$

SOLUTION

$$\begin{bmatrix} 1 & -3 & 2 & 12 \\ 2 & -5 & 5 & 14 \\ 1 & -2 & 3 & 20 \end{bmatrix} \xrightarrow[R_3 \to R_3 - R_1]{R_2 \to R_2 - 2R_1} \begin{bmatrix} 1 & -3 & 2 & 12 \\ 0 & 1 & 1 & -10 \\ 0 & 1 & 1 & 8 \end{bmatrix}$$

$$\xrightarrow{R_3 \to R_3 - R_2} \begin{bmatrix} 1 & -3 & 2 & 12 \\ 0 & 1 & 1 & -10 \\ 0 & 0 & 0 & 18 \end{bmatrix} \xrightarrow{\frac{1}{18}R_3} \begin{bmatrix} 1 & -3 & 2 & 12 \\ 0 & 1 & 1 & -10 \\ 0 & 0 & 0 & 1 \end{bmatrix}$$

This is in echelon form, so we may stop the Gaussian elimination process. Now if we translate the last row back into equation form, we get $0x + 0y + 0z = 1$, or $0 = 1$, which is false. No matter what values we pick for x, y, and z, the last equation will never be a true statement. This means the system *has no solution*. ■

A system that has no solution is said to be **inconsistent**. The procedure we used to show that the system in Example 1 is inconsistent works in general. If we use Gaussian elimination to change a system to echelon form, and if one of the equations we end up with is false, then the system is inconsistent. (The false equation will always have the form $0 = c$, where c is nonzero.)

The next example shows what happens when we apply Gaussian elimination to a system with infinitely many solutions.

EXAMPLE 2

Find the complete solution of the following system.

$$\begin{cases} -3x - 5y + 36z = 10 \\ -x + 7z = 5 \\ x + y - 10z = -4 \end{cases}$$

SOLUTION

$$\begin{bmatrix} -3 & -5 & 36 & 10 \\ -1 & 0 & 7 & 5 \\ 1 & 1 & -10 & -4 \end{bmatrix} \xrightarrow{R_1 \leftrightarrow R_3} \begin{bmatrix} 1 & 1 & -10 & -4 \\ -1 & 0 & 7 & 5 \\ -3 & -5 & 36 & 10 \end{bmatrix}$$

$$\xrightarrow[R_3 \to R_3 + 3R_1]{R_2 \to R_2 + R_1} \begin{bmatrix} 1 & 1 & -10 & -4 \\ 0 & 1 & -3 & 1 \\ 0 & -2 & 6 & -2 \end{bmatrix} \xrightarrow{R_3 \to R_3 + 2R_2} \begin{bmatrix} 1 & 1 & -10 & -4 \\ 0 & 1 & -3 & 1 \\ 0 & 0 & 0 & 0 \end{bmatrix}$$

The system is now in echelon form, so we stop using Gaussian elimination. Translating the last row back into an equation, we get

$$0x + 0y + 0z = 0$$

or

$$0 = 0$$

This equation is always true, no matter what x, y, and z are. Since the equation adds no new information about the variables, we can drop it from the system, which we now write in the form

$$\begin{bmatrix} 1 & 1 & -10 & -4 \\ 0 & 1 & -3 & 1 \end{bmatrix}$$

This corresponds to the system

$$\begin{cases} x + y - 10z = -4 \\ y - 3z = 1 \end{cases}$$

Neither of these equations determines a value for z, but we can use them to express the leading variables x and y in terms of z. From the last equation we get

$$y = 3z + 1$$

Substituting this value for y into the first equation gives us

$$x + (3z + 1) - 10z = -4$$
$$x - 7z + 1 = -4$$
$$x = 7z - 5$$

Since no value is determined for z, we can get a solution to the system by letting z be any real number and then using the above equations to calculate x and y. For example, if $z = 1$, then

$$x = 7z - 5 = 7(1) - 5 = 2$$

and

$$y = 3z + 1 = 3(1) + 1 = 4$$

Thus $(2, 4, 1)$ is a solution to the system. We would get a different solution if we let $z = 2$ because then

$$x = 7z - 5 = 7(2) - 5 = 9$$

and

$$y = 3z + 1 = 3(2) + 1 = 7$$

So (9, 7, 2) is also a solution. There are infinitely many solutions because z can be given any value. We write the complete solution as follows:

$$x = 7z - 5$$
$$y = 3z + 1$$
$$z = \text{any real number} \qquad \blacksquare$$

A system with infinitely many solutions is called **dependent**. In the complete solution to such a system, the variables that are *not* leading variables will be arbitrary, and the leading variables will *depend* on the arbitrary one(s). In Example 2, the leading variables x and y depended on (that is, were expressed in terms of) the nonleading variable z. If we use Gaussian elimination to convert a dependent system to echelon form and then discard any equations of the form $0 = 0$, we end up with a system that has fewer equations than variables. Example 2 ended up with only two equations in the three variables x, y, and z. In general, if we arrive at n equations in m variables ($m > n$) after this process, the complete solution will have $m - n$ arbitrary (nonleading) variables, and the n leading variables will be expressed in terms of these.

The following box summarizes what we have learned about systems with no solution, one solution, or infinitely many solutions.

SOLVING A SYSTEM IN ECHELON FORM

Suppose the matrix that represents a system of linear equations has been transformed by Gaussian elimination into echelon form.

1. If the echelon form contains a row that represents the equation $0 = c$, where c is nonzero, then the system has *no solution*.
2. If each of the variables in the echelon form of the system is a leading variable, then the system has *exactly one solution*, which we find using back substitution.
3. If not all of the variables in the echelon form are leading variables, then the system has *infinitely many solutions*. To solve the system, we use back substitution to express the leading variables in terms of the others.

Sometimes it is convenient to continue using Gaussian elimination on a matrix in echelon form to change each of the numbers *above* each leading entry to 0. The matrix is then said to be in *reduced echelon form*.

REDUCED ECHELON FORM OF A MATRIX

A matrix is in **reduced echelon form** if it is in echelon form and every number above and below each leading entry is 0.

The matrix on the left below is in reduced echelon form, but the one on the right is not, since the leading entries in the second and third rows do not have only 0's above them.

$$\begin{bmatrix} 1 & 3 & 0 & 0 & 0 \\ 0 & 0 & 1 & 0 & -3 \\ 0 & 0 & 0 & 1 & \frac{1}{2} \\ 0 & 0 & 0 & 0 & 0 \end{bmatrix} \qquad \begin{bmatrix} 1 & 3 & -\frac{1}{2} & 2 & 0 \\ 0 & 0 & 1 & 0 & -3 \\ 0 & 0 & 0 & 1 & \frac{1}{2} \\ 0 & 0 & 0 & 0 & 0 \end{bmatrix}$$

In the next example, we use the reduced echelon form to solve a system.

EXAMPLE 3

Find the complete solution of the system

$$\begin{cases} x + 2y - 3z - 4w = 10 \\ x + 3y - 3z - 4w = 15 \\ 2x + 2y - 6z - 8w = 10 \end{cases}$$

SOLUTION

$$\begin{bmatrix} 1 & 2 & -3 & -4 & 10 \\ 1 & 3 & -3 & -4 & 15 \\ 2 & 2 & -6 & -8 & 10 \end{bmatrix} \xrightarrow[\;R_3 \to R_3 - 2R_1\;]{R_2 \to R_2 - R_1} \begin{bmatrix} 1 & 2 & -3 & -4 & 10 \\ 0 & 1 & 0 & 0 & 5 \\ 0 & -2 & 0 & 0 & -10 \end{bmatrix}$$

$$\xrightarrow{R_3 \to R_3 + 2R_2} \begin{bmatrix} 1 & 2 & -3 & -4 & 10 \\ 0 & 1 & 0 & 0 & 5 \\ 0 & 0 & 0 & 0 & 0 \end{bmatrix} \xrightarrow{R_1 \to R_1 - 2R_2} \begin{bmatrix} 1 & 0 & -3 & -4 & 0 \\ 0 & 1 & 0 & 0 & 5 \\ 0 & 0 & 0 & 0 & 0 \end{bmatrix}$$

This is in reduced echelon form. Since the last row represents the equation $0 = 0$ we may discard it, and we end up with the system

$$\begin{bmatrix} 1 & 0 & -3 & -4 & 0 \\ 0 & 1 & 0 & 0 & 5 \end{bmatrix}$$

At this stage we have two equations in four unknowns, so the system is dependent.

The leading variables are x and y, and the arbitrary variables are z and w. From the second equation, $y = 5$, and from the first,

$$x - 3z - 4w = 0$$
$$x = 3z + 4w$$ ■

Note that we did not have to use back substitution in Example 3. If a system is in reduced echelon form, the solution can be determined directly, without back substitution.

Note also that z and w do *not* necessarily have to be the *same* real number in the solution for Example 3. We can choose arbitrary values for each if we wish to construct a specific solution to the system. For example, if we let $z = 1$ and we let $w = 2$, we get the solution $(11, 5, 1, 2)$. You should check that this does indeed satisfy all three of the original equations in Example 3.

Many of the linear systems that arise in practical problems turn out to be inconsistent or dependent. Both situations arise in the next two examples.

EXAMPLE 4

A biologist is performing an experiment on the effects of various combinations of vitamins. She wishes to feed each of her laboratory rabbits a diet that contains exactly 9 mg of niacin, 14 mg of thiamin, and 32 mg of riboflavin. She has available three different types of commercial rabbit pellets whose content of the relevant vitamins per ounce is given in the following table.

	Type A	Type B	Type C
Niacin (mg)	2	3	1
Thiamin (mg)	3	1	3
Riboflavin (mg)	8	5	7

How many ounces of each type of food should she give each rabbit daily to satisfy the experiment's requirements?

SOLUTION

If we let x represent the amount of type A to be fed to each rabbit, y the amount of type B, and z the amount of type C, then the daily requirements she has established lead to the linear equations

$$\begin{cases} 2x + 3y + \ z = \ 9 \\ 3x + \ y + 3z = 14 \\ 8x + 5y + 7z = 32 \end{cases}$$

We solve this system as follows.

$$
\begin{bmatrix} 2 & 3 & 1 & 9 \\ 3 & 1 & 3 & 14 \\ 8 & 5 & 7 & 32 \end{bmatrix}
\xrightarrow{R_2 \to R_2 - R_1}
\begin{bmatrix} 2 & 3 & 1 & 9 \\ 1 & -2 & 2 & 5 \\ 8 & 5 & 7 & 32 \end{bmatrix}
\xrightarrow[R_3 \to R_3 - 8R_2]{R_1 \to R_1 - 2R_2}
$$

$$
\begin{bmatrix} 0 & 7 & -3 & -1 \\ 1 & -2 & 2 & 5 \\ 0 & 21 & -9 & -8 \end{bmatrix}
\xrightarrow[R_1 \leftrightarrow R_2]{R_3 \to R_3 - 3R_1}
\begin{bmatrix} 1 & -2 & 2 & 5 \\ 0 & 7 & -3 & -1 \\ 0 & 0 & 0 & -5 \end{bmatrix}
$$

Since the last row translates into the equation $0 = -5$, which is false, we need go no further. The system has no solution, so no combination of the three foods will give the required vitamin combination. ■

EXAMPLE 5

Suppose that the biologist in Example 4 had specified 37 mg instead of 32 mg as the riboflavin requirement, but that all the other aspects of the experiment remained unchanged. Would there now be a combination of the three foods that would satisfy the requirements?

SOLUTION

The only change we need to make in the solution to Example 4 is to replace the 32 in the original system of equations by 37. If we then carry out the same row operations as before, we arrive at the matrix

$$
\begin{bmatrix} 1 & -2 & 2 & 5 \\ 0 & 7 & -3 & -1 \\ 0 & 0 & 0 & 0 \end{bmatrix}
$$

We continue using Gaussian elimination to put this into reduced echelon form.

$$
\xrightarrow{\frac{1}{7}R_2}
\begin{bmatrix} 1 & -2 & 2 & 5 \\ 0 & 1 & -\frac{3}{7} & -\frac{1}{7} \\ 0 & 0 & 0 & 0 \end{bmatrix}
\xrightarrow{R_1 \to R_1 + 2R_2}
\begin{bmatrix} 1 & 0 & \frac{8}{7} & \frac{33}{7} \\ 0 & 1 & -\frac{3}{7} & -\frac{1}{7} \\ 0 & 0 & 0 & 0 \end{bmatrix}
$$

The last equation now simply states that $0 = 0$ and so can be eliminated. The system has infinitely many solutions, with the leading variables x and y depending on the arbitrary variable z. The solution is

$$
\begin{cases} x = -\frac{8}{7}z + \frac{33}{7} \\ y = \frac{3}{7}z - \frac{1}{7} \\ z = \text{any real number} \end{cases}
$$

Because an amount of food cannot be negative, not every solution of the system provides a practical solution of the problem.

Since $y \geqslant 0$,

$$y = \tfrac{3}{7}z - \tfrac{1}{7} \geqslant 0$$
$$\tfrac{3}{7}z \geqslant \tfrac{1}{7}$$
$$z \geqslant \tfrac{1}{3}$$

Since $x \geqslant 0$,

$$x = -\tfrac{8}{7}z + \tfrac{33}{7} \geqslant 0$$
$$-\tfrac{8}{7}z \geqslant -\tfrac{33}{7}$$
$$z \leqslant \tfrac{33}{8}$$

This means that the solution to the problem would have to include the condition that the amount z of type C rabbit food used should be between $\tfrac{1}{3}$ oz and $\tfrac{33}{8}$ oz. ∎

EXERCISES 8.3

In Exercises 1–6 determine whether the matrix is in echelon form. If it is, determine whether it is in reduced echelon form.

1. $\begin{bmatrix} 1 & 0 & -3 \\ 0 & 1 & 5 \end{bmatrix}$ 2. $\begin{bmatrix} 1 & 3 & -3 \\ 0 & 1 & 5 \end{bmatrix}$

3. $\begin{bmatrix} 2 & 0 & 8 & 0 \\ 0 & 1 & 3 & 2 \\ 0 & 0 & 0 & 0 \end{bmatrix}$

4. $\begin{bmatrix} 1 & 0 & -7 & 0 \\ 0 & 1 & 3 & 0 \\ 0 & 0 & 0 & 1 \end{bmatrix}$

5. $\begin{bmatrix} 1 & 0 & 0 & 0 \\ 0 & 0 & 0 & 0 \\ 0 & 1 & 5 & 1 \end{bmatrix}$

6. $\begin{bmatrix} 1 & 0 & 0 & 1 \\ 0 & 1 & 0 & 2 \\ 0 & 0 & 1 & 3 \end{bmatrix}$

In Exercises 7–26 find the complete solution to each system of equations, or show that none exists.

7. $\begin{cases} x - y + 3z = 3 \\ 4x - 8y + 32z = 24 \\ 2x - 3y + 11z = 4 \end{cases}$ 8. $\begin{cases} -2x + 6y - 2z = -12 \\ x - 3y + 2z = 10 \\ -x + 3y + 2z = 6 \end{cases}$

9. $\begin{cases} x + 5y = 12 \\ 3x - 7y = 14 \\ 2x - 4y = 10 \end{cases}$ 10. $\begin{cases} 12x - 7y = 11 \\ -3x - 14y = 12 \\ 15x + 8y = 13 \end{cases}$

11. $\begin{cases} x - 3y = 1 \\ 3x - y = 5 \\ 4x - 8y = 3 \end{cases}$ 12. $\begin{cases} x + 2y + 3z = 7 \\ 3x + 2y + z = 21 \end{cases}$

13. $\begin{cases} x - y - z = 0 \\ 4x - 3y + 8z = 12 \end{cases}$ 14. $\begin{cases} 3x - 6y - 12z = 0 \\ -4x + 8y + 16z = 0 \end{cases}$

15. $\begin{cases} 2x - y + 5z = 12 \\ x + 4y - 2z = -3 \\ 8x + 5y + 11z = 30 \end{cases}$ 16. $\begin{cases} 3r + 2s - 3t = 10 \\ r - s - t = -5 \\ r + 4s - t = 20 \end{cases}$

17. $\begin{cases} 2x + y - 2z = 12 \\ -x - \tfrac{1}{2}y + z = -6 \\ 3x + \tfrac{3}{2}y - 3z = 18 \end{cases}$ 18. $\begin{cases} y - 5z = 7 \\ 3x + 2y = 12 \\ 3x + 10z = 80 \end{cases}$

19. $\begin{cases} x + y + z + w = 8 \\ y - w = 0 \\ 3x + 2y + z = 12 \\ -3x - 2y + z + 4w = 0 \end{cases}$

20. $\begin{cases} y - z + 2w = 0 \\ 3x + 2y + w = 0 \\ 2x + 4w = 12 \\ -2x - 2z + 5w = 6 \end{cases}$

21. $\begin{cases} 2x - y + 2z + w = 5 \\ -x + y + 4z - w = 3 \\ 3x - 2y - z = 0 \end{cases}$

22. $\begin{cases} 3t - u + v + 2w = 5 \\ t + u - v - w = 7 \\ 4t - 4u + 4v + 6w = 3 \end{cases}$

23. $\begin{cases} x - y \quad\ + w = 0 \\ 3x \qquad - z + 2w = 0 \\ x - 4y + z + 2w = 0 \end{cases}$

24. $\begin{cases} 3x_1 - 2x_2 + 4x_3 = -2 \\ x_1 - 2x_2 + x_3 = 0 \\ 4x_1 - 4x_2 + 5x_3 = -2 \\ \quad\ -4x_2 - x_3 = 2 \end{cases}$

25. $\begin{cases} 2x - y + z = 5 \\ 3x - 4y - 2z = 1 \\ x - 2y + 4z = 9 \\ 2x - 3y + 5z = 0 \end{cases}$

26. $\begin{cases} a + b + c + d + e = 2 \\ a \quad\ - c \quad\ + e = 2 \\ -2a + b \quad\ - d \quad\ = 0 \\ \quad\ 2b \qquad + 2e = 4 \end{cases}$

27. A nutritionist wishes to make a milk substitute by combining soya powder, ground millet, and nonfat dried milk powder with enough water to make 1 qt. She wants the mixture to contain 1.1 mg of thiamin, 3.1 mg of riboflavin, and 3.5 mg of niacin. The amounts of these nutrients per ounce of each substance are given in the following table.

	Soya powder	Ground millet	Dried milk
Thiamin (mg)	0.2	0.5	0.4
Riboflavin (mg)	0.2	2.0	1.4
Niacin (mg)	1.0	1.0	1.0

How many ounces of each food should she combine to satisfy her requirements for these nutrients? (Give all possible combinations.)

28. If the nutritionist of Exercise 27 decides she wants her product to have 1.2 mg of thiamin instead of 1.1 mg (without changing the other requirements), what combination of the three foods could she use?

29. A furniture factory makes wooden tables, chairs, and armoires. Each piece of furniture requires three production steps: cutting the wood, assembling, and finishing. The number of hours of each operation required to make a piece of furniture is given by the following table.

	Table	Chair	Armoire
Cutting (h)	$\frac{1}{2}$	1	1
Assembling (h)	$\frac{1}{2}$	$1\frac{1}{2}$	1
Finishing (h)	1	$1\frac{1}{2}$	2

The workers in the plant can provide 300 labor-hours of cutting, 400 h of assembling, and 590 h of finishing each week. How many tables, chairs, and armoires should be produced so that all available labor-hours are used? Or is this impossible?

30. Rework Exercise 29 assuming that one worker has been laid off, so that only 550 h of finishing labor are available each week.

31. I have some pennies, nickels, and dimes in my pocket. The total value of the coins is 72 cents, and the number of dimes is one-third of the total number of nickels and pennies. How many coins of each denomination do I have? [*Hint:* The number of each type of coin must be a nonnegative integer.]

32. A diagram of part of the network of streets in a city is shown in the figure, where the arrows indicate one-way streets. The numbers on the diagram show how many cars enter or leave this section of the city via the indicated street in a certain one-hour period. The variables x, y, z, and w represent the number of cars that travel along the portions of First, Second, Avocado, and Birch Streets shown in the figure during this period. Find x, y, z, and w, assuming that none of the cars involved in this problem stop or park on any of the streets shown in the diagram.

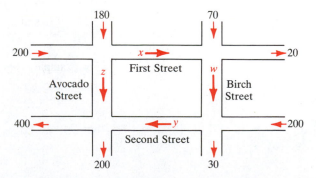

33. (a) Suppose that (x_0, y_0, z_0) and (x_1, y_1, z_1) are solutions of the system

$$\begin{cases} a_1 x + b_1 y + c_1 z = d_1 \\ a_2 x + b_2 y + c_2 z = d_2 \\ a_3 x + b_3 y + c_3 z = d_3 \end{cases}$$

Show that $\left(\dfrac{x_0 + x_1}{2}, \dfrac{y_0 + y_1}{2}, \dfrac{z_0 + z_1}{2} \right)$ is also a solution.

(b) Use the result of part (a) to prove that if the system has two different solutions, it has infinitely many.

SECTION 8.4
ALGEBRA OF MATRICES

Up to this point we have been using matrices simply as a notational convenience, to make our work in solving linear equations a little easier. Matrices have many other uses in mathematics and the sciences, and for most of these applications a knowledge of matrix algebra is essential. Like numbers, matrices can be added, subtracted, multiplied, and divided under certain circumstances, and in this section we learn how to perform these algebraic operations on matrices.

Recall that a matrix is simply a rectangular array of numbers enclosed between brackets. For example, let A be the matrix

$$\begin{bmatrix} -1 & 4 & 7 & 0 \\ 0 & 2 & 13 & 14 \\ \frac{1}{2} & 22 & 8 & -2 \end{bmatrix}$$

The **dimension** of a matrix is a pair of numbers that indicates how many rows and columns a matrix has. The matrix A is a 3×4 matrix because it has 3 rows and 4 columns. The individual numbers that make up a matrix are called its **entries**, and they are specified by their row and column position. In the above matrix A, the number 13 is the $(2, 3)$ entry, since it is in the second row and the third column. If the name of a matrix is A, we will often use the symbol a_{ij} to denote the (i, j) entry of the matrix. Thus for the above matrix, $a_{24} = 14$ and $a_{32} = 22$.

Two matrices are **equal** if they have the same dimension and their corresponding entries are equal. So

> $A = B$ if and only if both A and B have dimension $m \times n$ and $a_{ij} = b_{ij}$ for $i = 1, 2, \ldots, m$, and $j = 1, 2, \ldots, n$.

For example,

$$\begin{bmatrix} \sqrt{4} & 2^2 & 0 & e^0 \\ 0.5 & 1 & 0 & 1-1 \end{bmatrix} = \begin{bmatrix} 2 & 4 & 0 & 1 \\ \frac{1}{2} & \frac{2}{2} & 0 & 0 \end{bmatrix}$$

but

$$\begin{bmatrix} 1 & 2 \\ 3 & 4 \\ 5 & 6 \end{bmatrix} \neq \begin{bmatrix} 1 & 3 & 5 \\ 2 & 4 & 6 \end{bmatrix}$$

Addition and subtraction of matrices

Two matrices can be added or subtracted whenever they have the same dimension. (Otherwise their sum or difference is undefined.) If this is the case, we add or subtract the matrices by adding or subtracting corresponding entries. Thus we have the sum

$$\begin{bmatrix} 2 & -3 \\ 0 & 5 \\ 7 & -\frac{1}{2} \end{bmatrix} + \begin{bmatrix} 1 & 0 \\ -3 & 1 \\ 2 & 2 \end{bmatrix} = \begin{bmatrix} 3 & -3 \\ -3 & 6 \\ 9 & \frac{3}{2} \end{bmatrix}$$

because both the matrices being added have dimensions 3×2. The difference

$$\begin{bmatrix} 7 & -3 & 0 \\ 0 & 1 & 5 \end{bmatrix} - \begin{bmatrix} 6 & 0 & -6 \\ 8 & 1 & 9 \end{bmatrix} = \begin{bmatrix} 1 & -3 & 6 \\ -8 & 0 & -4 \end{bmatrix}$$

is also defined, since the matrices being subtracted are both 2×3. But the result of the operation

$$\begin{bmatrix} 7 & -3 & 0 \\ 0 & 1 & 5 \end{bmatrix} + \begin{bmatrix} 2 & -3 \\ 0 & 5 \\ 7 & -\frac{1}{2} \end{bmatrix}$$

is undefined, since we cannot take the sum of a 2×3 and a 3×2 matrix.

Multiplication of a matrix by a number

We can multiply a number times a matrix by multiplying every entry in the matrix by that number. For example,

$$5 \begin{bmatrix} 2 & -3 \\ 0 & 5 \\ 7 & -\frac{1}{2} \end{bmatrix} = \begin{bmatrix} 10 & -15 \\ 0 & 25 \\ 35 & -\frac{5}{2} \end{bmatrix}$$

Matrix multiplication

Multiplication of two matrices is not quite so easy to describe. We will see in later examples why taking the matrix product involves the following rather complex procedure.

First of all, the product AB (or $A \cdot B$) of two matrices A and B is defined only when the number of columns in A is equal to the number of rows in B. This means that if we write their dimensions side by side, the two inner numbers must match:

matrices	A	B
dimensions	$m \times n$	$n \times k$
	↑	↑
	columns in A	rows in B

If the dimensions of A and B match in this fashion, then the product AB will have dimension $m \times k$. Before describing the procedure for obtaining the elements of AB, we define the **inner product** of a row of A and a column of B.

If $[a_1 \quad a_2 \quad \cdots \quad a_n]$ is a row of A, and if $\begin{bmatrix} b_1 \\ b_2 \\ \vdots \\ b_n \end{bmatrix}$ is a column of B, then

their **inner product** is the number $a_1b_1 + a_2b_2 + \cdots + a_nb_n$.

For example,

$$[2 \quad -1 \quad 0 \quad 4] \cdot \begin{bmatrix} 5 \\ 4 \\ -3 \\ \frac{1}{2} \end{bmatrix} = 2 \cdot 5 + (-1) \cdot 4 + 0 \cdot (-3) + 4 \cdot \tfrac{1}{2} = 8$$

We now define the **product** AB of two matrices as follows.

THE PRODUCT OF TWO MATRICES

Suppose that A is an $m \times n$ matrix and B an $n \times k$ matrix. Then $C = AB$ is an $m \times k$ matrix, where c_{ij} is the inner product of the ith row of A and the jth column of B.

EXAMPLE 1

Let

$$A = \begin{bmatrix} 1 & 3 \\ -1 & 0 \end{bmatrix} \quad \text{and} \quad B = \begin{bmatrix} -1 & 5 & 2 \\ 0 & 4 & 7 \end{bmatrix}$$

Calculate, if possible, the products AB and BA.

SOLUTION

Since A has dimensions 2×2 and B has dimensions 2×3, the product AB will have dimension 2×3. We can thus write

$$AB = \begin{bmatrix} 1 & 3 \\ -1 & 0 \end{bmatrix} \begin{bmatrix} -1 & 5 & 2 \\ 0 & 4 & 7 \end{bmatrix} = \begin{bmatrix} ? & ? & ? \\ ? & ? & ? \end{bmatrix}$$

where the question marks must be filled in using the rule defining the entries of a matrix product. The $(1, 1)$ entry will be the inner product of the first row of A and the first column of B:

$$\begin{bmatrix} 1 & 3 \\ -1 & 0 \end{bmatrix} \begin{bmatrix} -1 & 5 & 2 \\ 0 & 4 & 7 \end{bmatrix} \qquad 1 \cdot (-1) + 3 \cdot 0 = -1$$

Olga Taussky-Todd *(b. 1906) is famous for her work in Number Theory. She is also one of the world's leaders in developing applications of Matrix Theory. She has successfully applied matrices to the study of aerodynamics. Taussky-Todd is professor (emerita) of mathematics at the California Institute of Technology in Pasadena, California.*

Similarly, we calculate the remaining entries as follows:

Entry:	*Inner Product of:*	*Value:*
$(1, 2)$	$\begin{bmatrix} 1 & 3 \\ -1 & 0 \end{bmatrix} \begin{bmatrix} -1 & 5 & 2 \\ 0 & 4 & 7 \end{bmatrix}$	$1 \cdot 5 + 3 \cdot 4 = 17$
$(1, 3)$	$\begin{bmatrix} 1 & 3 \\ -1 & 0 \end{bmatrix} \begin{bmatrix} -1 & 5 & 2 \\ 0 & 4 & 7 \end{bmatrix}$	$1 \cdot 2 + 3 \cdot 7 = 23$
$(2, 1)$	$\begin{bmatrix} 1 & 3 \\ -1 & 0 \end{bmatrix} \begin{bmatrix} -1 & 5 & 2 \\ 0 & 4 & 7 \end{bmatrix}$	$(-1) \cdot (-1) + 0 \cdot 0 = 1$
$(2, 2)$	$\begin{bmatrix} 1 & 3 \\ -1 & 0 \end{bmatrix} \begin{bmatrix} -1 & 5 & 2 \\ 0 & 4 & 7 \end{bmatrix}$	$(-1) \cdot 5 + 0 \cdot 4 = -5$
$(2, 3)$	$\begin{bmatrix} 1 & 3 \\ -1 & 0 \end{bmatrix} \begin{bmatrix} -1 & 5 & 2 \\ 0 & 4 & 7 \end{bmatrix}$	$(-1) \cdot 2 + 0 \cdot 7 = -2$

Thus we have

$$AB = \begin{bmatrix} -1 & 17 & 23 \\ 1 & -5 & -2 \end{bmatrix}$$

The product BA is not defined, however, because the dimensions are

$$2 \times 3 \quad \text{and} \quad 2 \times 2$$

The inner two numbers are not the same, so the rows and columns will not match up when we try to calculate the product. ■

The next example shows that even when both AB and BA are defined, they are not necessarily equal. This will prove that matrix multiplication is *not* commutative.

EXAMPLE 2

Let

$$A = \begin{bmatrix} 5 & 7 \\ -3 & 0 \end{bmatrix} \quad \text{and} \quad B = \begin{bmatrix} 1 & 2 \\ 9 & -1 \end{bmatrix}$$

Calculate the products AB and BA.

SOLUTION

Since both A and B are 2×2 matrices, both AB and BA are defined and are also 2×2 matrices.

$$AB = \begin{bmatrix} 5 & 7 \\ -3 & 0 \end{bmatrix} \begin{bmatrix} 1 & 2 \\ 9 & -1 \end{bmatrix} = \begin{bmatrix} 5 \cdot 1 + 7 \cdot 9 & 5 \cdot 2 + 7 \cdot (-1) \\ (-3) \cdot 1 + 0 \cdot 9 & (-3) \cdot 2 + 0 \cdot (-1) \end{bmatrix}$$

$$= \begin{bmatrix} 68 & 3 \\ -3 & -6 \end{bmatrix}$$

$$BA = \begin{bmatrix} 1 & 2 \\ 9 & -1 \end{bmatrix} \begin{bmatrix} 5 & 7 \\ -3 & 0 \end{bmatrix} = \begin{bmatrix} 1 \cdot 5 + 2 \cdot (-3) & 1 \cdot 7 + 2 \cdot 0 \\ 9 \cdot 5 + (-1) \cdot (-3) & 9 \cdot 7 + (-1) \cdot 0 \end{bmatrix}$$

$$= \begin{bmatrix} -1 & 7 \\ 48 & 63 \end{bmatrix}$$

This shows that, in general, $AB \neq BA$. In fact, in this example, AB and BA do not even have any entries in common. ∎

Although matrix multiplication is not commutative, it does obey the associative and distributive laws. That is, if $A, B, C,$ and D are matrices for which the products below are defined, we have

PROPERTIES OF MATRIX MULTIPLICATION

$A(BC) = (AB)C$ (associativity)

$A(B + C) = AB + AC,$ $(B + C)D = BD + CD$ (distributivity)

The next two examples give some indication of why mathematicians chose to define the matrix product in such an apparently bizarre fashion.

EXAMPLE 3

Show that the matrix equation

$$\begin{bmatrix} 1 & 2 & 4 \\ -1 & 1 & 2 \\ 2 & 3 & 3 \end{bmatrix} \begin{bmatrix} x \\ y \\ z \end{bmatrix} = \begin{bmatrix} 7 \\ 5 \\ 7 \end{bmatrix}$$

is equivalent to the system of equations in Example 1 of Section 8.2.

SOLUTION

If we perform the matrix multiplication on the left-hand side of the given equation, we get

$$\begin{bmatrix} x + 2y + 4z \\ -x + y + 2z \\ 2x + 3y + 3z \end{bmatrix} = \begin{bmatrix} 7 \\ 5 \\ 7 \end{bmatrix}$$

Since two matrices are equal if their corresponding entries are equal, this means that

$$x + 2y + 4z = 7$$
$$-x + y + 2z = 5$$
$$2x + 3y + 3z = 7$$

This is exactly the system of equations we had in Example 1 of Section 8.2. ■

The preceding example shows that our definition of matrix product allows us to express a system of linear equations as a single matrix equation in a natural way.

Arthur Cayley (1821–1895) *was an English mathematician who invented matrices and developed Matrix Theory. He practiced law until the age of 42, but his primary interest from adolescence was mathematics, and he published almost 200 papers on the subject in his spare time. In 1863 he accepted the offer of a professorship in mathematics at Cambridge, where he taught until his death. Cayley's work on matrices was of purely theoretical interest in his day, but in the twentieth century many of his results have found application in physics, the social sciences, business, and other fields.*

EXAMPLE 4

In a certain city the proportion of voters in each age group who registered as Democrats, Republicans, or Independents is given by the following matrix.

$$
\begin{array}{c}
\\
\text{Democrat} \\
\text{Republican} \\
\text{Independent}
\end{array}
\begin{array}{c}
Age \\
\begin{array}{ccc}
18\text{--}30 & 31\text{--}50 & \text{Over } 50
\end{array} \\
\begin{bmatrix}
0.30 & 0.60 & 0.50 \\
0.50 & 0.35 & 0.25 \\
0.20 & 0.05 & 0.25
\end{bmatrix}
\end{array} = A
$$

The next matrix gives the distribution, by age and sex, of the voting population of this city.

$$
Age
\begin{array}{c}
18\text{--}30 \\
31\text{--}50 \\
\text{Over } 50
\end{array}
\begin{array}{c}
\begin{array}{cc}
\text{Male} & \text{Female}
\end{array} \\
\begin{bmatrix}
5{,}000 & 6{,}000 \\
10{,}000 & 12{,}000 \\
12{,}000 & 15{,}000
\end{bmatrix}
\end{array} = B
$$

For the purposes of this problem, let us make the (highly unrealistic) assumption that within each age group, political preference is not related to gender. That is, the percentage of Democrat males in the 18–30 group, for example, is the same as the percentage of Democrat females in this group.
(a) Calculate the product AB.
(b) How many male Democrats are there in this city?
(c) How many female Republicans are there?

SOLUTION

(a) $$AB = \begin{bmatrix} 0.30 & 0.60 & 0.50 \\ 0.50 & 0.35 & 0.25 \\ 0.20 & 0.05 & 0.25 \end{bmatrix} \begin{bmatrix} 5{,}000 & 6{,}000 \\ 10{,}000 & 12{,}000 \\ 12{,}000 & 15{,}000 \end{bmatrix} = \begin{bmatrix} 13{,}500 & 16{,}500 \\ 9{,}000 & 10{,}950 \\ 4{,}500 & 5{,}550 \end{bmatrix}$$

(b) When we take the inner product of a row from A with a column from B, we are adding the number of people in each of the three age groups who belong to the category in question. For example, the $(2, 1)$ entry of AB (9,000) was obtained by taking the inner product of the "Republican" row from A with the "male" column from B. This number is therefore the total number of male Republicans in this city. We can label the rows and columns of AB as follows:

$$\begin{array}{c} \\ \text{Democrat} \\ \text{Republican} \\ \text{Independent} \end{array} \begin{array}{cc} \text{Male} & \text{Female} \\ \begin{bmatrix} 13{,}500 & 16{,}500 \\ 9{,}000 & 10{,}950 \\ 4{,}500 & 5{,}550 \end{bmatrix} \end{array} = AB$$

There are 13,500 male Democrats in this city.

(c) There are 10,950 female Republicans.   ∎

If you add the entries in the columns of matrix A in Example 4, you will see that in each case the sum is 1. (Can you see why this has to be true, given what the matrix is describing?) A matrix with this property is called **stochastic**. Stochastic matrices are studied extensively in statistics, where they arise frequently in situations like the one described in Example 4.

EXERCISES 8.4

In Exercises 1–21 the matrices A, B, C, D, E, F, and G are defined as follows:

$$A = \begin{bmatrix} 2 & -5 \\ 0 & 7 \end{bmatrix} \qquad B = \begin{bmatrix} 3 & \frac{1}{2} & 5 \\ 1 & -1 & 3 \end{bmatrix}$$

$$C = \begin{bmatrix} 2 & -\frac{5}{2} & 0 \\ 0 & 2 & -3 \end{bmatrix} \qquad D = \begin{bmatrix} 7 & 3 \end{bmatrix}$$

$$E = \begin{bmatrix} 0 & 0 & 0 & 0 & 0 \\ 0 & 0 & 0 & 0 & 0 \\ 0 & 0 & 0 & 0 & 0 \end{bmatrix} \qquad F = \begin{bmatrix} 1 & 0 & 0 \\ 0 & 1 & 0 \\ 0 & 0 & 1 \end{bmatrix}$$

$$G = \begin{bmatrix} 5 & -3 & 10 \\ 6 & 1 & 0 \\ -5 & 2 & 2 \\ 0 & 0 & 0 \end{bmatrix}$$

Carry out the algebraic operation in each exercise, or explain why it cannot be performed.

1. $B + C$ **2.** $B + F$ **3.** $C - B$

4. $5A$

5. $3B + 2C$

6. $C - 5A$

7. $2C - 6B$

8. DA

9. AD

10. BC

11. BF

12. GF

13. $(DA)B$

14. $D(AB)$

15. GE

16. A^2

17. A^3

18. $DB + DC$

19. B^2

20. F^2

21. $BF + FE$

22. What must be true about the dimensions of the matrices A and B if both products AB and BA are defined?

In Exercises 23–26 write the system of equations as a matrix equation. (See Example 3.)

23. $\begin{cases} 2x - 5y = 7 \\ 3x + 2y = 4 \end{cases}$

24. $\begin{cases} 6x - y + z = 12 \\ 2x \quad\ + z = 7 \\ \quad y - 2z = 4 \end{cases}$

25. $\begin{cases} 3x_1 + 2x_2 - x_3 + x_4 = 0 \\ x_1 \quad\quad - x_3 \quad\quad = 5 \\ \quad\quad 3x_2 + x_3 - x_4 = 4 \end{cases}$

26.
$$\begin{cases} x - y + z = 2 \\ 4x - 2y - z = 2 \\ x + y + 5z = 2 \\ -x - y - z = 2 \end{cases}$$

In Exercises 27–30 solve the matrix equation for the unknown matrix X, or explain why there is no solution. Here

$$A = \begin{bmatrix} 4 & 6 \\ 1 & 3 \end{bmatrix} \qquad B = \begin{bmatrix} 2 & 5 \\ 3 & 7 \end{bmatrix}$$

$$C = \begin{bmatrix} 2 & 3 \\ 1 & 0 \\ 0 & 2 \end{bmatrix} \qquad D = \begin{bmatrix} 10 & 20 \\ 30 & 20 \\ 10 & 0 \end{bmatrix}$$

27. $2X - A = B$

28. $5(X - C) = D$

29. $3X + B = C$

30. $A + D = 3X$

31. Let O represent the 2×2 **zero matrix**:

$$O = \begin{bmatrix} 0 & 0 \\ 0 & 0 \end{bmatrix}$$

If A and B are 2×2 matrices with $AB = O$, is it necessarily true that $A = O$ or $B = O$?

32. Prove that if A and B are 2×2 matrices, then

$$(A + B)^2 = A^2 + AB + BA + B^2$$

33. If A and B are 2×2 matrices, is it necessarily true that

$$(A + B)^2 \overset{?}{=} A^2 + 2AB + B^2$$

34. Let

$$A = \begin{bmatrix} 1 & 1 \\ 0 & 1 \end{bmatrix}$$

(a) Calculate A^2, A^3, and A^4.

(b) Find a general formula for A^n.

35. Let

$$A = \begin{bmatrix} 1 & 1 \\ 1 & 1 \end{bmatrix}$$

(a) Calculate A^2, A^3, and A^4.

(b) Find a general formula for A^n.

36. A small fast-food chain has restaurants in Santa Monica, Long Beach, and Anaheim. Only hamburgers, hot dogs, and milk shakes are sold by this chain. On a certain day, sales were distributed according to the following matrix.

Number of items sold

	Santa Monica	Long Beach	Anaheim	
Hamburgers	$\lceil$ 4000	1000	3500 $\rceil$	
Hot dogs	400	300	200	$= A$
Milk shakes	$\lfloor$ 700	500	900 $\rfloor$	

The price of each item is given by the matrix

Hamburger	Hot dog	Milk shake	
[$0.90	$0.80	$1.10]	$= B$

(a) Calculate the product BA.

(b) Interpret the entries in the product matrix BA.

37. A specialty car manufacturer has plants in Auburn, Biloxi, and Chattanooga. Three models are produced, with daily production given in the following matrix.

Cars produced each day

	Model K	Model R	Model W	
Auburn	$\lceil$ 12	10	0 $\rceil$	
Biloxi	4	4	20	$= A$
Chattanooga	$\lfloor$ 8	9	12 $\rfloor$	

Because of a wage increase, February profits are lower than January profits. The profit per car is tabulated in the matrix below.

	January	February	
Model K	$\lceil$ $1000	$500 $\rceil$	
Model R	$2000	$1200	$= B$
Model W	$\lfloor$ $1500	$1000 $\rfloor$	

(a) Calculate AB.

(b) Assuming all cars produced were sold, what was the daily profit in January from the Biloxi plant?

(c) What was the total daily profit (from all three plants) in February?

38. Let

$$A = \begin{bmatrix} 1 & 0 & 6 & -1 \\ 2 & \frac{1}{2} & 4 & 0 \end{bmatrix}$$

$$B = \begin{bmatrix} 1 & 7 & -9 & 2 \end{bmatrix}$$

$$C = \begin{bmatrix} 1 \\ 0 \\ -1 \\ -2 \end{bmatrix}$$

Determine which of the following products are defined, and calculate the ones that are:

ABC	ACB	BAC
BCA	CAB	CBA

SECTION 8.5
INVERSES OF MATRICES AND MATRIX EQUATIONS

We have seen in the preceding section that matrices can, when the dimensions are appropriate, be added, subtracted, and multiplied. In this section we will investigate division of matrices, which will allow us to solve equations that involve matrices.

First, we define **identity matrices**, which play the same role for matrix multiplication that the number 1 does for ordinary multiplication of numbers; that is, $1 \cdot a = a \cdot 1 = a$ for all numbers a. The term **main diagonal** in this definition refers to the entries of a square matrix whose row and column numbers are the same. (Note that these entries stretch diagonally down the matrix, from top left to bottom right.)

> The **identity matrix** I_n is the $n \times n$ matrix for which each main diagonal entry is a 1, and for which all other entries are 0.

Thus the 2×2, 3×3, and 4×4 identity matrices are, respectively,

$$I_2 = \begin{bmatrix} 1 & 0 \\ 0 & 1 \end{bmatrix} \qquad I_3 = \begin{bmatrix} 1 & 0 & 0 \\ 0 & 1 & 0 \\ 0 & 0 & 1 \end{bmatrix} \qquad I_4 = \begin{bmatrix} 1 & 0 & 0 & 0 \\ 0 & 1 & 0 & 0 \\ 0 & 0 & 1 & 0 \\ 0 & 0 & 0 & 1 \end{bmatrix}$$

Identity matrices behave like the number 1 in the sense that

$$A \cdot I_n = A \qquad \text{and} \qquad I_n \cdot B = B$$

whenever these products are defined. Thus multiplication by an identity of the appropriate size leaves a matrix unchanged. For example, one can verify by direct calculation that

$$\begin{bmatrix} 1 & 0 \\ 0 & 1 \end{bmatrix} \begin{bmatrix} 3 & 5 & 6 \\ -1 & 2 & 7 \end{bmatrix} = \begin{bmatrix} 3 & 5 & 6 \\ -1 & 2 & 7 \end{bmatrix}$$

or that

$$\begin{bmatrix} -1 & 7 & \frac{1}{2} \\ 12 & 1 & 3 \\ -2 & 0 & 7 \end{bmatrix} \begin{bmatrix} 1 & 0 & 0 \\ 0 & 1 & 0 \\ 0 & 0 & 1 \end{bmatrix} = \begin{bmatrix} -1 & 7 & \frac{1}{2} \\ 12 & 1 & 3 \\ -2 & 0 & 7 \end{bmatrix}$$

If A and B are $n \times n$ matrices, and if $AB = BA = I_n$, then we say that B is the **inverse** of A, and we write $B = A^{-1}$. The concept of the inverse of a matrix is analogous to that of the reciprocal of a real number. The following rule allows us to calculate the inverse of a 2×2 matrix.

INVERSE OF A 2 × 2 MATRIX

If $A = \begin{bmatrix} a & b \\ c & d \end{bmatrix}$ then $A^{-1} = \dfrac{1}{ad - bc} \begin{bmatrix} d & -b \\ -c & a \end{bmatrix}$

EXAMPLE 1

Let

$$A = \begin{bmatrix} 4 & 5 \\ 2 & 3 \end{bmatrix}$$

Find A^{-1} and verify that $AA^{-1} = A^{-1}A = I_2$.

SOLUTION

Using the rule, we get

$$A^{-1} = \frac{1}{4 \cdot 3 - 5 \cdot 2} \begin{bmatrix} 3 & -5 \\ -2 & 4 \end{bmatrix} = \frac{1}{2} \begin{bmatrix} 3 & -5 \\ -2 & 4 \end{bmatrix} = \begin{bmatrix} \frac{3}{2} & -\frac{5}{2} \\ -1 & 2 \end{bmatrix}$$

To verify that this is indeed the inverse of A, we calculate

$$AA^{-1} = \begin{bmatrix} 4 & 5 \\ 2 & 3 \end{bmatrix} \begin{bmatrix} \frac{3}{2} & -\frac{5}{2} \\ -1 & 2 \end{bmatrix} = \begin{bmatrix} 4 \cdot \frac{3}{2} + 5(-1) & 4\left(-\frac{5}{2}\right) + 5 \cdot 2 \\ 2 \cdot \frac{3}{2} + 3(-1) & 2\left(-\frac{5}{2}\right) + 3 \cdot 2 \end{bmatrix} = \begin{bmatrix} 1 & 0 \\ 0 & 1 \end{bmatrix}$$

and

$$A^{-1}A = \begin{bmatrix} \frac{3}{2} & -\frac{5}{2} \\ -1 & 2 \end{bmatrix} \begin{bmatrix} 4 & 5 \\ 2 & 3 \end{bmatrix} = \begin{bmatrix} \frac{3}{2} \cdot 4 + \left(-\frac{5}{2}\right)2 & \frac{3}{2} \cdot 5 + \left(-\frac{5}{2}\right)3 \\ (-1)4 + 2 \cdot 2 & (-1)5 + 2 \cdot 3 \end{bmatrix} = \begin{bmatrix} 1 & 0 \\ 0 & 1 \end{bmatrix}$$

■

The quantity $ad - bc$ that appears in the rule for calculating the inverse is called the **determinant** of the matrix. If the determinant is 0, then the matrix will not have an inverse (since we cannot divide by 0). In the next section we will learn how to calculate the determinant of a square matrix of any size, and how to use determinants to solve systems of equations.

■ INVERSES OF $n \times n$ MATRICES

For 3×3 and larger square matrices, the following technique provides the most efficient way to calculate the inverse. If A is an $n \times n$ matrix, we begin by constructing the $n \times 2n$ matrix which has the entries of A on the left and of the identity matrix I_n on the right:

$$\begin{bmatrix} a_{11} & a_{12} & \cdots & a_{1n} & 1 & 0 & \cdots & 0 \\ a_{21} & a_{22} & \cdots & a_{2n} & 0 & 1 & \cdots & 0 \\ \vdots & \vdots & \ddots & \vdots & \vdots & \vdots & \ddots & \vdots \\ a_{n1} & a_{n2} & \cdots & a_{nn} & 0 & 0 & \cdots & 1 \end{bmatrix}$$

We then use the elementary row operations on this new large matrix to change the left side into the identity matrix. The right side will be automatically transformed into A^{-1}. (We omit the proof of this fact.)

EXAMPLE 2

Find the inverse of the matrix

$$A = \begin{bmatrix} 1 & -2 & -4 \\ 2 & -3 & -6 \\ -3 & 6 & 15 \end{bmatrix}$$

and verify that $AA^{-1} = A^{-1}A = I_3$.

SOLUTION

We begin with the 3×6 matrix whose left half is A and whose right half is the identity matrix.

$$\begin{bmatrix} 1 & -2 & -4 & 1 & 0 & 0 \\ 2 & -3 & -6 & 0 & 1 & 0 \\ -3 & 6 & 15 & 0 & 0 & 1 \end{bmatrix}$$

We then transform the left half of this new matrix into the identity matrix by performing the following sequence of elementary row operations on the *entire* new matrix:

$$\xrightarrow[\begin{subarray}{l} R_2 \to R_2 - 2R_1 \\ R_3 \to R_3 + 3R_1 \end{subarray}]{} \begin{bmatrix} 1 & -2 & -4 & 1 & 0 & 0 \\ 0 & 1 & 2 & -2 & 1 & 0 \\ 0 & 0 & 3 & 3 & 0 & 1 \end{bmatrix}$$

$$\xrightarrow{\frac{1}{3}R_3} \begin{bmatrix} 1 & -2 & -4 & 1 & 0 & 0 \\ 0 & 1 & 2 & -2 & 1 & 0 \\ 0 & 0 & 1 & 1 & 0 & \frac{1}{3} \end{bmatrix}$$

$$\xrightarrow{R_1 \to R_1 + 2R_2} \begin{bmatrix} 1 & 0 & 0 & -3 & 2 & 0 \\ 0 & 1 & 2 & -2 & 1 & 0 \\ 0 & 0 & 1 & 1 & 0 & \frac{1}{3} \end{bmatrix}$$

$$\xrightarrow{R_2 \to R_2 - 2R_3} \begin{bmatrix} 1 & 0 & 0 & -3 & 2 & 0 \\ 0 & 1 & 0 & -4 & 1 & -\frac{2}{3} \\ 0 & 0 & 1 & 1 & 0 & \frac{1}{3} \end{bmatrix}$$

We have now transformed the left half of this matrix into the identity matrix. Note that to do this in as systematic a fashion as possible, we first changed the elements below the main diagonal to zeros, just as we would if we were doing Gaussian elimination. We then changed the main diagonal elements to ones by multiplying by the appropriate constant(s). Finally, we completed the process by changing the remaining entries on the left side to zeros. The right half is now A^{-1}.

$$A^{-1} = \begin{bmatrix} -3 & 2 & 0 \\ -4 & 1 & -\frac{2}{3} \\ 1 & 0 & \frac{1}{3} \end{bmatrix}$$

To verify this, we multiply

$$AA^{-1} = \begin{bmatrix} 1 & -2 & -4 \\ 2 & -3 & -6 \\ -3 & 6 & 15 \end{bmatrix} \begin{bmatrix} -3 & 2 & 0 \\ -4 & 1 & -\frac{2}{3} \\ 1 & 0 & \frac{1}{3} \end{bmatrix} = \begin{bmatrix} 1 & 0 & 0 \\ 0 & 1 & 0 \\ 0 & 0 & 1 \end{bmatrix}$$

and $\quad A^{-1}A = \begin{bmatrix} -3 & 2 & 0 \\ -4 & 1 & -\frac{2}{3} \\ 1 & 0 & \frac{1}{3} \end{bmatrix} \begin{bmatrix} 1 & -2 & -4 \\ 2 & -3 & -6 \\ -3 & 6 & 15 \end{bmatrix} = \begin{bmatrix} 1 & 0 & 0 \\ 0 & 1 & 0 \\ 0 & 0 & 1 \end{bmatrix}$ ■

The next example shows that not every square matrix has an inverse.

EXAMPLE 3

Try to find the inverse of the matrix

$$\begin{bmatrix} 2 & -3 & -7 \\ 1 & 2 & 7 \\ 1 & 1 & 4 \end{bmatrix}$$

SOLUTION

We proceed as follows.

$$\begin{bmatrix} 2 & -3 & -7 & 1 & 0 & 0 \\ 1 & 2 & 7 & 0 & 1 & 0 \\ 1 & 1 & 4 & 0 & 0 & 1 \end{bmatrix} \xrightarrow{R_1 \leftrightarrow R_2} \begin{bmatrix} 1 & 2 & 7 & 0 & 1 & 0 \\ 2 & -3 & -7 & 1 & 0 & 0 \\ 1 & 1 & 4 & 0 & 0 & 1 \end{bmatrix}$$

$$\xrightarrow[R_3 \to R_3 - R_1]{R_2 \to R_2 - 2R_1} \begin{bmatrix} 1 & 2 & 7 & 0 & 1 & 0 \\ 0 & -7 & -21 & 1 & -2 & 0 \\ 0 & -1 & -3 & 0 & -1 & 1 \end{bmatrix}$$

$$\xrightarrow{-\frac{1}{7}R_2} \begin{bmatrix} 1 & 2 & 7 & 0 & 1 & 0 \\ 0 & 1 & 3 & -\frac{1}{7} & \frac{2}{7} & 0 \\ 0 & -1 & -3 & 0 & -1 & 1 \end{bmatrix}$$

$$\xrightarrow[R_1 \to R_1 - 2R_2]{R_3 \to R_3 + R_2} \begin{bmatrix} 1 & 0 & 1 & \frac{2}{7} & \frac{3}{7} & 0 \\ 0 & 1 & 3 & -\frac{1}{7} & \frac{2}{7} & 0 \\ 0 & 0 & 0 & -\frac{1}{7} & -\frac{5}{7} & 1 \end{bmatrix}$$

At this point we would like to change the 0 in the (3, 3) position of this matrix to a 1, without changing the zeros in the (3, 1) and (3, 2) positions. But there is no way to accomplish this because no matter what multiple of rows 1 and/or 2 we add to row 3, we cannot change the third zero in row 3 without changing the first or second as well. Thus we cannot change the left half to the identity matrix. The original matrix does not have an inverse. ■

If we encounter a row of zeros on the left when trying to find an inverse, then the original matrix does not have an inverse.

MATRIX EQUATIONS

We saw in Section 8.4 that a system of linear equations can be written as a single matrix equation. For example, the system

$$\begin{cases} x + 2y + 4z = 7 \\ -x + y + 2z = 5 \\ 2x + 3y + 3z = 7 \end{cases}$$

is equivalent to the matrix equation

$$\begin{bmatrix} 1 & 2 & 4 \\ -1 & 1 & 2 \\ 2 & 3 & 3 \end{bmatrix} \begin{bmatrix} x \\ y \\ z \end{bmatrix} = \begin{bmatrix} 7 \\ 5 \\ 7 \end{bmatrix}$$

(See Example 3, Section 8.4.)

If we let

$$A = \begin{bmatrix} 1 & 2 & 4 \\ -1 & 1 & 2 \\ 2 & 3 & 3 \end{bmatrix} \quad X = \begin{bmatrix} x \\ y \\ z \end{bmatrix} \quad B = \begin{bmatrix} 7 \\ 5 \\ 7 \end{bmatrix}$$

then this matrix equation can be written

$$AX = B \tag{1}$$

This has the same form as, for example, the following simple real number equation:

$$3x = 12$$

We solve this latter equation by multiplying both sides by the reciprocal (or inverse) of 3:

$$\tfrac{1}{3}(3x) = \tfrac{1}{3}(12)$$
$$x = 4$$

Similarly, we can solve Equation 1 by multiplying both sides by the inverse of A (provided this inverse exists):

$$AX = B$$
$$A^{-1}(AX) = A^{-1}B$$
$$(A^{-1}A)X = A^{-1}B$$
$$I_3X = A^{-1}B$$
$$X = A^{-1}B \tag{2}$$

In this example,

$$A^{-1} = \frac{1}{9}\begin{bmatrix} 3 & -6 & 0 \\ -7 & 5 & 6 \\ 5 & -1 & -3 \end{bmatrix}$$

(Verify!), so that Equation 2 becomes

$$\begin{bmatrix} x \\ y \\ z \end{bmatrix} = \frac{1}{9}\begin{bmatrix} 3 & -6 & 0 \\ -7 & 5 & 6 \\ 5 & -1 & -3 \end{bmatrix}\begin{bmatrix} 7 \\ 5 \\ 7 \end{bmatrix}$$
$$= \frac{1}{9}\begin{bmatrix} -9 \\ 18 \\ 9 \end{bmatrix} = \begin{bmatrix} -1 \\ 2 \\ 1 \end{bmatrix}$$

Thus $x = -1$, $y = 2$, and $z = 1$ is the solution to the original system. (Compare this with the solution to Example 1 in Section 8.2.)

EXAMPLE 4

Solve the following matrix equation:

$$\begin{bmatrix} 2 & -5 \\ 3 & -6 \end{bmatrix} \begin{bmatrix} x \\ y \end{bmatrix} = \begin{bmatrix} 15 \\ 36 \end{bmatrix}$$

SOLUTION

Using the rule for calculating the inverse of a 2×2 matrix, we get

$$\begin{bmatrix} 2 & -5 \\ 3 & -6 \end{bmatrix}^{-1} = \frac{1}{2(-6) - (-5)3} \begin{bmatrix} -6 & -(-5) \\ -3 & 2 \end{bmatrix} = \frac{1}{3} \begin{bmatrix} -6 & 5 \\ -3 & 2 \end{bmatrix}$$

Multiplying both sides of the equation by this inverse matrix, we get

$$\begin{bmatrix} x \\ y \end{bmatrix} = \frac{1}{3} \begin{bmatrix} -6 & 5 \\ -3 & 2 \end{bmatrix} \begin{bmatrix} 15 \\ 36 \end{bmatrix} = \begin{bmatrix} 30 \\ 9 \end{bmatrix}$$

So $x = 30$ and $y = 9$. ■

EXAMPLE 5

A pet store owner feeds his hamsters and gerbils different mixtures of three types of rodent food pellets, which we will call brands A, B, and C. He wishes to be sure to feed his animals the correct amount of each brand to satisfy exactly their optimal daily requirements for protein, fat, and carbohydrates. Suppose that hamsters need 340 mg of protein, 280 mg of fat, and 440 mg of carbohydrates, and gerbils need 480 mg of protein, 360 mg of fat, and 680 mg of carbohydrates each day. How many grams of each food should the storekeeper feed his hamsters and gerbils daily if the amounts of these nutrients in one gram of each brand are given in the following table?

	Brand A	Brand B	Brand C
Protein (mg)	10	0	20
Fat (mg)	10	20	10
Carbohydrates (mg)	5	10	30

SOLUTION

If we let x_1, x_2, and x_3 be the grams of brands A, B, and C, respectively, that the hamsters should eat, and if we let y_1, y_2, and y_3 be the corresponding amounts for the gerbils, then we want to solve the matrix equations

$$\begin{bmatrix} 10 & 0 & 20 \\ 10 & 20 & 10 \\ 5 & 10 & 30 \end{bmatrix} \begin{bmatrix} x_1 \\ x_2 \\ x_3 \end{bmatrix} = \begin{bmatrix} 340 \\ 280 \\ 440 \end{bmatrix} \tag{3}$$

and
$$\begin{bmatrix} 10 & 0 & 20 \\ 10 & 20 & 10 \\ 5 & 10 & 30 \end{bmatrix} \begin{bmatrix} y_1 \\ y_2 \\ y_3 \end{bmatrix} = \begin{bmatrix} 480 \\ 360 \\ 680 \end{bmatrix} \qquad (4)$$

Since the coefficient matrix on the left is the same in both of these equations, we can solve each one by multiplying both sides by the inverse of this matrix. We therefore begin by finding this inverse.

$$\begin{bmatrix} 10 & 0 & 20 & 1 & 0 & 0 \\ 10 & 20 & 10 & 0 & 1 & 0 \\ 5 & 10 & 30 & 0 & 0 & 1 \end{bmatrix} \xrightarrow{2R_3} \begin{bmatrix} 10 & 0 & 20 & 1 & 0 & 0 \\ 10 & 20 & 10 & 0 & 1 & 0 \\ 10 & 20 & 60 & 0 & 0 & 2 \end{bmatrix}$$

$$\xrightarrow[R_3 \to R_3 - R_1]{R_2 \to R_2 - R_1} \begin{bmatrix} 10 & 0 & 20 & 1 & 0 & 0 \\ 0 & 20 & -10 & -1 & 1 & 0 \\ 0 & 20 & 40 & -1 & 0 & 2 \end{bmatrix}$$

$$\xrightarrow{R_3 \to R_3 - R_2} \begin{bmatrix} 10 & 0 & 20 & 1 & 0 & 0 \\ 0 & 20 & -10 & -1 & 1 & 0 \\ 0 & 0 & 50 & 0 & -1 & 2 \end{bmatrix}$$

$$\xrightarrow{\frac{1}{5}R_3} \begin{bmatrix} 10 & 0 & 20 & 1 & 0 & 0 \\ 0 & 20 & -10 & -1 & 1 & 0 \\ 0 & 0 & 10 & 0 & -\frac{1}{5} & \frac{2}{5} \end{bmatrix}$$

$$\xrightarrow[R_1 \to R_1 - 2R_3]{R_2 \to R_2 + R_3} \begin{bmatrix} 10 & 0 & 0 & 1 & \frac{2}{5} & -\frac{4}{5} \\ 0 & 20 & 0 & -1 & \frac{4}{5} & \frac{2}{5} \\ 0 & 0 & 10 & 0 & -\frac{1}{5} & \frac{2}{5} \end{bmatrix}$$

$$\xrightarrow{\frac{1}{10}R_1, \frac{1}{20}R_2, \frac{1}{10}R_3} \begin{bmatrix} 1 & 0 & 0 & 0.10 & 0.04 & -0.08 \\ 0 & 1 & 0 & -0.05 & 0.04 & 0.02 \\ 0 & 0 & 1 & 0 & -0.02 & 0.04 \end{bmatrix}$$

So
$$\begin{bmatrix} 10 & 0 & 20 \\ 10 & 20 & 10 \\ 5 & 10 & 30 \end{bmatrix}^{-1} = \frac{1}{100} \begin{bmatrix} 10 & 4 & -8 \\ -5 & 4 & 2 \\ 0 & -2 & 4 \end{bmatrix}$$

and if we now multiply both sides of Equations 3 and 4 by this inverse matrix, we get

$$\begin{bmatrix} x_1 \\ x_2 \\ x_3 \end{bmatrix} = \frac{1}{100} \begin{bmatrix} 10 & 4 & -8 \\ -5 & 4 & 2 \\ 0 & -2 & 4 \end{bmatrix} \begin{bmatrix} 340 \\ 280 \\ 440 \end{bmatrix} = \begin{bmatrix} 10 \\ 3 \\ 12 \end{bmatrix}$$

and
$$\begin{bmatrix} y_1 \\ y_2 \\ y_3 \end{bmatrix} = \frac{1}{100} \begin{bmatrix} 10 & 4 & -8 \\ -5 & 4 & 2 \\ 0 & -2 & 4 \end{bmatrix} \begin{bmatrix} 480 \\ 360 \\ 680 \end{bmatrix} = \begin{bmatrix} 8 \\ 4 \\ 20 \end{bmatrix}$$

This means that each hamster should be fed 10 g of brand A, 3 g of brand B, and 12 g of brand C, whereas each gerbil should be fed 8 g of brand A, 4 g of brand B, and 20 g of brand C daily.   ∎

Since a lot of work is usually involved in finding the inverse of a 3×3 or larger matrix, the method used in Example 5 is actually useful only when we are solving several systems of equations with the same coefficient matrix. However, if we have access to a calculator or computer program that calculates matrix inverses, then this becomes the preferred method under all circumstances.

EXERCISES 8.5

In Exercises 1 and 2 find the inverse of the given matrix and verify that $A^{-1}A = AA^{-1} = I_2$ and $B^{-1}B = BB^{-1} = I_3$.

1. $A = \begin{bmatrix} 7 & 4 \\ 3 & 2 \end{bmatrix}$

2. $B = \begin{bmatrix} 1 & 3 & 2 \\ 0 & 2 & 2 \\ -2 & -1 & 0 \end{bmatrix}$

In Exercises 3–18 find the inverse of the matrix, if it exists.

3. $\begin{bmatrix} 3 & 7 \\ 2 & 5 \end{bmatrix}$

4. $\begin{bmatrix} 3 & 5 \\ 4 & 7 \end{bmatrix}$

5. $\begin{bmatrix} 2 & 5 \\ -5 & -13 \end{bmatrix}$

6. $\begin{bmatrix} -7 & 4 \\ 8 & -5 \end{bmatrix}$

7. $\begin{bmatrix} 6 & -3 \\ -8 & 4 \end{bmatrix}$

8. $\begin{bmatrix} \frac{1}{2} & \frac{1}{3} \\ 5 & 4 \end{bmatrix}$

9. $\begin{bmatrix} 0.4 & -1.2 \\ 0.3 & 0.6 \end{bmatrix}$

10. $\begin{bmatrix} 4 & 2 & 3 \\ 3 & 3 & 2 \\ 1 & 0 & 1 \end{bmatrix}$

11. $\begin{bmatrix} 2 & 4 & 1 \\ -1 & 1 & -1 \\ 1 & 4 & 0 \end{bmatrix}$

12. $\begin{bmatrix} 5 & 7 & 4 \\ 3 & -1 & 3 \\ 6 & 7 & 5 \end{bmatrix}$

13. $\begin{bmatrix} 1 & 2 & 3 \\ 4 & 5 & -1 \\ 1 & -1 & -10 \end{bmatrix}$

14. $\begin{bmatrix} 2 & 1 & 0 \\ 1 & 1 & 4 \\ 2 & 1 & 2 \end{bmatrix}$

15. $\begin{bmatrix} 0 & -2 & 2 \\ 3 & 1 & 3 \\ 1 & -2 & 3 \end{bmatrix}$

16. $\begin{bmatrix} 3 & -2 & 0 \\ 5 & 1 & 1 \\ 2 & -2 & 0 \end{bmatrix}$

17. $\begin{bmatrix} 1 & 2 & 0 & 3 \\ 0 & 1 & 1 & 1 \\ 0 & 1 & 0 & 1 \\ 1 & 2 & 0 & 2 \end{bmatrix}$

18. $\begin{bmatrix} 1 & 0 & 1 & 0 \\ 0 & 1 & 0 & 1 \\ 1 & 1 & 1 & 0 \\ 1 & 1 & 1 & 1 \end{bmatrix}$

In Exercises 19–26 solve the system of equations by converting to a matrix equation and using the inverse of the coefficient matrix, as in Example 4. Use the inverses from Exercises 3–6, 11, 12, 16, and 17.

19. $\begin{cases} 3x + 7y = 4 \\ 2x + 5y = 0 \end{cases}$

20. $\begin{cases} 3x + 5y = 10 \\ 4x + 7y = 20 \end{cases}$

21. $\begin{cases} 2x + 5y = 2 \\ -5x - 13y = 20 \end{cases}$

22. $\begin{cases} -7x + 4y = 0 \\ 8x - 5y = 100 \end{cases}$

23. $\begin{cases} 2x + 4y + z = 7 \\ -x + y - z = 0 \\ x + 4y = -2 \end{cases}$

24. $\begin{cases} 5x + 7y + 4z = 1 \\ 3x - y + 3z = 1 \\ 6x + 7y + 5z = 1 \end{cases}$

25. $\begin{cases} 3x - 2y = 6 \\ 5x + y + z = 12 \\ 2x - 2y = 18 \end{cases}$

26. $\begin{cases} x + 2y + 3w = 0 \\ y + z + w = 1 \\ y + w = 2 \\ x + 2y + 2w = 3 \end{cases}$

In Exercises 27 and 28 solve the matrix equation by multiplying both sides by the appropriate inverse matrix.

27. $\begin{bmatrix} 3 & -2 \\ -4 & 3 \end{bmatrix} \begin{bmatrix} x & y & z \\ u & v & w \end{bmatrix} = \begin{bmatrix} 1 & 0 & -1 \\ 2 & 1 & 3 \end{bmatrix}$

28. $\begin{bmatrix} 0 & -2 & 2 \\ 3 & 1 & 3 \\ 1 & -2 & 3 \end{bmatrix} \begin{bmatrix} x & u \\ y & v \\ z & w \end{bmatrix} = \begin{bmatrix} 3 & 6 \\ 6 & 12 \\ 0 & 0 \end{bmatrix}$

29. A nutritionist is studying the effects of the nutrients folic acid, choline, and inositol. He has three different types of food available, which contain the following amounts of these nutrients per ounce:

	Type A	Type B	Type C
Folic acid (mg)	3	1	3
Choline (mg)	4	2	4
Inositol (mg)	3	2	4

(a) Find the inverse of the matrix

$$\begin{bmatrix} 3 & 1 & 3 \\ 4 & 2 & 4 \\ 3 & 2 & 4 \end{bmatrix}$$

and use it to solve the remaining parts of this problem.

(b) How many ounces of each food should the nutritionist feed his laboratory rats if he wants their diet to contain 10 mg of folic acid, 14 mg of choline, and 13 mg of inositol?

(c) How much of each food should be given to supply 9 mg of folic acid, 12 mg of choline, and 10 mg of inositol?

(d) Is there any combination of these foods that will supply 2 mg of folic acid, 4 mg of choline, and 11 mg of inositol?

30. Refer to Exercise 29. Suppose it is found that food C has been improperly labeled, and actually contains 4 mg of folic acid, 6 mg of choline, and 5 mg of inositol per ounce. Would it still be possible to use matrix inversion to solve (b), (c), and (d) of Exercise 29? Why or why not?

In Exercises 31–34 find the inverse of the matrix.

31. $\begin{bmatrix} 1 & 2e^x \\ e^{-x} & 1 \end{bmatrix}$

32. $\begin{bmatrix} x & 1 \\ -1 & 1/x \end{bmatrix}$

33. $\begin{bmatrix} e^x & -e^{2x} \\ e^{2x} & e^{3x} \end{bmatrix}$

34. $\begin{bmatrix} 1 & e^x & 0 \\ e^x & -e^{2x} & 0 \\ 0 & 0 & 2 \end{bmatrix}$

35. A matrix that has an inverse is called **invertible**. Find two 2×2 invertible matrices whose sum is not invertible.

36. Find the inverse of the matrix

$$\begin{bmatrix} a & 0 & 0 & 0 \\ 0 & b & 0 & 0 \\ 0 & 0 & c & 0 \\ 0 & 0 & 0 & d \end{bmatrix}$$

where $abcd \neq 0$.

SECTION 8.6
DETERMINANTS AND CRAMER'S RULE

If a matrix is **square** (that is, if it has the same number of rows as columns) then we can assign to it a number called its **determinant**. Determinants can be used to solve matrix equations, as we will see later in this section. They are also useful in determining whether a matrix has an inverse, without our actually going through the process of trying to find its inverse.

We denote the determinant of a square matrix A by the symbol $|A|$, and we begin by defining $|A|$ for the simplest cases. If A is a 1×1 matrix, then it has only one entry, and we define the determinant to be the value of that entry; that is, if $A = [a]$, then $|A| = a$. If A is a 2×2 matrix, then

$$A = \begin{bmatrix} a & b \\ c & d \end{bmatrix}$$

and we define the determinant of A to be

$$|A| = \begin{vmatrix} a & b \\ c & d \end{vmatrix} = ad - bc$$

EXAMPLE 1

Evaluate $|A|$ for $A = \begin{bmatrix} 6 & -3 \\ 2 & 3 \end{bmatrix}$

SOLUTION

$$\begin{vmatrix} 6 & -3 \\ 2 & 3 \end{vmatrix} = 6 \cdot 3 - (-3)2 = 18 - (-6) = 24 \qquad \blacksquare$$

Emmy Noether (1882–1935) *was one of the foremost mathematicians of the early twentieth century. Her ground-breaking work in abstract algebra provided much of the foundation for this field, and her work in Invariant Theory was essential in the development of Einstein's theory of general relativity. Although women were not allowed to study at German universities at the time, she audited courses unofficially and went on to receive a doctorate at Erlangen* summa cum laude, *despite the opposition of the academic senate, which declared that women students would "overthrow all academic order." She subsequently taught mathematics at Göttingen, Moscow, and Frankfurt. In 1933 she left Germany to escape Nazi persecution, accepting a position at Bryn Mawr College in suburban Philadelphia. She lectured there and at the Institute for Advanced Study in Princeton, New Jersey, until her untimely death in 1935.*

Note that we can think of the evaluation of a 2×2 determinant as a "cross-product" operation. We take the product of the diagonal from top left to bottom right, and subtract the product from top right to bottom left.

To define the concept of determinant for an arbitrary $n \times n$ matrix, we must first introduce the following terminology.

Let A be an $n \times n$ matrix.

1. The **minor** M_{ij} of the element a_{ij} is the determinant of the matrix obtained by deleting the ith row and jth column of A.

2. The **cofactor** A_{ij} of the element a_{ij} is

$$A_{ij} = (-1)^{i+j}M_{ij}$$

For example, if A is the matrix

$$\begin{bmatrix} 2 & 3 & -1 \\ 0 & 2 & 4 \\ -2 & 5 & 6 \end{bmatrix}$$

then M_{12} is the determinant of the matrix obtained by deleting the first row and second column from A. Thus

$$M_{12} = \begin{vmatrix} 2 & 3 & -1 \\ 0 & 2 & 4 \\ -2 & 5 & 6 \end{vmatrix} = \begin{vmatrix} 0 & 4 \\ -2 & 6 \end{vmatrix} = 0(6) - 4(-2) = 8$$

so

$$A_{12} = (-1)^{1+2}M_{12} = -8$$

Similarly,

$$M_{33} = \begin{vmatrix} 2 & 3 & -1 \\ 0 & 2 & 4 \\ -2 & 5 & 6 \end{vmatrix} = \begin{vmatrix} 2 & 3 \\ 0 & 2 \end{vmatrix} = 2 \cdot 2 - 3 \cdot 0 = 4$$

so

$$A_{33} = (-1)^{3+3} M_{33} = 4$$

Note that the cofactor of a_{ij} is just the minor of a_{ij} multiplied by either 1 or -1, depending on whether $i + j$ is even or odd. Thus in a 3×3 matrix we obtain the cofactor of any element by prefixing its minor with the sign obtained from the following checkerboard pattern.

$$\begin{bmatrix} + & - & + \\ - & + & - \\ + & - & + \end{bmatrix}$$

We are now ready to define the determinant of any square matrix.

THE DETERMINANT OF A SQUARE MATRIX

If A is an $n \times n$ matrix, then the **determinant** of A is obtained by multiplying each element of the first row by its cofactor, and then adding the results. In symbols,

$$|A| = \begin{vmatrix} a_{11} & a_{12} & \cdots & a_{1n} \\ a_{21} & a_{22} & \cdots & a_{2n} \\ \vdots & \vdots & \ddots & \vdots \\ a_{n1} & a_{n2} & \cdots & a_{nn} \end{vmatrix} = a_{11}A_{11} + a_{12}A_{12} + \cdots + a_{1n}A_{1n}$$

EXAMPLE 2

Evaluate the determinant of the matrix

$$A = \begin{bmatrix} 2 & 3 & -1 \\ 0 & 2 & 4 \\ -2 & 5 & 6 \end{bmatrix}$$

SOLUTION

$$|A| = 2\begin{vmatrix} 2 & 4 \\ 5 & 6 \end{vmatrix} - 3\begin{vmatrix} 0 & 4 \\ -2 & 6 \end{vmatrix} + (-1)\begin{vmatrix} 0 & 2 \\ -2 & 5 \end{vmatrix}$$

$$= 2(2 \cdot 6 - 4 \cdot 5) - 3[0 \cdot 6 - 4(-2)] - [0 \cdot 5 - 2(-2)]$$

$$= -16 - 24 - 4$$

$$= -44$$

■

In our definition of the determinant, we used the cofactors of elements in the first row only. This is sometimes called **expanding the determinant by the first row**. In fact, *we can expand the determinant by any row or column in the same way, and obtain the same result*. Although we will not prove this, the next example illustrates this principle.

EXAMPLE 3

Expand the determinant of the matrix A in Example 2 by the second row and by the third column, and show that the value obtained is the same in each case.

SOLUTION

Expanding by the second row, we get

$$|A| = \begin{vmatrix} 2 & 3 & -1 \\ 0 & 2 & 4 \\ -2 & 5 & 6 \end{vmatrix} = -0\begin{vmatrix} 3 & -1 \\ 5 & 6 \end{vmatrix} + 2\begin{vmatrix} 2 & -1 \\ -2 & 6 \end{vmatrix} - 4\begin{vmatrix} 2 & 3 \\ -2 & 5 \end{vmatrix}$$

$$= 0 + 2[2 \cdot 6 - (-1)(-2)] - 4[2 \cdot 5 - 3(-2)]$$

$$= 0 + 20 - 64$$

$$= -44$$

Expanding by the third column gives

$$|A| = -1\begin{vmatrix} 0 & 2 \\ -2 & 5 \end{vmatrix} - 4\begin{vmatrix} 2 & 3 \\ -2 & 5 \end{vmatrix} + 6\begin{vmatrix} 2 & 3 \\ 0 & 2 \end{vmatrix}$$

$$= -[0 \cdot 5 - 2(-2)] - 4[2 \cdot 5 - 3(-2)] + 6(2 \cdot 2 - 3 \cdot 0)$$

$$= -4 - 64 + 24$$

$$= -44$$

In both cases we obtained the same value for the determinant as when we expanded by the first row in Example 2.   ■

The following principle allows us to determine whether a square matrix has an inverse, without actually calculating the inverse. This is one of the most important uses of the determinant in matrix algebra, and is the reason for the name *determinant*.

INVERTIBILITY CRITERION | If A is a square matrix, then A has an inverse if and only if $|A| \neq 0$.

Although we will not prove this fact, we have already seen (in the preceding section) why it is true in the case of 2×2 matrices.

EXAMPLE 4

Show that the matrix A has no inverse, where

$$A = \begin{bmatrix} 1 & 2 & 0 & 4 \\ 0 & 0 & 0 & 3 \\ 5 & 6 & 2 & 6 \\ 2 & 4 & 0 & 9 \end{bmatrix}$$

SOLUTION

We begin by calculating the determinant of A. Since all but one of the elements of the second row is zero, we need to calculate only the cofactor of the 3 if we expand the determinant by the second row.

$$|A| = -0 \cdot A_{21} + 0 \cdot A_{22} - 0 \cdot A_{23} + 3 \cdot A_{24} = 3A_{24}$$

$$= 3 \begin{vmatrix} 1 & 2 & 0 \\ 5 & 6 & 2 \\ 2 & 4 & 0 \end{vmatrix} \qquad \text{(expand this by the third column)}$$

$$= 3(-2) \begin{vmatrix} 1 & 2 \\ 2 & 4 \end{vmatrix}$$

$$= 3(-2)(1 \cdot 4 - 2 \cdot 2) = 0$$

Since the determinant of A is zero, A cannot have an inverse, by the Invertibility Criterion.   ■

The preceding example shows that if we expand a determinant about a row or column that contains many zeros, our work is considerably reduced because we do not have to evaluate the cofactors of the elements that are zero. The following principle enables us in many cases to simplify the process of finding a determinant by introducing zeros into it without changing its value.

ROW AND COLUMN TRANSFORMATIONS OF A DETERMINANT

If A is a square matrix, and if the matrix B is obtained from A by adding a multiple of one row to another, or a multiple of one column to another, then $|A| = |B|$.

EXAMPLE 5

Find the determinant of the matrix A. Does it have an inverse?

$$A = \begin{bmatrix} 8 & 2 & -1 & -4 \\ 3 & 5 & -3 & 11 \\ 24 & 6 & 1 & -12 \\ 2 & 2 & 7 & -1 \end{bmatrix}$$

SOLUTION

If we subtract three times row 1 from row 3, that will change all but one of the elements of row 3 to zeros:

$$\begin{bmatrix} 8 & 2 & -1 & -4 \\ 3 & 5 & -3 & 11 \\ 0 & 0 & 4 & 0 \\ 2 & 2 & 7 & -1 \end{bmatrix}$$

This new matrix has the same determinant as A, and if we expand its determinant by the third row, we get

$$|A| = 4 \begin{vmatrix} 8 & 2 & -4 \\ 3 & 5 & 11 \\ 2 & 2 & -1 \end{vmatrix}$$

Now adding two times column 3 to column 1 in this determinant gives us

$$|A| = 4 \begin{vmatrix} 0 & 2 & -4 \\ 25 & 5 & 11 \\ 0 & 2 & -1 \end{vmatrix} \qquad \text{(expand this by the first column)}$$

$$= 4(-25) \begin{vmatrix} 2 & -4 \\ 2 & -1 \end{vmatrix}$$

$$= 4(-25)[2(-1) - (-4)2] = -600$$

Since the determinant of A is not zero, A does have an inverse. ∎

CRAMER'S RULE

The solutions of linear equations can sometimes be expressed using determinants. To illustrate, let us try to solve the following pair of linear equations for the variable x.

$$\begin{cases} ax + by = r \\ cx + dy = s \end{cases}$$

If $d \neq 0$, then we can eliminate the variable y from the first equation by multiplying the second equation by b/d and then subtracting it from the first, which gives

$$ax - \left(\frac{b}{d}\right)cx = r - \left(\frac{b}{d}\right)s$$

If we now multiply both sides of the equation by d and factor x from the left, we get

$$(ad - bc)x = rd - bs$$

Assuming that $ad - bc \neq 0$, we can now solve this equation for x, obtaining

$$x = \frac{rd - bs}{ad - bc}$$

The numerator and denominator of this fraction look like the determinants of 2×2 matrices. In fact, we can write the solution for x as

$$x = \frac{\begin{vmatrix} r & b \\ s & d \end{vmatrix}}{\begin{vmatrix} a & b \\ c & d \end{vmatrix}}$$

Using the same sort of technique, we can solve the original pair of equations for y, to get

$$y = \frac{\begin{vmatrix} a & r \\ c & s \end{vmatrix}}{\begin{vmatrix} a & b \\ c & d \end{vmatrix}}$$

Notice that the denominator in each case is the determinant of the coefficient matrix, which we will call D. The numerator in the solution for x is the determinant of the matrix obtained from D by replacing the coefficients of x by r and s, respectively. Similarly, in the solution for y the numerator is the determinant of the matrix obtained from D by replacing the coefficients of y by r and s. Thus if we define

$$D = \begin{bmatrix} a & b \\ c & d \end{bmatrix} \qquad D_x = \begin{bmatrix} r & b \\ s & d \end{bmatrix} \qquad D_y = \begin{bmatrix} a & r \\ c & s \end{bmatrix}$$

we can write the solution of the system as

$$x = \frac{|D_x|}{|D|} \qquad \text{and} \qquad y = \frac{|D_y|}{|D|}$$

This pair of formulas is known as **Cramer's Rule**, and it can be used to solve any pair of linear equations in two unknowns in which the determinant of the coefficient matrix is not zero.

EXAMPLE 6

Use Cramer's Rule to solve the system

$$\begin{cases} 2x + 6y = -1 \\ x + 8y = 2 \end{cases}$$

For this system, we have

$$|D| = \begin{vmatrix} 2 & 6 \\ 1 & 8 \end{vmatrix} = 2 \cdot 8 - 6 \cdot 1 = 10$$

$$|D_x| = \begin{vmatrix} -1 & 6 \\ 2 & 8 \end{vmatrix} = (-1)8 - 6 \cdot 2 = -20$$

and

$$|D_y| = \begin{vmatrix} 2 & -1 \\ 1 & 2 \end{vmatrix} = 2 \cdot 2 - (-1)1 = 5$$

The solution is

$$x = \frac{|D_x|}{|D|} = \frac{-20}{10} = -2$$

and

$$y = \frac{|D_y|}{|D|} = \frac{5}{10} = \frac{1}{2}$$ ■

Cramer's Rule can be extended to apply to any system of n linear equations in n variables in which the determinant of the coefficient matrix is not zero. As we saw in the preceding section, any such system can be written in matrix form as

$$\begin{bmatrix} a_{11} & a_{12} & \cdots & a_{1n} \\ a_{21} & a_{22} & \cdots & a_{2n} \\ \vdots & \vdots & \ddots & \vdots \\ a_{n1} & a_{n2} & \cdots & a_{nn} \end{bmatrix} \begin{bmatrix} x_1 \\ x_2 \\ \vdots \\ x_n \end{bmatrix} = \begin{bmatrix} b_1 \\ b_2 \\ \vdots \\ b_n \end{bmatrix}$$

By analogy with what we did in the case of two equations in two unknowns, we let D be the coefficient matrix in the above system, and we let D_{x_i} be the matrix obtained by replacing the ith column of D by the numbers $b_1, b_2, \ldots, b_n$ that appear to the right of the equal sign in the system. The solution of the system is then given by the following rule:

CRAMER'S RULE

$$x_1 = \frac{|D_{x_1}|}{|D|}, \quad x_2 = \frac{|D_{x_2}|}{|D|}, \quad \ldots, \quad x_n = \frac{|D_{x_n}|}{|D|}$$

EXAMPLE 7

Use Cramer's Rule to solve the system

$$\begin{cases} 2x - 3y + 4z = 1 \\ x \quad\quad + 6z = 0 \\ 3x - 2y \quad\quad = 5 \end{cases}$$

SOLUTION

First we evaluate the determinants that appear in Cramer's Rule.

$$|D| = \begin{vmatrix} 2 & -3 & 4 \\ 1 & 0 & 6 \\ 3 & -2 & 0 \end{vmatrix} = -38 \qquad |D_x| = \begin{vmatrix} 1 & -3 & 4 \\ 0 & 0 & 6 \\ 5 & -2 & 0 \end{vmatrix} = -78$$

$$|D_y| = \begin{vmatrix} 2 & 1 & 4 \\ 1 & 0 & 6 \\ 3 & 5 & 0 \end{vmatrix} = -22 \qquad |D_z| = \begin{vmatrix} 2 & -3 & 1 \\ 1 & 0 & 0 \\ 3 & -2 & 5 \end{vmatrix} = 13$$

Now we use Cramer's Rule to get the solution:

$$x = \frac{|D_x|}{|D|} = \frac{-78}{-38} = \frac{39}{19}$$

$$y = \frac{|D_y|}{|D|} = \frac{-22}{-38} = \frac{11}{19}$$

$$z = \frac{|D_z|}{|D|} = \frac{13}{-38} = -\frac{13}{38}$$

If we had solved the system in Example 7 using Gaussian elimination, our work would have involved matrices whose elements are fractions with fairly large denominators. Thus in cases like Examples 6 and 7, Cramer's Rule provides an efficient method for solving systems of linear equations. But in systems with more than three equations, evaluating the various determinants involved is usually a long and tedious task. Moreover, the rule does not apply if $|D| = 0$ or if D is not a square matrix. So Cramer's Rule is a useful alternative to Gaussian elimination in only a limited set of situations.

EXERCISES 8.6

In Exercises 1–8 find the determinant of the given matrix, if it exists.

1. $\begin{bmatrix} 2 & 3 \\ -1 & 0 \end{bmatrix}$ **2.** $\begin{bmatrix} -4 & 6 \\ 2 & -5 \end{bmatrix}$ **3.** $[6]$

4. $\begin{bmatrix} 4 \\ 2 \end{bmatrix}$ **5.** $[3 \quad -1]$ **6.** $[0]$

7. $\begin{bmatrix} \frac{1}{2} & \frac{1}{3} \\ 1 & \frac{2}{3} \end{bmatrix}$ **8.** $\begin{bmatrix} \frac{3}{5} & -0.6 \\ \frac{1}{3} & 4 \end{bmatrix}$

In Exercises 9–14 evaluate the given minor and cofactor using the matrix

$$A = \begin{bmatrix} 1 & 0 & \frac{1}{2} \\ -3 & 5 & 2 \\ 0 & 0 & 4 \end{bmatrix}$$

9. M_{11}, A_{11} **10.** M_{33}, A_{33}

11. M_{12}, A_{12} **12.** M_{13}, A_{13}

13. M_{23}, A_{23} **14.** M_{32}, A_{32}

In Exercises 15–20 find the determinant of the matrix. Determine whether the matrix has an inverse, but do not try to find the inverse.

15. $\begin{bmatrix} 1 & 3 & 7 \\ 2 & 0 & -1 \\ 0 & 2 & 6 \end{bmatrix}$ **16.** $\begin{bmatrix} -2 & -\frac{3}{2} & \frac{1}{2} \\ 2 & 4 & 0 \\ \frac{1}{2} & 2 & 1 \end{bmatrix}$

17. $\begin{bmatrix} 30 & 0 & 20 \\ 0 & -10 & -20 \\ 40 & 0 & 10 \end{bmatrix}$ **18.** $\begin{bmatrix} 1 & 2 & 5 \\ -2 & -3 & 2 \\ 3 & 5 & 3 \end{bmatrix}$

19. $\begin{bmatrix} 1 & 3 & 3 & 0 \\ 0 & 2 & 0 & 1 \\ -1 & 0 & 0 & 2 \\ 1 & 6 & 4 & 1 \end{bmatrix}$ **20.** $\begin{bmatrix} 1 & 2 & 0 & 2 \\ 3 & -4 & 0 & 4 \\ 0 & 1 & 6 & 0 \\ 1 & 0 & 2 & 0 \end{bmatrix}$

In Exercises 21–24 evaluate the determinant. Use row or column operations whenever possible to simplify your work.

21. $\begin{vmatrix} 0 & 0 & 4 & 6 \\ 2 & 1 & 1 & 3 \\ 2 & 1 & 2 & 3 \\ 3 & 0 & 1 & 7 \end{vmatrix}$ **22.** $\begin{vmatrix} -2 & 3 & -1 & 7 \\ 4 & 6 & -2 & 3 \\ 7 & 7 & 0 & 5 \\ 3 & -12 & 4 & 0 \end{vmatrix}$

23. $\begin{vmatrix} 1 & 2 & 3 & 4 & 5 \\ 0 & 2 & 4 & 6 & 8 \\ 0 & 0 & 3 & 6 & 9 \\ 0 & 0 & 0 & 4 & 8 \\ 0 & 0 & 0 & 0 & 5 \end{vmatrix}$ **24.** $\begin{vmatrix} 2 & -1 & 6 & 4 \\ 7 & 2 & -2 & 5 \\ 4 & -2 & 10 & 8 \\ 6 & 1 & 1 & 4 \end{vmatrix}$

25. Let

$$B = \begin{bmatrix} 4 & 1 & 0 \\ -2 & -1 & 1 \\ 4 & 0 & 3 \end{bmatrix}$$

(a) Evaluate $|B|$ by expanding by the second row.
(b) Evaluate $|B|$ by expanding by the third column.
(c) Do your results in parts (a) and (b) agree?

26. Consider the system

$$\begin{cases} x + 2y + 6z = 5 \\ -3x - 6y + 5z = 8 \\ 2x + 6y + 9z = 7 \end{cases}$$

(a) Verify that $x = -1$, $y = 0$, $z = 1$ is a solution of the system.

(b) Find the determinant of the coefficient matrix.
(c) Without solving the system, determine whether there are any other solutions.
(d) Can Cramer's Rule be used to solve this system? Why or why not?

In Exercises 27–42 use Cramer's Rule to solve the system.

27. $\begin{cases} 2x - y = -9 \\ x + 2y = 8 \end{cases}$ **28.** $\begin{cases} 6x + 12y = 33 \\ 4x + 7y = 20 \end{cases}$

29. $\begin{cases} x - 6y = 3 \\ 3x + 2y = 1 \end{cases}$ **30.** $\begin{cases} \frac{1}{2}x + \frac{1}{3}y = 1 \\ \frac{1}{4}x - \frac{1}{6}y = -\frac{3}{2} \end{cases}$

31. $\begin{cases} 0.4x + 1.2y = 0.4 \\ 1.2x + 1.6y = 3.2 \end{cases}$ **32.** $\begin{cases} 10x - 17y = 21 \\ 20x - 31y = 39 \end{cases}$

33. $\begin{cases} x - y + 2z = 0 \\ 3x + z = 11 \\ -x + 2y = 0 \end{cases}$ **34.** $\begin{cases} 5x - 3y + z = 6 \\ 4y - 6z = 22 \\ 7x + 10y = -13 \end{cases}$

35. $\begin{cases} 2x_1 + 3x_2 - 5x_3 = 1 \\ x_1 + x_2 - x_3 = 2 \\ 2x_2 + x_3 = 8 \end{cases}$ **36.** $\begin{cases} -2a + c = 2 \\ a + 2b - c = 9 \\ 3a + 5b + 2c = 22 \end{cases}$

37. $\begin{cases} \frac{1}{3}x - \frac{1}{5}y + \frac{1}{2}z = \frac{7}{10} \\ -\frac{2}{3}x + \frac{2}{5}y + \frac{3}{2}z = \frac{11}{10} \\ x - \frac{4}{5}y + z = \frac{9}{5} \end{cases}$ **38.** $\begin{cases} 2x - y = 5 \\ 5x + 3z = 19 \\ 4y + 7z = 17 \end{cases}$

39. $\begin{cases} 3y + 5z = 4 \\ 2x - z = 10 \\ 4x + 7y = 0 \end{cases}$ **40.** $\begin{cases} 2x - 5y = 4 \\ x + y - z = 8 \\ 3x + 5z = 0 \end{cases}$

41. $\begin{cases} 3r - s + 3t = 7 \\ 4r + 5s - 2t = 0 \\ 9r + s + t = 0 \end{cases}$ **42.** $\begin{cases} \theta + \phi + \psi = 2 \\ 2\theta - \phi + \psi = 4 \\ \theta - 3\phi + 2\psi = 0 \end{cases}$

43. (a) Show that the equation

$$\begin{vmatrix} x_1 & y_1 & 1 \\ x_2 & y_2 & 1 \\ x & y & 1 \end{vmatrix} = 0$$

is an equation for the line that passes through the points (x_1, y_1) and (x_2, y_2).

(b) Use the result of part (a) to find an equation for the line that passes through the points $(20, 50)$ and $(-10, 25)$.

44. Evaluate the determinant

$$\begin{vmatrix} a & a & a & a & a \\ 0 & a & a & a & a \\ 0 & 0 & a & a & a \\ 0 & 0 & 0 & a & a \\ 0 & 0 & 0 & 0 & a \end{vmatrix}$$

In Exercises 45–48 solve for x.

45. $\begin{vmatrix} x & 12 & 13 \\ 0 & x-1 & 23 \\ 0 & 0 & x-2 \end{vmatrix} = 0$

46. $\begin{vmatrix} x & 1 & 1 \\ 1 & 1 & x \\ x & 1 & x \end{vmatrix} = 0$

47. $\begin{vmatrix} 1 & 0 & x \\ x^2 & 1 & 0 \\ x & 0 & 1 \end{vmatrix} = 0$

48. $\begin{vmatrix} a & b & x-a \\ x & x+b & x \\ 0 & 1 & 1 \end{vmatrix} = 0$

SECTION 8.7
NONLINEAR SYSTEMS

Up to this point we have been studying systems of *linear* equations. As we have seen, mathematicians have developed several techniques for handling such systems. In calculus and the sciences, however, one often encounters systems of nonlinear equations as well, so we study them in this section. Unfortunately, there are no general techniques for nonlinear systems like the ones we have been applying to linear systems. We have to approach each problem on an individual basis and solve it using whatever ad hoc method or combination of methods happens to work in that particular situation.

The technique we use most often is simple substitution. If the system we are dealing with consists of a linear equation and a quadratic polynomial in two variables (as in Example 1), then this method always gives us the complete solution.

EXAMPLE 1

Find all solutions of the following system:

$$\begin{cases} x^2 + y^2 = 100 \\ 3x - y = 10 \end{cases}$$

SOLUTION

The graph of the first equation is a circle and the graph of the second is a line (see Figure 1). The figure shows that the graphs intersect in two points, so the system has two solutions. We solve the system by solving for y in the second equation and then substituting into the first.

$$y = 3x - 10$$
$$x^2 + (3x - 10)^2 = 100$$
$$x^2 + (9x^2 - 60x + 100) = 100$$
$$10x^2 - 60x = 0$$
$$10x(x - 6) = 0$$
$$x = 0 \quad \text{or} \quad x = 6$$

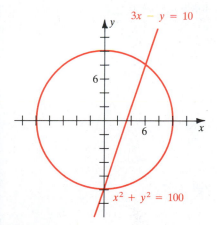

Figure 1

If $x = 0$, then $y = 3(0) - 10 = -10$, and if $x = 6$, then $y = 3(6) - 10 = 8$. Thus the solutions are $(0, -10)$ and $(6, 8)$.   ■

EXAMPLE 2

Solve the system:

$$\begin{cases} x^2 + 2y^2 = 11 \\ 3x^2 + 4y = 23 \end{cases}$$

SOLUTION

Here we could solve for y in the second equation and substitute into the first just as in Example 1. But this would lead to a fourth-degree equation in x involving fractions, which might be difficult to solve. Instead, we multiply the first equation by 3 and subtract the second from it to eliminate the x:

$$3x^2 + 6y^2 = 33$$
$$\underline{3x^2 + 4y\ = 23}$$
$$6y^2 - 4y\ = 10$$

We now solve this new equation by factoring:

$$6y^2 - 4y - 10 = 0$$
$$2(3y^2 - 2y - 5) = 0$$
$$2(y + 1)(3y - 5) = 0$$
$$y = -1 \quad \text{or} \quad y = \tfrac{5}{3}$$

We can now solve for the corresponding values of x by substituting into either of the two original equations. If $y = -1$, then using the second equation, we get

$$3x^2 + 4(-1) = 23$$
$$3x^2 = 27$$
$$x^2 = 9$$
$$x = 3 \quad \text{or} \quad x = -3$$

If $y = \tfrac{5}{3}$, then

$$3x^2 + 4\left(\tfrac{5}{3}\right) = 23$$
$$3x^2 = 23 - \tfrac{20}{3} = \tfrac{49}{3}$$
$$x^2 = \tfrac{49}{9}$$
$$x = \tfrac{7}{3} \quad \text{or} \quad x = -\tfrac{7}{3}$$

This means that there are four solutions:

$$(3, -1) \qquad (-3, -1) \qquad \left(\tfrac{7}{3}, \tfrac{5}{3}\right) \qquad \left(-\tfrac{7}{3}, \tfrac{5}{3}\right) \qquad \blacksquare$$

Some nonlinear systems are really just disguised versions of linear ones, like the system in the next example. We can solve these using any of the available linear methods.

EXAMPLE 3

Solve the system

$$\begin{cases} 4x^3 + 6y^2 = 22 \\ 5x^3 + 8y^2 = 32 \end{cases}$$

SOLUTION

If we let $u = x^3$ and $v = y^2$, then the system becomes linear:

$$\begin{cases} 4u + 6v = 22 \\ 5u + 8v = 32 \end{cases}$$

We can solve this using Gaussian elimination, matrix inversion, Cramer's Rule, or any method that works on linear systems. Omitting the details of the calculations, we have

$$u = -8 \qquad \text{and} \qquad v = 9$$

Since $u = x^3$ and $v = y^2$, this gives us

$$x = -2 \qquad \text{and} \qquad y = \pm 3$$

so the solutions of the original system are

$$(-2, 3) \qquad \text{and} \qquad (-2, -3) \qquad \blacksquare$$

EXAMPLE 4

Solve the system

$$\begin{cases} 3x^2 + 2y^2 + 15x = 0 \\ \qquad\qquad xy + y^2 = 0 \end{cases}$$

SOLUTION

We begin by tackling the second equation, since it looks more manageable. Factoring, we get

$$(x + y)y = 0$$

so

$$x = -y \qquad \text{or} \qquad y = 0$$

We now substitute each of these possibilities into the first equation, to see what the consequences are. If $x = -y$, then

$$3(-y)^2 + 2y^2 + 15(-y) = 0$$
$$5y^2 - 15y = 0$$
$$5y(y - 3) = 0$$
$$y = 0 \quad \text{or} \quad y = 3$$

Since these values for y were obtained from the assumption that $x = -y$, we get the following solutions of the original system:

$$(0, 0) \quad \text{and} \quad (-3, 3)$$

Now we check what happens in the first equation if $y = 0$.

$$3x^2 + 2(0)^2 + 15x = 0$$
$$3x^2 + 15x = 0$$
$$3x(x + 5) = 0$$
$$x = 0 \quad \text{or} \quad x = -5$$

This leads to the solutions $(0, 0)$ and $(-5, 0)$. We have already found the first of these, so the solutions are

$$(0, 0) \qquad (-3, 3) \qquad (-5, 0) \qquad \blacksquare$$

The next example involves three equations and three variables.

EXAMPLE 5

Solve the system

$$\begin{cases} x + yz = 0 \\ y + 4xz = 0 \\ x^2 + y^2 = 20 \end{cases}$$

SOLUTION

If we solve for x in the first equation and substitute into the second, we get

$$y + 4(-yz)z = 0$$
$$y(1 - 4z^2) = 0$$
$$y = 0 \quad \text{or} \quad z^2 = \tfrac{1}{4}$$

Substituting $y = 0$ into the first equation leads to $x = 0$, but this does not satisfy the third equation. So it must be true that $z^2 = \tfrac{1}{4}$, or $z = \pm\tfrac{1}{2}$. Thus $y = \mp 2x$ (from the second equation), and the third equation becomes

$$x^2 + 4x^2 = 20$$
$$5x^2 = 20$$
$$x^2 = 4$$
$$x = \pm 2$$

This leads to four possible solutions:

$$\left(2, 4, -\tfrac{1}{2}\right) \quad \left(2, -4, \tfrac{1}{2}\right) \quad \left(-2, 4, \tfrac{1}{2}\right) \quad \left(-2, -4, -\tfrac{1}{2}\right) \quad \blacksquare$$

We now consider an application that arises from a problem in geometry.

EXAMPLE 6

A right triangle has an area of 120 ft^2 and a perimeter of 60 ft. Find the lengths of its sides.

SOLUTION

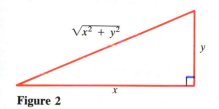

$\sqrt{x^2 + y^2}$

y

x

Figure 2

Let x and y be the lengths of the sides adjacent to the right angle. Then by the Pythagorean Theorem, the hypotenuse has length $\sqrt{x^2 + y^2}$. (See Figure 2.) Since the area is 120 ft^2, we have

$$\tfrac{1}{2}xy = 120$$

or
$$xy = 240 \tag{1}$$

Also, since the perimeter is 60 ft,

$$x + y + \sqrt{x^2 + y^2} = 60 \tag{2}$$

We simplify Equation 2 as follows:

$$\sqrt{x^2 + y^2} = 60 - x - y$$
$$\left(\sqrt{x^2 + y^2}\right)^2 = (60 - x - y)^2$$
$$x^2 + y^2 = 3600 + x^2 + y^2 - 120x - 120y + 2xy$$
$$120x + 120y = 3600 + 2xy$$
$$60x + 60y = 1800 + xy \tag{3}$$

Thus we must solve the system that combines Equations 1 and 3:

$$\begin{cases} xy = 240 \\ 60x + 60y = 1800 + xy \end{cases}$$

Adding these equations (to eliminate the xy term) and simplifying, we get

$$60x + 60y = 2040$$
$$x + y = 34$$
$$y = 34 - x$$

Substituting this into Equation 1 gives

$$x(34 - x) = 240$$
$$x^2 - 34x + 240 = 0$$
$$(x - 24)(x - 10) = 0$$

So either $x = 24$ and $y = 10$, or $x = 10$ and $y = 24$. In either case, the hypotenuse is

$$\sqrt{(10)^2 + (24)^2} = \sqrt{676} = 26$$

The sides of the triangle are 10 ft, 24 ft, and 26 ft long.   ■

USING GRAPHING DEVICES TO SOLVE NONLINEAR SYSTEMS

Graphing devices are sometimes useful in solving systems of equations that involve just two variables. Note that with most graphing devices, any equation must first be expressed in terms of one or more functions of the form $y = f(x)$ before the calculator can graph it. Not all equations can be readily expressed in this way, so not all systems can be solved by this technique.

EXAMPLE 7

Find all solutions of the following system, correct to one decimal.

$$\begin{cases} \dfrac{x^2}{12} + \dfrac{y^2}{7} = 1 \\ \qquad y = 3x^2 - 6x + \tfrac{1}{2} \end{cases}$$

SOLUTION

The graph of the first equation is an ellipse with x-intercepts $x = \pm\sqrt{12} \approx \pm 3.46$ and with y-intercepts $y = \pm\sqrt{7} \approx \pm 2.65$. We therefore select the viewing rectangle $[-4, 4]$ by $[-3, 3]$ to be sure that our graph contains the entire ellipse. Solving for y in terms of x, we get

$$\frac{y^2}{7} = 1 - \frac{x^2}{12}$$

$$y^2 = 7\left(1 - \frac{x^2}{12}\right)$$

$$y = \pm\sqrt{7\left(1 - \frac{x^2}{12}\right)}$$

To graph the entire ellipse, we must graph both of the functions

$$y = \sqrt{7[1 - (x^2/12)]} \qquad \text{and} \qquad y = -\sqrt{7[1 - (x^2/12)]}$$

The graph of the second equation in the system is a parabola. We graph it in the same viewing rectangle as the ellipse (see Figure 3). There are two intersection

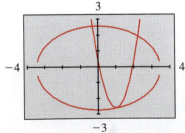

Figure 3
$\dfrac{x^2}{12} + \dfrac{y^2}{7} = 1$, $y = 3x^2 - 6x + \tfrac{1}{2}$

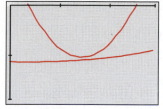

Figure 4

points on the top half of the ellipse. We zoom in and use the cursor to locate them. Their approximate coordinates are $(-0.3, 2.6)$ and $(2.2, 2.0)$. There also appears to be an intersection point on the bottom of the ellipse, near the vertex of the parabola. However, when we zoom in we see that the parabola and the ellipse come close to each other here but do not intersect (see Figure 4). Thus the system has only two solutions; correct to the nearest tenth, they are

$$(-0.3, 2.6) \qquad \text{and} \qquad (2.2, 2.0) \qquad \blacksquare$$

EXERCISES 8.7

In Exercises 1–24 find all real solutions (x, y) of the system of equations.

1. $\begin{cases} x^2 + y^2 = 8 \\ x + y = 0 \end{cases}$

2. $\begin{cases} x^2 + y = 9 \\ x - y + 3 = 0 \end{cases}$

3. $\begin{cases} y + x^2 = 4x \\ y + 4x = 16 \end{cases}$

4. $\begin{cases} x - y^2 = 0 \\ y - x^2 = 0 \end{cases}$

5. $\begin{cases} x - 2y = 2 \\ y^2 - x^2 = 2x + 4 \end{cases}$

6. $\begin{cases} y = 4 - x^2 \\ y = x^2 - 4 \end{cases}$

7. $\begin{cases} x - y = 4 \\ xy = 12 \end{cases}$

8. $\begin{cases} x^3 - y = 0 \\ -2x + y = 4 \end{cases}$

9. $\begin{cases} 3x^2 - y^2 = 11 \\ x^2 + 4y^2 = 8 \end{cases}$

10. $\begin{cases} xy = 24 \\ 2x^2 - y^2 + 4 = 0 \end{cases}$

11. $\begin{cases} x^2 y = 16 \\ x^2 + 4y + 16 = 0 \end{cases}$

12. $\begin{cases} 2x^2 + 4y = 13 \\ x^2 - y^2 = \frac{7}{2} \end{cases}$

13. $\begin{cases} x + \sqrt{y} = 0 \\ y^2 - 4x^2 = 12 \end{cases}$

14. $\begin{cases} \sqrt{x} - 2\sqrt{y} = 1 \\ 2x - 4(y + \sqrt{y}) = 2 \end{cases}$

15. $\begin{cases} x^2 + y^2 = 9 \\ x^2 - y^2 = 1 \end{cases}$

16. $\begin{cases} x^2 + 2y^2 = 2 \\ 2x^2 - 3y = 15 \end{cases}$

17. $\begin{cases} 2x^2 - 8y^3 = 19 \\ 4x^2 + 16y^3 = 34 \end{cases}$

18. $\begin{cases} x^4 - y^3 = 17 \\ 3x^4 + 5y^3 = 53 \end{cases}$

19. $\begin{cases} \dfrac{2}{x} - \dfrac{3}{y} = 1 \\ -\dfrac{4}{x} + \dfrac{7}{y} = 1 \end{cases}$

20. $\begin{cases} \dfrac{4}{x^2} + \dfrac{6}{y^4} = \dfrac{7}{2} \\ \dfrac{1}{x^2} - \dfrac{2}{y^4} = 0 \end{cases}$

21. $\begin{cases} 3\sqrt{x} + 5\sqrt{y} = 19 \\ 2\sqrt{x} + 7\sqrt{y} = 20 \end{cases}$

22. $\begin{cases} 2\sqrt{x} - \sqrt{y} = 3 \\ 4\sqrt{x} + 3\sqrt{y} = 1 \end{cases}$

23. $\begin{cases} x^2 - xy + 2y^2 = 8 \\ x^3 - xy^2 = 0 \end{cases}$

24. $\begin{cases} xy - 3x = 0 \\ x^3 - y + 11 = 0 \end{cases}$

In Exercises 25–30 find all real solutions (x, y, z) of the system of equations.

25. $\begin{cases} x - y = 2 \\ y + z = 0 \\ x^2 + y^2 + z^2 = 4 \end{cases}$

26. $\begin{cases} xy + z = 0 \\ yz + x = 0 \\ x^2 + y^2 = 2 \end{cases}$

27. $\begin{cases} x^2 + yz = 0 \\ y + xz = 2 \\ xyz = 1 \end{cases}$

28. $\begin{cases} xy - xz = 0 \\ y + yz = 2 \\ x^2 + y^2 = 5 \end{cases}$

29. $\begin{cases} x^2 + y + z = 0 \\ 2x^2 - y + 3z = -4 \\ y^2 + yz = 0 \end{cases}$

30. $\begin{cases} x^2 + y^3 + z^4 = 4 \\ 3x^2 - y^3 - z^4 = 12 \\ 2x^2 - 3y^3 + 2z^4 = 13 \end{cases}$

In Exercises 31–38 use a graphing device to find all real solutions (x, y) of the system of equations.

31. $\begin{cases} x^2 + y^2 = 25 \\ x + 3y = 2 \end{cases}$

32. $\begin{cases} x^2 + y^2 = 17 \\ x^2 - 2x + y^2 = 13 \end{cases}$

33. $\begin{cases} \dfrac{x^2}{9} + \dfrac{y^2}{18} = 1 \\ y = -x^2 + 6x - 2 \end{cases}$

34. $\begin{cases} x^2 - y^2 = 3 \\ y = x^2 - 2x - 8 \end{cases}$

35. $\begin{cases} x^4 + 16y^4 = 32 \\ x^2 + 2x + y = 0 \end{cases}$

36. $\begin{cases} x^2 - 2x - y^2 = 0 \\ x - y^5 = 0 \end{cases}$

37. $\begin{cases} y = e^x + e^{-x} \\ y = 5 - x^2 \end{cases}$

38. $\begin{cases} \ln x + \ln y = 3 \\ (\ln x)^2 - \ln y = 0 \end{cases}$

39. A right triangle has a perimeter of 40 cm and an area of 60 cm². What are the lengths of its sides?

40. A rectangle has an area of 180 cm² and a perimeter of 54 cm. What are its dimensions?

41. A right triangle has an area of 54 in². The product of the lengths of the three sides is 1620 in³. What are the lengths of its sides?

42. A right triangle has an area of 84 ft² and a hypotenuse 25 ft long. What are the lengths of its other two sides?

43. The perimeter of a rectangle is 70 and its diagonal is 25. Find its length and width.

44. A circular piece of sheet metal has a diameter of 20 in. The edges are to be cut off to form a rectangle of area 160 in² (see the figure). What are the dimensions of the rectangle?

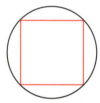

45. Find an equation for the line that passes through the points of intersection of the circles $x^2 + y^2 = 25$ and $x^2 - 3x + y^2 + y = 30$.

46. (a) For what value of k does the system

$$\begin{cases} y = x^2 \\ y = x + k \end{cases}$$

have exactly one solution?

(b) Graph both equations in the system on the same set of axes, using the value of k you chose in part (a).

(c) Based on your graph, how many solutions will the system have if k is smaller than your value from part (a)? How many will there be if k is larger?

In Exercises 47–50 find all real solutions of the system.

47. $\begin{cases} x - y = 3 \\ x^3 - y^3 = 387 \end{cases}$ [*Hint:* Factor the left side of the second equation.]

48. $\begin{cases} x^2 + xy + xz = 1 \\ xy + y^2 + yz = 3 \\ xz + yz + z^2 = 5 \end{cases}$ [*Hint:* Add the equations and factor the result.]

49. $\begin{cases} 2^x + 2^y = 10 \\ 4^x + 4^y = 68 \end{cases}$

50. $\begin{cases} \log x + \log y = \frac{3}{2} \\ 2 \log x - \log y = 0 \end{cases}$

SECTION 8.8
SYSTEMS OF INEQUALITIES

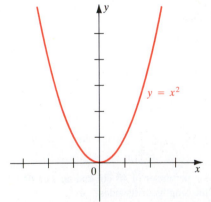

Figure 1

In this section we study systems of inequalities in two variables from a graphical point of view. First we consider the graph of a single inequality. We already know that the graph of $y = x^2$, for example, is the parabola in Figure 1. If we replace the equal sign by the symbol $\geq$, we obtain the **inequality**

$$y \geq x^2$$

Its graph consists of not just the parabola in Figure 1, but also every point whose y-coordinate is *larger* than x^2. We indicate the solution in Figure 2 by shading the points *above* the parabola.

Similarly, the graph of $y \leq x^2$ in Figure 3 consists of all points on and *below* the parabola, whereas the graphs of $y > x^2$ and $y < x^2$ do not include the points on the parabola itself, as indicated by the broken curves in Figures 4 and 5.

The graph of an inequality, in general, consists of a region in the plane whose boundary is in the graph of the equation obtained by replacing the inequality sign ($\geq$, $\leq$, $>$, or $<$) with an equal sign. To determine which side of this graph gives the solution set of the inequality, we need to check only **test points**, as illustrated in the next example.

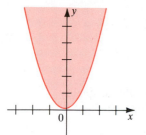

Figure 2
$y \geq x^2$

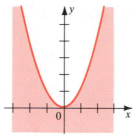

Figure 3
$y \leq x^2$

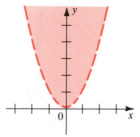

Figure 4
$y > x^2$

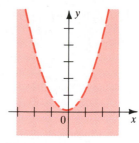

Figure 5
$y < x^2$

EXAMPLE 1

Graph the following inequalities:

(a) $x^2 + y^2 < 25$ (b) $x + 2y \geq 5$

SOLUTION

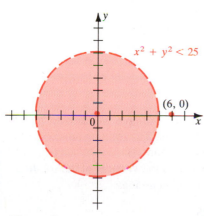

Figure 6

(a) The graph of $x^2 + y^2 = 25$ is a circle of radius 5 centered at the origin. The points on the circle itself do not satisfy the inequality, since it is of the form $<$, so we graph the circle with a broken curve in Figure 6.

To determine whether the inside or the outside of the circle satisfies the inequality, we use the test points $(0, 0)$ on the inside and $(6, 0)$ on the outside. (Note that *any* points inside and outside the circle can serve as test points. We chose these for simplicity.)

Check: Does $(0, 0)$ satisfy $x^2 + y^2 < 25$?

$$0^2 + 0^2 \overset{?}{<} 25$$
$$0 \overset{?}{<} 25 \quad \textit{Yes}$$

Check: Does $(6, 0)$ satisfy $x^2 + y^2 < 25$?

$$6^2 + 0^2 \overset{?}{<} 25$$
$$36 \overset{?}{<} 25 \quad \textit{No}$$

Thus the graph of $x^2 + y^2 < 25$ consists of the points inside the circle only (see Figure 6).

(b) The graph of $x + 2y = 5$ is the line shown in Figure 7. We use the test points $(0, 0)$ and $(5, 5)$ on opposite sides of the line.

Check: Does $(0, 0)$ satisfy $x + 2y \geq 5$?

$$(0) + 2(0) \overset{?}{\geq} 5$$
$$0 \overset{?}{\geq} 5 \quad \textit{No}$$

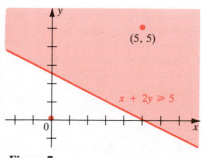

Figure 7

Check: Does $(5, 5)$ satisfy $x + 2y \geq 5$?

$$(5) + 2(5) \overset{?}{\geq} 5$$
$$15 \overset{?}{\geq} 5 \qquad \textit{Yes}$$

Our check shows that the points *above* the line satisfy the inequality.

Alternatively, we could put the inequality into slope-intercept form and graph it directly:

$$x + 2y \geq 5$$
$$2y \geq -x + 5$$
$$y \geq -\tfrac{1}{2}x + \tfrac{5}{2}$$

From this form, we see that the graph includes the points whose y-coordinates are *larger* than those on the line $y = -\tfrac{1}{2}x + \tfrac{5}{2}$; that is, the graph consists of the points on or *above* this line as in Figure 7. ∎

EXAMPLE 2

Graph the solution set of the pair of inequalities

$$\begin{cases} x^2 + y^2 < 25 \\ x + 2y \geq 5 \end{cases}$$

SOLUTION

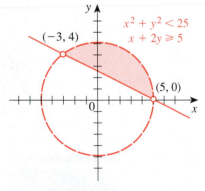

Figure 8

These are the two inequalities of Example 1. In this example we wish to graph those points that simultaneously satisfy both inequalities. The solution thus consists of the intersection of the graphs in Example 1 (see Figure 8).

The points $(-3, 4)$ and $(5, 0)$ in Figure 8 are called the **vertices** of the solution set. They are obtained by simultaneously solving the *equations*

$$\begin{cases} x^2 + y^2 = 25 \\ x + 2y = 5 \end{cases}$$

We solve this system of equations by substitution. Solving for x in the second equation gives $x = 5 - 2y$, and substituting this into the first equation gives

$$(5 - 2y)^2 + y^2 = 25$$
$$25 - 20y + 5y^2 = 25$$
$$-20y + 5y^2 = 0$$
$$-5y(4 - y) = 0$$

Thus $y = 0$ or $y = 4$. When $y = 0$, we have $x = 5 - 2(0) = 5$, and when $y = 4$, we have $x = 5 - 2(4) = -3$. So the points of intersection of these curves are $(5, 0)$ and $(-3, 4)$.

Note that in this case the vertices are not part of the solution set, since they do not satisfy the inequality $x^2 + y^2 < 25$. They simply show where the "corners" of the solution set lie. ∎

An inequality is **linear** if it can be put into one of the following forms:

$$ax + by \geq c \qquad ax + by \leq c \qquad ax + by > c \qquad ax + by < c$$

In the next example we graph the solution set of a system of linear inequalities.

EXAMPLE 3

Graph the solution set of the system

$$\begin{cases} x + 3y \leq 12 \\ x + y \leq 8 \\ x \geq 3 \\ y \geq 0 \end{cases}$$

SOLUTION

In Figure 9 we first graph the lines given by the equations that correspond to each of the inequalities. The shaded region is the set of points that satisfy all four inequalities simultaneously. The coordinates of the vertices are obtained by simultaneously solving the equations of the lines that intersect at that vertex. For example, the vertex $(6, 2)$ lies on both the lines

$$x + 3y = 12$$
and $\qquad\qquad x + y = 8$

In this case the vertices *are* part of the solution set. ∎

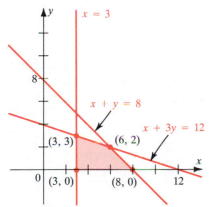

Figure 9

EXAMPLE 4

A manufacturer of insulating materials produces two different brands: Foamboard and Plastiflex. Each cubic yard of Foamboard weighs 8 lb, and each cubic yard of Plastiflex weighs 24 lb. The products are moved from the plant to the loading dock on carts that have a maximum capacity of 25 yd^3 and 432 lb. Find a system of inequalities that describes all possible combinations of Foamboard and Plastiflex that can be carried on such a cart. Graph the solution set of this system.

SOLUTION

First we let

$$x = \text{number of cubic yards of Foamboard on a cart}$$
$$y = \text{number of cubic yards of Plastiflex on a cart}$$

Since a cart can carry no more than 25 yd^3, we have

$$x + y \leq 25$$

In addition, the total weight cannot exceed 432 lb. Since there are $8x$ pounds of Foamboard and $24y$ pounds of Plastiflex on the cart, this means that

$$8x + 24y \leq 432$$

Dividing both sides of this inequality by 8 simplifies it to

$$x + 3y \leq 54$$

Finally, negative amounts would be meaningless in this context, so

$$x \geq 0 \qquad \text{and} \qquad y \geq 0$$

Thus the possible amounts of material that a cart can hold are given by the system

$$\begin{cases} x + y \leq 25 \\ x + 3y \leq 54 \\ x \geq 0 \\ y \geq 0 \end{cases}$$

The graph is shown in Figure 10.   ■

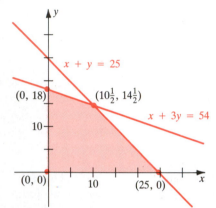

Figure 10

When a region in the plane can be covered by a (sufficiently large) circle, it is said to be **bounded**. A region that is not bounded is called **unbounded**. For example, the regions graphed in Figures 6, 8, 9, and 10 are bounded, while those in Figures 2, 3, 4, 5, and 7 are unbounded. Unbounded regions cannot be "fenced in"—they extend infinitely far in some direction.

EXERCISES 8.8

In Exercises 1–12 graph the inequality.

1. $y > -1$

2. $x \leq 4$

3. $y \geq 2x + 2$

4. $y < -x + 5$

5. $2x - y \leq 8$

6. $3x + 4y + 12 > 0$

7. $4x + 5y < 25$

8. $-x^2 + y \geq 10$

9. $y > x^3 + 1$

10. $x - y^2 \leq 0$

11. $x^2 + y^2 \geq 5$

12. $xy < 4$

In Exercises 13–34 graph the solutions of the system of inequalities. In each case, find the coordinates of all vertices, and determine whether the solution set is bounded.

13. $\begin{cases} x + y \leq 4 \\ \quad\ y \geq x \end{cases}$

14. $\begin{cases} 2x + 3y > 12 \\ 3x - \ y < 21 \end{cases}$

15. $\begin{cases} y < \frac{1}{4}x + 2 \\ y \geq 2x - 5 \end{cases}$

16. $\begin{cases} x - y > 0 \\ 4 + y \leq 2x \end{cases}$

17. $\begin{cases} x \geq 0, y \geq 0 \\ 3x + 5y \leq 15 \\ 3x + 2y \leq 9 \end{cases}$

18. $\begin{cases} \qquad x > 2 \\ \qquad y < 12 \\ 2x - 4y > 8 \end{cases}$

19. $\begin{cases} y < 9 - x^2 \\ y \geq x + 3 \end{cases}$

20. $\begin{cases} x \geq y^2 \\ x + y \geq 6 \end{cases}$

21. $\begin{cases} x^2 + y^2 \leq 4 \\ x - y > 0 \end{cases}$

22. $\begin{cases} x > 0, \ y > 0 \\ x + y < 10 \\ x^2 + y^2 > 9 \end{cases}$

23. $\begin{cases} x^2 - y \leq 0 \\ 2x^2 + y \leq 12 \end{cases}$

24. $\begin{cases} x^2 + y^2 < 9 \\ 2x + y^2 \geq 1 \end{cases}$

25. $\begin{cases} x + 2y \leq 14 \\ 3x - y \geq 0 \\ x - y \geq 2 \end{cases}$

26. $\begin{cases} y < x + 6 \\ 3x + 2y \geq 12 \\ x - 2y \leq 2 \end{cases}$

27. $\begin{cases} x \geq 0, \ y \geq 0 \\ x \leq 5 \\ x + y \leq 7 \\ x + 2y \geq 4 \end{cases}$

28. $\begin{cases} x \geq 0, \ y \geq 0 \\ y \leq 4 \\ 2x + y \leq 8 \\ 20x + 3y \leq 66 \end{cases}$

29. $\begin{cases} y > x + 1 \\ x + 2y \leq 12 \\ x + 1 > 0 \end{cases}$

30. $\begin{cases} x + y > 12 \\ y < \frac{1}{2}x - 6 \\ 3x + y < 6 \end{cases}$

31. $\begin{cases} x^2 + y^2 \leq 8 \\ x \geq 2 \\ y \geq 0 \end{cases}$

32. $\begin{cases} x^2 - y \geq 0 \\ x + y < 6 \\ x - y < 6 \end{cases}$

33. $\begin{cases} x^2 + y^2 < 9 \\ x + y > 0 \\ x \leq 0 \end{cases}$

34. $\begin{cases} y \geq x^3 \\ y \leq 2x + 4 \\ x + y \geq 0 \end{cases}$

35. A publishing company publishes a total of no more than 100 books every year. At least 20 of these are nonfiction, but the company always publishes at least as much fiction as nonfiction. Find a system of inequalities that describes the possible number of fiction and nonfiction books the company can produce each year consistent with these policies. Graph the solution set.

36. A man and his daughter manufacture unfinished tables and chairs. Each table requires 3 h of sawing and 1 h of assembly. Each chair requires 2 h of sawing and 2 h of assembly. The two of them can do a total of up to 12 h of sawing and 8 h of assembly work each day. Find a system of inequalities that describes all possible combinations of tables and chairs that they can make daily. Graph the solution set.

SECTION 8.9
APPLICATION: LINEAR PROGRAMMING

Linear programming is a mathematical technique used to determine the optimal allocation of resources in business, the military, and other areas of human endeavor. For example, a manufacturer who makes several different products from the same raw materials can use linear programming to tell how much of each he or she should produce to maximize profits. This technique is probably the most important practical application of systems of linear inequalities. In 1975 Leonid Kantorovich and T.C. Koopmans won the Nobel Prize in economics for their work in the development of this subject.

Although linear programming can be applied to very complex problems with hundreds or even thousands of variables, we will consider only a few simple examples to which the graphical methods of the preceding section can be applied. We introduce the technique with a typical problem.

EXAMPLE 1

A small shoe manufacturer makes two different styles of shoes: oxfords and loafers. Two machines are used in the process: a cutting machine and a sewing machine. Each type of shoe requires 15 min per pair on the cutting machine. Oxfords require 10 min of sewing per pair, and loafers require 20 min of sewing per pair. Because the manufacturer can hire only one operator for each machine, each is available for just 8 h per day. If the profit on each pair of oxfords is $15 and on each pair of loafers is $20, how many pairs of each type should be produced per day for maximum profit?

SOLUTION

First we organize the information given into a table. To be consistent, we convert all times to hours.

	Oxfords	Loafers	Time available
Time on cutting machine (h)	$\frac{1}{4}$	$\frac{1}{4}$	8
Time on sewing machine (h)	$\frac{1}{6}$	$\frac{1}{3}$	8
Profit	$15	$20	

Let
$$x = \text{number of pairs of oxfords made daily}$$
$$y = \text{number of pairs of loafers made daily}$$

The total number of cutting hours needed is then $\frac{1}{4}x + \frac{1}{4}y$. Since only 8 h are available on the cutting machine, we have

$$\tfrac{1}{4}x + \tfrac{1}{4}y \leq 8$$

Similarly, by considering the amount of time needed and available on the sewing machine, we get

$$\tfrac{1}{6}x + \tfrac{1}{3}y \leq 8$$

Since we cannot produce a negative number of shoes, we also have

$$x \geq 0 \quad \text{and} \quad y \geq 0$$

Thus x and y must satisfy the system of inequalities

$$\begin{cases} \tfrac{1}{4}x + \tfrac{1}{4}y \leq 8 \\ \tfrac{1}{6}x + \tfrac{1}{3}y \leq 8 \\ \qquad\quad x \geq 0 \\ \qquad\quad y \geq 0 \end{cases}$$

If we multiply the first inequality by 4 and the second by 6, we obtain the simplified system

Linear programming is used by the telephone industry to determine the most efficient way to route telephone calls. The routing decisions have to be made very rapidly by computer so as not to keep callers waiting on the line for the telephone connection. Since the data in such problems are huge, a very fast method for solving linear programming problems is essential. In 1984 the 28-year-old mathematician Narendra Karmarkar, working at Bell Labs in Murray Hill, New Jersey, discovered just such a method. His idea is so ingenious and his method so fast that the discovery caused a sensation in the mathematical world. His technique is now also used by airlines in scheduling passengers, flight personnel, fuel, baggage, and maintenance workers so as to minimize costs. Although mathematical discoveries rarely make the news, this one was reported in Time, *Dec. 3, 1984.*

$$\begin{cases} x + y \le 32 \\ x + 2y \le 48 \\ x \ge 0 \\ y \ge 0 \end{cases}$$

The solution of this system (with vertices labeled) is graphed in Figure 1.

We wish to determine which values for x and y give maximum profit. The only values that satisfy the restrictions of the problem are the ones that correspond to points of the shaded region in Figure 1. This is called the **feasible region** for the problem. Since each pair of oxfords provides $15 profit and each pair of loafers $20, the total profit will be

$$P = 15x + 20y$$

As x or y increases, profit will increase as well. Thus it seems reasonable that the maximum profit will occur at a point on one of the outside edges of the feasible region, where it is impossible to increase x or y without going outside the region. In fact, it can be shown that the maximum value will occur at a vertex. This means that we need to check the profit only at the vertices.

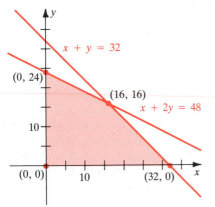

$x + y = 32$

$(0, 24)$

$(16, 16)$

$x + 2y = 48$

10

$(0, 0)$ 10 $(32, 0)$

Figure 1

Vertex	$P = 15x + 20y$
$(0, 0)$	0
$(0, 24)$	$15(0) + 20(24) = \$480$
$(16, 16)$	$15(16) + 20(16) = \$560 \leftarrow$ maximum profit
$(32, 0)$	$15(32) + 20(0) = \$480$

The largest value of P occurs at the point $(16, 16)$, where $P = \$560$. Thus the manufacturer should make 16 pairs of oxfords and 16 pairs of loafers, for a maximum daily profit of $560.   ■

All the linear programming problems that we consider follow the pattern of this example. Each problem involves two variables, and certain restrictions described in the problem lead to a system of linear inequalities that involve these variables. The graph of this system is called the **feasible region**. We consider only bounded feasible regions. The function we are trying to maximize or minimize is called the **objective function**. This function will always attain its largest and smallest values at the **vertices** of the feasible region, so checking its value at all vertices gives the solution to the problem.

EXAMPLE 2

A car dealer has warehouses in Millville and Trenton and has dealerships in Camden and Atlantic City. Every car sold at the dealerships must be delivered from one of the warehouses. On a certain day the Camden salespeople sell 10 cars and the Atlantic City sales staff sell 12. The Millville warehouse has 15 cars available and the Trenton warehouse has 10. It costs $50 to ship a car from Millville to Camden, $40 from Millville to Atlantic City, $60 from Trenton to Camden, and $55 from Trenton to Atlantic City. How many cars should be moved from each warehouse to each dealership to fill the orders at minimum cost?

SOLUTION

The first step is to organize the given information. Rather than construct a table, we draw a diagram to show the flow of cars from the warehouses to the dealerships (see Figure 2). The diagram shows the number of cars available or required at each location, and the cost of shipping between locations.

Since there are four possible routes, there seem to be four variables here. But if we let x be the number of cars to be shipped from Millville to Camden, then $10 - x$ cars would have to be shipped from Trenton to Camden, since the Camden dealership needs 10 cars in all. Similarly, if y cars are shipped from Millville to Atlantic City, $12 - y$ would be shipped from Trenton to Atlantic City.

We now derive the inequalities that define the feasible region. First of all, the number of cars shipped on each route cannot be negative, so

$$x \geq 0$$
$$10 - x \geq 0$$
$$y \geq 0$$
$$12 - y \geq 0$$

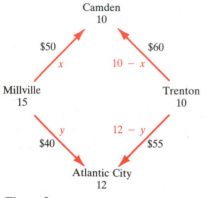

Figure 2

Second, the total number of cars shipped from each warehouse cannot exceed the number of cars available there, so

$$x + y \leq 15$$
$$(10 - x) + (12 - y) \leq 10$$

Simplifying the latter inequality, we get

$$22 - x - y \leq 10$$
$$-x - y \leq -12$$
$$x + y \geq 12$$

The inequalities $10 - x \geq 0$ and $12 - y \geq 0$ can be rewritten as $x \leq 10$ and $y \leq 12$, respectively. Thus the feasible region is described by the system of inequalities

$$\begin{cases} x \geq 0, \ y \geq 0 \\ x \leq 10, \ y \leq 12 \\ x + y \leq 15 \\ x + y \geq 12 \end{cases}$$

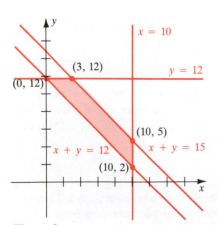

Figure 3

The feasible region is graphed in Figure 3.

From Figure 2, we see that the total cost of shipping the cars is

$$C = 50x + 40y + 60(10 - x) + 55(12 - y)$$
$$= 50x + 40y + 600 - 60x + 660 - 55y$$
$$= 1260 - 10x - 15y$$

This is the objective function. We check its value at each vertex.

Vertex	$C = 1260 - 10x - 15y$
(0, 12)	$1260 - 10(0) - 15(12) = \$1080$
(3, 12)	$1260 - 10(3) - 15(12) = \$1050$ ← minimum cost
(10, 5)	$1260 - 10(10) - 15(5) = \$1085$
(10, 2)	$1260 - 10(10) - 15(2) = \$1130$

The lowest cost is incurred at the point (3, 12). Thus the dealer should ship

3 cars from Millville to Camden

12 cars from Millville to Atlantic City

7 cars from Trenton to Camden

0 cars from Trenton to Atlantic City   ■

In the 1940s mathematicians developed matrix methods for solving linear programming problems that involve more than two variables. These methods were first used by the Allied military in World War II to solve supply problems similar to (but of course much more complicated than) Example 2. Improving these matrix methods is an active and exciting area of current mathmematical research.

EXERCISES 8.9

In Exercises 1–4 you are given a set of inequalities describing a feasible region and an objective function. Graph the feasible region, determine the coordinates of its vertices, and then find the maximum and minimum values of the objective function on the feasible region by checking its values at the vertices.

1. $\begin{cases} x \geqslant 0, \ y \geqslant 0 \\ 2x + \ y \leqslant 10 \\ 2x + 4y \leqslant 28 \end{cases}$

   $P = 140 - x + 3y$

2. $\begin{cases} x \geqslant 0, \ y \geqslant 0 \\ x \leqslant 10, \ y \leqslant 20 \\ x + \ y \geqslant 5 \\ x + 2y \leqslant 18 \end{cases}$

   $Q = 70x + 82y$

3. $\begin{cases} x \geqslant 3, \ y \geqslant 4 \\ 2x + \ y \leqslant 24 \\ 2x + 3y \leqslant 36 \end{cases}$

   $R = 12 + x + 4y$

4. $\begin{cases} x \leqslant 36, \ y \leqslant 40 \\ x \leqslant 2y \\ 2x + y \geqslant 60 \end{cases}$

   $S = 200 + 3x - y$

5. A furniture manufacturer makes wooden tables and chairs. The production process involves two basic types of labor: carpentry and finishing. Making a table requires 2 h of carpentry and 1 h of finishing, whereas making a chair requires 3 h of carpentry and $\frac{1}{2}$ h of finishing. The profit per table is \$35 and per chair is \$20. The manufacturer's employees can supply a maximum of 108 h of carpentry

work and 20 h of finishing work per day. How many tables and chairs should be made each day to maximize profits?

6. A housing contractor has subdivided a farm into 100 building lots. He has designed two types of homes for these lots: colonial and ranch style. A colonial requires $30,000 of capital and provides a profit of $4000. A ranch-style house requires $40,000 of capital and provides an $8000 profit. If he has $3.6 million of capital at hand, how many houses of each type should he build for maximum profit? Will any of the lots be left vacant?

7. A trucker is planning to carry citrus fruit from Florida to Montreal. Each crate of oranges is 4 ft^3 in volume and weighs 80 lb. Each crate of grapefruit has a volume of 6 ft^3 and weighs 100 lb. Her truck has a maximum capacity of 300 ft^3 and can carry no more than 5600 lb. Moreover, she is not permitted to carry more crates of grapefruit than crates of oranges. If she makes a profit of $2.50 on each crate of oranges and $4 on each crate of grapefruit, how many crates of each type of fruit should she carry for maximum profit?

8. A manufacturer of calculators produces two types: a standard type and a scientific model. Long-term demand for the two types mandates that the company manufacture at least 100 standard and 80 scientific calculators each day. However, because of limitations on production capacity, no more than 200 standard and 170 scientific calculators can be made daily. To satisfy a shipping contract, a total of at least 200 calculators must be shipped every day.
  (a) If it costs $5 to produce a standard calculator and $7 for a scientific one, how many of each type should be made each day to minimize costs?
  (b) If each standard calculator results in a $2 loss, but each scientific one produces a $5 profit, how many of each type should be made each day to maximize net profits?

9. An electronics discount chain has a sale on a certain brand of stereo. The chain has stores in Santa Monica and El Toro and warehouses in Long Beach and Pasadena. To satisfy rush orders, 15 sets must be shipped from the warehouses to the Santa Monica store, and 19 must be shipped to the El Toro store. The cost of shipping a set from Long Beach to Santa Monica is $5; from Long Beach to El Toro, $6; from Pasadena to Santa Monica, $4; and from Pasadena to El Toro, $5.50. If the Long Beach warehouse has 24 sets and the Pasadena warehouse has 18, how many sets should be shipped from each

warehouse to each store to fill the orders at a minimum shipping cost?

10. A man owns two building supply stores on the east and west sides of a city. Two customers order some $\frac{1}{2}$-inch plywood. Customer A needs 50 sheets and customer B needs 70 sheets. The east-side store has 80 sheets and the west-side store has 45 sheets of this plywood in stock. Delivery costs per sheet are: $0.50 from the east-side store to customer A, $0.60 from the east-side store to customer B, $0.40 from the west-side store to A, and $0.55 from the west-side store to B. How many sheets should be shipped from each store to each customer to minimize delivery costs?

11. Refer to Example 4 of Section 8.8. Suppose that a worker can load 12 yd^3 of Foamboard per minute and 10 yd^3 of Plastiflex per minute. If each cubic yard of Foamboard brings $0.50 profit and each cubic yard of Plastiflex $0.70, how should the carts be loaded to provide maximum profit per minute for each cartload?

12. A confectioner sells two different types of nut mixtures. Each package of the standard mixture contains 100 g of cashews and 200 g of peanuts and sells for $1.95. Each package of the deluxe mixture contains 150 g of cashews and 50 g of peanuts and sells for $2.25. The confectioner has 15 kg of cashews and 20 kg of peanuts available. Based on past sales statistics, he wishes to have at least as many standard as deluxe packages available. How many bags of each type of mixture should he package to maximize his revenue?

13. A furniture manufacturer has factories in two locations. The Vancouver factory can produce 20 sofas, 20 chairs, and 35 ottomans each day, whereas the Seattle factory can produce 50 sofas, 25 chairs, and 25 ottomans each day. It costs $3000 per day to operate the Vancouver factory and $4000 per day to operate the Seattle factory. An order for 400 sofas, 300 chairs, and 375 ottomans is received. The order is to be filled in no more than 30 days. How many days should each factory be operated to fill the order at minimum cost?

14. A biologist wishes to feed laboratory rabbits a mixture of two different types of foods. Type I contains 8 g of fat, 12 g of carbohydrate, and 2 g of protein per ounce, whereas type II contains 12 g of fat, 12 g of carbohydrate, and 1 g of protein per ounce. Type I costs $0.20 per ounce and type II costs $0.30 per ounce. Each rabbit is to receive a daily minimum of 24 g of fat, 36 g of carbohydrate, and 4 g of protein, but should get no more than 5 oz of food per day. How many ounces of each type of food should be fed to each rabbit daily to satisfy the requirements at minimum cost?

15. A woman wishes to invest $12,000 in three types of bonds: municipal bonds paying 7% interest per annum, bank investment certificates paying 8%, and high-risk bonds paying 12%. For tax reasons, she wants the amount invested in municipal bonds to be at least three times as much as the amount invested in bank certificates. To keep her level of risk manageable, she will invest no more than $2000 in high-risk bonds. How much should she invest in each type of bond to maximize her annual interest yield? [*Hint:* Let $x =$ amount in municipal bonds and $y =$ amount in bank certificates. Then the amount in high-risk bonds will be $12,000 - x - y$.]

16. Refer to Exercise 15. Suppose the investor decides to increase to $3000 the maximum amount she will allow herself to invest in high-risk bonds but leaves the other conditions unchanged. By how much will her maximum possible interest yield increase?

17. A small software company publishes computer games and educational and utility software. Their policy is to market a total of 36 new programs each year, with at least 4 of these being games. The number of utility programs pub-

lished is never more than twice the number of educational programs. On the average, the company can expect to make an annual profit of $5000 on each computer game, $8000 on each educational program, and $6000 on each utility program. How many of each type of program should they publish annually for maximum profit?

18. All parts of this problem refer to the following feasible region and objective function.

$$\begin{cases} x \geqslant 0 \\ x \geqslant y \\ x + 2y \leqslant 12 \\ x + y \leqslant 10 \end{cases}$$
$$P = x + 4y$$

(a) Graph the feasible region.

(b) On your graph from part (a), sketch the graphs of the linear equations obtained by setting P equal to 40, 36, 32, and 28.

(c) If we continue to decrease the value of P, at which vertex of the feasible region will these lines first touch the feasible region?

(d) Verify that the maximum value of P on the feasible region occurs at the vertex you chose in part (c).

SECTION 8.10
APPLICATION: PARTIAL FRACTIONS

The process of adding and subtracting fractions by first writing them with a common denominator is familiar for both numerical fractions and rational functions. Thus it is easy to add

$$\frac{1}{3} + \frac{1}{4} = \frac{4}{12} + \frac{3}{12} = \frac{4 + 3}{12} = \frac{7}{12}$$

or to subtract

$$\frac{1}{x - 1} - \frac{1}{x + 1} = \frac{x + 1}{(x + 1)(x - 1)} - \frac{x - 1}{(x + 1)(x - 1)} = \frac{2}{x^2 - 1}$$

The ancient Egyptians used a number notation that required them to write all fractions as sums of reciprocals of whole numbers. For them it was therefore important to know how to reverse this process—for example, to be able to write 7/12 as the sum of the more elementary fractions 1/3 and 1/4. For us this skill is of little use in the context of numerical fractions. For rational functions, however, the process opposite to "bringing to a common denominator" will turn out to be

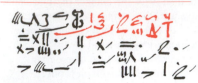

*The oldest known mathematical document is the **Rhind papyrus**, an Egyptian scroll written in 1650 B.C. by the scribe Ahmes, who writes that it is an exact copy of one from two hundred years earlier. Ahmes claims that his papyrus contains "a thorough study of all things, insight into all that exists, knowledge of all obscure secrets." Actually, the document contains rules for doing arithmetic including multiplication and division of fractions and several exercises with solutions. One exercise reads: A heap and its seventh make nineteen; how large is the heap?*

very important in your study of calculus. You will have to know, for example, how to split up $2/(x^2 - 1)$ into the difference of the simpler functions $1/(x - 1)$ and $1/(x + 1)$. These simpler functions are called **partial fractions**, and since the process of finding them involves solving linear equations, we study it in this section.

Let r be the rational function

$$r(x) = \frac{P(x)}{Q(x)}$$

where the degree of P is less than the degree of Q. It can be shown, using advanced algebra techniques, that every polynomial with real coefficients can be factored completely into linear and irreducible quadratic factors; that is, factors of the form $ax + b$ and $ax^2 + bx + c$, where a, b, and c are real numbers. For instance,

$$x^4 - 1 = (x^2 - 1)(x^2 + 1) = (x - 1)(x + 1)(x^2 + 1)$$

After we have factored the denominator Q of r completely, we will be able to express $r(x)$ as a sum of **partial fractions** of the form

$$\frac{A}{(ax + b)^i} \quad \text{and} \quad \frac{Ax + B}{(ax^2 + bx + c)^j}$$

This sum is called the **partial fraction decomposition** of r. We now explain the details in the four cases that occur.

Case 1: The Denominator $Q(x)$ Is a Product of Distinct Linear Factors

This means that we can write

$$Q(x) = (a_1x + b_1)(a_2x + b_2) \cdots (a_nx + b_n)$$

where no factor is repeated. In this case the partial fraction decomposition of r takes the form

$$r(x) = \frac{P(x)}{Q(x)} = \frac{A_1}{a_1x + b_1} + \frac{A_2}{a_2x + b_2} + \cdots + \frac{A_n}{a_nx + b_n}$$

where the constants $A_1, A_2, \ldots, A_n$ are determined as in the following example.

EXAMPLE 1

Find the partial fraction decomposition of

$$\frac{5x + 7}{x^3 + 2x^2 - x - 2}$$

SOLUTION

The denominator factors as follows.

$$x^3 + 2x^2 - x - 2 = x^2(x + 2) - (x + 2) = (x^2 - 1)(x + 2)$$
$$= (x - 1)(x + 1)(x + 2)$$

This gives us the partial fraction decomposition

$$\frac{5x + 7}{x^3 + 2x^2 - x - 2} = \frac{A}{x - 1} + \frac{B}{x + 1} + \frac{C}{x + 2}$$

Multiplying both sides by the common denominator $(x - 1)(x + 1)(x + 2)$, we get

$$5x + 7 = A(x + 1)(x + 2) + B(x - 1)(x + 2) + C(x - 1)(x + 1)$$
$$= A(x^2 + 3x + 2) + B(x^2 + x - 2) + C(x^2 - 1)$$
$$= (A + B + C)x^2 + (3A + B)x + (2A - 2B - C)$$

If two polynomials are equal, their coefficients are equal. Thus, since $5x + 7$ has no x^2 term, we have that $A + B + C = 0$. Similarly, by comparing the coefficients of x, we see that $3A + B = 5$, and by comparing constant terms, we get that $2A - 2B - C = 7$. This leads to the following system of linear equations for A, B, and C.

$$\begin{cases} A + B + C = 0 \\ 3A + B \phantom{{}+ C} = 5 \\ 2A - 2B - C = 7 \end{cases}$$

We solve this system using the methods developed in Section 8.2.

$$\begin{bmatrix} 1 & 1 & 1 & 0 \\ 3 & 1 & 0 & 5 \\ 2 & -2 & -1 & 7 \end{bmatrix} \xrightarrow[\substack{R_2 \to R_2 - 3R_1 \\ R_3 \to R_3 - 2R_1}]{} \begin{bmatrix} 1 & 1 & 1 & 0 \\ 0 & -2 & -3 & 5 \\ 0 & -4 & -3 & 7 \end{bmatrix} \xrightarrow[\substack{R_3 \to R_3 - 2R_2 \\ -R_2}]{}$$

$$\begin{bmatrix} 1 & 1 & 1 & 0 \\ 0 & 2 & 3 & -5 \\ 0 & 0 & 3 & -3 \end{bmatrix} \xrightarrow[\substack{R_2 \to R_2 - R_3 \\ \frac{1}{3}R_3}]{} \begin{bmatrix} 1 & 1 & 1 & 0 \\ 0 & 2 & 0 & -2 \\ 0 & 0 & 1 & -1 \end{bmatrix}$$

Thus we see that $C = -1$, $B = -1$, and $A = 2$, so the required partial fraction decomposition is

$$\frac{5x + 7}{x^3 + 2x^2 - x - 2} = \frac{2}{x - 1} + \frac{-1}{x + 1} + \frac{-1}{x + 2} \qquad \blacksquare$$

The same method of attack works in each of the remaining cases. We set up the partial fraction decomposition with the unknown constants $A, B, C, \ldots$. We then multiply both sides of the resulting equation by the common denominator, simplify the right-hand side of the equation, and equate coefficients. This gives a set of linear equations that will always have a unique solution (provided the partial fraction decomposition has been set up correctly).

Case 2: The Denominator $Q(x)$ Is a Product of Linear Factors, Some of Which Are Repeated

Suppose the complete factorization of $Q(x)$ contains the linear factor $ax + b$ repeated k times; that is, $(ax + b)^k$ is a factor of $Q(x)$. Then corresponding to each such factor the partial fraction decomposition for $P(x)/Q(x)$ will contain

$$\frac{A_1}{ax + b} + \frac{A_2}{(ax + b)^2} + \cdots + \frac{A_k}{(ax + b)^k}$$

EXAMPLE 2

Find the partial fraction decomposition of

$$\frac{x^2 + 1}{x(x - 1)^3}$$

SOLUTION

Because the factor $x - 1$ is repeated three times in the denominator, the partial fraction decomposition is

$$\frac{x^2 + 1}{x(x - 1)^3} = \frac{A}{x} + \frac{B}{x - 1} + \frac{C}{(x - 1)^2} + \frac{D}{(x - 1)^3}$$

Multiplying both sides by the common denominator $x(x - 1)^3$ gives

$$\begin{aligned} x^2 + 1 &= A(x - 1)^3 + Bx(x - 1)^2 + Cx(x - 1) + Dx \\ &= A(x^3 - 3x^2 + 3x - 1) + B(x^3 - 2x^2 + x) + C(x^2 - x) + Dx \\ &= (A + B)x^3 + (-3A - 2B + C)x^2 + (3A + B - C + D)x - A \end{aligned}$$

Equating coefficients, we get the equations

$$\begin{cases} A + B & = 0 \\ -3A - 2B + C & = 1 \\ 3A + B - C + D & = 0 \\ -A & = 1 \end{cases}$$

If we rearrange these equations by putting the last one in the first position, we can easily see (without having to use matrix techniques) that the solution to the system is $A = -1$, $B = 1$, $C = 0$, and $D = 2$, so

$$\frac{x^2 + 1}{x(x - 1)^3} = \frac{-1}{x} + \frac{1}{x - 1} + \frac{2}{(x - 1)^3} \qquad \blacksquare$$

Case 3: The Denominator $Q(x)$ Has Irreducible Quadratic Factors, None of Which Is Repeated

If the complete factorization of $Q(x)$ contains the quadratic factor $ax^2 + bx + c$ (which cannot be factored further), then corresponding to this the partial fraction decomposition of $P(x)/Q(x)$ will have a term of the form

$$\frac{Ax + B}{ax^2 + bx + c}$$

EXAMPLE 3

Find the partial fraction decomposition of

$$\frac{2x^2 - x + 4}{x^3 + 4x}$$

SOLUTION

Since $x^3 + 4x = x(x^2 + 4)$, which cannot be factored further, we write

$$\frac{2x^2 - x + 4}{x^3 + 4x} = \frac{A}{x} + \frac{Bx + C}{x^2 + 4}$$

Multiplying by $x(x^2 + 4)$, we get

$$2x^2 - x + 4 = A(x^2 + 4) + (Bx + C)x$$
$$= (A + B)x^2 + Cx + 4A$$

Equating coefficients gives the equations

$$\begin{cases} A + B = 2 \\ \quad\quad C = -1 \\ 4A = 4 \end{cases}$$

and so $A = 1$, $B = 1$, and $C = -1$. The required partial fraction decomposition is

$$\frac{2x^2 - x + 4}{x^3 + 4x} = \frac{1}{x} + \frac{x - 1}{x^2 + 4}$$ ■

Case 4: The Denominator $Q(x)$ Has a Repeated Irreducible Quadratic Factor

If the complete factorization of $Q(x)$ contains the factor $(ax^2 + bx + c)^k$, where $ax^2 + bx + c$ cannot be factored further, then corresponding to this the partial fraction decomposition of $P(x)/Q(x)$ will have the terms

$$\frac{A_1x + B_1}{ax^2 + bx + c} + \frac{A_2x + B_2}{(ax^2 + bx + c)^2} + \cdots + \frac{A_kx + B_k}{(ax^2 + bx + c)^k}$$

EXAMPLE 4

Write out the form of the partial fraction decomposition of

$$\frac{x^5 - 3x^2 + 12x - 1}{x^3(x^2 + x + 1)(x^2 + 2)^3}$$

SOLUTION

$$\frac{x^5 - 3x^2 + 12x - 1}{x^3(x^2 + x + 1)(x^2 + 2)^3}$$

$$= \frac{A}{x} + \frac{B}{x^2} + \frac{C}{x^3} + \frac{Dx + E}{x^2 + x + 1} + \frac{Fx + G}{x^2 + 2} + \frac{Hx + I}{(x^2 + 2)^2} + \frac{Jx + K}{(x^2 + 2)^3} \quad ■$$

In order to find the values of $A, B, C, D, E, F, G, H, I, J,$ and K in Example 4, we would have to solve a system of 11 linear equations. Although certainly possible, this would involve a great deal of work!

It is important to note that the techniques we have described in this section apply only to rational functions $P(x)/Q(x)$ in which the degree of P is less than the degree of Q. If this is not the case, we must first use long division to divide Q into P.

EXAMPLE 5

Find the partial fraction decomposition of

$$\frac{2x^4 + 4x^3 - 2x^2 + x + 7}{x^3 + 2x^2 - x - 2}$$

SOLUTION

Since the degree of the numerator is larger than the degree of the denominator, we use long division to obtain

$$\frac{2x^4 + 4x^3 - 2x^2 + x + 7}{x^3 + 2x^2 - x - 2} = 2x + \frac{5x + 7}{x^3 + 2x^2 - x - 2}$$

The remainder term now satisfies the requirement that the degree of the numerator is less than the degree of the denominator. At this point we would proceed as in Example 1 to obtain the decomposition

$$\frac{2x^4 + 4x^3 - 2x^2 + x + 7}{x^3 + 2x^2 - x - 2} = 2x + \frac{2}{x - 1} + \frac{-1}{x + 1} + \frac{-1}{x + 2} \quad ■$$

EXERCISES 8.10

In Exercises 1–26 find the partial fraction decomposition of the given rational function.

1. $\dfrac{4}{x^2 - 4}$

2. $\dfrac{2x + 1}{x^2 + x - 2}$

3. $\dfrac{x + 14}{x^2 - 2x - 8}$

4. $\dfrac{8x - 3}{2x^2 - x}$

5. $\dfrac{x}{8x^2 - 10x + 3}$

6. $\dfrac{7x - 3}{x^3 + 2x^2 - 3x}$

7. $\dfrac{9x^2 - 9x + 6}{2x^3 - x^2 - 8x + 4}$

8. $\dfrac{-3x^2 - 3x + 27}{(x + 2)(2x^2 + 3x - 9)}$

9. $\dfrac{x^2 + 1}{x^3 + x^2}$

10. $\dfrac{3x^2 + 5x - 13}{(3x + 2)(x^2 - 4x + 4)}$

11. $\dfrac{2x}{4x^2 + 12x + 9}$

12. $\dfrac{x - 4}{(2x - 5)^2}$

13. $\dfrac{4x^2 - x - 2}{x^4 + 2x^3}$

14. $\dfrac{x^3 - 2x^2 - 4x + 3}{x^4}$

15. $\dfrac{-10x^2 + 27x - 14}{(x - 1)^3(x + 2)}$

16. $\dfrac{-2x^2 + 5x - 1}{x^4 - 2x^3 + 2x - 1}$

17. $\dfrac{3x^3 + 22x^2 + 53x + 41}{(x + 2)^2(x + 3)^2}$

18. $\dfrac{3x^2 + 12x - 20}{x^4 - 8x^2 + 16}$

19. $\dfrac{x - 3}{x^3 + 3x}$

20. $\dfrac{3x^2 - 2x + 8}{x^3 - x^2 + 2x - 2}$

21. $\dfrac{2x^3 + 7x + 5}{(x^2 + x + 2)(x^2 + 1)}$

22. $\dfrac{x^2 + x + 1}{2x^4 + 3x^2 + 1}$

23. $\dfrac{x^4 + x^3 + x^2 - x + 1}{x(x^2 + 1)^2}$

24. $\dfrac{2x^2 - x + 8}{(x^2 + 4)^2}$

25. $\dfrac{x^5 - 2x^4 + x^3 + x + 5}{x^3 - 2x^2 + x - 2}$

26. $\dfrac{x^5 - 3x^4 + 3x^3 - 4x^2 + 4x + 12}{(x - 2)^2(x^2 + 2)}$

In Exercises 27 and 28 give the form of the partial fraction decomposition of the given rational function (as in Example 4).

27. $\dfrac{x^3 + x + 1}{x(2x - 5)^3(x^2 + 2x + 5)^2}$

28. $\dfrac{1}{(x^6 - 1)(x^4 - 1)}$

29. Determine A and B in terms of a and b:

$$\frac{ax + b}{x^2 - 1} = \frac{A}{x - 1} + \frac{B}{x + 1}$$

30. Determine A, B, C, and D in terms of a and b:

$$\frac{ax^3 + bx^2}{(x^2 + 1)^2} = \frac{Ax + B}{x^2 + 1} + \frac{Cx + D}{(x^2 + 1)^2}$$

CHAPTER 8 REVIEW

Define, state, or discuss each of the following.

1. System of equations

2. Linear equations

3. Gaussian elimination

4. Triangular form

5. Elementary row operations

6. Matrix

7. Echelon form

8. Reduced echelon form

9. Leading variables

10. Inconsistent system of equations

11. Dependent system of equations

12. Addition and subtraction of matrices

13. Product of matrices

14. Identity matrix

15. Inverse of a matrix

16. Minor

17. Cofactor

18. Determinant

19. Invertibility criterion

20. Row and column transformations of a determinant

21. Cramer's Rule

22. Nonlinear system of equations

23. System of inequalities

24. Bounded and unbounded regions

25. Vertex

26. Linear programming

27. Feasible region

28. Objective function

29. Partial fraction decomposition

REVIEW EXERCISES

In Exercises 1–6 solve the system of equations and graph the lines.

1. $\begin{cases} 2x + 4y = 16 \\ 4x - y = 5 \end{cases}$

2. $\begin{cases} y = 4x + 4 \\ x = 3y + 10 \end{cases}$

3. $\begin{cases} 2x - 7y = 28 \\ y = \frac{2}{7}x - 4 \end{cases}$

4. $\begin{cases} 6x - 8y = 15 \\ -\frac{3}{2}x + 2y = -4 \end{cases}$

5. $\begin{cases} 2x - y = 1 \\ x + 3y = 10 \\ 3x + 4y = 15 \end{cases}$

6. $\begin{cases} 2x + 5y = 9 \\ -x + 3y = 1 \\ 7x - 2y = 14 \end{cases}$

In Exercises 7–14 find the complete solution of the system using Gaussian elimination, or show that there is no solution.

7. $\begin{cases} x - 2y + z = 0 \\ 3x - y + 2z = 0 \\ 4x - 9y = 21 \end{cases}$

8. $\begin{cases} 2x - y + 3z = 2 \\ 4x - 9z = 2 \\ 3x + 2y + 6z = 18 \end{cases}$

9. $\begin{cases} x - 2y + 3z = 1 \\ 2x - y + z = 3 \\ 2x - 7y + 11z = 2 \end{cases}$

10. $\begin{cases} x + y + z + w = 2 \\ 2x - 3z = 5 \\ x - 2y + 4w = 9 \\ x + y + 2z + 3w = 5 \end{cases}$

11. $\begin{cases} x - 3y + z = 4 \\ 4x - y + 15z = 5 \end{cases}$

12. $\begin{cases} 2x - 3y + 4z = 3 \\ 4x - 5y + 9z = 13 \\ 2x + 7z = 0 \end{cases}$

13. $\begin{cases} -x + 4y + z = 8 \\ 2x - 6y + z = -9 \\ x - 6y - 4z = -15 \end{cases}$

14. $\begin{cases} x - z + w = 2 \\ 2x + y - 2w = 12 \\ 3y + z + w = 4 \\ x + y - z = 10 \end{cases}$

15. A man invests his savings in two accounts, one paying 6% interest per annum and the other paying 7%. He has twice as much invested in the 7% account as in the 6% account, and his annual interest income is $600. How much does he have invested in each account?

16. Find the values of *a*, *b*, and *c* if the parabola

$$y = ax^2 + bx + c$$

is to pass through the points $(1, 0)$, $(-1, -4)$, and $(2, 11)$.

In Exercises 17–28, let

$$A = \begin{bmatrix} 2 & 0 & -1 \end{bmatrix} \qquad B = \begin{bmatrix} 1 & 2 & 4 \\ -2 & 1 & 0 \end{bmatrix}$$

$$C = \begin{bmatrix} \frac{1}{2} & 3 \\ 2 & \frac{3}{2} \\ -2 & 1 \end{bmatrix} \qquad D = \begin{bmatrix} 1 & 4 \\ 0 & -1 \\ 2 & 0 \end{bmatrix}$$

$$E = \begin{bmatrix} 2 & -1 \\ -\frac{1}{2} & 1 \end{bmatrix} \qquad F = \begin{bmatrix} 4 & 0 & 2 \\ -1 & 1 & 0 \\ 7 & 5 & 0 \end{bmatrix}$$

$$G = \begin{bmatrix} 5 \end{bmatrix}$$

Carry out the operation indicated in each exercise, or explain why it cannot be performed.

17. $A + B$

18. $C - D$

19. $2C + 3D$

20. $5B - 2C$

21. GA

22. AG

23. BC

24. CB

25. BF

26. FC

27. $(C + D)E$

28. $F(2C - D)$

In Exercises 29–34 find the determinant and, if possible, the inverse of the matrix.

29. $\begin{bmatrix} 1 & 4 \\ 2 & 9 \end{bmatrix}$

30. $\begin{bmatrix} 2 & 2 \\ 1 & -3 \end{bmatrix}$

31. $\begin{bmatrix} 4 & -12 \\ -2 & 6 \end{bmatrix}$

32. $\begin{bmatrix} 2 & 4 & 0 \\ -1 & 1 & 2 \\ 0 & 3 & 2 \end{bmatrix}$

33. $\begin{bmatrix} 3 & 0 & 1 \\ 2 & -3 & 0 \\ 4 & -2 & 1 \end{bmatrix}$

34. $\begin{bmatrix} 1 & 0 & 0 & 1 \\ 0 & 2 & 0 & 2 \\ 0 & 0 & 3 & 3 \\ 0 & 0 & 0 & 4 \end{bmatrix}$

In Exercises 35 and 36 express the system of linear equations as a matrix equation. Then solve the matrix equation by multiplying both sides by the inverse of the coefficient matrix.

35. $\begin{cases} 12x - 5y = 10 \\ 5x - 2y = 17 \end{cases}$

36. $\begin{cases} 2x + y + 5z = \frac{1}{3} \\ x + 2y + 2z = \frac{1}{4} \\ x \qquad + 3z = \frac{1}{6} \end{cases}$

In Exercises 37–40 solve the system using Cramer's Rule.

37. $\begin{cases} 2x + 7y = 13 \\ 6x + 16y = 30 \end{cases}$

38. $\begin{cases} 12x - 11y = 140 \\ 7x + 9y = 20 \end{cases}$

39. $\begin{cases} 2x - y + 5z = 0 \\ -x + 7y \qquad = 9 \\ 5x + 4y + 3z = -9 \end{cases}$

40. $\begin{cases} 3x + 4y - z = 10 \\ x \qquad - 4z = 20 \\ 2x + y + 5z = 30 \end{cases}$

In Exercises 41–44 find all solutions of the system.

41. $\begin{cases} x^2 + y^2 + 6y = 0 \\ x - 2y = 3 \end{cases}$

42. $\begin{cases} x^2 + y^2 = 10 \\ x^2 + 2y^2 - 7y = 0 \end{cases}$

43. $\begin{cases} 3x^4 + \dfrac{4}{y} = 50 \\ x^4 - \dfrac{8}{y} = 12 \end{cases}$

44. $\begin{cases} x^2 + yz = 0 \\ y^2 + xz = 0 \\ x^2 + xy + y^2 = 3 \end{cases}$

In Exercises 45–48 graph the solution set of the system of inequalities. Find the coordinates of all vertices, and determine if the solution set is bounded or unbounded.

45. $\begin{cases} x^2 + y^2 < 9 \\ x + y < 0 \end{cases}$

46. $\begin{cases} y - x^2 \geq 4 \\ y < 20 \end{cases}$

47. $\begin{cases} x \geq 0, \ y \geq 1 \\ x + 2y \leq 12 \\ y \leq x + 4 \end{cases}$

48. $\begin{cases} x \geq 4 \\ x + y \geq 24 \\ x \leq 2y + 12 \end{cases}$

49. Find the maximum and minimum values of the function $P = 3x + 4y$ on the region described by the inequalities in Exercise 47.

50. (a) Find the minimum value of the function $Q = 60 + 3x + 5y$ on the region described by the inequalities in Exercise 48.
(b) Explain why Q has no maximum value on this region.

51. A farmer wishes to plant oats and barley on 400 acres of his land. The land can produce 40 bushels of oats or 50 bushels of barley per acre. After harvest, the farmer will have to store the grain for several months in order to get the best price for it, and he has facilities to store no more than 18,000 bushels of grain.
(a) If he can get $2.05 per bushel for oats and $1.80 for barley, how many acres of each grain should he plant for maximum revenue?
(b) If instead the price for oats is $1.20 and for barley is $1.60 per bushel, how many acres of each should he plant to maximize his revenue?

52. A woman wishes to invest $12,000, some in a high-risk stock with an expected annual dividend of 15%, some in long-term bonds yielding 10% interest per annum, and the remainder in a bank money market account paying an annual yield of 6% interest. She wishes to put at least $4000 into the money market account. Also, the amount invested in the high-risk stock should be no more than half the total of her other two investments. How much should she put in each investment to maximize her annual interest and dividend yield?

In Exercises 53 and 54 solve for x, y, and z in terms of a, b, and c.

53. $\begin{cases} -x + y + z = a \\ x - y + z = b \\ x + y - z = c \end{cases}$

54. $\begin{cases} ax + by + cz = a - b + c \\ bx + by + cz = c \\ cx + cy + cz = c \end{cases}$
$(a \neq b, \ b \neq c, \ c \neq 0)$

55. For what values of k do the three lines
$$x + y = 12$$
$$kx - y = 0$$
$$y - x = 2k$$
have a common point of intersection?

56. For what value of k does the system
$$\begin{cases} kx + y + z = 0 \\ x + 2y + kz = 0 \\ -x \qquad + 3z = 0 \end{cases}$$
have infinitely many solutions?

 Use a graphing device to solve each of the systems in Exercises 57–60, correct to the nearest hundredth.

57. $\begin{cases} 0.32x + 0.43y = 0 \\ 7x - 12y = 341 \end{cases}$

58. $\begin{cases} \sqrt{12}x - 3\sqrt{2}y = 660 \\ 7137x + 3931y = 20{,}000 \end{cases}$

59. $\begin{cases} x - y^2 = 10 \\ \quad x = \frac{1}{22}y + 12 \end{cases}$ **60.** $\begin{cases} y = 5^x + x \\ y = x^5 + 5 \end{cases}$

In Exercises 61–64 find the partial fraction decomposition of the rational function.

61. $\dfrac{3x + 1}{x^2 - 2x - 15}$ **62.** $\dfrac{8}{x^3 - 4x}$

63. $\dfrac{2x - 4}{x(x - 1)^2}$ **64.** $\dfrac{x + 6}{x^3 - 2x^2 + 4x - 8}$

CHAPTER 8 TEST

1. An airplane takes $2\frac{1}{2}$ h to make a trip of 600 km against the wind. It takes 50 min to travel 300 km with the wind. Find the speed of the wind and the speed of the airplane in still air.

In Problems 2–5 find all solutions of the system. Determine whether the system is linear or nonlinear. If it is linear, state whether it is inconsistent, dependent, or neither.

2. $\begin{cases} 2x - 5y = 9 \\ 7x + 6y = 8 \end{cases}$

3. $\begin{cases} 3x - y + z = 5 \\ x \quad\quad - 4z = 7 \\ x - y + 9z = -8 \end{cases}$

4. $\begin{cases} 2x - y + z = 0 \\ 3x + 2y - 3z = 1 \\ x - 4y + 5z = -1 \end{cases}$ **5.** $\begin{cases} 2x^2 + y^2 = 6 \\ 3x^2 - 4y = 11 \end{cases}$

In Problems 6–13, let

$$A = \begin{bmatrix} 2 & 3 \\ 2 & 4 \end{bmatrix}$$

$$B = \begin{bmatrix} 1 & 6 \\ 0 & 1 \\ -1 & 5 \end{bmatrix} \qquad C = \begin{bmatrix} 1 & 0 & 4 \\ -1 & 1 & 2 \\ 0 & 1 & 3 \end{bmatrix}$$

Find the following, or explain why the indicated operation cannot be performed.

6. $A + B$ **7.** AB **8.** $BA - 3B$

9. CBA **10.** A^{-1} **11.** B^{-1}

12. $|B|$ **13.** $|C|$

14. Write a matrix equation equivalent to the given system of linear equations. Find the inverse of the coefficient matrix, and use it to solve the system.

$$\begin{cases} 3x - 5y = 51 \\ 2x + 3y = 64 \end{cases}$$

15. Solve using Cramer's Rule:

$$\begin{cases} 2x \quad\quad - z = 14 \\ 3x - y + 5z = 0 \\ 4x + 2y + 3z = -2 \end{cases}$$

16. Only one of the following matrices has an inverse. Find the determinant of each matrix, and use the determinants to identify the one that has an inverse. Then find the inverse.

$$A = \begin{bmatrix} 1 & 4 & 1 \\ 0 & 2 & 0 \\ 1 & 0 & 1 \end{bmatrix} \qquad B = \begin{bmatrix} 1 & 4 & 0 \\ 0 & 2 & 0 \\ -3 & 0 & 1 \end{bmatrix}$$

17. Graph the following system of inequalities, indicating the coordinates of the vertices:

$$\begin{cases} x^2 - 2x - y + 5 \leq 0 \\ \quad\quad\quad\quad y \leq 5 + 2x \end{cases}$$

18. Find the partial fraction decomposition of the rational function

$$\frac{4x - 1}{(x - 1)^2(x + 2)}$$

19. A farmer grows wheat and barley on 200 acres of land. He can borrow no more than $10,000 at the beginning of the season for production costs, and has no other financial resources available. It costs $60 per acre to grow wheat and $40 per acre to grow barley, and the land yields 50 bushels of wheat or 40 bushels of barley per acre. The market predictions for the fall harvest indicate that he can make a profit of $2.50 per bushel of wheat and $2.00 per bushel of barley. How many acres of each grain should he plant to maximize his profits?

20. Use a graphing calculator to find all solutions of the following system, correct to two decimals.

$$\begin{cases} 2x^2 + y^2 = 16 \\ \quad\quad y = x^4 - 4x^3 + 6x^2 - 4x \end{cases}$$

FOCUS ON PROBLEM SOLVING

It is sometimes useful to attack a problem by first **working backward**. This means that we assume the conclusion and work backward step by step until we arrive at something that is given or known. Then we may be able to reverse the steps in the argument and proceed forward from the given to the conclusion.

We have already used this procedure in solving equations. For example, in solving the equation $2x + 7 = 23$ we assume that x is a number that satisfies $2x + 7 = 23$ and work backward. We subtract 7 from each side of the equation and then divide each side by 2 to get $x = 8$. Each of these steps can be reversed, so we have solved the problem.

EXAMPLE 1

How is it possible to bring up from a river exactly 6 gallons of water when you only have two containers, a 9-gallon pail and a 4-gallon pail?

SOLUTION

If you work this problem in the forward direction you might be lucky and discover, out of the many possibilities for proceeding, a correct solution. But it is more systematic to work backward.

Imagine that we have 6 gal of water in the 9-gal pail, together with the full 4-gal pail. We could fill the larger pail from the smaller one, leaving just 1 gal in the smaller pail. Then we could empty the large pail and pour the 1 gal of water into it. Finally we could exactly fill the larger pail by adding 8 gal of water using the smaller pail twice.

Now, by reversing the procedure, we have the solution to the problem. Start with a full 9-gal pail. Use it to fill the 4-gal pail, then empty the 4-gal pail. Again fill the 4-gal pail from the larger one and empty the smaller one. This leaves 1 gal in the larger pail. Transfer it to the smaller pail and fill the larger one. Use the larger pail to fill the smaller one. This leaves 6 gal in the larger pail. ■

In the next example we combine the technique of working backward with the method of taking cases.

EXAMPLE 2

Solve the equation $\left|2x - |3x + 1|\right| = 1$.

SOLUTION

We assume that x is a number such that $\left|2x - |3x + 1|\right| = 1$. It follows from Property 4 of absolute values that there are two cases:

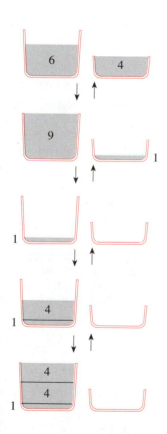

Work backward

Take cases
$$2x - |3x + 1| = 1 \qquad \text{or} \qquad 2x - |3x + 1| = -1$$

$$|3x + 1| = 2x - 1 \qquad\qquad |3x + 1| = 2x + 1$$

Take cases again

$3x + 1 = 2x - 1$ or $3x + 1 = 1 - 2x$ $\qquad$ $3x + 1 = 2x + 1$ or $3x + 1 = -2x - 1$

$\quad x = -2 \qquad\qquad 5x = 0 \qquad\qquad\qquad x = 0 \qquad\qquad 5x = -2$

$\qquad\qquad\qquad\qquad x = 0 \qquad\qquad\qquad\qquad\qquad\qquad x = -\frac{2}{5}$

Thus working backward gives three potential solutions: $x = 0$, $x = -2$, and $x = -\frac{2}{5}$. But is it possible to reverse the steps in each case? Let us check by trying to verify that these are indeed solutions:

Try $x = 0$: $\quad \big|2(0) - |3(0) + 1|\big| = |0 - 1| = 1 \qquad\qquad$ 0 is a solution

Try $x = -2$: $\quad \big|2(-2) - |3(-2) + 1|\big| = |-4 - 5| = 9 \qquad$ -2 is not a solution

Try $x = -\frac{2}{5}$: $\quad \big|2\big(-\frac{2}{5}\big) - |3\big(-\frac{2}{5}\big) + 1|\big| = \big|-\frac{4}{5} - \frac{1}{5}\big| = 1 \qquad$ $-\frac{2}{5}$ is a solution

Therefore the only solutions are 0 and $-\frac{2}{5}$. ∎

PROBLEMS

1. Three containers hold 19 gal, 13 gal, and 7 gal, respectively. The 19-gallon container is empty. The other two are full. How can you measure out 10 gal using no other container?

2. Justin, Sasha, and Vanessa each have a bag of marbles. First Justin gives Sasha and Vanessa each as many marbles as they already have. Then Sasha gives Justin and Vanessa as many marbles as they now have. Finally, Vanessa gives Justin and Sasha as many marbles as they now have. Everyone ends up with 16 marbles. How many did each person have to begin with?

In Problems 3–6 solve the equation.

3. $\big|5 - |x - 1|\big| = 3$ $\qquad\qquad\qquad\qquad$ 4. $\big||2x + 1| + 5\big| = 10$

5. $\big|4x - |x + 1|\big| = 3$ $\qquad\qquad\qquad\qquad$ 6. $\big||3x + 1| - x\big| = 2$

7. Draw the graph of the equation $|x| + |y| = 1 + |xy|$.

8. Sketch the region in the plane consisting of all points (x, y) such that

$$|x| + |y| \leq 1$$

9. Sketch the region in the plane consisting of all points (x, y) such that

$$|x - y| + |x| - |y| \leq 2$$

10. If x and y are positive real numbers, prove that

$$\sqrt{xy} \le \frac{x + y}{2}$$

11. Solve the following system of equations:

$$\begin{cases} 3x + 5y + z + 7t = -8 \\ 4x + 8y + 2z + 6t = -8 \\ 6x + 2y + 8z + 4t = 8 \\ 7x + y + 5z + 3t = 8 \end{cases}$$

[*Hint:* Instead of immediately using the methods of Chapter 8, look for a shortcut.]

12. Find the complete solution of the following system of equations:

$$\begin{cases} x + y + z = 2 \\ x^2 + y^2 + z^2 = 2 \\ xy = z^2 \end{cases}$$

9 TOPICS IN ANALYTIC GEOMETRY

In studying the procedures of geometric thought we may hope to reach what is most essential in the human mind.

HENRI POINCARÉ

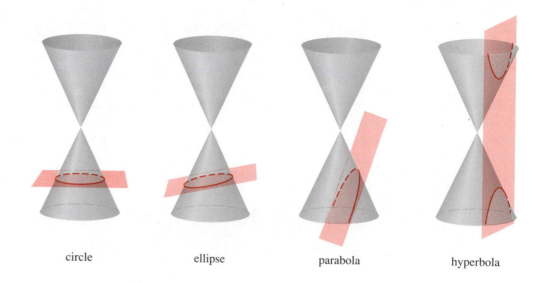

circle   ellipse   parabola   hyperbola

In this chapter we study the geometry of **conic sections** (or simply **conics**). Conic sections are the curves formed by the intersection of a plane with a pair of circular cones. These curves can have four basic shapes, called **circles**, **ellipses**, **parabolas**, and **hyperbolas**, as illustrated in the figure.

The ancient Greeks studied these curves because they considered the geometry of conic sections to be very beautiful. The mathematician Apollonius (262–190 B.C.) wrote a definitive eight-volume work on the subject. In more modern times, conics were found to be useful as well as beautiful. Galileo discovered in 1590 that the path of a missile shot upward at an angle is a parabola. In 1609, Kepler found that the planets move in elliptical orbits around the sun. In 1668, Newton was the first to build a reflecting telescope, whose principle is based on the properties of parabolas and hyperbolas. In this century, many further applications of conic sections have been developed. One important application is the LORAN radio navigation system, which uses the intersection points of hyperbolas to pinpoint the location of ships and aircraft.

In addition to studying conics, we learn in this chapter about two other ways of describing points and curves in the Cartesian plane: polar coordinates and parametric equations. Both of these topics require a thorough understanding of trigonometry.

SECTION 9.1
PARABOLAS

We have seen that the graph of the equation $y = ax^2 + bx + c$ is a U-shaped curve called a **parabola** that opens either upward or downward, depending on whether the sign of a is positive or negative (see Figure 1). The lowest or highest point of the parabola is called the **vertex**, and the parabola is symmetric about its **axis**.

In this section we study parabolas from a more geometric (rather than algebraic) point of view. We begin with the geometric definition of a parabola.

Figure 1
$y = ax^2 + bx + c$

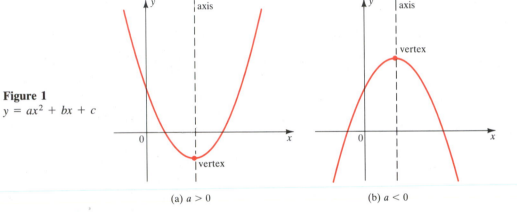

(a) $a > 0$ (b) $a < 0$

Archimedes (287–212 B.C.) *was the greatest mathematician of the ancient world. He was born and lived in Syracuse, a Greek colony on Sicily, a generation after Euclid (see page 38). He was renowned as a mechanical genius for his many engineering inventions, including pulleys to lift heavy ships and the spiral screw used for transporting water to higher levels. He is said to have used parabolic mirrors to concentrate the rays of the sun to set fire to the Roman ships attacking Syracuse. King Hieron II of Syracuse once suspected a goldsmith of keeping part of the gold intended for the king's crown and replacing it with an equal weight of silver. The king asked Archimedes for advice. While* deep in thought at a public bath, *Archimedes discovered the solution to the king's problem when he noticed that his body's volume was the same as the volume of water it displaced from the tub. As the story is told, he immediately ran home naked shouting "Eureka, eureka!" ["I have found it, I have found it!"]. This incident attests to his enormous powers of concentration. In spite of his engineering prowess, Archimedes was most proud of his mathematical discoveries. These included the formulas for the volume of the sphere: $V = \frac{4}{3}\pi r^3$; the surface area of a sphere: $S = 4\pi r^2$; and a careful analysis of the properties of parabolas and other conics.*

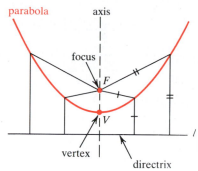

parabola

axis

focus

F

V

vertex

directrix

Figure 2

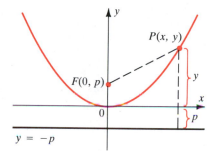

y

$P(x, y)$

$F(0, p)$

y

x

0

p

$y = -p$

Figure 3

A **parabola** is the set of points in a plane equidistant from a fixed point F (called the **focus**) and a fixed line l (called the **directrix**).

This definition is illustrated in Figure 2. Note that the vertex V of the parabola lies halfway between the focus and the directrix, and that the axis of symmetry is the line that runs through the focus perpendicular to the directrix.

In this section we restrict our attention to parabolas that are situated with the vertex at the origin and that have a vertical or horizontal axis of symmetry. (Parabolas in more general positions will be considered in Sections 9.4 and 9.5.) If the focus of such a parabola is the point $F(0, p)$, then the axis of symmetry must be vertical and the directrix has the equation $y = -p$. (See Figure 3, which illustrates the case $p > 0$.)

If $P(x, y)$ is any point on the parabola, then the distance from P to the focus F (using the Distance Formula) is

$$\sqrt{x^2 + (y - p)^2}$$

and the distance from P to the directrix is

$$|y - (-p)| = |y + p|$$

By the definition of a parabola, these two distances must be equal:

$$\sqrt{x^2 + (y - p)^2} = |y + p|$$

Squaring both sides and simplifying, we get

$$x^2 + (y - p)^2 = |y + p|^2 = (y + p)^2$$
$$x^2 + y^2 - 2py + p^2 = y^2 + 2py + p^2$$
$$x^2 - 2py = 2py$$
$$x^2 = 4py$$

PARABOLA WITH VERTICAL AXIS

The equation of a parabola with focus $F(0, p)$ and directrix $y = -p$ is

$$x^2 = 4py$$

If $p > 0$, then the parabola opens upward, but if $p < 0$, it opens downward (see Figure 4). If x is replaced by $-x$, the equation remains unchanged, so the graph is symmetric about the y-axis.

Figure 4

(a) $x^2 = 4py$ $p > 0$

(b) $x^2 = 4py$ $p < 0$

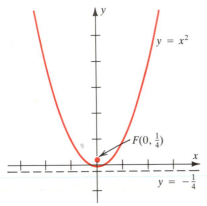

Figure 5

EXAMPLE 1

Find the focus and directrix of the parabola $y = x^2$ and sketch the graph.

SOLUTION

Comparing the equation $y = x^2$ with the general equation $x^2 = 4py$, we see that $4p = 1$, so $p = \frac{1}{4}$. Thus the focus is $\left(0, \frac{1}{4}\right)$ and the directrix is $y = -\frac{1}{4}$. The graph is shown in Figure 5.   ■

Reflecting the graph in Figure 3 about the diagonal line $y = x$ has the effect of interchanging the role of x and y. Thus, using the same sort of argument as before, we can show the following:

Galileo Galilei (1564–1642) was born in Pisa, Italy. He began his university education by studying medicine, but later abandoned this in favor of science and mathematics. At the age of 25 he demonstrated that light objects fall at the same rate as heavier ones, by dropping cannonballs of various sizes from the Leaning Tower of Pisa. This contradicted the then-accepted view of Aristotle that heavier objects fall more quickly. He also showed that the distance an object falls is proportional to the square of the time it has been falling, and from this was able to prove that the path of a projectile is a parabola.

Galileo was a gifted inventor. He constructed the first telescope, and using it, discovered sunspots and the moons of Jupiter. His advocacy of the Copernican view that the earth revolves around the sun led to his being called before the Inquisition. By then an old man, almost blind from observing the sun through his telescope, he was forced to recant his views, but he is said to have muttered under his breath "the earth nevertheless does move." Galileo revolutionized science by expressing scientific principles in the language of mathematics. He said, "The great book of nature is written in mathematical symbols."

**PARABOLA WITH
HORIZONTAL AXIS**

> The equation of a parabola with focus $F(p, 0)$ and directrix $x = -p$ is
>
> $$y^2 = 4px$$

In this case the x-axis is the axis of symmetry, and if $p > 0$, the parabola opens to the right, whereas if $p < 0$, it opens to the left (see Figure 6).

Figure 6

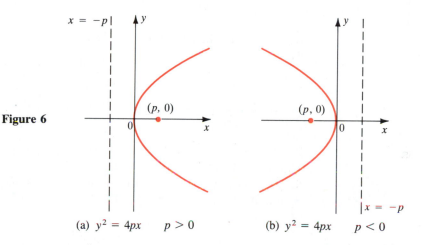

(a) $y^2 = 4px$ $p > 0$ (b) $y^2 = 4px$ $p < 0$

EXAMPLE 2

Find the focus and directrix of the parabola $6x + y^2 = 0$, and sketch the graph.

SOLUTION

We first write the equation as $y^2 = -6x$. Comparing this with the general equation $y^2 = 4px$, we see that $-6 = 4p$, so $p = -\frac{3}{2}$. Thus the focus is $\left(-\frac{3}{2}, 0\right)$ and the directrix is $x = \frac{3}{2}$. The graph is shown in Figure 7.

Figure 7

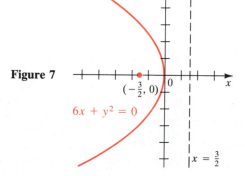

$\left(-\frac{3}{2}, 0\right)$

$6x + y^2 = 0$

$x = \frac{3}{2}$

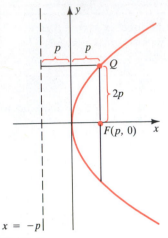

Figure 8

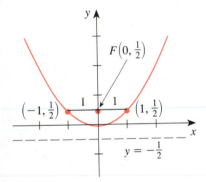

Figure 9
$y = \frac{1}{2}x^2$

Figure 10
Parabolic reflector

The coordinates of the focus can be used to help us estimate the "width" of a parabola when sketching its graph. The line segment that runs through the focus perpendicular to the axis, with endpoints on the parabola, is called the **latus rectum**, and its length is the **focal diameter** of the parabola (see Figure 8). From the figure we can see that the distance from an endpoint Q of the latus rectum to the directrix is $|2p|$. Thus the distance from Q to the focus must be $|2p|$ as well (by the definition of a parabola), and so the focal diameter is $|4p|$.

EXAMPLE 3

Find the focus, directrix, and focal diameter of the parabola $y = \frac{1}{2}x^2$, and sketch the graph.

SOLUTION

We first put the equation in the form $x^2 = 4py$:

$$x^2 = 2y$$

Thus $4p = 2$, so the focal diameter is 2. Also $p = \frac{1}{2}$, so the focus is $\left(0, \frac{1}{2}\right)$ and the directrix is $y = -\frac{1}{2}$. The latus rectum extends one unit to the left and to the right of the focus, since the focal diameter is 2 (see Figure 9). ■

Parabolas have an important property that makes them useful as reflectors for lamps and telescopes. Light from a source placed at the focus of a surface with parabolic cross section (see Figure 10) will be reflected in such a way that it travels parallel to the axis of the parabola. Thus a parabolic reflector will concentrate the light into a beam of parallel rays. Conversely, light approaching the reflector in rays parallel to its axis of symmetry will be reflected to the focus. This principle, which can be proved using calculus, is used in the construction of reflecting telescopes.

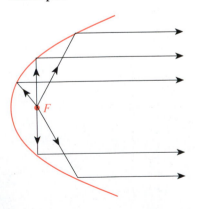

EXERCISES 9.1

In Exercises 1–12 find the focus, directrix, and focal diameter of the given parabola, and sketch its graph.

1. $y^2 = 4x$

2. $x^2 = y$

3. $x^2 = 9y$

4. $y^2 = 3x$

5. $y = 5x^2$

6. $y = -2x^2$

7. $x = -8y^2$

8. $x = \frac{1}{2}y^2$

9. $x^2 + 6y = 0$

10. $x - 7y^2 = 0$

11. $5x + 3y^2 = 0$

12. $8x^2 + 12y = 0$

In Exercises 13–20 find an equation for each parabola sketched.

13.

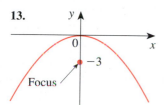

14.

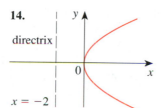

15.

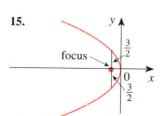

16.

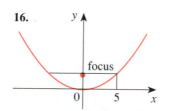

17.

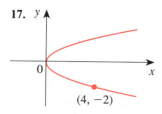

18.

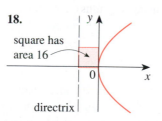

19.

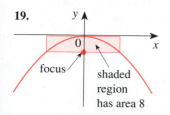

20.

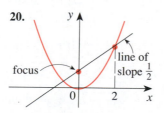

In Exercises 21–26 find an equation for the parabola described. The vertex of the parabola is to be the origin.

21. Its directrix is $x = 3$.

22. Its focus is $\left(0, -\frac{1}{2}\right)$.

23. Its focus is on the positive x-axis, two units away from the directrix.

24. Its focal diameter is 7 and its focus is on the negative y-axis.

25. Its directrix has y-intercept 6.

26. It opens upward with the focus five units from the vertex.

27. A lamp with a parabolic reflector is shown in the figure. The bulb is at the focus and the focal diameter is 12 cm.

(a) Find an equation of the parabola.

(b) Find the diameter $|CD|$ of the opening 20 cm from the vertex.

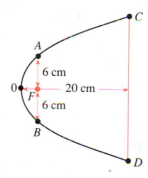

28. A reflector for a radiotelescope is parabolic in cross section, with the receiver at the focus. The reflector is 1 ft deep and 20 ft wide from rim to rim (see the figure). How far is the receiver from the vertex of the parabolic reflector?

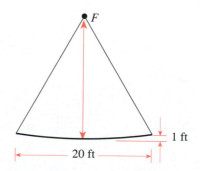

SECTION 9.2
ELLIPSES

An ellipse is an oval curve that looks like an elongated circle. More precisely, we have the following definition:

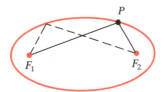

Figure 1

> An **ellipse** is the set of all points in the plane the sum of whose distances from two fixed points F_1 and F_2 is a constant (see Figure 1). These two fixed points are called the **foci** (plural of **focus**) of the ellipse.

The definition suggests a simple method for drawing an ellipse. Place a sheet of paper on a drawing board and insert thumbtacks at the two points that are to be the foci of the ellipse. Attach the ends of a string to the thumbtacks, as shown in Figure 2. With the point of a pencil, hold the string taut. Then carefully move the pencil around the foci, keeping the string taut at all times. The pencil will trace out an ellipse, since the sum of the distances from its point to the foci will always equal the length of the string, which is constant.

If the string is only slightly longer than the distance between the foci, then the ellipse traced out will be elongated in shape as in Figure 3(a), but if the foci are close together relative to the length of the string, the ellipse will be almost circular as in Figure 3(b).

Figure 2

Figure 3

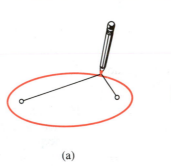

(a)

(b)

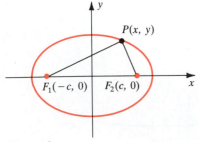

Figure 4

To obtain the simplest possible equation for an ellipse, we place the foci on the x-axis at $F_1(-c, 0)$ and $F_2(c, 0)$, so that the origin is halfway between them (see Figure 4). For later convenience, we let the sum of the distances from a point on the ellipse to the foci be $2a$. Then if $P(x, y)$ is any point on the ellipse, we have

$$|PF_1| + |PF_2| = 2a$$

so, from the Distance Formula,

$$\sqrt{(x + c)^2 + y^2} + \sqrt{(x - c)^2 + y^2} = 2a$$

Johannes Kepler (1571–1630) was
the first to give a correct description
of the motion of the planets. The cos-
mology of his time postulated complicated systems of circles moving on
circles to describe these motions.
Kepler sought a simpler and more
harmonious description. As official
astronomer at the imperial court in
Prague, he studied the astronomical
observations of the Danish astronomer
Tycho Brahe, whose data was at the
time the most accurate available.
After numerous attempts and failures
at a theory, Kepler made the momen-
tous discovery that the orbits of the
planets are elliptical. Two of his three
great laws of planetary motion are:
(1) the orbit of each planet is an
ellipse with the sun at one focus and
(2) the line segment that joins the sun
to a planet sweeps out equal areas in
equal times. His formulation of these
laws is perhaps the most impressive
deduction from empirical data in the
history of science.

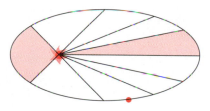

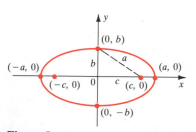

Figure 5
$$\frac{x^2}{a^2} + \frac{y^2}{b^2} = 1 \qquad a > b$$

or

$$\sqrt{(x - c)^2 + y^2} = 2a - \sqrt{(x + c)^2 + y^2}$$

Squaring both sides and multiplying, we get

$$x^2 - 2cx + c^2 + y^2 = 4a^2 - 4a\sqrt{(x + c)^2 + y^2} + (x^2 + 2cx + c^2 + y^2)$$

which simplifies to

$$4a\sqrt{(x + c)^2 + y^2} = 4a^2 + 4cx$$

Dividing both sides by 4 and squaring again, we get

$$a^2[(x + c)^2 + y^2] = (a^2 + cx)^2$$
$$a^2x^2 + 2a^2cx + a^2c^2 + a^2y^2 = a^4 + 2a^2cx + c^2x^2$$
$$(a^2 - c^2)x^2 + a^2y^2 = a^2(a^2 - c^2)$$

Since the sum of the distances from P to the foci must be larger than the distance
between the foci, we have that $2a > 2c$, or $a > c$. Thus $a^2 - c^2 > 0$, and we can
divide both sides of the preceding equation by $a^2(a^2 - c^2)$ to get

$$\frac{x^2}{a^2} + \frac{y^2}{a^2 - c^2} = 1$$

For convenience, let $b^2 = a^2 - c^2$ (where $b > 0$). Since $b^2 < a^2$, it follows that
$b < a$. The preceding equation then becomes

$$\frac{x^2}{a^2} + \frac{y^2}{b^2} = 1 \qquad a > b$$

This is the equation of an ellipse. To graph it, we need to know the x- and
y-intercepts. Setting $y = 0$, we get

$$\frac{x^2}{a^2} = 1$$

so $x^2 = a^2$ and $x = \pm a$. Thus the ellipse crosses the x-axis at $(a, 0)$ and $(-a, 0)$.
These points are called the **vertices** of the ellipse, and the segment that joins them
is called the **major axis**. Its length is $2a$.

Similarly, if we set $x = 0$, we get $y = \pm b$, so the ellipse crosses the y-axis at
$(0, b)$ and $(0, -b)$. The segment that joins these points is called the **minor axis**
and has length $2b$. Note that $2a > 2b$, so the major axis is longer than the minor
axis.

In Section 1.9 we studied several tests that detect symmetry in a graph. If we
replace x by $-x$ or y by $-y$ in the ellipse equation, it remains unchanged. Thus
the ellipse is symmetric about both the x- and y-axes, and hence about the origin
as well. For this reason, the origin is called the **center** of the ellipse. The complete
graph is shown in Figure 5.

We saw earlier in this section (Figure 3) that if $2a$ is only slightly greater than $2c$, the ellipse is long and thin, while if $2a$ is much greater than $2c$, the ellipse is almost circular. We define the **eccentricity** of the ellipse to be the ratio

$$e = \frac{c}{a}$$

Thus, if e is close to 1, c is almost equal to a, and the ellipse is elongated in shape, but if e is close to 0, the ellipse is close to a circle in shape. (For any ellipse, $0 < e < 1$.) The eccentricity is a measure of how "stretched out" the ellipse is.

We summarize the preceding discussion as follows:

ELLIPSE WITH HORIZONTAL MAJOR AXIS

The graph of the equation

$$\frac{x^2}{a^2} + \frac{y^2}{b^2} = 1 \qquad a > b$$

is an ellipse with foci $(\pm c, 0)$, where $c^2 = a^2 - b^2$. The center is $(0, 0)$ and the vertices are $(\pm a, 0)$. The major axis is horizontal and has length $2a$, and the minor axis is vertical and has length $2b$. The eccentricity is $e = c/a$. (See Figure 5.)

EXAMPLE 1

Find the foci, vertices, eccentricity, and the lengths of the major and minor axes for the following ellipse and sketch the graph:

$$\frac{x^2}{9} + \frac{y^2}{4} = 1$$

SOLUTION

Here $a^2 = 9$ and $b^2 = 4$, so $c^2 = a^2 - b^2 = 9 - 4 = 5$. Thus we have

foci: $\left(\pm\sqrt{5}, 0\right)$

vertices: $(\pm 3, 0)$

eccentricity: $\sqrt{5}/3 \approx 0.745$

length of major axis: 6

length of minor axis: 4

The graph is shown in Figure 6.   ■

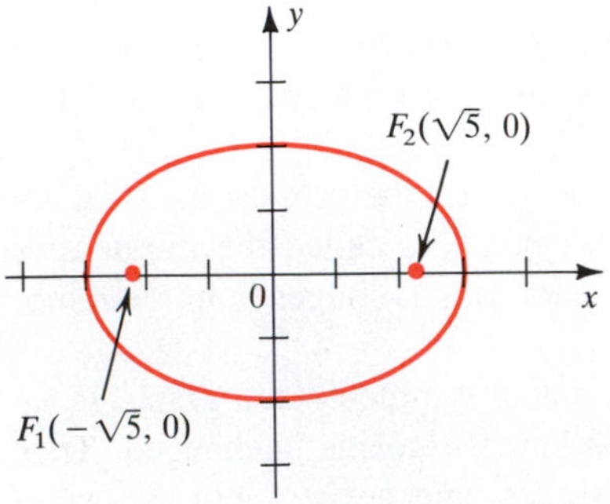

Figure 6
$\dfrac{x^2}{9} + \dfrac{y^2}{4} = 1$

EXAMPLE 2

The vertices of an ellipse are $(\pm 4, 0)$ and the eccentricity is $\frac{1}{2}$. Find its equation and sketch the graph.

SOLUTION

Since the vertices are $(\pm 4, 0)$, we have that $a = 4$. Also, $e = c/a = \frac{1}{2}$, so

$$\frac{c}{4} = \frac{1}{2}$$
$$c = 2$$

Finally, $c^2 = a^2 - b^2$, so $2^2 = 4^2 - b^2$, and hence

$$b = \sqrt{4^2 - 2^2}$$
$$= 2\sqrt{3}$$

Thus the equation of the ellipse is

$$\frac{x^2}{16} + \frac{y^2}{12} = 1$$

Its graph is shown in Figure 7.

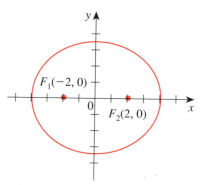

Figure 7
$$\frac{x^2}{16} + \frac{y^2}{12} = 1$$

If the foci of the ellipse are placed on the y-axis at $(0, \pm c)$ rather than on the x-axis, then the roles of x and y are reversed in the discussion leading up to the equation for a horizontal ellipse. Thus we have the following description of such ellipses.

ELLIPSE WITH VERTICAL MAJOR AXIS

The graph of the equation

$$\frac{x^2}{b^2} + \frac{y^2}{a^2} = 1 \qquad a > b$$

is an ellipse with foci $(0, \pm c)$, where $c^2 = a^2 - b^2$. The center is $(0, 0)$ and the vertices are $(0, \pm a)$. The major axis is vertical and has length $2a$, and the minor axis is horizontal and has length $2b$. The eccentricity is $e = c/a$. (See Figure 8.)

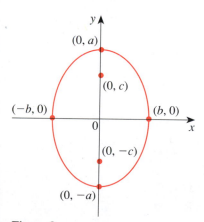

Figure 8
$$\frac{x^2}{b^2} + \frac{y^2}{a^2} = 1 \qquad a > b$$

EXAMPLE 3

Find the foci of the ellipse $16x^2 + 9y^2 = 144$ and sketch the graph.

SOLUTION

Dividing through by 144, we get

$$\frac{x^2}{9} + \frac{y^2}{16} = 1$$

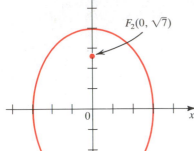
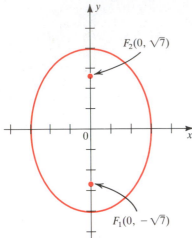

Figure 9
$16x^2 + 9y^2 = 144$

Since $16 > 9$, this is an ellipse with its foci on the y-axis, and with $a = 4$ and $b = 3$. We have

$$c^2 = a^2 - b^2 = 16 - 9 = 7$$
$$c = \sqrt{7}$$

Thus the foci are $\left(0, \pm\sqrt{7}\right)$. The graph is shown in Figure 9.   ■

EXAMPLE 4

Find an equation for the ellipse with foci $(0, \pm 2)$ and vertices $(0, \pm 3)$.

SOLUTION

For this ellipse, $c = 2$ and $a = 3$. Therefore $b^2 = a^2 - c^2 = 9 - 4 = 5$, so the equation of the ellipse is

$$\frac{x^2}{5} + \frac{y^2}{9} = 1$$ ■

Like parabolas, ellipses have an interesting reflection property that has a number of practical consequences. If a light source is placed at one focus of a reflecting surface with elliptical cross sections, then all the light will be reflected off the surface to the other focus, as shown in Figure 10. This principle, which works for sound waves as well as for light, is used in *lithotripsy*, the new treatment for removing kidney stones. The patient sits in a tub with elliptical cross sections and is placed so that the kidney is at one focus. High-intensity sound waves generated at the other focus are reflected to the stone and destroy it with minimal damage to surrounding tissue. The patient is spared the trauma of surgery and recovers within days instead of weeks.

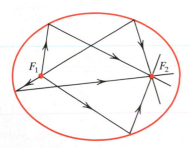

Figure 10

EXERCISES 9.2

In Exercises 1–14 find the vertices, foci, and eccentricity of the ellipse. Determine the lengths of the major and minor axes, and sketch the graph.

1. $\dfrac{x^2}{25} + \dfrac{y^2}{9} = 1$

2. $\dfrac{x^2}{16} + \dfrac{y^2}{25} = 1$

3. $9x^2 + 4y^2 = 36$

4. $4x^2 + 25y^2 = 100$

5. $x^2 + 4y^2 = 16$

6. $4x^2 + y^2 = 16$

7. $2x^2 + y^2 = 3$

8. $5x^2 + 6y^2 = 30$

9. $x^2 + 4y^2 = 1$

10. $9x^2 + 4y^2 = 1$

11. $\frac{1}{2}x^2 + \frac{1}{8}y^2 = \frac{1}{4}$

12. $x^2 = 4 - 2y^2$

13. $y^2 = 1 - 2x^2$

14. $20x^2 + 4y^2 = 5$

In Exercises 15–26 find an equation for the ellipse that satisfies the given conditions.

15. Foci $(\pm 4, 0)$, vertices $(\pm 5, 0)$

16. Foci $(0, \pm 3)$, vertices $(0, \pm 5)$

17. Length of major axis 4, length of minor axis 2, foci on y-axis

18. Length of major axis 6, length of minor axis 4, foci on x-axis

19. Foci $(0, \pm 2)$, length of minor axis 6

20. Foci $(\pm 5, 0)$, length of major axis 12

21. Endpoints of major axis $(\pm 10, 0)$, distance between foci 6

22. Endpoints of minor axis $(0, \pm 3)$, distance between foci 8

23. Length of major axis 10, foci on x-axis, ellipse passes through point $(\sqrt{5}, 2)$

24. Eccentricity $\frac{1}{9}$, foci $(0, \pm 2)$

25. Eccentricity 0.8, foci $(\pm 1.5, 0)$

26. Eccentricity $\sqrt{3}/2$, foci on y-axis, length of major axis 4

In Exercises 27 and 28 find the intersection points of the pair of ellipses. Sketch the graphs of each pair of equations on the same coordinate axes and show the points of intersection.

27. $\begin{cases} 4x^2 + y^2 = 4 \\ 4x^2 + 9y^2 = 36 \end{cases}$

28. $\begin{cases} \dfrac{x^2}{16} + \dfrac{y^2}{9} = 1 \\ \dfrac{x^2}{9} + \dfrac{y^2}{16} = 1 \end{cases}$

29. The planets move around the sun in elliptical orbits with the sun at one focus. The point in the orbit at which the planet is closest to the sun is called **perihelion**, and the point at which it is farthest is called **aphelion**. These points are the vertices of the orbit. The earth's distance from the sun is 147,000,000 km at perihelion and 153,000,000 km at aphelion. Find an equation for the earth's orbit. (Place the origin at the center of the orbit, with the sun on the x-axis.)

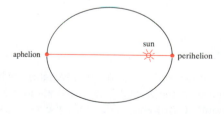

30. With an eccentricity of 0.25, Pluto's orbit is the most eccentric in the solar system. The length of the minor axis of its orbit is approximately 10,000,000,000 km. Find an equation for Pluto's orbit. (See Exercise 29.)

31. For an object in an elliptical orbit around the moon, the points in the orbit that are closest to and farthest from the center of the moon are called **perilune** and **apolune**, respectively. These are the vertices of the orbit. The center of the moon is at one of the foci of the orbit. The *Apollo 11* spacecraft was placed in a lunar orbit with perilune at 68 mi and apolune at 195 mi above the surface of the moon. Assuming the moon is a sphere of radius 1075 mi, find an equation for the orbit of *Apollo 11*.

32. If $k > 0$, the following equation represents an ellipse:

$$\frac{x^2}{k} + \frac{y^2}{4 + k} = 1$$

Show that all the ellipses represented by this equation have the same foci, no matter what the value of k is.

33. A "sunburst" window above a doorway is constructed in the shape of the top half of an ellipse (see the figure). The window is 20 in. tall at its highest point and 80 in. wide at the bottom. How tall is the window at a point 25 in. from the center of the base?

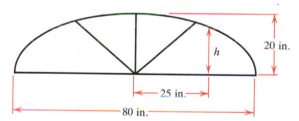

34. The **ancillary circle** of an ellipse is the circle with radius equal to half the length of the minor axis and center the same as the ellipse (see the figure). The ancillary circle is thus the largest circle that can fit within an ellipse.

(a) Find an equation for the ancillary circle of the ellipse $x^2 + 4y^2 = 16$.

(b) For the ellipse and ancillary circle of part (a), show that if (s, t) is a point on the ancillary circle, then $(2s, t)$ is a point on the ellipse.

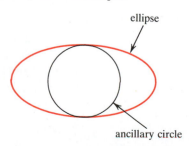

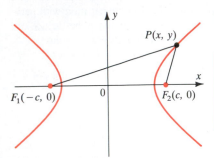

Figure 1

P is on the hyperbola if
$$||PF_1| - |PF_2|| = 2a.$$

Although ellipses and hyperbolas have completely different shapes, their definitions and equations are similar. Instead of using a *sum* of distances from two fixed foci, as in the case of an ellipse, we use the *difference* to define a hyperbola.

> A **hyperbola** is the set of all points in the plane the difference of whose distances from two fixed points F_1 and F_2 is a constant (see Figure 1). The points F_1 and F_2 are called the **foci** of the hyperbola.

As in the case of the ellipse, we get the simplest equation for the hyperbola by placing the foci on the x-axis at $(\pm c, 0)$, as shown in Figure 1. From the definition, if $P(x, y)$ lies on the hyperbola, either $|PF_1| - |PF_2|$ or $|PF_2| - |PF_1|$ must equal some positive constant, which we call $2a$. Thus we have

$$|PF_1| - |PF_2| = \pm 2a$$

or $\qquad \sqrt{(x + c)^2 + y^2} - \sqrt{(x - c)^2 + y^2} = \pm 2a$

Proceeding as we did in the case of the ellipse in Section 9.2, we simplify this to

$$(c^2 - a^2)x^2 - a^2y^2 = a^2(c^2 - a^2) \qquad (1)$$

If we consider the triangle PF_1F_2 in Figure 1, we see that

$$||PF_1| - |PF_2|| < 2c$$

so $\qquad\qquad 2a < 2c \qquad \text{or} \qquad a < c$

Thus $c^2 - a^2 > 0$, so we can write $b^2 = c^2 - a^2$. We then simplify Equation 1 to

$$\frac{x^2}{a^2} - \frac{y^2}{b^2} = 1 \qquad (2)$$

If we replace x by $-x$ or y by $-y$ in Equation 2, the equation remains unchanged, so the hyperbola is symmetric about both the x- and y-axes and about the origin. The x-intercepts are $\pm a$, and the points $(a, 0)$ and $(-a, 0)$ are the **vertices** of the hyperbola. But setting $x = 0$ in Equation 2 leads to $-y^2 = b^2$, which is impossible, so the hyperbola has no y-intercept. Furthermore, rewriting the equation, we get

$$\frac{x^2}{a^2} = \frac{y^2}{b^2} + 1 \geqslant 1$$

so $x^2 \geq a^2$, and hence $x \geq a$ or $x \leq -a$. This means that the hyperbola consists of two parts, called its **branches**. The segment joining the two vertices on the separate branches is called the **transverse axis** of the hyperbola and the origin is called its **center**.

As a further guide to graphing the hyperbola, we solve for y in Equation 2:

$$y = \pm \frac{b}{a}\sqrt{x^2 - a^2}$$

Considering only the portion of the hyperbola in the first quadrant (so that $x > 0$ and $y \geq 0$), we have

$$y = \frac{b}{a}\sqrt{x^2 - a^2}$$

$$= \frac{b}{a}x\sqrt{1 - \frac{a^2}{x^2}}$$

As x becomes larger and larger, a^2/x^2 becomes smaller and smaller, so that $\sqrt{1 - a^2/x^2}$ approaches $\sqrt{1} = 1$ in value. Thus we see that as x gets larger, the line $y = (b/a)x$ and the hyperbola get closer together. By symmetry, the same sort of situation exists in the other quadrants, and we see that the hyperbola approaches the lines

$$y = \pm \frac{b}{a}x$$

These lines are called the **asymptotes** of the hyperbola and are a useful guide in graphing it. (Recall that we discussed asymptotes for rational functions in Section 3.9.)

A convenient way to locate the asymptotes and graph the hyperbola is to first plot the vertices $(a, 0)$ and $(-a, 0)$ and the points $(0, b)$ and $(0, -b)$. We then draw horizontal and vertical segments through these points to construct a rectangle as shown in Figure 2. We call this rectangle the **central box** of the hyperbola. The slopes of the diagonals of the central box are $\pm b/a$, so by extending them we obtain the asymptotes $y = \pm(b/a)x$. Finally, we use the asymptotes as a guide in sketching the hyperbola.

We summarize the main results of our discussion as follows:

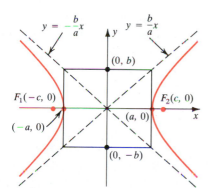

Figure 2

Hyperbola $\dfrac{x^2}{a^2} - \dfrac{y^2}{b^2} = 1$ with

asymptotes $y = \pm \dfrac{b}{a}x$

HYPERBOLA WITH HORIZONTAL TRANSVERSE AXIS

The graph of the equation

$$\frac{x^2}{a^2} - \frac{y^2}{b^2} = 1 \tag{3}$$

is a hyperbola with vertices $(\pm a, 0)$, asymptotes $y = \pm(b/a)x$, and foci $(\pm c, 0)$, where $c^2 = a^2 + b^2$. (See Figure 2.)

EXAMPLE 1

Find the foci, vertices, and asymptotes of the following hyperbola, and sketch the graph:

$$9x^2 - 16y^2 = 144$$

SOLUTION

First we divide both sides of the equation by 144 to put it into the form of Equation 3:

$$\frac{x^2}{16} - \frac{y^2}{9} = 1$$

This means that $a = 4$, $b = 3$, and $c = \sqrt{16 + 9} = 5$. The vertices are $(\pm 4, 0)$, the foci are $(\pm 5, 0)$, and the asymptotes are $y = \pm\frac{3}{4}x$. After drawing the central box and asymptotes, we complete the sketch of the hyperbola (Figure 3). ∎

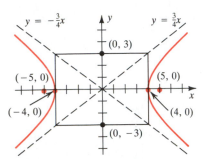

Figure 3
$9x^2 - 16y^2 = 144$

Placing the foci of the hyperbola on the y-axis rather than on the x-axis has the effect of reversing the roles of x and y in the derivation of Equation 3. This leads to the following result, which is illustrated in Figure 4.

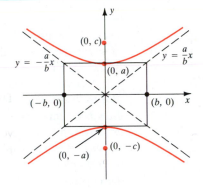

Figure 4
Hyperbola $\dfrac{y^2}{a^2} - \dfrac{x^2}{b^2} = 1$ with asymptotes $y = \pm\dfrac{a}{b}x$

HYPERBOLA WITH VERTICAL TRANSVERSE AXIS

The graph of the equation

$$\frac{y^2}{a^2} - \frac{x^2}{b^2} = 1 \qquad (4)$$

is a hyperbola with vertices $(0, \pm a)$, asymptotes $y = \pm(a/b)x$, and foci $(0, \pm c)$, where $c^2 = a^2 + b^2$.

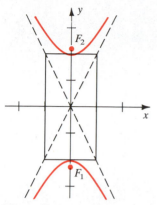

Figure 5
$\dfrac{y^2}{4} - x^2 = 1$

EXAMPLE 2

Find the equation and the foci of the hyperbola with vertices $(0, \pm 2)$ and asymptotes $y = \pm 2x$. Sketch the graph.

SOLUTION

Since the foci are on the y-axis, the equation will take the form of Equation 4, with $a = 2$. From the asymptote equation, we see that $a/b = 2$. Thus $b = a/2 = 1$, and $c^2 = a^2 + b^2 = 2^2 + 1^2 = 5$. The foci are $\left(0, \pm\sqrt{5}\right)$ and the equation of the hyperbola is

$$\frac{y^2}{4} - x^2 = 1$$

The graph is shown in Figure 5.   ■

EXAMPLE 3

Find the foci, vertices, and asymptotes of the following hyperbola, and sketch the graph:

$$x^2 - 9y^2 + 9 = 0$$

SOLUTION

We begin by writing the equation in the standard form for hyperbolas:

$$x^2 - 9y^2 = -9 \qquad \text{or} \qquad y^2 - \frac{x^2}{9} = 1$$

This is in the form of Equation 4, so the hyperbola has its foci and vertices on the y-axis. We have $a^2 = 1$ and $b^2 = 9$, so $c = \sqrt{a^2 + b^2} = \sqrt{10}$. Thus the foci are $\left(0, \pm\sqrt{10}\right)$ and the vertices are $(0, \pm 1)$. Since $a = 1$ and $b = 3$, the asymptotes are $y = \pm\frac{1}{3}x$. The graph is shown in Figure 6.   ■

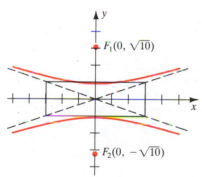

Figure 6
$x^2 - 9y^2 + 9 = 0$

In the LORAN (LOng RAnge Navigation) system, hyperbolas are used on board ships and aircraft to determine their location. In Figure 7, radio stations at A and B transmit signals simultaneously for reception by the ship at P. The onboard computer converts the time difference in reception of these signals into a distance difference $|PA| - |PB|$. By the definition of a hyperbola, this locates the ship on one branch of a hyperbola with foci at A and B (sketched in black in the figure). The same procedure is carried out with two other radio stations at C and D, and this locates the ship on a second hyperbola (shown in color in Figure 7). The coordinates of the intersection point of these two hyperbolas, which can be calculated precisely by the computer, give the location of P.

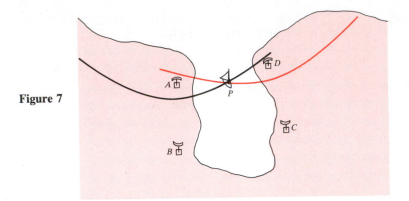

Figure 7

EXERCISES 9.3

In Exercises 1–12 find the vertices, foci, and asymptotes of the hyperbola, and sketch the graph.

1. $\dfrac{x^2}{4} - \dfrac{y^2}{16} = 1$

2. $\dfrac{y^2}{9} - \dfrac{x^2}{16} = 1$

3. $y^2 - \dfrac{x^2}{25} = 1$

4. $\dfrac{x^2}{2} - y^2 = 1$

5. $x^2 - y^2 = 1$

6. $9x^2 - 4y^2 = 36$

7. $25y^2 - 9x^2 = 225$

8. $x^2 - y^2 + 4 = 0$

9. $x^2 - 4y^2 - 8 = 0$

10. $x^2 - 2y^2 = 3$

11. $4y^2 - x^2 = 1$

12. $9x^2 - 16y^2 = 1$

In Exercises 13–24 find an equation for the hyperbola that satisfies the given conditions.

13. Foci $(\pm 5, 0)$, vertices $(\pm 3, 0)$

14. Foci $(0, \pm 10)$, vertices $(0, \pm 8)$

15. Foci $(0, \pm 2)$, vertices $(0, \pm 1)$

16. Foci $(\pm 6, 0)$, vertices $(\pm 2, 0)$

17. Vertices $(\pm 1, 0)$, asymptotes $y = \pm 5x$

18. Vertices $(0, \pm 6)$, asymptotes $y = \pm\frac{1}{3}x$

19. Foci $(0, \pm 8)$, asymptotes $y = \pm\frac{1}{2}x$

20. Vertices $(0, \pm 6)$, hyperbola passes through $(-5, 9)$

21. Asymptotes $y = \pm x$, hyperbola passes through $(5, 3)$

22. Foci $(\pm 3, 0)$, hyperbola passes through $(4, 1)$

23. Foci $(\pm 5, 0)$, length of transverse axis 6

24. Foci $(0, \pm 1)$, length of transverse axis 1

25. (a) Show that the asymptotes of the hyperbola $x^2 - y^2 = 5$ are perpendicular to each other.

(b) Find an equation for the hyperbola with vertices $(\pm c, 0)$ and with asymptotes perpendicular to each other.

26. The hyperbolas

$$\frac{x^2}{a^2} - \frac{y^2}{b^2} = 1 \qquad \text{and} \qquad \frac{x^2}{a^2} - \frac{y^2}{b^2} = -1$$

are said to be **conjugate** to each other.

(a) Show that the hyperbolas

$$x^2 - 4y^2 + 16 = 0 \qquad \text{and} \qquad 4y^2 - x^2 + 16 = 0$$

are conjugate to each other, and graph them on the same coordinate axes.

(b) What do the hyperbolas of part (a) have in common?

(c) Show that any pair of conjugate hyperbolas have the relationship you discovered in part (b).

27. Derive Equation 1 in the text from the equation that precedes it.

28. (a) For the hyperbola

$$\frac{x^2}{9} - \frac{y^2}{16} = 1$$

determine the values of a, b, and c, and find the coordinates of the foci F_1 and F_2.

(b) Show that the point $P\left(5, \frac{16}{3}\right)$ lies on this hyperbola.

(c) Find $|PF_1|$ and $|PF_2|$.

(d) Verify that the difference between $|PF_1|$ and $|PF_2|$ is $2a$.

29. Refer to Figure 7 in the text. Suppose that the radio stations at A and B are 500 mi apart, and that the ship at P receives Station A's signal 2640 microseconds (μs) before it receives the signal from B.

(a) Assuming that radio signals travel at 980 ft/μs, find $|PA| - |PB|$.

(b) Find an equation for the branch of the hyperbola indicated in black in the figure. (Place A and B on the y-axis with the origin halfway between them. Use miles as the unit of distance.)

(c) If A is due north of B, and if P is due east of A, how far is P from A?

30. Some comets, such as Halley's comet, are a permanent part of the solar system, travelling in elliptical orbits around the sun. Others pass through the solar system only once, following a hyperbolic path with the sun at a focus. The figure shows the path of such a comet. Find an equation for the path, assuming that the closest the comet comes to the sun is 2×10^9 mi and that the path the comet was taking before it neared the solar system is at a right angle to the path it continues on after leaving the solar system.

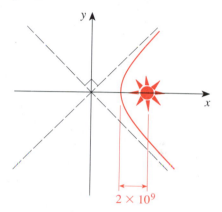

2×10^9

■ ■ **SECTION 9.4**
■ ■ **SHIFTED CONICS**

In the preceding sections we have studied parabolas with vertices at the origin, and ellipses and hyperbolas with centers at the origin. We restricted ourselves to these cases because the equations then have the simplest possible form. In this section we consider conics whose vertices and centers are not necessarily at the origin, and we determine how this affects their equations.

If we replace x by $x - h$ in any equation, the graph of the new equation is simply the old graph shifted to the right h units if h is positive, or to the left $-h$ units if h is negative. Similarly, the effect of replacing y by $y - k$ is to shift the graph upward k units (or downward $-k$ units if k is negative). In Section 2.5 we applied these principles to the graphs of functions. Here we apply them to the conics.

For example, consider the ellipse with equation

$$\frac{x^2}{a^2} + \frac{y^2}{b^2} = 1$$

(See Figure 1.) If we shift it so that its center is at the point (h, k) instead of at the origin, then its equation becomes

$$\frac{(x - h)^2}{a^2} + \frac{(y - k)^2}{b^2} = 1$$

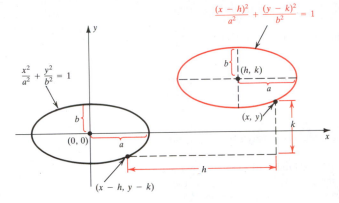

Figure 1
Shifted ellipse

EXAMPLE 1

Sketch the graph of the following ellipse, and determine the coordinates of the foci:

$$\frac{(x + 1)^2}{4} + \frac{(y - 2)^2}{9} = 1 \tag{1}$$

SOLUTION

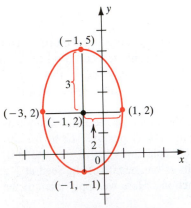

Figure 2
$\dfrac{(x + 1)^2}{4} + \dfrac{(y - 2)^2}{9} = 1$

From the preceding discussion, we see that this ellipse has the same shape as the ellipse

$$\frac{x^2}{4} + \frac{y^2}{9} = 1 \tag{2}$$

but that its center is at the point $(-1, 2)$ instead of at the origin. In fact, the graph of Equation 1 is the graph of Equation 2, shifted to the left 1 unit and upward 2 units.

Since $a^2 = 9$ and $b^2 = 4$, we have $c^2 = 9 - 4 = 5$, so $c = \sqrt{5}$. Thus the foci of the ellipse of Equation 2 occur at $(0, \pm\sqrt{5})$. This means we obtain the foci of the ellipse of Equation 1 by shifting these points 1 unit to the left and 2 units upward, to get

$$\left(-1, 2 + \sqrt{5}\right) \quad \text{and} \quad \left(-1, 2 - \sqrt{5}\right)$$

The same shifts are used to find the location of the vertices, as shown in Figure 2.   ■

This shifting technique can be applied to parabolas and hyperbolas as well. The results are summarized in Figures 3 and 4.

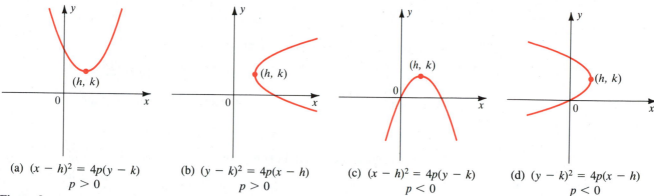

(a) $(x - h)^2 = 4p(y - k)$
$p > 0$

(b) $(y - k)^2 = 4p(x - h)$
$p > 0$

(c) $(x - h)^2 = 4p(y - k)$
$p < 0$

(d) $(y - k)^2 = 4p(x - h)$
$p < 0$

Figure 3
Shifted parabolas

Figure 4
Shifted hyperbolas

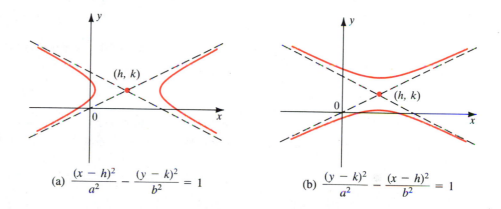

(a) $\dfrac{(x - h)^2}{a^2} - \dfrac{(y - k)^2}{b^2} = 1$

(b) $\dfrac{(y - k)^2}{a^2} - \dfrac{(x - h)^2}{b^2} = 1$

EXAMPLE 2

Determine the vertex, focus, and directrix and sketch the graph of the parabola

$$x^2 - 4x = 8y - 28$$

SOLUTION

We complete the square in x to put this equation into one of the forms in Figure 3.

$$x^2 - 4x + 4 = 8y - 28 + 4$$
$$(x - 2)^2 = 8y - 24$$
$$(x - 2)^2 = 8(y - 3)$$

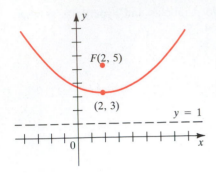

Figure 5
$x^2 - 4x = 8y - 28$

This is a parabola that opens upward, with vertex at $(2, 3)$. Since $4p = 8$, we have $p = 2$, so the focus is 2 units above the vertex and the directrix is 2 units below the vertex. Thus the focus is $(2, 5)$ and the directrix is $y = 1$. The graph is shown in Figure 5. ∎

EXAMPLE 3

Show that the following equation represents a hyperbola:

$$9x^2 - 72x - 16y^2 - 32y = 16$$

Find its center, vertices, foci, and asymptotes, and sketch its graph.

SOLUTION

We first complete the square in both x and y:

$$9(x^2 - 8x \quad\quad) - 16(y^2 + 2y \quad\quad) = 16$$
$$9(x^2 - 8x + 16) - 16(y^2 + 2y + 1) = 16 + 9 \cdot 16 - 16 \cdot 1$$
$$9(x - 4)^2 - 16(y + 1)^2 = 144$$

Now we divide both sides of the equation by 144 to get

$$\frac{(x - 4)^2}{16} - \frac{(y + 1)^2}{9} = 1 \tag{3}$$

This is a hyperbola with center $(4, -1)$ and with a horizontal transverse axis. Its graph will have the same shape as the unshifted hyperbola

$$\frac{x^2}{16} - \frac{y^2}{9} = 1 \tag{4}$$

Since $a^2 = 16$ and $b^2 = 9$, we have $a = 4$, $b = 3$, and $c = \sqrt{a^2 + b^2} = \sqrt{16 + 9} = 5$. Thus the foci lie 5 units to the left and to the right of the center, and the vertices lie 4 units on either side of the center.

$$\text{foci:} \quad (9, -1) \text{ and } (-1, -1)$$
$$\text{vertices:} \quad (8, -1) \text{ and } (0, -1)$$

The asymptotes of the unshifted hyperbola of Equation 4 are $y = \pm\frac{3}{4}x$, so the asymptotes of the hyperbola of Equation 3 are

$$y + 1 = \pm\frac{3}{4}(x - 4)$$
$$y + 1 = \pm\frac{3}{4}x \mp 3$$
$$y = \frac{3}{4}x - 4 \quad \text{and} \quad y = -\frac{3}{4}x + 2$$

The graph is shown in Figure 6.

Figure 6
$9x^2 - 72x - 16y^2 - 32y = 16$

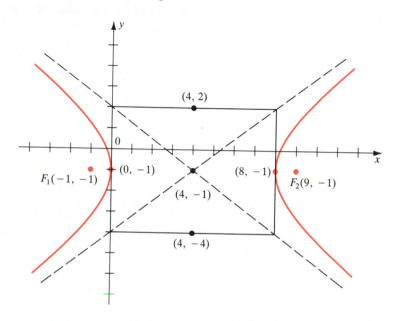

If we multiply out and simplify the equations of any of the shifted conics illustrated in Figures 1, 3, and 4, we will always obtain an equation of the form

$$Ax^2 + Cy^2 + Dx + Ey + F = 0$$

where A and C are not both zero. Conversely, if we begin with an equation of this form, we can complete the square in x and y to see which type of conic section the equation represents. In some cases, called the **degenerate cases**, the graph of the equation turns out to be just a pair of lines or a single point, or the equation may have no graph at all. The next example illustrates such a case.

EXAMPLE 4

Sketch the graph of the equation $9x^2 - y^2 + 18x + 6y = 0$.

SOLUTION

Because the coefficients of x^2 and y^2 have opposite sign, this equation looks as if it should represent a hyperbola, like the equation of Example 3. To see if this is in fact the case, we complete the square:

$$9(x^2 + 2x \quad\quad) - (y^2 - 6y \quad\quad) = 0$$
$$9(x^2 + 2x + 1) - (y^2 - 6y + 9) = 9 - 9 = 0$$
$$9(x + 1)^2 - (y - 3)^2 = 0$$

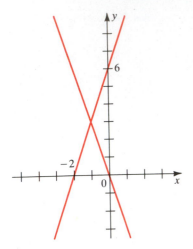

Figure 7
$9x^2 - y^2 + 18x + 6y = 0$

For this to fit the form of the equation of a hyperbola, we would need a nonzero constant to the right of the equal sign. In fact, further analysis shows that this is the equation of a pair of intersecting lines:

$$(y - 3)^2 = 9(x + 1)^2$$
$$y - 3 = \pm 3(x + 1)$$
$$y = 3x + 6 \qquad \text{or} \qquad y = -3x$$

These lines are graphed in Figure 7. ∎

Because the equation in Example 4 looked at first glance like the equation of a hyperbola but in fact turned out to represent simply a pair of lines, we refer to its graph as a **degenerate hyperbola**. Degenerate ellipses and parabolas can also arise when we complete the square in an equation that seems to represent a conic. For example, the equation

$$4x^2 + y^2 - 8x + 2y + 6 = 0$$

looks as if it should represent an ellipse, because the coefficients of x^2 and y^2 have the same sign. But completing the square leads to

$$4(x - 1)^2 + (y + 1)^2 = -1$$

which has no solution at all (since the sum of two squares cannot be negative). This equation is therefore degenerate.

To summarize, we have the following theorem.

GENERAL EQUATION OF A CONIC SECTION

The graph of the equation

$$Ax^2 + Cy^2 + Dx + Ey + F = 0$$

where A and C are not both zero, is a conic or a degenerate conic. In the nondegenerate cases, the graph is

1. a parabola if A or C is zero
2. an ellipse if A and C have the same sign (or a circle if $A = C$)
3. a hyperbola if A and C have opposite sign

EXERCISES 9.4

In Exercises 1–4 find the center, foci, and vertices of the ellipse, and determine the lengths of the major and minor axes. Then sketch the graph.

1. $\dfrac{(x-2)^2}{9} + \dfrac{(y-1)^2}{4} = 1$

2. $\dfrac{(x-3)^2}{16} + (y+3)^2 = 1$

3. $\dfrac{x^2}{9} + \dfrac{(y+5)^2}{25} = 1$

4. $\dfrac{(x+2)^2}{4} + y^2 = 1$

In Exercises 5–8 find the vertex, focus, and directrix of the parabola, and sketch the graph.

5. $(x-3)^2 = 8(y+1)$

6. $(y+5)^2 = -6x + 12$

7. $-4\left(x + \frac{1}{2}\right)^2 = y$

8. $y^2 = 16x - 8$

In Exercises 9–12 find the center, foci, vertices, and asymptotes of the hyperbola. Then sketch the graph.

9. $\dfrac{(x+1)^2}{9} - \dfrac{(y-3)^2}{16} = 1$

10. $(x-8)^2 - (y+6)^2 = 1$

11. $y^2 - \dfrac{(x+1)^2}{4} = 1$

12. $\dfrac{(y-1)^2}{25} - (x+3)^2 = 1$

In Exercises 13–24 complete the square to determine whether the equation represents an ellipse, a parabola, a hyperbola, or a degenerate conic. Then sketch the graph of the equation. If the graph is an ellipse, find the center, foci, vertices, and lengths of the major and minor axes. If it is a parabola, find the vertex, focus, and directrix. If it is a hyperbola, find the center, foci, vertices, and asymptotes. If the equation has no graph, explain why.

13. $9x^2 - 36x + 4y^2 = 0$

14. $y^2 = 4(x + 2y)$

15. $x^2 - 4y^2 - 2x + 16y = 20$

16. $x^2 + 6x + 12y + 9 = 0$

17. $4x^2 + 25y^2 - 24x + 250y + 561 = 0$

18. $2x^2 + y^2 = 2y + 1$

19. $16x^2 - 9y^2 - 96x + 288 = 0$

20. $4x^2 - 4x - 8y + 9 = 0$

21. $x^2 + 16 = 4(y^2 + 2x)$

22. $x^2 - y^2 = 10(x - y) + 1$

23. $3x^2 + 4y^2 - 6x - 24y + 39 = 0$

24. $x^2 + 4y^2 + 20x - 40y + 300 = 0$

25. What must the value of F be if the graph of the equation

$$4x^2 + y^2 + 4(x - 2y) + F = 0$$

is **(a)** an ellipse? **(b)** a single point? **(c)** the empty set?

26. Find an equation for the ellipse that shares a vertex and a focus with the parabola $x^2 + y = 100$ and that has its other focus at the origin.

■■
■ **SECTION 9.5**
■■ **ROTATION OF AXES**

In Section 9.4 we studied conics with equations of the form

$$Ax^2 + Cy^2 + Dx + Ey + F = 0$$

We saw that the graph is always an ellipse, parabola, or hyperbola with horizontal or vertical axes (except in the degenerate cases). In this section we study the most

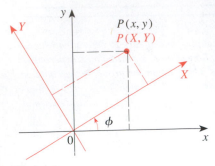

Figure 1

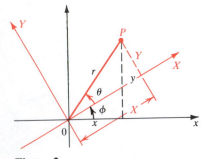

Figure 2

general second-degree equation

$$Ax^2 + Bxy + Cy^2 + Dx + Ey + F = 0 \qquad (1)$$

We will see that by rotating the coordinate axes through an appropriate angle, we can eliminate the term Bxy and then use our knowledge of conic sections to analyze the graph.

In Figure 1 the x- and y-axes have been rotated through an acute angle ϕ about the origin to produce a new pair of axes, which we call the X- and Y-axes. A point P that has coordinates (x, y) in the old system has coordinates (X, Y) in the new system. If we let r denote the distance of P from the origin and let θ be the angle that the segment OP makes with the new X-axis, then we can see from Figure 2 (by considering the two right triangles in the figure) that

$$X = r \cos \theta \qquad\qquad Y = r \sin \theta$$
$$x = r \cos(\theta + \phi) \qquad\qquad y = r \sin(\theta + \phi)$$

Using the addition formula for cosine, we see that

$$\begin{aligned}
x &= r \cos(\theta + \phi) \\
&= r(\cos \theta \cos \phi - \sin \theta \sin \phi) \\
&= (r \cos \theta) \cos \phi - (r \sin \theta) \sin \phi \\
&= X \cos \phi - Y \sin \phi
\end{aligned}$$

Similarly, we can apply the addition formula for sine to the expression for y to obtain $y = X \sin \phi + Y \cos \phi$. By treating these equations for x and y as a system of linear equations in the variables X and Y (see Exercise 27), we obtain expressions for X and Y in terms of x and y, as detailed in the following box.

ROTATION OF AXES FORMULAS

Suppose the x- and y-axes in a coordinate plane are rotated through the acute angle ϕ to produce the X- and Y-axes, as shown in Figure 1. Then the coordinates (x, y) and (X, Y) of a point in the xy- and the XY-planes are related as follows:

$$x = X \cos \phi - Y \sin \phi \qquad\qquad X = x \cos \phi + y \sin \phi$$
$$y = X \sin \phi + Y \cos \phi \qquad\qquad Y = -x \sin \phi + y \cos \phi$$

EXAMPLE 1

If the coordinate axes are rotated through 30°, find the XY-coordinates of the point with xy-coordinates $(2, -4)$.

SOLUTION

Using the Rotation of Axes Formulas with $x = 2$, $y = -4$, and $\phi = 30°$, we get

$$X = 2 \cos 30° + (-4) \sin 30° = 2\left(\frac{\sqrt{3}}{2}\right) - 4\left(\frac{1}{2}\right) = \sqrt{3} - 2$$

$$Y = -2 \sin 30° + (-4) \cos 30° = -2\left(\frac{1}{2}\right) - 4\left(\frac{\sqrt{3}}{2}\right) = -1 - 2\sqrt{3}$$

The XY-coordinates are $\left(-2 + \sqrt{3}, -1 - 2\sqrt{3}\right)$. ∎

EXAMPLE 2

Show by rotating the coordinate axes through $45°$ that the graph of the equation $xy = 2$ is a hyperbola.

SOLUTION

We use the Rotation of Axes Formulas with $\phi = 45°$ to obtain

$$x = X \cos 45° - Y \sin 45° = \frac{X}{\sqrt{2}} - \frac{Y}{\sqrt{2}}$$

$$y = X \sin 45° + Y \cos 45° = \frac{X}{\sqrt{2}} + \frac{Y}{\sqrt{2}}$$

Substituting these expressions into the original equation gives

$$\left(\frac{X}{\sqrt{2}} - \frac{Y}{\sqrt{2}}\right)\left(\frac{X}{\sqrt{2}} + \frac{Y}{\sqrt{2}}\right) = 2$$

$$\frac{X^2}{2} - \frac{Y^2}{2} = 2$$

$$\frac{X^2}{4} - \frac{Y^2}{4} = 1$$

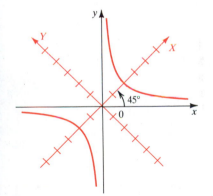

Figure 3
$xy = 2$

We recognize this as a hyperbola with vertices $(\pm 2, 0)$ in the XY-coordinate system. Its asymptotes are $Y = \pm X$, which correspond to the coordinate axes in the xy-system (see Figure 3). ∎

The method illustrated in Example 2 can be used in any equation of the form of Equation 1 to eliminate the term that involves the product xy. We can always choose an angle ϕ so that when the coordinate axes are rotated through this angle, the equation in the new coordinate system does not involve the product XY. In

Example 2 the angle ϕ was given to us, but to see what angle works in any given situation, we perform the following analysis.

After rotating the coordinate axes through an angle ϕ, we substitute the values of x and y from the Rotation of Axes Formulas into Equation 1 to see what the equation becomes in XY-coordinates:

Equation 1

$$Ax^2 + Bxy + Cy^2 + Dx + Ey + F = 0$$

Substituting for x and y using Rotation of Axes Formulas

$$A(X \cos \phi - Y \sin \phi)^2 + B(X \cos \phi - Y \sin \phi)(X \sin \phi + Y \cos \phi)$$
$$+ C(X \sin \phi + Y \cos \phi)^2 + D(X \cos \phi - Y \sin \phi)$$
$$+ E(X \sin \phi + Y \cos \phi) + F = 0$$

If we multiply this out and collect like terms, we obtain an equation of the form

$$A'X^2 + B'XY + C'Y^2 + D'X + E'Y + F' = 0 \qquad (2)$$

where the coefficients A', B', C', $\ldots$ are expressed in terms of the original coefficients A, B, C, $\ldots$, $\sin \phi$, and $\cos \phi$. In particular, the coefficient B' of XY is

$$B' = 2(C - A) \sin \phi \cos \phi + B(\cos^2\phi - \sin^2\phi)$$

(See Exercise 28.) To eliminate the XY-term we would like to choose ϕ so that $B' = 0$, that is,

$$2(C - A) \sin \phi \cos \phi + B(\cos^2\phi - \sin^2\phi) = 0$$
$$B(\cos^2\phi - \sin^2\phi) = 2(A - C) \sin \phi \cos \phi$$
$$\frac{\cos^2\phi - \sin^2\phi}{2 \sin \phi \cos \phi} = \frac{A - C}{B}$$

Using the double-angle formulas for cosine and sine, we can rewrite this as

$$\frac{\cos 2\phi}{\sin 2\phi} = \frac{A - C}{B} \qquad \text{or} \qquad \cot 2\phi = \frac{A - C}{B}$$

This proves the following theorem.

SIMPLIFYING THE GENERAL CONIC EQUATION

To eliminate the xy-term in the general conic equation

$$Ax^2 + Bxy + Cy^2 + Dx + Ey + F = 0$$

rotate the axes through the acute angle ϕ that satisfies

$$\cot 2\phi = \frac{A - C}{B}$$

EXAMPLE 3

Identify and sketch the curve

$$6\sqrt{3}\,x^2 + 6xy + 4\sqrt{3}\,y^2 = 21\sqrt{3}$$

SOLUTION

To eliminate the xy-term, we rotate the axes thorough an angle ϕ that satisfies

$$\cot 2\phi = \frac{A - C}{B} = \frac{6\sqrt{3} - 4\sqrt{3}}{6} = \frac{1}{\sqrt{3}}$$

Thus $2\phi = 60°$ and hence $\phi = 30°$. With this value of ϕ, we get

$$x = X\left(\frac{\sqrt{3}}{2}\right) - Y\left(\frac{1}{2}\right)$$

$$y = X\left(\frac{1}{2}\right) + Y\left(\frac{\sqrt{3}}{2}\right)$$

Substituting these values for x and y into the given equation leads to

$$6\sqrt{3}\left(\frac{X\sqrt{3}}{2} - \frac{Y}{2}\right)^2 + 6\left(\frac{X\sqrt{3}}{2} - \frac{Y}{2}\right)\left(\frac{X}{2} + \frac{Y\sqrt{3}}{2}\right) + 4\sqrt{3}\left(\frac{X}{2} + \frac{Y\sqrt{3}}{2}\right)^2 = 21\sqrt{3}$$

Multiplying this out and simplifying, we get

$$7\sqrt{3}\,X^2 + 3\sqrt{3}\,Y^2 = 21\sqrt{3}$$

or, after dividing through by $21\sqrt{3}$,

$$\frac{X^2}{3} + \frac{Y^2}{7} = 1$$

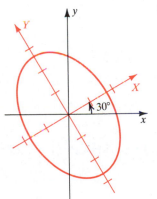

Figure 4
$6\sqrt{3}\,x^2 + 6xy + 4\sqrt{3}\,y^2 = 21\sqrt{3}$

This is the equation of an ellipse in the XY-coordinate system. The foci lie on the Y-axis. Because $a^2 = 7$ and $b^2 = 3$, the length of the major axis is $2\sqrt{7}$ and the length of the minor axis is $2\sqrt{3}$. The ellipse is sketched in Figure 4.   ■

In the preceding example we were able to determine ϕ without difficulty, since we remembered that $\cot 60° = 1/\sqrt{3}$. In general, finding ϕ is not quite so easy. As the next example illustrates, the following half-angle formulas, which are valid for $0 < \phi < \pi/2$, are useful in determining ϕ (see Section 7.4):

$$\cos \phi = \sqrt{\frac{1 + \cos 2\phi}{2}} \qquad \sin \phi = \sqrt{\frac{1 - \cos 2\phi}{2}}$$

EXAMPLE 4

Identify and sketch the curve

$$64x^2 + 96xy + 36y^2 - 15x + 20y - 25 = 0$$

SOLUTION

This is in the form of Equation 1 with $A = 64$, $B = 96$, and $C = 36$. Thus

$$\cot 2\phi = \frac{A - C}{B} = \frac{64 - 36}{96} = \frac{7}{24}$$

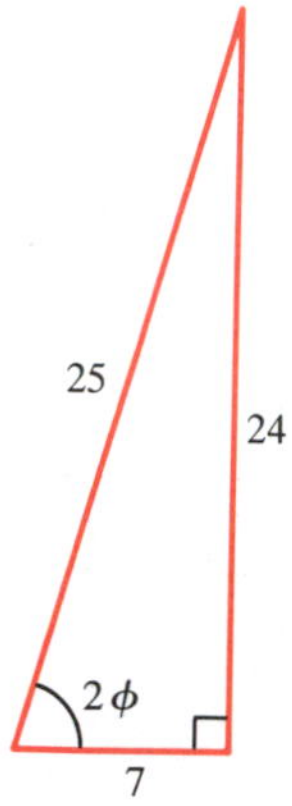

In Figure 5 we sketch a triangle with $\cot 2\phi = \frac{7}{24}$. We see that

$$\cos 2\phi = \frac{7}{25}$$

so, using the half-angle formulas, we get

$$\cos \phi = \sqrt{\frac{1 + \frac{7}{25}}{2}} = \sqrt{\frac{16}{25}} = \frac{4}{5}$$

and

$$\sin \phi = \sqrt{\frac{1 - \frac{7}{25}}{2}} = \sqrt{\frac{9}{25}} = \frac{3}{5}$$

The Rotation of Axes Formulas then give:

$$x = \tfrac{4}{5}X - \tfrac{3}{5}Y \qquad \text{and} \qquad y = \tfrac{3}{5}X + \tfrac{4}{5}Y$$

Substituting into the given equation, we have

$$64\left(\tfrac{4}{5}X - \tfrac{3}{5}Y\right)^2 + 96\left(\tfrac{4}{5}X - \tfrac{3}{5}Y\right)\left(\tfrac{3}{5}X + \tfrac{4}{5}Y\right)$$

$$+ 36\left(\tfrac{3}{5}X + \tfrac{4}{5}Y\right)^2 - 15\left(\tfrac{4}{5}X - \tfrac{3}{5}Y\right) + 20\left(\tfrac{3}{5}X + \tfrac{4}{5}Y\right) - 25 = 0$$

Multiplying out and collecting like terms, we get

$$100X^2 + 25Y - 25 = 0$$

which simplifies to

$$-4X^2 = Y - 1$$

We recognize this as the equation of a parabola that opens along the negative Y-axis and has vertex $(0, 1)$ in XY-coordinates. Since $4p = -4$, we have $p = -1$, so the focus is $(0, 0)$ and the directrix is $Y = 2$. Using

$$\phi = \cos^{-1}\tfrac{4}{5} \approx 37°$$

we sketch the graph in Figure 6.

■

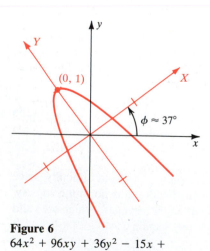

Figure 5

Figure 6
$64x^2 + 96xy + 36y^2 - 15x + 20y - 25 = 0$

EXERCISES 9.5

In Exercises 1–6 determine the XY-coordinates of the given point if the coordinate axes are rotated through the indicated angle.

1. $(1, 1)$, $\phi = 45°$ **2.** $(-2, 1)$, $\phi = 30°$

3. $\left(3, -\sqrt{3}\right)$, $\phi = 60°$ **4.** $(2, 0)$, $\phi = 15°$

5. $(0, 2)$, $\phi = 55°$ **6.** $\left(\sqrt{2}, 4\sqrt{2}\right)$, $\phi = 45°$

In Exercises 7–10 determine the equation of the given conic in XY-coordinates when the coordinate axes are rotated through the indicated angle.

7. $y = (x - 1)^2$, $\phi = 45°$

8. $x^2 - y^2 = 2y$, $\phi = \cos^{-1}\frac{3}{5}$

9. $x^2 + 2\sqrt{3}\,xy - y^2 = 4$, $\phi = 30°$

10. $xy = x + y$, $\phi = \pi/4$

In Exercises 11–24 use rotation of axes to identify and sketch the graph of each equation.

11. $xy = 8$ **12.** $xy + 4 = 0$

13. $x^2 + 2xy + y^2 + x - y = 0$

14. $13x^2 + 6\sqrt{3}\,xy + 7y^2 = 16$

15. $x^2 + 2\sqrt{3}\,xy - y^2 + 2 = 0$

16. $21x^2 + 10\sqrt{3}\,xy + 31y^2 = 144$

17. $11x^2 - 24xy + 4y^2 + 20 = 0$

18. $25x^2 - 120xy + 144y^2 - 156x - 65y = 0$

19. $\sqrt{3}\,x^2 + 3xy = 3$

20. $153x^2 + 192xy + 97y^2 = 225$

21. $2\sqrt{3}\,x^2 - 6xy + \sqrt{3}\,x + 3y = 0$

22. $9x^2 - 24xy + 16y^2 = 100(x - y - 1)$

23. $52x^2 + 72xy + 73y^2 = 40x - 30y + 75$

24. $(7x + 24y)^2 = 600x - 175y + 25$

25. (a) Use rotation of axes to show that the following equation represents a hyperbola:

$$7x^2 + 48xy - 7y^2 - 200x - 150y + 600 = 0$$

(b) Find the XY- and xy-coordinates of the center, vertices, and foci.

(c) Find the equations of the asymptotes in XY- and xy-coordinates.

26. (a) Use rotation of axes to show that the following equation represents a parabola:

$$2\sqrt{2}\,(x + y)^2 = 7x + 9y$$

(b) Find the XY- and xy-coordinates of the vertex and focus.

(c) Find the equation of the directrix in XY- and xy-coordinates.

27. Solve the equations

$$x = X \cos \phi - Y \sin \phi$$
$$y = X \sin \phi + Y \cos \phi$$

for X and Y in terms of x and y.   [*Hint:* To begin, multiply the first equation by cos ϕ and the second by sin ϕ, and then add the two equations to solve for X.]

28. Suppose that a rotation through the angle ϕ changes Equation 1 to Equation 2 (in the text). Show the following:
(a) $B' = 2(C - A) \sin \phi \cos \phi + B(\cos^2\phi - \sin^2\phi)$
(b) $A + C = A' + C'$

29. For Equation 1 in the text, the quantity

$$B^2 - 4AC$$

is called the **discriminant** of the equation. Suppose that a rotation through the angle ϕ changes Equation 1 to Equation 2 (in the text). Show that Equation 1 and Equation 2 have the same discriminant.

30. Use Exercise 29 to show that, except in the degenerate cases, Equation 1 represents an ellipse (or a circle) if the discriminant is negative, a parabola if the discriminant is zero, and a hyperbola if the discriminant is positive.

31. Use Exercise 30 to determine the type of curve in each of Exercises 11–24.

32. Show that the graph of the equation

$$\sqrt{x} + \sqrt{y} = 1$$

is a part of a parabola by rotating the axes through an

angle of 45°. [*Hint:* First convert the equation to one that does not involve radicals.]

33. Let Z, Z', and R be the matrices

$$Z = \begin{bmatrix} x \\ y \end{bmatrix}, \qquad Z' = \begin{bmatrix} X \\ Y \end{bmatrix},$$

$$R = \begin{bmatrix} \cos \phi & -\sin \phi \\ \sin \phi & \cos \phi \end{bmatrix}$$

Show that the Rotation of Axes Formulas can be written as

$$Z = RZ' \qquad \text{and} \qquad Z' = R^{-1}Z$$

SECTION 9.6
POLAR COORDINATES

A coordinate system is a method for specifying the location of a point in the plane. Up to now we have been dealing with the rectangular (or Cartesian) coordinate system, which describes locations using a rectangular grid. Using rectangular coordinates is like describing a location in a city by saying that, for example, it is at the corner of 48th Street and 7th Avenue. But we might also describe this same location by saying that it is 3 miles northwest of city hall. Instead of specifying the location with respect to a grid of streets and avenues, we can describe it by giving its distance and direction from a fixed reference point.

The *polar coordinate system* uses distances and directions to specify the location of points in the plane. To set up this system, we first choose a fixed point O called the **origin** (or **pole**). We then draw a ray (half-line) starting at O, called the **polar axis**. This axis is usually drawn horizontally to the right of O and coincides with the x-axis in rectangular coordinates. Now let P be any point in the plane. Let r be the distance from P to the origin, and let θ be the angle between the polar axis and the segment OP, as shown in Figure 1. Then the ordered pair (r, θ) uniquely specifies the location of P. We write $P(r, \theta)$ and refer to r and θ as the **polar coordinates** of P. We use the convention that θ is positive if measured in a counterclockwise direction from the polar axis, and negative if measured in a clockwise direction. It is customary to use radian measure for θ. If $r = 0$, then $P = O$ no matter what value θ has, so $(0, \theta)$ represents the pole for any value of θ.

Because the angles $\theta + 2n\pi$ (for $n = \pm 1, \pm 2, \pm 3, \ldots$) all have the same terminal side as the angle θ, each point has infinitely many representations in polar coordinates. For example, $(2, \pi/3)$, $(2, 7\pi/3)$, and $(2, -5\pi/3)$ all represent the same point, as illustrated in Figure 2. Moreover, we also allow r to take on negative

Figure 1

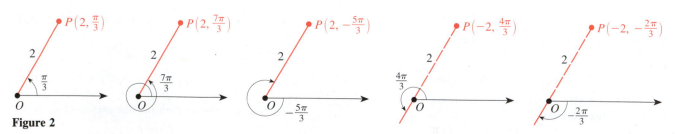

Figure 2

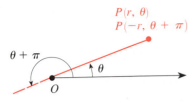

Figure 3

values, with the understanding that if $-r$ is negative, then the point $P(-r, \theta)$ lies r units away from the origin in the direction *opposite* to that given by θ. Thus the point P graphed in Figure 2 can also be described by the coordinates $(-2, 4\pi/3)$ or $(-2, -2\pi/3)$. By this convention, the coordinates (r, θ) and $(-r, \theta + \pi)$ represent the same point (see Figure 3). In fact, any point $P(r, \theta)$ can also be represented by

$$P(r, \theta + 2n\pi) \qquad \text{and} \qquad P(-r, \theta + (2n + 1)\pi)$$

for any integer n.

EXAMPLE 1

Plot the points whose polar coordinates are given:

(a) $(1, 3\pi/4)$ (b) $(3, -\pi/6)$ (c) $(3, 3\pi)$ (d) $(-4, \pi/4)$

SOLUTION

The points are plotted in Figure 4. Note that the point in part (d) lies 4 units from the origin along the angle $5\pi/4$, because the given value of r is negative.

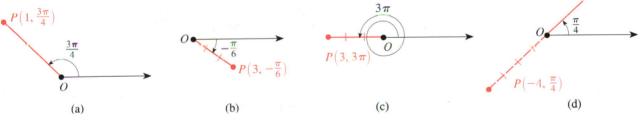

Figure 4

    (a)   (b)   (c)   (d)

Situations often arise in which we need to consider polar and rectangular coordinates simultaneously. The connection between the two systems is illustrated in Figure 5, where the polar axis coincides with the positive x-axis. Although we have pictured the case where $r > 0$ and θ is acute, the following discussion holds for any angle θ and for any nonzero value of r. From the figure, we see that

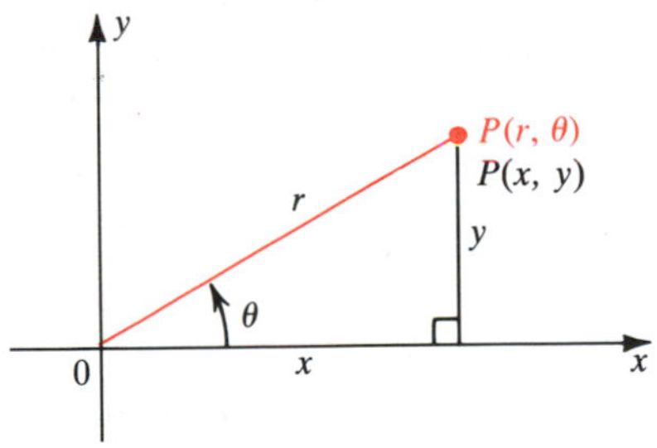

$$\cos \theta = \frac{x}{r} \qquad \text{and} \qquad \sin \theta = \frac{y}{r}$$

Figure 5

Thus we have

$$x = r \cos \theta \qquad \text{and} \qquad y = r \sin \theta$$

Note that these equations also hold at the origin, since $r = 0$ and $x = y = 0$ there.

From Figure 5 we also see that $\tan \theta = y/x$ and that $x^2 + y^2 = r^2$ by the Pythagorean Theorem. This proves the formulas in the following box, which relate polar and rectangular coordinates.

RELATIONSHIP BETWEEN POLAR AND RECTANGULAR COORDINATES

(a) To change from polar to rectangular coordinates, use the formulas

$$x = r \cos \theta \qquad \text{and} \qquad y = r \sin \theta$$

(b) To change from rectangular to polar coordinates, use the formulas

$$\tan \theta = \frac{y}{x} \quad (x \neq 0) \qquad \text{and} \qquad r^2 = x^2 + y^2$$

EXAMPLE 2

Find rectangular coordinates for the point that has polar coordinates $(4, 2\pi/3)$.

SOLUTION

Since $r = 4$ and $\theta = 2\pi/3$, we have

$$x = r \cos \theta = 4 \cos \frac{2\pi}{3} = 4 \cdot \left(-\frac{1}{2}\right) = -2$$

$$y = r \sin \theta = 4 \sin \frac{2\pi}{3} = 4 \cdot \frac{\sqrt{3}}{2} = 2\sqrt{3}$$

Thus the point has rectangular coordinates $\left(-2, 2\sqrt{3}\right)$. ■

EXAMPLE 3

Find polar coordinates for the point that has rectangular coordinates $(2, -2)$.

SOLUTION

Using $x = 2$, $y = -2$, we get

$$r^2 = x^2 + y^2 = 2^2 + (-2)^2 = 8$$

so $r = 2\sqrt{2}$ or $-2\sqrt{2}$. Also

$$\tan \theta = \frac{y}{x} = \frac{-2}{2} = -1$$

so $\theta = 3\pi/4$ or $-\pi/4$. Since the point $(2, -2)$ lies in quadrant IV (see Figure 6), we can represent it in polar coordinates as $\left(2\sqrt{2}, -\pi/4\right)$ or $\left(-2\sqrt{2}, 3\pi/4\right)$. ■

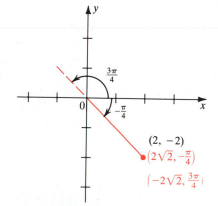

(2, −2)
$\left(2\sqrt{2}, -\frac{\pi}{4}\right)$
$\left(-2\sqrt{2}, \frac{3\pi}{4}\right)$

Figure 6

   Note that the equations relating polar and rectangular coordinates do not uniquely determine r or θ. When we use these equations to find the polar coordinates of a point, we must be careful that the values we choose for r and θ give us a point in the correct quadrant, as we saw in Example 3.

GRAPHS OF POLAR EQUATIONS

The **graph of a polar equation** $r = f(\theta)$ consists of all points P that have at least one polar representation (r, θ) whose coordinates satisfy the equation. In the next two examples, we see that circles centered at the origin and lines that pass through the origin have particularly simple equations in polar coordinates.

EXAMPLE 4

Sketch the graph of the equation $r = 3$.

SOLUTION

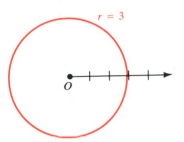

Figure 7

The graph consists of all points whose r-coordinate is 3, that is, all points that are 3 units away from the origin. Therefore this is the circle of radius 3 centered at the origin, as shown in Figure 7.   ■

In general, the graph of the equation $r = a$ is a circle of radius $|a|$ centered at the origin. Squaring both sides of this equation, we get $r^2 = a^2$, and since $r^2 = x^2 + y^2$, the equivalent equation in rectangular coordinates is

$$x^2 + y^2 = a^2$$

EXAMPLE 5

Sketch the graph of the equation $\theta = \pi/3$ and express the equation in rectangular coordinates.

SOLUTION

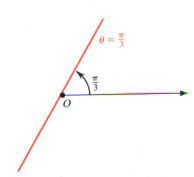

Figure 8

The graph consists of all points whose θ-coordinate is $\pi/3$. This is the straight line that passes through the origin and makes an angle of $\pi/3$ with the polar axis (see Figure 8). Note that the points $(r, \pi/3)$ on the line with $r > 0$ lie in quadrant I, whereas those with $r < 0$ lie in quadrant III. If the point (x, y) lies on this line, then

$$\frac{y}{x} = \tan \theta = \tan \frac{\pi}{3} = \sqrt{3}$$

Thus the rectangular equation of this line is $y = \sqrt{3}\, x$.   ■

To sketch a polar curve whose graph is not so obvious as the ones in the preceding examples, we rely on two techniques. One technique is to plot points calculated for sufficiently many values of θ and then join them in a continuous curve. This is what we did when we first learned to graph functions in rectangular coordinates. The other technique is to convert the polar equation into rectangular coordinates in the hope that the resulting equation is one that we recognize from our previous work. Both methods are used in the next example.

EXAMPLE 6

(a) Sketch the curve with polar equation $r = 2 \sin \theta$.

(b) Convert the equation of part (a) to rectangular coordinates.

SOLUTION

(a) We first use the equation to determine the polar coordinates of several points on the curve. The results are shown in the following table:

θ	0	$\pi/6$	$\pi/4$	$\pi/3$	$\pi/2$	$2\pi/3$	$3\pi/4$	$5\pi/6$	π
$r = 2 \sin \theta$	0	1	$\sqrt{2}$	$\sqrt{3}$	2	$\sqrt{3}$	$\sqrt{2}$	1	0

We plot these points in Figure 9 and then join them to sketch the curve. The graph appears to be a circle. We have used values of θ only between 0 and π, since the same points (this time expressed with negative r-coordinates) would be obtained if we allowed θ to range from π to 2π.

Figure 9
$r = 2 \sin \theta$

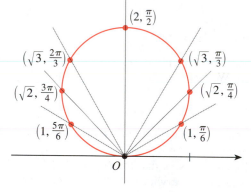

(b) If we multiply both sides of the equation of this curve by r, we get

$$r^2 = 2r \sin \theta$$

Using the formulas that relate polar and rectangular coordinates, we replace r^2 by $x^2 + y^2$ and $r \sin \theta$ by y to obtain the rectangular equation

$$x^2 + y^2 = 2y$$

Thus $x^2 + y^2 - 2y = 0$, and we complete the square in y to get

$$x^2 + (y - 1)^2 = 1$$

which is a circle of radius 1 centered at the point (0, 1). ∎

Using the method from part (b) of the preceding example, we can show the following:

---

1. The equation

$$r = 2a \sin \theta$$

represents a circle of radius $|a|$ centered at the point that has polar coordinates $(a, \pi/2)$.

2. The equation

$$r = 2a \cos \theta$$

represents a circle of radius $|a|$ centered at the point that has polar coordinates $(a, 0)$.

EXAMPLE 7

(a) Sketch the curve $r = 2 + 2 \cos \theta$.
(b) Convert the equation of part (a) to rectangular coordinates.

SOLUTION

(a) We tabulate the values of r for some convenient values of θ, as in Example 6.

θ	0	$\pi/6$	$\pi/4$	$\pi/3$	$\pi/2$	$2\pi/3$	$3\pi/4$	$5\pi/6$	π
$r = 2 + 2 \cos \theta$	4	$2 + \sqrt{3}$	$2 + \sqrt{2}$	3	2	1	$2 - \sqrt{2}$	$2 - \sqrt{3}$	0

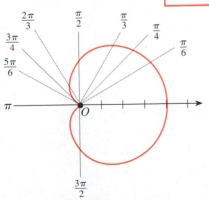

As θ increases from 0 to π, we see that r decreases from 4 to 0. Plotting the points in this table gives us the upper half of the curve shown in Figure 10. We obtain the lower half of this curve by noting that replacing θ by $-\theta$ in the equation leaves it unchanged, since $\cos(-\theta) = \cos \theta$. Thus for every point (r, θ) on the upper half of the curve, there is a corresponding point $(r, -\theta)$ below the polar axis. This means that the lower half is simply a mirror image of the upper half, and we complete the graph using symmetry.

(b) We first multiply both sides of the equation by r:

$$r^2 = 2r + 2r \cos \theta$$

Using $r^2 = x^2 + y^2$ and $y = r \cos \theta$, we can convert two of the three terms in the equation into rectangular coordinates, but to eliminate the remaining r

Figure 10
$r = 2 + 2 \cos \theta$

requires more work:

$$x^2 + y^2 = 2r + 2y$$
$$x^2 + y^2 - 2y = 2r$$
$$(x^2 + y^2 - 2y)^2 = 4r^2 \qquad \text{(squaring both sides)}$$
$$(x^2 + y^2 - 2y)^2 = 4(x^2 + y^2) \qquad \text{(since } r^2 = x^2 + y^2)$$

The rectangular equation is in this case much more complicated than the polar equation, so the polar form is the more useful one. ■

The curve in Figure 10 is called a **cardioid** because it is heart-shaped. In general, any equation of the form

$$r = a(1 \pm \cos \theta) \qquad \text{or} \qquad r = a(1 \pm \sin \theta)$$

represents a cardioid.

We saw in Example 7 that exploiting symmetry often saves a lot of work when sketching polar curves. We list three tests for symmetry. Figure 11 shows why these tests work.

TESTS FOR SYMMETRY

1. If a polar equation is unchanged when we replace θ by $-\theta$, the graph is symmetric about the polar axis [Figure 11(a)].
2. If the equation is unchanged when we replace r by $-r$, then graph is symmetric about the origin [Figure 11(b)].
3. If the equation is unchanged when we replace θ by $\pi - \theta$, the graph is symmetric about the vertical line $\theta = \pi/2$ (the y-axis) [Figure 11(c)].

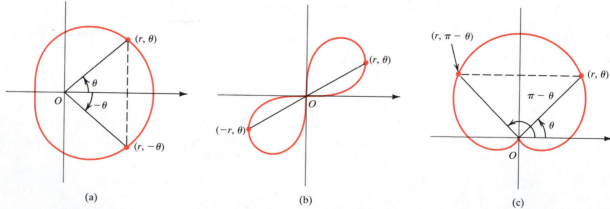

(a)   (b)   (c)

Figure 11

EXAMPLE 8

Sketch the curve $r = \cos 2\theta$.

SOLUTION

Instead of plotting points as we did in Examples 6 and 7, we reason as follows. First we sketch the graph of $r = \cos 2\theta$ in *rectangular* coordinates, with θ plotted along the horizontal axis and r on the vertical axis. The graph is shown in Figure 12. From this graph we can find at a glance the values of r that correspond to increasing values of θ. As θ increases from 0 to $\pi/4$, Figure 12 shows that r decreases from 1 to 0, and so we draw the corresponding portion of the polar curve in Figure 13 (indicated by ①). As θ increases from $\pi/4$ to $\pi/2$, the value of r goes from 0 to -1. This means that the distance from the origin increases from 0 to 1, but instead of being in quadrant I, this portion of the polar curve (indicated by the ②) lies on the opposite side of the origin in quadrant III. The remainder of the curve is drawn in a similar fashion, with the arrows and numbers indicating the order in which the portions are traced out. The resulting curve has four petals and is called a **four-leafed rose**.

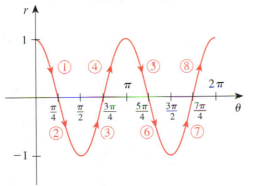

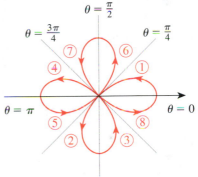

Figure 12
$r = \cos 2\theta$ in rectangular coordinates

Figure 13
Four-leafed rose $r = \cos 2\theta$ in polar coordinates

In general, the graphs of equations of the form

$$r = a \cos n\theta \qquad \text{or} \qquad r = a \sin n\theta$$

are n-leafed roses if n is odd and $2n$-leafed roses if n is even (as in Example 8).

EXERCISES 9.6

In Exercises 1–6 plot the point that has the given polar coordinates. Then give two other polar coordinate representations of each point, one with $r < 0$ and the other with $r > 0$.

1. $(3, \pi/2)$ **2.** $(2, 3\pi/4)$ **3.** $(-1, 7\pi/6)$

4. $(-2, -\pi/3)$ **5.** $(-5, 0)$ **6.** $(3, 1)$

In Exercises 7–12 find the rectangular coordinates for the point whose polar coordinates are given.

7. $(4, \pi/6)$ **8.** $(6, 2\pi/3)$ **9.** $\left(\sqrt{2}, -\pi/4\right)$

10. $(-1, 5\pi/2)$ **11.** $(5, 5\pi)$ **12.** $(0, 13\pi)$

In Exercises 13–18 convert the given rectangular coordinates to polar coordinates with $r > 0$ and $0 \le \theta < 2\pi$.

13. $(-1, 1)$ **14.** $\left(3\sqrt{3}, -3\right)$ **15.** $\left(\sqrt{8}, \sqrt{8}\right)$

16. $\left(-\sqrt{6}, -\sqrt{2}\right)$ **17.** $(3, 4)$ **18.** $(1, -2)$

In Exercises 19–24 convert the equation to polar form.

19. $x = y$ **20.** $x^2 + y^2 = 9$ **21.** $y = x^2$

22. $y = 5$ **23.** $x = 4$ **24.** $x^2 - y^2 = 1$

In Exercises 25–38 convert the polar equation to rectangular coordinates.

25. $r = 7$ **26.** $\theta = \pi$

27. $r \cos \theta = 6$ **28.** $r = 6 \cos \theta$

29. $r^2 = \tan \theta$ **30.** $r^2 = \sin 2\theta$

31. $r = \dfrac{1}{\sin \theta - \cos \theta}$ **32.** $r = \dfrac{1}{1 + \sin \theta}$

33. $r = 1 + \cos \theta$ **34.** $r = \dfrac{4}{1 + 2 \sin \theta}$

35. $r = 2 \sec \theta$ **36.** $r = 2 - \cos \theta$

37. $\sec \theta = 2$ **38.** $\cos 2\theta = 1$

In Exercises 39–60 sketch the curve whose polar equation is given.

39. $r = 3$ **40.** $r = -1$

41. $\theta = -\pi/2$ **42.** $\theta = 5\pi/6$

43. $r = 6 \sin \theta$ **44.** $r = \cos \theta$

45. $r = -2 \cos \theta$ **46.** $r = 2 \sin \theta + 2 \cos \theta$

47. $r = 2 - 2 \cos \theta$ **48.** $r = 1 + \sin \theta$

49. $r = -3(1 + \sin \theta)$ **50.** $r = \cos \theta - 1$

51. $r = \theta, \ \theta \ge 0$ (spiral)

52. $r\theta = 1, \ \theta > 0$ (reciprocal spiral)

53. $r = \sin 2\theta$ (four-leafed rose)

54. $r = 2 \cos 3\theta$ (three-leafed rose)

55. $r^2 = \cos 2\theta$ (lemniscate)

56. $r^2 = 4 \sin 2\theta$ (lemniscate)

57. $r = 2 + \sin \theta$ (limaçon)

58. $r = 1 - 2 \cos \theta$ (limaçon)

59. $r = 2 + \sec \theta$ (conchoid)

60. $r = \sin \theta \tan \theta$ (cissoid)

*In Exercises 61–64 we consider **polar equations of conics**. It can be shown that any polar equation of the form*

$$r = \frac{ed}{1 \pm e \cos \theta} \quad or \quad r = \frac{ed}{1 \pm e \sin \theta} \quad (e > 0)$$

is a conic section. It is a parabola if $e = 1$, an ellipse if $0 < e < 1$, and a hyperbola if $e > 1$. In each exercise, (a) identify the conic and (b) convert the equation to rectangular coordinates.

61. $r = \dfrac{1}{1 - \cos \theta}$ **62.** $r = \dfrac{2}{1 + 2 \sin \theta}$

63. $r = \dfrac{4}{1 + 2 \sin \theta}$ **64.** $r = \dfrac{12}{1 - 4 \cos \theta}$

65. (a) Show that the distance between the points whose polar coordinates are (r_1, θ_1) and (r_2, θ_2) is

$$\sqrt{r_1^2 + r_2^2 - 2r_1 r_2 \cos(\theta_2 - \theta_1)}$$

[*Hint:* Use the Law of Cosines.]

(b) Find the distance between the points whose polar coordinates are $(3, 3\pi/4)$ and $(-1, 7\pi/6)$.

66. Show that the graph of $r = a \cos \theta + b \sin \theta$ is a circle, and find its center and radius.

SECTION 9.7
PARAMETRIC EQUATIONS

So far we have described a curve by giving an equation that the coordinates of all points on the curve must satisfy. For example, we know that the equation $y = x^2$ represents a parabola in rectangular coordinates and that $r = \sin \theta$ represents a circle in polar coordinates. We now study another method for describing a curve in the plane, which in many situations turns out to be more useful and natural than either rectangular or polar equations. In this method, the x- and y-coordinates of points on the curve are given separately as functions of an additional variable t, called the **parameter**:

$$x = f(t) \qquad y = g(t)$$

These are called **parametric equations** for the curve. Substituting a value of t into each equation determines the coordinates of a point (x, y). As t varies, the point $(x, y) = (f(t), g(t))$ varies and traces out the curve. If we think of t as representing time, then as t increases, we can imagine a particle at $(x, y) = (f(t), g(t))$ moving along the curve.

EXAMPLE 1

Sketch the curve defined by the parametric equations

$$x = t^2 - 3t \qquad y = t - 1$$

Eliminate the parameter t to obtain a single equation for the curve in the variables x and y.

SOLUTION

For every value of t we get a point on the curve. For example, if $t = 0$, then $x = 0$ and $y = -1$, so the corresponding point is $(0, -1)$. In Figure 1 we plot the points (x, y) determined by the values of t shown in the following table.

t	-2	-1	0	1	2	3	4	5
x	10	4	0	-2	-2	0	4	10
y	-3	-2	-1	0	1	2	3	4

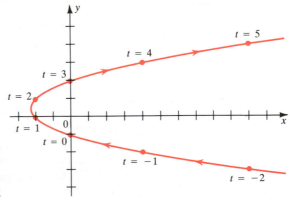

Figure 1

As t increases, a particle whose position is given by the parametric equations moves along the curve in the direction of the arrows. The curve seems to be a parabola. We can confirm this by eliminating the parameter t from the parametric equations and reducing them to a single equation as follows. First we solve for t in the second equation to get $t = y + 1$. Substituting this into the first equation, we get

$$x = (y + 1)^2 - 3(y + 1) = y^2 - y - 2$$

The curve is the parabola $x = y^2 - y - 2$.   ■

Notice that we would obtain the same graph as in Example 1 from the parametrization

$$x = t^2 - t - 2 \qquad y = t$$

because the points on this curve also satisfy the equation $x = y^2 - y - 2$. But the same value of t produces different points on the curve in these two parametrizations. For example, when $t = 0$, the particle that traces out the curve in Figure 1 is at $(0, -1)$, whereas in the parametrization of the preceding equations, the particle is already at $(-2, 0)$ when $t = 0$. Thus a parametrization contains more information than just the curve being parametrized. It also indicates *how* that curve is being traced out.

EXAMPLE 2

Describe and graph the curve represented by the parametric equations

$$x = \cos t \qquad y = \sin t \qquad 0 \leqslant t \leqslant 2\pi$$

SOLUTION

To identify the curve, we eliminate the parameter. Since $\cos^2 t + \sin^2 t = 1$ and since $x = \cos t$ and $y = \sin t$ for every point (x, y) on the curve, we have

$$x^2 + y^2 = (\cos t)^2 + (\sin t)^2 = 1$$

This means that all points on the curve satisfy the equation $x^2 + y^2 = 1$, so the graph is a circle of radius 1 centered at the origin. As t increases from 0 to 2π, the point given by the parametric equations starts at $(1, 0)$ and moves counterclockwise once around the circle, as shown in Figure 2. Notice that the parameter t can be interpreted as the angle shown in the figure.   ■

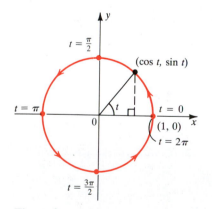

Figure 2

EXAMPLE 3

Find parametric equations for the line of slope 3 that passes through the point (2, 6).

SOLUTION

Let us start at the point (2, 6) and move up and to the right along this line. Because the line has slope 3, for every one unit we move to the right, we must move up three units. In other words, if we increase the x-coordinate by t units, we must correspondingly increase the y-coordinate by $3t$ units. This leads to the parametric equations

$$x = 2 + t \qquad y = 6 + 3t$$

To confirm that these equations give the desired line, we eliminate the parameter by solving for t in the first equation and substituting into the second, to get

$$y = 6 + 3(x - 2) = 3x$$

Thus the slope-intercept form of the equation of this line is $y = 3x$, which is a line of slope 3 that does pass through (2, 6) as required. The graph is shown in Figure 3.   ■

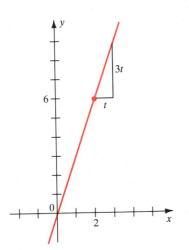

Figure 3

EXAMPLE 4

Sketch the curve with parametric equations

$$x = \sin t \qquad y = 2 - \cos^2 t$$

SOLUTION

To eliminate the parameter, we first use the trigonometric identity $\cos^2 t = 1 - \sin^2 t$ to change the second equation to

$$y = 2 - \cos^2 t = 2 - (1 - \sin^2 t) = 1 + \sin^2 t$$

Now we can substitute $\sin t = x$ from the first equation to get

$$y = 1 + x^2$$

and so the point (x, y) moves along the parabola $y = 1 + x^2$. However, since $-1 \le \sin t \le 1$, we have $-1 \le x \le 1$, so the parametric equations represent only the part of the parabola between $x = -1$ and $x = 1$. Since $\sin t$ is periodic, the point $(x, y) = (\sin t, 2 - \cos^2 t)$ moves back and forth infinitely often along the parabola between the points $(-1, 2)$ and $(1, 2)$ as shown in Figure 4.   ■

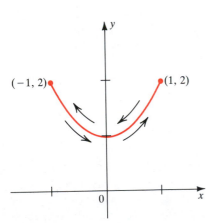

Figure 4

EXAMPLE 5

As a circle rolls along a straight line, the curve traced out by a fixed point P on the circumference of the circle is called a **cycloid** (see Figure 5). If the circle has radius a and rolls along the x-axis, with one position of the point P being at the origin, find parametric equations for the cycloid.

Figure 5

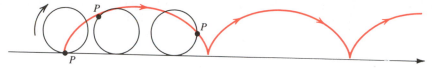

SOLUTION

Figure 6 shows the circle and the point P after the circle has rolled through an angle θ (in radians). The distance $|OT|$ that the circle has rolled must be the same as the length of the arc PT, which, by the arc length formula, is $a\theta$ (see Section 6.1). This means that the center of the circle is $C(a\theta, a)$.

Let the coordinates of P be (x, y). Then from Figure 6 (which illustrates the case $0 < \theta < \pi/2$), we see that

$$x = |OT| - |PQ| = a\theta - a \sin \theta = a(\theta - \sin \theta)$$
$$y = |TC| - |QC| = a - a \cos \theta = a(1 - \cos \theta)$$

so parametric equations for the cycloid are

$$x = a(\theta - \sin \theta) \qquad y = a(1 - \cos \theta) \qquad \blacksquare$$

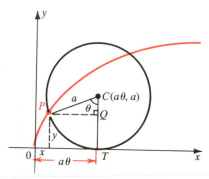

Figure 6

The cycloid has a number of interesting physical properties. It is the "curve of quickest descent" in the following sense. Let us choose two points P and Q not directly above each other, and join them with a wire. Suppose we allow a bead to slide down the wire under the influence of gravity (ignoring the resulting friction). Of all possible shapes that the wire can be bent into, the bead will slide from P to Q the fastest when the shape is half of an arch of an inverted cycloid (see Figure 7). The cycloid is also the "curve of equal descent" in the sense that no matter where we place a bead B on a cycloid-shaped wire, it takes the same time

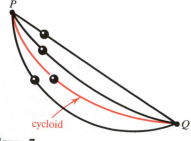

cycloid

Figure 7

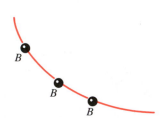

Figure 8

to slide to the bottom (see Figure 8). These rather surprising properties of the cycloid were proved (using calculus) in the 17th century by several mathematicians and physicists, including Johann Bernoulli, Blaise Pascal, and Christiaan Huygens.

USING GRAPHING DEVICES TO GRAPH PARAMETRIC CURVES

Most graphing calculators and computer graphing programs can be used to graph parametric equations. Such devices are particularly useful when sketching complicated curves that would be difficult to graph by hand.

EXAMPLE 6

Use a graphing device to sketch the following parametric curves. Discuss their similarities and differences.

(a) $x = \sin 2t$ (b) $x = \sin 3t$

    $y = 2 \cos t$ $y = 2 \cos t$

SOLUTION

In both parts (a) and (b), the graph will lie inside the rectangle given by $-1 \le x \le 1$, $-2 \le y \le 2$, since both the sine and the cosine of any number will be between -1 and 1. Thus we may use the viewing rectangle $[-1.5, 1.5]$ by $[-2.5, 2.5]$.

(a) Since $2 \cos t$ is periodic with period 2π (see Section 5.3), and since $\sin 2t$ has period π, letting t vary over the interval $0 \le t \le 2\pi$ will give us the complete graph, which is shown in Figure 9(a).

(b) Again, letting t take on values between 0 and 2π gives the complete graph shown in Figure 9(b).

Both graphs are *closed curves*, which means that they form loops with no starting or ending point, and both graphs cross over themselves. However, the graph in Figure 9(a) has two loops, like a figure eight, whereas the graph in Figure 9(b) has three loops. ∎

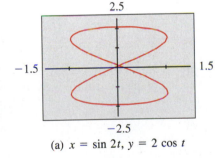

(a) $x = \sin 2t$, $y = 2 \cos t$

(b) $x = \sin 3t$, $y = 2 \cos t$

Figure 9

The curves graphed in Example 6 are called Lissajous figures. A **Lissajous figure** is the graph of a pair of parametric equations of the form

$$x = A \sin \omega_1 t$$
$$y = B \cos \omega_2 t$$

where A, B, ω_1, and ω_2 are real constants. Since $\sin \omega_1 t$ and $\cos \omega_2 t$ are both between -1 and 1, a Lissajous figure will lie inside the rectangle determined by $-A \le x \le A$, $-B \le y \le B$. This fact can be used to choose a viewing rectangle when graphing a Lissajous figure, as in Example 6.

Recall from Section 9.6 that rectangular coordinates (x, y) and polar coordinates (r, θ) are related by the equations $x = r \cos \theta$, $y = r \sin \theta$. Thus we can graph the polar equation $r = f(\theta)$ by changing it to parametric form as follows:

$$x = r \cos \theta = f(\theta) \cos \theta$$

[since $r = f(\theta)$]

$$y = r \sin \theta = f(\theta) \sin \theta$$

Replacing θ by the standard parametric variable t, we have the following result.

POLAR EQUATIONS IN PARAMETRIC FORM

The graph of the polar equation $r = f(\theta)$ is the same as the graph of the parametric equations

$$x = f(t) \cos t$$
$$y = f(t) \sin t$$

EXAMPLE 7

Use a graphing device to sketch a graph of the polar equation $r = \ln \theta$, where $1 \leq \theta \leq 10\pi$.

SOLUTION

We first put the equation in parametric form:

$$x = \ln t \cos t$$
$$y = \ln t \sin t$$

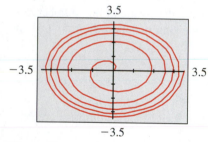

Figure 10
$x = \ln t \cos t, y = \ln t \sin t$

Since $\ln 10\pi \approx 3.45$, we use the viewing rectangle $[-3.5, 3.5]$ by $[-3.5, 3.5]$, and we let t vary from 1 to $10\pi \approx 31.42$. The resulting graph shown in Figure 10 is a *logarithmic spiral*. ∎

EXERCISES 9.7

In Exercises 1–22, (a) sketch the curve represented by the parametric equations, and (b) find a rectangular-coordinate equation for the curve by eliminating the parameter.

1. $x = 2t$, $y = t + 6$

2. $x = 6t - 4$, $y = 3t$, $t \geq 0$

3. $x = t^2$, $y = t - 2$, $2 \leq t \leq 4$

4. $x = 2t + 1$, $y = \left(t + \frac{1}{2}\right)^2$

5. $x = \sqrt{t}$, $y = 1 - t$

6. $x = t^2$, $y = t^4 + 1$

7. $x = \dfrac{1}{t}$, $y = t + 1$

8. $x = t + 1$, $y = \dfrac{t}{t + 1}$

9. $x = 4t^2$, $y = 8t^3$

10. $x = |t|$, $y = \left|1 - |t|\right|$

11. $x = 2 \sin t$, $y = 2 \cos t$, $0 \leq t \leq \pi$

12. $x = 2 \cos t$, $y = 3 \sin t$, $0 \leq t \leq 2\pi$

13. $x = \sin^2 t,\quad y = \sin^4 t$ 14. $x = \sin^2 t,\quad y = \cos t$

15. $x = \cos t,\quad y = \cos 2t$ 16. $x = \cos 2t,\quad y = \sin 2t$

17. $x = \sec t,\quad y = \tan t,\quad 0 \le t < \pi/2$

18. $x = \cot t,\quad y = \csc t,\quad 0 < t < \pi$

19. $x = e^t,\quad y = e^{-t}$

20. $x = e^{2t},\quad y = e^t,\quad t \ge 0$

21. $x = \cos^2 t,\quad y = \sin^2 t$

22. $x = \cos^3 t,\quad y = \sin^3 t,\quad 0 \le t \le 2\pi$

In Exercises 23–26 find parametric equations for the line with the given properties.

23. Slope $\frac{1}{2}$, passing through $(4, -1)$

24. Slope -2, passing through $(-10, -20)$

25. Passing through $(6, 7)$ and $(7, 8)$

26. Passing through $(12, 7)$ and the origin

27. Find parametric equations for the circle $x^2 + y^2 = a^2$.

28. Find parametric equations for the ellipse

$$\frac{x^2}{a^2} + \frac{y^2}{b^2} = 1$$

29. Show by eliminating the parameter θ that the following parametric equations represent a hyperbola:

$$x = a \tan \theta \qquad y = b \sec \theta$$

30. Show that the following parametric equations represent a part of the hyperbola of Exercise 29:

$$x = a\sqrt{t} \qquad y = b\sqrt{t + 1}$$

In Exercises 31–34 sketch the curve given by the parametric equations.

31. $x = t \cos t,\quad y = t \sin t,\quad t \ge 0$

32. $x = \sin t,\quad y = \sin 2t$

33. $x = \dfrac{3t}{1 + t^3},\quad y = \dfrac{3t^2}{1 + t^3}$

34. $x = \cot t,\quad y = 2 \sin^2 t,\quad 0 < t < \pi$

35. If a projectile is fired with an initial speed of v_0 ft/s at an angle α above the horizontal, then its position after t sec-

onds is given by the parametric equations

$$x = (v_0 \cos \alpha)t \qquad y = (v_0 \sin \alpha)t - 16t^2$$

(where x and y are measured in feet). Show that the path of the projectile is a parabola by eliminating the parameter t.

36. Referring to Exercise 35, suppose that a gun fires a bullet into the air with an initial speed of 2048 ft/s at an angle of 30° to the horizontal.
   (a) After how many seconds will the bullet hit the ground?
   (b) How far from the gun will the bullet hit the ground?
   (c) What is the maximum height attained by the bullet?

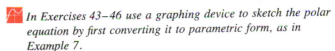 *In Exercises 37–42 use a graphing device to sketch the curve represented by the given parametric equations.*

37. $x = \sin t,\quad y = 2 \cos 3t$

38. $x = 2 \sin t,\quad y = \cos 4t$

39. $x = 3 \sin 5t,\quad y = 5 \cos 3t$

40. $x = \sin 4t,\quad y = \cos 3t$

41. $x = \sin(\cos t),\quad y = \cos t^{3/2},\quad 0 \le t \le 2\pi$

42. $x = 2 \cos t + \cos 2t,\quad y = 2 \sin t - \sin 2t$

In Exercises 43–46 use a graphing device to sketch the polar equation by first converting it to parametric form, as in Example 7.

43. $r = e^{\theta/12},\quad 0 \le \theta \le 4\pi$

44. $r = \sin \theta + 2 \cos \theta$

45. $r = \dfrac{4}{2 - \cos \theta}$ 46. $r = 2^{\sin \theta}$

47. In Example 5, suppose that the point P that traces out the curve lies not on the edge of the circle, but rather at a fixed point inside the rim, at a distance b from the center (where $b < a$). The curve traced out by P is called a **curtate cycloid** (or **trochoid**). Show that parametric equations for the curtate cycloid are

$$x = a\theta - b \sin \theta \qquad y = a - b \cos \theta$$

Sketch the graph.

48. In Exercise 47 if the point P lies *outside* the circle at a distance b from the center (where $b > a$), then the curve traced out by P is called a **prolate cycloid**. Show that

parametric equations for the prolate cycloid are the same as the equations for the curtate cycloid, and sketch the graph for the case where $a = 1$ and $b = 2$.

49. A circle C of radius b rolls on the inside of a larger circle of radius a centered at the origin. Let P be a fixed point on the smaller circle, with initial position at the point $(a, 0)$, as shown in the figure. The curve traced out by P is called a **hypocycloid**.

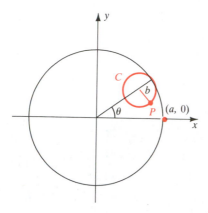

(a) Show that parametric equations for the hypocycloid are

$$x = (a - b) \cos \theta + b \cos\left(\frac{a - b}{b}\theta\right)$$

$$y = (a - b) \sin \theta - b \sin\left(\frac{a - b}{b}\theta\right)$$

(b) If $a = 4b$, the hypocycloid is called an **astroid**. Show that in this case the parametric equations can be reduced to

$$x = a \cos^3\theta \qquad y = a \sin^3\theta$$

Sketch the curve and eliminate the parameter to obtain an equation for the astroid in rectangular coordinates.

50. If the circle C of Exercise 49 rolls on the *outside* of the larger circle, the curve traced out by P is called an **epicycloid**. Find parametric equations for the epicycloid.

51. In the figure, the circle of radius a is stationary and, for every θ, the point P is the midpoint of the segment QR. The curve traced out by P for $0 < \theta < \pi$ is called the **longbow curve**. Find parametric equations for this curve.

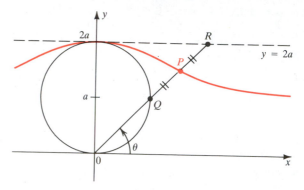

52. A string is wound around a circle and then unwound while being held taut. The curve traced out by the point P at the end of the string is called the **involute** of the circle, as shown in the figure. If the circle has radius a and is centered at the origin, and if the initial position of P is at $(a, 0)$, show that parametric equations for the involute in terms of the parameter θ are

$$x = a(\cos \theta + \theta \sin \theta) \qquad y = a(\sin \theta - \theta \cos \theta)$$

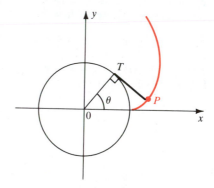

53. Eliminate the parameter θ in the parametric equations for the cycloid (Example 5) to obtain a rectangular coordinate equation for the section of the curve given by $0 \leqslant \theta \leqslant \pi$.

CHAPTER 9 REVIEW

Define, state, or discuss each of the following.

1. Conic section

2. Parabola

3. Focus, directrix, vertex, and focal diameter of a parabola

4. Ellipse

5. Foci of an ellipse

6. Major and minor axes of an ellipse

7. Eccentricity of an ellipse

8. Hyperbola

9. Foci of a hyperbola

10. Asymptotes of a hyperbola

11. Shifted conics

12. The equation $Ax^2 + Cy^2 + Dx + Ey + F = 0$

13. Rotation of axes

14. The equation $Ax^2 + Bxy + Cy^2 + Dx + Ey + F = 0$

15. Polar coordinates

16. Graphs of polar equations

17. Parametric equations and their graphs

18. Cycloid

REVIEW EXERCISES

In Exercises 1–4 find the vertex, focus, and directrix of the parabola, and sketch the graph.

1. $x^2 + 8y = 0$

2. $2x - y^2 = 0$

3. $x - y^2 + 4y - 2 = 0$

4. $2x^2 + 6x + 5y + 10 = 0$

In Exercises 5–8 find the center, vertices, foci, and the lengths of the major and minor axes of the ellipse, and sketch the graph.

5. $x^2 + 4y^2 = 16$

6. $9x^2 + 4y^2 = 1$

7. $4x^2 + 9y^2 = 36y$

8. $2x^2 + y^2 = 2 + 4(x - y)$

In Exercises 9–12 find the center, vertices, foci, and asymptotes of the hyperbola, and sketch the graph.

9. $x^2 - 2y^2 = 16$

10. $x^2 - 4y^2 + 16 = 0$

11. $9y^2 + 18y = x^2 + 6x + 18$

12. $y^2 = x^2 + 6y$

In Exercises 13–24 determine the type of curve represented by the equation. Find the foci and vertices (if any), and sketch the graph.

13. $\dfrac{x^2}{12} + y = 1$

14. $\dfrac{x^2}{12} + \dfrac{y^2}{144} = \dfrac{y}{12}$

15. $x^2 - y^2 + 144 = 0$

16. $x^2 + 6x = 9y^2$

17. $4x^2 + y^2 = 8(x + y)$

18. $3x^2 - 6(x + y) = 10$

19. $x = y^2 - 16y$

20. $2x^2 + 4 = 4x + y^2$

21. $2x^2 - 12x + y^2 + 6y + 26 = 0$

22. $36x^2 - 4y^2 - 36x - 8y = 31$

23. $9x^2 + 8y^2 - 15x + 8y + 27 = 0$

24. $x^2 + 4y^2 = 4x + 8$

In Exercises 25–32 find an equation for the conic section with the given properties.

25. The parabola with focus $F(0, 1)$ and directrix $y = -1$

26. The ellipse with center $C(0, 4)$, foci $F_1(0, 0)$ and $F_2(0, 8)$, and major axis of length 10

27. The hyperbola with vertices $V(0, \pm 2)$ and asymptotes $y = \pm \frac{1}{2}x$

28. The hyperbola with center $C(2, 4)$, foci $F_1(2, 7)$ and $F_2(2, 1)$, and vertices $V_1(2, 6)$ and $V_2(2, 2)$

29. The ellipse with foci $F_1(1, 1)$ and $F_2(1, 3)$, and with one vertex on the x-axis

30. The parabola with vertex $V(5, 5)$ and directrix the y-axis

31. The ellipse with vertices $V_1(7, 12)$ and $V_2(7, -8)$, and containing the point $P(1, 8)$

32. The parabola with vertex $V(-1, 0)$, horizontal axis of symmetry, and crossing the y-axis at $y = 2$

In Exercises 33–36 use rotation of axes to determine what type of conic is represented by the equation, and sketch the graph.

33. $x^2 + 4xy + y^2 = 1$

34. $5x^2 - 6xy + 5y^2 - 8x + 8y - 8 = 0$

35. $7x^2 - 6\sqrt{3}xy + 13y^2 - 4\sqrt{3}x - 4y = 0$

36. $9x^2 + 24xy + 16y^2 = 25$

In Exercises 37–44 sketch the curve whose polar equation is given. Express the equation in rectangular coordinates.

37. $r = 3 + 3\cos\theta$

38. $r = 3\sin\theta$

39. $r = 2\sin 2\theta$

40. $r = 4\cos 3\theta$

41. $r^2 = \sec 2\theta$

42. $r^2 = 4\sin 2\theta$

43. $r = \sin\theta + \cos\theta$

44. $r = \dfrac{4}{2 + \cos\theta}$

In Exercises 45–48 sketch the parametric curve, and eliminate the parameter to find an equation in rectangular coordinates that all points on the curve satisfy.

45. $x = 1 - t^2, \quad y = 1 + t$

46. $x = t^2 - 1, \quad y = t^2 + 1$

47. $x = 1 + \cos t, \quad y = 1 - \sin t, \quad 0 \leq t \leq \pi/2$

48. $x = \dfrac{1}{t} + 2, \quad y = \dfrac{2}{t^2}, \quad 0 < t \leq 2$

In Exercises 49 and 50 use a graphing device to sketch the parametric curve.

49. $x = \cos 2t, \quad y = \sin 3t$

50. $x = \sin(t + \cos 2t), \quad y = \cos(t + \sin 3t)$

51. The curves C, D, E, and F are defined parametrically as follows, where the parameter t takes on all real values unless otherwise stated:

$$C: \quad x = t, \quad y = t^2$$
$$D: \quad x = \sqrt{t}, \quad y = t, \quad t \geq 0$$
$$E: \quad x = \sin t, \quad y = 1 - \cos^2 t$$
$$F: \quad x = e^t, \quad y = e^{2t}$$

(a) Show that the points on all four of these curves satisfy the same rectangular coordinate equation.

(b) Sketch the graph of each curve and explain how the curves differ from one another.

52. In the figure the point P is the midpoint of the segment QR and $0 \leq \theta < \pi/2$. Using θ as the parameter, find a parametric representation for the curve traced out by P.

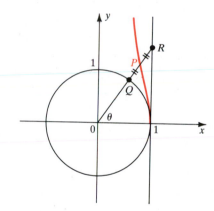

CHAPTER 9 TEST

1. Find the focus and directrix of the parabola $x^2 = -6y$, and sketch the graph.

2. Find the vertices, foci, and the lengths of the major and minor axes for the ellipse

$$\frac{x^2}{8} + \frac{y^2}{12} = 1$$

Sketch the graph.

3. Find the vertices, foci, and asymptotes of the hyperbola

$$\frac{y^2}{49} - \frac{x^2}{36} = 1$$

Sketch the graph.

In Problems 4–6 sketch the graph of the equation.

4. $16x^2 + 36y^2 - 96x + 36y + 9 = 0$

5. $9x^2 - 8y^2 + 36x + 64y = 92$

6. $2x + y^2 + 8y + 8 = 0$

7. Find an equation for the hyperbola with foci $(0, \pm 5)$ and with asymptotes $y = \pm \frac{3}{4}x$.

8. Find an equation for the parabola with focus $(2, 4)$ and with directrix the x-axis.

9. (a) Use rotation of axes to eliminate the xy-term in the equation

$$5x^2 + 4xy + 2y^2 = 18$$

(b) Sketch the graph of the equation.
(c) Find the coordinates of the vertices of this conic (in the xy-coordinate system).

10. Graph the polar equation $r = 2 + \cos \theta$.

11. Convert the polar equation $r = 2 \cos \theta - 4 \sin \theta$ to rectangular coordinates and identify the graph.

12. (a) Sketch the graph of the parametric curve

$$x = 3 \sin \theta + 3 \qquad y = 2 \cos \theta \qquad 0 \le \theta \le \pi$$

(b) Eliminate the parameter θ in part (a) to obtain an equation for this curve in rectangular coordinates.

FOCUS ON PROBLEM SOLVING

When faced with a new problem, it is useful to **try to recognize something familiar**. You may be able to relate some aspect of the problem to previous knowledge, maybe even to a previously solved problem.

The following example illustrates this principle. It also illustrates other aspects of the four-stage problem-solving strategy outlined on pages 100–105.

EXAMPLE

Express the hypotenuse h of a right triangle in terms of its area A and its perimeter P.

SOLUTION

Understand the problem

Let us first sort out the information by identifying the unknown quantity and the data.

$$\text{Unknown: } h$$

$$\text{Given quantities: } A, P$$

Draw a diagram
Connect the given with the unknown
Introduce something extra

It helps to draw a diagram and we do so in Figure 1.

In order to connect the given quantities to the unknown, we introduce two extra variables a and b, which are the lengths of the other two sides of the triangle. This enables us to express the given condition, which is that the triangle is right-angled, by the Pythagorean Theorem:

$$h^2 = a^2 + b^2$$

The other connections among the variables come by writing expressions for the area and perimeter:

$$A = \tfrac{1}{2}ab \qquad P = a + b + h$$

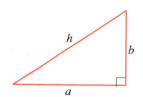

Figure 1

Since A and P are given, notice that we now have three equations in the three unknowns a, b, and h:

$$h^2 = a^2 + b^2 \qquad (1)$$

$$A = \tfrac{1}{2}ab \qquad (2)$$

$$P = a + b + h \qquad (3)$$

Relate to the familiar

Although we have the correct number of equations, they are not easy to solve in a straightforward fashion. But if we use the problem-solving strategy of trying to recognize something familiar, then we can solve these equations by an easier method. Look at the right sides of Equations 1, 2, and 3. Do these expressions remind you of anything familiar? Notice that they contain the ingredients of one of the Special Product Formulas from Section 1.3:

$$(a + b)^2 = a^2 + 2ab + b^2$$

Using this idea we express $(a + b)^2$ in two ways. From Equations 1 and 2 we have

$$(a + b)^2 = (a^2 + b^2) + 2ab = h^2 + 4A$$

From Equation 3 we have

$$(a + b)^2 = (P - h)^2 = P^2 - 2Ph + h^2$$

Thus
$$h^2 + 4A = P^2 - 2Ph + h^2$$
$$2Ph = P^2 - 4A$$
$$h = \frac{P^2 - 4A}{2P}$$

This is the required expression. ■

PROBLEMS

Don't feel bad if you don't solve these problems right away. Problems 7 and 8 were sent to Albert Einstein by his friend Wertheimer. Einstein (and his friend Bucky) enjoyed the problems and wrote back to Wertheimer. Here is part of Einstein's reply:

Your letter gave us a lot of amusement. The first intelligence-test fooled both of us (Bucky and me). Only on working it out did I notice that no time is available for the downhill run! Mr. Bucky was also taken in by the second example, but I was not. Such drolleries show us how stupid we are!

(See Mathematical Intelligencer, Spring 1990, page 41)

1. The sum of two numbers is 4 and their product is 1. Find the sum of their cubes.

2. Draw the graph of the equation

$$x^2y - y^3 - 5x^2 + 5y^2 = 0$$

without making a table of values.

3. In a right triangle, the hypotenuse has length 5 cm and another side has length 3 cm. What is the length of the altitude that is perpendicular to the hypotenuse?

4. The perimeter of a right triangle is 60 cm and the altitude perpendicular to the hypotenuse is 12 cm. Find the lengths of the three sides.

5. Write 116 as a sum of four perfect squares.

6. How many integers are there from one to one million (inclusive) that are either perfect squares or perfect cubes (or both)?

7. An old broken-down car has to travel a 2-mile route, uphill and down. Because it is so old, it can take the first mile—the ascent—no faster than at an average speed of 15 mi/h. How fast does it have to cover the second mile—on the descent it can go faster, of course—in order to achieve an average speed of 30 mi/h?

8. An amoeba propagates by simple division; it takes three minutes for each split. I put such an amoeba into a glass container with a nutrient fluid; the rate of multiplication rises, of course; it takes one hour until the vessel is full of amoebas. How long would it take to fill the vessel likewise, if I start not with one amoeba, but two?

10

SEQUENCES AND SERIES

A mathematician, like a painter or poet, is a maker of patterns.

G. H. HARDY

Certainly let us learn proving, but also let us learn guessing.

GEORGE POLYA

In this chapter we study sequences and series of numbers. Roughly speaking, a sequence is a list of numbers written in a specific order and a series is what one gets by adding the numbers in a sequence. Sequences and series have many theoretical and practical uses. Among other applications, we consider how series are used to calculate the value of an annuity.

Section 10.7 introduces a special kind of proof called mathematical induction. In Section 10.8 we use mathematical induction to prove a formula for expanding $(a + b)^n$ for any natural number n.

SECTION 10.1
SEQUENCES

A *sequence* is a set of numbers written in a specific order:

$$a_1, a_2, a_3, a_4, \ldots, a_n, \ldots$$

The number a_1 is called the *first term*, a_2 is the *second term*, and in general a_n is the *nth term*. Since for every natural number n there is a corresponding number a_n, we can define a sequence as a function.

> A **sequence** is a function f whose domain is the set of natural numbers. The values $f(1), f(2), f(3), \ldots$ are called the **terms** of the sequence.

We usually write a_n instead of the function notation $f(n)$ for the value of the function at the number n.

Here is a simple example of a sequence:

$$2, 4, 6, 8, 10, \ldots$$

The dots indicate that the sequence continues indefinitely. When we write a sequence in this way we are saying that it is clear what the subsequent terms of the sequence are. This sequence consists of even numbers. To be more accurate, however, we need to specify a procedure for finding *all* the terms of the sequence. This can be done by giving a formula for the *nth* term a_n of the sequence. In this case,

$$a_n = 2n$$

Another way to write this sequence is to use function notation:

$$a(n) = 2n$$

so $a(1) = 2$, $a(2) = 4$, $a(3) = 6, \ldots$

and the sequence can be written as

$$
\begin{array}{cccccc}
2, & 4, & 6, & 8, & \ldots, & 2n, & \ldots \\
\uparrow & \uparrow & \uparrow & \uparrow & & \uparrow & \\
\text{1st} & \text{2nd} & \text{3rd} & \text{4th} & & \text{nth} & \\
\text{term} & \text{term} & \text{term} & \text{term} & & \text{term} &
\end{array}
$$

Notice how the formula $a_n = 2n$ gives all the terms of the sequence. For instance, substituting 1, 2, 3, and 4 for n gives the first four terms:

$$a_1 = 2 \cdot 1 = 2 \qquad a_2 = 2 \cdot 2 = 4$$

$$a_3 = 2 \cdot 3 = 6 \qquad a_4 = 2 \cdot 4 = 8$$

To find the 103rd term of this sequence we use $n = 103$ to get

$$a_{103} = 2 \cdot 103 = 206$$

EXAMPLE 1

Find the first five terms and the 100th term of the sequences defined by the following formulas:

(a) $a_n = 2n - 1$ (b) $c_n = n^2 - 1$ (c) $t_n = \dfrac{n}{n+1}$ (d) $r_n = \dfrac{(-1)^n}{2^n}$

SOLUTION

(a) Using the formula we have

$$a_1 = 2(1) - 1 = 1 \qquad a_2 = 2(2) - 1 = 3 \qquad a_3 = 2(3) - 1 = 5$$
$$a_4 = 2(4) - 1 = 7 \qquad a_5 = 2(5) - 1 = 9 \quad \text{and} \quad a_{100} = 2(100) - 1 = 199$$

The sequence can be written as

$$1, 3, 5, 7, 9, \ldots, 2n - 1, \ldots$$

This is the sequence of odd numbers.

(b) We have

$$c_1 = 1^2 - 1 = 0 \qquad c_2 = 2^2 - 1 = 3 \qquad c_3 = 3^2 - 1 = 8$$
$$c_4 = 4^2 - 1 = 15 \qquad c_5 = 5^2 - 1 = 24 \quad \text{and} \quad c_{100} = 100^2 - 1 = 9999$$

This sequence can be written as

$$0, 3, 8, 15, 24, \ldots, n^2 - 1, \ldots$$

(c) From the formula for t_n we get $t_1 = \frac{1}{2}$, $t_2 = \frac{2}{3}$, $t_3 = \frac{3}{4}$, $t_4 = \frac{4}{5}$, $t_5 = \frac{5}{6}$, and $t_{100} = \frac{100}{101}$. This sequence can be written as follows:

$$\frac{1}{2}, \frac{2}{3}, \frac{3}{4}, \frac{4}{5}, \frac{5}{6}, \ldots, \frac{n}{n+1}, \ldots$$

(d) $r_1 = -\frac{1}{2}$, $r_2 = \frac{1}{4}$, $r_3 = -\frac{1}{8}$, $r_4 = \frac{1}{16}$, $r_5 = -\frac{1}{32}$, and $r_{100} = 1/2^{100}$. (The number 2^{100} has 31 digits, so we will not write it here.) The sequence can be written as follows:

$$-\frac{1}{2}, \frac{1}{4}, -\frac{1}{8}, \frac{1}{16}, -\frac{1}{32}, \ldots, \frac{(-1)^n}{2^n}, \ldots \qquad \blacksquare$$

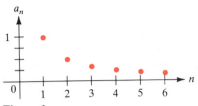

Figure 1

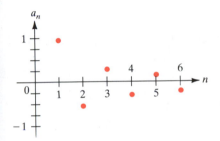

Figure 2

Notice the presence of $(-1)^n$ in this last sequence. It has the effect of making successive terms of the sequence alternately positive and negative.

It is often useful to picture a sequence by sketching its graph. Since a sequence is a function whose domain is the natural numbers, we can draw its graph in the Cartesian plane. For instance, the graph of the sequence

$$1, \frac{1}{2}, \frac{1}{3}, \frac{1}{4}, \frac{1}{5}, \frac{1}{6}, \cdots \frac{1}{n}, \cdots$$

is shown in Figure 1. Compare this to the graph of

$$1, -\frac{1}{2}, \frac{1}{3}, -\frac{1}{4}, \frac{1}{5}, -\frac{1}{6}, \cdots, \frac{(-1)^{n+1}}{n}, \cdots$$

shown in Figure 2. Note that the graph of a sequence consists of isolated points which are *not* connected.

Some sequences do not have simple defining formulas like those of Example 1. It may happen that the nth term of a sequence depends on some or all of the terms preceding it. Sequences defined in this way are called **recursive.** Here are two examples.

EXAMPLE 2

Find the first five terms of the sequence defined by

$$a_n = 3(a_{n-1} + 2)$$

and $a_1 = 1$.

SOLUTION

The defining formula for this sequence is recursive. It allows us to find the nth term a_n if we know the preceding term a_{n-1}. Thus we can find the second term from the first term, the third term from the second term, the fourth term from the third term, and so on. Since we are given the first term $a_1 = 1$, we can proceed as follows:

$$a_2 = 3(a_1 + 2) = 3(1 + 2) = 9$$

$$a_3 = 3(a_2 + 2) = 3(9 + 2) = 33$$

$$a_4 = 3(a_3 + 2) = 3(33 + 2) = 105$$

$$a_5 = 3(a_4 + 2) = 3(105 + 2) = 321$$

Thus the first five terms of this sequence are

$$1, 9, 33, 105, 321, \ldots$$

Notice that in order to find the 100th term of the sequence in Example 2 we must first find all the preceding 99 terms.

EXAMPLE 3

Find the first 11 terms of the sequence defined recursively by

$$F_n = F_{n-1} + F_{n-2}$$

where $F_1 = 1$ and $F_2 = 1$.

SOLUTION

To find F_n we need to find the two preceding terms F_{n-1} and F_{n-2}. Since we are given F_1 and F_2 we proceed as follows:

$$F_3 = F_2 + F_1 = 1 + 1 = 2$$
$$F_4 = F_3 + F_2 = 2 + 1 = 3$$
$$F_5 = F_4 + F_3 = 3 + 2 = 5$$

It is clear what is happening here. Each term is simply the sum of the two terms that precede it, so we can easily write down as many terms as we please. Here are the first 11 terms:

$$1, 1, 2, 3, 5, 8, 13, 21, 34, 55, 89, \ldots$$

■

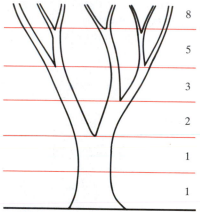

Figure 3

The Fibonacci sequence in the branching of a tree.

A prime number is a natural number with no factor other than 1 and itself. By convention, 1 is not considered prime.

The sequence in Example 3 is called the **Fibonacci sequence**, named after the 13th century Italian mathematician who used it to solve a problem about the breeding of rabbits (see Exercise 37). It has found numerous other applications in nature. In fact, so many phenomena behave like the Fibonacci sequence that there is a mathematical journal (the *Fibonacci Quarterly*) devoted entirely to its properties. Figures 3 and 4 show two applications.

The sequences we have considered so far are defined by either a formula or a recursive procedure. But not all sequences can be defined in this way. For example, there is no known formula that produces the sequence of prime numbers:

$$2, 3, 5, 7, 11, 13, 17, 19, 23, \ldots$$

If we let a_n be the digit in the nth decimal place of the number π, we get the sequence

$$1, 4, 1, 5, 9, 2, 6, 5, 4, \ldots$$

Again there is no simple formula for finding the terms of this sequence.

Finding patterns is a very important part of mathematics. Consider a sequence

$$1, 4, 9, 16, \ldots$$

Fibonacci (1175–1250) was born in Pisa, Italy, and educated in North Africa. He traveled widely in the Mediterranean area and learned the various methods then used for writing numbers. On returning to Pisa in 1202 Fibonacci advocated the use of the Hindu-Arabic decimal system, the one we use today, over the Roman numeral system used in Europe in his time. His most famous book Liber Abaci *is devoted to expounding the advantages of the Hindu-Arabic numerals. In fact, multiplication and division were so complicated using Roman numerals that it required a college degree to master these skills. Interestingly, in 1299 the city of Florence outlawed the use of the decimal system for merchants and other businesses, allowing numbers to be written only using Roman numerals or words. One can only speculate about the reasons for this.*

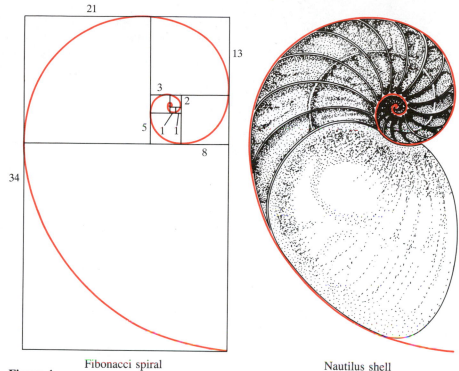

Fibonacci spiral   Nautilus shell

Figure 4

Can you detect a pattern in these numbers? In other words, can you define a sequence whose first four terms are these numbers? The answer to this question seems easy; these numbers are the squares of the numbers 1, 2, 3, 4. Thus the sequence we are looking for is defined by $a_n = n^2$. We point out, however, that this is not the only sequence whose first four terms are 1, 4, 9, 16. In other words the answer to our problem is not unique (see Exercises 31 and 32). In the next example we are interested in finding an *obvious* sequence whose first few terms agree with the given ones.

EXAMPLE 4

Find the *n*th term of a sequence whose first few terms are given.
(a) $\frac{1}{2}, \frac{3}{4}, \frac{5}{6}, \frac{7}{8}, \ldots$ (b) $-2, 4, -8, 16, -32, \ldots$

SOLUTION

(a) We notice that the numerators of these fractions are the odd numbers and the denominators are the even numbers. Even numbers are of the form $2n$ and odd numbers are of the form $2n - 1$ (an odd number differs from an even number by 1). So a sequence that has these numbers for its first four terms is given by

$$a_n = \frac{2n - 1}{2n}$$

(b) These numbers are powers of 2 and since they alternate in sign, a sequence that agrees with these terms is given by

$$a_n = (-1)^n 2^n$$

It is a good idea to check that these formulas do indeed generate the given terms. ∎

EXERCISES 10.1

In Exercises 1–12 find the first four terms as well as the 1000th term of each sequence.

1. $a_n = \dfrac{1}{n+2}$

2. $a_n = n^3 + n^2 + n + 1$

3. $a_n = \dfrac{(-1)^n}{n^2}$

4. $a_n = \dfrac{1}{n^2}$

5. $a_n = 1 + (-1)^n$

6. $a_n = 1 - \dfrac{1}{2^n}$

7. $a_n = (-2)^n$

8. $a_n = \dfrac{n^2}{n^3 + 1}$

9. $a_n = (-1)^{n+1}\dfrac{n}{n+1}$

10. $a_n = 3n - 2$

11. $a_n = (-1)^n\dfrac{2n+1}{\sqrt{n+1}}$

12. $a_n = 3$

In Exercises 13–20 a sequence is defined recursively. Find the first five terms of the sequence.

13. $a_n = 2(a_{n-1} - 2)$ and $a_1 = 3$

14. $a_n = \dfrac{a_{n-1}}{2}$ and $a_1 = -8$

15. $a_n = 2a_{n-1} + 1$ and $a_1 = 1$

16. $a_n = (a_{n-1})^2$ and $a_1 = 2$

17. $a_n = a_{n-1} - a_{n-2}$ and $a_1 = 0,\ a_2 = 1$

18. $a_n = \dfrac{1}{1 + a_{n-1}}$ and $a_1 = 1$

19. $a_n = a_{n-1} + a_{n-2}$ and $a_1 = 1,\ a_2 = 2$

20. $a_n = a_{n-1} + a_{n-2} + a_{n-3}$ and $a_1 = a_2 = a_3 = 1$

In Exercises 21–30 find the nth term of a sequence whose first several terms are given.

21. $2, 4, 8, 16, \ldots$

22. $-\frac{1}{3}, \frac{1}{9}, -\frac{1}{27}, \frac{1}{81}, \ldots$

23. $1, 4, 7, 10, \ldots$

24. $5, -25, 125, -625, \ldots$

25. $1, \frac{3}{4}, \frac{5}{9}, \frac{7}{16}, \frac{9}{25}, \ldots$

26. $\frac{3}{4}, \frac{4}{5}, \frac{5}{6}, \frac{6}{7}, \ldots$

27. $r, r^2, r^3, r^4, \ldots$

28. $a, a + d, a + 2d, a + 3d, \ldots$

29. $0, 2, 0, 2, 0, 2, \ldots$

30. $1, \frac{1}{2}, 3, \frac{1}{4}, 5, \frac{1}{6}, \ldots$

Exercises 31 and 32 explain why a finite number of terms do not uniquely determine a sequence.

31. (a) Show that the first four terms of the sequence $a_n = n^2$ are

$$1, 4, 9, 16, \ldots$$

(b) Show that the first four terms of the sequence $a_n = n^2 + (n - 1)(n - 2)(n - 3)(n - 4)$ are also

$$1, 4, 9, 16, \ldots$$

but that the sequences do not agree from the fifth term on.

(c) Find a sequence whose first six terms are the same as those of $a_n = n^2$ but whose succeeding terms differ from this sequence.

32. Find two different sequences that begin

$$2, 4, 8, 16, \ldots$$

33. Find a formula for the nth term of the sequence

$$\sqrt{2},\ \sqrt{2\sqrt{2}},\ \sqrt{2\sqrt{2\sqrt{2}}},\ \sqrt{2\sqrt{2\sqrt{2\sqrt{2}}}},\ \ldots$$

[Hint: Write each term as a power of 2.]

34. Find the first 100 terms of the sequence defined by

$$a_{n+1} = \begin{cases} \dfrac{a_n}{2} & \text{if } a_n \text{ is an even number} \\[2mm] 3a_n + 1 & \text{if } a_n \text{ is an odd number} \end{cases}$$

and $a_1 = 11$.

35. Repeat Exercise 34 with $a_1 = 25$.

36. Find the first ten terms of the sequence defined by

$$a_n = a_{n-a_{n-1}} + a_{n-a_{n-2}}$$

(a) where $a_1 = 1$, $a_2 = 1$.
(b) where $a_1 = 1$, $a_2 = 2$.

37. Fibonacci posed the following problem: Suppose that rabbits live forever and that every month each pair produces a new pair that becomes productive at age 2 months. If we start with one newborn pair, how many pairs of rabbits will there be in the nth month? Show that the answer is F_n, where F_n is the nth term of the Fibonacci sequence.

SECTION 10.2
ARITHMETIC AND GEOMETRIC SEQUENCES

In this section we study two special kinds of sequences: arithmetic sequences, whose terms are generated by successively adding a fixed constant, and geometric sequences, whose terms are generated by successively multiplying by a fixed constant.

ARITHMETIC SEQUENCES

Perhaps the simplest way to generate a sequence is to start with a number a and add to it a fixed constant d over and over again.

An **arithmetic sequence** is a sequence of the form

$$a, a + d, a + 2d, a + 3d, a + 4d, \ldots$$

The number a is the **first term**, and d is the **common difference** of the sequence.

The number d is called the common difference because any two consecutive terms of an arithmetic sequence differ by d.

EXAMPLE 1

(a) If $a = 2$ and $d = 3$ we get the arithmetic sequence

$$2, 2 + 3, 2 + 6, 2 + 9, \ldots$$

or

$$2, 5, 8, 11, \ldots$$

Any two consecutive terms of this sequence differ by $d = 3$.

(b) Consider the sequence

$$9, 4, -1, -6, -11, \ldots$$

Here the common difference is $d = -5$. Notice that the terms of an arithmetic sequence decrease if the common difference is negative. ∎

It is easy to find a formula for the nth term of an arithmetic sequence.

$$\text{the 1st term is} \quad a + 0d$$
$$\text{the 2nd term is} \quad a + 1d$$
$$\text{the 3rd term is} \quad a + 2d$$
$$\text{the 4th term is} \quad a + 3d$$
$$\vdots \qquad\qquad \vdots$$

Continuing in this manner we get the following formula.

THE nTH TERM OF AN ARITHMETIC SEQUENCE

> The nth term of the arithmetic sequence $a, a + d, a + 2d, a + 3d, \ldots$ is
> $$a_n = a + (n - 1)d$$

For example, the nth terms of the sequences in parts (a) and (b) of Example 1 are

$$a_n = 2 + 3(n - 1)$$

and

$$a_n = 9 - 5(n - 1)$$

An arithmetic sequence is completely determined by the first term a and the common difference d. Thus if we know the first two terms of an arithmetic sequence, we can find a formula for the nth term, as the following example shows.

EXAMPLE 2

Find the first six terms as well as the 300th term of the arithmetic sequence

$$13, 7, \ldots$$

SOLUTION

Since the first term is 13, we have $a = 13$. The common difference is $d = 7 - 13 = -6$. Thus the nth term of this sequence is

$$a_n = 13 - 6(n - 1)$$

From this we find the first six terms:

$$13, 7, 1, -5, -11, -17, \ldots$$

The 300th term is $a_{300} = 13 - 6(299) = -1781$. ∎

In fact an arithmetic sequence is completely determined by any two of its terms. Here is an example.

EXAMPLE 3

The 11th term of an arithmetic sequence is 52 and the 19th term is 92. Find the 1000th term.

SOLUTION

To find the nth term of this sequence we need to find a and d in the formula

$$a_n = a + (n - 1)d$$

From this formula, we get

$$a_{11} = a + (11 - 1)d = a + 10d$$
$$a_{19} = a + (19 - 1)d = a + 18d$$

Since $a_{11} = 52$ and $a_{19} = 92$, we get the two equations:

$$\begin{cases} 52 = a + 10d \\ 92 = a + 18d \end{cases}$$

Solving this system for a and d, we get $a = 2$ and $d = 5$. (Verify this.) Thus the nth term of this sequence is

$$a_n = 2 + 5(n - 1)$$

The 1000th term is $a_{1000} = 2 + 5(999) = 4997$.   ■

GEOMETRIC SEQUENCES

Another simple way of generating a sequence is to start with a number a and repeatedly multiply by a fixed nonzero constant r.

A **geometric sequence** is a sequence of the form

$$a, \ ar, \ ar^2, \ ar^3, \ ar^4, \ \ldots$$

The number a is the **first term**, and r is the **common ratio** of the sequence.

The number r is called the common ratio because the ratio of any two consecutive terms of the sequence is r.

EXAMPLE 4

(a) If $a = 3$ and $r = 2$ we get the geometric sequence

$$3, 3 \cdot 2, 3 \cdot 2^2, 3 \cdot 2^3, 3 \cdot 2^4, \ldots$$

or $\qquad\qquad\qquad$ $3, 6, 12, 24, 48, \ldots$

Notice that the ratio of any two consecutive terms is $r = 2$.

(b) The sequence

$$2, -10, 50, -250, 1250, \ldots$$

is a geometric sequence with $a = 2$ and $r = -5$. Notice that when r is negative the terms of the sequence alternate in sign.

(c) The sequence

$$1, \tfrac{1}{3}, \tfrac{1}{9}, \tfrac{1}{27}, \tfrac{1}{81}, \ldots$$

is a geometric sequence with $a = 1$ and $r = \tfrac{1}{3}$.

Notice that when $0 < r < 1$ the terms of the sequence decrease, and when $r > 1$ the terms increase. (What happens if $r = 1$?) ■

To find a formula for the nth term of a geometric sequence we consider the pattern generated by the first few terms:

$$\begin{aligned}
&\text{the 1st term is} & ar^0 \\
&\text{the 2nd term is} & ar^1 \\
&\text{the 3rd term is} & ar^2 \\
&\text{the 4th term is} & ar^3 \\
&\qquad\vdots & \vdots
\end{aligned}$$

Since this pattern continues, we get the following formula.

THE nth TERM OF A GEOMETRIC SEQUENCE

> The nth term of the geometric sequence $a, ar, ar^2, ar^3, ar^4, \ldots$ is
>
> $$a_n = ar^{n-1}$$

Thus the nth terms of the sequences in parts (a), (b), and (c) of Example 4 are, respectively,

$$a_n = 3(2)^{n-1}$$
$$a_n = 2(-5)^{n-1}$$
$$a_n = 1\left(\tfrac{1}{3}\right)^{n-1}$$

Geometric sequences occur naturally. Here is a simple example. Suppose that when a ball is dropped its elasticity is such that it bounces up one-third of the distance it has fallen. If this ball is dropped from a height of 2 m, it bounces up to a height of $2\left(\frac{1}{3}\right) = \frac{2}{3}$ m. On its second bounce it returns to a height of $\left(\frac{2}{3}\right)\left(\frac{1}{3}\right) = \frac{2}{9}$ m, and so on (see Figure 1). Thus the height h_n that the ball reaches on its nth bounce is given by the geometric sequence

$$h_n = \frac{2}{3}\left(\frac{1}{3}\right)^{n-1} = 2\left(\frac{1}{3}\right)^n$$

Figure 1

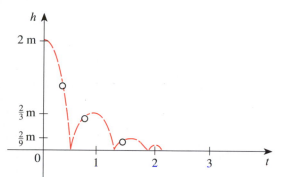

We can find the nth term of a geometric sequence if we know any two terms, as the following examples show.

EXAMPLE 5

Find the eighth term of the geometric sequence 5, 15, 45, ...

SOLUTION

To find a formula for the nth term of this sequence we need to find a and r. Clearly $a = 5$. To find r we find the ratio of any two consecutive terms. For instance,

$$r = \frac{45}{15} = 3$$

Thus

$$a_n = 5(3)^{n-1}$$

The eighth term is

$$a_8 = 5(3)^{8-1} = 5(3)^7 = 10{,}935$$ ∎

EXAMPLE 6

The third term of a geometric series is $\frac{63}{4}$ and the sixth term is $\frac{1701}{32}$. Find the fifth term.

SOLUTION

Since this series is geometric, its nth term is given by the formula

$$a_n = ar^{n-1}$$

Thus

$$a_3 = ar^{3-1} = ar^2$$

and

$$a_6 = ar^{6-1} = ar^5$$

From the values we are given for these two terms we get the system of equations:

$$\begin{cases} \frac{63}{4} = ar^2 & (1) \\ \frac{1701}{32} = ar^5 & (2) \end{cases}$$

One way to solve this system is to divide Equation 2 by Equation 1 to get

$$\frac{ar^5}{ar^2} = \frac{\frac{1701}{32}}{\frac{63}{4}}$$

So

$$r^3 = \frac{27}{8} \quad \text{or} \quad r = \frac{3}{2}$$

Substituting for r in Equation 1 gives

$$\frac{63}{4} = a\left(\frac{3}{2}\right)^2 \quad \text{or} \quad a = 7$$

It follows that the nth term of this sequence is

$$a_n = 7\left(\frac{3}{2}\right)^{n-1}$$

and so the fifth term is

$$a_5 = 7\left(\frac{3}{2}\right)^{5-1} = 7\left(\frac{3}{2}\right)^4 = \frac{567}{16}$$

EXERCISES 10.2

In Exercises 1–8 determine the common difference, the fifth term, the nth term and the 100th term for the given arithmetic sequence.

1. 4, 9, 14, 19, . . .

2. 11, 8, 5, 2, . . .

3. $-12, -8, -4, 0, \ldots$

4. $\frac{7}{6}, \frac{5}{3}, \frac{13}{6}, \frac{8}{3}, \ldots$

5. 25, 26.5, 28, 29.5, . . .

6. 15, 12.3, 9.6, 6.9, . . .

7. $2, 2 + s, 2 + 2s, 2 + 3s, \ldots$

8. $-t, -t + 3, -t + 6, -t + 9, \ldots$

In Exercises 9–18 determine the common ratio, the fifth term, and the nth term for the given geometric sequence.

9. 2, 6, 18, 54, . . .

10. $7, \frac{14}{3}, \frac{28}{9}, \frac{56}{27}, \ldots$

11. $0.3, -0.09, 0.027, -0.0081, \ldots$

12. $1, \sqrt{2}, 2, 2\sqrt{2}, \ldots$

13. $144, -12, 1, -\frac{1}{12}, \ldots$

14. $-8, -2, -\frac{1}{2}, -\frac{1}{8}, \ldots$

15. $3, 3^{5/3}, 3^{7/3}, 27, \ldots$

16. $t, \frac{t^2}{2}, \frac{t^3}{4}, \frac{t^4}{8}, \ldots$

17. $1, s^{2/7}, s^{4/7}, s^{6/7}, \ldots$

18. $5, 5^{c+1}, 5^{2c+1}, 5^{3c+1}, \ldots$

In Exercises 19–30 the first four terms of a sequence are given. In each exercise determine whether the given terms can be the terms of an arithmetic sequence, a geometric sequence, or neither. Find the next term if the sequence is arithmetic or geometric.

19. $5, -3, 5, -3, \ldots$

20. $\frac{1}{3}, 1, \frac{5}{3}, \frac{7}{3}, \ldots$

21. $\sqrt{3}, 3, 3\sqrt{3}, 9, \ldots$

22. $1, -1, 1, -1, \ldots$

23. $2, -1, \frac{1}{2}, 2, \ldots$

24. $-3, 1, 5, 8, \ldots$

25. $x - 1, x, x + 1, x + 2, \ldots$

26. $\dfrac{\sqrt{2}}{\sqrt{2} + 1}, \dfrac{2}{\sqrt{2} + 1}, \dfrac{4}{\sqrt{2} + 2}, \dfrac{4}{\sqrt{2} + 1}, \ldots$

27. $16, 8, 4, 1, \ldots$

28. $-3, -\frac{3}{2}, 0, \frac{3}{2}, \ldots$

29. $1, \frac{3}{2}, 2, \frac{5}{2}, \ldots$

30. $\sqrt{5}, \sqrt[3]{5}, \sqrt[6]{5}, 1, \ldots$

31. The tenth term of an arithmetic sequence is $\frac{55}{2}$ and the second term is $\frac{7}{2}$. Find the first term.

32. The 12th term of an arithmetic sequence is 32 and the fifth term is 18. Find the 20th term.

33. The 100th term of an arithmetic sequence is 98 and the common difference is 2. Find the first three terms.

34. The first term of a geometric sequence is 8, and the second term is 4. Find the fifth term.

35. The first term of a geometric sequence is 3 and the third term is $\frac{4}{3}$. Find the fifth term.

36. The first term of a geometric sequence is 1 and the fifth term is 7^4. Find the common ratio, assuming it is positive.

37. The common ratio in a geometric sequence is $\frac{2}{5}$ and the fourth term is $\frac{5}{2}$. Find the third term.

38. The common ratio in a geometric sequence is $\frac{3}{2}$ and the fifth term is 1. Find the first three terms.

39. Which term of the arithmetic sequence $1, 4, 7, \ldots$ is 88?

40. Which term of the geometric sequence $2, 6, 18, \ldots$ is 118,098?

41. The first term of an arithmetic sequence is 1 and the common difference is 4. Is 11,937 a term of this sequence? If so, which term is it?

42. The second term and the fifth term of a geometric sequence are 10 and 1250, respectively. Is 31,250 a term of this sequence? If so, which term is it?

43. A ball is dropped from a height of 80 ft. The elasticity of this ball is such that it rebounds three-fourths of the distance it has fallen. How high does the ball go on the fifth bounce? Find a formula for how high the ball goes on the nth bounce.

44. The size of a bacteria culture increases by 8% every hour and there are 5000 bacteria present initially. How many bacteria will be present at the end of 5 h? Find a formula for the number of bacteria present after n hours.

45. Suppose that the value of a certain machine depreciates 15% each year. What is the value of the machine after 6 years if its original cost is $12,500?

46. A certain radioactive substance decays so that at the end of each year there is only 0.85% as much as there was at the beginning of the year. If there was originally 1 kg of the substance, find the amount that remains after 5 years. Find a formula for the amount that remains after n years.

47. A truck radiator holds 5 gallons and is filled with water. A gallon of water is removed from the radiator and replaced with a gallon of antifreeze; then a gallon of the mixture is removed from the radiator and again replaced by a gallon of antifreeze. This process is repeated indefinitely. How much water remains in the tank after this process is repeated three times? five times? n times?

48. Show that the sequence defined recursively by

$$a_{n+1} = \frac{3a_n + 1}{3} \qquad \text{and} \qquad a_1 = c$$

is an arithmetic sequence and find the common difference.

49. Show that the sequence defined recursively by $a_{n+1} = 3a_n$ and $a_1 = c$ is a geometric sequence and find the common ratio.

50. If $a_1, a_2, a_3, \ldots$ is a geometric sequence with common ratio r, show that the sequence

$$\frac{1}{a_1}, \frac{1}{a_2}, \frac{1}{a_3}, \ldots$$

is also a geometric sequence and find the common ratio.

51. If $a_1, a_2, a_3, \ldots$ is a geometric sequence with common ratio r, show that the sequence

$$a_1^2, a_2^2, a_3^2, \ldots$$

is also a geometric sequence and find the common ratio.

52. If $a_1, a_2, a_3, \ldots$ is a geometric sequence with common ratio $r > 0$ and $a_1 > 0$, show that the sequence

$$\log a_1, \log a_2, \log a_3, \ldots$$

is an arithmetic sequence and find the common difference.

53. If $a_1, a_2, a_3, \ldots$ is an arithmetic sequence with common difference d, show that the sequence

$$10^{a_1}, 10^{a_2}, 10^{a_3}, \ldots$$

is a geometric sequence and find the common ratio.

54. Show that a right triangle whose sides are in arithmetic progression is similar to a 3-4-5 triangle.

55. If the sum of three consecutive terms in an arithmetic sequence is 15 and their product is 80, find the three terms. [*Hint:* Let x denote the middle term.]

56. The sum of 5 consecutive terms of an arithmetic sequence is 260. Find the middle term.

57. The first three terms of an arithmetic sequence are $x - 1$, $x + 1$, and $3x + 3$. Find x.

58. If the product of three consecutive terms in a geometric sequence is 216 and their sum is 21, find the three terms. [*Hint:* Let x denote the middle term.]

59. Let $a_1, a_2, a_3, \ldots$ be a geometric sequence with positive terms that satisfy

$$a_n = a_{n+1} + a_{n+2}$$

Find the common ratio.

60. If the numbers $a_1, a_2, \ldots, a_n$ form an arithmetic sequence, then $a_2, a_3, \ldots, a_{n-1}$ are called *arithmetic means* between a_1 and a_n. Insert three arithmetic means between 2 and 14. (Notice that if only one arithmetic mean is inserted between two numbers, then it is their average.)

61. If the numbers $a_1, a_2, \ldots, a_n$ form a geometric sequence, then $a_2, a_3, \ldots, a_{n-1}$ are called *geometric means* between a_1 and a_n. Insert three geometric means between 5 and 80.

62. A sequence is called *harmonic* if the reciprocals of the terms of the sequence form an arithmetic sequence. Determine if the sequence

$$1, \tfrac{3}{5}, \tfrac{3}{7}, 3, \ldots$$

is harmonic.

63. The *harmonic mean* of two numbers is the reciprocal of the arithmetic mean (see Exercise 60) of the reciprocals of the two numbers. Find the harmonic mean of 3 and 5.

SECTION 10.3
SERIES

In this section we are interested in adding the terms of a sequence. For example, we might want to add the first 100 terms of the sequence

$$1, 2, 3, 4, \ldots$$

to find

$$1 + 2 + 3 + 4 + \cdots + 100$$

Before tackling these problems we need a more compact way of writing such long sums.

SIGMA NOTATION

Given a sequence

$$a_1, a_2, a_3, a_4, \ldots$$

we can write the sum of the first n terms using **sigma notation**. This notation derives its name from the Greek letter Σ (capital sigma, corresponding to our S for sum) and is used as follows:

$$\sum_{k=1}^{n} a_k = a_1 + a_2 + a_3 + a_4 + \cdots + a_n$$

The left side of this expression is read "The sum of a_k from $k = 1$ to $k = n$." The letter k is called the **index of summation** or the **summation variable** and the idea is to replace k in the expression after the sigma by the integers $1, 2, 3, \ldots, n$, and add the resulting expressions, arriving at the right side. For example, the sum of the squares of the first five integers can be written

$$\sum_{k=1}^{5} k^2 = 1^2 + 2^2 + 3^2 + 4^2 + 5^2 = 55$$

Although we often use the letter k for the index of summation, any other letter can be used without affecting the sum. So this last sum can be written

$$\sum_{j=1}^{5} j^2 = 1^2 + 2^2 + 3^2 + 4^2 + 5^2$$

The index of summation need not start at 1. For example,

$$\sum_{j=3}^{7} j^2 = 3^2 + 4^2 + 5^2 + 6^2 + 7^2$$

The following examples illustrate these points.

EXAMPLE 1

Find the following sums:

(a) $\displaystyle\sum_{k=1}^{3} k^3(k - 1)$ (b) $\displaystyle\sum_{j=3}^{5} \frac{1}{j}$ (c) $\displaystyle\sum_{i=5}^{10} i$ (d) $\displaystyle\sum_{i=1}^{6} 2$

SOLUTION

(a) $\displaystyle\sum_{k=1}^{3} k^3(k - 1) = 1^3(1 - 1) + 2^3(2 - 1) + 3^3(3 - 1) = 0 + 8 + 54 = 62$

(b) $\displaystyle\sum_{j=3}^{5} \frac{1}{j} = \frac{1}{3} + \frac{1}{4} + \frac{1}{5} = \frac{47}{60}$

(c) $\displaystyle\sum_{i=5}^{10} i = 5 + 6 + 7 + 8 + 9 + 10 = 45$

(d) $\displaystyle\sum_{i=1}^{6} 2 = 2 + 2 + 2 + 2 + 2 + 2 = 12$ ■

EXAMPLE 2

Write the following sums using sigma notation:
(a) $1^3 + 2^3 + 3^3 + 4^3 + 5^3 + 6^3 + 7^3$
(b) $\sqrt{3} + \sqrt{4} + \sqrt{5} + \cdots + \sqrt{77}$

SOLUTION

(a) We can write

$$1^3 + 2^3 + 3^3 + 4^3 + 5^3 + 6^3 + 7^3 = \sum_{k=1}^{7} k^3$$

(b) A natural way to write this sum is

$$\sqrt{3} + \sqrt{4} + \sqrt{5} + \cdots + \sqrt{77} = \sum_{k=3}^{77} \sqrt{k}$$

However, there is no unique way of writing a sum in sigma notation. We also could write this last sum as

$$\sqrt{3} + \sqrt{4} + \sqrt{5} + \cdots + \sqrt{77} = \sum_{k=0}^{74} \sqrt{k + 3}$$

or

$$\sqrt{3} + \sqrt{4} + \sqrt{5} + \cdots + \sqrt{77} = \sum_{k=1}^{75} \sqrt{k + 2}$$ ■

The following properties of sums are natural consequences of properties of the real numbers.

PROPERTIES OF SUMS

Let $a_1, a_2, a_3, a_4, \ldots$ and $b_1, b_2, b_3, b_4, \ldots$ be sequences. Then for every positive integer n and any real number c,

1. $\displaystyle\sum_{k=1}^{n} (a_k + b_k) = \sum_{k=1}^{n} a_k + \sum_{k=1}^{n} b_k$

2. $\displaystyle\sum_{k=1}^{n} (a_k - b_k) = \sum_{k=1}^{n} a_k - \sum_{k=1}^{n} b_k$

3. $\displaystyle\sum_{k=1}^{n} ca_k = c\left(\sum_{k=1}^{n} a_k\right)$

To prove Property 1 we write out the left side of the equation to get

$$\sum_{k=1}^{n} (a_k + b_k) = (a_1 + b_1) + (a_2 + b_2) + (a_3 + b_3) + \cdots + (a_n + b_n)$$

Since addition is commutative and associative, we can rearrange the terms on the right side to read

$$\sum_{k=1}^{n} (a_k + b_k) = (a_1 + a_2 + a_3 + \cdots + a_n) + (b_1 + b_2 + b_3 + \cdots + b_n)$$

Rewriting the right side using sigma notation gives Property 1. Property 2 is proved in a similar manner. To prove Property 3 we use the Distributive Law.

$$\sum_{k=1}^{n} ca_k = ca_1 + ca_2 + ca_3 + \cdots + ca_n$$

$$= c(a_1 + a_2 + a_3 + \cdots + a_n)$$

$$= c\left(\sum_{k=1}^{n} a_k\right)$$

■ SERIES

When we add some of the terms of a sequence we get a *series* (or a *finite series*). For example,

$$\sum_{k=1}^{1,000,000} a_k$$

is the series that consists of the first one million terms of the sequence $a_1, a_2, a_3, a_4, \ldots$ added together.

DEFINITION OF A SERIES

Let $a_1, a_2, a_3, a_4, \ldots$ be a sequence. A sum of the form

$$a_1 + a_2 + a_3 + a_4 + \cdots + a_N$$

is called a **series**. The number a_1 is called the **first term** of the series, a_2 the **second term**, and so on. The number that the series adds to is called the **sum** of the series.

In this and the next section we find the sums of series that consist of many terms. For example, suppose we want to find the sum of the series

$$\sum_{k=1}^{1000} \left(\frac{1}{k} - \frac{1}{k+1}\right)$$

Since this series has a thousand terms it would take a long time to write down all the terms and add them together. We need a better way of doing this. So, we start by adding a few terms of the series and try to detect a pattern as we add more and more terms.

To do this we need some notation. We write S_1 for the first term of a series, S_2 for the sum of the first two terms, S_3 for the sum of the first three terms, and so on. These are called *partial sums* because we are only partially adding the terms of the series.

THE PARTIAL SUMS OF A SERIES

For the series

$$a_1 + a_2 + a_3 + a_4 + \cdots + a_N$$

the **partial sums** are

$$S_1 = a_1$$
$$S_2 = a_1 + a_2$$
$$S_3 = a_1 + a_2 + a_3$$
$$S_4 = a_1 + a_2 + a_3 + a_4$$
$$\vdots$$
$$S_n = a_1 + a_2 + a_3 + \cdots + a_n$$
$$\vdots$$

S_1 is called the **first partial sum**, S_2 is the **second partial sum**, and so on. S_n is called the **nth partial sum**.

The sequence $S_1, S_2, S_3, \ldots, S_n, \ldots$ is called the **sequence of partial sums**.

EXAMPLE 3

Find the sum of the series $\displaystyle\sum_{k=1}^{1000}\left(\frac{1}{k} - \frac{1}{k+1}\right)$.

SOLUTION

We begin by finding the first few partial sums of this series.

$$S_1 = \left(1 - \frac{1}{2}\right) \qquad\qquad\qquad\qquad\qquad = 1 - \frac{1}{2}$$

$$S_2 = \left(1 - \frac{1}{2}\right) + \left(\frac{1}{2} - \frac{1}{3}\right) \qquad\qquad\qquad = 1 - \frac{1}{3}$$

$$S_3 = \left(1 - \frac{1}{2}\right) + \left(\frac{1}{2} - \frac{1}{3}\right) + \left(\frac{1}{3} - \frac{1}{4}\right) \qquad = 1 - \frac{1}{4}$$

$$S_4 = \left(1 - \frac{1}{2}\right) + \left(\frac{1}{2} - \frac{1}{3}\right) + \left(\frac{1}{3} - \frac{1}{4}\right) + \left(\frac{1}{4} - \frac{1}{5}\right) = 1 - \frac{1}{5}$$

Do we detect a pattern here? Of course, we have

$$S_n = 1 - \frac{1}{n+1}$$

It is now easy to find the sum of as many terms of this series as we please. For instance the sum of the first 100 terms is

$$S_{100} = 1 - \frac{1}{101} = \frac{100}{101}$$

The sum of the series is the sum of all 1000 terms. Thus the sum of this series is

$$S_{1000} = 1 - \frac{1}{1001} = \frac{1000}{1001}$$ ■

EXAMPLE 4

Find the sum of the series $\displaystyle\sum_{k=1}^{100} \frac{1}{2^k}$.

SOLUTION

We first find a formula for the nth partial sum of this series. We do this by writing down the first few partial sums and trying to see a pattern:

$$S_1 = \frac{1}{2} \qquad\qquad = \frac{1}{2}$$

$$S_2 = \frac{1}{2} + \frac{1}{4} \qquad\qquad = \frac{3}{4}$$

$$S_3 = \frac{1}{2} + \frac{1}{4} + \frac{1}{8} \qquad = \frac{7}{8}$$

$$S_4 = \frac{1}{2} + \frac{1}{4} + \frac{1}{8} + \frac{1}{16} = \frac{15}{16}$$

Notice that in the value of each partial sum the denominator is a power of 2 and the numerator is one less than the denominator. In general,

$$S_n = \frac{2^n - 1}{2^n} = 1 - \frac{1}{2^n}$$

Now we see that the sum of the given series is

$$S_{100} = 1 - \frac{1}{2^{100}}$$ ■

EXERCISES 10.3

In Exercises 1–12 find the given sum.

1. $\displaystyle\sum_{k=3}^{6} k^2$

2. $\displaystyle\sum_{i=1}^{4} \frac{1}{i}$

3. $\displaystyle\sum_{k=0}^{5} (7 - 2k)$

4. $\displaystyle\sum_{j=1}^{100} (-1)^j$

5. $\displaystyle\sum_{i=1}^{8} [1 + (-1)^i]$

6. $\displaystyle\sum_{i=4}^{12} 10$

✓7. $\displaystyle\sum_{k=1}^{5} 2^{k-1}$

8. $\displaystyle\sum_{j=1}^{10} \frac{3}{j+2}$

9. $\displaystyle\sum_{m=0}^{4} (-3)^{m+2}$

10. $\displaystyle\sum_{i=1}^{3} i2^{i}$

11. $\displaystyle\sum_{m=3}^{5} (2^m + m^2)$

12. $\displaystyle\sum_{k=1}^{1} k^{100}$

In Exercises 13–20 write the given sum without using sigma notation.

✓13. $\displaystyle\sum_{k=1}^{5} \sqrt{k}$

14. $\displaystyle\sum_{i=0}^{4} \frac{2i-1}{2i+1}$

15. $\displaystyle\sum_{k=0}^{6} \sqrt{k+4}$

16. $\displaystyle\sum_{k=3}^{100} x^{k}$

17. $\displaystyle\sum_{i=1}^{8} ix^{i+1}$

18. $\displaystyle\sum_{k=6}^{9} k(k+3)$

19. $\displaystyle\sum_{j=1}^{n} (-1)^{j+1} x^{j}$

20. $\displaystyle\sum_{j=1}^{8} \frac{x^{j}}{j^{2}}$

In Exercises 21–30 write the given sum using sigma notation.

21. $1 + 2 + 3 + 4 + \cdots + 100$

22. $2 + 4 + 6 + \cdots + 2n$

23. $\dfrac{1}{2\ln 2} - \dfrac{1}{3\ln 3} + \dfrac{1}{4\ln 4}$
$\qquad - \dfrac{1}{5\ln 5} + \cdots + \dfrac{1}{100\ln 100}$

24. $1 + x + x^2 + x^3 + \cdots + x^{100}$

25. $1 - \dfrac{x}{3} + \dfrac{x^2}{9} - \dfrac{x^3}{27} + \dfrac{x^4}{81} - \dfrac{x^5}{243}$

26. $\dfrac{10}{15} + \dfrac{11}{16} + \dfrac{12}{17} + \cdots + \dfrac{100}{105}$

27. $1 - 2x + 3x^2 - 4x^3 + 5x^4 + \cdots + 100x^{99}$

28. $\dfrac{1}{1\cdot 2} + \dfrac{1}{2\cdot 3} + \dfrac{1}{3\cdot 4} + \cdots + \dfrac{1}{999\cdot 1000}$

29. $1\cdot 2\cdot 3 + 2\cdot 3\cdot 4 + 3\cdot 4\cdot 5 + \cdots + 97\cdot 98\cdot 99$

30. $\dfrac{\sqrt{1}}{1^2} + \dfrac{\sqrt{2}}{2^2} + \dfrac{\sqrt{3}}{3^2} + \cdots + \dfrac{\sqrt{n}}{n^2}$

In Exercises 31–34 find the first six partial sums S_1, S_2, S_3, S_4, S_5, S_6 of the given series.

31. $1 + 3 + 5 + 7 + \cdots + 1001$

32. $1^2 + 2^2 + 3^2 + \cdots + 600^2$

33. $\displaystyle\sum_{k=1}^{100} \frac{1}{3^k}$

34. $\displaystyle\sum_{j=1}^{20} (-1)^j$

In Exercises 35–41 find a formula for the nth partial sum S_n of the given series and then find the sum of the series. (See Examples 3 and 4.)

35. $\displaystyle\sum_{k=1}^{1000} \left(\frac{1}{k+1} - \frac{1}{k+2} \right)$

36. $\displaystyle\sum_{k=1}^{100} \left(\frac{1}{2k-1} - \frac{1}{2k+1} \right)$

37. $\displaystyle\sum_{k=1}^{20} \frac{2}{3^k}$

38. $\displaystyle\sum_{j=1}^{20} \frac{4}{5^j}$

39. $\displaystyle\sum_{i=1}^{99} \left(\sqrt{i} - \sqrt{i+1} \right)$

40. $\displaystyle\sum_{k=1}^{20} (2^{k-1} - 2^k)$

41. $\displaystyle\sum_{k=1}^{999999} \log\!\left(\frac{k}{k+1} \right)$ [*Hint:* Use a property of logarithms to write the kth term as a difference.]

42. Let $a_1, a_2, a_3, \ldots$ be a sequence.
  (a) Show that

$$\sum_{k=1}^{n} [a_k - a_{k+1}] = a_1 - a_{n+1}$$

A series of this form is called a **telescoping series**.
  (b) Which of the series in Exercises 35–41 are telescoping?

SECTION 10.4
ARITHMETIC AND GEOMETRIC SERIES

In this section we find formulas for the sums of series whose terms form an arithmetic or geometric sequence.

◼ ARITHMETIC SERIES

An **arithmetic series** is a series whose terms form an arithmetic sequence. We will find a formula for the nth partial sum of an arithmetic series.

Let us begin with a simple example. Suppose we want to find the sum of the numbers 1, 2, 3, 4, . . . , 100, that is,

$$\sum_{k=1}^{100} k$$

When the famous mathematician Carl Friedrich Gauss was a schoolboy his teacher asked the class this question and expected that it would keep the students busy for a long time. But Gauss answered the question almost immediately. His idea was that, since we are adding numbers that are produced according to a fixed pattern, there must also be a pattern (or formula) for finding the sum. He started by writing the numbers from 1 to 100 and below them the same numbers in reverse order. Writing S for the sum and adding corresponding terms gives

$$
\begin{array}{rccccccc}
S = & 1 + & 2 + & 3 + \cdots + & 98 + & 99 + & 100 \\
S = & 100 + & 99 + & 98 + \cdots + & 3 + & 2 + & 1 \\
\hline
2S = & 101 + & 101 + & 101 + \cdots + & 101 + & 101 + & 101
\end{array}
$$

It follows that $2S = 100(101) = 10{,}100$ and so $S = 5050$.

Of course the sequence of natural numbers 1, 2, 3, . . . is an arithmetic sequence (with $a = 1$ and $d = 1$) and the method used for summing the first 100 terms of this series can be used to find a formula for the nth partial sum of any arithmetic series. We want to find the sum of the first n terms of the arithmetic sequence whose terms are $a_k = a + (k - 1)d$, that is, we want to find

$$S_n = \sum_{k=1}^{n} [a + (k - 1)d]$$
$$= a + (a + d) + (a + 2d) + (a + 3d) + \cdots + [a + (n - 1)d]$$

Using Gauss's method we write

$$
\begin{array}{rccccc}
S_n = & a & + & (a + d) & + \cdots + & [a + (n - 2)d] + & [a + (n - 1)d] \\
S_n = & [a + (n - 1)d] + & [a + (n - 2)d] + \cdots + & (a + d) & + & a \\
\hline
2S_n = & [2a + (n - 1)d] + & [2a + (n - 1)d] + \cdots + & [2a + (n - 1)d] + & [2a + (n - 1)d]
\end{array}
$$

There are n identical terms on the right side of this equation, so

$$2S_n = n[2a + (n - 1)d] \qquad \text{or} \qquad S_n = \frac{n}{2}[2a + (n - 1)d]$$

Notice that $a_n = a + (n - 1)d$ is the last term of this series. So we can write

$$S_n = \frac{n}{2}[a + a + (n - 1)d] = n\left(\frac{a + a_n}{2}\right)$$

This last formula says that the sum of the first n terms of an arithmetic series is the average of the first and last terms multiplied by n, the number of terms in the series. We summarize these results.

SUM OF AN ARITHMETIC SERIES

The sum S_n of the first n terms of an arithmetic series

$$S_n = a + (a + d) + (a + 2d) + (a + 3d) + \cdots + [a + (n - 1)d]$$

is given by

$$S_n = \frac{n}{2}[2a + (n - 1)d] \tag{1}$$

or

$$S_n = n\left(\frac{a + a_n}{2}\right) \tag{2}$$

EXAMPLE 1

Find the sum of the first 50 odd numbers.

SOLUTION

The odd numbers form an arithmetic sequence whose nth term is $a_n = 2n - 1$, so the 50th odd number is $a_{50} = 2(50) - 1 = 99$. Substituting in Formula 2 for the sum of an arithmetic series, we get

$$S_{50} = 50\left(\frac{a + a_{50}}{2}\right) = 50\left(\frac{1 + 99}{2}\right) = 50 \cdot 50 = 2500 \qquad \blacksquare$$

EXAMPLE 2

Find the sum of the first 40 terms of the arithmetic sequence

$$3, 7, 11, 15, \ldots$$

SOLUTION

For this arithmetic sequence $a = 3$ and $d = 4$. Using Formula 1 for the sum of an arithmetic series, we get

$$S_{40} = \frac{40}{2}[2(3) + (40 - 1)4] = 20(6 + 156) = 3240 \qquad \blacksquare$$

EXAMPLE 3

An arithmetic series has first term 5 and 50th term 103. How many terms of this series must be added to get 572?

SOLUTION

We first find the common difference. Since, for an arithmetic sequence, $a_n = a + (n - 1)d$, we get

$$a_{50} = a + (50 - 1)d$$

so

$$103 = 5 + 49d$$

Solving for d gives $d = 2$.

We are also asked to find n when $S_n = 572$. Using Formula 1 for the sum of an arithmetic series and substituting for S_n, a, and d gives

$$572 = \frac{n}{2}[2 \cdot 5 + (n - 1)2]$$

Solving for n we have

$$572 = 5n + n(n - 1)$$
$$n^2 + 4n - 572 = 0$$

so

$$(n - 22)(n + 26) = 0$$

This gives $n = 22$ or $n = -26$. But since n is a *number* of terms in a sequence we must have $n = 22$. ∎

GEOMETRIC SERIES

A **geometric series** is a series whose terms form a geometric sequence. So adding the first n terms of the geometric sequence

$$a, ar, ar^2, ar^3, ar^4, \ldots, ar^{n-1}, \ldots$$

we get the geometric series

$$S_n = \sum_{k=1}^{n} ar^{k-1} = a + ar + ar^2 + ar^3 + ar^4 + \cdots + ar^{n-1}$$

To find a formula for S_n we multiply S_n by r and subtract from S_n to get

$$S_n = a + ar + ar^2 + ar^3 + ar^4 + \cdots + ar^{n-1}$$
$$\underline{rS_n = \quad ar + ar^2 + ar^3 + ar^4 + \cdots + ar^{n-1} + ar^n}$$
$$S_n - rS_n = a - ar^n$$

So $\quad S_n(1 - r) = a(1 - r^n) \quad$ or $\quad S_n = \dfrac{a(1 - r^n)}{1 - r} \quad (r \neq 1)$

SUM OF A GEOMETRIC SERIES

The sum S_n of the first n terms of a geometric series

$$S_n = a + ar + ar^2 + ar^3 + ar^4 + \cdots + ar^{n-1} \qquad (r \neq 1)$$

is given by

$$S_n = a\frac{1 - r^n}{1 - r}$$

EXAMPLE 4

Find the sum of the first five terms of the geometric sequence

$$1, 0.7, 0.49, 0.343, \ldots$$

SOLUTION

The sum is a geometric series with $a = 1$ and $r = 0.7$. Using the formula for the sum of a geometric series with $n = 5$, we get

$$S_5 = 1\frac{1 - (0.7)^5}{1 - 0.7} = 2.7731$$

Thus the sum of the first 5 terms of this sequence is 2.7731. ∎

EXAMPLE 5

Find the sum of the series $\displaystyle\sum_{k=1}^{5} 7\left(-\frac{2}{3}\right)^k$.

SOLUTION

This is a geometric series with first term $a = 7\left(-\frac{2}{3}\right) = -\frac{14}{3}$ and common ratio $r = -\frac{2}{3}$, and there are five terms. Thus by the formula for the sum of a geometric series we have

$$S_5 = -\frac{14}{3}\frac{\left[1 - \left(-\frac{2}{3}\right)^5\right]}{1 - \left(-\frac{2}{3}\right)} = -\frac{14}{3}\frac{1 + \frac{32}{243}}{\frac{5}{3}} = -\frac{770}{243}$$
∎

EXERCISES 10.4

In Exercises 1–6 find the sum S_n of the arithmetic series that satisfies the given conditions.

1. $a = 4$, $d = 2$, $n = 20$ **2.** $a = 100$, $d = -5$, $n = 8$

3. $a_1 = 55$, $d = 12$, $n = 10$

4. $a_2 = 8$, $a_5 = 9.5$, $n = 15$

5. $a_3 = 980$, $a_{10} = 910$, $n = 5$

6. $a_4 = 21$, $d = 3$, $n = 10$

In Exercises 7–12 find the sum S_n of the geometric series that satisfies the given conditions.

7. $a = 5$, $r = 2$, $n = 6$ **8.** $a = \frac{2}{3}$, $r = \frac{1}{3}$, $n = 4$

9. $a_3 = 28$, $a_6 = 224$, $n = 6$

10. $a_2 = \frac{10}{3}$, $a_4 = \frac{40}{27}$, $r < 0$, $n = 4$

11. $a_3 = 0.18$, $r = 0.3$, $n = 5$

12. $a_2 = 0.12$, $a_5 = 0.00096$, $n = 4$

In Exercises 13–18 find the sum of the arithmetic series.

13. $1 + 5 + 9 + \cdots + 401$

14. $-3 + \left(-\frac{3}{2}\right) + 0 + \frac{3}{2} + 3 + \cdots + 30$

15. $0.7 + 2.7 + 4.7 + \cdots + 56.7$

16. $-10 - 9.9 - 9.8 - \cdots - 0.1$

17. $\sum_{k=0}^{10} (3 + 0.25k)$ **18.** $\sum_{n=0}^{20} (1 - 2n)$

In Exercises 19–24 find the sum of the geometric series.

19. $1 + 3 + 9 + \cdots + 2187$

20. $1 - \frac{1}{2} + \frac{1}{4} - \frac{1}{8} + \cdots - \frac{1}{512}$

21. $0.7 + 0.49 + 0.343 + \cdots + 0.16807$

22. $1 - \sqrt{2} + 2 - 2\sqrt{2} + \cdots + 32$

23. $\sum_{k=0}^{10} 3\left(\frac{1}{2}\right)^k$ **24.** $\sum_{j=0}^{5} 7\left(\frac{3}{2}\right)^j$

In Exercises 25–36 determine whether the series is arithmetic or geometric and find its sum.

25. $4 + 2.4 + 1.44 + \cdots + 0.5184$

26. $2 + 5 + 8 + \cdots + 32$

27. $1 - x + x^2 - x^3 + \cdots + x^{20}$

28. $1 - \sqrt{3} + 3 - 3\sqrt{3} + \cdots + 243$

29. $2 + 4 + 6 + \cdots + 1000$

30. $\sqrt{5} + 2\sqrt{5} + 3\sqrt{5} + \cdots + 100\sqrt{5}$

31. $\frac{1}{2} + 1 + \frac{3}{2} + \cdots + 64$ **32.** $\frac{1}{2} + 1 + 2 + \cdots + 64$

33. $\sum_{i=0}^{8} \left(1 + \sqrt{2}i\right)$ **34.** $\sum_{k=0}^{8} 2\left(\sqrt{3}\right)^k$

35. $\sum_{n=0}^{8} 5^{n/3}$ **36.** $\sum_{i=0}^{8} \sqrt{5} \cdot 2^i$

37. An arithmetic sequence has first term $a = 5$ and common difference $d = 2$. How many terms of this sequence must be added to get 2700?

38. A geometric sequence has first term $a = 1$ and common ratio $r = -\frac{1}{2}$. How many terms of this sequence must be added to get $\frac{341}{512}$?

39. The sum of the first four terms of a geometric series is 50 and the common ratio is $r = \frac{1}{2}$. Find the first term.

40. The sum of the first ten terms of an arithmetic series is 100 and the first term is 1. Find the tenth term.

41. The sum of the first 20 terms of an arithmetic series is 155 and the first term is 3. Find the common difference.

42. The sum of the first and 20th terms of an arithmetic sequence is 182. Find the sum of the first 20 terms.

43. The second term in a geometric series is $\frac{14}{3}$ and the fifth term is $\frac{112}{81}$. Find the sum of the first four terms.

44. The common ratio in a certain geometric sequence is $r = 0.2$ and the sum of the first four terms is 1248. Find the first term.

45. Find the product of the numbers

$$10^{1/10}, \ 10^{2/10}, \ 10^{3/10}, \ 10^{4/10}, \ldots, \ 10^{19/10}$$

46. Find the sum of the first ten terms of the sequence

$$a + b, \ a^2 + 2b, \ a^3 + 3b, \ a^4 + 4b, \ldots$$

47. A very patient women wishes to become a billionaire. She decides on a simple scheme. She puts aside 1 cent the first day, 2 cents the second day, 4 cents the third day, and so on, doubling the number of cents she puts aside each day. How much money will she have at the end of 30 days? How many days will it take for this woman to realize her wish?

48. A ball is dropped from a height of 9 ft. The elasticity of the ball is such that it always bounces back one-third of the distance from which it falls.
   (a) Find the total distance the ball has traveled at the instant it hits the ground for the fifth time.
   (b) Find a formula for the total distance the ball has traveled at the instant it hits the ground for the nth time.

49. When an object is allowed to fall freely near the surface of the earth the gravitational pull is such that it falls 16 ft in the first second, 48 ft in the next second, 80 ft in the next second, and so on.
   (a) Find the total distance the ball falls in 6 seconds.

(b) Find a formula for the distance the ball falls in n seconds.

50. In the well-known song "The Twelve Days of Christmas," a person gives his sweetheart k gifts on the kth day for each of the 12 days of Christmas. The person also repeats each gift identically on each subsequent day. Thus on the twelfth day the sweetheart receives a gift for the first day, 2 gifts for the second, 3 gifts for the third, and so on. Show that the number of gifts on the 12th day is an arithmetic series and find its sum.

51. The following is a well known children's rhyme:

As I was going to St. Ives
I met a man with seven wives;
Every wife had seven sacks;
Every sack had seven cats;
Every cat had seven kits;
Kits, cats, sacks, and wives,
How many were going to St. Ives?

Assuming that the entire group is actually going to St. Ives, show that the answer to the question in the children's rhyme is the sum of a geometric series and find its sum.

SECTION 10.5
ANNUITIES AND INSTALLMENT BUYING

Many financial transactions involve payments that are made at regular intervals. For example, if you deposit $100 each month in an interest-bearing account, what will the value of the account be at the end of five years? If you borrow $100,000 to buy a house, how much will the monthly payments be in order to pay off the loan in 30 years? These questions involve the sum of a series of numbers and we use the results of the preceding section to answer them here.

 THE AMOUNT OF AN ANNUITY

An **annuity** is a sum of money that is paid in regular equal payments. Although the word annuity suggests annual (or yearly) payments, they can be made semiannually, quarterly, monthly, or at other regular intervals. Payments are usually made at the end of the payment interval. The **amount of an annuity** is the sum of all the individual payments from the time of the first payment until the last payment is made, together with all the interest. We denote this sum by A_f (the subscript f here is used to denote *final* amount).

EXAMPLE 1

An investor deposits $400 every December 15 and June 15 for ten years into an account that bears interest at the rate of 8% per year compounded semiannually. How much will she have in the account at the time of the last payment?

SOLUTION

We need to find the amount of an annuity consisting of 20 semiannual payments of $400 each. Since the interest rate is 8% per year compounded semiannually, the interest rate per time period is $i = 0.04$. Notice that the first payment is in the account for 19 time periods, the second for 18 time periods, and so on. The last payment receives no interest. The situation can be illustrated by the time line in Figure 1.

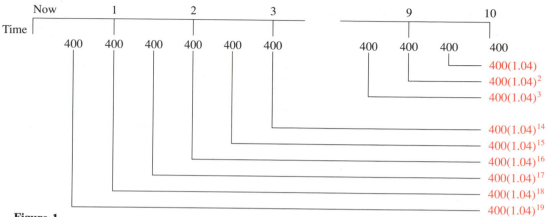

Figure 1

The amount A_f of the annuity is the sum of these 20 amounts. Thus

$$A_f = 400 + 400(1.04) + 400(1.04)^2 + \cdots + 400(1.04)^{19}$$

But this is a geometric series with $a = 400$, $r = 1.04$, and $n = 20$, so

$$A_f = 400\frac{1 - (1.04)^{20}}{1 - 1.04} \approx 11{,}911.23$$

Thus the amount of the annuity after ten years is \$11,911.23. ■

In general, the regular annuity payment is called the **periodic rent** and is denoted by R. We also let i denote the interest rate per time period and n the number of payments. (We always assume that the time period in which interest is compounded is equal to the time between payments.) By the same reasoning as in Example 1, we see that the amount A_f of an annuity is

$$A_f = R + R(1 + i) + R(1 + i)^2 + \cdots + R(1 + i)^{n-1}$$

Since this is a geometric series with n terms where $a = R$ and $r = 1 + i$, the formula for the sum of a geometric series gives

$$A_f = R\frac{1 - (1 + i)^n}{1 - (1 + i)} = R\frac{1 - (1 + i)^n}{-i} = R\frac{(1 + i)^n - 1}{i}$$

AMOUNT OF AN ANNUITY

The amount A_f of an annuity consisting of n regular equal payments of size R with interest rate i per time period is given by

$$A_f = R\frac{(1 + i)^n - 1}{i}$$

EXAMPLE 2

How much money should be invested every month at 12% per year compounded monthly in order to have $4000 in 18 months?

SOLUTION

In this problem $i = 0.12/12 = 0.01$, $A_f = 4000$, and $n = 18$. We need to find the amount R of each payment. By the formula for the amount of an annuity

$$4000 = R\frac{(1 + 0.01)^{18} - 1}{0.01}$$

Solving for R we get

$$R = \frac{4000(0.01)}{(1 + 0.01)^{18} - 1} \approx \frac{40}{1.196147 - 1} \approx 203.928$$

Thus the monthly investment should be $203.93. ◼

THE PRESENT VALUE OF AN ANNUITY

If you are to receive $10,000 five years from now it will be worth much less than getting $10,000 right now. This is because of the interest you can accumulate during that time. What smaller amount would you be willing to accept *now* instead of receiving $10,000 in five years? This is the amount of money that, together with interest, will be worth $10,000 in five years. The amount we are looking for here is called the *discounted value* or *present value*. If the interest rate is 8% compounded quarterly, then the interest per time period is $i = 0.08/4 = 0.02$ and there are $4 \times 5 = 20$ time periods. If we let PV denote the present value, then by the formula for compound interest (Section 4.2) we have

$$10,000 = PV(1 + i)^n = PV(1 + 0.02)^{20}$$

so

$$PV = 10,000(1 + 0.02)^{-20} \approx 6729.713$$

Thus, in this situation, the present value of $10,000 is $6729.71. The same reasoning leads to the following general formula for present value.

$$PV = A(1 + i)^{-n}$$

Similarly, the **present value of an annuity** is the amount A_p that must be invested now at the interest rate i per time period in order to provide n payments each of amount R. Clearly A_p is the sum of the present values of each individual payment (see Exercise 20). Another way of finding A_p is to notice that A_p is the present value of A_f:

$$A_p = A_f(1 + i)^{-n} = R\frac{(1 + i)^n - 1}{i}(1 + i)^{-n} = R\frac{1 - (1 + i)^{-n}}{i}$$

**THE PRESENT VALUE
OF AN ANNUITY**

The present value A_p of an annuity consisting of n regular equal payments of size R with interest rate i per time period is given by

$$A_p = R\frac{1 - (1 + i)^{-n}}{i}$$

EXAMPLE 3

A person wins $10,000,000 in the California lottery. But the amount is paid in equal yearly installments of half a million dollars for 20 years. What is the present value of his winnings? We assume that he can get 10% interest compounded annually.

SOLUTION

The winnings are paid as an annuity and we need to find its present value. Here $i = 0.1$, $R = \$500,000$, and $n = 20$. Thus

$$A_p = 500,000\frac{1 - (1 + 0.1)^{-20}}{0.1} \approx 4,256,781.859$$

This means that the winner really won only $4,256,781.86 if it were paid to him immediately. ∎

INSTALLMENT BUYING

When you buy a house or a car by installment, the payments you make are an annuity whose present value is the amount of the loan.

EXAMPLE 4

A student wishes to buy a car. He can afford to pay $200 per month but has no down payment. If he can make these payments for four years and if the interest rate is 12%, what price car can he buy?

SOLUTION

The payments the student makes constitute an annuity whose present value is the price of the car (which is also the amount of the loan, in this case). Here we have $i = 0.12/12 = 0.01$, $R = 200$, $n = 12 \times 4 = 48$. Thus

$$A_p = R\frac{1 - (1 + i)^{-n}}{i} = 200\frac{1 - (1 + 0.01)^{-48}}{0.01} \approx 7594.792$$

Thus the student can buy a car worth $7594.79. ∎

EXAMPLE 5

A couple borrows $100,000 at 9% interest as a mortgage loan on a house. They expect to make monthly payments for 30 years to repay the loan. What is the size of each payment?

SOLUTION

The payments form an annuity with present value $A_p = \$100,000$. Also, $i = 0.09/12 = 0.0075$, and $n = 12 \times 30 = 360$. We are looking for the amount R of each payment. From the formula

$$A_p = R\frac{1 - (1 + i)^{-n}}{i}$$

we get

$$R = \frac{iA_p}{1 - (1 + i)^{-n}}$$

In our case,

$$R = \frac{(0.0075)(100,000)}{1 - (1 + 0.0075)^{-360}} \approx 804.622$$

Thus the monthly payments are $804.62. ■

We now give an example that illustrates the use of graphing devices in solving problems related to installment buying.

EXAMPLE 6

A car dealer sells a new car for $18,000. He also offers to sell the same car for payments of $405 per month for five years. What interest rate is this car dealer charging?

SOLUTION

The payments form an annuity with present value $A_p = \$18,000$, $R = 405$, and $n = 12 \times 5 = 60$. To find the interest rate we must solve the equation

$$R = \frac{iA_p}{1 - (1 + i)^{-n}}$$

for i. A little experimentation will convince you that it is not possible to algebraically solve this equation for i. So to find i we use a graphing device to graph R as a function of the interest rate x and then use the graph to find the interest rate corresponding to the value of R we are interested in ($405 in this case). Since $i = x/12$, we graph the function

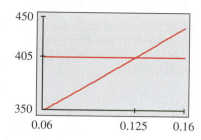

$$R(x) = \frac{\frac{x}{12}(18{,}000)}{1 - \left(1 + \frac{x}{12}\right)^{-60}}$$

in the viewing rectangle $[0.06, 0.16] \times [350, 450]$. We also graph the horizontal line $R(x) = 405$ in the same viewing rectangle. Then by moving the cursor to the point of intersection of the two graphs we find that the corresponding x-value is approximately 0.125. Thus the interest rate is about $12\frac{1}{2}\%$ per year. ∎

EXERCISES 10.5

1. Find the amount of an annuity that consists of 20 annual payments of $5000 each into an account that pays interest of 12% per year.

2. Find the amount of an annuity that consists of 20 semi-annual payments of $500 each into an account that pays 6% interest per year compounded semiannually.

3. Find the amount of an annuity that consists of 16 quarterly payments of $300 each into an account that pays interest of 8% per year compounded quarterly.

4. How much money should be invested every quarter at 10% per year compounded quarterly in order to have $5,000 in 2 years?

5. How much money should be invested monthly at 6% per year compounded monthly in order to have $2,000 in 8 months?

6. What is the present value of an annuity that consists of 20 semiannual payments of $1000 at the interest rate of 9% compounded semiannually?

7. How much money must be invested now at 9% per year compounded semiannually to provide an annuity of 20 payments of $200 every 6 months, the first payment being in 6 months?

8. A 55-year-old man deposits $50,000 to set up an annuity with an insurance company at 8% per year compounded semiannually until age 65. If he is to be paid twice per year, how large is each payment he receives?

9. A woman wants to borrow $12,000 in order to buy a car. She wants to repay the loan by monthly installments for four years. If the interest rate on this loan is $10\frac{1}{2}\%$ compounded monthly, what is the amount of each payment?

10. What is the monthly payment on a 30-yr mortgage of $80,000 at 12% interest? What is the monthly payment on this same mortgage if it is to be repaid over a 15-yr period?

11. What is the monthly payment on a 30-yr mortgage of $100,000 at 8% interest compounded monthly? What is the total amount paid on this loan over the 30-yr period?

12. A couple can afford to pay $650 per month toward buying a house. If the mortgage rates are 9% and they intend to secure a 30-yr mortgage, how much can they borrow?

13. A couple secure a 30-yr loan of $100,000 at $9\frac{3}{4}\%$ compounded monthly toward buying a house.
    (a) What is the amount of the monthly payments?
    (b) What is the total amount that will be paid by the couple over the 30-yr period?
    (c) If, instead of buying a house, the couple deposits the monthly payments in an account that pays $9\frac{3}{4}\%$ interest compounded monthly, how much will be in the account at the end of the 30-yr period?

14. Jane agrees to buy a car by putting a down payment of $2000 and making payments of $220 per month for three years. If the interest rate is 8% compounded monthly, what is the actual purchase price of the car?

15. Mike buys a ring for his fiancee by paying $30 per month for one year. If the interest rate is 10% compounded monthly, what is the price of the ring?

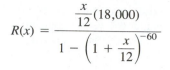 16. Janet decides to buy a $12,500 car by paying $420 a month for three years. Assuming that interest is compounded monthly, what interest rate is she paying on the loan?

17. John buys a stereo system that is advertised for $640. He agrees to pay $32 per month for two years. Assuming that interest is compounded monthly, what interest rate is he paying?

18. A woman purchases a $2000 diamond ring by making a down payment of $200 and paying monthly installments of $88 for two years. Assuming that interest is compounded monthly, what interest rate is she paying?

19. An item at a department store is priced at $189.99 and can be bought by making 20 payments of $10.50. Find the interest rate assuming that interest is compounded monthly.

20. (a) Draw a time line as in Example 1 to show that the present value of an annuity is the sum of the present values of each payment; that is,

$$A_p = \frac{R}{1 + i} + \frac{R}{(1 + i)^2} + \frac{R}{(1 + i)^3} + \cdots + \frac{R}{(1 + i)^n}$$

(b) Use part (a) to derive the formula for A_p given in the text.

SECTION 10.6
INFINITE GEOMETRIC SERIES

So far we have been discussing series with a finite number of terms. Let us write down a series with an infinite number of terms

$$a_1 + a_2 + a_3 + a_4 + \cdots$$

The dots mean that we are to continue the addition indefinitely. A series of this kind is called an **infinite series**.

WHAT IS AN INFINITE SERIES?

What meaning can we attach to the sum of infinitely many numbers? It seems at first that it is not possible to add infinitely many numbers and arrive at a finite number. But consider the following problem. You have a cake and you want to eat it by first eating half the cake, then eating half of what remains, then again eating half of what remains. This process can continue indefinitely since at each stage some of the cake remains. (See Figure 1.)

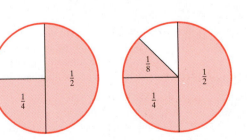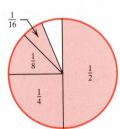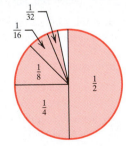

Figure 1

Does this mean that it is impossible to eat all of the cake? Of course not. Let us write down what you have eaten from this cake:

$$\frac{1}{2} + \frac{1}{4} + \frac{1}{8} + \frac{1}{16} + \cdots + \frac{1}{2^n} + \cdots$$

This is an infinite series and we would like to note two things about it. First, from Figure 1 it is clear that no matter how many terms of this series we add, the total will never exceed 1. Secondly, the more terms of this series that we add, the closer the sum is to 1 (see Figure 1). This suggests that the number 1 can be written as the sum of infinitely many smaller numbers:

$$1 = \frac{1}{2} + \frac{1}{4} + \frac{1}{8} + \frac{1}{16} + \cdots + \frac{1}{2^n} + \cdots$$

To make this more precise, let us look at the partial sums of this series:

$$S_1 = \frac{1}{2} \qquad\qquad = \frac{1}{2}$$

$$S_2 = \frac{1}{2} + \frac{1}{4} \qquad\qquad = \frac{3}{4}$$

$$S_3 = \frac{1}{2} + \frac{1}{4} + \frac{1}{8} \qquad = \frac{7}{8}$$

$$S_4 = \frac{1}{2} + \frac{1}{4} + \frac{1}{8} + \frac{1}{16} = \frac{15}{16}$$

$$\vdots$$

and in general (see Example 4 of Section 10.3),

$$S_n = 1 - \frac{1}{2^n}$$

As n gets larger and larger, we are adding more and more of the terms of this series. Intuitively, as n gets larger, S_n gets closer to the sum of the series. Now notice that as n gets large $1/2^n$ gets closer and closer to 0. Thus S_n gets close to $1 - 0 = 1$. Using the notation of Section 3.9 we can write

$$S_n \to 1 \qquad \text{as} \qquad n \to \infty$$

In general, if S_n gets close to a finite number S as n gets large, we say that S is the **sum of the infinite series**.

INFINITE GEOMETRIC SERIES

We call an infinite series of the form

$$a + ar + ar^2 + ar^3 + ar^4 + \cdots + ar^{n-1} + \cdots$$

an **infinite geometric series**. We can apply the reasoning used earlier to find the sum of an infinite geometric series. The nth partial sum of such a series is given by the formula for the sum of a geometric series in Section 10.4:

$$S_n = a\frac{1 - r^n}{1 - r} \qquad r \neq 1$$

It can be shown that if $|r| < 1$, then r^n gets close to 0 as n gets large (you can easily convince yourself of this using a calculator). It follows that S_n gets close to $a/(1 - r)$ as n gets large, or

$$S_n \to \frac{a}{1 - r} \qquad \text{as} \qquad n \to \infty$$

Thus the sum of this infinite geometric series is $a/(1 - r)$. We summarize this result.

SUM OF AN INFINITE GEOMETRIC SERIES

If $|r| < 1$, then the infinite geometric series

$$a + ar + ar^2 + ar^3 + ar^4 + \cdots + ar^{n-1} + \cdots$$

has the sum

$$S = \frac{a}{1 - r}$$

EXAMPLE 1

Find the sum of the infinite geometric series

$$2 + \frac{2}{5} + \frac{2}{25} + \frac{2}{125} + \cdots + \frac{2}{5^n} + \cdots$$

SOLUTION

We use the formula for the sum of an infinite geometric series. In this case $a = 2$ and $r = \frac{1}{5}$. Thus the sum of this infinite series is

$$S = \frac{2}{1 - \frac{1}{5}} = \frac{5}{2}$$

EXAMPLE 2

Find the fraction that represents the rational number $2.3\overline{51}$.

SOLUTION

This repeating decimal can be written as a series:

$$\frac{23}{10} + \frac{51}{1000} + \frac{51}{100,000} + \frac{51}{10,000,000} + \frac{51}{1,000,000,000} + \cdots$$

The terms of this series after the first term form an infinite geometric series with

$$a = \frac{51}{1000} \quad \text{and} \quad r = \frac{1}{100}$$

Thus the sum of this part of the series is

$$S = \frac{\frac{51}{1000}}{1 - \frac{1}{100}} = \frac{\frac{51}{1000}}{\frac{99}{100}} = \frac{51}{1000} \cdot \frac{100}{99} = \frac{51}{990}$$

So

$$2.3\overline{51} = \frac{23}{10} + \frac{51}{990} = \frac{2328}{990} = \frac{388}{165} \quad \blacksquare$$

EXERCISES 10.6

In Exercises 1–8 find the sum of the given infinite geometric series.

1. $1 - \dfrac{1}{3} + \dfrac{1}{9} - \dfrac{1}{27} + \cdots$

2. $\dfrac{2}{5} + \dfrac{4}{25} + \dfrac{8}{125} + \cdots$

3. $\dfrac{1}{3^6} + \dfrac{1}{3^8} + \dfrac{1}{3^{10}} + \dfrac{1}{3^{12}} + \cdots$

4. $3 - \dfrac{3}{2} + \dfrac{3}{4} - \dfrac{3}{8} + \cdots$

5. $-\dfrac{100}{9} + \dfrac{10}{3} - 1 + \dfrac{3}{10} - \cdots$

6. $\dfrac{1}{\sqrt{2}} + \dfrac{1}{2} + \dfrac{1}{2\sqrt{2}} + \dfrac{1}{4} + \cdots$

7. $5^{4/3} - 5^{5/3} + 5^{6/3} - 5^{7/3} + \cdots$

8. $\dfrac{1}{1 + \sqrt{2}} - 1 - \dfrac{1}{1 - \sqrt{2}} - \cdots$

In Exercises 9–14 express each repeating decimal as a fraction.

9. $0.777\ldots$

10. $0.2\overline{53}$

11. $0.030303\ldots$

12. $2.11\overline{25}$

13. $0.\overline{112}$

14. $0.123123123\ldots$

15. The elasticity of a ball is such that it rebounds two-thirds the distance which it falls. If this ball is dropped from a distance of 12 ft from the ground, use an infinite geometric series to approximate the total distance the ball travels before it stops bouncing.

16. A certain ball rebounds to half the height from which it is dropped. If this ball is dropped from a distance of 1 m from the ground, use an infinite geometric series to approximate the total distance the ball travels before it comes to rest.

17. If the ball in Exercise 16 is dropped from a height of 8 ft then the time required for its first complete bounce (from the instant the ball first touches the ground till the next time it touches the ground) is 1 s. Each subsequent complete bounce requires $1/\sqrt{2}$ as long as the complete bounce preceding it. Use an infinite geometric series to estimate the time required for the ball to stop bouncing from the instant it first touches the ground.

18. The midpoints of the sides of a square of side 1 are joined to form a new square. This procedure is repeated for each new square. (See the figure.)
(a) Find the sum of the areas of all the squares.
(b) Find the sum of the perimeters of all the squares.

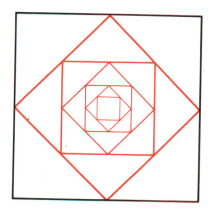

19. A circular disk of radius R is cut out of paper as shown in part (a) of the figure. Two disks of radius $\frac{1}{2}R$ are cut out of paper and placed on top of the first disk [part (b)], then four disks of radius $\frac{1}{4}R$ are placed on these two disks [part (c)]. Assuming that this process can be repeated indefinitely, find the total area of all the disks.

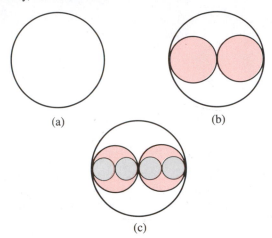

(a)   (b)

(c)

20. An **annuity in perpetuity** is one that continues forever. Such annuities are useful in setting up scholarship funds to ensure that the award continues.

(a) Draw a time line (as in Example 1 of Section 10.5) to show that the amount of money to be invested now (A_p) at interest rate i per time period in order to set up an annuity in perpetuity of amount R per time period is

$$A_p = \frac{R}{1+i} + \frac{R}{(1+i)^2} + \frac{R}{(1+i)^3}$$
$$+ \cdots + \frac{R}{(1+i)^n} + \cdots$$

(b) Find the sum of the infinite series in part (a) to show that

$$A_p = \frac{R}{i}$$

21. How much money must be invested now at 10% per year compounded annually to provide an annuity of $5000 every year in perpetuity? The first payment is due in one year. (Refer to Exercise 20.)

22. How much money must be invested now at 8% per year compounded *quarterly* in order to provide an annuity of $3000 per year in perpetuity? The first payment is due in one year. (Refer to Exercise 20.)

SECTION 10.7
MATHEMATICAL INDUCTION

There are two aspects to mathematics—discovery and proof—and both are of equal importance. It is necessary to discover something before attempting to prove it, and we can only be certain of its truth once it has been proved. In this section we look more carefully at the relationship between these two parts of mathematics.

CONJECTURE AND PROOF

Let us try a simple experiment. We add up more and more of the odd numbers as follows:

$$1 = 1$$
$$1 + 3 = 4$$
$$1 + 3 + 5 = 9$$
$$1 + 3 + 5 + 7 = 16$$
$$1 + 3 + 5 + 7 + 9 = 25$$

What do you notice about the numbers on the right side of these equations? They are in fact all perfect squares. These equations say that

The sum of the first 1 odd number is 1^2.

The sum of the first 2 odd numbers is 2^2.

The sum of the first 3 odd numbers is 3^2.

The sum of the first 4 odd numbers is 4^2.

The sum of the first 5 odd numbers is 5^2.

This leads naturally to the following question: Is it true that for every natural number n, the sum of the first n odd numbers is n^2? Could this remarkable property be true? We could try a few more numbers and find that the pattern persists for the first 6, 7, 8, 9, and 10 odd numbers. At this point we feel quite sure that this is always true, so we make a conjecture:

The sum of the first n odd numbers is n^2.

Since we know that the nth odd number is $2n - 1$, we can write this statement more precisely as:

$$1 + 3 + 5 + \cdots + (2n - 1) = n^2$$

It is important to realize that this is still a conjecture. We cannot conclude that a property is true for all numbers (there are infinitely many) by checking a finite number of cases. To see this more clearly, suppose someone tells us that he has added up the first trillion odd numbers and found out that they do *not* add up to one trillion squared. What would you tell this person? It would be silly to tell him that you are sure it is true because you have already checked it out for the first five cases. You could, however, take out paper and pencil and start checking it out yourself, but this task would probably take up the rest of your life. The tragedy would be that after completing this task you would still not be sure of the truth of the conjecture. Do you see why?

Herein lies the power of mathematical proof. A **proof** is a clear argument that demonstrates the truth of a statement beyond doubt. We consider here a special kind of proof called mathematical induction that will help us prove statements like the one we were just considering. But before we do this let us try another experiment. Consider the polynomial

$$p(n) = n^2 - n + 41$$

Let us find some of the values of $p(n)$ for natural numbers n:

$p(1) = 41$ $p(2) = 43$ $p(3) = 47$

$p(4) = 53$ $p(5) = 61$ $p(6) = 71$

$p(7) = 83$ $p(8) = 97$ $p(9) = 113$

We notice this time that all the values that we have calculated are prime numbers. We might want to conclude at this point that all the values of this polynomial are prime. But let's try a few more values:

$$p(10) = 131 \quad \text{(prime)} \qquad p(11) = 151 \quad \text{(prime)} \qquad p(12) = 173 \quad \text{(prime)}$$
$$p(13) = 197 \quad \text{(prime)} \qquad p(14) = 223 \quad \text{(prime)} \qquad p(15) = 251 \quad \text{(prime)}$$

At this point we are getting tired of calculating more values. We make the following conjecture:

> For every natural number n, $p(n)$ is a prime number.

If you try values for n from 16 to 40 you will find that again $p(n)$ is prime. But our conjecture is too hasty. It is easily seen that $p(41) = 41^2 - 41 + 41 = 41^2$ is not prime. This is the first n for which $p(n)$ is not prime! So our conjecture is false.

This illustrates clearly that we cannot be certain of the truth of a statement by checking out special cases. We need a convincing argument to determine the truth of a statement—a proof.

■ MATHEMATICAL INDUCTION

We consider a special kind of proof called **mathematical induction**. Here is how it works. Suppose we have a statement that says something about all natural numbers n. Let's call this statement P. For example, we could consider the statement

> P: For every natural number n, the sum of the first n odd numbers is n^2.

Since this statement is about *all* natural numbers, it contains infinitely many statements; we will call them $P(1)$, $P(2)$, ...

> $P(1)$: The sum of the first 1 odd number is 1^2.
>
> $P(2)$: The sum of the first 2 odd numbers is 2^2.
>
> $P(3)$: The sum of the first 3 odd numbers is 3^2.
>
> $\vdots$ $\qquad\qquad$ $\vdots$

How can we prove all of these statements at once? Mathematical induction is a clever way of doing just that.

The crux of the idea is this: Suppose that we can prove that whenever one of these statements is true, the one following it in the list is also true. In other words,

> For every k, if $P(k)$ is true then $P(k + 1)$ is true.

This is called the **induction step** because it leads us from the truth of one statement to the next. Now, suppose that we can also prove that

> $P(1)$ is true.

The induction step now leads us through the following chain of statements:

$$P(1) \text{ is true, so } P(2) \text{ is true.}$$
$$P(2) \text{ is true, so } P(3) \text{ is true.}$$
$$P(3) \text{ is true, so } P(4) \text{ is true.}$$
$$\vdots \qquad\qquad \vdots$$

So we see that if both of these statements are proved, then statement P is proved for all n. We summarize this important method of proof.

PRINCIPLE OF MATHEMATICAL INDUCTION

For each natural number n, let $P(n)$ be a statement depending on n. Suppose that the following two conditions are satisfied.

1. $P(1)$ is true.

2. For every natural number k, if $P(k)$ is true, then $P(k + 1)$ is true.

Then $P(n)$ is true for all natural numbers n.

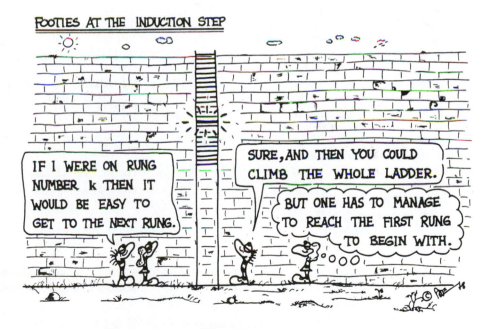

FOOTIES AT THE INDUCTION STEP

IF I WERE ON RUNG NUMBER k THEN IT WOULD BE EASY TO GET TO THE NEXT RUNG.

SURE, AND THEN YOU COULD CLIMB THE WHOLE LADDER.

BUT ONE HAS TO MANAGE TO REACH THE FIRST RUNG TO BEGIN WITH.

To apply this principle there are two steps:

Step 1   Prove that $P(1)$ is true.

Step 2   Assume that $P(k)$ is true and use this assumption to prove that $P(k + 1)$ is true.

Blaise Pascal (1623–1662) is considered one of the most versatile minds in modern history. He was a writer and philosopher as well as a gifted mathematician and physicist. Among his contributions that appear in this book are the theory of probability, Pascal's triangle, and the Principle of Mathematical Induction. Pascal's father, himself a mathematician, believed that his son should not study mathematics until he was 15 or 16. But at age 12 Blaise insisted on learning geometry, and proved most of its elementary theorems himself. At 19 he invented the first mechanical adding machine. In 1647, after writing a major treatise on the conic sections, he abruptly abandoned mathematics because he felt that his intense studies were contributing to his ill health. He devoted himself instead to frivolous recreations such as gambling, but this only served to pique his interest in probability. In 1654 he miraculously survived a carriage accident in which his horses ran off a bridge. Taking this to be a sign from God, he entered a monastery, where he pursued theology and philosophy, writing his famous Pensées. He also continued his mathematical research. He valued faith and intuition more than reason as the source of truth, declaring that "the heart has its own reasons, which reason cannot know."

Notice that in Step 2 we do not prove that $P(k)$ is true. We only show that *if $P(k)$ is true, then* $P(k + 1)$ is also true. The assumption that $P(k)$ is true is called the **induction hypothesis**.

We use mathematical induction to prove that the conjecture we made at the beginning of this section is true.

EXAMPLE 1

Prove that for all natural numbers n, $1 + 3 + 5 + \cdots + (2n - 1) = n^2$.

SOLUTION

Let $P(n)$ denote the statement

$$1 + 3 + 5 + \cdots + (2n - 1) = n^2 \tag{1}$$

Step 1 We need to show that $P(1)$ is true. But $P(1)$ is simply the statement that $1 = 1^2$, which is of course true.

Step 2 We assume that $P(k)$ is true. Thus our induction hypothesis is that

$$1 + 3 + 5 + \cdots + (2k - 1) = k^2 \tag{2}$$

We need to show that $P(k + 1)$ is true, that is,

$$1 + 3 + 5 + \cdots + (2k - 1) + [2(k + 1) - 1] = (k + 1)^2$$

[We get $P(k + 1)$ by substituting $k + 1$ for each n in Equation 1.] To show this let us add the quantity $[2(k + 1) - 1]$ to both sides of Equation 2 to get

$$\begin{aligned} 1 + 3 + 5 + \cdots + (2k - 1) + [2(k + 1) - 1] &= k^2 + [2(k + 1) - 1] \\ &= k^2 + [2k + 2 - 1] \\ &= k^2 + 2k + 1 \\ &= (k + 1)^2 \end{aligned}$$

Thus $P(k + 1)$ follows from $P(k)$ and this completes the induction step.

Having proved Steps 1 and 2 we conclude from the Principle of Mathematical Induction that $P(n)$ is true for all natural numbers n. ∎

EXAMPLE 2

Prove that for every natural number n,

$$1 + 2 + 3 + \cdots + n = \frac{n(n + 1)}{2}$$

SOLUTION

We let $P(n)$ denote the statement $1 + 2 + 3 + \cdots + n = n(n + 1)/2$. We want to show that $P(n)$ is true for all natural numbers n.

Step 1 We need to show that $P(1)$ is true. But $P(1)$ says that

$$1 = \frac{1(1 + 1)}{2}$$

which is clearly true.

Step 2 Assume that $P(k)$ is true, that is,

$$1 + 2 + 3 + \cdots + k = \frac{k(k + 1)}{2} \tag{3}$$

This is our induction hypothesis. We want to use this to show that $P(k + 1)$ is true. The statement $P(k + 1)$ is

$$1 + 2 + 3 + \cdots + k + (k + 1) = \frac{(k + 1)[(k + 1) + 1]}{2} \tag{4}$$

In other words we want to derive Equation 4 from Equation 3. Since the left sides of these two equations differ by the quantity $k + 1$, let us try adding $k + 1$ to both sides of Equation 3 and manipulating the right side to arrive at Equation 4. Thus

$$
\begin{aligned}
1 + 2 + 3 + \cdots + k + (k + 1) &= \frac{k(k + 1)}{2} + (k + 1) \\
&= (k + 1)\left(\frac{k}{2} + 1\right) \\
&= (k + 1)\left(\frac{k + 2}{2}\right) \\
&= \frac{(k + 1)[(k + 1) + 1]}{2}
\end{aligned}
$$

But this is exactly $P(k + 1)$. Thus we have shown that if $P(k)$ is true, then $P(k + 1)$ is true. This completes the induction step.

Steps 1 and 2 together show that $P(n)$ is true for all n by the Principle of Mathematical Induction. ■

It might happen that a statement $P(n)$ is false for the first few natural numbers, but true from some number on. For example, we may want to prove that $P(n)$ is true for $n \geq 5$. Notice that if we prove that $P(5)$ is true, then this together with the induction step would imply the truth of $P(5), P(6), P(7), \ldots$. The next example illustrates this point.

EXAMPLE 3

Prove that $4n < 2^n$ for all $n \geq 5$.

SOLUTION

Let $P(n)$ denote the statement $4n < 2^n$.

Step 1 $P(5)$ is the statement that $4 \cdot 5 < 2^5$ or $20 < 32$, which is true.

Step 2   Assume that $P(k)$ is true. Thus our induction hypothesis is

$$4k < 2^k \tag{5}$$

We want to use this to show that $P(k + 1)$ is true, that is,

$$4(k + 1) < 2^{k+1} \tag{6}$$

So, let us start by adding 4 to both sides of (5). (The motivation for this is that we are trying to derive (6), and the left side of (6) is $4k + 4$.) This gives

$$4k + 4 < 2^k + 4 \tag{7}$$

Now since $k \geq 5$, it follows that $4 < 2^k$. From this and (7) we get

$$4(k + 1) = 4k + 4 < 2^k + 4 < 2^k + 2^k = 2 \cdot 2^k = 2^{k+1}$$

or $4(k + 1) < 2^{k+1}$

But this is exactly (6). Thus we have derived $P(k + 1)$ from $P(k)$ and this completes the induction step.

Having proved Steps 1 and 2, we conclude that $P(n)$ is true for all natural numbers $n \geq 5$.   ■

EXERCISES 10.7

In Exercises 1–9 use mathematical induction to prove that the given formula is true for all natural numbers n.

✓**1.** $5 + 8 + 11 + \cdots + (3n + 2) = \dfrac{n(3n + 7)}{2}$

2. $1^2 + 2^2 + 3^2 + \cdots + n^2 = \dfrac{n(n + 1)(2n + 1)}{6}$

✓**3.** $1 \cdot 2 + 2 \cdot 3 + 3 \cdot 4 + \cdots + n(n + 1)$
$$= \dfrac{n(n + 1)(n + 2)}{3}$$

4. $1 \cdot 3 + 2 \cdot 4 + 3 \cdot 5 + \cdots + n(n + 2)$
$$= \dfrac{n(n + 1)(2n + 7)}{6}$$

✓**5.** $1^3 + 2^3 + 3^3 + \cdots + n^3 = \dfrac{n^2(n + 1)^2}{4}$

6. $1^3 + 3^3 + 5^3 + \cdots + (2n - 1)^3 = n^2(2n^2 - 1)$

7. $2^3 + 4^3 + 6^3 + \cdots + (2n)^3 = 2n^2(n + 1)^2$

8. $\dfrac{1}{1\cdot 2}+\dfrac{1}{2\cdot 3}+\dfrac{1}{3\cdot 4}+\cdots+\dfrac{1}{n(n+1)}=\dfrac{n}{(n+1)}$

9. $1\cdot 2+2\cdot 2^2+3\cdot 2^3+4\cdot 2^4+\cdots+n\cdot 2^n$
$$=2[1+(n-1)2^n]$$

10. Prove that $n<2^n$ for all natural numbers n.

11. Prove that if $x>-1$, then $(1+x)^n\ge 1+nx$ for all natural numbers n.

12. Prove that $(n+1)^2<2n^2$ for all natural numbers $n\ge 3$.

13. Show that n^2-n+41 is odd for all natural numbers n.

14. Show that n^3-n+3 is divisible by 3 for all natural numbers n.

15. Show that 8^n-3^n is divisible by 5 for all natural numbers n.

16. Show that $3^{2n}-1$ is divisible by 8 for every natural number n.

17. Let $a_{n+1}=3a_n$ and $a_1=5$. Show that $a_n=5\cdot 3^{n-1}$ for all natural numbers n.

18. A sequence is defined recursively by $a_{n+1}=3a_n-8$ and $a_1=4$. Find an explicit formula for a_n and prove that the formula you found is true using mathematical induction.

19. Show that $x-y$ is a factor of x^n-y^n for all natural numbers n.
[*Hint:* $x^{k+1}-y^{k+1}=x^k(x-y)+(x^k-y^k)y$]

20. Show that $x+y$ is a factor of $x^{2n-1}+y^{2n-1}$ for all natural numbers n.

21. Determine whether each of the following statements is true or false. If the statement is true, prove it. If it is false, give an example where it fails.
(a) $p(n)=n^2-n+11$ is prime for all n.
(b) $n^2>n$ for all $n\ge 2$.
(c) $2^{2n+1}+1$ is divisible by 3 for all $n\ge 1$.
(d) $n^3\ge(n+1)^2$ for all $n\ge 2$.
(e) n^3-n is divisible by 3 for all $n\ge 2$.
(f) n^3-6n^2+11n is divisible by 6 for all $n\ge 1$

In Exercises 22–26, F_n denotes the nth term of the Fibonacci sequence discussed in Section 10.1. Use mathematical induction to prove the given statements.

22. F_{3n} is even for all natural numbers n.

23. $F_1+F_2+F_3+\cdots+F_n=F_{n+2}-1$

24. $F_1^2+F_2^2+F_3^2+\cdots+F_n^2=F_nF_{n+1}$

25. If $a_{n+2}=a_{n+1}\cdot a_n$ and $a_1=a_2=2$, then $a_n=2^{F_n}$ for all natural numbers n.

26. For all $n\ge 2$,
$$\begin{bmatrix}1&1\\1&0\end{bmatrix}^n=\begin{bmatrix}F_{n+1}&F_n\\F_n&F_{n-1}\end{bmatrix}$$

27. Let a_n be the nth term of the sequence defined recursively by
$$a_{n+1}=\frac{1}{1+a_n}$$
and $a_1=1$. Find a formula for a_n in terms of the Fibonacci numbers F_n. Prove that the formula you found is valid for all natural numbers n.

28. Let F_n be the nth term of the Fibonacci sequence. Find and prove an inequality relating n and F_n for natural numbers n.

29. Find and prove an inequality relating $100n$ and n^3.

30. What is wrong with the following "proof" by mathematical induction that all girls have blond hair? Let $P(n)$ denote the statement: In any group of n girls, if one of them has blond hair, then they all do.

Step 1 The statement is clearly true for $n=1$.

Step 2 Suppose that $P(k)$ is true. We show that $P(k+1)$ is true. Consider a group of $k+1$ girls, one of whom has blond hair; we call her *Ex*. Remove a girl, call her *Es*, from the group. Now we have a group of k girls, one of whom has blond hair, and by our induction hypothesis all k girls have blond hair. Now put *Es* back in the group and remove another girl. Again by our induction hypothesis all k girls in this group have blond hair, and so *Es* also has blond hair. It follows that all $k+1$ girls in the group have blond hair and this completes the induction step.

Since everyone knows at least one girl with blond hair, it follows that all girls have blond hair.

SECTION 10.8
THE BINOMIAL THEOREM

An expression of the form $a + b$ is called a **binomial**. Although in principle it is easy to raise $a + b$ to any power, raising it to very high powers would be a tedious task. In this section we find a formula that gives the expansion of $(a + b)^n$ for any natural number n.

We discover this formula by finding a pattern for the successive powers of $a + b$. So let us look at some special cases:

$$(a + b)^1 = a + b$$
$$(a + b)^2 = a^2 + 2ab + b^2$$
$$(a + b)^3 = a^3 + 3a^2b + 3ab^2 + b^3$$
$$(a + b)^4 = a^4 + 4a^3b + 6a^2b^2 + 4ab^3 + b^4$$
$$(a + b)^5 = a^5 + 5a^4b + 10a^3b^2 + 10a^2b^3 + 5ab^4 + b^5$$
$$\vdots$$

The following simple patterns emerge for the expansion of $(a + b)^n$:

1. There are $n + 1$ terms, the first being a^n and the last b^n.
2. The exponents of a decrease by 1 from term to term while the exponents of b increase by 1.
3. The sum of the exponents of a and b in each term is n.

For instance, notice how the exponents of a and b behave in the expansion of $(a + b)^5$.

The exponents of a decrease:

$$(a + b)^5 = a^{⑤} + 5a^{④}b^1 + 10a^{③}b^2 + 10a^{②}b^3 + 5a^{①}b^4 + b^5$$

The exponents of b increase:

$$(a + b)^5 = a^5 + 5a^4b^{①} + 10a^3b^{②} + 10a^2b^{③} + 5a^1b^{④} + b^{⑤}$$

With these observations we can write the form of the expansion of $(a + b)^n$ for any natural number n. For example, putting a question mark for the missing coefficients, we have

$$(a + b)^8 = a^8 + ?a^7b + ?a^6b^2 + ?a^5b^3 + ?a^4b^4 + ?a^3b^5 + ?a^2b^6 + ?ab^7 + b^8$$

To complete the expansion we need to determine these coefficients. To find a pattern, let us write the coefficients in the expansion of $(a + b)^n$ for the first few values of n in a triangular array as shown below. This array is called **Pascal's triangle**.

$$
\begin{array}{ll}
(a+b)^0 & \qquad\qquad 1 \\
(a+b)^1 & \qquad\quad 1 \quad 1 \\
(a+b)^2 & \qquad 1 \quad 2 \quad 1 \\
(a+b)^3 & \quad 1 \quad 3 \quad 3 \quad 1 \\
(a+b)^4 & 1 \quad 4 \quad 6 \quad 4 \quad 1 \\
(a+b)^5 & 1 \quad 5 \quad 10 \quad 10 \quad 5 \quad 1
\end{array}
$$

The row corresponding to $(a + b)^0$ is called the zeroth row and is added for the purposes of symmetry. The key observation about Pascal's triangle is the following:

KEY PROPERTY OF PASCAL'S TRIANGLE

Every entry (other than a 1) is the sum of the two entries diagonally above it.

From this property it is easy to find any row of Pascal's triangle from the row above it. For instance we find the sixth and seventh rows starting with the fifth row:

$$
\begin{array}{ll}
(a+b)^5 & \quad 1 \quad 5 \quad 10 \quad 10 \quad 5 \quad 1 \\
(a+b)^6 & 1 \quad 6 \quad 15 \quad 20 \quad 15 \quad 6 \quad 1 \\
(a+b)^7 & 1 \quad 7 \quad 21 \quad 35 \quad 35 \quad 21 \quad 7 \quad 1
\end{array}
$$

To see why this property holds let us consider the following expansions:

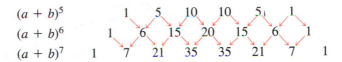

$$(a+b)^5 = \qquad a^5 + 5a^4b + 10a^3b^2 + 10a^2b^3 + 5ab^4 + b^5$$

$$(a+b)^6 = a^6 + 6a^5b + 15a^4b^2 + 20a^3b^3 + 15a^2b^4 + 6ab^5 + b^6$$

We arrive at the expansion of $(a + b)^6$ by multiplying $(a + b)^5$ by $(a + b)$. Now notice, for instance, that the circled term in the expansion of $(a + b)^6$ is obtained via this multiplication from the two circled terms above it. (In fact we get this term when the two terms above it are multiplied by b and a, respectively.) Thus its coefficient is the sum of the coefficients of these two terms. This is the observation we will use at the end of this section in proving the Binomial Theorem.

Having found these patterns, we can now easily obtain the expansion of any binomial, at least to relatively small powers.

EXAMPLE 1

Find the expansion of $(a + b)^7$ using Pascal's triangle.

SOLUTION

The first term in the expansion is a^7 and the last term is b^7. Using the fact that the exponent of a decreases by 1 from term to term and that of b increases by 1 from term to term, we have

$$(a + b)^7 = a^7 + ?a^6b + ?a^5b^2 + ?a^4b^3 + ?a^3b^4 + ?a^2b^5 + ?ab^6 + b^7$$

The appropriate coefficients appear in the seventh row of Pascal's triangle. Thus

$$(a + b)^7 = a^7 + 7a^6b + 21a^5b^2 + 35a^4b^3 + 35a^3b^4 + 21a^2b^5 + 7ab^6 + b^7$$

EXAMPLE 2

Use Pascal's triangle to expand $(2 - 3x)^5$.

SOLUTION

We find the expansion of $(a + b)^5$ and then substitute 2 for a and $-3x$ for b. Using Pascal's triangle for the coefficients, we get

$$(a + b)^5 = a^5 + 5a^4b + 10a^3b^2 + 10a^2b^3 + 5ab^4 + b^5$$

Substituting $a = 2$ and $b = -3x$ gives

$$(2 - 3x)^5 = (2)^5 + 5(2)^4(-3x) + 10(2)^3(-3x)^2 + 10(2)^2(-3x)^3 + 5(2)(-3x)^4 + (-3x)^5$$
$$= 32 - 240x + 720x^2 - 1080x^3 + 810x^4 - 243x^5$$

Although Pascal's triangle is useful in finding the binomial expansion for reasonably small values of n, it is not practical to use it for finding $(a + b)^n$ for large values of n. The reason is that the method we use for finding the successive rows of Pascal's triangle is recursive. Thus to find the 100th row of this triangle we must first find all the preceding rows.

We need to examine the pattern in the coefficients more carefully to get a formula that will allow us to calculate directly any coefficient in the binomial expansion. Such a formula exists, and the rest of this section is devoted to finding and proving it. However, to state this formula we need some notation, which we discuss next.

THE BINOMIAL COEFFICIENTS AND PASCAL'S TRIANGLE

The product of the first n natural numbers is called n **factorial** and is denoted by $n!$. Thus

$$n! = 1 \cdot 2 \cdot 3 \cdot \cdots \cdot (n - 1)n$$

For example,

$$3! = 1 \cdot 2 \cdot 3 = 6$$

$$6! = 1 \cdot 2 \cdot 3 \cdot 4 \cdot 5 \cdot 6 = 720$$

$$11! = 1 \cdot 2 \cdot 3 \cdot 4 \cdot 5 \cdot 6 \cdot 7 \cdot 8 \cdot 9 \cdot 10 \cdot 11 = 39{,}916{,}800$$

We also define

$$0! = 1$$

This definition of 0! makes many formulas involving factorials shorter and simpler to write (the binomial formula is a notable example here). We now use factorials to define the binomial coefficients.

THE BINOMIAL COEFFICIENTS

Let n and r be nonnegative integers with $r \leq n$. The **binomial coefficient** is denoted by $\binom{n}{r}$ and is defined by

$$\binom{n}{r} = \frac{n!}{r!(n-r)!}$$

EXAMPLE 3

(a) $\binom{9}{4} = \dfrac{9!}{4!(9-4)!} = \dfrac{9!}{4!5!} = \dfrac{1 \cdot 2 \cdot 3 \cdot 4 \cdot 5 \cdot 6 \cdot 7 \cdot 8 \cdot 9}{(1 \cdot 2 \cdot 3 \cdot 4)(1 \cdot 2 \cdot 3 \cdot 4 \cdot 5)}$

$$= \frac{6 \cdot 7 \cdot 8 \cdot 9}{1 \cdot 2 \cdot 3 \cdot 4} = 126$$

(b) $\binom{100}{3} = \dfrac{100!}{3!(100-3)!} = \dfrac{1 \cdot 2 \cdot 3 \cdot \cdots \cdot 97 \cdot 98 \cdot 99 \cdot 100}{(1 \cdot 2 \cdot 3)(1 \cdot 2 \cdot 3 \cdot \cdots \cdot 97)}$

$$= \frac{98 \cdot 99 \cdot 100}{1 \cdot 2 \cdot 3} = 161{,}700$$

(c) $\binom{100}{97} = \dfrac{100!}{97!(100-97)!} = \dfrac{1 \cdot 2 \cdot 3 \cdot \cdots \cdot 97 \cdot 98 \cdot 99 \cdot 100}{(1 \cdot 2 \cdot 3 \cdot \cdots \cdot 97)(1 \cdot 2 \cdot 3)}$

$$= \frac{98 \cdot 99 \cdot 100}{1 \cdot 2 \cdot 3} = 161{,}700$$

Although the binomial coefficient $\binom{n}{r}$ is defined in terms of a fraction, all the results of this example are natural numbers. In fact, the binomial coefficient $\binom{n}{r}$ is always a natural number (see Exercise 58). Notice that the binomial coefficients in parts (b) and (c) of Example 3 are equal. This is a special case of the relation

$$\binom{n}{r} = \binom{n}{n-r}$$

which you are asked to prove in Exercise 54.

To see the connection between the binomial coefficients and the binomial expansion of $(a + b)^n$, let us calculate the following binomial coefficients:

$$\binom{5}{2} = \frac{5!}{2!(5-2)!} = 10$$

$$\binom{5}{0} = 1 \qquad \binom{5}{1} = 5 \qquad \binom{5}{2} = 10 \qquad \binom{5}{3} = 10 \qquad \binom{5}{4} = 5 \qquad \binom{5}{5} = 1$$

Notice that these are precisely the entries in the fifth row of Pascal's triangle. In fact it is true that, for all natural numbers n, the nth row of Pascal's triangle has the entries

$$\binom{n}{0} \qquad \binom{n}{1} \qquad \binom{n}{2} \quad \cdots \quad \binom{n}{n-1} \qquad \binom{n}{n}$$

So we can write Pascal's triangle as follows:

$$\binom{0}{0}$$

$$\binom{1}{0} \qquad \binom{1}{1}$$

$$\binom{2}{0} \qquad \binom{2}{1} \qquad \binom{2}{2}$$

$$\binom{3}{0} \qquad \binom{3}{1} \qquad \binom{3}{2} \qquad \binom{3}{3}$$

$$\binom{4}{0} \qquad \binom{4}{1} \qquad \binom{4}{2} \qquad \binom{4}{3} \qquad \binom{4}{4}$$

$$\binom{5}{0} \qquad \binom{5}{1} \qquad \binom{5}{2} \qquad \binom{5}{3} \qquad \binom{5}{4} \qquad \binom{5}{5}$$

$$\binom{n}{0} \qquad \binom{n}{1} \qquad \binom{n}{2} \qquad \cdots \qquad \binom{n}{n-1} \qquad \binom{n}{n}$$

Pascal's triangle appears in this Chinese document by Chu Shi-kie dated 1303. The title reads "The Old Method Chart of the Seven Multiplying Squares." The triangle was rediscovered by Pascal (see p. 690).

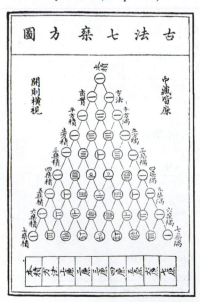

To show that this pattern actually holds we need to show that any entry in this version of Pascal's triangle is the sum of the two entries diagonally above it. In other words, we need to show that the entries satisfy the key property of Pascal's triangle. Stated in terms of the binomial coefficients, this property says:

KEY PROPERTY OF THE BINOMIAL COEFFICIENTS

For any nonnegative integers r and k with $r \leq k$,

$$\binom{k}{r-1} + \binom{k}{r} = \binom{k+1}{r} \qquad (1)$$

Notice that the two terms on the left side of this equation are adjacent entries in the kth row of Pascal's triangle, and the term on the right side is the entry diagonally below them in the $(k + 1)$st row. Thus this equation is a restatement of the key property of Pascal's triangle in terms of the binomial coefficients. A proof of this identity is outlined in Exercise 57.

■ THE BINOMIAL THEOREM

We are now ready to state the Binomial Theorem.

THE BINOMIAL THEOREM

$$(a + b)^n = \binom{n}{0}a^n + \binom{n}{1}a^{n-1}b + \binom{n}{2}a^{n-2}b^2 + \cdots + \binom{n}{n-1}ab^{n-1} + \binom{n}{n}b^n$$

We prove this theorem at the end of this section. First we show some of its applications.

EXAMPLE 4

Use the Binomial Theorem to expand $(x + y)^4$.

SOLUTION

By the Binomial Theorem,

$$(x + y)^4 = \binom{4}{0}x^4 + \binom{4}{1}x^3y + \binom{4}{2}x^2y^2 + \binom{4}{3}xy^3 + \binom{4}{4}y^4$$

Verify that

$$\binom{4}{0} = 1 \qquad \binom{4}{1} = 4 \qquad \binom{4}{2} = 6 \qquad \binom{4}{3} = 4 \qquad \binom{4}{4} = 1$$

It follows that

$$(x + y)^4 = x^4 + 4x^3y + 6x^2y^2 + 4xy^3 + y^4$$

■

EXAMPLE 5

Use the Binomial Theorem to expand $\left(\sqrt{x} - 1\right)^8$.

SOLUTION

We first find the expansion of $(a + b)^8$ and then substitute $\sqrt{x}$ for a and -1 for b. Using the Binomial Theorem we have

$$(a + b)^8 = \binom{8}{0}a^8 + \binom{8}{1}a^7b + \binom{8}{2}a^6b^2 + \binom{8}{3}a^5b^3 + \binom{8}{4}a^4b^4$$
$$+ \binom{8}{5}a^3b^5 + \binom{8}{6}a^2b^6 + \binom{8}{7}ab^7 + \binom{8}{8}b^8$$

Verify that

$$\binom{8}{0} = 1 \quad \binom{8}{1} = 8 \quad \binom{8}{2} = 28 \quad \binom{8}{3} = 56 \quad \binom{8}{4} = 70$$
$$\binom{8}{5} = 56 \quad \binom{8}{6} = 28 \quad \binom{8}{7} = 8 \quad \binom{8}{8} = 1$$

So

$$(a + b)^8 = a^8 + 8a^7b + 28a^6b^2 + 56a^5b^3 + 70a^4b^4 + 56a^3b^5$$
$$+ 28a^2b^6 + 8ab^7 + b^8$$

Performing the substitutions $a = x^{1/2}$ and $b = -1$ gives

$$\left(\sqrt{x} - 1\right)^8 = (x^{1/2})^8 + 8(x^{1/2})^7(-1) + 28(x^{1/2})^6(-1)^2 + 56(x^{1/2})^5(-1)^3$$
$$+ 70(x^{1/2})^4(-1)^4 + 56(x^{1/2})^3(-1)^5 + 28(x^{1/2})^2(-1)^6$$
$$+ 8(x^{1/2})(-1)^7 + (-1)^8$$

This simplifies to

$$\left(\sqrt{x} - 1\right)^8 = x^4 - 8x^{7/2} + 28x^3 - 56x^{5/2} + 70x^2 - 56x^{3/2} + 28x - 8x^{1/2} + 1$$
■

The Binomial Theorem can be used to find particular terms of a binomial expansion without having to find the entire expansion.

GENERAL TERM OF THE BINOMIAL EXPANSION

The term that contains a^r in the expansion of $(a + b)^n$ is

$$\binom{n}{n - r}a^rb^{n-r}$$

EXAMPLE 6

Find the term that contains x^5 in the expansion of $(2x + y)^{20}$.

SOLUTION

The term that contains x^5 is given by the formula for the general term with $a = 2x$, $b = y$, $n = 20$, and $r = 5$. So this term is

$$\binom{20}{15}a^5 b^{15} = \frac{20!}{15!(20-15)!}(2x)^5 y^{15} = \frac{20!}{15!\,5!}32x^5 y^{15} = 496{,}128x^5 y^{15} \qquad \blacksquare$$

EXAMPLE 7

Find the coefficient of x^8 in the expansion of $\left(x^2 + \dfrac{1}{x}\right)^{10}$.

SOLUTION

Notice that both x^2 and $1/x$ are powers of x. So the power of x in each term of the expansion is determined by both terms of the binomial. To find the required coefficient we first find the general term in the expansion. By the formula for the general term with $a = x^2$, $b = 1/x$, and $n = 10$, the general term is

$$\binom{10}{10-r}(x^2)^r\left(\frac{1}{x}\right)^{10-r} = \binom{10}{10-r}(x^2)^r(x^{-1})^{10-r} = \binom{10}{10-r}x^{3r-10}$$

Thus the term that contains x^8 is the term where

$$3r - 10 = 8$$

or

$$r = 6$$

So the required coefficient is that of the sixth term and is given by

$$\binom{10}{10-4} = \binom{10}{6} = 210 \qquad \blacksquare$$

PROOF OF THE BINOMIAL THEOREM

We now give a proof of the Binomial Theorem using mathematical induction.

Proof   Let $P(n)$ denote the statement

$$(a + b)^n = \binom{n}{0}a^n + \binom{n}{1}a^{n-1}b + \binom{n}{2}a^{n-2}b^2 + \cdots + \binom{n}{n-1}ab^{n-1} + \binom{n}{n}b^n$$

Step 1 We show that $P(1)$ is true. But $P(1)$ is just the statement

$$(a + b)^1 = \binom{1}{0}a^1 + \binom{1}{1}b^1 = 1a + 1b = a + b$$

which is certainly true.

Step 2 We assume that $P(k)$ is true and show that $P(k + 1)$ is true. The statement $P(k)$ reads

$$(a + b)^k = \binom{k}{0}a^k + \binom{k}{1}a^{k-1}b + \binom{k}{2}a^{k-2}b^2 + \cdots + \binom{k}{k-1}ab^{k-1} + \binom{k}{k}b^k$$

Multiplying both sides of this equation by $(a + b)$ and collecting like terms gives:

$$(a + b)^{k+1} = (a + b)\left[\binom{k}{0}a^k + \binom{k}{1}a^{k-1}b + \binom{k}{2}a^{k-2}b^2 + \cdots + \binom{k}{k-1}ab^{k-1} + \binom{k}{k}b^k\right]$$

$$= a\left[\binom{k}{0}a^k + \binom{k}{1}a^{k-1}b + \binom{k}{2}a^{k-2}b^2 + \cdots + \binom{k}{k-1}ab^{k-1} + \binom{k}{k}b^k\right]$$

$$+ b\left[\binom{k}{0}a^k + \binom{k}{1}a^{k-1}b + \binom{k}{2}a^{k-2}b^2 + \cdots + \binom{k}{k-1}ab^{k-1} + \binom{k}{k}b^k\right]$$

$$= \binom{k}{0}a^{k+1} + \binom{k}{1}a^k b + \binom{k}{2}a^{k-1}b^2 + \cdots + \binom{k}{k-1}a^2 b^{k-1} + \binom{k}{k}ab^k$$

$$+ \binom{k}{0}a^k b + \binom{k}{1}a^{k-1}b^2 + \binom{k}{2}a^{k-2}b^3 + \cdots + \binom{k}{k-1}ab^k + \binom{k}{k}b^{k+1}$$

$$= \binom{k}{0}a^{k+1} + \left[\binom{k}{0} + \binom{k}{1}\right]a^k b + \left[\binom{k}{1} + \binom{k}{2}\right]a^{k-1}b^2 + \cdots + \left[\binom{k}{k-1} + \binom{k}{k}\right]ab^k + \binom{k}{k}b^{k+1}$$

Using the key property of the binomial coefficients, we can write each of the expressions in square brackets as a single binomial coefficient. Also, writing the first and last coefficients as $\binom{k+1}{0}$ and $\binom{k+1}{k+1}$ (these are both equal to 1 by Exercise 52) gives

$$(a + b)^{k+1} = \binom{k+1}{0}a^{k+1} + \binom{k+1}{1}a^k b + \binom{k+1}{2}a^{k-1}b^2 + \cdots + \binom{k+1}{k}ab^k + \binom{k+1}{k+1}b^{k+1}$$

But this last equation is precisely $P(k + 1)$ and this completes the induction step.

Having proved steps 1 and 2, we conclude by the Principle of Mathematical Induction that the theorem is true for all natural numbers n. ∎

EXERCISES 10.8

In Exercises 1–12 use Pascal's triangle to expand the expression.

1. $(x + y)^7$ **2.** $(2x + 1)^4$ **3.** $\left(x + \dfrac{1}{x}\right)^4$

4. $(x - y)^5$ **5.** $\left(3 - \sqrt{3}\right)^5$ **6.** $\left(\sqrt{a} + \sqrt{b}\right)^6$

7. $(x^2y - 1)^5$ **8.** $\left(1 + \sqrt{2}\right)^6$ **9.** $(2x - 3y)^3$

10. $(1 + x^3)^3$ **11.** $\left(\dfrac{1}{x} - \sqrt{x}\right)^5$ **12.** $\left(2 + \dfrac{x}{2}\right)^5$

In Exercises 13–18 evaluate the given expression.

13. $\dbinom{5}{2}$ **14.** $\dbinom{10}{6}$ **15.** $\dbinom{100}{98}$ **16.** $\dbinom{10}{5}$

17. $\dbinom{5}{0} + \dbinom{5}{1} + \dbinom{5}{2} + \dbinom{5}{3} + \dbinom{5}{4} + \dbinom{5}{5}$

18. $\dbinom{5}{0} - \dbinom{5}{1} + \dbinom{5}{2} - \dbinom{5}{3} + \dbinom{5}{4} - \dbinom{5}{5}$

In Exercises 19–22 use the Binomial Theorem to expand the given expression.

19. $(x + 2y)^4$ **20.** $(1 - x)^5$

21. $\left(1 + \dfrac{1}{x}\right)^6$ **22.** $(2A + B^2)^4$

23. Find the first three terms in the expansion of $(x + 2y)^{20}$.

24. Find the first four terms in the expansion of $(x^{1/2} + 1)^{30}$.

25. Find the last two terms in the expansion of $(a^{2/3} + a^{1/3})^{25}$.

26. Find the first three terms in the expansion of $\left(x + \dfrac{1}{x}\right)^{40}$.

27. Find the middle term in the expansion of $(x^2 + 1)^{18}$.

28. Find the fifth term in the expansion of $(ab - 1)^{20}$.

29. Find the 24th term in the expansion of $(a + b)^{25}$.

30. Find the 28th term in the expansion of $(A - B)^{30}$.

31. Find the 100th term in the expansion of $(1 + y)^{100}$.

32. Find the second term in the expansion of $\left(x^2 - \dfrac{1}{x}\right)^{25}$.

33. Find the term containing x^4 in the expansion of $(x + 2y)^{10}$.

34. Find the term containing y^3 in the expansion of $\left(\sqrt{2} + y\right)^{12}$.

35. Find the term containing b^8 in the expansion of $(a + b^2)^{12}$.

36. Find the term containing a in the expansion of $\left(\sqrt{a} + \dfrac{1}{\sqrt{a}}\right)^{10}$.

37. Find the term that does not contain x in the expansion of $\left(8x + \dfrac{1}{2x}\right)^8$.

38. Find the term that does not contain m in the expansion of $(mn + m^{-4})^5$.

39. Find the term containing c^7 in the expansion of $\left(2c + \sqrt{c}\right)^8$.

40. Find the coefficient of r^{-5} in the expansion of
$$\left(\dfrac{r^2}{4} - \dfrac{4}{r^3}\right)^5$$

In Exercises 41–44 find the number of distinct terms in the expansion of the given expression.

41. $\left(x + \dfrac{1}{x}\right)^{20}$ **42.** $\left(x^2 + \dfrac{1}{x}\right)^6$

43. $(a^2 - 2ab + b^2)^5$

44. $[(x + y)^2(x - y)^2]^3$

In Exercises 45–48 simplify the given expression.

45. $x^4 + 4x^3y + 6x^2y^2 + 4xy^3 + y^4$

46. $(x - 1)^5 + 5(x - 1)^4 + 10(x - 1)^3 + 10(x - 1)^2 + 5(x - 1) + 1$

47. $8a^3 + 12a^2b + 6ab^2 + b^3$

48. $x^8 + 4x^6y + 6x^4y^2 + 4x^2y^3 + y^4$

49. Expand $(a^2 + a + 1)^4$.
[*Hint:* $a^2 + a + 1 = a^2 + (a + 1)$]

50. Show that $(1.01)^{100} > 2$.
[*Hint:* Note that $(1.01)^{100} = (1 + 0.01)^{100}$ and use the Binomial Theorem to show that the sum of the first two terms of the expansion is greater than 2.]

51. Which is larger: $(100!)^{101}$ or $(101!)^{100}$?

52. Show that $\binom{n}{0} = 1$ and $\binom{n}{n} = 1$.

53. Show that $\binom{n}{1} = \binom{n}{n-1} = n$.

54. Show that $\binom{n}{r} = \binom{n}{n-r}$ for $0 \le r \le n$.

55. (a) Show that $\binom{n}{0} + \binom{n}{1} + \binom{n}{2} + \cdots + \binom{n}{n} = 2^n$.

[*Hint:* $2^n = (1 + 1)^n$]

(b) Give a counting argument that explains the equality in part (a).

56. Show that $\binom{n}{0} - \binom{n}{1} + \binom{n}{2} - \cdots + (-1)^k\binom{n}{k} + $

$\cdots + (-1)^n\binom{n}{n} = 0$.   [*Hint:* $0 = 1 - 1$]

57. In this exercise we prove the identity

$$\binom{n}{r-1} + \binom{n}{r} = \binom{n+1}{r}$$

(a) Write out the left side of this equation as the sum of two fractions.

(b) Show that a common denominator of the expression you found in part (a) is $r!(n - r + 1)!$.

(c) Add the two fractions using the common denominator in part (b), simplify the numerator, and notice that the resulting expression is equal to the right side of the equation.

58. Prove that $\binom{n}{r}$ is an integer for all n and for $0 \le r \le n$.

[*Suggestion:* Use induction to show that the statement is true for all n, and use Exercise 57 for the induction step.]

CHAPTER 10 REVIEW

Define, state, or discuss each of the following.

1. Sequence
2. Recursive sequence
3. Fibonacci sequence
4. Arithmetic sequence
5. Common difference
6. Geometric sequence
7. Common ratio
8. Sigma notation
9. Series
10. Arithmetic series
11. Sum of an arithmetic series
12. Geometric series
13. Sum of a geometric series
14. Annuity
15. Infinite series
16. Sum of an infinite geometric series
17. Mathematical induction
18. Pascal's triangle
19. Key property of Pascal's triangle
20. Binomial coefficients
21. The Binomial Theorem
22. General term of the binomial expansion

REVIEW EXERCISES

In Exercises 1–6 find the first four terms as well as the tenth term of the sequence with the given nth term.

1. $a_n = \dfrac{n^2}{n+1}$ **2.** $a_n = (-1)^n\dfrac{2^n}{n}$ **3.** $a_n = \dfrac{(-1)^n + 1}{n^3}$ **4.** $a_n = \dfrac{n(n+1)}{2}$ **5.** $a_n = \dfrac{(2n)!}{2^n n!}$ **6.** $a_n = \binom{n+1}{2}$

In Exercises 7–12 a sequence is defined recursively. Find the first seven terms of the sequence.

7. $a_n = a_{n-1} + 2n - 1, \quad a_1 = 1$

8. $a_n = \dfrac{a_{n-1}}{n}, \quad a_1 = 1$

9. $a_n = a_{n-1} + 2a_{n-2}, \quad a_1 = 1, a_2 = 3$

10. $a_n = \sqrt{3a_{n-1}}, \quad a_1 = \sqrt{3}$

11. $a_n = (a_{n-1} - 1)!, \quad a_1 = 3$

12. $a_n = \dbinom{n+1}{a_{n-1}}, \quad a_1 = 1$

In Exercises 13–24 the first four terms of a sequence are given. In each case determine whether the given terms can be the terms of an arithmetic sequence, a geometric sequence, or neither. For those sequences that are arithmetic or geometric, find the fifth term.

13. $5, 5.5, 6, 6.5, \ldots$

14. $1, -\frac{3}{2}, 2, -\frac{5}{2}, \ldots$

15. $\sqrt{2}, 2\sqrt{2}, 3\sqrt{2}, 4\sqrt{2}, \ldots$

16. $\sqrt{2}, 2, 2\sqrt{2}, 4, \ldots$

17. $t - 3, t - 2, t - 1, t, \ldots$

18. $t^3, t^2, t, 1, \ldots$

19. $\dfrac{3}{4}, \dfrac{1}{2}, \dfrac{1}{3}, \dfrac{2}{9}, \ldots$

20. $\dfrac{a}{c}, 1, \dfrac{c}{a}, \left(\dfrac{c}{a}\right)^2, \ldots$

21. $\ln a, \ln 2a, \ln 3a, \ln 4a, \ldots$

22. $a, 1, \dfrac{1}{a}, \dfrac{1}{a^2}, \ldots$

23. $a, abc^3, ab^2c^6, ab^3c^9, \ldots$

24. $a, a + b^2, a + 2b^2, a + 3b^2, \ldots$

25. Show that $3, 6i, -12, -24i, \ldots$ is a geometric sequence and find the common ratio.

26. Find the nth term of the geometric sequence $2, 2 + 2i, 4i, -4 + 4i, -8, \ldots$

27. The sixth term of an arithmetic sequence is 17 and the fourth term is 11. Find the second term.

28. The 20th term of an arithmetic sequence is 96 and the common difference is 5. Find the nth term.

29. The third term of a geometric sequence is 9 and the common ratio is $\frac{3}{2}$. Find the fifth term.

30. The second term of a geometric sequence is 10 and the fifth term is $\frac{1250}{27}$. Find the nth term.

31. The frequencies of musical notes measured in cycles per second form a geometric sequence. Middle C has a frequency of 256 and C an octave higher has a frequency of 512. Find the frequency of C two octaves below middle C.

32. A person has two parents, four grandparents, eight great-grandparents, and so on. How many ancestors does a person have 15 generations back?

33. A certain type of bacteria divides every five seconds. If a petri dish contains three of these bacteria, how many bacteria are in the dish at the end of one minute?

34. If $a_1, a_2, a_3, \ldots$ and $b_1, b_2, b_3, \ldots$ are arithmetic sequences, show that $a_1 + b_1, a_2 + b_2, a_3 + b_3, \ldots$ is also an arithmetic sequence.

35. If $a_1, a_2, a_3, \ldots$ and $b_1, b_2, b_3, \ldots$ are geometric sequences, show that $a_1b_1, a_2b_2, a_3b_3, \ldots$ is also a geometric sequence.

36. (a) If $a_1, a_2, a_3, \ldots$ is an arithmetic sequence, is the sequence $a_1 + 2, a_2 + 2, a_3 + 2, \ldots$ arithmetic?
(b) If $a_1, a_2, a_3, \ldots$ is a geometric sequence, is the sequence $5a_1, 5a_2, 5a_3, \ldots$ geometric?

37. Find the values of x for which the sequence $6, x, 12, \ldots$ is
(a) arithmetic   (b) geometric

38. Find the values of x and y for which the sequence $2, x, y, 17, \ldots$ is
(a) arithmetic   (b) geometric

In Exercises 39–42 find the indicated sum.

39. $\displaystyle\sum_{k=3}^{6} (k + 1)^2$

40. $\displaystyle\sum_{i=1}^{4} \dfrac{2i}{2i - 1}$

41. $\displaystyle\sum_{k=1}^{6} (k + 1)2^{k-1}$

42. $\displaystyle\sum_{m=1}^{5} 3^{m-2}$

In Exercises 43–46 write the indicated sum without using sigma notation. Do not evaluate.

43. $\displaystyle\sum_{k=1}^{10} (k - 1)^2$

44. $\displaystyle\sum_{j=2}^{100} \dfrac{1}{j - 1}$

45. $\displaystyle\sum_{k=1}^{50} \dfrac{3^k}{2^{k+1}}$

46. $\displaystyle\sum_{n=1}^{10} n^2 2^n$

In Exercises 47–50 write the given sum using sigma notation. Do not evaluate.

47. $3 + 6 + 9 + 12 + \cdots + 99$

48. $1^2 + 2^2 + 3^2 + \cdots + 100^2$

49. $1 \cdot 2^3 + 2 \cdot 2^4 + 3 \cdot 2^5 + 4 \cdot 2^6 + \cdots + 100 \cdot 2^{102}$

50. $\dfrac{1}{1 \cdot 2} + \dfrac{1}{2 \cdot 3} + \dfrac{1}{3 \cdot 4} + \cdots + \dfrac{1}{999 \cdot 1000}$

In Exercises 51–58 determine whether the series is arithmetic or geometric and find its sum.

51. $1 + 0.9 + (0.9)^2 + \cdots + (0.9)^5$

52. $3 + 3.7 + 4.4 + \cdots + 10$

53. $1 - \sqrt{5} + 5 - 5\sqrt{5} + \cdots + 625$

54. $\sqrt{5} + 2\sqrt{5} + 3\sqrt{5} + \cdots + 100\sqrt{5}$

55. $\frac{1}{3} + \frac{2}{3} + 1 + \frac{4}{3} + \cdots + 33$

56. $a + abc^3 + ab^2c^6 + ab^3c^9 + \cdots + ab^8c^{24}$

57. $\displaystyle\sum_{n=0}^{6} 3(-4)^n$ **58.** $\displaystyle\sum_{k=0}^{8} 7(5)^{k/2}$

59. The first term of an arithmetic sequence is $a = 7$ and the common difference is $d = 3$. How many terms of this sequence must be added to get 325?

60. A geometric sequence has first term $a = 81$ and common ratio $r = -\frac{2}{3}$. How many terms of this sequence must be added to get 55?

61. The sum of the first eight terms of an arithmetic series is 100 and the first term is 2. Find the tenth term.

62. The sum of the first three terms of a geometric series is 52 and the common ratio is $r = 3$. Find the first term.

63. A city has a population of 100,000. If the population is increasing at the rate of 10% per year, what will be the population of this city in 10 years? Find a formula for the population of the city after n years.

64. Refer to Exercise 32. What is the total number of ancestors of a person in 15 generations?

65. Find the amount of an annuity consisting of 16 annual payments of $1000 each into an account that pays 8% interest per year compounded annually.

66. How much money should be invested every quarter at 12% per year compounded quarterly in order to have $10,000 in one year?

67. What are the monthly payments on a mortgage of $60,000 at 9% interest if (a) the loan is to be repaid in 30 years? (b) the loan is to be repaid in 15 years?

In Exercises 68–71 find the sum of the given infinite geometric series.

68. $1 - \dfrac{2}{5} + \dfrac{4}{25} - \dfrac{8}{125} + \cdots$

69. $0.1 + 0.01 + 0.001 + 0.0001 + \cdots$

70. $1 + \dfrac{1}{3^{1/2}} + \dfrac{1}{3} + \dfrac{1}{3^{3/2}} + \cdots$

71. $a + ab^2 + ab^4 + ab^6 + \cdots$

In Exercises 72–75 use mathematical induction to prove that the given formula is true for all natural numbers n.

72. $1 + 4 + 7 + \cdots + (3n - 2) = \dfrac{n(3n - 1)}{2}$

73. $1^4 + 2^4 + 3^4 + \cdots + n^4$
$$= \dfrac{n(n + 1)(2n + 1)(3n^2 + 3n - 1)}{30}$$

74. $\dfrac{1}{1 \cdot 3} + \dfrac{1}{3 \cdot 5} + \dfrac{1}{5 \cdot 7} + \cdots + \dfrac{1}{(2n - 1)(2n + 1)}$
$$= \dfrac{n}{2n + 1}$$

75. $\left(1 + \dfrac{1}{1}\right)\left(1 + \dfrac{1}{2}\right)\left(1 + \dfrac{1}{3}\right) \cdots \left(1 + \dfrac{1}{n}\right) = n + 1$

76. Show that $7^n - 1$ is divisible by 6 for all natural numbers n.

77. Show that $11^{n+2} + 12^{2n+1}$ is divisible by 133 for every natural number n.

78. Let $a_{n+1} = 3a_n + 4$ and $a_1 = 4$. Show that $a_n = 2 \cdot 3^n - 2$ for all natural numbers n.

79. Prove that the Fibonacci number F_{4n} is divisible by 3 for all natural numbers n.

80. Find and prove an inequality that relates 2^n and $n!$.

In Exercises 81–84 evaluate the given expression.

81. $\dbinom{5}{2}\dbinom{5}{3}$ **82.** $\dbinom{10}{2} + \dbinom{10}{6}$

83. $\displaystyle\sum_{k=0}^{5} \dbinom{5}{k}$ **84.** $\displaystyle\sum_{k=0}^{8} \dbinom{8}{k}\dbinom{8}{8 - k}$

In Exercises 85 and 86 expand the given expression.

85. $(1 - x^2)^6$ **86.** $(2x + y)^4$

87. Find the 20th term in the expansion of $(a + b)^{22}$.

88. Find the first three terms in the expansion of $(b^{-2/3} + b^{1/3})^{20}$.

89. Find the coefficient of s^5 in the expansion of

$$\left(\frac{s^3}{2} - \frac{2}{s^2}\right)^5$$

90. Find the term containing A^6 in the expansion of $(A + 3B)^{10}$.

In Exercises 91 and 92 simplify the given expression.

91. $(a + 1)^3 - 3(a + 1)^2 + 3(a + 1) - 1$

92. $x^5 + 5x^4y + 10x^3y^2 + 10x^2y^3 + 5xy^4 + y^5$

CHAPTER 10 TEST

1. Find the tenth term of the sequence whose nth term is

$$a_n = \frac{n}{1 - n^2}$$

2. A sequence is defined recursively by

$$a_{n+2} = (a_n)^2 - a_{n+1}.$$

If $a_1 = 1$ and $a_2 = 1$, find a_5.

3. Find the 30th term in the arithmetic sequence 80, 76, 72, ...

4. The second term of a geometric sequence is 125 and the fifth term is 1. Is $\frac{1}{5}$ a term of this sequence? If so, which term is it?

5. Determine whether each of the following statements is true or false. If it is true, prove it. If it is false, give an example where it fails.

6. Find the sum of the infinite geometric series

$$1 + \frac{1}{2^{1/2}} + \frac{1}{2} + \frac{1}{2^{3/2}} + \cdots$$

9. Use mathematical induction to prove that for all natural numbers n

$$1^2 + 2^2 + 3^2 + \cdots + n^2 = \frac{n(n + 1)(2n + 1)}{6}$$

10. Write the following expression without using sigma notation and then find the sum.

(a) If $a_1, a_2, a_3, \ldots$ is an arithmetic sequence, then the sequence $a_1^2, a_2^2, a_3^2, \ldots$ is also arithmetic.

(b) If $a_1, a_2, a_3, \ldots$ is a geometric sequence, then the sequence $a_1^2, a_2^2, a_3^2, \ldots$ is also geometric.

6. (a) Write the formula for the sum of a (finite) arithmetic series.

(b) The first term of an arithmetic series is 10 and the tenth term is 2. Find the sum of the first ten terms.

(c) Find the common difference and the 100th term of the series in part (b).

7. (a) Write the formula for the sum of a (finite) geometric series.

(b) Find the sum of the geometric series

$$\frac{1}{3} + \frac{2}{3^2} + \frac{2^2}{3^3} + \frac{2^3}{3^4} + \cdots + \frac{2^9}{3^{10}}$$

(a) $\displaystyle\sum_{n=1}^{5} (1 - n^2)$ **(b)** $\displaystyle\sum_{n=3}^{6} (-1)^n 2^{n-2}$

11. Write the expansion of $(a + b)^n$ using sigma notation.

12. Expand $(2x + y^2)^5$.

13. Find the term that contains a^3 in the expansion of $(2a + b)^{100}$.

14. Find the term that does not contain x in the expansion of

$$\left(2x + \frac{1}{x}\right)^{10}$$

FOCUS ON PROBLEM SOLVING

The solutions to many of the problems of mathematics involve **finding patterns**. The algebraic formulas we have found in this book are compact ways of describing a pattern. For example, the familiar equation $(a + b)^2 = a^2 + 2ab + b^2$ gives the pattern for multiplying the sum of two numbers. Another example we have encountered is the pattern for the sum of the first n odd numbers:

$$1 + 3 + 5 + \cdots + (2n - 1) = n^2$$

How do we discover patterns? In many cases, a good way to start is to experiment with the problem at hand. How can we be sure a pattern always holds? One way is to use mathematical induction, but there are other ways. In the example we give here we find a pattern for adding the cubes of the first n natural numbers. We then find a geometrical interpretation of this sum that shows the pattern always holds. This startling connection between number patterns and geometrical patterns is credited to the 11th century mathematician Abu Bekr Mohammed ibn Alhusain Al Karchi.

THE GNOMONS OF AL KARCHI

We prove the beautiful formula

$$1^3 + 2^3 + 3^3 + \cdots + n^3 = (1 + 2 + 3 + \cdots + n)^2$$

But first, here is how this formula was discovered. The sum of the first n natural numbers is called a *triangular number*. The first few triangular numbers are 1, $1 + 2 = 3$, $1 + 2 + 3 = 6$, $1 + 2 + 3 + 4 = 10$, and in general (as we have already shown) $1 + 2 + 3 + \cdots + n = \frac{1}{2}n(n + 1)$. The name "triangular" comes from the following pictures:

| 1 | $1 + 2 = 3$ | $1 + 2 + 3 = 6$ | $1 + 2 + 3 + 4 = 10$ | $1 + 2 + 3 + 4 + 5 = 15$ |

The sequence of triangular numbers

$$1, 3, 6, 10, 15, \ldots$$

is so well known that any mathematician would instantly recognize it. Now let's look at the sums of the cubes:

$$1^3 = 1, \qquad 1^3 + 2^3 = 9, \qquad 1^3 + 2^3 + 3^3 = 36,$$
$$1^3 + 2^3 + 3^3 + 4^3 = 100, \qquad 1^3 + 2^3 + 3^3 + 4^3 + 5^3 = 225, \ldots$$

We get the sequence

$$1, 9, 36, 100, 225, \ldots$$

It doesn't take long to notice that the numbers in this sequence are the squares of the triangular numbers. It appears that the sum of the cubes of the first n natural numbers equals the square of the sum of the first n natural numbers.

To show that this pattern always holds, Al Karchi sketches the following diagram:

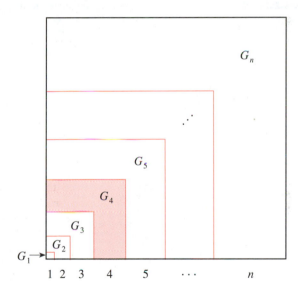

Each region G_n in the shape of an inverted L is called a "gnomon." Let us find the area of the gnomon G_4. This area is

Area $= 4^2 + 2[4 \times (1 + 2 + 3)]$
$\phantom{\text{Area }} = 64$

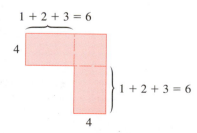

Similarly, the areas of the gnomons G_1, G_2, G_3, G_4, G_5, . . . are 1, 8, 27, 64, 125, ... These are the cubes of the natural numbers. The pattern persists here also, since the area of the nth gnomon G_n is n^3, as the following calculation shows:

$$n^2 + 2[n \times (1 + 2 + \cdots + (n - 1))] = n^2 + 2n\frac{(n - 1)n}{2}$$
$$= n^2 + n^3 - n^2 = n^3$$

Now comes Al Karchi's punchline: The first n gnomons form a square of side $1 + 2 + 3 + \cdots + n$ and so the sum of the areas of these gnomons equals the area of the square. That is, $1^3 + 2^3 + \cdots + n^3 = (1 + 2 + \cdots + n)^2$.

PROBLEMS

1. Use the diagram shown to find and prove a formula for the sum of the first n odd numbers.

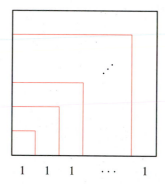

$$1 \quad 1 \quad 1 \quad \cdots \quad 1$$

2. (a) Find the product $\left(1 - \frac{1}{2}\right)\left(1 - \frac{1}{3}\right)\left(1 - \frac{1}{4}\right) \cdots \left(1 - \frac{1}{200}\right)$.

(b) Find the product $\left(1 - \frac{1}{4}\right)\left(1 - \frac{1}{9}\right)\left(1 - \frac{1}{16}\right) \cdots \left(1 - \frac{1}{n^2}\right)$.

3. Prove that

$$\frac{1}{\sin 2} + \frac{1}{\sin 4} + \frac{1}{\sin 8} + \cdots + \frac{1}{\sin 2^n} = \cot 1 - \cot 2^n$$

4. Prove that

$$\frac{n^5}{5} + \frac{n^4}{2} + \frac{n^3}{3} - \frac{n}{30}$$

is an integer for all natural numbers n.

5. Find a formula for the sums

$$S_n = 1 \cdot 1! + 2 \cdot 2! + 3 \cdot 3! + \cdots + n \cdot n!$$

and prove that your formula holds for all n.

6. Find a formula for the sums

$$S_n = \frac{1}{2!} + \frac{2}{3!} + \frac{3}{4!} + \cdots + \frac{n}{(n+1)!}$$

and prove that your formula holds for all n.

7. Prove that the number of people who have shaken hands an odd number of times is an even number.

8. An $n \times n$ magic square consists of the numbers from 1 to n^2 arranged in a square in such a way that the sum of the numbers in any row, column, or diagonal is the same. Let us call this constant sum S. Find a formula for S for an $n \times n$ magic square.

9. Consider a 6×6 grid as shown in the figure.

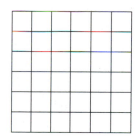

(a) Find the number of squares of all sizes in this grid. Generalize your result to an $n \times n$ grid.

(b) Find the number of rectangles of all sizes in this grid. Generalize your result to an $n \times n$ grid.

10. (a) How many zeros are at the end of the number 100!?

(b) If $S = 1! + 2! + 3! + \cdots + 99!$, what is the last digit in the value of S?

The earliest known magic square is a 3×3 magic square from ancient China. According to Chinese legend it appeared on the back of a turtle emerging from the river Lo.

4	9	2
3	5	7
8	1	6

Here is an example of a 4×4 magic square:

16	3	2	13
5	10	11	8
9	6	7	12
4	15	14	1

ANSWERS TO ODD-NUMBERED EXERCISES AND CHAPTER TESTS

CHAPTER 1

Exercises 1.1 (page 9)

1. $3x + 3y$ **3.** $8m$ **5.** $-8y$ **7.** $-5x + 10y$
9. False   **11.** True   **13.** True   **15.** True
17. True   **19.** $x > 0$ **21.** $t < 4$ **23.** $a \geq \pi$
25. $-5 < x < \frac{1}{3}$ **27.** $|p - 3| \leq 5$
29. $\{1, 2, 3, 4, 5, 6, 8\}$ **31.** $\{8\}$
33. $\{1, 2, 3, 4, 5, 6, 7, 8, 9, 10\}$
35. $\{x \mid -2 \leq x < 4\}$ **37.** $\{x \mid -1 < x \leq 5\}$

39. $-3 < x < 0$

41. $2 \leq x < 8$

43. $-1 \leq x \leq 1$

45. $x \geq 2$

47. $x \leq -2$

49. $(-\infty, 1]$

51. $[1, 2]$

53. $(-2, 1]$

55. $(-1, \infty)$

57.

59.

61.

63.

65. 100   **67.** 4   **69.** π **71.** $3 - \sqrt{3}$ **73.** -1
75. 10   **77.** 15   **79.** 26   **81.** 19

Exercises 1.2 (page 19)

1. 64   **3.** $\frac{625}{8}$ **5.** 16   **7.** 100,000   **9.** -2
11. $\frac{1}{2}$ **13.** 0.02   **15.** 10   **17.** 6   **19.** 3
21. 2187   **23.** $-\frac{1}{5}$ **25.** 256   **27.** $\frac{125}{512}$ **29.** 3

31. False **33.** False **35.** True **37.** 2^7 **39.** 2^{36}
41. 2^{-6} **43.** 2^{-6} **45.** t^5 **47.** $6x^7y^5$ **49.** $16x^{10}$
51. $4/b^2$ **53.** $64r^7s$ **55.** $648y^7$ **57.** x^3/y
59. y^2z^9/x^5 **61.** $s^3/(q^7r^6)$ **63.** $x^{13/15}$ **65.** $16b^{9/10}$
67. $1/(c^{2/3}d)$ **69.** $y^{1/2}$ **71.** $32x^{12}/y^{16/15}$
73. $x^{15}/y^{15/2}$ **75.** $4a^2/(3b^{1/3})$ **77.** $3t^{25/6}/s^{1/2}$
79. $5\sqrt{3}$ **81.** $2\sqrt{5}$ **83.** $|x|$ **85.** $x\sqrt[5]{y}$
87. $a^{6/5}b^{7/5}$ **89.** $|xy^3|$ **91.** $2\sqrt[6]{x}$ **93.** x^{a+b+c}
95. $x^{5/3}$ **97.** $x^{5/2}$ **99.** x^{n+5} **101.** $x^{3/4}$
103. $\sqrt{6}/6$ **105.** $\sqrt{15}/10$ **107.** $\sqrt{3xy}/(3y)$
109. $(3 + \sqrt{5})/4$ **111.** $2(\sqrt{a} - 1)/(a - 1)$
113. $1/[2(\sqrt{2} - 1)]$ **115.** $46/(7 + \sqrt{3})$ **117.** No
119. Yes **121.** No **123.** 1.5×10^8 km **125.** $7^{1/4}$
127. (a)

Exercises 1.3 (page 27)

1. $-8x - 4$ **3.** $3x^2 - x + 7$
5. $x^3 + 3x^2 - 6x + 11$ **7.** $-t^4 + t^3 - t^2 - 10t + 5$
9. $x^{3/2} - x$ **11.** $y^{7/3} - y^{1/3}$ **13.** $2x^2 + 9x - 18$
15. $21t^2 - 29t + 10$ **17.** $3x^2 + 5xy - 2y^2$
19. $1 - 4y + 4y^2$ **21.** $x - y$
23. $2x^3 - 7x^2 + 7x - 5$ **25.** $x^3 + x^2 - 2x$
27. $x^4 - x^3 + 2x^2 - x + 1$ **29.** $30y^4 + y^5 - y^6$
31. $4x^4 + 12x^2y^2 + 9y^4$ **33.** $x^4 - a^4$
35. $1 + 3a^3 + 3a^6 + a^9$ **37.** $a - 1/b^2$
39. $x^5 + x^4 - 3x^3 + 3x - 2$ **41.** $1 - x^{2/3} + x^{4/3} - x^2$
43. $1 - 2b^2 + b^4$ **45.** $3x^4y^4 + 7x^3y^5 - 6x^2y^3 - 14xy^4$
47. $2x(1 + 6x^2)$ **49.** $3y^3(2y - 5)$ **51.** $(x + 6)(x + 1)$
53. $(x - 4)(x + 2)$ **55.** $(y - 3)(y - 5)$
57. $(2x + 3)(x + 1)$ **59.** $9(x - 2)(x + 2)$
61. $(3x + 2)(2x - 3)$ **63.** $(4x - 3)(2x + 5)$
65. $(t + 1)(t^2 - t + 1)$ **67.** $(2t - 3)^2$ **69.** $x(x + 1)^2$
71. $(2x + y)^2$ **73.** $x^2(x + 3)(x - 1)$
75. $(2x - 5)(4x^2 + 10x + 25)$
77. $(x^2 + 2)(x - 1)(x + 1)$ **79.** $(y - 2)(y + 2)(y - 3)$
81. $(2x^2 + 1)(x + 2)$
83. $(x - y)(x + y)(x^2 + xy + y^2)(x^2 - xy + y^2)$
85. $(s + t^3)(s^2 - st^3 + t^6)$ **87.** $(x^4 + 1)(x - 1)(x + 1)$
89. $x^{1/2}(x - 1)(x + 1)$ **91.** $x^{-3/2}(x + 1)^2$
93. $(x^2 + 3)(x^2 + 1)^{-1/2}$ **99.** $(x^2 - x + 2)(x^2 + x + 2)$

Exercises 1.4 (page 34)

1. $\dfrac{x + 1}{x + 3}$ **3.** $\dfrac{-y}{y + 1}$ **5.** $\dfrac{x(2x + 3)}{2x - 3}$ **7.** $\dfrac{1}{t^2 + 9}$
9. $\dfrac{x + 4}{x + 1}$ **11.** $\dfrac{(2x + 1)(2x - 1)}{(x + 5)^2}$ **13.** $x^2(x + 1)$

15. $\dfrac{x}{yz}$ **17.** $\dfrac{3x + 7}{(x - 3)(x + 5)}$ **19.** $\dfrac{1}{(x + 1)(x + 2)}$
21. $\dfrac{3x + 2}{(x + 1)^2}$ **23.** $\dfrac{u^2 + 3u + 1}{u + 1}$ **25.** $\dfrac{2x + 1}{x^2(x + 1)}$
27. $\dfrac{2x + 7}{(x + 3)(x + 4)}$ **29.** $\dfrac{x - 2}{x^2 - 9}$ **31.** $\dfrac{5x - 6}{x(x - 1)}$
33. $\dfrac{-5}{(x + 1)(x + 2)(x - 3)}$ **35.** $-xy$ **37.** $\dfrac{c}{c - 2}$
39. $\dfrac{3x + 7}{x^2 + 2x - 1}$ **41.** $\dfrac{y - x}{xy}$ **43.** $\dfrac{-1}{a(a + h)}$
45. $\dfrac{-3}{(2 + x)(2 + x + h)}$ **47.** $\dfrac{1}{\sqrt{1 - x^2}}$
49. $\dfrac{r - s}{t(\sqrt{r} - \sqrt{s})}$ **51.** $\dfrac{-1}{\sqrt{x}\sqrt{x + h}(\sqrt{x} + \sqrt{x + h})}$
53. $\dfrac{x + 1}{\sqrt{x^2 + x + 1} - x}$ **55.** True **57.** False
59. False **61.** True **63.** False

Exercises 1.5 (page 45)

1. Yes **3.** Yes **5.** Yes, $x = 9$ **7.** No **9.** 4
11. 12 **13.** $-\frac{3}{4}$ **15.** $\frac{32}{9}$ **17.** $-\frac{1}{3}$ **19.** -20
21. $\frac{29}{2}$ **23.** No solution **25.** No solution **27.** ± 11
29. $\pm 2\sqrt{2}$ **31.** $-3, 4$ **33.** $-5, -\frac{2}{3}$ **35.** $-1 \pm \sqrt{3}$
37. $\dfrac{-1 \pm \sqrt{5}}{2}$ **39.** $-3, 5$ **41.** $-60, 24$
43. $\dfrac{-1 \pm \sqrt{7}}{3}$ **45.** $\dfrac{1 \pm \sqrt{5}}{4}$ **47.** No real solution
49. $\dfrac{-5 \pm \sqrt{13}}{2}$ **51.** $\dfrac{\sqrt{5} \pm 1}{2}$ **53.** $-50, 100$
55. -4 **57.** 5 **59.** $\pm\sqrt{3}$ **61.** 0, 1, 2
63. $\pm\sqrt{2}, 4$ **65.** 4 **67.** 4 **69.** 21
71. No real solution **73.** $-1, 0, 3$
75. $\pm 2\sqrt{2}, \pm 3\sqrt{3}$ **77.** $-\frac{1}{2}$ **79.** 5.061
81. $0.259, -0.248$ **83.** $R = \dfrac{PV}{nT}$ **85.** $R_1 = \dfrac{RR_2}{R_2 - R}$
87. $r = \pm\sqrt{3V/(\pi h)}$ **89.** $b = \pm\sqrt{c^2 - a^2}$
91. all real k **93.** None **95.** One
97. Two solutions **99.** $k = \pm 20$ **101.** 1.79 in

Exercises 1.6 (page 55)

1. $3x + 3$ **3.** $4w + 6$ **5.** $d/55$ **7.** $7.5c$
9. 49 m $\times$ 56 m **11.** 43 \$5 bills and 31 \$10 bills
13. 5 quarters, 10 dimes, 15 nickels **15.** 40 yr
17. \$4500 at 9% and \$1500 at 8% **19.** $1\frac{1}{2}$ h
21. 450 mi **23.** 4 h **25.** 15 mi/h and 30 mi/h
27. 71, 73, 75 **29.** 200 mL **31.** 0.6 L **33.** 3 h

35. 120 ft **37.** 19, 36 **39.** 12 in × 24 in
41. 13 in × 13 in **43.** 120 ft × 126 ft
45. (a) After 1 s and $1\frac{1}{2}$ s (b) Never (c) 25 ft
(d) After $1\frac{1}{4}$ s (e) After $2\frac{1}{2}$ s
47. (a) After 17 yr, on Jan. 1, 2009
(b) After 18.612 yr, on Aug. 12, 2010 **49.** 6 km/h
51. Irene 3 h, Henry $4\frac{1}{2}$ h **53.** 215,000 mi
55. 16 mi, no

Exercises 1.7 (page 66)

1. $(-\infty, 5]$

3. $(-2, \infty)$

5. $[-1, \infty)$

7. $(3, \infty)$

9. $(-\infty, -1]$

11. $(2, 6)$

13. $(0, 1]$

15. $\left[-1, \frac{1}{2}\right)$

17. $[2, 3]$

19. $\left[\frac{7}{3}, \infty\right)$

21. $(-\infty, 0) \cup \left(\frac{1}{4}, \infty\right)$

23. $(-\infty, 1) \cup (2, \infty)$

25. $[-2, 4]$

27. $\left[-1, \frac{1}{2}\right]$

29. $(-\infty, -1) \cup (4, \infty)$

31. $(-\infty, -3) \cup (6, \infty)$

33. $(-2, 2)$

35. $(-\infty, \infty)$

37. $(-\infty, -2) \cup [1, \infty)$

39. $(-4, -3)$

41. $(-\infty, 5) \cup [16, \infty)$

43. $(-2, 0) \cup (2, \infty)$

45. $(-\infty, -1] \cup [1, \infty)$

47. $[-2, -1) \cup (0, 1]$

49. $68 \leq F \leq 86$
51. (a) $-\frac{1}{3}P + \frac{560}{3}$ (b) From \$215 to \$290
53. $0 \leq t \leq 3$
55. $-\frac{4}{3} \leq x \leq \frac{4}{3}$ **57.** $(-\infty, -2) \cup (7, \infty)$
63. $x \geq \dfrac{c(a + b)}{ab}$ **65.** $x > \dfrac{c - b}{a}$

Exercises 1.8 (page 72)

1. $|x - 3| = \begin{cases} x - 3 & \text{if } x \geq 3 \\ 3 - x & \text{if } x < 3 \end{cases}$

3. $|3x - 10| = \begin{cases} 3x - 10 & \text{if } x \geq \frac{10}{3} \\ 10 - 3x & \text{if } x < \frac{10}{3} \end{cases}$

5. $|x^2 + 1| = x^2 + 1$ **7.** $3|x + 3|$ **9.** $\frac{1}{2}|x - 5|$
11. $x^2 + 9$ **13.** $\pm\frac{3}{2}$ **15.** 3.99, 4.01
17. $-4, -\frac{2}{5}$ **19.** $-\frac{3}{2}, -\frac{1}{4}$ **21.** $(-3, 3)$
23. $(3, 5)$ **25.** $(-\infty, -7] \cup [-3, \infty)$ **27.** $[1.3, 1.7]$
29. $(-4, 8)$ **31.** $(-6.001, -5.999)$
33. $[-4, -1] \cup [1, 4]$ **35.** $\left(-\frac{15}{2}, -7\right) \cup \left(-7, -\frac{13}{2}\right)$
37. $\left(\frac{1}{2}, \infty\right)$ **39.** $(-1, \infty)$

Exercises 1.9 (page 83)

1. (a)

3. (a)

(b) 5 (c) $\left(\frac{5}{2}, 3\right)$ (b) $\sqrt{74}$ (c) $\left(\frac{5}{2}, \frac{1}{2}\right)$

5. (a)

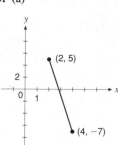

(b) $2\sqrt{37}$ (c) $(3, -1)$

7.

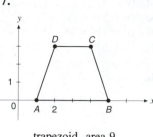

trapezoid, area 9

21.

23. A
29. $(0, -4)$
31. $\left(1, \frac{7}{2}\right)$

33. $1, -1$, no symmetry

35. $\frac{5}{3}, -5$, no symmetry

9.

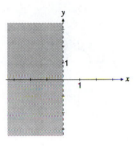

11.

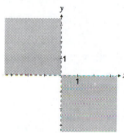

37. $\pm 1, 1$, symmetry about y-axis

39. $0, 0$, symmetry about y-axis

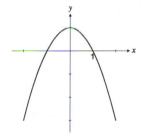

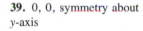

13.

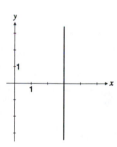

15.

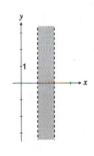

41. no intercept, symmetry about origin

43. $0, 0$, no symmetry

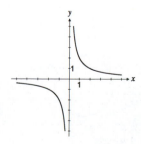

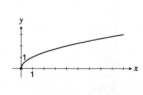

17.

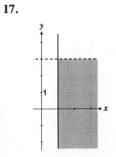

19.

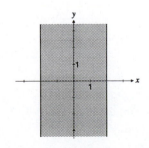

45. ±3, ±3, symmetry about both axes and origin

47. ±2, 2, symmetry about y-axis

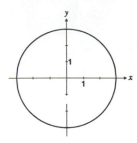

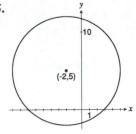

61. $(x - 3)^2 + (y + 1)^2 = 25$ **63.** $x^2 + y^2 = 65$
65. $(x - 2)^2 + (y - 3)^2 = 13$
67. $(x - 7)^2 + (y + 3)^2 = 9$ **69.** (2, 5), 4
71. $\left(-\frac{1}{2}, 0\right), \frac{1}{2}$ **73.** $\left(\frac{1}{4}, -\frac{1}{4}\right), \sqrt{5/8}$

75.

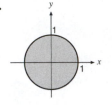

77.

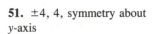

49. 0, 0, symmetry about y-axis

51. ±4, 4, symmetry about y-axis

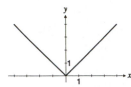

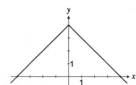

79.

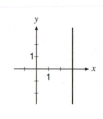

81.

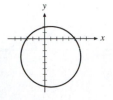

83. 12π

85. $a^2 + b^2 > 4c, \left(-\dfrac{a}{2}, -\dfrac{b}{2}\right), \frac{1}{2}\sqrt{a^2 + b^2 - 4c}$

53. 0, 0, symmetry about origin

55.

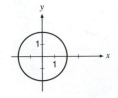

Exercises 1.10 (page 93)

1. 2   **3.** $-\frac{9}{2}$ **5.** $x + y - 4 = 0$
7. $6x - y - 15 = 0$ **9.** $2x - 3y + 19 = 0$
11. $5x + y - 11 = 0$ **13.** $3x - y - 2 = 0$
15. $3x - y - 3 = 0$ **17.** $y = 5$
19. $x + 2y + 11 = 0$ **21.** $5x - 2y + 1 = 0$

57.

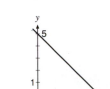

59.

23. (a)

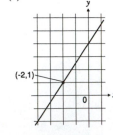

25. −1, 5

(b) $3x - 2y + 8 = 0$

27. $-\frac{1}{3}$, 0

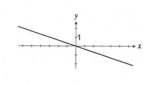

29. $\frac{3}{2}$, 3

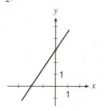

31. $\frac{3}{4}$, -3

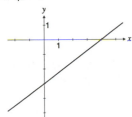

33. $-\frac{3}{4}$, $\frac{1}{4}$

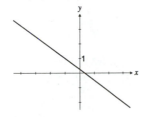

39. $x - y - 3 = 0$

41. (b) $4x - 3y - 24 = 0$

43. (a)

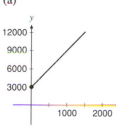

(b) The slope represents production costs per toaster; the y-intercept represents monthly fixed costs.

45. (a) $C = \frac{1}{4}d + 260$ (b) $635

(c)

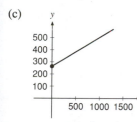

The slope represents cost per mile.

(d) The y-intercept represents annual fixed costs.

47. (a) $P = 0.434d + 15$, where P is pressure in lb/in² and d is depth in feet.   (b) 196 ft

Review Exercises for Chapter 1 (page 96)

1. $-1 < x \le 3$

3. $(2, \infty)$

5. 6   **7.** $\frac{1}{72}$ **9.** $\frac{1}{6}$ **11.** 11   **13.** x^{-2}
15. x^{4m+2} **17.** $12x^5y^4$ **19.** $9x^3$ **21.** x^2y^2
23. $x\left(2 - \sqrt{x}\right)/(4 - x)$ **25.** $4r^{5/2}/s^7$
27. $(x - 2)(x + 5)$ **29.** $(4t + 3)(t - 4)$
31. $(5 - 4t)(5 + 4t)$
33. $(x - 1)(x^2 + x + 1)(x + 1)(x^2 - x + 1)$
35. $x^{-1/2}(x - 1)^2$ **37.** $(x - 2)(4x^2 + 3)$
39. $\sqrt{x^2 + 2}(x^4 + 2x^3 + 5x^2 + 4x + 4)$
41. $3(2x^2 - 7x + 1)$ **43.** $4a^4 - 4a^2b + b^2$
45. $x^3 - 6x^2 + 11x - 6$ **47.** $2x^{3/2} + x - x^{1/2}$
49. $\dfrac{x - 3}{2x + 3}$ **51.** $\dfrac{3(x + 3)}{x + 4}$ **53.** $\dfrac{x + 1}{x - 4}$
55. $\dfrac{x + 1}{(x - 1)(x^2 + 1)}$ **57.** $\dfrac{1}{x + 1}$ **59.** $-\dfrac{1}{2x}$
61. $6x + 3h - 5$ **63.** No   **65.** Yes   **67.** No
69. 9   **71.** $\frac{2}{33}$ **73.** $-1, \frac{1}{2}$ **75.** $0, \pm\frac{5}{2}$
77. $\dfrac{-2 \pm \sqrt{7}}{3}$ **79.** $\dfrac{3 \pm \sqrt{6}}{3}$ **81.** ± 3 **83.** 1
85. 20 lb raisins, 30 lb nuts
87. $\frac{1}{4}\left(\sqrt{329} - 3\right) \approx 3.78$ mi/h   **89.** 12 cm, 16 cm
91. $(-3, -1]$

93. $(-\infty, -6) \cup (2, \infty)$

95. $[-4, -1)$

97. $(-\infty, -2) \cup (2, 4]$

99. $[2, 8]$

101. $(-\infty, -1] \cup [0, \infty)$

103. (a) $\left[-3, \frac{8}{3}\right]$ (b) $(0, 1)$

105. (b) 5   (c) $\left(\frac{3}{2}, 2\right)$ (d) $y = \frac{4}{3}x$ (e) $x^2 + y^2 = 25$
107. (b) $\sqrt{53}$ (c) $\left(\frac{3}{2}, -2\right)$ (d) $y = -\frac{2}{7}x - \frac{11}{7}$
(e) $(x - 5)^2 + (y + 3)^2 = 53$

109.

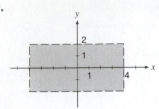

111. *B*   **113.** $(x - 2)^2 + (y + 5)^2 = 2$
115. $\left(x - \frac{1}{2}\right)^2 + \left(y - \frac{11}{2}\right)^2 = \frac{17}{2}$
117. Circle, center $(-1, 3)$, radius 1   **119.** No graph
121. $2x - 3y = 16$ **123.** $x + 5y = 0$

125. No symmetry   **127.** No symmetry

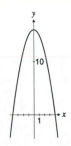   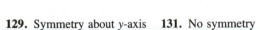

129. Symmetry about *y*-axis   **131.** No symmetry

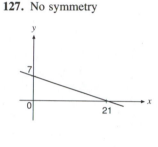

133. Symmetry about both   **135.** Symmetry about origin
axes and origin

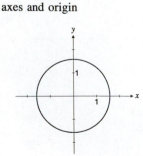   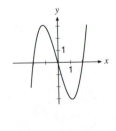

Chapter 1 Test (page 99)

1.

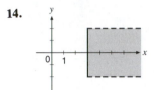

2. (a) 16 (b) $\frac{1}{16}$ (c) 27 (d) $\frac{1}{8}$ **3.** $x^{b/2}$
4. (a) $8\sqrt{2}$ (b) $54a^6b^{14}$ (c) y^{32}/x^8 (d) $(x + 2)/(x - 2)$
(e) $2(x - 1)/(x^2 - 4)$ **5.** (a) $-3 - 7x$
(b) $2x^2 - 7x - 15$ (c) $x - y$
(d) $9t^2 + 24t + 16$ (e) $8 - 12x^2 + 6x^4 - x^6$
6. (a) $(3x - 5)(3x + 5)$ (b) $(3x + 5)(2x - 1)$
(c) $(x^2 - 3)(x - 4)$ (d) $x(x + 3)(x^2 - 3x + 9)$
(e) $3x^{-1/2}(x - 1)(x - 2)$ **7.** $x\left(\sqrt{x} + 2\right)/(x - 4)$
8. (a) $-\frac{4}{3}$ (b) 0   **9.** 120 mi   **10.** (a) $-3, 4$
(b) $\dfrac{-1 \pm \sqrt{5}}{2}$

11. 50 ft by 120 ft   **12.** (a) $\left(-\frac{5}{2}, 3\right]$ (b) $(0, 1) \cup (2, \infty)$
(c) $(1, 5)$ (d) $[-4, -1)$ **13.** 41 °F to 50 °F

14.

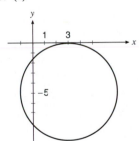

15. 26; $(4, 8)$; $(x - 4)^2 + (y - 8)^2 = 169$

16. (a)   (b)

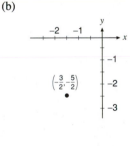

17. $3x - y = 15$

PRINCIPLES OF PROBLEM SOLVING (page 104)

1. (a) $7920\pi \approx 24{,}881$ mi   (b) $2\pi \approx 6.3$ ft
3. A 40% discount   **5.** 2
7. $1729 = 1^3 + 12^3 = 9^3 + 10^3$ **9.** No   **11.** No

CHAPTER 2

Exercises 2.1 (page 111)

1. $0, 12, \frac{3}{4}, 7 - 3\sqrt{5}, a^2 - 3a + 2, a^2 + 3a + 2,$
$a^2 + 2ab + b^2 - 3a - 3b + 2$
3. $-\frac{1}{3}, -3, (1 - \pi)/(1 + \pi), (1 - a)/(1 + a),$
$(2 - a)/a, (1 + a)/(1 - a)$
5. $-4, 10, 3\sqrt{2}, 5 + 7\sqrt{2}, 2x^2 - 3x - 4, 2x^2 + 7x + 1,$
$4x^2 + 6x - 8, 8x^2 + 6x - 4$
7. $1 + 2a, 2 + 2a + 2h, 1 + 2a + 2h, 2$
9. $3 - 5a + 4a^2, 6 - 5a + 4a^2 - 5h + 4h^2,$
$3 - 5a - 5h + 4a^2 + 8ah + 4h^2, -5 + 8a + 4h$
11. $15 + 8h, 8x + 8h - 1, 8$
13. $1/x, 1/(2 + h), 1/(x + h), -1/[x(x + h)]$

15.

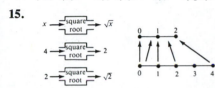

17. $[-1, 5], [-2, 10]$ **19.** $[-2, 3], [-6, 14]$

21. $(-\infty, \infty), (-\infty, 2]$ **23.** $\left[\frac{5}{2}, \infty\right), [0, \infty)$

25. $[-1, 1], [3, 4]$ **27.** $\{x \mid x \neq -4\}$
29. $\{x \mid x \neq \pm 1\}$ **31.** $(-\infty, \infty)$ **33.** $\left(-\infty, \frac{5}{2}\right]$
35. $(-\infty, \infty)$ **37.** $[-2, 3) \cup (3, \infty)$ **39.** $(10, \infty)$
41. $[0, 1]$ **43.** $\left(-\infty, -\frac{1}{2}\right] \cup \left[\frac{1}{2}, \infty\right)$
45. $(-\infty, 0] \cup [6, \infty)$ **47.** $[0, \pi)$

Exercises 2.2 (page 120)

1. (a) $1, -1, 3, 4$ (b) $[-3, 4]$ (c) $[-1, 4]$
(d) Decreasing on $[-3, 0], [3, 4]$; increasing on $[0, 3]$
3. (a), (c)   **5.** Function, domain $[-3, 2]$, range $[-2, 2]$
7. Not a function
9. (a)

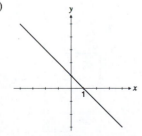

(b) R (c) Decreasing on R

11. (a)

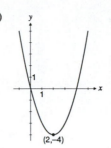

(b) R
(c) Decreasing on
$(-\infty, 2]$, increasing on $[2, \infty)$

15. (a)

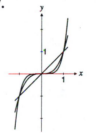

17.

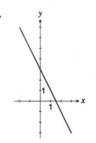

21.

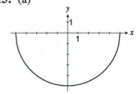

25.

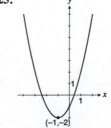

13. (a)

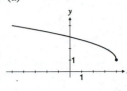

(b) $(-\infty, 4]$
(c) Decreasing on $(-\infty, 4]$

(b) $[-5, 5]$ (c) Decreasing on
$[-5, 0]$, increasing on $[0, 5]$

19.

23.

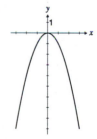

27.

29.

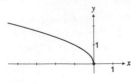

31.

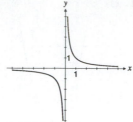

49.

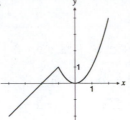

51.

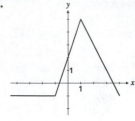

33.

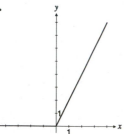

35.

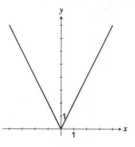

53. $f(x) = -\frac{7}{6}x - \frac{4}{3}, \ -2 \leqslant x \leqslant 4$

55. $f(x) = 1 - \sqrt{-x}$

57. Even

59. Neither

61. Odd

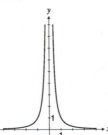 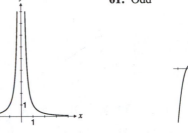

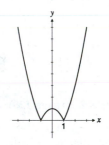

37.

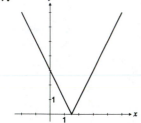

39.

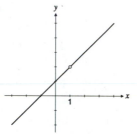

63. Neither

65. (a)

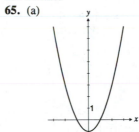

(b)

41.

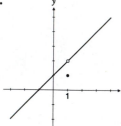

43.

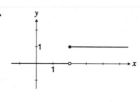

Exercises 2.3 (page 129)

1. (d)   **3.** (c)

5.   **7.**

45.

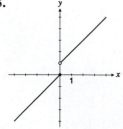

47.

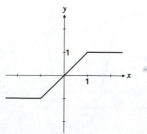

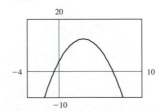 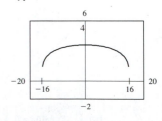

9.

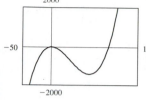

11.

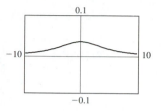

13.

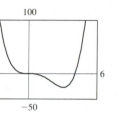

15.

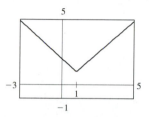

17.

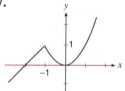

19.

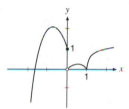

21.

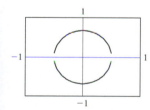

23. (a)

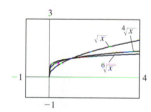

(b)

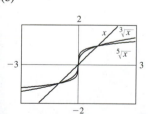

(c)

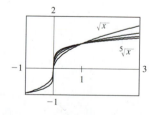

(d) Graphs of even roots are similar to $\sqrt{x}$, graphs of odd roots are similar to $\sqrt[3]{x}$. As n increases the graph of $y = \sqrt[n]{x}$ becomes steeper near 0 and flatter when $x > 1$.

25. g

27. (a)

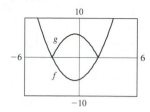

(b) The graph of g is obtained from that of f by reflecting the part below the x-axis, about the x-axis.

29.

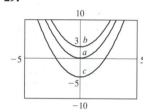

The graph in part (b) is obtained by shifting the graph in part (a), 3 units upward.
The graph in part (c) is obtained by shifting the graph in part (a), 5 units downward.

31.

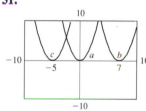

The graph in part (b) is obtained by shifting the graph in part (a), 7 units to the right.
The graph in part (c) is obtained by shifting the graph in part (a), 5 units to the left.

Exercises 2.4 (page 135)

1.

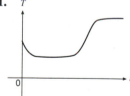

3.

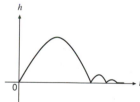

5.

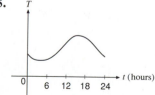

7. $A = 10x - x^2, 0 < x < 10$
9. $A = \sqrt{3}x^2/4, x > 0$ **11.** $r = \sqrt{A/\pi}, A > 0$
13. $A = x^2 + (48/x), x > 0$
15. $A = 15x - (\pi + 4)x^2/8, x > 0$
17. $A = 2x(1200 - x), 0 < x < 1200$
19. $d = 25t, t \geq 0$

21.

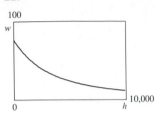

23.

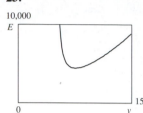

25. $R = kt$ **27.** $v = k/z$ **29.** $y = ks/t$
31. $z = k\sqrt{y}$ **33.** $y = 18x$ **35.** $M = 15x/y$
37. $W = 360/r^2$ **39.** (a) $F = kx$ (b) 8 (c) 32 N
41. (a) $C = \frac{1}{8}pn$ (b) \$57,500
43. (a) $R = kL/d^2, k = \frac{7}{2400} = 0.00291\overline{6}$ (b) $\approx 137\ \Omega$

Exercises 2.5 (page 143)

1. Shift 10 units downward   **3.** Shift 1 unit left
5. Stretch vertically by a factor 8
7. Stretch vertically by a factor 6 and reflect in x-axis
9. Shift 2 units right and 3 units downward
11. Shrink vertically by a factor 2 and shift 9 units upward

13. (a)

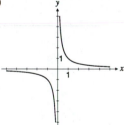

(b) (i)

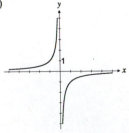

(ii)

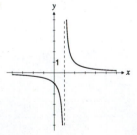

(iii)

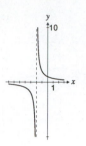

(iv)

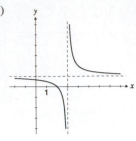

15.

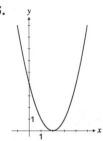

17.

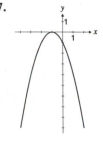

19.

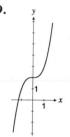

21.

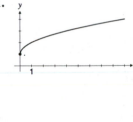

23.

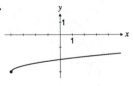

25.

27.

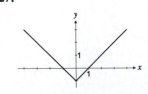

29.

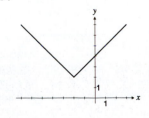

31.

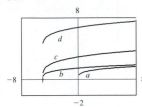

33.

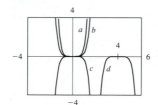

35. (a) (i)

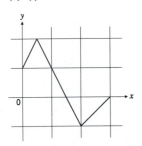

(ii)

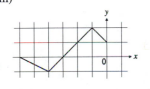

(iii)   (iv)

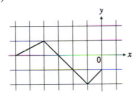

(b) (i) Shrink horizontally by a factor of a.
   (ii) Stretch horizontally by a factor of a.
   (iii) Reflect in the y-axis.

37.

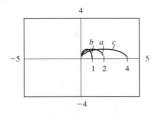

For part (b), shrink part (a) horizontally by a factor of 2.
For part (c), stretch part (a) horizontally by a factor of 2.

Exercises 2.6 (page 150)

1.

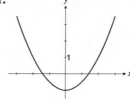

3.

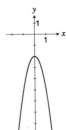

Vertex $(0, -1)$; x-intercepts $\pm\sqrt{2}$; y-intercept -1

Vertex $(0, -2)$; no x-intercept; y-intercept -2

5.

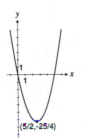

7.

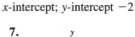
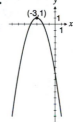

Vertex $\left(\frac{5}{2}, -\frac{25}{4}\right)$; x-intercepts 0, 5; y-intercept 0

Vertex $(-3, 1)$; x-intercepts $-4, -2$; y-intercept -8

9.

11.

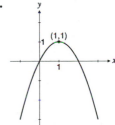

Vertex $(5, 7)$; no x-intercept; y-intercept 57

Maximum $f(1) = 1$

13.

15.

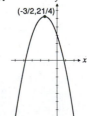

Minimum $f(-1) = -2$

Maximum $f\left(-\frac{3}{2}\right) = \frac{21}{4}$

17.

19.

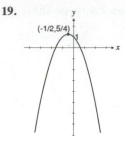

Minimum $g(2) = 1$

Maximum $h\left(-\frac{1}{2}\right) = \frac{5}{4}$

21. Minimum $f\left(-\frac{1}{2}\right) = \frac{3}{4}$ **23.** Maximum $f\left(-\frac{25}{7}\right) = \frac{1325}{7}$
25. $f(x) = 2x^2 - 4x$ **27.** $(-\infty, \infty)$, $(-\infty, 1]$
29. 25 ft   **31.** ± 50 **33.** $-12, -12$
35. 600 ft by 1200 ft   **37.** 14,062.5 ft^2
41. $f(x) = x^2 + x - 3$

Exercises 2.7 (page 154)

1. (a) -4.01 (b) -4.011025
3. Local maximum ≈ 0.38 when $x \approx -0.58$; local minimum ≈ -0.38 when $x \approx 0.58$
5. Local maximum 0 when $x = 0$;
local minimum ≈ -13.61 when $x \approx -1.71$;
local minimum ≈ -73.32 when $x \approx 3.21$
7. Local maximum ≈ 5.66 when $x \approx 4.00$
9. Local maximum ≈ 0.38 when $x \approx -1.73$;
local minimum ≈ -0.38 when $x \approx 1.73$ **11.** 7.5 mi/h
13. Height ≈ 1.44 ft, length and width of base ≈ 2.88 ft
15. Width ≈ 8.40 ft, straight height ≈ 4.20 ft
17. Width ≈ 3.27, height ≈ 5.33

Exercises 2.8 (page 158)

1. $(f + g)(x) = x^2 + 5$, $(-\infty, \infty)$;
$(f - g)(x) = x^2 - 2x - 5$, $(-\infty, \infty)$;
$(fg)(x) = x^3 + 4x^2 - 5x$, $(-\infty, \infty)$;
$(f/g)(x) = (x^2 - x)/(x + 5)$, $(-\infty, -5) \cup (-5, \infty)$
3. $(f + g)(x) = \sqrt{1 + x} + \sqrt{1 - x}$, $[-1, 1]$;
$(f - g)(x) = \sqrt{1 + x} - \sqrt{1 - x}$, $[-1, 1]$;
$(fg)(x) = \sqrt{1 - x^2}$, $[-1, 1]$;
$(f/g)(x) = \sqrt{(1 + x)/(1 - x)}$, $[-1, 1)$
5. $(f + g)(x) = \sqrt{x} + \sqrt[3]{x}$, $[0, \infty)$;
$(f - g)(x) = \sqrt{x} - \sqrt[3]{x}$, $[0, \infty)$;
$(fg)(x) = x^{5/6}$, $[0, \infty)$; $(f/g)(x) = \sqrt[6]{x}$, $(0, \infty)$
7. $\{x \mid -3 \leq x \leq 4, \quad x \neq \pm\sqrt{2}\} =$
$\left[-3, -\sqrt{2}\right) \cup \left(-\sqrt{2}, \sqrt{2}\right) \cup \left(\sqrt{2}, 4\right]$

9.

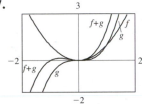

11.

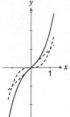

13.

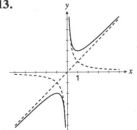

15.

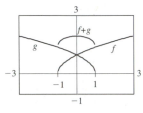

17.

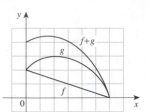

Exercises 2.9 (page 161)

1. 1   **3.** 16   **5.** -11 **7.** -29 **9.** $1 - 3x^2$
11. $9x - 20$ **13.** 4   **15.** 5   **17.** 4
19. $(f \circ g)(x) = 8x + 1$, $(-\infty, \infty)$; $(g \circ f)(x) = 8x + 11$,
$(-\infty, \infty)$; $(f \circ f)(x) = 4x + 9$, $(-\infty, \infty)$;
$(g \circ g)(x) = 16x - 5$, $(-\infty, \infty)$
21. $(f \circ g)(x) = 3(6x^2 + 7x + 2)$, $(-\infty, \infty)$;
$(g \circ f)(x) = 6x^2 - 3x + 2$, $(-\infty, \infty)$;
$(f \circ f)(x) = 8x^4 - 8x^3 + x$, $(-\infty, \infty)$;
$(g \circ g)(x) = 9x + 8$, $(-\infty, \infty)$
23. $(f \circ g)(x) = \sqrt{x^2 - 1}$, $(-\infty, -1] \cup [1, \infty)$;
$(g \circ f)(x) = x - 1$, $[1, \infty)$;
$(f \circ f)(x) = \sqrt{\sqrt{x - 1} - 1}$, $[2, \infty)$;
$(g \circ g)(x) = x^4$, $(-\infty, \infty)$
25. $(f \circ g)(x) = \sqrt[3]{1 - \sqrt{x}}$, $[0, \infty)$;
$(g \circ f)(x) = 1 - \sqrt[6]{x}$, $[0, \infty)$; $(f \circ f)(x) = \sqrt[9]{x}$, $(-\infty, \infty)$;
$(g \circ g)(x) = 1 - \sqrt{1 - \sqrt{x}}$, $[0, 1]$
27. $(f \circ g)(x) = (3x - 4)/(3x - 2)$, $x \neq 2$, $x \neq \frac{2}{3}$;
$(g \circ f)(x) = -(x + 2)/(3x)$, $x \neq -\frac{1}{2}$, $x \neq 0$;
$(f \circ f)(x) = (5x + 4)/(4x + 5)$, $x \neq -\frac{1}{2}$, $x \neq -\frac{5}{4}$;
$(g \circ g)(x) = x/(4 - x)$, $x \neq 2$, $x \neq 4$

29. $(f \circ g \circ h)(x) = \sqrt{x - 1} - 1$
31. $(f \circ g \circ h)(x) = (\sqrt{x} - 5)^4 + 1$
33. $g(x) = x - 9, f(x) = x^5$
35. $g(x) = x^2, f(x) = x/(x + 4)$
37. $f(x) = |x|, g(x) = 1 - x^3$
39. $h(x) = x^2, g(x) = x + 1, f(x) = 1/x$
41. $f(x) = x^9, g(x) = 4 + x, h(x) = \sqrt[3]{x}$
43. $A(t) = 3600\pi t^2$ **45.** $g(x) = x^2 + x - 1$ **47.** Yes

Exercises 2.10 (page 168)

1. No   **3.** Yes   **5.** No   **7.** Yes   **9.** Yes
11. No   **13.** No   **15.** Yes   **17.** 2   **19.** 1
21. $f^{-1}(x) = \frac{1}{4}(x - 7)$
23. $f^{-1}(x) = (5x - 1)/(2x + 3)$
25. $f^{-1}(x) = \frac{1}{5}(x^2 - 2), x \geq 0$ **27.** $f^{-1}(x) = \frac{1}{5}(3 - x)$
29. $f^{-1}(x) = \sqrt{4 - x}$ **31.** $f^{-1}(x) = (x - 4)^3$

33. (a), (b)

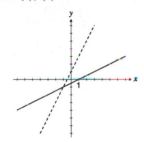

(c) $f^{-1}(x) = \frac{1}{2}(x - 1)$

35. (a), (b)

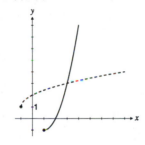

(c) $f^{-1}(x) = x^2 - 2x,$
$x \geq 1$

37. (a), (b)

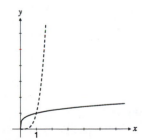

(c) $f^{-1}(x) = \sqrt[4]{x}$

41. $m \neq 0, f^{-1}(x) = \dfrac{x - b}{m}$

Review Exercises for Chapter 2 (page 169)

1. $3, 1 + 2\sqrt{2}, 1 + \sqrt{a}, 1 + \sqrt{-x - 1}, 1 + \sqrt{x^2 - 1},$
$x + 2\sqrt{x - 1}$

3. (b), (c) are graphs of functions, (c) is one-to-one
5. $(-\infty, \infty), [4, \infty)$ **7.** $(-1, \infty)$
9. $\{x \mid x \neq -\frac{1}{2}, x \neq 3\}$

11.

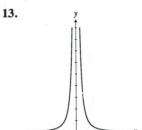

13.

15.

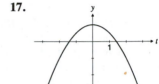

17.

19.

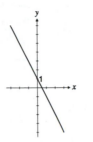

(3,-3)

21.

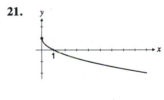

23.

25.

27.

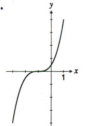

29. (c)

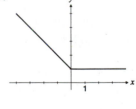
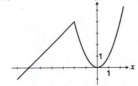

31.

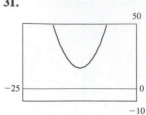

33.

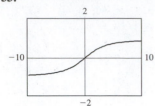

35. $[-2.1, 0.2] \cup [1.9, \infty)$
37. (a) Shift 8 units upward   (b) Shift 8 units left
(c) Stretch vertically by a factor 2, then shift 1 unit upward
(d) Shift 2 units right and 2 units downward
(e) Reflect in x-axis   (f) Reflect in line $y = x$
39. $f\left(-\frac{1}{2}\right) = \frac{5}{4}$
41. Local maximum ≈ 3.79 when $x \approx 0.46$;
local minimum ≈ 2.81 when $x \approx -0.46$
43. (a) Neither   (b) Odd   (c) Even   (d) Neither
45. $(f \circ g)(x) = -3x^2 + 6x - 1, R$;
$(g \circ f)(x) = -9x^2 + 12x - 3, R$; $(f \circ f)(x) = 9x - 4, R$;
$(g \circ g)(x) = -x^4 + 4x^3 - 6x^2 + 4x, R$
47. $(f \circ g \circ h)(x) = 1 + \sqrt{x}$ **49.** Yes   **51.** No
53. No   **55.** $f^{-1}(x) = \frac{1}{3}(x + 2)$

57. (a), (b)

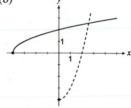

(c) $f^{-1}(x) = \sqrt{x + 4}$

59. $M = 8z$ **61.** (a) $I = k/d^2$ (b) 64,000 (c) 160 candles
63. $A = b\sqrt{4 - b}$
65. (a) $A(x) = (x^2/16) + (\sqrt{3}/36)(10 - x)^2, 0 \le x \le 10$
(b) $40\sqrt{3}/(9 + 4\sqrt{3}) \approx 4.35$ m
67. Square with side 10 m

Chapter 2 Test (page 172)

1. (a), (b) are graphs of functions, (a) is one-to-one
2. $(2, \infty)$

3. (a)

(b)

4. (a)

(b) $f(4) = 12$

5. (a) $-3, 3$
(b)

6. Shift 3 units right, reflect in x-axis, then shift 2 units upward
7. $(f \circ g)(x) = 4x^2 - 8x + 2, (g \circ f)(x) = 2x^2 + 4x - 5$
8. (a) $f^{-1}(x) = 3 - x^2,$ (b)
$x \ge 0$

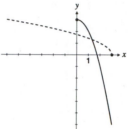

9. (a) $A = 400x - 2x^2$ (b) 100 ft
10. (a)

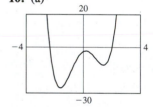

(b) No
(c) Local minimum ≈ -27.18
when $x \approx -1.61$.
Local maximum ≈ -2.55
when $x \approx 0.18$.
Local minimum ≈ -11.93
when $x \approx 1.43$.
(d) $[-27.18, \infty)$

FOCUS ON PROBLEM SOLVING
(page 175)

1. 9   **3.** $f_n(x) = x^{2^{n+1}}$
5. 15,999,999,999,992,000,000,000,001

7.

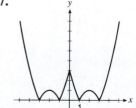

9.

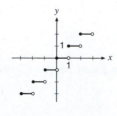

11.

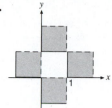

13.

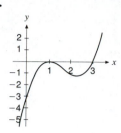

15.

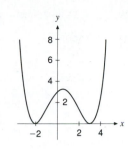

17.

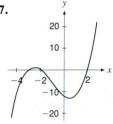

19.

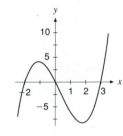

CHAPTER 3

Exercises 3.1 (page 182)

1.

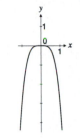

3.

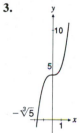

21.

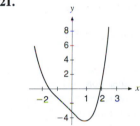

23.

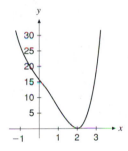

5.

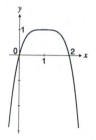

7.

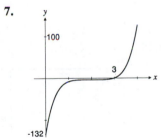

25.

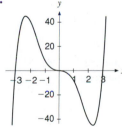

9.

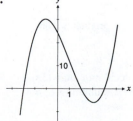

11.

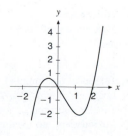

27. (a)

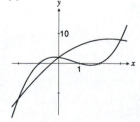

(b) Three (c) (0, 2), (3, 8), (−2, −12)

29. (a) $V(x) = 2x^2(18 - x)$

(b)

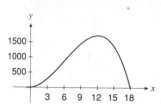

(c) $0 < x < 18$

Exercises 3.2 (page 187)

1.

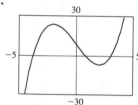

x-intercepts: 3, 0.79, −3.79
y-intercept: 9
local maximum:
(−2, 25)
local minimum: (2, −7)

3.

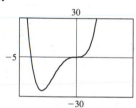

x-intercepts: −4, 0
y-intercept: 0
local minimum:
(−3, −27)

5.

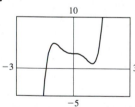

x-intercept: −1.42
y-intercept: 3
local maximum: (−1, 5)
local minimum: (1, 1)

7. Local maximum: (−0.33, 0.19);
local minimum: (1.00, −1.00)
9. Local maximum: (0, 4): local minima: (−1.58, −2.25),
(1.58, −2.25)
11. No extrema
13. Local maximum: (−0.50, 0); local minimum (2.50, −54)
15. Local maximum: (0.44, 0.33);
local minima: (1.09, −1.15), (1.12, −3.36)

17.

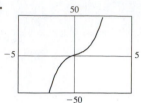

$y \to \infty$ as $x \to \infty$
$y \to -\infty$ as $x \to -\infty$

19.

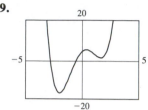

$y \to \infty$ as $x \to \pm\infty$

21.

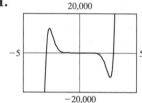

$y \to \infty$ as $x \to \infty$
$y \to -\infty$ as $x \to -\infty$

23.

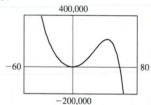

$y \to -\infty$ as $x \to \infty$
$y \to \infty$ as $x \to -\infty$

25.

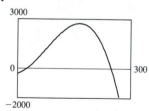

(a) 26 blenders
(b) Maximum profit is $3276.22 when producing 166
blenders.

27. (a) Local maximum (1.8, 2.1); local minimum (3.6, −0.6)
(b) Local maximum (1.8, 7.1); local minimum (3.6, 4.4)
29. (a) 3 x-intercepts, 2 local extrema
(b) 1 x-intercept, no local extrema

Exercises 3.3 (page 195)

In answers 1–25, the first polynomial given is the quotient and the second is the remainder.

1. $4x + 5$, 12   **3.** $x^2 + 5$, 13
5. $x^4 - x^3 + 2x^2 - 2x + 3$, -3
7. $x^5 + 5x^4 + 10x^3 + 10x^2 + 5x + 1$, 0
9. $x + 2$, $8x - 1$ **11.** $3x + 1$, $7x - 5$
13. $x^3 + 1$, 0   **15.** $2x^2 - 1$, $4x^2 + 2x$
17. 0, $6x - 10$ **19.** $2x^2 + 4x$, 1
21. $x^{100} + x^{99} + x^{98} + \cdots + x^3 + x^2 + x + 1$, 0
23. $\frac{1}{2}x + \frac{1}{2}$, -1 **25.** $\frac{1}{2}x^3 - \frac{3}{4}x^2 - \frac{3}{8}x + \frac{5}{16}$, $\frac{37}{16}$
27. 17   **29.** 1.92   **31.** 20   **33.** -273
35. 100   **37.** $\frac{49}{64}$ **39.** 5   **45.** $-1 \pm \sqrt{6}$
47. $-\frac{3}{2}x^3 + 3x^2 + \frac{15}{2}x - 9$
49. $x^4 - 4x^3 - 7x^2 + 22x + 24$ **51.** No
53. $\dfrac{-1 \pm \sqrt{13}}{3}$

Exercises 3.4 (page 205)

1. ± 1, ± 7
3. ± 1, ± 3, ± 5, ± 15, $\pm\frac{1}{2}$, $\pm\frac{3}{2}$, $\pm\frac{5}{2}$, $\pm\frac{15}{2}$, $\pm\frac{1}{4}$, $\pm\frac{3}{4}$, $\pm\frac{5}{4}$, $\pm\frac{15}{4}$
5. 3 or 1 positive, 1 negative; 4 or 2 real
7. 1 positive, 1 negative; 2 real
9. 2 or 0 positive, 0 negative; 3 or 1 real (since 0 is a root)
11. 5, 3, or 1 positive, 2 or 0 negative; 7, 5, 3, or 1 real
17. 3, -1 **19.** 3, -1 **21.** -2, $\frac{3}{2}$ **23.** 1
25. $-2, 2, 3$ **27.** $-1, 2, 3$ **29.** $1, 2, 4$ **31.** $-6, 1$
33. -1 **35.** ± 1 **37.** $-2, 1, 2$
39. $1, -1, -2, -4$ **41.** ± 2, $\pm\frac{3}{2}$ **43.** $-\frac{3}{2}, \frac{1}{2}, 1$
45. -1, $\pm\frac{1}{2}$ **47.** $-2, \frac{1}{2}, \pm 1$ **49.** $\pm\frac{1}{2}$, $\sqrt{5}$
51. $-2, 1, 3, 4$ **53.** $-1, \frac{3}{4}, 4$ **55.** ± 1 **65.** $2, -1$

Exercises 3.5 (page 211)

1. 1.3   **3.** 0.79   **5.** -0.54 **7.** 0.26   **9.** 4.18
11. 0.62   **13.** $2, -2 \pm\sqrt{2}$ **15.** $\frac{1}{2}, -1, -1.47$
17. $-0.88, 1.35, 2.53$ **19.** $1, -2, \dfrac{-1 \pm \sqrt{5}}{2}$
21. $\frac{1}{2}, -0.71$ **23.** 2.626 ft $\times$ 3.808 ft   **25.** 2.76 m

Exercises 3.6 (page 217)

1. $-2, -1, 1$ **3.** $-\frac{3}{2}, -1, 1, 4$ **5.** $-1, \frac{5}{2}$
7. -2 **9.** 0.67   **11.** $-1.00, 1.50$
13. $-0.93, 1.11, 5.82$ **15.** $-2.00, -1.24, 2.00, 3.24$
17. -1.50 **19.** $-1.41, 1.41, 3.00$ **21.** 11.3 ft
23. 10 in. $\times$ 10 in. $\times$ 4 in., 13.8 in. $\times$ 13.8 in. $\times$ 2.1 in.

Exercises 3.7 (page 224)

1. Real part 5, imaginary part -6
3. Real part 2, imaginary part $\dfrac{\sqrt{3}}{2}$
5. Real part $\sqrt{2} - 2$, imaginary part $\sqrt{3} - \sqrt{5}$
7. $10 - i$ **9.** $\frac{11}{2} + \frac{9}{2}i$ **11.** $-19 + 4i$
13. $8 + 24i$ **15.** $13 - i$ **17.** $-33 - 56i$
19. $-i$ **21.** $-\frac{1}{2} + \frac{1}{2}i$ **23.** $-5 + 12i$
25. $-\frac{8}{13} + \frac{12}{13}i$ **27.** $-i$ **29.** 1   **31.** $5i$
33. -6 **35.** $\left(3 + 5\sqrt{3}\right) + \left(3\sqrt{15} - \sqrt{5}\right)i$
37. $\dfrac{2 + 7\sqrt{2}}{15} + \dfrac{\sqrt{7} - 2\sqrt{14}}{15}i$ **39.** $-\dfrac{i}{6}$

41. 3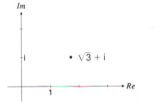

43. $\sqrt{29}$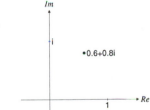

45. 2

47. 1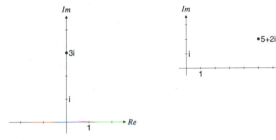

49.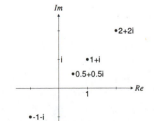

51.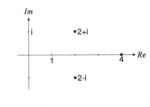

53.

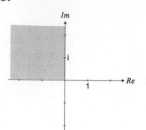

55.

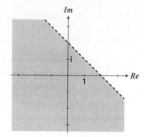

57.

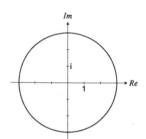

59.

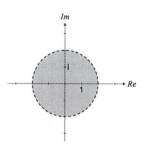

Exercises 3.8 (page 232)

1. $\pm 2i$ **3.** $\dfrac{1 \pm i\sqrt{3}}{2}$ **5.** $-2 \pm 2i$

7. $\dfrac{5 \pm i\sqrt{23}}{6}$ **9.** $4 \pm i$ **11.** $\dfrac{-3 \pm i\sqrt{3}}{2}$

13. $0, i$ **15.** $x^3 - 2x^2 + x - 2$

17. $x^4 - 8x^3 + 39x^2 - 62x + 50$

19. $2x^4 - 8x^3 + 16x^2 - 16x + 8$ **21.** $2, \pm 2i$

23. $\pm 2i, \pm i\sqrt{2}/2$ **25.** $\pm 1, \pm i$ **27.** $-2, 1 \pm i\sqrt{3}$

29. $\pm 2, \pm 2i$ **31.** $\pm 3, \dfrac{\pm 3 \pm 3\sqrt{3}i}{2}$ **33.** $\pm i\sqrt{5}$

35. $-2, \pm 2i$ **37.** $1, \dfrac{1 \pm i\sqrt{3}}{2}$ **39.** $2, \dfrac{1 \pm i\sqrt{3}}{2}$

41. $-2, 1, \pm 3i$

43. $(x + 3)\left(x - \dfrac{3 + 3\sqrt{3}i}{2}\right)\left(x - \dfrac{3 - 3\sqrt{3}i}{2}\right)$

45. $(x - 2)(x + 2)\left[x + \left(1 + i\sqrt{3}\right)\right]\left[x + \left(1 - i\sqrt{3}\right)\right] \cdot$
$\left[x - \left(1 + i\sqrt{3}\right)\right]\left[x - \left(1 - i\sqrt{3}\right)\right]$

47. $2\left(x + \tfrac{3}{2}\right)\left(x + 1 - i\sqrt{2}\right)\left(x + 1 + i\sqrt{2}\right)$

51. $-22i$

53. (a) $x^4 - 2x^3 + 3x^2 - 2x + 2$
(b) $x^2 - (1 + 2i)x - 1 + i$

Exercises 3.9 (page 244)

1. x-intercept 6; y-intercept -6
3. x-intercept 0; y-intercept 0
5. no x-intercept, no y-intercept
7. Vertical: $x = -3$; horizontal: $y = 0$
9. Vertical: $x = 3$, $x = -2$; horizontal: $y = 1$
11. Horizontal: $y = 0$
13. Vertical: $x = 1$; slant: $y = x + 1$
15. Vertical: $x = -3$

17.

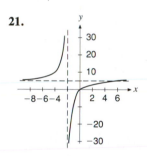

19.

21.

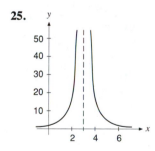

23.

25.

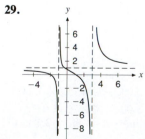

27.

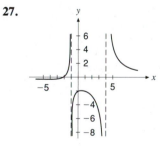

29.

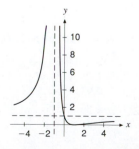

31.

33.

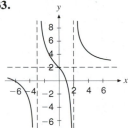

35.

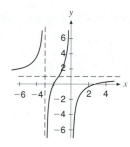

37.

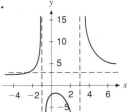

39.

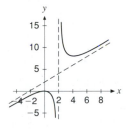

41.

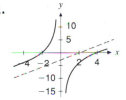

43.

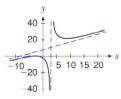

45.

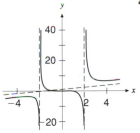

47.

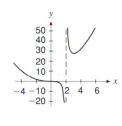

51.

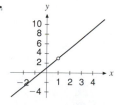

53.

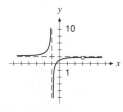

55. $y = \dfrac{x^2 - 5x + 6}{x^2 + 3x - 4}$ **57.** $y = \dfrac{6x^2 - 15x}{2x - 1}$

Exercises 3.10 (page 248)

1. Vertical: $x = -2$; horizontal: $y = 3$

3. Vertical: $x = -3$, $x = 3$; horizontal: $y = 3$

5.

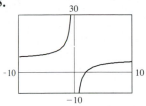

Vertical: $x = 0$
horizontal: $y = 7$
x-intercept: 2
y-intercept: none
no extrema

7.

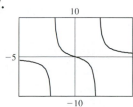

Vertical: $x = -2$, $x = 2$
horizontal: $y = 0$
x-intercept: 0
y-intercept: 0
no extrema

9.

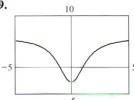

Vertical: none
horizontal: $y = 6$
x-intercept: ± 1
y-intercept: -3
local minimum $(0, -3)$

11.

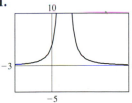

Vertical: $x = 1$
horizontal: $y = 0$
x-intercept: none
y-intercept: 4
no extrema

13.

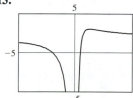

Vertical: $x = 0$
horizontal: $y = 2$
x-intercepts: -2, $\frac{1}{2}$
y-intercept: none
local maximum
$(1.33, 3.13)$

15.

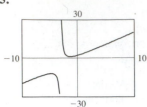

Vertical: $x = -3$

17.

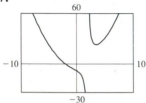

Vertical: $x = 2$

19.

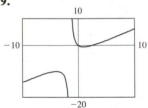

Vertical: $x = -1.5$
x-intercepts: 0, 2.5
y-intercept: 0
local maximum:
    $(-3.9, -10.4)$
local minimum: $(0.9, -0.6)$
end behavior: $y = x - 4$

21.

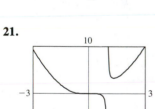

Vertical: $x = 1$
x-intercept: 0
y-intercept: 0
local minimum: $(1.4, 3.1)$
end behavior: $y = x^2$

23.

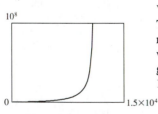

Vertical: $v \approx 11{,}000$
The vertical asymptote
represents the escape
velocity from the earth's
gravitational pull:
$11{,}000$ m/s ≈ 1900 mi/h

25. $y = \dfrac{2x^2}{(x - 1)(x - 4)}$ [Other answers are also possible.]

Exercises 3.11 (page 255)

1. $(-\infty, -4) \cup (5, \infty)$

3. $\left(-\infty, \dfrac{-3 - 3\sqrt{5}}{2}\right] \cup \left[\dfrac{-3 + 3\sqrt{5}}{2}, \infty\right)$ **5.** $[-4, 4]$

7. $\left[-4, \frac{1}{3}\right]$ **9.** $(-2, 4)$ **11.** $\left[-2, \frac{7}{3}\right]$ **13.** All $x \in R$

15. $[-1 - 2\sqrt{2}, -1 + 2\sqrt{2}]$ **17.** $(-\infty, 1) \cup \left(\frac{7}{2}, \infty\right)$

19. $(-\infty, -4] \cup [-2, 1] \cup [3, \infty)$ **21.** $[-1, 1]$

23. $(-\infty, -3) \cup \left(\frac{1}{2}, 3\right)$ **25.** $(-\infty, -3] \cup [1, \infty)$

27. $(-3.53, 3.53)$ **29.** $(0, 2) \cup (3, \infty)$

31. $\left[-\frac{1}{3}, \frac{1}{2}\right] \cup [1, \infty)$ **33.** $(1, 10)$

35. $\left(-7, \frac{5}{2}\right] \cup (5, \infty)$

37. $(-\infty, -1 - \sqrt{3}) \cup [0, -1 + \sqrt{3})$

39. $(-\infty, -3) \cup \left(-\frac{2}{3}, 1\right) \cup (3, \infty)$ **41.** $(-4, 3]$

43. $\left[-8, -\frac{5}{2}\right)$ **45.** $\left(0, \dfrac{3 - \sqrt{3}}{2}\right] \cup \left(1, \dfrac{3 + \sqrt{3}}{2}\right]$

47. $(-\infty, -2) \cup (-1, 1) \cup (1, \infty)$ **49.** $[-2, 0) \cup (1, 3]$

51. $\left(-3, -\frac{1}{2}\right) \cup (2, \infty)$ **53.** $(0, 1]$

55. $(-\infty, -2) \cup (5, \infty)$ **57.** $\left(-\frac{1}{2}, 0\right) \cup \left(\frac{1}{2}, \infty\right)$

59. $[-2, 3]$ **61.** $(-\infty, -1] \cup [1, \infty)$

63. $(-\infty, a] \cup [b, c] \cup [d, \infty)$ **65.** $[-2, 1] \cup [3, \infty)$

67. $(-\infty, -1.37) \cup (0.37, 1.00)$ **69.** $(0, 1.6)$

Review Exercises for Chapter 3 (page 256)

1.

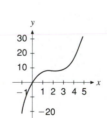

3.

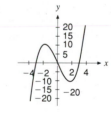

5.

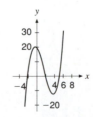

7.

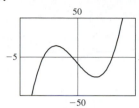

x-int: −3, −0.5, 3
y-int: −9
local maximum:
  (−1.9, 15.1)
local minimum:
  (1.6, −27.1)
$y \to \infty$ as $x \to \infty$,
$y \to -\infty$ as $x \to -\infty$

9.

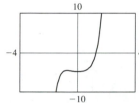

x-int: 1.3; y-int: −5
local maximum:
  (−0.7, −4.7)
local minimum: (0, −5)
$y \to \infty$ as $x \to \infty$,
$y \to -\infty$ as $x \to -\infty$

*In answers 11–17, the first polynomial given is the quotient
and the second is the remainder.*

11. $x^2 + 2x + 7$, 10   **13.** $x - 3$, −9

15. $x^3 - 5x^2 + 4$, −5

17. $x^3 + \left(\sqrt{3} + 1\right)x^2 + \left(\sqrt{3} + 1\right)x + \sqrt{3}$, 2

19. 3   **23.** 8   **25.** ±1, ±2, ±3, ±6, ±18   **27.** 1.62

29. 1.33   **31.** $4x^3 - 18x^2 + 14x + 12$

33. No; since the complex conjugates of imaginary zeros
will also be zeros, the polynomial would have 8 zeros,
contradicting the requirement that it have degree 4.

35. −3, 1, 5   **37.** $-1 \pm 2i$, −2 (multiplicity 2)

39. ± 2, 1 (multiplicity 3)   **41.** ±2, $\pm 1 \pm i\sqrt{3}$

43. 1, 3, $\dfrac{-1 \pm i\sqrt{7}}{2}$ **45.** $-1, 2, 2 \pm i\sqrt{2}$ **47.** 1.57

49. $x = -0.5, 3$ **51.** $x \approx -0.24, 4.24$

53.

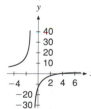

55.

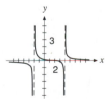

57.

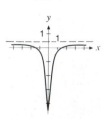

59.

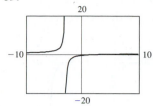

x-int: 3; y-int: −0.5
vertical: $x = -3$
horizontal: $y = 0.5$
no local extrema

61.

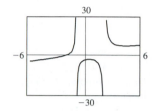

x-int: −2; y-int: −4
vertical: $x = -1$, $x = 2$
slant: $y = x + 1$
local maximum:
  (0.425, −3.599)
local minimum:
  (4.216, 7.175)

63. $(-\infty, -1] \cup \left[\frac{3}{2}, \infty\right)$ **65.** $(-3, 3)$ **67.** $\left(-2, \frac{1}{7}\right]$

69. $\left[-3, \frac{3}{8}\right]$ **71.** [0.74, 1.94]

Chapter 3 Test (page 259)

1.

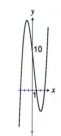

2. (a) ±1, ±2, ±3, ±4, ±6, ±9, ±12, ±18, ±36,
$\pm\frac{1}{2}, \pm\frac{3}{2}, \pm\frac{9}{2}$ (b) 4, 2, or 0 positive real roots; 0 negative
real roots   (d) $\frac{1}{2}$, 1, $\frac{3}{2}$, 2, 3, 4, $\frac{9}{2}$, 6   (e) $\frac{3}{2}$, 2 (double), 3
(f) $2\left(x - \frac{3}{2}\right)(x - 2)^2(x - 3)$
3. $x^5 + x^4 + 2x^3 + 10x^2 + 13x + 5$
4. (a) P and Q: by Rational Roots Theorem;
R: by Descartes' Rule   (b) No; Descartes' Rule
(c) Two (one positive, one negative) by Descartes' Rule
(d) Only possible rational roots are 1 and −1, neither of
which is a root

5. 1.2   **6.** 4, $\frac{1}{2}$, $2 \pm \sqrt{3}$ **7.** ±2i, $\dfrac{-1 \pm i\sqrt{3}}{2}$

8. (a) *r*, *u*,   (b) *s*   (c) *s*   (d)

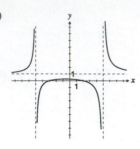

9. $(-\infty, -1] \cup \left(\frac{5}{2}, 3\right]$ **10.** $\left(-1 - \sqrt{5}, -1 + \sqrt{5}\right)$

11. (a)

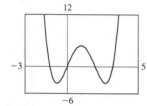

x-intercepts:
   $-1.24, 0, 2, 3.24$
local maximum: $(1, 5)$
local minima:
   $(-0.73, -4), (2.73, -4)$

(b) $(-\infty, -1.24] \cup [0, 2] \cup [3.24, \infty)$

FOCUS ON PROBLEM SOLVING
(page 261)

1. Yes; possible only for odd *n*
3. *Hint:* Consider another monk making the trip from the top at the same time the first monk leaves the bottom.
5. $x^3 - 6x^2 + 12x - 6$; none   **7.** 11 cm

CHAPTER 4

Exercises 4.1 (page 269)

1.

3.

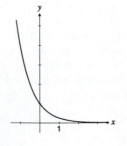

5.

7. $R, (-\infty, 0), y = 0$

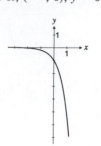

9. $R, (-3, \infty), y = -3$

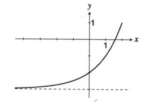

11. $R, (4, \infty), y = 4$

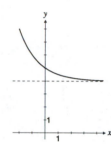

13. $R, (0, \infty), y = 0$

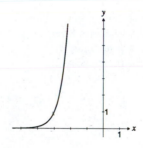

15. $R, (-\infty, 0), y = 0$

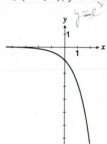

17. $R, (-1, \infty), y = -1$

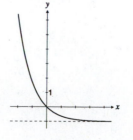

19. $R, (0, \infty), y = 0$

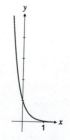

21. R, $(-\infty, 5)$, $y = 5$

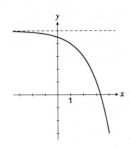

23. R, $[1, \infty)$, no asymptote

(iii)

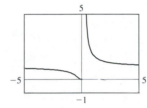

f ultimately grows much more quickly than g

(b) 1.2, 22.4

43. -0.57 **45.** 0.36

47.

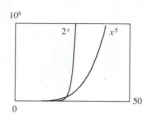

Vertical asymptote
$x = 0$
horizontal asymptote
$y = 1$

25.

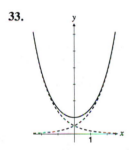

27. ± 1
29. $0, \frac{4}{3}$
31. $-\sqrt{2} < x < \sqrt{2}$

33.

39. (a)

(b)

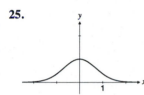

41. (a) (i)

(ii)

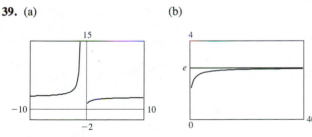

49. Local maximum ≈ 0.37 when $x \approx 1.00$.
51. Local minimum ≈ 0.69 when $x \approx 0.37$.
53. (a) Increasing on $(-\infty, 0.50]$, decreasing on $[0.50, \infty)$
(b) $(0, 1.78]$

Exercises 4.2 (page 276)

1. (a) $(1500)2^{2t}$ (b) 2381   (c) $(1500)2^{48} \approx 4.22 \times 10^{17}$
3. (a) $(10,000)3^t$ (b) $(10,000)3^{2.5} \approx 1.56 \times 10^5$
5. (a) 6.6 billion   (b) 46.9 billion
7. $(150)2^{-t/1590}$, 97.0 mg
9. (a) 12 g (b) 7.135 g   (c) 0.315 g
11. (a) \$16,288.95   (b) \$26,532.98   (c) \$43,219.42
13. (a) \$4,615.87   (b) \$4,658.91   (c) \$4,697.04
(d) \$4,703.11   (e) \$4,704.68   (f) \$4,704.93
(g) \$4,704.94   **15.** \$7,678.96
17. (a) $m(t) = (300)2^{-t/140}$ (b) ≈ 82 days
19. About 8.33 yr

Exercises 4.3 (page 283)

1. $2^6 = 64$ **3.** $10^{-2} = 0.01$ **5.** $8^3 = 512$
7. $a^c = b$ **9.** $\log_2 8 = 3$ **11.** $\log_{10} 0.0001 = -4$
13. $\log_4 0.125 = -\frac{3}{2}$ **15.** $\log_r t = s$ **17.** 4
19. 3   **21.** 1   **23.** -3 **25.** $\frac{1}{2}$ **27.** -1
29. $-\frac{2}{3}$ **31.** 37   **33.** 4   **35.** $\sqrt{5}$ **37.** 1024
39. $\frac{95}{3}$ **41.** 2   **43.** $\log_5 16$ **45.** $1 - \log_2 3$
47. e^{10} **49.** $3 - e^2$ **51.** $\frac{1}{12} \ln 17$ **53.** 3, -2
55. 3^{16}

57.

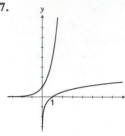

59. $(4, \infty)$, R, $x = 4$

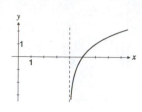

79.

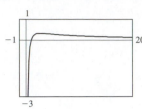

Domain $= (0, \infty)$
Vertical asymptote
 $x = 0$
Horizontal asymptote
 $y = 0$
Local maximum
 ≈ 0.37 when $x \approx 2.72$

81. f grows more slowly than g.   **83.** 2.21
85. 0.00, 1.14   **87.** About 265 mi
89. $\log_{10} 2 < x < \log_{10} 5$ **91.** (a) $(1, \infty)$
(b) $f^{-1}(x) = 10^{2^x}$ **93.** $\log_2 3$

Exercises 4.4 (page 289)

1. $\log_2 x + \log_2(x - 1)$ **3.** $23 \log 7$
5. $\log_2 A + 2 \log_2 B$ **7.** $\log_3 x + \frac{1}{2} \log_3 y$
9. $\frac{1}{3} \log_5(x^2 + 1)$ **11.** $\frac{1}{2}(\ln a + \ln b)$
13. $3 \log x + 4 \log y - 6 \log z$
15. $\log_2 x + \log_2(x^2 + 1) - \frac{1}{2} \log_2(x^2 - 1)$
17. $\ln x + \frac{1}{2}(\ln y - \ln z)$ **19.** $\frac{1}{4} \log(x^2 + y^2)$
21. $\frac{1}{3}[\log(x^2 + 4) - \log(x^2 + 1) - 2 \log(x^3 - 7)]$
23. $\frac{1}{2} \ln x + 4 \ln z - \frac{1}{3} \ln(y^2 + 6y + 17)$ **25.** $\frac{3}{2}$
27. 1   **29.** 3   **31.** $\ln 8$ **33.** 16   **35.** $\log_3 160$
37. $\log_2(AB/C^2)$ **39.** $\log[x^4(x - 1)^2/\sqrt[3]{x^2 + 1}]$
41. $\ln[5x^2(x^2 + 5)^3]$

43. $\log\left[\sqrt[3]{2x + 1} \sqrt{(x - 4)/(x^4 - x^2 - 1)}\right]$ **45.** No
47. Yes   **49.** No   **51.** Yes   **53.** Yes   **55.** $\frac{3}{2}$
57. 5   **59.** 5   **61.** $\frac{13}{12}$ **63.** 6   **65.** 6.584963
67. 0.943028   **69.** -43.067656 **71.** 2.149159
73. 6.212567   **75.** -2.946865 **77.** 2.807355
79. 2.182658

81.

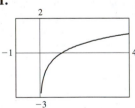

87. $2 < x < 4$ or
 $7 < x < 9$

61. $(-\infty, 0)$, R, $x = 0$

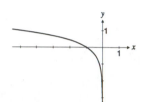

63. $(0, \infty)$, R, $x = 0$

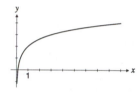

65. $(0, \infty)$, R, $x = 0$

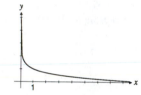

67. $(0, \infty)$, $[0, \infty)$, $x = 0$

69. $\left(-\frac{2}{5}, \infty\right)$ **71.** $(-\infty, -1) \cup (1, \infty)$ **73.** $(0, 2)$

75.

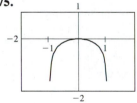

Domain $= (-1, 1)$
Vertical asymptotes $x = 1$,
 $x = -1$
Local maximum value 0
 when $x = 0$

77.

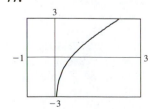

Domain $= (0, \infty)$
Vertical asymptote
 $x = 0$
No maximum or
 minimum

Exercises 4.5 (page 298)

1. (a) 2.3   (b) 3.5   (c) 8.3   **3.** (a) 10^{-3} M
(b) 3.2×10^{-7} M   **5.** $4.8 \leq$ pH ≤ 6.4
7. 42.5 min   **9.** 43.5 min   **11.** 2103   **13.** 16 yr
15. 149 h   **17.** (a) 137 °F   (b) 116 min   **19.** 5 yr

21. 8.15 yr   **23.** $\log 20 \approx 1.3$ **25.** Twice as intense
27. 6.3×10^{-3} watts/m^2 **29.** $t = -\frac{5}{13} \ln\left(1 - \frac{13}{60}I\right)$
31. (b) 106 dB

Review Exercises for Chapter 4 (page 300)

1. R, $(0, \infty)$, $y = 0$ **3.** R, $(-\infty, 5)$, $y = 5$

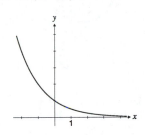

5. $(1, \infty)$, R, $x = 1$ **7.** $(0, \infty)$, R, $x = 0$

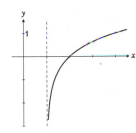

9. R, $(-1, \infty)$, $y = -1$ **11.** $(0, \infty)$, R, $x = 0$

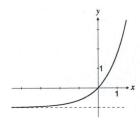

13. $\left(-\infty, \frac{1}{2}\right)$ **15.** $2^{10} = 1024$ **17.** $10^y = x$
19. $\log_2 64 = 6$ **21.** $\log 74 = x$ **23.** 7   **25.** 45
27. 6   **29.** -3 **31.** $\frac{1}{2}$ **33.** 2   **35.** 92
37. $\frac{2}{3}$ **39.** $\log A + 2 \log B + 3 \log C$
41. $\frac{1}{2}[\ln(x^2 - 1) - \ln(x^2 + 1)]$
43. $2 \log_5 x + \frac{3}{2} \log_5(1 - 5x) - \frac{1}{2} \log_5(x^3 - x)$
45. $\log 96$ **47.** $\log_2[(x - y)^{3/2}/(x^2 + y^2)^2]$
49. $\log\left[(x^2 - 4)/\sqrt{x^2 + 4}\right]$ **51.** -15

53. $\frac{1}{3}(5 - \log_5 26)$ **55.** $\frac{4}{3} \ln 10$ **57.** 3
59. $-4, 2$ **61.** 0.430618   **63.** 2.303600

65.

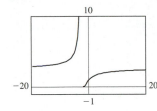

Vertical asymptote
$x = -2$
Horizontal asymptote
$y = 2.72$
No maximum or
minimum

67.

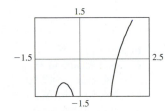

Vertical asymptotes
$x = -1$, $x = 0$, $x = 1$
Local maximum $\approx$
-0.41 when $x \approx -0.58$

69. 2.42   **71.** $0.16 < x < 3.15$
73. Increasing on $(-\infty, 0]$ and $[1.10, \infty)$, decreasing on $[0, 1.10]$
75. 1.953445   **77.** $\log_4 258$
79. 7.9, basic   **81.** (a) 5278   (b) 58 min
83. (a) 9.97 mg   (b) 1.39×10^5 yr
85. (a) $16,081.15   (b) $16,178.18   (c) $16,197.65
(d) $16,198.31   **87.** 8.0

Chapter 4 Test (page 302)

1.

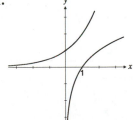

2. $(3, \infty)$, R, $x = 3$ **3.** (a) $\frac{2}{3}$ (b) 2
4. $\frac{1}{2}[\log(x^2 - 1) - 3 \log x - 5 \log(y^2 + 1)]$
5. $\ln\left[x\sqrt{3 - x^4}/(x^2 + 1)^2\right]$ **6.** (a) $\log_2 10$ (b) $3 - e^4$

7. (a) 20 min   (b) $(1000)2^{3t}$ (c) 22,627   (d) 1.3 h
(e)

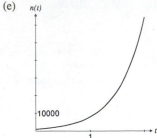

8. $\left(-4, \frac{8}{5}\right)$

9. (a)

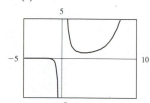

(b) $x = 0$, $y = 0$
(c) Local minimum $\approx$
    0.74 when $x \approx 3.00$
(d) $(-\infty, 0) \cup [0.74, \infty)$
(e) -0.85, 0.96, 9.92

FOCUS ON PROBLEM SOLVING

(page 305)

3. 686   **5.** 5   **7.** *Hint:* Use contradiction
9. Infinitely far

CHAPTER 5

Exercises 5.1 (page 316)

7. $P\left(\frac{3}{5}, \frac{4}{5}\right)$ **9.** $P\left(\frac{2}{3}, -\sqrt{5}/3\right)$ **11.** $P\left(\sqrt{2}/3, -\sqrt{7}/3\right)$

13. Quadrant I   **15.** Quadrant II

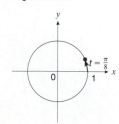   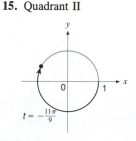

17. Quadrant III   **19.** Quadrant III

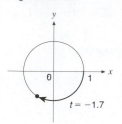   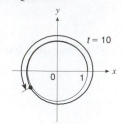

21. $\left(\sqrt{2}/2, \sqrt{2}/2\right)$ **23.** $\left(\frac{1}{2}, -\sqrt{3}/2\right)$ **25.** $(-1, 0)$
27. $\left(-\frac{1}{2}, \sqrt{3}/2\right)$ **29.** $\left(-\sqrt{2}/2, -\sqrt{2}/2\right)$
31. (a) $\left(-\frac{3}{5}, \frac{4}{5}\right)$ (b) $\left(\frac{3}{5}, -\frac{4}{5}\right)$ (c) $\left(-\frac{3}{5}, -\frac{4}{5}\right)$ (d) $\left(\frac{3}{5}, \frac{4}{5}\right)$
33. (a) $\pi/4$ (b) $\pi/3$ (c) $\pi/3$ (d) $\pi/6$ **35.** (a) $\pi/5$
(b) $\pi/6$ (c) $\pi/3$ (d) $\pi/6$ **37.** (a) $\pi/4$
(b) $\left(-\sqrt{2}/2, \sqrt{2}/2\right)$ **39.** (a) $\pi/3$ (b) $\left(-\frac{1}{2}, -\sqrt{3}/2\right)$
41. (a) $\pi/4$ (b) $\left(-\sqrt{2}/2, -\sqrt{2}/2\right)$ **43.** (a) $\pi/6$
(b) $\left(-\sqrt{3}/2, -\frac{1}{2}\right)$ **45.** (a) $\pi/3$ (b) $\left(\frac{1}{2}, \sqrt{3}/2\right)$
47. (a) $\pi/3$ (b) $\left(-\frac{1}{2}, -\sqrt{3}/2\right)$ **49.** $(0.5, 0.8)$
51. $(0.5, -0.9)$

Exercises 5.2 (page 325)

1. (a) 0   (b) 1   **3.** (a) 0   (b) -1 **5.** (a) 1   (b) -1
7. (a) $\sqrt{2}/2$ (b) $-\sqrt{2}/2$ **9.** (a) $\frac{1}{2}$ (b) $-\frac{1}{2}$
11. (a) $\frac{1}{2}$ (b) $-2\sqrt{3}/3$ **13.** (a) $\sqrt{3}/3$ (b) $-\sqrt{3}/3$
15 (a) 2   (b) $-2\sqrt{3}/3$ **17.** (a) $\sqrt{2}/2$ (b) $\sqrt{2}$
19. (a) -1 (b) -1
21. $\sin 0 = 0$, $\cos 0 = 1$, $\tan 0 = 0$, $\sec 0 = 1$, others undefined
23. $\sin \pi = 0$, $\cos \pi = -1$, $\tan \pi = 0$, $\sec \pi = -1$, others undefined   **25.** $\frac{4}{5}, \frac{3}{5}, \frac{4}{3}$ **27.** $-\sqrt{13}/7, \frac{6}{7}, -\sqrt{13}/6$
29. $\frac{9}{41}, \frac{40}{41}, \frac{9}{40}$ **31.** $-\frac{12}{13}, -\frac{5}{13}, \frac{12}{5}$ **33.** (a) 0.8
(b) 0.84147   **35.** (a) 0.9   (b) 0.93204   **37.** (a) 1
(b) 1.02964   **39.** (a) -0.6 (b) -0.57482
41. Positive   **43.** Negative   **45.** II   **47.** II
49. $\sin t = \sqrt{1 - \cos^2 t}$ **51.** $\tan t = (\sin t)/\sqrt{1 - \sin^2 t}$
53. $\sec t = -\sqrt{1 + \tan^2 t}$ **55.** $\tan t = \sqrt{1 - \sec^2 t}$
57. $\tan^2 t = (\sin^2 t)/(1 - \sin^2 t)$
59. $(1 - \sin^2 t)/\sec t = \cos^3 t$
61. $\cos t = -\frac{4}{5}$, $\tan t = -\frac{3}{4}$, $\csc t = \frac{5}{3}$, $\sec t = -\frac{5}{4}$,
$\cot t = -\frac{4}{3}$
63. $\sin t = -\frac{3}{5}$, $\cos t = \frac{4}{5}$, $\csc t = -\frac{5}{3}$, $\sec t = \frac{5}{4}$,
$\cot t = -\frac{4}{3}$
65. $\sin t = -\sqrt{3}/2$, $\cos t = \frac{1}{2}$, $\tan t = -\sqrt{3}$,
$\csc t = -2\sqrt{3}/3$, $\cot t = -\sqrt{3}/3$

67. $\cos t = -\sqrt{15}/4$, $\tan t = \sqrt{15}/15$, $\csc t = -4$, $\sec t = -4\sqrt{15}/15$, $\cot t = \sqrt{15}$ **69.** Odd   **71.** Odd
73. Even   **75.** Neither

Exercises 5.3 (page 335)

1.

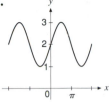

3.

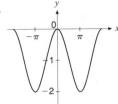

5.

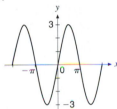

7.

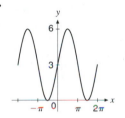

9.

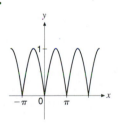

11. $2\pi/3$, 3

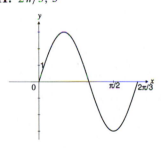

13. 4π, 10

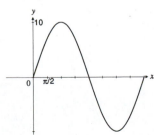

15. 6π, 1

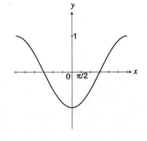

17. $\frac{2}{3}$, 3

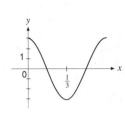

19. 3π, 1

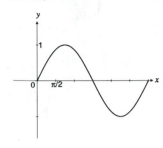

21. 1, 2π, $\pi/2$

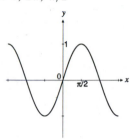

23. 2, 2, $\frac{1}{3}$

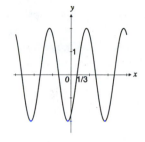

25. 2, 2π, $\pi/2$

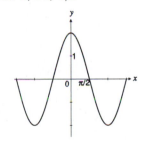

27. 5, $2\pi/3$, $\pi/12$

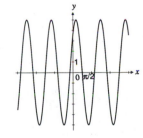

29. 2, 3π, $\pi/4$

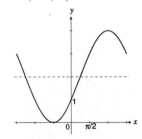

31. 3, 2, $-\frac{1}{2}$

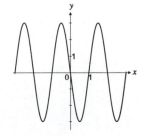

33. 2, π, $\pi/6$

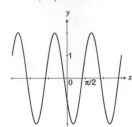

35. 1, $2\pi/3$, $-\pi/3$

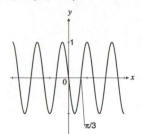

13. 2π

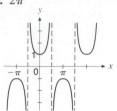

15. π

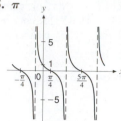

37. (a) 4, 2π, 0   (b) $y = 4 \sin x$ **39.** (a) 3, 4π, 0
(b) $y = 3 \sin \frac{1}{2}x$ **41.** (a) $\frac{1}{2}$, π, $-\pi/3$
(b) $y = \frac{1}{2} \sin 2(x + \pi/3)$ **43.** Period 2
45. Not periodic   **47.** 2π **49.** π **51.** 2π **53.** π

17. 2π

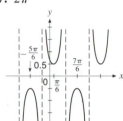

19. $\pi/2$

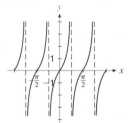

Exercises 5.4 (page 344)

1. π

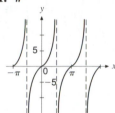

3. π

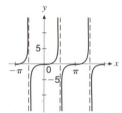

5. π

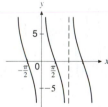

7. 2π

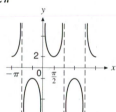

21. 1

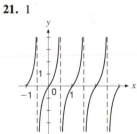

23. π

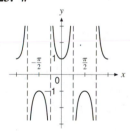

9. 2π

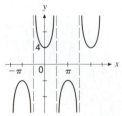

11. π

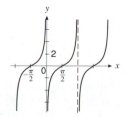

25. π

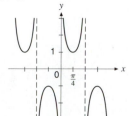

27. $\pi/3$

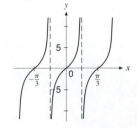

29. $2\pi/3$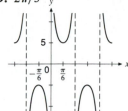

31. $\pi/2$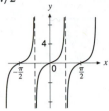

45. 2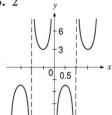

47. $\pi/2$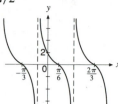

33. $\pi/2$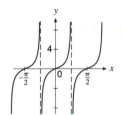

35. $\pi/2$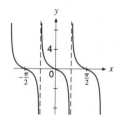

49. $2\pi/3$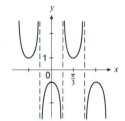

Exercises 5.5 (page 349)

37. 2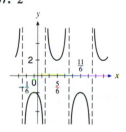

39. 2π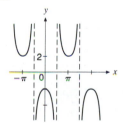

1.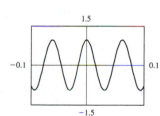

3.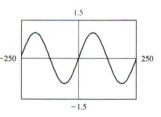

41. $2\pi/3$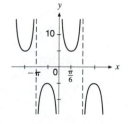

43. $3\pi/2$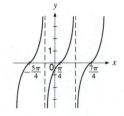

5.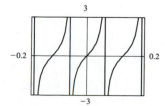

7.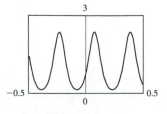

9.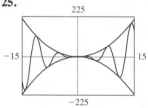

11.

13. Maximum value 1.76 when $x \approx 0.94$, minimum value -1.76 when $x \approx -0.94$ (The same maximum and minimum values occur at infinitely many other values of x.)

15. Maximum value 6.97 when $x \approx 5.24$, minimum value -0.68 when $x \approx 1.04$

17. Maximum value 3.00 when $x \approx 1.57$, minimum value -1.00 when $x \approx -1.57$ (The same maximum and minimum values occur at infinitely many other values of x.)

19. 1.16   **21.** 0.34, 2.80   **23.** 0.00, 1.90

25.

$y = x^2 \sin x$ is a sine curve that lies between the graphs of $y = x^2$ and $y = -x^2$

27.

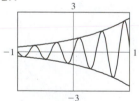

$y = e^x \sin 5\pi x$ is a sine curve that lies between the graphs of $y = e^x$ and $y = -e^x$

29.

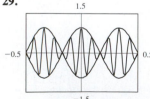

$y = \cos 3\pi x \cos 21\pi x$ is a cosine curve that lies between the graphs of $y = \cos 3\pi x$ and $y = -\cos 3\pi x$

31. (a)

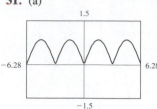

(b) Period π (c) Even

33. (a)

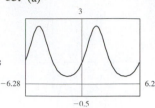

(b) Period 2π (c) Neither

35. (a)

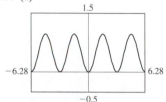

(b) Period π (c) Even

37. (a) Odd   (b) 0, $\pm 2\pi$, $\pm 4\pi$, $\pm 6\pi$, . . .

(c)

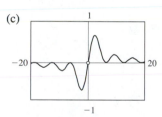

(d) $f(t)$ approaches 0   (e) $f(t)$ approaches 0

Exercises 5.6 (page 357)

1. (a) 4, $\frac{1}{4}$, 4   (b) 0   **3.** (a) 0.3, π, $1/\pi$ (b) 0.3

(c)

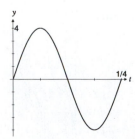

(c)

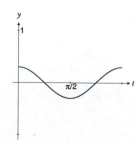

5. (a) 1000, 1, 1   (b) 0   **7.** (a) 1, 2, $\frac{1}{2}$ (b) 0

(c)

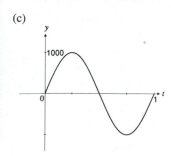

(c)

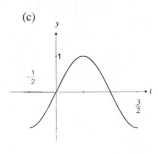

9. (a) 8, $\frac{2}{3}$, $\frac{3}{2}$ (b) 4

(c)

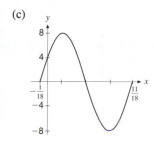

11. $y = 6 \sin 4\pi t$ **13.** $y = 2 \sin 6\pi t$
15. $y = -6 \sin(\pi t/6)$ **17.** $y = 8 \sin 20\pi t$
19. $y = 5 \cos 2\pi t$ **21.** $y = 11 + 10 \sin(\pi t/10)$
23. $y = 155 \sin 120\pi t$ **25.** $E(t) = 310 \cos 200\pi t$, 219 V
27. $R = 20 + 1.5 \sin(2\pi t/5.4)$, R in millions of miles
29. (a) $P(t) = 10 \sin 40\pi t$

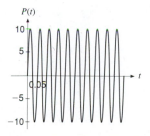

(b) The period decreases. The period increases.
31. $y = 10e^{-2.1t} \cos 10\pi t$ **33.** (a) $y = 6e^{-2.8t} \cos 4\pi t$
(b) 0.9 s   **35.** (a) 0.8   (b) $y = 3e^{-0.8t} \cos 330\pi t$

Review Exercises for Chapter 5 (page 361)

1. (b) $\frac{1}{2}$, $\sqrt{3}/2$, $\sqrt{3}/3$ **3.** (a) $\pi/6$ (b) $\left(\sqrt{3}/2, \frac{1}{2}\right)$
(c) $\sin t = \frac{1}{2}$, $\cos t = \sqrt{3}/2$, $\tan t = \sqrt{3}/3$, $\csc t = 2$,

$\sec t = 2\sqrt{3}/3$, $\cot t = \sqrt{3}$ **5.** (a) $\pi/4$
(b) $\left(\sqrt{2}/2, \sqrt{2}/2\right)$ (c) $\sin t = \sqrt{2}/2$, $\cos t = \sqrt{2}/2$,
$\tan t = 1$, $\csc t = \sqrt{2}$, $\sec t = \sqrt{2}$, $\cot t = 1$
 7. (a) $\sqrt{2}/2$ (b) $-\sqrt{2}/2$
 9. (a) 0.89121   (b) 0.45360   **11.** (a) 0   (b) Undefined
13. (a) Undefined   (b) 0   **15.** (a) 0.41421   (b) 2.41421
17. $(\sin t)/(1 - \sin^2 t)$ **19.** $(\sin t)/\sqrt{1 - \sin^2 t}$
21. $\tan t = -\frac{5}{12}$, $\csc t = \frac{13}{5}$, $\sec t = -\frac{13}{12}$, $\cot t = -\frac{12}{5}$
23. $\sin t = 2\sqrt{5}/5$, $\cos t = -\sqrt{5}/5$, $\tan t = -2$,
$\sec t = -\sqrt{5}$ **25.** $\left(16 - \sqrt{17}\right)/4$ **27.** $2 + \sqrt{3}$

29. (a) 10, 4π, 0   **31.** (a) 1, 4π, 0
(b)   (b)

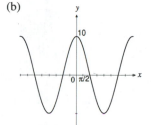 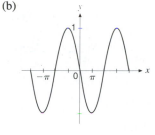

33. (a) 3, π, 1   **35.** (a) 1, 4, $-\frac{1}{3}$
(b)   (b)

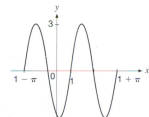 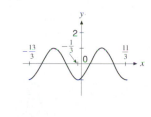

37. $y = 5 \sin 4x$ **39.** $y = \frac{1}{2} \sin 2\pi\left(x + \frac{1}{3}\right)$

41. π **43.** π

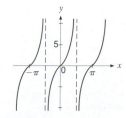 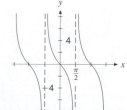

45. π

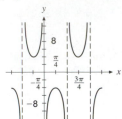

47. 2π

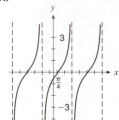

49. (a)

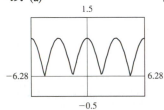

(b) Period π (c) Even

51. (a)

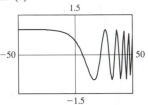

(b) Not periodic (c) Neither

(b) Not periodic (c) Even

53. (a)

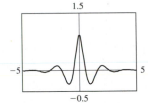

55.

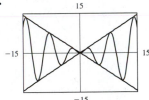

$y = x \sin x$ is a sine function whose graph lies between those of $y = x$ and $y = -x$

57.

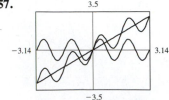

The graphs are related by graphical addition.

59. 1.76, -1.76 **61.** 0.30, 2.84
63. (a) Odd (b) 0, $\pm\pi$, $\pm2\pi$, ...
(c)

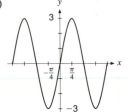

(d) $f(t)$ approaches 0   (e) $f(t)$ approaches 0

65. $y = 5e^{-0.3t} \cos 4\pi t$ **67.** $y = -50 \cos 8\pi t$

Chapter 5 Test (page 363)

1. $-\frac{12}{13}$ **2.** (a) $\frac{3}{5}$ (b) $-\frac{4}{5}$ (c) $\frac{3}{4}$ (d) $\frac{5}{4}$
3. $\tan t = -(\sin t)/\sqrt{1 - \sin^2 t}$ **4.** $-\frac{2}{15}$

5. (a) 3, π, 0   **6.** (a) 1, 4π, $\pi/3$
(b)   (b)

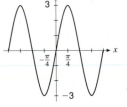

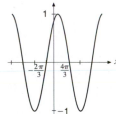

7. π **8.** $\pi/2$

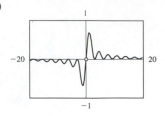

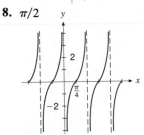

9. $y = 2 \sin 2(x + \pi/3)$ **10.** $y = 4 \cos(\pi t/6)$
11. $y = 5 \sin 4\pi t$
12. (a)

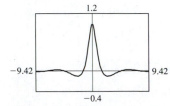

(b) Even   (c) Not periodic
(d) Minimum value -0.11 when $x \approx \pm 2.54$, maximum value 1 when $x = 0$
13. (a)

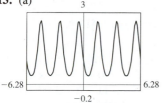

(b) Neither   (c) Period $2\pi/3$
(d) Minimum value 0.37, maximum value 2.72 (Both values are attained at infinitely many values of x.)
14. (a)

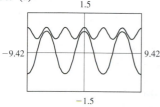

(b) y_1 has period π, y_2 has period 2π
(c) $\sin(\cos x) < \cos(\sin x)$

FOCUS ON PROBLEM SOLVING
(page 366)

1. 63   **3.** 3   **5.** (a) None (b) 20   **7.** 0

CHAPTER 6

Exercises 6.1 (page 375)

1. $2\pi/9 \approx 0.698$ rad   **3.** $2\pi/5 \approx 1.257$ rad
5. $\pi/4 \approx 0.785$ rad   **7.** $17\pi/4 \approx 13.352$ rad
9. $\pi/5 \approx 0.628$ rad   **11.** $-630°$
13. $360°/\pi \approx 114.6°$ **15.** $40°$ **17.** $36°$
19. $660°, 1020°, -60°, -420°$
21. $11\pi/4, 19\pi/4, -5\pi/4, -13\pi/4$
23. $332°, 692°, -28°, -748°$ **25.** $9.8, 16.1, -2.8, -9.1$
27. $137°, 857°, -223°, -583°$ **29.** Yes   **31.** No
33. No   **35.** Yes   **37.** $13°$ **39.** $63°$ **41.** $280°$
43. $2\pi/5$ **45.** π **47.** $\pi/4$ **49.** $55\pi/9 \approx 19.2$
51. 4   **53.** 4 mi   **55.** 2 rad $\approx 114.6°$
57. $36/\pi \approx 11.459$ m   **59.** $330\pi \approx 1037$ mi
61. 1.6 million mi   **63.** 1.15 mi   **65.** 50 m^2
67. 4 m   **69.** $\frac{5}{32}$ rad   **71.** 6 cm^2

Exercises 6.2 (page 385)

1. $\sin \theta = \frac{4}{5}$, $\cos \theta = \frac{3}{5}$, $\tan \theta = \frac{4}{3}$, $\csc \theta = \frac{5}{4}$, $\sec \theta = \frac{5}{3}$, $\cot \theta = \frac{3}{4}$
3. $\sin \theta = \frac{4}{7}$, $\cos \theta = \sqrt{33}/7$, $\tan \theta = 4\sqrt{33}/33$, $\csc \theta = \frac{7}{4}$, $\sec \theta = 7\sqrt{33}/33$, $\cot \theta = \sqrt{33}/4$
5. $\sin \theta = 5\sqrt{34}/34$, $\cos \theta = 3\sqrt{34}/34$, $\tan \theta = \frac{5}{3}$, $\csc \theta = \sqrt{34}/5$, $\sec \theta = \sqrt{34}/3$, $\cot \theta = \frac{3}{5}$
7. (a) $2\sqrt{13}/13$, $2\sqrt{13}/13$ (b) $\frac{2}{3}, \frac{2}{3}$
(c) $\sqrt{13}/3$, $\sqrt{13}/3$ **9.** (a) $\frac{5}{13}, \frac{5}{13}$ (b) $\frac{5}{12}, \frac{5}{12}$ (c) $\frac{13}{12}, \frac{13}{12}$
11. $\frac{25}{2}$ **13.** $13\sqrt{3}/2$ **15.** 16.51658
17. $x = 28 \cos \theta$, $y = 28 \sin \theta$
19. $x = 4 \tan \theta$, $y = 4 \sec \theta$
21. $\cos \theta = \frac{4}{5}$, $\tan \theta = \frac{3}{4}$, $\csc \theta = \frac{5}{3}$, $\sec \theta = \frac{5}{4}$, $\cot \theta = \frac{4}{3}$
23. $\sin \theta = \sqrt{2}/2$, $\cos \theta = \sqrt{2}/2$, $\tan \theta = 1$, $\csc \theta = \sqrt{2}$, $\sec \theta = \sqrt{2}$
25. $\sin \theta = 4\sqrt{3}/7$, $\cos \theta = \frac{1}{7}$, $\tan \theta = 4\sqrt{3}$, $\csc \theta = 7\sqrt{3}/12$, $\cot \theta = \sqrt{3}/12$ **27.** $\dfrac{1 + \sqrt{3}}{2}$
29. 1   **31.** $\frac{1}{2}$ **33.** $\frac{1}{2}$ **35.** 0.26, 0.97, 0.27
37.   **39.**
41. 1026 ft   **43.** (a) 2100 mi   (b) No   **45.** 19 ft
47. $38.7°$ **49.** 345 ft   **51.** 415 ft, 152 ft
53. 2570 ft   **55.** 5808 ft   **57.** 91.7 million mi
59. 3960 mi   **61.** 230.9   **63.** 63.7   **65.** $\sin \theta \cos \theta$
67. $a = \sin \theta$, $b = \tan \theta$, $c = \sec \theta$, $d = \cos \theta$

Exercises 6.3 (page 398)

1. (a) $45°$ (b) $35°$ (c) $1°$ **3.** (a) $50°$ (b) $50°$
(c) $26°$ **5.** (a) $2\pi/5$ (b) $\pi/6$ (c) $\pi/3$ **7.** (a) 1.4
(b) $7 - 2\pi \approx 0.72$ (c) 0.7   **9.** $\frac{1}{2}$ **11.** $\sqrt{2}/2$
13. $-\sqrt{3}/2$ **15.** 1   **17.** $-\sqrt{3}/2$ **19.** $\sqrt{3}/3$
21. 1   **23.** -2 **25.** $-\sqrt{3}/2$ **27.** $\frac{1}{2}$
29. $-\sqrt{3}/3$ **31.** $-\sqrt{2}$ **33.** $\sqrt{2}/2$
35. Undefined   **37.** $-\frac{1}{2}$ **39.** Negative   **41.** Positive
43. III   **45.** IV   **47.** $\tan \theta = -\dfrac{\sqrt{1 - \cos^2\theta}}{\cos \theta}$
49. $\cos \theta = \sqrt{1 - \sin^2\theta}$ **51.** $\sec \theta = -\sqrt{1 + \tan^2\theta}$
53. $\sin \theta = -\dfrac{\sqrt{\sec^2\theta - 1}}{\sec \theta}$
55. $\cos \theta = -\frac{4}{5}$, $\tan \theta = -\frac{3}{4}$, $\csc \theta = \frac{5}{3}$, $\sec \theta = -\frac{5}{4}$, $\cot \theta = -\frac{4}{3}$

57. $\sin \theta = -\frac{3}{5}$, $\cos \theta = \frac{4}{5}$, $\csc \theta = -\frac{5}{3}$, $\sec \theta = \frac{5}{4}$,
$\cot \theta = -\frac{4}{3}$
59. $\sin \theta = \frac{1}{2}$, $\cos \theta = \sqrt{3}/2$, $\tan \theta = \sqrt{3}/3$,
$\sec \theta = 2\sqrt{3}/3$, $\cot \theta = \sqrt{3}$
61. $\sin \theta = 3\sqrt{5}/7$, $\tan \theta = -3\sqrt{5}/2$, $\csc \theta = 7\sqrt{5}/15$,
$\sec \theta = -\frac{7}{2}$, $\cot \theta = -2\sqrt{5}/15$ **63.** (a) $\sqrt{3}/2$, $\sqrt{3}$
(b) $\frac{1}{2}$, $\sqrt{3}/4$ (c) $\frac{3}{4}$, 0.88967 **65.** 19.1
67. $25\sqrt{3} \approx 43.3$ **69.** $(4\pi/3) - \sqrt{3} \approx 2.46$
71. (a) $\frac{1}{2}\sqrt{2 - \sqrt{3}}$, $\frac{1}{2}\sqrt{2 + \sqrt{3}}$
(b) $\frac{1}{2}\sqrt{2 - \sqrt{2}}$, $\frac{1}{2}\sqrt{2 + \sqrt{2}}$
(c) $\frac{1}{2}\sqrt{2 - \sqrt{2 + \sqrt{3}}}$, $\frac{1}{2}\sqrt{2 + \sqrt{2 + \sqrt{3}}}$

Exercises 6.4 (page 405)

1. 21.5 **3.** 134.6 **5.** $44°$
7. $\angle C = 114°$, $a \approx 51$, $b \approx 24$
9. $\angle C = 62°$, $a \approx 200$, $b \approx 242$

11. $\angle B = 85°$, $a \approx 5$, $c \approx 9$ **13.** $\angle A = 100°$, $a \approx 89$,
$c \approx 71$

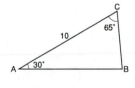

15. $\angle B \approx 30°$, $\angle C \approx 40°$, $c \approx 19$ **17.** No solution
19. $\angle A_1 \approx 125°$, $\angle C_1 \approx 30°$, $a_1 \approx 49$; $\angle A_2 \approx 5°$,
$\angle C_2 \approx 150°$, $a_2 \approx 5.6$ **21.** No solution   **23.** 219 ft
25. (a) 1018 mi   (b) 1017 mi   **27.** 155 ft
29. (a) $d\dfrac{\sin \alpha}{\sin(\beta - \alpha)}$ (c) 2350 ft.   **31.** $48.2°$

Exercises 6.5 (page 412)

1. 13 **3.** $29.89°$ **5.** 15
7. $\angle A \approx 39.3°$, $\angle B \approx 20.7°$, $c \approx 24.6$
9. $\angle A \approx 48°$, $\angle B \approx 79°$, $c \approx 3.2$
11. $\angle A \approx 50°$, $\angle B \approx 73°$, $\angle C \approx 57°$
13. $\angle A_1 \approx 83.6°$, $\angle C_1 \approx 56.4°$, $a_1 \approx 193$; $\angle A_2 \approx 16.4°$,
$\angle C_2 \approx 123.6$, $a_2 \approx 54.9$ **15.** No such triangle   **17.** 2

19. 25 **21.** $84.6°$ **23.** 24 **25.** 2.30 mi
27. 23.1 mi   **29.** 2179 mi   **31.** $96°$ **33.** 211 ft
35. 3835 ft   **37.** 3.85 cm^2 **39.** 14.3 m
41. $\$165,554$

Review Exercises for Chapter 6 (page 415)

1. (a) $7\pi/18 \approx 1.22$ (b) $7\pi/3 \approx 7.33$
(c) $-4\pi/3 \approx -4.19$ (d) $-2\pi/9 \approx -0.70$ **3.** (a) $630°$
(b) $-60°$ (c) $315°$ (d) $(378/\pi)° \approx 120.3°$ **5.** 8 m
7. 82 ft   **9.** 0.619 rad $\approx 35.4°$ **11.** $18{,}151$ ft^2
13. $\sin \theta = 5/\sqrt{74}$, $\cos \theta = 7/\sqrt{74}$, $\tan \theta = \frac{5}{7}$,
$\csc \theta = \sqrt{74}/5$, $\sec \theta = \sqrt{74}/7$, $\cot \theta = \frac{7}{5}$
15. $x \approx 3.83$, $y \approx 3.21$ **17.** $x \approx 2.92$, $y \approx 3.11$
19.

21. $a = \cot \theta$, $b = \csc \theta$ **23.** 48 m   **25.** 1076 mi
27. $-\sqrt{2}/2$ **29.** 1 **31.** $-\sqrt{3}/3$ **33.** $-\sqrt{2}/2$
35. $2\sqrt{3}/3$ **37.** $-\sqrt{3}$
39. $\sin \theta = \frac{12}{13}$, $\cos \theta = -\frac{5}{13}$, $\tan \theta = -\frac{12}{5}$, $\csc \theta = \frac{13}{12}$,
$\sec \theta = -\frac{13}{5}$, $\cot \theta = -\frac{5}{12}$ **41.** $60°$
43. $\tan \theta = \sqrt{1 - \cos^2\theta}/\cos \theta$
45. $\tan^2\theta = \sin^2\theta/(1 - \sin^2\theta)$
47. $\sin \theta = \sqrt{7}/4$, $\cos \theta = \frac{3}{4}$, $\csc \theta = 4\sqrt{7}/7$,
$\cot \theta = 3\sqrt{7}/7$
49. $\cos \theta = -\frac{4}{5}$, $\tan \theta = -\frac{3}{4}$, $\csc \theta = \frac{5}{3}$, $\sec \theta = -\frac{5}{4}$,
$\cot \theta = -\frac{4}{3}$ **51.** $-\sqrt{5}/5$ **53.** 1 **55.** 5.32
57. 148.07 **59.** 77.82 **61.** 77.3 mi   **63.** 3.87 mi
65. 119.2 m   **67.** 14.98

Chapter 6 Test (page 419)

1. $5\pi/3$, $-\pi/10$, $5\pi/4$ **2.** $150°$, $-495°$, $137.5°$
3. (a) $\sqrt{2}/2$ (b) $\sqrt{3}/3$ (c) 2 (d) 1
4. $(26 + 6\sqrt{13})/39$ **5.** $a = \cot \theta$, $b = 1 - \sin \theta$
6. $(4 - 3\sqrt{2})/4$ **7.** $-\frac{13}{12}$ **8.** $\tan \theta = -\sqrt{\sec^2\theta - 1}$
9. 19.6 ft   **10.** 9.1 **11.** 250.5 **12.** 7.9
13. 19.5 **14.** 15.3 m^2 **15.** 24.3 m   **16.** $129.9°$
17. 44.9 **18.** 548 ft

FOCUS ON PROBLEM SOLVING
(page 422)

1. Yes   **3.** No   **5.** Yes   **7.** (b) 12 cm   (c) Plane
9. $b = a[1 + \sin(\theta/2)]/[1 - \sin(\theta/2)]$
11. $4\pi + 6\sqrt{3} \approx 23$ cm

CHAPTER 7

Exercises 7.1 (page 431)

1. $\sin x$ **3.** 1   **5.** $\sec A \csc A$ **7.** $\sec u$
9. $\tan x$ **11.** $\sin y$ **13.** $\sin^2 x$ **15.** $\sec x$
17. $2 \sec u$ **19.** $\cos^2 x$ **21.** $\cos \theta$ **23.** $\cot t$
111. $\tan \theta$ **113.** $\tan \theta$ **115.** $a \cos \theta$ **123.** yes
125. no   **127.** no
129.

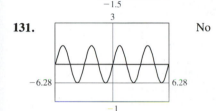

Yes

131.

No

Exercises 7.2 (page 441)

1. $\dfrac{\pi}{3} + 2k\pi, \dfrac{2\pi}{3} + 2k\pi$

3. $\dfrac{\pi}{6} + 2k\pi, \dfrac{\pi}{4} + k\pi, \dfrac{11\pi}{6} + 2k\pi$

5. $\dfrac{\pi}{6} + k\pi, \dfrac{5\pi}{6} + k\pi$ **7.** $\dfrac{\pi}{2} + k\pi$

9. $\dfrac{\pi}{6} + 2k\pi, \dfrac{5\pi}{6} + 2k\pi$ **11.** No solution

13. $\dfrac{\pi}{6} + \dfrac{k\pi}{3}$ **15.** $\dfrac{\pi}{3} + 2k\pi, \dfrac{5\pi}{3} + 2k\pi$

17. $\dfrac{\pi}{6}, \dfrac{3\pi}{4}, \dfrac{5\pi}{6}, \dfrac{7\pi}{4}$ **19.** $\dfrac{\pi}{4}, \dfrac{3\pi}{4}, \dfrac{5\pi}{4}, \dfrac{7\pi}{4}$

21. $\dfrac{\pi}{6}, \dfrac{\pi}{4}, \dfrac{5\pi}{6}, \dfrac{7\pi}{6}, \dfrac{5\pi}{4}, \dfrac{11\pi}{6}$ **23.** $\dfrac{\pi}{3}, \dfrac{2\pi}{3}, \dfrac{4\pi}{3}, \dfrac{5\pi}{3}$

25. $\dfrac{\pi}{6}, \dfrac{\pi}{4}, \dfrac{5\pi}{6}, \dfrac{5\pi}{4}$ **27.** $\dfrac{\pi}{4}, \dfrac{3\pi}{4}, \dfrac{5\pi}{4}, \dfrac{7\pi}{4}$

29. $\dfrac{1}{4}, \dfrac{3}{4}, \dfrac{5}{4}, \dfrac{7}{4}$ **31.** $0, \pi$ **33.** No solution   **35.** $\dfrac{3\pi}{2}$

37. $\dfrac{\pi}{4}, \dfrac{\pi}{3}, \dfrac{2\pi}{3}, \dfrac{5\pi}{4}, \dfrac{4\pi}{3}, \dfrac{5\pi}{3}$ **39.** $0, \dfrac{\pi}{2}, \pi, \dfrac{3\pi}{2}$

41. $\dfrac{\pi}{2} - 1, \dfrac{3\pi}{2} - 1$ **43.** $\dfrac{7\pi}{6}, \dfrac{11\pi}{6}$ **45.** $\dfrac{\pi}{2}, \dfrac{3\pi}{2}$

47. $\dfrac{\pi}{3}, \dfrac{5\pi}{3}$

49. $((2k + 1)\pi, -2)$ **51.** $\left(\dfrac{\pi}{3} + k\pi, \sqrt{3}\right)$

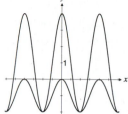

53. $\left(\dfrac{\pi}{3} + 2k\pi, \dfrac{3}{2}\right), \left(\dfrac{5\pi}{3} + 2k\pi, \dfrac{3}{2}\right)$ **55.** $\dfrac{9}{2}$

57. $0, \pm 0.95$ **59.** 1.92   **61.** ± 0.71

Exercises 7.3 (page 448)

1. $(\sqrt{6} - \sqrt{2})/4$ **3.** $-2 - \sqrt{3}$ **5.** $(\sqrt{2} - \sqrt{6})/4$
7. $\sqrt{2}/2$ **9.** $\sqrt{3}$ **11.** $-\dfrac{56}{65}, -\dfrac{16}{65}$ **13.** 0, 1
47. $\dfrac{\pi}{3}, \dfrac{5\pi}{3}$ **49.** $0, \dfrac{\pi}{2}, \pi, \dfrac{3\pi}{2}$ **55.** $\tan \gamma = \dfrac{17}{6}$

57. (a)

$$\sin^2\left(x + \dfrac{\pi}{4}\right) + \sin^2\left(x - \dfrac{\pi}{4}\right) = 1$$

Exercises 7.4 (page 457)

1. (a) $2 \sin 44° \cos 44°$
(b) $1 - 2 \sin^2(\pi/8)$ or $2 \cos^2(\pi/8) - 1$ or
$\cos^2(\pi/8) - \sin^2(\pi/8)$ (c) $2 \sin 2\theta \cos 2\theta$
3. (a) $(1 - \cos 26°)/\sin 26°$ or $\sin 26°/(1 + \cos 26°)$
(b) $\sqrt{(1 - \cos 56°)/2}$
(c) $\left(1 - \cos \dfrac{\theta}{4}\right)/\sin \dfrac{\theta}{4}$ or $\sin \dfrac{\theta}{4}/\left(1 + \cos \dfrac{\theta}{4}\right)$
5. (a) $\sin 36°$ (b) $\sin 6\theta$ **7.** (a) $\cos 68°$ (b) $\cos 10\theta$
9. (a) $\tan 4°$ (b) $\tan 2\theta$ **11.** $\dfrac{1}{2}\sqrt{2 - \sqrt{3}}$
13. $\dfrac{1}{2}\sqrt{2 - \sqrt{2 + \sqrt{3}}}$ **15.** $\dfrac{1}{2}\sqrt{2 + \sqrt{2 + \sqrt{2}}}$
17. $\dfrac{1}{2}\sqrt{2 - \sqrt{3}}$ **19.** $\dfrac{120}{169}, \dfrac{119}{169}, \dfrac{120}{119}$ **21.** $-\dfrac{24}{25}, -\dfrac{7}{25}, \dfrac{24}{7}$

23. $\frac{24}{25}, \frac{7}{25}, \frac{24}{7}$ **25.** $\sqrt{10}/10, 3\sqrt{10}/10, \frac{1}{3}$

27. $\sqrt{(3 + 2\sqrt{2})/6}, \sqrt{(3 - 2\sqrt{2})/6}, 3 + 2\sqrt{2}$

29. $\sqrt{6}/6, -\sqrt{30}/6, -\sqrt{5}/5$

49. (a)

$$\frac{\sin 3x}{\sin x} - \frac{\cos 3x}{\cos x} = 2$$

51. $\frac{\pi}{6}, \frac{\pi}{2}, \frac{5\pi}{6}, \frac{3\pi}{2}$ **53.** $\frac{\pi}{2}, \frac{7\pi}{6}, \frac{11\pi}{6}$ **55.** 0

57. (a)

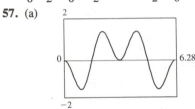

(b) 0, 1.57, 3.14, 4.71   (c) 0, $\frac{\pi}{2}, \pi, \frac{3\pi}{2}$

59. (a)

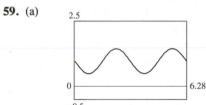

(b) None   (c) None

Exercises 7.5 (page 464)

1. $\frac{1}{2}(\sin 5x - \sin x)$ **3.** $\frac{3}{2}(\cos 11x + \cos 3x)$

5. $(\sqrt{2} + \sqrt{3})/2$ **7.** $-5 - 2\sqrt{6}$ **9.** $2 \sin 4x \cos x$

11. $2 \sin 5x \sin x$ **13.** $-2 \cos \frac{9}{2}x \sin \frac{5}{2}x$

15. $2 \cos 10\pi x \cos \pi x$ **17.** $\sqrt{6}/2$ **19.** $\sqrt{2}/2$

29. $k\pi/2$

31. $\frac{\pi}{9} + \frac{2k\pi}{3}, \frac{\pi}{2} + k\pi, \frac{5\pi}{9} + \frac{2k\pi}{3}$

33. $\frac{\pi}{8} + \frac{k\pi}{4}, \frac{\pi}{6} + 2k\pi, \frac{5\pi}{6} + 2k\pi$

35. $\frac{\pi}{4} + \frac{k\pi}{2}, \frac{\pi}{6} + 2k\pi, \frac{5\pi}{6} + 2k\pi$

43. $\sqrt{2} \sin\left(x + \frac{\pi}{4}\right)$ **45.** $6 \sin \pi\left(x + \frac{1}{3}\right)$

47. $2 \sin 2\left(x + \frac{\pi}{12}\right)$

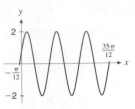

49. $0, \frac{\pi}{6}, \frac{2\pi}{3}, \frac{5\pi}{6}, \frac{4\pi}{3}, \frac{3\pi}{2}$ **51.** $\frac{2}{3}, \frac{5}{3}, \frac{8}{3}, \frac{11}{3}, \frac{14}{3}, \frac{17}{3}$

53. (a)

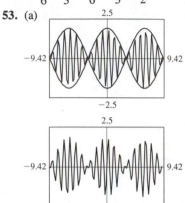

The graph of $y = f(x)$ lies between the two other graphs.

Exercises 7.6 (page 474)

1. (a) $\frac{\pi}{6}$ (b) $\frac{\pi}{3}$ (c) Not defined   **3.** (a) $\frac{\pi}{4}$ (b) $\frac{\pi}{4}$

(c) $-\frac{\pi}{4}$ **5.** (a) $\frac{\pi}{2}$ (b) 0 (c) π **7.** (a) $\frac{\pi}{6}$

(b) $-\frac{\pi}{6}$ (c) Not defined   **9.** (a) 0.87696 (b) 2.09601

(c) 1.50510   **11.** $\frac{1}{2}$ **13.** $\frac{\sqrt{3}}{3}$ **15.** $\frac{1}{2}$ **17.** $\frac{\sqrt{3}}{2}$

19. $\frac{\pi}{3}$ **21.** $\frac{\pi}{3}$ **23.** $\frac{\pi}{3}$ **25.** $\frac{4}{5}$ **27.** $\frac{12}{13}$ **29.** $\frac{24}{25}$

31. $\frac{\sqrt{3}}{2}$ **33.** 1   **35.** 0   **37.** $\frac{\sqrt{10} - 3\sqrt{30}}{20}$

39. $\sqrt{1 - x^2}$ **41.** $\frac{x}{\sqrt{1 - x^2}}$ **43.** $\frac{1 - x^2}{1 + x^2}$

45. $\frac{1 - \sqrt{1 - x^2}}{x}$ **47.** $\frac{x\sqrt{1 - x^2} - 1}{\sqrt{1 + x^2}}$

51. $\theta = \tan^{-1}\frac{50}{s}$ **53.** $\theta = \tan^{-1}\frac{d}{2}$

55. (a) $\tan^{-1}2, \tan^{-1}(-1)$ (b) 1.10715, -0.78540

57. (a) $\tan^{-1}1$ (b) 0.78540   **59.** (a) $\sin^{-1}\frac{1}{3}, \pi - \sin^{-1}\frac{1}{3}$

(b) 0.33984, 2.80176

65. (a)

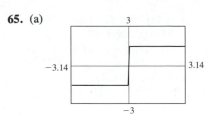

$$\tan^{-1}x + \tan^{-1}\frac{1}{x} = \begin{cases} \pi/2, & x > 0 \\ -\pi/2, & x < 0 \end{cases}$$

67. (a) 0.71   **(b)** $\dfrac{\sqrt{2}}{2}$

Exercises 7.7 (page 483)

1. $\sqrt{2}\left(\cos\dfrac{\pi}{4} + i\sin\dfrac{\pi}{4}\right)$ **3.** $2\left(\cos\dfrac{7\pi}{4} + i\sin\dfrac{7\pi}{4}\right)$

5. $4\left(\cos\dfrac{11\pi}{6} + i\sin\dfrac{11\pi}{6}\right)$

7. $\sqrt{2}\left(\cos\dfrac{3\pi}{2} + i\sin\dfrac{3\pi}{2}\right)$

9. $5\sqrt{2}\left(\cos\dfrac{\pi}{4} + i\sin\dfrac{\pi}{4}\right)$

11. $8\left(\cos\dfrac{11\pi}{6} + i\sin\dfrac{11\pi}{6}\right)$ **13.** $20(\cos\pi + i\sin\pi)$

15. $5\left[\cos\left(\tan^{-1}\frac{4}{3}\right) + i\sin\left(\tan^{-1}\frac{4}{3}\right)\right]$

17. $3\sqrt{2}\left(\cos\dfrac{3\pi}{4} + i\sin\dfrac{3\pi}{4}\right)$ **19.** $8\left(\cos\dfrac{\pi}{6} + i\sin\dfrac{\pi}{6}\right)$

21. $\sqrt{5}\left[\cos\left(\tan^{-1}\frac{1}{2}\right) + i\sin\left(\tan^{-1}\frac{1}{2}\right)\right]$

23. $2\left(\cos\dfrac{\pi}{4} + i\sin\dfrac{\pi}{4}\right)$

25. $z_1 = 2\left(\cos\dfrac{\pi}{6} + i\sin\dfrac{\pi}{6}\right)$

$z_2 = 2\left(\cos\dfrac{\pi}{3} + i\sin\dfrac{\pi}{3}\right)$

$z_1z_2 = 4\left(\cos\dfrac{\pi}{2} + i\sin\dfrac{\pi}{2}\right)$

$\dfrac{z_1}{z_2} = \cos\dfrac{\pi}{6} - i\sin\dfrac{\pi}{6}$

$\dfrac{1}{z_1} = \dfrac{1}{2}\left(\cos\dfrac{\pi}{6} - i\sin\dfrac{\pi}{6}\right)$

27. $z_1 = 4\left(\cos\dfrac{11\pi}{6} + i\sin\dfrac{11\pi}{6}\right)$

$z_2 = \sqrt{2}\left(\cos\dfrac{3\pi}{4} + i\sin\dfrac{3\pi}{4}\right)$

$z_1z_2 = 4\sqrt{2}\left(\cos\dfrac{7\pi}{12} + i\sin\dfrac{7\pi}{12}\right)$

$\dfrac{z_1}{z_2} = 2\sqrt{2}\left(\cos\dfrac{13\pi}{12} + i\sin\dfrac{13\pi}{12}\right)$

$\dfrac{1}{z_1} = \dfrac{1}{4}\left(\cos\dfrac{11\pi}{6} - i\sin\dfrac{11\pi}{6}\right)$

29. $z_1 = 5\sqrt{2}\left(\cos\dfrac{\pi}{4} + i\sin\dfrac{\pi}{4}\right)$

$z_2 = 4(\cos 0 + i\sin 0)$

$z_1z_2 = 20\sqrt{2}\left(\cos\dfrac{\pi}{4} + i\sin\dfrac{\pi}{4}\right)$

$\dfrac{z_1}{z_2} = \dfrac{5\sqrt{2}}{4}\left(\cos\dfrac{\pi}{4} + i\sin\dfrac{\pi}{4}\right)$

$\dfrac{1}{z_1} = \dfrac{\sqrt{2}}{10}\left(\cos\dfrac{\pi}{4} - i\sin\dfrac{\pi}{4}\right)$

31. $z_1 = 20(\cos\pi + i\sin\pi)$

$z_2 = 2\left(\cos\dfrac{\pi}{6} + i\sin\dfrac{\pi}{6}\right)$

$z_1z_2 = 40\left(\cos\dfrac{7\pi}{6} + i\sin\dfrac{7\pi}{6}\right)$

$\dfrac{z_1}{z_2} = 10\left(\cos\dfrac{5\pi}{6} + i\sin\dfrac{5\pi}{6}\right)$

$\dfrac{1}{z_1} = \dfrac{1}{20}(\cos\pi - i\sin\pi)$

33. $z_1 = 3\sqrt{2}\left(\cos\dfrac{3\pi}{4} + i\sin\dfrac{3\pi}{4}\right)$

$z_2 = 2\sqrt{2}\left(\cos\dfrac{7\pi}{4} + i\sin\dfrac{7\pi}{4}\right)$

$z_1z_2 = 12\left(\cos\dfrac{\pi}{2} + i\sin\dfrac{\pi}{2}\right)$

$\dfrac{z_1}{z_2} = \dfrac{3}{2}(\cos\pi - i\sin\pi)$

$\dfrac{1}{z_1} = \dfrac{\sqrt{2}}{6}\left(\cos\dfrac{3\pi}{4} - i\sin\dfrac{3\pi}{4}\right)$

35. -1024 **37.** $512\left(-\sqrt{3} + i\right)$ **39.** -1

41. 4096 **43.** $8(-1 + i)$ **45.** $\dfrac{1}{2048}\left(-\sqrt{3} - i\right)$

47. $2\sqrt{2}\left(\cos\dfrac{\pi}{12} + i\sin\dfrac{\pi}{12}\right),$

$2\sqrt{2}\left(\cos\dfrac{13\pi}{12} + i\sin\dfrac{13\pi}{12}\right)$

49. $3\left(\cos\dfrac{3\pi}{8} + i\sin\dfrac{3\pi}{8}\right), 3\left(\cos\dfrac{7\pi}{8} + i\sin\dfrac{7\pi}{8}\right),$

$3\left(\cos\dfrac{11\pi}{8} + i\sin\dfrac{11\pi}{8}\right), 3\left(\cos\dfrac{15\pi}{8} + i\sin\dfrac{15\pi}{8}\right)$

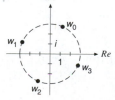

51. $\pm 1, \pm i, \dfrac{\sqrt{2}}{2}(\pm 1 \pm i)$ **53.** $\dfrac{\sqrt{3}}{2} + \dfrac{1}{2}i, -\dfrac{\sqrt{3}}{2} + \dfrac{1}{2}i, -i$

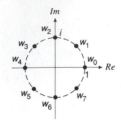

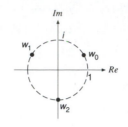

55. $\dfrac{\sqrt{2}}{2}(\pm 1 \pm i)$

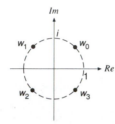

57. $\dfrac{\sqrt{2}}{2}(\pm 1 \pm i)$

59. $2\left(\cos \dfrac{\pi}{18} + i \sin \dfrac{\pi}{18}\right), 2\left(\cos \dfrac{13\pi}{18} + i \sin \dfrac{13\pi}{18}\right),$

$2\left(\cos \dfrac{25\pi}{18} + i \sin \dfrac{25\pi}{18}\right)$ **61.** $\pm 1, \pm\dfrac{1}{2} \pm \dfrac{\sqrt{3}}{2}i$

63. $-3i, -i$

Exercises 7.8 (page 492)

1. **3.**

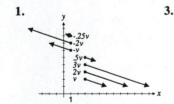

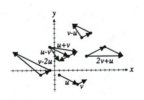

5. $\langle 5, 7 \rangle$ **7.** $\langle -4, -3 \rangle$ **9.** $\langle 0, 2 \rangle$
11. $\langle 4, 14 \rangle, \langle -9, -3 \rangle, \langle 5, 8 \rangle, \langle -6, 17 \rangle$
13. $\langle 0, -2 \rangle, \langle 6, 0 \rangle, \langle -2, -1 \rangle, \langle 8, -3 \rangle$
15. $4\mathbf{i}, -9\mathbf{i} + 6\mathbf{j}, 5\mathbf{i} - 2\mathbf{j}, -6\mathbf{i} + 8\mathbf{j}$
17. $\sqrt{5}, \sqrt{13}, 2\sqrt{5}, \frac{1}{2}\sqrt{13}, \sqrt{26}, \sqrt{10}, \sqrt{5} - \sqrt{13}$
19. $\sqrt{101}, 2\sqrt{2}, 2\sqrt{101}, \sqrt{2}, \sqrt{73}, \sqrt{145}, \sqrt{101} - 2\sqrt{2}$
21. $20\sqrt{3}\mathbf{i} + 20\mathbf{j}$ **23.** $-\dfrac{\sqrt{2}}{2}\mathbf{i} - \dfrac{\sqrt{2}}{2}\mathbf{j}$
25. $4 \cos 10°\mathbf{i} + 4 \sin 10°\mathbf{j} \approx 3.94\mathbf{i} + 0.69\mathbf{j}$
27. $15\sqrt{3}, -15$ **29.** $5, 53.13°$ **31.** $13, 157.38°$

33. $2, 60°$ **35.** (a) $425\mathbf{i} + 40\mathbf{j}$ (b) 427 mi/h, N84.6°E
37. 794 mi/h, N26.6°W   **39.** (a) $20\mathbf{i} + 17.32\mathbf{j}$
(b) 26.5 mi/h, N49.1°E   **41.** 7.4 mi/h, 22.8 mi/h
43. (a) $\langle 5, -3 \rangle$ (b) $\langle -5, 3 \rangle$ **45.** (a) $-4\mathbf{j}$ (b) $4\mathbf{j}$
47. (a) $\langle -7.57, 10.61 \rangle$ (b) $\langle 7.57, -10.61 \rangle$
49. $\mathbf{T_1} \approx -56.5\mathbf{i} + 67.4\mathbf{j}, \mathbf{T_2} \approx 56.5\mathbf{i} + 32.6\mathbf{j}$

Exercises 7.9 (page 500)

1. (a) 2 (b) $45°$ **3.** (a) 13 (b) $56°$ **5.** (a) -1
(b) $97°$ **7.** (a) $5\sqrt{3}$ (b) $30°$ **9.** Yes   **11.** No
13. Yes   **15.** 9 **17.** -5 **19.** $-\frac{12}{5}$ **21.** -24
23. -28 **25.** 25 **27.** 16 ft-lb   **29.** 8660 ft-lb
31. 1164 lb   **33.** $23.6°$

Review Exercises for Chapter 7 (page 502)

23. (a)   (b) Yes

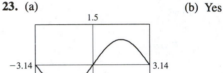
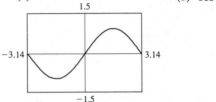

25. (a)   (b) No

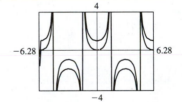

27. (a) $2 \sin^2 3x + \cos 6x = 1$

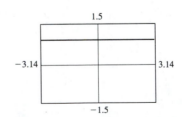

29. $0, \pi$ **31.** $\dfrac{\pi}{6}, \dfrac{5\pi}{6}$ **33.** $\dfrac{\pi}{3}, \dfrac{5\pi}{3}$ **35.** $\dfrac{2\pi}{3}, \dfrac{4\pi}{3}$

37. $\dfrac{\pi}{3}, \dfrac{2\pi}{3}, \dfrac{3\pi}{4}, \dfrac{4\pi}{3}, \dfrac{5\pi}{3}, \dfrac{7\pi}{4}$

39. $\dfrac{\pi}{6}, \dfrac{\pi}{2}, \dfrac{5\pi}{6}, \dfrac{7\pi}{6}, \dfrac{3\pi}{2}, \dfrac{11\pi}{6}$ **41.** $\dfrac{\pi}{6}$ **43.** 1.18

45. $\frac{1}{2}\sqrt{2 + \sqrt{3}}$ **47.** $\sqrt{2} - 1$ **49.** $\sqrt{2}/2$

51. $\sqrt{2}/2$ **53.** $\dfrac{\sqrt{2} + \sqrt{3}}{4}$ **55.** $2\dfrac{\sqrt{10} + 1}{9}$

57. $2\dfrac{2\sqrt{5} + \sqrt{2}}{8 - \sqrt{10}}$ **59.** $\sqrt{(3 + 2\sqrt{2})/6}$ **61.** $\pi/3$

63. $\frac{1}{2}$ **65.** $2/\sqrt{21}$ **67.** $\frac{7}{9}$ **69.** $\theta = \cos^{-1}\dfrac{x}{3}$

71. $s = 7920 \cos^{-1}\left(\dfrac{3960}{h + 3960}\right)$

73. $4\sqrt{2}\left(\cos\dfrac{\pi}{4} + i\sin\dfrac{\pi}{4}\right)$

75. $\sqrt{34}\left[\cos\left(\tan^{-1}\frac{3}{5}\right) + i\sin\left(\tan^{-1}\frac{3}{5}\right)\right]$

77. $\sqrt{2}\left(\cos\dfrac{3\pi}{4} + i\sin\dfrac{3\pi}{4}\right)$ **79.** $8\left(-1 + i\sqrt{3}\right)$

81. $-\frac{1}{32}\left(1 + i\sqrt{3}\right)$ **83.** $\pm 2\sqrt{2}(1 - i)$

85. $\pm 1, \pm\frac{1}{2} \pm \dfrac{\sqrt{3}}{2}i$

87. $\langle 6, 4\rangle, \langle -10, 2\rangle, \langle -4, 6\rangle, \langle -22, 7\rangle$ **89.** $\sqrt{5}$ **91.** 3

93. 3   **95.** $3\mathbf{i} - 4\mathbf{j}$ **97.** $(10, -2)$

99. (a) $(0.41\mathbf{i} + 5.08\mathbf{j}) \times 10^4$ (b) 5.10×10^4 lb, N4.6°E

101. Orthogonal   **103.** Not orthogonal, 45°   **105.** -6

Chapter 7 Test (page 505)

2. (a) $\dfrac{2\pi}{3}, \dfrac{4\pi}{3}$ (b) $\dfrac{\pi}{6}, \dfrac{\pi}{2}, \dfrac{5\pi}{6}, \dfrac{3\pi}{2}$ **3.** $\tan\theta$

4. (a) $\frac{1}{2}$ (b) $\frac{1}{2}\sqrt{2 - \sqrt{3}}$ **5.** (a) 0   (b) $\frac{1}{2}\left(3 - \sqrt{5}\right)$

6. $-\frac{33}{65}$ **7.** (a) $\frac{1}{2}(\sin 8x - \sin 2x)$ (b) $-2\cos\frac{7}{2}x\sin\frac{3}{2}x$

8. -2

9.

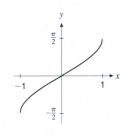

Domain R

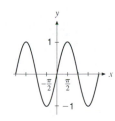

Domain $[-1, 1]$

10. (a) $\theta = \tan^{-1}\dfrac{x}{4}$ (b) $\theta = \cos^{-1}\dfrac{3}{x}$ **11.** (a) $\frac{40}{41}$

(b) $\dfrac{\pi}{3}$ **12.** (a) $2\left(\cos\dfrac{\pi}{3} + i\sin\dfrac{\pi}{3}\right)$ (b) -512

13. $-8, \sqrt{3} + i$

14. $-3i, 3\left(\pm\dfrac{\sqrt{3}}{2} + \dfrac{1}{2}i\right)$

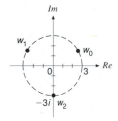

15. (a) $-6\mathbf{i} + 10\mathbf{j}$ (b) $2\sqrt{34}$ (c) $\left\langle -3/\sqrt{34}, 5/\sqrt{34}\right\rangle$

16. (a) $\langle 19, -3\rangle$ (b) $5\sqrt{2}$ (c) 0   (d) Yes   **17.** 45°

18. (a) $14\mathbf{i} + 6\sqrt{3}\,\mathbf{j}$ (b) 17.4 mi/h, N53.4°E   **19.** 90

FOCUS ON PROBLEM SOLVING
(page 508)

1. $-5 \leqslant x \leqslant 0$ **3.** $9, -\frac{7}{3}$

5. $n\pi - \dfrac{\pi}{4} \leqslant x \leqslant n\pi + \dfrac{\pi}{4}$, where n is any integer

7. $A = 2, B = 1, C = 9, D = 7, E = 8$ **9.** 14

13. $\dfrac{\pi}{2}$

17. *Hint:* First divide the eggs into three groups of three.

CHAPTER 8

Exercises 8.1 (page 517)

1. $(4, 0)$ **3.** Parallel

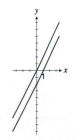

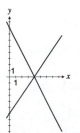

5. $\left(-\frac{5}{2}, 4\right)$

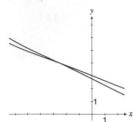

7. $(2, 1)$ **9.** $\left(2, \frac{1}{2}\right)$ **11.** $(10, -9)$ **13.** $(0, -2)$
15. $\left(x, \frac{1}{15}x + \frac{2}{5}\right)$ **17.** $(-1, 4)$ **19.** $(-3, -7)$
21. $(12, -5)$ **23.** $\left(\frac{1}{5}, \frac{2}{5}\right)$ **25.** No solution
27. $\left(\frac{7}{3}, 7\right)$ **29.** $(-1, 2)$ **31.** $\left(\frac{5}{2}, 4\right)$ **33.** $\left(\frac{2}{3}, -\frac{1}{6}\right)$
35. $\left(-\dfrac{1}{a-1}, \dfrac{1}{a-1}\right)$ **37.** $\left(\dfrac{1}{a+b}, \dfrac{1}{a+b}\right)$
39. 22, 12   **41.** 5 dimes, 9 quarters
43. Plane's speed 120 mi/h, wind speed 30 mi/h
45. Run 5 mi/h, cycle 20 mi/h
47. 200 g of A, 40 g of B   **49.** 25%, 10%
51. 25 mi/h, 5:24 P.M.   **53.** 72
55. $y = -5x^2 + 17x$ **57.** $(3.87, 2.74)$
59. $(61.00, 20.00)$ **61.** $(1.89, -0.28)$

Exercises 8.2 (page 527)

1. Linear   **3.** Not linear   **5.** Not linear

7. $\begin{cases} 2x + 3y = 1 \\ 4x + 2y = 3 \end{cases}$ **9.** $\begin{cases} y = 0 \\ x + z = 0 \\ -2y + 2z = 7 \end{cases}$ **11.** $(1, 3, 2)$

13. $(1, 0, 1)$ **15.** $(-1, 5, 0)$ **17.** $(10, 3, -2)$
19. $(2, 0, -2)$ **21.** $\left(\frac{1}{3}, 1, -\frac{1}{2}\right)$ **23.** $(1, 3, 0, -2)$
25. $(1, -1, 2, -2, 0)$ **27.** $(1, -3, 7)$
29. $(1, -1, -3)$ **31.** 2 VitaMax, 1 Vitron, 2 VitaPlus
33. 11 pennies, 4 nickels, 5 dimes, 10 quarters
35. Standard $40, deluxe $55, first class $90
37. $y = 2x^2 - 5x + 6$ **39.** $(1, -2, 3)$

Exercises 8.3 (page 536)

1. Reduced echelon form   **3.** Not in echelon form
5. Not in echelon form   **7.** No solution
9. $x = 7, y = 1$ **11.** No Solution
13. $x = 12 - 11z, y = 12 - 12z, z =$ any number
15. $x = -2z + 5, y = z - 2, z =$ any number
17. $x = -\frac{1}{2}y + z + 6, y =$ any number, $z =$ any number
19. $x = 2, y = w, z = -2w + 6, w =$ any number

21. $x = -12w + 20, y = -19w + 31, z = 2w - 2,$
$w =$ any number   **23.** $x = \frac{1}{3}z - \frac{2}{3}w, y = \frac{1}{3}z + \frac{1}{3}w,$
$z =$ any number, $w =$ any number   **25.** No solution
27. $x, y,$ and z are oz of soya, millet, and milk powder,
respectively: $x = -\frac{1}{3}z + \frac{13}{6}, y = -\frac{2}{3}z + \frac{4}{3}, 0 \leqslant z \leqslant 2$
29. Impossible   **31.** 7 pennies, 5 nickels, 4 dimes

Exercises 8.4 (page 544)

1. $\begin{bmatrix} 5 & -2 & 5 \\ 1 & 1 & 0 \end{bmatrix}$ **3.** $\begin{bmatrix} -1 & -3 & -5 \\ -1 & 3 & -6 \end{bmatrix}$
5. $\begin{bmatrix} 13 & -\frac{7}{2} & 15 \\ 3 & 1 & 3 \end{bmatrix}$ **7.** $\begin{bmatrix} -14 & -8 & -30 \\ -6 & 10 & -24 \end{bmatrix}$

9. Impossible   **11.** $\begin{bmatrix} 3 & \frac{1}{2} & 5 \\ 1 & -1 & 3 \end{bmatrix}$

13. $[28 \quad 21 \quad 28]$ **15.** $\begin{bmatrix} 0 & 0 & 0 & 0 & 0 \\ 0 & 0 & 0 & 0 & 0 \\ 0 & 0 & 0 & 0 & 0 \\ 0 & 0 & 0 & 0 & 0 \end{bmatrix}$

17. $\begin{bmatrix} 8 & -335 \\ 0 & 343 \end{bmatrix}$ **19.** Impossible   **21.** Impossible

23. $\begin{bmatrix} 2 & -5 \\ 3 & 2 \end{bmatrix}\begin{bmatrix} x \\ y \end{bmatrix} = \begin{bmatrix} 7 \\ 4 \end{bmatrix}$

25. $\begin{bmatrix} 3 & 2 & -1 & 1 \\ 1 & 0 & -1 & 0 \\ 0 & 3 & 1 & -1 \end{bmatrix}\begin{bmatrix} x_1 \\ x_2 \\ x_3 \\ x_4 \end{bmatrix} = \begin{bmatrix} 0 \\ 5 \\ 4 \end{bmatrix}$ **27.** $\begin{bmatrix} 3 & \frac{11}{2} \\ 2 & 5 \end{bmatrix}$

29. Impossible   **31.** No   **33.** No

35. (a) $A^2 = \begin{bmatrix} 2 & 2 \\ 2 & 2 \end{bmatrix}, A^3 = \begin{bmatrix} 4 & 4 \\ 4 & 4 \end{bmatrix}, A^4 = \begin{bmatrix} 8 & 8 \\ 8 & 8 \end{bmatrix}$

(b) $A^n = \begin{bmatrix} 2^{n-1} & 2^{n-1} \\ 2^{n-1} & 2^{n-1} \end{bmatrix}$ **37.** (a) $\begin{bmatrix} 32,000 & 18,000 \\ 42,000 & 26,800 \\ 44,000 & 26,800 \end{bmatrix}$

(b) $42,000   (c) $71,600

Exercises 8.5 (page 554)

1. $\begin{bmatrix} 1 & -2 \\ -\frac{3}{2} & \frac{7}{2} \end{bmatrix}$ **3.** $\begin{bmatrix} 5 & -7 \\ -2 & 3 \end{bmatrix}$ **5.** $\begin{bmatrix} 13 & 5 \\ -5 & -2 \end{bmatrix}$

7. No inverse   **9.** $\begin{bmatrix} 1 & 2 \\ -\frac{1}{2} & \frac{2}{3} \end{bmatrix}$ **11.** $\begin{bmatrix} -4 & -4 & 5 \\ 1 & 1 & -1 \\ 5 & 4 & -6 \end{bmatrix}$

13. No inverse   **15.** $\begin{bmatrix} -\frac{9}{2} & -1 & 4 \\ 3 & 1 & -3 \\ \frac{7}{2} & 1 & -3 \end{bmatrix}$

17. $\begin{bmatrix} 0 & 0 & -2 & 1 \\ -1 & 0 & 1 & 1 \\ 0 & 1 & -1 & 0 \\ 1 & 0 & 0 & -1 \end{bmatrix}$ **19.** $x = 20, y = -8$

21. $x = 126, y = -50$ **23.** $x = -38, y = 9, z = 47$

25. $x = -12, y = -21, z = 93$ **27.** $\begin{bmatrix} 7 & 2 & 3 \\ 10 & 3 & 5 \end{bmatrix}$

29. (a) $\begin{bmatrix} 0 & 1 & -1 \\ -2 & \frac{3}{2} & 0 \\ 1 & -\frac{3}{2} & 1 \end{bmatrix}$ (b) 1 oz A, 1 oz B, 2 oz C

(c) 2 oz A, 0 oz B, 1 oz C   (d) No

31. $\begin{bmatrix} -1 & 2e^x \\ e^{-x} & -1 \end{bmatrix}$ **33.** $\frac{1}{2}\begin{bmatrix} e^{-x} & e^{-2x} \\ -e^{-2x} & e^{-3x} \end{bmatrix}$

35. For example, $\begin{bmatrix} 1 & 0 \\ 0 & 1 \end{bmatrix}$ and $\begin{bmatrix} -1 & 0 \\ 0 & -1 \end{bmatrix}$. (There are infinitely many possible answers.)

Exercises 8.6 (page 563)

1. 3   **3.** 6   **5.** No determinant   **7.** 0
9. 20, 20   **11.** $-12, 12$ **13.** 0, 0
15. -6, has an inverse   **17.** 5000, has an inverse
19. -4, has an inverse   **21.** -18 **23.** 120
25. -2 **27.** $(-2, 5)$ **29.** $(0.6, -0.4)$
31. $(4, -1)$ **33.** $(4, 2, -1)$ **35.** $(1, 3, 2)$
37. $(0, -1, 1)$ **39.** $\left(\frac{189}{29}, -\frac{108}{29}, \frac{88}{29}\right)$
41. $\left(-\frac{49}{80}, \frac{77}{40}, \frac{287}{80}\right)$ **43.** (b) $5x - 6y = -200$
45. 0, 1, 2   **47.** 1, -1

Exercises 8.7 (page 571)

1. $(2, -2), (-2, 2)$ **3.** $(4, 0)$ **5.** $(-2, -2)$
7. $(6, 2), (-2, -6)$
9. $(2, 1), (2, -1), (-2, 1), (-2, -1)$
11. No solution   **13.** $\left(-\sqrt{6}, 6\right)$
15. $\left(\sqrt{5}, 2\right), \left(\sqrt{5}, -2\right), \left(-\sqrt{5}, 2\right), \left(-\sqrt{5}, -2\right)$
17. $\left(3, -\frac{1}{2}\right), \left(-3, -\frac{1}{2}\right)$ **19.** $\left(\frac{1}{5}, \frac{1}{3}\right)$ **21.** $(9, 4)$
23. $(0, 2), (0, -2), (2, 2), (-2, -2), \left(\sqrt{2}, -\sqrt{2}\right),$
$\left(-\sqrt{2}, \sqrt{2}\right)$ **25.** $(2, 0, 0), \left(\frac{2}{3}, -\frac{4}{3}, \frac{4}{3}\right)$
27. $(-1, 1, -1)$ **29.** $(2, 0, -4), (-2, 0, -4), (0, 1, -1)$
31. $(-4.51, 2.17), (4.91, -0.97)$
33. $(1.23, 3.87), (-0.35, -4.21)$

35. $(-2.30, -0.70), (0.48, -1.19)$
37. $(1.19, 3.59), (-1.19, 3.59)$
39. 8 cm, 15 cm, 17 cm   **41.** 9 in., 12 in., 15 in.
43. 20, 15   **45.** $y = 3x + 5$ **47.** $(8, 5), (-5, -8)$
49. $(1, 3), (3, 1)$

Exercises 8.8 (page 576)

1. 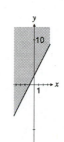   **3.**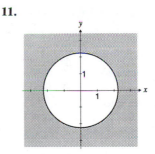

5.   **7.**

9. 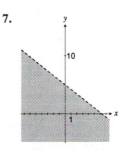   **11.**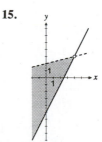

13.   **15.**

not bounded   not bounded

17.

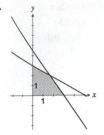

bounded

19.

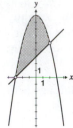

bounded

33.

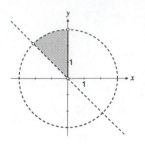

bounded

21.

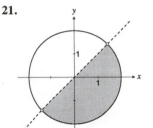

bounded

23.

bounded

35. x = number of fiction books
    y = number of nonfiction books $\begin{cases} x + y \leqslant 100 \\ 20 \leqslant y, \, x \geqslant y \\ x \geqslant 0, \, y \geqslant 0 \end{cases}$

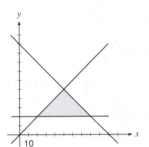

25.

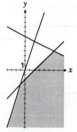

not bounded

27.

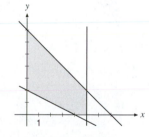

bounded

Exercises 8.9 (page 581)

1.

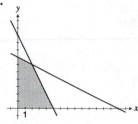

3.

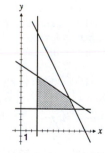

maximum 161   maximum 55
minimum 135   minimum 31

29.

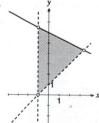

bounded

31.

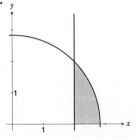

bounded

5. 3 tables, 34 chairs
7. 30 grapefruit crates, 30 orange crates
9. 15 Pasadena to Santa Monica, 3 Pasadena to El Toro,
0 Long Beach to Santa Monica, 16 Long Beach to El Toro
11. $10\frac{1}{2}$ yd^3 Foamboard, $14\frac{1}{2}$ yd^3 Plastiflex
13. 10 days Vancouver, 4 days Seattle
15. $7500 in municipal bonds, $2500 in bank certificates,
$2000 in high-risk bonds
17. 4 games, 32 educational, 0 utility

Exercises 8.10 (page 589)

1. $\dfrac{1}{x-2} - \dfrac{1}{x+2}$ **3.** $\dfrac{3}{x-4} - \dfrac{2}{x+2}$

5. $\dfrac{-\frac{1}{2}}{2x-1} + \dfrac{\frac{3}{2}}{4x-3}$ **7.** $\dfrac{2}{x-2} + \dfrac{3}{x+2} - \dfrac{1}{2x-1}$

9. $\dfrac{2}{x+1} - \dfrac{1}{x} + \dfrac{1}{x^2}$ **11.** $\dfrac{1}{2x+3} - \dfrac{3}{(2x+3)^2}$

13. $\dfrac{2}{x} - \dfrac{1}{x^3} - \dfrac{2}{x+2}$

15. $\dfrac{4}{x+2} - \dfrac{4}{x-1} + \dfrac{2}{(x-1)^2} + \dfrac{1}{(x-1)^3}$

17. $\dfrac{3}{x+2} - \dfrac{1}{(x+2)^2} - \dfrac{1}{(x+3)^2}$ **19.** $\dfrac{x+1}{x^2+3} - \dfrac{1}{x}$

21. $\dfrac{2x-5}{x^2+x+2} + \dfrac{5}{x^2+1}$

23. $\dfrac{1}{x^2+1} - \dfrac{x+2}{(x^2+1)^2} + \dfrac{1}{x}$

25. $x^2 + \dfrac{3}{x-2} - \dfrac{x+1}{x^2+1}$

27. $\dfrac{A}{x} + \dfrac{B}{2x-5} + \dfrac{C}{(2x-5)^2} + \dfrac{D}{(2x-5)^3} + \dfrac{Ex+F}{x^2+2x+5}$
$+ \dfrac{Gx+H}{(x^2+2x+5)^2}$

29. $A = \dfrac{a+b}{2},\ B = \dfrac{a-b}{2}$

Review Exercises for Chapter 8 (page 590)

1. (2, 3)

3. x = any number
$y = \frac{2}{7}x - 4$

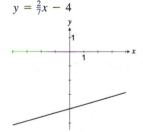

5. No solution

7. (3, −1, −5)
9. No solution

11. $x = -4z + 1,\ y = -z - 1,\ z$ = any number
13. $x = 6 - 5z,\ y = \frac{1}{2}(7 - 3z),\ z$ = any number
15. \$3000 at 6%, \$6000 at 7%   **17.** Impossible

19. $\begin{bmatrix} 4 & 18 \\ 4 & 0 \\ 2 & 2 \end{bmatrix}$ **21.** $[10 \quad 0 \quad -5]$ **23.** $\begin{bmatrix} -\frac{7}{2} & 10 \\ 1 & -\frac{9}{2} \end{bmatrix}$

25. $\begin{bmatrix} 30 & 22 & 2 \\ -9 & 1 & -4 \end{bmatrix}$ **27.** $\begin{bmatrix} -\frac{1}{2} & \frac{11}{2} \\ \frac{15}{4} & -\frac{3}{2} \\ -\frac{1}{2} & 1 \end{bmatrix}$

29. $1, \begin{bmatrix} 9 & -4 \\ -2 & 1 \end{bmatrix}$ **31.** 0, no inverse

33. $-1, \begin{bmatrix} 3 & 2 & -3 \\ 2 & 1 & -2 \\ -8 & -6 & 9 \end{bmatrix}$ **35.** (65, 154)   **37.** $\left(\frac{1}{5}, \frac{9}{5}\right)$

39. $\left(-\frac{87}{26}, \frac{21}{26}, \frac{39}{26}\right)$ **41.** $(-3, -3), \left(\frac{9}{5}, -\frac{3}{5}\right)$
43. (2, 2), (−2, 2)

45.

bounded

47.

bounded

49. Maximum 34, minimum 4
51. (a) 200 acres oats, 200 acres barley
(b) No oats, 360 acres barley
53. $x = \dfrac{b+c}{2},\ y = \dfrac{a+c}{2},\ z = \dfrac{a+b}{2}$
55. 2, 3   **57.** (21.41, −15.93)
59. (11.94, −1.39), (12.07, 1.44)   **61.** $\dfrac{2}{x-5} + \dfrac{1}{x+3}$
63. $\dfrac{-4}{x} + \dfrac{4}{x-1} + \dfrac{-2}{(x-1)^2}$

Chapter 8 Test (page 592)

1. Wind speed 60 km/h, airplane 300 km/h
2. (2, −1); linear neither inconsistent nor dependent

3. No solution; linear, inconsistent

4. $x = \frac{1}{7}(z + 1)$, $y = \frac{1}{7}(9z + 2)$,
  $z =$ any number; linear, dependent

5. $(\pm 1, -2)$, $\left(\pm\frac{5}{3}, -\frac{2}{3}\right)$; nonlinear

6. Incompatible dimensions   **7.** Incompatible dimensions

8. $\begin{bmatrix} 11 & 9 \\ 2 & 1 \\ 11 & 2 \end{bmatrix}$ **9.** $\begin{bmatrix} 46 & 95 \\ 4 & 11 \\ 26 & 55 \end{bmatrix}$ **10.** $\begin{bmatrix} 2 & -\frac{3}{2} \\ -1 & 1 \end{bmatrix}$

11. B is not square   **12.** B is not square   **13.** -3

14. $\left(\frac{473}{19}, \frac{90}{19}\right)$ **15.** $(5, -5, -4)$

16. $|A| = 0$, $|B| = 2$, $B^{-1} = \begin{bmatrix} 1 & -2 & 0 \\ 0 & \frac{1}{2} & 0 \\ 3 & -6 & 1 \end{bmatrix}$

17.

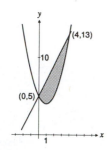

18. $\dfrac{1}{x - 1} + \dfrac{1}{(x - 1)^2} - \dfrac{1}{x + 2}$

19. He should grow $166\frac{2}{3}$ acres of wheat and no barley.

20. $(-0.49, 3.93)$, $(2.34, 2.24)$

FOCUS ON PROBLEM SOLVING
(page 594)

3. $-7, -1, 3, 9$ **5.** $-\frac{2}{3}, \frac{4}{3}$

7.

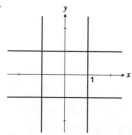

9.

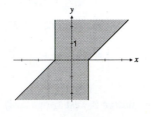

11. $x = z = 1$, $y = t = -1$

CHAPTER 9

Exercises 9.1 (page 603)

1. Focus $F(1, 0)$; directrix
  $x = -1$;
  focal diameter 4

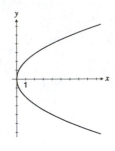

3. $F\left(0, \frac{9}{4}\right)$; $y = -\frac{9}{4}$; 9

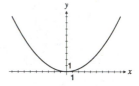

5. $F\left(0, \frac{1}{20}\right)$; $y = -\frac{1}{20}$; $\frac{1}{5}$ **7.** $F\left(-\frac{1}{32}, 0\right)$; $x = \frac{1}{32}$; $\frac{1}{8}$

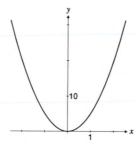

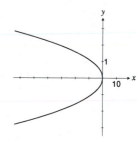

9. $F\left(0, -\frac{3}{2}\right)$; $y = \frac{3}{2}$; 6   **11.** $F\left(-\frac{5}{12}, 0\right)$; $x = \frac{5}{12}$; $\frac{5}{3}$

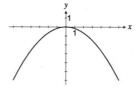

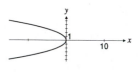

13. $x^2 = -12y$ **15.** $y^2 = -3x$ **17.** $x = y^2$

19. $x^2 = -4\sqrt{2}y$ **21.** $y^2 = -12x$ **23.** $y^2 = 4x$

25. $x^2 = -24y$ **27.** (a) $y^2 = 12x$ (b) $8\sqrt{15} \approx 31$ cm

Exercises 9.2 (page 608)

1. Vertices $V(\pm 5, 0)$; foci $F(\pm 4, 0)$; eccentricity $\frac{4}{5}$; major axis 10, minor axis 6

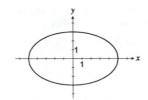

15. $\dfrac{x^2}{25} + \dfrac{y^2}{9} = 1$ **17.** $x^2 + \dfrac{y^2}{4} = 1$

19. $\dfrac{x^2}{9} + \dfrac{y^2}{13} = 1$ **21.** $\dfrac{x^2}{100} + \dfrac{y^2}{91} = 1$

23. $\dfrac{x^2}{25} + \dfrac{y^2}{5} = 1$ **25.** $\dfrac{64x^2}{225} + \dfrac{64y^2}{81} = 1$

27. $(0, \pm 2)$

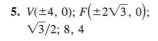

3. $V(0, \pm 3)$; $F(0, \pm\sqrt{5})$; $\sqrt{5}/3$; 6, 4

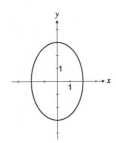

5. $V(\pm 4, 0)$; $F(\pm 2\sqrt{3}, 0)$; $\sqrt{3}/2$; 8, 4

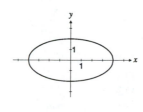

29. $\dfrac{x^2}{2.250 \times 10^{16}} + \dfrac{y^2}{2.249 \times 10^{16}} = 1$

31. $\dfrac{x^2}{1,453,200} + \dfrac{y^2}{1,449,200} = 1$

33. $5\sqrt{39}/2 \approx 15.6$ in

7. $V(0, \pm\sqrt{3})$; $F(0, \pm\sqrt{3/2})$; $1/\sqrt{2}$; $2\sqrt{3}$, $\sqrt{6}$

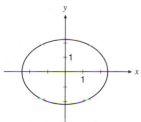

9. $V(\pm 1, 0)$; $F(\pm\sqrt{3}/2, 0)$; $\sqrt{3}/2$; 2, 1

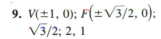

Exercises 9.3 (page 614)

1. Vertices $V(\pm 2, 0)$; foci $F(\pm 2\sqrt{5}, 0)$; asymptotes $y = \pm 2x$

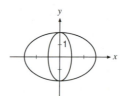

3. $V(0, \pm 1)$; $F(0, \pm\sqrt{26})$; $y = \pm\frac{1}{5}x$

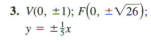

11. $V(0, \pm\sqrt{2})$; $F(0, \pm\sqrt{3/2})$; $\sqrt{3}/2$; $2\sqrt{2}$, $\sqrt{2}$

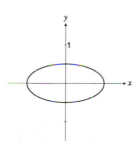

13. $V(0, \pm 1)$; $F(0, \pm 1/\sqrt{2})$; $1/\sqrt{2}$; 2, $\sqrt{2}$

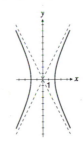

5. $V(\pm 1, 0)$; $F(\pm\sqrt{2}, 0)$; $y = \pm x$

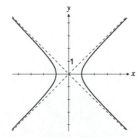

7. $V(0, \pm 3)$; $F(0, \pm\sqrt{34})$; $y = \pm\frac{3}{5}x$

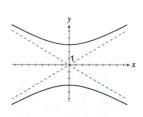

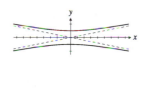

9. $V(\pm 2\sqrt{2}, 0)$;
$F(\pm\sqrt{10}, 0)$;
$y = \pm\frac{1}{2}x$

11. $V(0, \pm\frac{1}{2})$; $F(0, \pm\sqrt{5}/2)$;
$y = \pm\frac{1}{2}x$

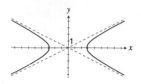

9. Center $C(-1, 3)$;
foci $F_1(-6, 3)$, $F_2(4, 3)$;
vertices $V_1(-4, 3)$,
$V_2(2, 3)$;
asymptotes
$y = \pm\frac{4}{3}(x + 1) + 3$

11. $C(-1, 0)$; $F(-1, \pm\sqrt{5})$;
$V(-1, \pm 1)$;
$y = \pm\frac{1}{2}(x + 1)$

13. $\dfrac{x^2}{9} - \dfrac{y^2}{16} = 1$ **15.** $y^2 - \dfrac{x^2}{3} = 1$

17. $x^2 - \dfrac{y^2}{25} = 1$ **19.** $\dfrac{5y^2}{64} - \dfrac{5x^2}{256} = 1$

21. $\dfrac{x^2}{16} - \dfrac{y^2}{16} = 1$ **23.** $\dfrac{x^2}{9} - \dfrac{y^2}{16} = 1$

25. (b) $x^2 - y^2 = \dfrac{c^2}{2}$ **29.** (a) 490 mi

(b) $\dfrac{y^2}{60,025} - \dfrac{x^2}{2475} = 1$ (c) 10.1 mi

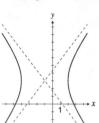

13. Ellipse; $C(2, 0)$;
$F(2, \pm\sqrt{5})$;
$V(2, \pm 3)$;
major axis 6, minor axis 4

15. Hyperbola; $C(1, 2)$;
$F_1(-\frac{3}{2}, 2)$, $F_2(\frac{7}{2}, 2)$;
$V(1 \pm \sqrt{5}, 2)$;
asymptotes
$y = \pm\frac{1}{2}(x - 1) + 2$

Exercises 9.4 (page 621)

1. Center $C(2, 1)$;
foci $F(2 \pm \sqrt{5}, 1)$;
vertices $V_1(-1, 1)$,
$V_2(5, 1)$;
major axis 6, minor axis 4

3. $C(0, -5)$; $F_1(0, -1)$,
$F_2(0, -9)$;
$V_1(0, 0)$, $V_2(0, -10)$;
axes 10, 6

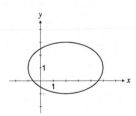

17. Ellipse; $C(3, -5)$;
$F(3 \pm \sqrt{21}, -5)$;
$V_1(-2, -5)$, $V_2(8, -5)$;
major axis 10,
minor axis 4

19. Hyperbola; $C(3, 0)$;
$F(3, \pm 5)$; $V(3, \pm 4)$;
asymptotes $y = \pm\frac{4}{3}(x - 3)$

5. Vertex $V(3, -1)$;
focus $F(3, 1)$; directrix
$y = -3$

7. $V(-\frac{1}{2}, 0)$; $F(-\frac{1}{2}, -\frac{1}{16})$;
$y = \frac{1}{16}$

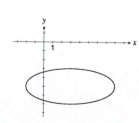

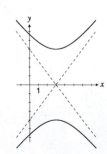

21. Degenerate conic (pair of lines), $y = \pm\frac{1}{2}(x - 4)$

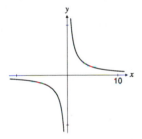

23. point $(1, 3)$

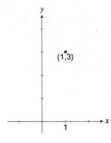

25. (a) $F < 17$ (b) $F = 17$ (c) $F > 17$

Exercises 9.5 (page 627)

1. $\left(\sqrt{2}, 0\right)$ **3.** $\left(0, -2\sqrt{3}\right)$ **5.** $(1.6383, 1.1472)$
7. $X^2 + Y^2 - 2XY - 3\sqrt{2}X + \sqrt{2}Y + 2 = 0$
9. $X^2 - Y^2 = 2$
11. Hyperbola, $X^2 - Y^2 = 16$

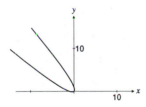

13. Parabola, $Y = \sqrt{2}X^2$

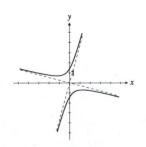

15. Hyperbola, $Y^2 - X^2 = 1$

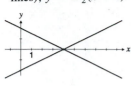

17. Hyperbola, $\dfrac{X^2}{4} - Y^2 = 1$

19. Hyperbola, $3X^2 - Y^2 = 2\sqrt{3}$

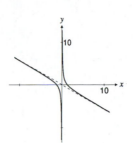

21. Hyperbola, $(X - 1)^2 - 3Y^2 = 1$

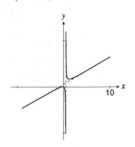

23. Ellipse, $X^2 + \dfrac{(Y + 1)^2}{4} = 1$

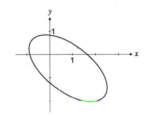

25. (a) $(X - 5)^2 - Y^2 = 1$

(b) XY-coordinates: $C(5, 0)$; $V_1(6, 0)$, $V_2(4, 0)$; $F\left(5 \pm \sqrt{2}, 0\right)$;
xy-coordinates: $C(4, 3)$; $V_1\left(\frac{24}{5}, \frac{18}{5}\right)$, $V_2\left(\frac{16}{5}, \frac{12}{5}\right)$; $F\left(4 \pm \frac{4}{5}\sqrt{2}, 3 \pm \frac{3}{5}\sqrt{2}\right)$

(c) $Y = \pm(X - 5)$; $7x + y - 25 = 0$, $x - 7y + 25 = 0$

Exercises 9.6 (page 636)

1.

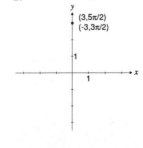

$\left(-3, \dfrac{3\pi}{2}\right), \left(3, \dfrac{5\pi}{2}\right)$

3.

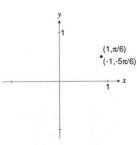

$\left(-1, -\dfrac{5\pi}{6}\right), \left(1, \dfrac{\pi}{6}\right)$

5.

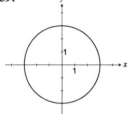

(5,π)
(-5,2π)

$(-5, 2\pi), (5, \pi)$

7. $\left(2\sqrt{3}, 2\right)$ **9.** $(1, -1)$ **11.** $(-5, 0)$

13. $\left(\sqrt{2}, \dfrac{3\pi}{4}\right)$ **15.** $\left(4, \dfrac{\pi}{4}\right)$ **17.** $\left(5, \tan^{-1}\dfrac{4}{3}\right)$

19. $\theta = \dfrac{\pi}{4}$ **21.** $r = \tan\theta \sec\theta$ **23.** $r = 4\sec\theta$

25. $x^2 + y^2 = 49$ **27.** $x = 6$ **29.** $x^2 + y^2 = \dfrac{y}{x}$

31. $y - x = 1$ **33.** $x^2 + y^2 = (x^2 + y^2 - x)^2$

35. $x = 2$ **37.** $y = \pm\sqrt{3}\,x$

39.

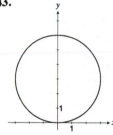

41.

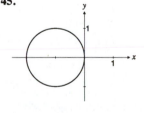

43.

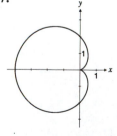

45.

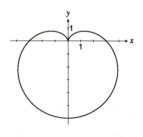

47.

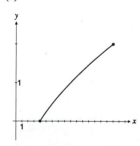

49.

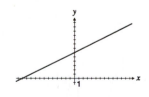

51.

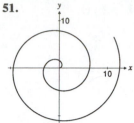

53.

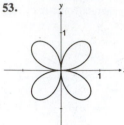

55.

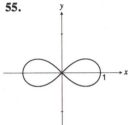

57.

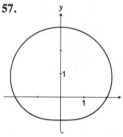

59.

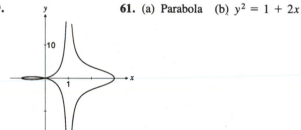

61. (a) Parabola   (b) $y^2 = 1 + 2x$

63. (a) Hyperbola   (b) $x^2 - 3y^2 + 16y - 16 = 0$

65. (b) $\sqrt{10 + 6\cos(5\pi/12)} \approx 3.40$

Exercises 9.7 (page 642)

1. (a)

3. (a)

(b) $x - 2y + 12 = 0$ (b) $x = (y + 2)^2$

5. (a)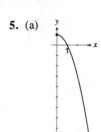

(b) $x = \sqrt{1 - y}$

7. (a)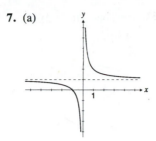

(b) $y = \dfrac{1}{x} + 1$

9. (a)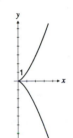

(b) $x^3 = y^2$

11. (a)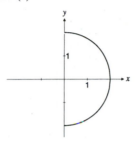

(b) $x^2 + y^2 = 4$

13. (a)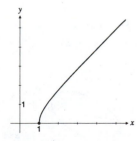

(b) $y = x^2$

15. (a)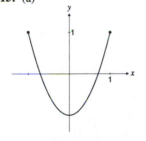

(b) $y = 2x^2 - 1$

17. (a)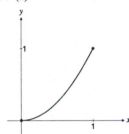

(b) $x^2 - y^2 = 1$

19. (a)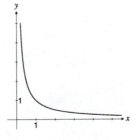

(b) $xy = 1$

21. (a)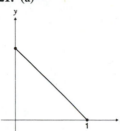

(b) $x + y = 1$

23. $x = 4 + t,\ y = -1 + \frac{1}{2}t$
25. $x = 6 + t,\ y = 7 + t$
27. $x = a \cos t,\ y = a \sin t$

31.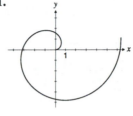

33.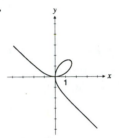

37.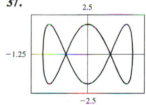

39.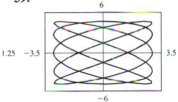

41.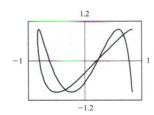

43.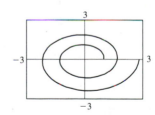

45.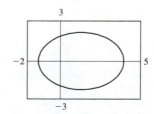

47.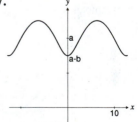

49. (b)

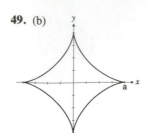

$$x^{2/3} + y^{2/3} = a^{2/3}$$

51. $x = a(\sin \theta \cos \theta + \cot \theta)$, $y = a(1 + \sin^2\theta)$

53. $y = a - a \cos\left(\dfrac{x + \sqrt{2ay - y^2}}{a}\right)$

Review Exercises for Chapter 9 (page 645)

1. Vertex $V(0, 0)$;
focus $F(0, -2)$;
directrix $y = 2$

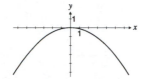

3. $V(-2, 2)$; $F\left(-\frac{7}{4}, 2\right)$;
directrix $x = -\frac{9}{4}$

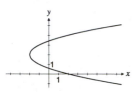

5. Center $C(0, 0)$;
vertices $V(\pm 4, 0)$;
foci $F\left(\pm 2\sqrt{3}, 0\right)$;
major axis 8, minor axis 4

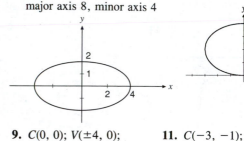

7. $C(0, 2)$; $V(\pm 3, 2)$;
$F\left(\pm\sqrt{5}, 2\right)$;
axes 6, 4

9. $C(0, 0)$; $V(\pm 4, 0)$;
$F\left(\pm 2\sqrt{6}, 0\right)$;
asymptotes $y = \pm\dfrac{1}{\sqrt{2}}x$

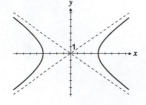

11. $C(-3, -1)$;
$V\left(-3, -1 \pm \sqrt{2}\right)$;
$F\left(-3, -1 \pm 2\sqrt{5}\right)$;
asymptotes $y = \frac{1}{3}x$,
$y = -\frac{1}{3}x - 2$

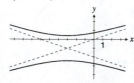

13. Parabola; $F(0, -2)$;
$V(0, 1)$

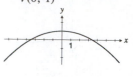

15. Hyperbola; $F\left(0, \pm 12\sqrt{2}\right)$;
$V(0, \pm 12)$

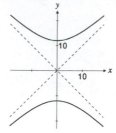

17. Ellipse; $F\left(1, 4 \pm \sqrt{15}\right)$;
$V\left(1, 4 \pm 2\sqrt{5}\right)$

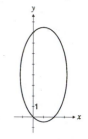

19. Parabola; $F\left(-\frac{255}{4}, 8\right)$;
$V(-64, 8)$

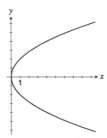

21. Ellipse;
$$F\left(3, -3 \pm \dfrac{1}{\sqrt{2}}\right);$$
$V_1(3, -4)$, $V_2(3, -2)$

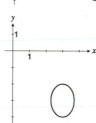

23. has no graph

25. $x^2 = 4y$

27. $\dfrac{y^2}{4} - \dfrac{x^2}{16} = 1$

29. $\dfrac{(x-1)^2}{3} + \dfrac{(y-2)^2}{4} = 1$

31. $\dfrac{4(x-7)^2}{225} + \dfrac{(y-2)^2}{100} = 1$

33. Hyperbola; $3X^2 - Y^2 = 1$

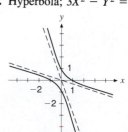

35. Ellipse,
$(X - 1)^2 + 4Y^2 = 1$

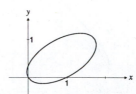

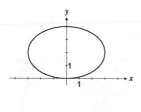

37. $(x^2 + y^2 - 3x)^2 = 9(x^2 + y^2)$

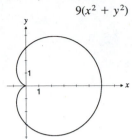

39. $(x^2 + y^2)^3 = 16x^2y^2$

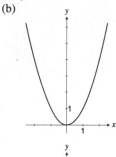

41. $x^2 - y^2 = 1$

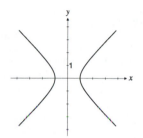

43. $x^2 + y^2 = x + y$

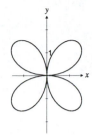

45. $x = 2y - y^2$

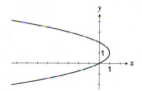

47. $(x - 1)^2 + (y - 1)^2 = 1$

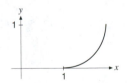

49.

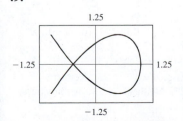

51. (a) $y = x^2$
(b)

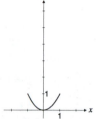

The curves are different parts of the parabola $y = x^2$.

Chapter 9 Test (page 647)

1. Focus $F\left(0, -\frac{3}{2}\right)$; directrix $y = \frac{3}{2}$

2. Vertices $V(0, \pm 2\sqrt{3})$; foci $F(0, \pm 2)$; major axis $4\sqrt{3}$, minor axis $4\sqrt{2}$

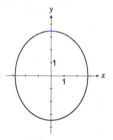

3. Vertices $V(0, \pm 7)$; foci $F(0, \pm\sqrt{85})$; asymptotes $y = \pm\frac{7}{6}x$

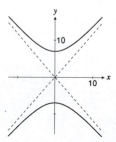

4. $\dfrac{(x-3)^2}{9} + \dfrac{\left(y+\frac{1}{2}\right)^2}{4} = 1$

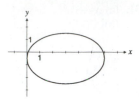

5. $9(x+2)^2 - 8(y-4)^2 = 0$

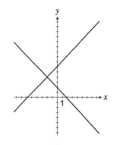

6. $(y+4)^2 = -2(x-4)$

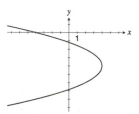

7. $\dfrac{y^2}{9} - \dfrac{x^2}{16} = 1$ **8.** $x^2 - 4x - 8y + 20 = 0$

9. (a) $\dfrac{X^2}{3} + \dfrac{Y^2}{18} = 1$

(b)

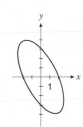

(c) $\left(-3\sqrt{2/5}, 6\sqrt{2/5}\right), \left(3\sqrt{2/5}, -6\sqrt{2/5}\right)$

10.

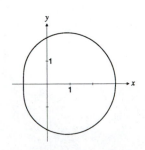

11. $(x-1)^2 + (y+2)^2 = 5$, circle

12. (a)

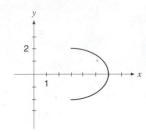

(b) $\dfrac{(x-3)^2}{9} + \dfrac{y^2}{4} = 1$

FOCUS ON PROBLEM SOLVING
(page 649)

1. 52   **3.** 2.4 cm   **5.** $116 = 1^2 + 3^2 + 5^2 + 9^2$
7. Impossible

CHAPTER 10

Exercises 10.1 (page 656)

1. $1/3, 1/4, 1/5, 1/6; 1/1002$
3. $-1, 1/4, -1/9, 1/16; 1/1{,}000{,}000$ **5.** $0, 2, 0, 2; 2$
7. $-2, 4, -8, 16; 2^{1000}$
9. $1/2, -2/3, 3/4, -4/5; -1000/1001$
11. $-3/\sqrt{2}, 5/\sqrt{3}, -7/2, 9/\sqrt{5}; 2001/\sqrt{1001}$
13. $3, 2, 0, -4, -12$ **15.** $1, 3, 7, 15, 31$
17. $0, 1, 1, 0, -1$ **19.** $1, 2, 3, 5, 8$ **21.** 2^n
23. $3n - 2$ **25.** $(2n-1)/n^2$ **27.** r^n
29. $1 + (-1)^n$
31. (c) $n^2 + (n-1)(n-2)(n-3)(n-4)(n-5)(n-6)$
33. $2^{(2^n-1)/2^n}$
35. 25, 76, 38, 19, 58, 29, 88, 44, 22, 11, 34, 17, 52, 26, 13, 40, 20, 10, 5, 16, 8, 4, 2, 1, 4, 2, 1, . . .

Exercises 10.2 (page 662)

1. $5, 24, 4 + 5(n-1), 499$
3. $4, 4, -12 + 4(n-1), 384$
5. $1.5, 31, 25 + 1.5(n-1), 173.5$
7. $s, 2 + 4s, 2 + (n-1)s, 2 + 99s$
9. $3, 162, 2 \cdot 3^{n-1}$ **11.** $-0.3, 0.00243, (0.3)(-0.3)^{n-1}$
13. $-1/12, 1/144, 144(-1/12)^{n-1}$
15. $3^{2/3}, 3^{11/3}, 3^{(2n+1)/3}$ **17.** $s^{2/7}, s^{8/7}, s^{2(n-1)/7}$
19. Neither   **21.** Geometric, $9\sqrt{3}$ **23.** Neither
25. Arithmetic, $x + 3$ **27.** Neither   **29.** Arithmetic, 3

31. $\frac{1}{2}$ **33.** $-100, -98, -96$ **35.** $\frac{16}{27}$ **37.** $\frac{25}{4}$
39. 30th   **41.** Yes, 2985th   **43.** 19 ft   **45.** \$4714.37
47. $64/25, 1024/625, 5(4/5)^n$ **49.** 3   **51.** r^2
53. $r = 10^d$ **55.** 2, 5, 8   **57.** 0
59. $(\sqrt{5} - 1)/2$ **61.** 10, 20, 40   **63.** 15/4

Exercises 10.3 (page 669)

1. 86   **3.** 12   **5.** 8   **7.** 31   **9.** 549
11. 106   **13.** $\sqrt{1} + \sqrt{2} + \sqrt{3} + \sqrt{4} + \sqrt{5}$
15. $\sqrt{4} + \sqrt{5} + \sqrt{6} + \sqrt{7} + \sqrt{8} + \sqrt{9} + \sqrt{10}$
17. $1 \cdot x^2 + 2 \cdot x^3 + \cdots + 8 \cdot x^9$
19. $x - x^2 + x^3 - \cdots + (-1)^{n+1}x^n$
21. $\sum_{k=1}^{100} k$ **23.** $\sum_{k=2}^{100} \frac{(-1)^k}{k \ln k}$ **25.** $\sum_{k=0}^{5}(-1)^k\frac{x^k}{3^k}$
27. $\sum_{k=1}^{100}(-1)^{k+1}kx^{k-1}$ **29.** $\sum_{k=1}^{97}k(k + 1)(k + 2)$
31. 1, 4, 9, 16, 25, 36   **33.** $\frac{1}{3}, \frac{4}{9}, \frac{13}{27}, \frac{40}{81}, \frac{121}{243}, \frac{364}{729}$
35. $S_n = \frac{1}{2} - \frac{1}{n + 2}, \frac{250}{501}$ **37.** $S_n = 1 - \frac{1}{3^n}, 1 - \frac{1}{3^{20}}$
39. $S_n = 1 - \sqrt{n + 1}, -9$ **41.** $S_n = -\log(n + 1), -6$

Exercises 10.4 (page 674)

1. 460   **3.** 1090   **5.** 4900   **7.** 315   **9.** 441
11. 2.8502   **13.** 20,301   **15.** 832.3   **17.** 46.75
19. 3280   **21.** 1.94117   **23.** 5.997070313
25. Geometric, 9.2224   **27.** Geometric, $(1 + x^{21})/(1 + x)$
29. Arithmetic, 250,500   **31.** Arithmetic, 4128
33. Arithmetic, $9\left(1 + 4\sqrt{2}\right)$
35. Geometric, $124/\left(\sqrt[5]{5} - 1\right)$ **37.** 50   **39.** 80/3
41. 1/2   **43.** 455/27   **45.** 10^{19}
47. \$10,737,418.23, 37 days
49. (a) 576 ft   (b) $16n^2$ ft   **51.** 2801

Exercises 10.5 (page 681)

1. \$360,262.21   **3.** \$5591.79   **5.** \$245.66
7. \$2601.59   **9.** \$307.24   **11.** \$733.76, \$264,153.60
13. (a) \$859.15   (b) \$309,294.00   (c) \$1,841,519.29
15. \$341.24   **17.** 18%   **19.** 12%

Exercises 10.6 (page 685)

1. 3/4   **3.** 1/648   **5.** $-1000/117$
7. $5^{4/3}/(1 + 5^{1/3})$ **9.** 7/9   **11.** 1/33
13. 112/999   **15.** 60   **17.** $2 + \sqrt{2}$ s   **19.** $2\pi R^2$
21. \$50,000

Exercises 10.8 (page 713)

1. $x^7 + 7x^6y + 21x^5y^2 + 35x^4y^3 + 35x^3y^4 + 21x^2y^5 + 7xy^6 + y^7$
3. $x^4 + 4x^2 + 6 + \dfrac{4}{x^2} + \dfrac{1}{x^4}$ **5.** $1188 - 684\sqrt{3}$
7. $x^{10}y^5 - 5x^8y^4 + 10x^6y^3 - 10x^4y^2 + 5x^2y - 1$
9. $8x^3 - 36x^2y + 54xy^2 - 27y^3$
11. $\dfrac{1}{x^5} - \dfrac{5}{x^{7/2}} + \dfrac{10}{x^2} - \dfrac{10}{x^{1/2}} + 5x - x^{5/2}$ **13.** 10
15. 4950   **17.** 32
19. $x^4 + 8x^3y + 24x^2y^2 + 32xy^3 + 16y^4$
21. $1 + \dfrac{6}{x} + \dfrac{15}{x^2} + \dfrac{20}{x^3} + \dfrac{15}{x^4} + \dfrac{6}{x^5} + \dfrac{1}{x^6}$
23. $x^{20} + 40x^{19}y + 760x^{18}y^2$ **25.** $25a^{26/3} + a^{25/3}$
27. $48,620x^{18}$ **29.** $300a^2b^{23}$ **31.** $100y^{99}$
33. $13,440x^4y^6$ **35.** $495a^8b^8$ **37.** 17,920
39. $1792c^7$ **41.** 21   **43.** 11   **45.** $(x + y)^4$
47. $(2a + b)^3$
49. $a^8 + 4a^7 + 10a^6 + 16a^5 + 19a^4 + 16a^3 + 10a^2 + 4a + 1$ **51.** $(101!)^{100}$

Review Exercises for Chapter 10 (page 714)

1. 1/2, 4/3, 9/4, 16/5; 100/11
3. 0, 1/4, 0, 1/32; 1/500
5. 1, 3, 15, 105; 654,729,075
7. 1, 4, 9, 16, 25, 36, 49   **9.** 1, 3, 5, 11, 21, 43, 85
11. 3, 2, 1, 1, 1, 1, 1   **13.** Arithmetic, 7
15. Arithmetic, $5\sqrt{2}$ **17.** Arithmetic, $t + 1$
19. Geometric, 4/27   **21.** Neither
23. Geometric, ab^4c^{12} **25.** $2i$ **27.** 5   **29.** 81/4
31. 64   **33.** 12,288   **37.** (a) 9   (b) $\pm6\sqrt{2}$
39. 126   **41.** 384   **43.** $0^2 + 1^2 + 2^2 + \cdots + 9^2$
45. $\dfrac{3}{2^2} + \dfrac{3^2}{2^3} + \dfrac{3^3}{2^4} + \cdots + \dfrac{3^{50}}{2^{51}}$
47. $\sum_{k=1}^{33} 3k$ **49.** $\sum_{k=1}^{100} k2^{k+2}$ **51.** 4.68559
53. $(1 + 5^{9/2})/(1 + 5^{1/2}) \approx 432.17$ **55.** Arithmetic, 1650
57. Geometric, 9831   **59.** 13
61. 29   **63.** (a) 259,374   (b) $100,000(1.1)^n$
65. \$30,324.28   **67.** (a) \$482.77   (b) \$608.56
69. 1/9   **71.** $a/(1 - b^2)$ **81.** 100   **83.** 32
85. $1 - 6x^2 + 15x^4 - 20x^6 + 15x^8 - 6x^{10} + x^{12}$
87. $1540a^3b^{19}$ **89.** 5   **91.** a^3

Chapter 10 Test (page 717)

1. $-10/99$ **2.** -1 **3.** -36 **4.** Yes, 6th term

5. (a) False   (b) True

6. (a) $S_n = \dfrac{n}{2}[2a + (n-1)d]$ or $S_n = n\left(\dfrac{a + a_n}{2}\right)$ (b) 60

(c) $-8/9$, -78 **7.** (a) $S_n = \dfrac{a(1 - r^n)}{1 - r}$

(b) $58{,}025/59{,}049$ **8.** $2 + \sqrt{2}$ **10.** (a) -50 (b) 10

11. $(a + b)^n = \displaystyle\sum_{k=0}^{n} \binom{n}{k} a^{n-k} b^k$

12. $32x^5 + 80x^4y^2 + 80x^3y^4 + 40x^2y^6 + 10xy^8 + y^{10}$

13. $1{,}293{,}600\, a^3 b^{97}$ **14.** 8064

FOCUS ON PROBLEM SOLVING
(page 720)

1. $1 + 3 + 5 + \cdots + (2n - 1) = n^2$ **5.** $(n + 1)! - 1$

7. *Hint:* Use mathematical induction on the number of handshakes.

9. (a) 91; $n(n + 1)(2n + 1)/6$ (b) 441; $[n(n + 1)/2]^2$

INDEX

TO THE STUDENT

Many students experience difficulty with *precalculus* mathematics in calculus courses. Calculus requires that you understand and remember precalculus topics. For this reason, it may be helpful to retain this text as a reference in your calculus course. It has been written with this purpose in mind.

SOME REFERENCES TO CALCULUS TOPICS IN THIS TEXT

GRAPHS OF THE TRIGONOMETRIC FUNCTIONS

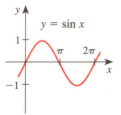

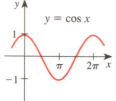

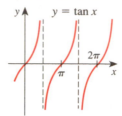

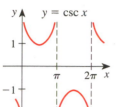

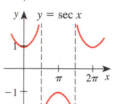

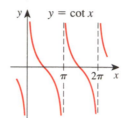

SINE AND COSINE CURVES

$y = a \sin k(x - b) \quad (k > 0)$

$y = a \cos k(x - b) \quad (k > 0)$

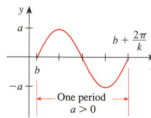

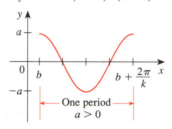

amplitude: $|a|$ period: $2\pi/k$ phase shift: b

GRAPHS OF THE INVERSE TRIGONOMETRIC FUNCTIONS

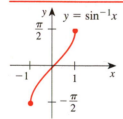

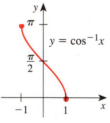

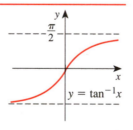

FUNDAMENTAL IDENTITIES

$$\sec x = \frac{1}{\cos x} \qquad \csc x = \frac{1}{\sin x} \qquad \tan x = \frac{\sin x}{\cos x} \qquad \cot x = \frac{1}{\tan x}$$

$$\sin^2 x + \cos^2 x = 1 \qquad 1 + \tan^2 x = \sec^2 x \qquad 1 + \cot^2 x = \csc^2 x$$

$$\sin(-x) = -\sin x \qquad \cos(-x) = \cos x \qquad \tan(-x) = -\tan x$$

COFUNCTION IDENTITIES

$$\sin\left(\frac{\pi}{2} - x\right) = \cos x \qquad \cos\left(\frac{\pi}{2} - x\right) = \sin x$$

$$\tan\left(\frac{\pi}{2} - x\right) = \cot x \qquad \cot\left(\frac{\pi}{2} - x\right) = \tan x$$

$$\sec\left(\frac{\pi}{2} - x\right) = \csc x \qquad \csc\left(\frac{\pi}{2} - x\right) = \sec x$$

REDUCTION IDENTITIES

$$\sin(x + \pi) = -\sin x \qquad \sin\left(x + \frac{\pi}{2}\right) = \cos x$$

$$\cos(x + \pi) = -\cos x \qquad \cos\left(x + \frac{\pi}{2}\right) = -\sin x$$

$$\tan(x + \pi) = \tan x \qquad \tan\left(x + \frac{\pi}{2}\right) = -\cot x$$

THE LAW OF SINES

$$\frac{\sin A}{a} = \frac{\sin B}{b} = \frac{\sin C}{c}$$

THE LAW OF COSINES

$$a^2 = b^2 + c^2 - 2bc \cos A$$
$$b^2 = a^2 + c^2 - 2ac \cos B$$
$$c^2 = a^2 + b^2 - 2ab \cos C$$

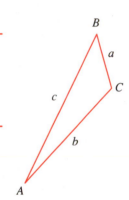

ADDITION AND SUBTRACTION FORMULAS

$$\sin(x + y) = \sin x \cos y + \cos x \sin y$$
$$\sin(x - y) = \sin x \cos y - \cos x \sin y$$
$$\cos(x + y) = \cos x \cos y - \sin x \sin y$$
$$\cos(x - y) = \cos x \cos y + \sin x \sin y$$
$$\tan(x + y) = \frac{\tan x + \tan y}{1 - \tan x \tan y} \qquad \tan(x - y) = \frac{\tan x - \tan y}{1 + \tan x \tan y}$$

DOUBLE-ANGLE FORMULAS

$$\sin 2x = 2 \sin x \cos x$$

$$\cos 2x = \cos^2 x - \sin^2 x$$
$$= 2 \cos^2 x - 1$$
$$= 1 - 2 \sin^2 x$$

$$\tan 2x = \frac{2 \tan x}{1 - \tan^2 x}$$